BIODIVERSITY IN MANAGED LANDSCAPES

BIODIVERSITY IN MANAGED LANDSCAPES

Theory and Practice

Edited by

Robert C. Szaro

David W. Johnston

New York Oxford
OXFORD UNIVERSITY PRESS
1996

Oxford University Press

Oxford New York
Athens Auckland Bangkok Bombay
Calcutta Cape Town Dar es Salaam Delhi
Florence Hong Kong Istanbul Karachi
Kuala Lumpur Madras Madrid Melbourne
Mexico City Nairobi Paris Singapore
Taipei Tokyo Toronto

and associated companies in
Berlin Ibadan

Published by Oxford University Press, Inc.,
198 Madison Avenue, New York, New York 10016

Library of Congress Cataloging-in-Publication Data
Biodiversity in managed landscapes : theory and practice / edited by Robert C. Szaro, David W. Johnston.
p. cm.
Based on a symposium held in Sacramento, California, July 13–17, 1992.
Includes bibliographical references and index.
ISBN 0-19-507958-2
1. Biological diversity conservation—Congresses. 2. Ecosystem management—Congresses.
3. Biological diversity conservation—Government policy—Congresses. I. Szaro, Robert C.
II. Johnston, David W.
QH75.A1B524 1996
333.95'16—dc20 95-35423

9 8 7 6 5 4 3 2 1

Printed in the United States of America
on acid free paper

This book is dedicated to the memory of Alwyn H. Gentry and Edward T. LaRoe who devoted their lives to the conservation of biodiversity.

FOREWORD

"Where do we go from here" so far as dealing with biodiversity in management is concerned. Fulfilling that task requires knowing where we are in terms of knowledge and professional and public understanding of biodiversity. Based on my experience over the last few years, I have the impression that for many people dealing with the preservation of biodiversity in the real world can be likened to eating small Jello squares with chop sticks. The intent is clear enough but there remains a lot to be worked out in practice.

As a first step it seems essential to describe what biodiversity is. The definition that emerged seems simple enough—to scientists. Biodiversity can be described as the sum of extant life forms with concerns ranging from diversity of alleles to individuals to populations to communities to ecosystems and the inherent processes at each level of concern. And, it came clear that the path to salvation is to retain as much biodiversity as possible over the long term. What is there in the objective of conservation of biodiversity that contributes to the emergence of biodiversity retention as the newest "Holy Grail" for conservation and other associated brands of biologists?

It seems to me that the emergence of preserving biodiversity as a unifying concept for biologists and the public is the present culmination of a trail that reaches back to Darwin's underlying theory of evolution—i.e., the preservation of biodiversity retains the optimum cultural medium in which the ongoing process of evolution thrives best. The preservation of biodiversity allows biologists of all ilks a way to heed the voices of our prophets who cried out in the intellectual wilderness; Aldo Leopold's pleas to be careful in our tinkering with the earth to save all the cogs and wheels is the foremost example. The concept of retaining biodiversity has attributes of a worthy Holy Grail—it is bright and shining and new, mysterious, potentially unifying, and sufficiently complex to accommodate an almost infinite number of quests.

Yet, the concept is familiar enough to allow some comfort that the quest is a worthy one. This seems significant, for as T.S. Eliot observed, "We shall not cease from exploration And the end of our exploring will be to arrive at where we started And know the place for the first time." So, we have arrived through a long journey, back at the idea of evolution and the preservation of the cultural medium to allow that process, that unifying concept, its fullest potential. So, some biologists begin again our course of exploration and adventure. Somewhere in the backs of our minds many of us suspect or feel that the retention of biodiversity is somehow central to the welfare of the earth and its inhabitants.

Several years ago, the Endangered Species Committee ("The God Squad") was convened by Secretary of Interior Manuel Lujan at the request of Cy Jamison, Director of the Bureau of Land Management (BLM). The purpose of their deliberation was to determine if 44 timber sales proposed by the BLM and determined by the U.S. Fish and

Wildlife Service to represent jeopardy to the welfare of the northern spotted owl, a "threatened" sub-species, should be allowed to proceed because of anticipated economic impact of preserving the owl. During that hearing, the attorney representing the timber interests announced to the press that, in the course of the hearing, he would "defrock the high priest of the cult of biology." Several biologists spent a day or so each on the witness stand in the course of that action. Whether or not they were defrocked, frocked, or otherwise folded, mutilated, and spindled is not the point. The point is that they were attacked on the very basis of biological theory that underlaid the accepted conservation strategy with the intent to establish that there is a "cult of biology" that is a threat to people and their welfare.

This may well have been a precursor of actions to come and provides the biological community with a view of the future. It may be well at this point to take time "to see ourselves as others see us." It is not my intent to blaspheme, but let me describe what runs through my head as this issue unfolds.

Charles Darwin brought back what became "holy" tablets from his retreat to the Galapagos that described the theory of evolution that has come to underlie all of the fields of biology. After some wandering in the intellectual wilderness the theory of evolution became the foundation of biology and the "cult of biology" was formed. Many orders came into being—geneticists, ecologists, taxonomists, botanists, zoologists, etc., etc., who devoted themselves to the development of knowledge and theory and its encasement in peer reviewed journals and scholarly treatises. From this base arose new action orders dedicated to bringing this knowledge into a useful state for application in the manipulation of the earth for human benefit. Among these new action orders were agronomists, foresters, aguaculturists, wildlife biologists, and many others.

This reductionist approach, both in the development of basic knowledge and in application in management, became accepted *modus operandi*. Each of these disciplines raised their towers of understanding, of knowledge, of practice, one upon or next to the others. And, each tower was some different in some way from the others. With time, the sounds emanating from each of the towers became more different still. For those striving for coherence and for means of integrating science from whatever source of managing renewable natural resources including retention of biodiversity it was indeed, collectively, a tower of Babel.

Then, the science of ecology began its slow development in the first third of the 20th century. Ecology came to be called,"the subversive science" in that as revealed that, as Barry Commoner put it, everything is connected to everything else and there is no free lunch. This meant that some new integrative approaches of applying scientifically derived information were required to, if not replace, augment the tribal rites of the separate orders of knowledge and practice. From this intellectual ferment arose an idea that just now seems to be coming to fruition—the extension of ethics outside the confines of human relationships between humans and to the land itself. It is at this confluence of the two merging streams of ecology and ethics that we find ourselves today. In these newly mingling streams lies much promise and much danger.

We must look for some unifying mechanism to bring useful order out of seeming chaos. To do that requires a focus on a worthy objective and new language, an *Esperanza*, to facilitate communication. The unifying concept that has emerged has been biodiversity and its preservation and a new applicable *Esperanza* could be contained in the concepts and approaches evolving in the fabricated field of conservation biology. These may, together, constitute a viable means of providing a unification of ideas and knowledge from myriad disciplinary orders in search of a new vision and a new mission.

But, again in stepping back and "seeing ourselves as others see us" let me describe some additional impressions. The new order in the evolution of biological disciplines is developing a new language where comprehension separates the washed from the unwashed. Meetings on biodiversity have been complete with a latin service—"*in situ*, *ex situ*, *alpha*, *beta*, and *gamma*," amen. I have seen transparency and transparency projected that reveals hieroglyphics and other mysterious writings worthy of ancient tombs of the Pharaohs—boxes and arrows and "black boxes" of great wonderment.

High priests swoop in from afar (no doubt on great silver winged birds) to impart wisdom and guidance and insights and then quickly depart to spread the word elsewhere. They talk of world institutes, concerned with biodiversity, of national institutes, and of unifying and coordinating infrastructures and with appropriate staff and resources that transcended boundaries of land ownerships, states and even nations. There are intimations of new covenants between humans and the earth. They even try to impart a vision—a necessary thing, for without vision the people perish.

I am left with the feeling of temptation for those who struggle in the trenches of natural resource management to form themselves into "cargo cults" that struggle day to day scanning the sky for the return of the great silver birds bearing wisdom, institutes, and resources. Indeed, the big silver birds may return—heavily laden with goodies, or much more likely, not.

I have watched the faces of the foot soldiers in the rag tag army that deals with natural resource management day in and day out. I have visited with the grunts at opportune moments as they watch and listen and evaluate the chants for biodiversity conservation. They, in general, have not been encouraged by what they hear. Their comments ranged from complaints about unnecessary complexity, unintelligible presentations, expressions of joy for the institutes and mandates to come, and concerns over the likelihood of extreme political resistance and construction of barriers to achievement of any biodiversity retention. I am reminded of the scene from the *Wizard of Oz* in which Dorothy, Toto, Scarecrow, Tin Woodsman, and Cowardly Lion were traveling more deeply into the gloom of the Enchanted Forest. The further they went into the gloom of the Forest the more frightened they became. They began to chant together, softly at first, "lions and tigers and bears" and, then, louder, "Lions and Tigers and Bears!" and, then, louder, "LIONS AND TIGERS AND BEARS!!" This continued until they were essentially paralyzed by fear.

The goals and concepts applicable to the retention of biodiversity emanating from the Tower of Babel seemed increasingly complex and convoluted. I vaguely remember that some philosopher said (as reported by columnist George Wills) that road-blocks to achievement could be divided into messes and problems. Messes, he said, were too difficult given present circumstances, to clean up. Messes were to be endured. Problems, on the other hand, can be defined, addressed and solved.

I suggest that the foot soldiers leave the messes to the gurus and high priests that come and go on the big silver birds to contemplate and concentrate on the solution of problems with the tools and resources at hand. Forget about the full latin service and drop back on the folk mass suitable for however many instruments and voices are available. Start thinking and being concerned about biodiversity at an appropriate intellectual/philosophical level. Is, intuitively, the preservation of biodiversity a good guide for a general conservation strategy? The answer that emerges again and again is yes. Why? It makes good sense for a number of reasons. Then, we can ask the question—what can the individual or agency do, here and now to come to grips with the "problem" of the place and time? It really doesn't take super sophisticated science nor does it require the blessing of the Cardinals to address problems at the local or regional level.

The word "biodiversity" now regularly appears in the press and in conversations in Congress and even in Presidential politics. That's the good news. The bad news is that it is quite clear that the use of the word is far more prevalent than understanding of the concept. You know there is a long way to go in education of the Nation's leaders when a Congressman asks you if a certain action is apt to be "good for *the* biodiversity?"

If you are convinced that retention of biodiversity is a worthy goal, don't wait for the high priests and gurus to point out *the* way. Do the best you can with what you have at your disposal. Don't wait, start now. Don't be intimidated. More "big picture" thinking will be forthcoming as "biodiversity retention," "ecosystem management," "new forestry," "land-scape scale management," "new perspective's in forestry," etc., move from concept to context. But, in the meantime, be aware and do what you can.

There are repeated calls for inventory and monitoring to a degree that boggles my mind. I am, admittedly, not overly enamored with intensive inventory and monitoring as an expenditure of scarce resources. This is not because the information would not be valuable but because experience tells me that such efforts have, historically, fared badly both in terms of maintaining adequate funding and agency commitment and in making good on any use of the information gathered.

This reservation goes hand in hand with the repeated calls for "getting our stuff together" from the top down. Well, certainly that is quite important over the longer term. At any rate, such things are more the purview of those who walk the halls of power, compose and put through legislation, ply the skies in the big silver birds, and appear to us all on our TV screens and in the press. Messes and their resolution are their purview. Bless them all. For the rest of us, we can respond when called upon but our immediate best contribution is dealing with problems.

At the moment in the United States, our concerns with the retention of biodiversity seem to be subsumed into the attention to the activities occurring in the name of the Endangered Species Act. The controversy swirling around the plans for the conservation of the northern spotted owl, red-cockaded woodpecker, grizzly bear, and the various runs of races of salmon disguises the fact that these are really issues that swish around the retention of biodiversity. To cast these issues, for example, as "jobs vs. owls" trivializes the real issue - the retention of biodiversity. There is a need to move beyond the Endangered Species Act to a broader concept.

I heard two definitions concerning the utility of conservation biology at the conference on biodiversity held in July 1992 in Sacramento, CA. It is well to dwell on those descriptions. They are very different and quite apt to yield different results as our concerns over biodiversity manifest themselves in management applications, laws and regulations, and legal action.

The first of these descriptions was made by Hal Salwasser as follows:

> The long term utility of conservation biology depends on its validity as an objective synthesis of different scientific disciplines and how it applies ecological theory and empirical facts to the practical resolution of complex and difficult choices in conservation policy, *not* on how it co-mingles scientific theory with the value system of its practitioners or aids in the political agenda of particular interest groups.

Contrast that to this statement by Michael Soulé:

> Conservation biology is the mission-oriented servant of radical movements in the cause to save biological diversity.

Which matches your view? There is a big difference.

So, now we return to the beginning, and ask—where do we go from here? A logical first move is to synthesize the applicable information on the principles and theory surrounding biodiversity. This book answers that need by bringing together and synthesizing relevant principles, practices, and policies for biodiversity conservation. Of course, we need more research (what else would a research scientist say?) of both the applied and basic nature.

In ending, I want to emphasize that people, too, are part of ecosystems. People, more than any other factor, influence the degree of biodiversity that will be retained over the next centuries. Satisfying human needs or controlling human demands or both is certainly one key to preservation of biodiversity. I noted a quote in the paper during the last presidential election that emphasizes the point. The sometime Presidential Candidate, Ross Perot, when asked about the controversy that swirls around the economic and social costs of instituting the conservation plan for the northern spotted owl commented that:

> If our children are hungry we will cut every last tree. And, we won't worry about spotted owls except maybe to eat them.

Believe me, it doesn't take sophisticated science nor an intricate mathematical model to surface that truth.

Jack Ward Thomas, Chief
USDA Forest Service
Washington, D.C.

ACKNOWLEDGMENTS

Most of the chapters in this book were originally presented at the Symposium "Biodiversity in Managed Landscapes" held in Sacramento, California, July 13-17, 1992. This milestone symposium was coordinated with The Biodiversity Conservation Strategy Programme organized by the World Conservation Union (IUCN), the United Nations Environment Programme (UNEP), and the World Resources Institute (WRI). It was co-sponsored and funded by a variety of agencies and environmental groups including the USDA Forest Service, USDI Fish and Wildlife Service, USDI National Park Service, USDI Bureau of Land Management, US Department of Defense, US Agency for International Development, Environmental Protection Agency, Smithsonian Institution, World Resources Institute, Ecological Society of America, Environmental Law Institute, International Union of Forestry Research Organizations (IUFRO), World Conservation Union (IUCN), The Nature Conservancy, Society of American Foresters, American Forest Council, Forest Policy Center, Wilderness Society, Center for Natural Lands Management, California Department of Fish and Game, and The Wildlife Society. The Pacific Southwest Research Station of the USDA Forest Service hosted the meeting.

We thank the following members of the symposium organizing committee for their help in putting together the original program: Peter Boice, Department of Defense; Carol Hamilton, USDI Bureau of Land Management; Ted LaRoe, USDI Fish and Wildlife Service; Peter Jutro and Timothy Titus, Environmental Protection Agency; Michael Ruggiero and John Dennis, USDI National Park Service; Thomas Ledig and Constance Millar, USDA Forest Service. Special thanks to the local arrangements committee: Susan Bates, Thomas Ledig, M. Thompson Conkle, Clair Kinlaw, Constance Millar, and Robert Westfall.

We also want to express our thanks to the chapter reviewers for their help in improving this volume: Peter Ashton, Keith Aubry, George Barrowclough, Donald Barry, Charles Bonham, Dan Botkin, William Brown, Stephen Brush, William Burch, Walter Courtenay, Jr., Thomas Crow, Richard Cunningham, Frank Davis, John Day, Jr., James Deacon, D. Decker, B. Driver, F. Dominic Dottavio, K. Elliot, Donald Falk, William Gregg, Jr., Vernon Gilbert, Jr., Dennis Glick, Gary Hartshorn, Carolyn Hunsaker, Malcolm Hunter, Wm. Robert Irvin, David Jablonski, Roy Jemison, Michael Johnson, Rachael Kaplan, Roger Kasperson, Ross Kiester, Russell Lande, Peter Lesica, Hiram Li, Thomas Ledig, Jane Lubchenco, Ariel Lugo, Gary Machlis, Lynn Maguire, Bruce Marcot, Eric Menges, William Michener, Richard Miller, Will Moir, Michael Morrison, Gary Mullins, Robert Naiman, W. Nehlson, Frederic Nichols, Bryan Norton, Barry Noon, Reed Noss, Mike O'Connell, Robert O'Neill, Gordon Orians, David Orr, Bruce Pavlik, Ajith Perera, George

Peterson, E.C. Pielou, David Pimentel, Edwin Pister, Richard Primack, Alan Randall, Martin Raphael, Walter Reid, Willis Rietveld, Michael Ruggiero, Mark Sagoff, Hal Salwasser, Christine Schonewald-Cox, Herbert Schroeder, J. Michael Scott, Owen Sexton, Mark Shaffer, Lee Simons, Clifford Smith, Robert Smythe, Carol Spears, Thomas Spies, Charles Stone, Allan Stoner, Jr., James Sweeney, Jay Stauffer, Jr., Charlotte Taylor, Stanley Temple, Monica Turner, Charles Van Sickle, Richard Walsh, Kenneth Watt, Christopher Whelan, John Wiens, Robert Wiese, David Wilcove, R. Gerald Wright, and John Zasada.

Fritz Knopf and Thomas Ledig provided substantial editorial help in the early stages of this publication. We thank Vince Dong, Anne Harrison, and Kathy Stewart for their help in preparing the camera-ready copy of the text. Special thanks are also given to Kirk Jensen for his patience in seeing this effort through to its completion. Finally, we recognize all the authors for their patience and responsiveness in refining their contributions.

CONTRIBUTORS

Michael J. Bean
Environmental Defense Fund
1875 Connecticut Avenue NW
Suite 1016
Washington, D.C. 20009
U.S.A.

Peter Bridgewater
Australian National Parks and Wildlife Service
GPO Box 636
Canberra, ACT 2601
Australia

Sandra Brown
U.S. Environmental Protection Agency
Environmental Research Laboratory
200 SW 35th Street
Corvallis, Oregon 97333
U.S.A.

Marybeth Buechner
Biology Department
Consumnes River College
8401 Center Parkway
Sacramento, California 95823-5799
U.S.A.

John R. Busby
Australian National Parks and Wildlife Service
GPO Box 636
Canberra, ACT 2601
Australia

Aram Calhoun
Department of Plant Biology & Pathology
University of Maine
Orono, Maine 04469
U.S.A.

James A. Caplan
USDA Forest Service
P.O. Box 96090
Washington, D.C. 20090-6090
U.S.A.

Charles W. Cartwright
USDA Forest Service
Southwestern Region, Federal Building
517 Gold Ave., S.W.
Albuquerque, New Mexico 87102
U.S.A.

Tim W. Clark
School of Forestry and Environmental Studies
Yale University
205 Prospect Street
New Haven, Connecticut 06511
U.S.A.

Walter R. Courtenay, Jr.
Department of Biological Sciences
Florida Atlantic University
Boca Raton, Florida 33431-0991
U.S.A.

Wendy Craik
Great Barrier Reef Marine Park Authority
P.O. Box 1379
Townsville, Queensland 4810
Australia

L.E.M. Crowley
School of Biological Sciences
Macquarie University, NSW 2109
C.J.M. Glover, South Australian Museum
North Terrace, Adelaide, SA 5000
Australia

Blair Csuti
Idaho Cooperative Fish and Wildlife Research Unit
College of Forestry, Wildlife and Range Sciences
University of Idaho
Moscow, Idaho 83843
U.S.A.

Kenneth A. Dahlberg
Department of Political Science
Western Michigan University
Kalamazoo, Michigan 49008
U.S.A.

Francisco Dallmeier
Smithsonian Institution
1100 Jefferson Drive, S.W.
I.C. 3123
Washington, D.C. 20560
U.S.A.

Frank W. Davis
Department of Geography
University of California - Santa Barbara
Santa Barbara, California 93106
U.S.A.

Gary E. Davis
National Park Service
Channel Islands National Park
1901 Spinnaker Drive
Ventura, Californai 93001
U.S.A.

Kerry J. Dawson
School of Environmental Design
University of Georgia
Athens, Georgia 30602
U.S.A.

John G. Dennis
Wildife and Vegetation Division
National Park Service
P.O. Box 37127
Washington, DC 20013-7127
U.S.A.

Arlene T. Doyle
Western Washington University
Huxley College of Environmental Studies
Bellingham, Washington 98225-9181
U.S.A.

Douglas H. Erwin
Department of Paleobiology
NHB 121
National Museum of Natural History
Washington, D.C. 20560
U.S.A.

Deborah J. Forester
Department of Forest Resources
University of Idaho
Moscow, Idaho 83843
U.S.A.

Jerry F. Franklin
College of Forest Resources AR-10
University of Washington
Seattle, Washington 98195
U.S.A.

Jiragorn Gajaseni
Department of Biology
Faculty of Science
Chulalongkorn University
Bangkok 10300
Thailand

Robert H. Gardner
Appalachian Environmental Laboratory
Gunter Hall
Frostburg, Maryland 21532
U.S.A.

Alwyn H. Gentry
Missouri Botanical Garden
P.O. Box 299
St. Louis, Missouri 63166-0299
U.S.A.

Dennis Glick
Greater Yellowstone Coalition
P.O. Box 1874
Bozeman, Montana 59715
U.S.A.

Jeffrey H. Gove
USDA Forest Service
Northeastern Forest Experiment Station
P.O. Box 640
Durham, New Hampshire 03824
U.S.A.

Andrew J. Hansen
Biology Department
Montana State University
Bozeman, Montana 59717
U.S.A.

Lauren L. Hicks
Plum Creek
999 3rd Ave
Suite 2300
Seattle, Washington 98104
U.S.A.

Stephen D. Hopper
Kings Park and Botanic Garden
West Perth, Western Australia 6005
Australia

Patrick Hounihan
Forest Science Department
Oregon State University
Corvallis, Oregon 97331
U.S.A.

Malcolm L. Hunter, Jr.
Department of Wildlife
College of Forest Resources
240 Nutting Hall, University of Maine
Orono, ME 04469-0125
U.S.A.

Gordon Brent Ingram
Department of Forest Resources Management
University of British Columbia and 1230 Hamilton Street #204
Vancouver V6B 2S8
Canada

Walter Ivantsoff
School of Biological Sciences
Macquarie University, NSW 2109
C.J.M. Glover, South Australian Museum
North Terrace, Adelaide, SA 5000
Australia

Robert E. Jenkins, Jr.
2200 Wilson Blvd. #102-218
Arlington, Virginia 22201
U.S.A.

K. Bruce Jones
EPA/ EMSL-LV/EAD
P.O. Box 93478
Las Vegas, Nevada 89193-3478
U.S.A.

Carl F. Jordan
School of Ecology
The University of Georgia
Athens, Georgia 30602
U.S.A.

Graeme Kelleher
Great Barrier Reef Marine Park Authority
P.O. Box 1379
Townsville, Queensland 4810
Australia

Winifred B. Kessler
University of Northern British Columbia
Prince George, British Columbia V2L 4V3
Canada

Russell Lande
Department of Biology
University of Oregon
Eugene, Oregon 97403
U.S.A.

Javier Lobón-Cervia
Department of Evolutionary Ecology
National Museum of Natural Sciences (CSIC)
José Gutierrez Abascal, 2
28006 Madrid
Spain

Lloyd L. Loope
National Park Service
Haleakala National Park
P.O. Box 369
Makawao, Maui, Hawaii 96768
U.S.A.

Ariel E. Lugo
International Institute of Tropical Forestry
USDA Forest Service
Call Box 25000
Guadalcanal Street, Estacion Experimental
Rio Piedras, Puerto Rico 00928-2500

Gary E. Machlis
Department of Forest Resources
University of Idaho
Moscow, Idaho 83843
U.S.A.

Bruce G. Marcot
USDA Forest Service
Pacific Northwest Research Station
P.O. Box 3890
Portland, Oregon 97208
U.S.A.

Rodrigo Matta-Machado
School of Ecology
The University of Georgia
Athens, Georgia 30602
U.S.A.

Jeffrey A. McNeely
IUCN - The World Conservation Union
1196 Gland
Switzerland

Gary K. Meffe
Savannah River Ecology Lab.
Drawer E
Aiken, South Carolina 29802
U.S.A.

Peter B. Moyle
Department of Wildlife & Fisheries Biolgoy
University of California, Davis
Davis, California 95616-5270
U.S.A.

Kenton R. Miller
World Resources Institute
1709 New York Avenue, N.W.
Washington, D.C. 20006
U.S.A.

Gary W. Mullins
School of Natural Resources
The Ohio State University
2021 Coffey Road
Columbus, Ohio 43210
U.S.A.

Dennis D. Murphy
Center for Conservation Biology
Stanford University
Stanford, California 94305-5020
U.S.A.

Gene Namkoong
Department of Forest Science
Faculty of Forestry, McMillan Building
193-2357 Main Hall
Vancouver, British Columbia V6T-1Z4
Canada

Reed F. Noss
Department of Fisheries and Wildlife
Oregon State University
Corvallis, Oregon 97331
U.S.A.

Michael A. O'Connell
The Nature Conservancy
2699 Lee Road, Suite 500
Winter Park, Florida 32789
U.S.A.

Robert V. O'Neill
Environmental Sciences Division
Oak Ridge National Laboratory
P.O. Box 2008, MS: 6038
Oak Ridge, Tennessee 37831-6038
U.S.A.

Ganapati P. Patil
Center for Statistical Ecology and Environmental Statistics
Department of Statistics
The Pennslyvania State University
421 Classroom Building
University Park, Pennsylvania 16802-2112
U.S.A.

Bruce M. Pavlik
Mills College
Department of Botany
Oakland, California 94613
U.S.A.

Scott M. Pearson
Biology Department
Mars Hill College
Mars Hill, North Carolina 28754
U.S.A.

Walter V. Reid
World Resources Institute
1709 New York Avenue, NW
Washington, D.C. 20006
U.S.A.

Brett R. Riddle
University of Nevada Las Vegas
Department of Biological Sciences
4905 Maryland Pk.
Las Vegas, Nevada 89154
U.S.A.

John N. Rinne
Rocky Mountain Forest and Range Experiment Station
Southwestern Forestry Sciences Complex
Northern Arizona University
2500 S. Pine Knoll Dr.
Flagstaff, Arizona 86001
U.S.A.

James A. Rochelle
Weyerhauser Company
WTC 2H2
Takoma, Washington 98477
U.S.A.

Michael A. Ruggiero
National Biological Survey
1849 C Street NW
3660 - MIB
Washington, D.C. 20240
U.S.A.

Hal Salwasser
School of Forestry
The University of Montana
Missoula, Montana 59812
U.S.A.

Raymond Sauvajot
Ecology Graduate Group
Division of Environmental Studies
University of California Davis
Davis, California 95616
U.S.A.

J. Michael Scott
Idaho Coop. Fish & Wild. Research Unit
College of For., Wild. & Range Sciences
University of Idaho
Moscow, Idaho 83843-9637
U.S.A.

Thomas A. Spies
USDA Forest Service
Pacific Northwest Research Station
Forestry Sciences Laboratory
3200 SW Jefferson Way
Corvallis, Oregon 97331
U.S.A.

Peter A. Stine
USDI Fish & Wildlife Service
Carlsbad Field Office
Ecological Services
2730 Loker Avenue West
Carlsbad, California 92008
U.S.A.

Charles P. Stone
National Park Service
Hawaii Volcanoes National Park
Box 52
Hawaii National Park, Hawaii 96718
U.S.A.

Lawrence Stritch
USDA Forest Service
Shawnee National Forest
901 S. Commercial St.
Harrisburg, Illinois 62946
U.S.A.

Robert C. Szaro
USDA Forest Service
Research - 1NW
P.O. Box 96090
Washington, D.C. 20090-6090
U.S.A.

Charles Taillie
Center for Statistical Ecology and Environmental Statistics
Department of Statistics
The Pennslyvania State University
421 Classroom Building
University Park, Pennsylvania 16802-2112
U.S.A.

Alan R. Templeton
Department of Biology
Washington University
St. Louis, Missouri 63130
U.S.A.

Monica G. Turner
Department of Zoology
Birge Hall
430 Lincoln Drive
University of Wisconsin
Madison, Wisconsin 53706
U.S.A.

John D. Varley
P.O. Box 168
Yellowstone National Park, Wyoming 82190
U.S.A.

Dan W. Walton
Australian National Parks and Wildlife Service
GPO Box 636
Canberra, ACT 2601
Australia

Michael D. Watson
School of Natural Resources
The Ohio State University
2021 Coffey Road
Columbus, Ohio 43210
U.S.A.

Neil E. West
Department of Range Science
Utah State University
Logan, Utah 84322-5230
U.S.A.

Jack E. Williams
Intermountain Research Station
316 Myrtle Street
Boise, Idaho 83702
U.S.A.

Mary F. Willson
USDA Forest Service
Forestry Sciences Laboratory
2770 Sherwood Lane (Suite 2B)
Juneau, AK 99801
U.S.A.

Wim J. Wolff
DLO - Institute for Forestry and Nature Research
P.O. Box 23
6700 AA Wageningen
The Netherlands

CONTENTS

II. INVENTORY, MONITORING, AND ASSESSMENT

III. STRATEGIES FOR MAINTAINING BIODIVERSITY

IV. MAINTAINING BIODIVERSITY ON A LANDSCAPE SCALE

V. POLICY AND SOCIAL CONSIDERATIONS

VI. SUMMARY

INTRODUCTION

Efforts to develop the framework Convention on Biological Diversity were launched by the United Nations Environment Programme (UNEP) in May 1989 when the Governing Council of the UNEP unanimously adopted a resolution to begin negotiations on an international convention to conserve biodiversity. The primary tenet behind the development of the Convention was that biodiversity is fundamental to human life, and, as such, it provides the basis for ecosystems, for the regulation of water and the atmosphere, and for agricultural production. Recognition was also growing that the values of biodiversity should be more fully recognized so that the costs of conserving biodiversity and sustainably using its components are seen as better economic and environmental alternatives. The experts involved agreed that the Biological Diversity Convention should include measures and obligations for networks of protected areas, inventories and classification of species and habitats, training of scientific and management personnel, and land-use planning and species protection that safeguard biodiversity.

A growing global consciousness on environmental issues culminated with the United Nations Conference on the Environment and Development (UNCED or also referred to as the "Earth Summit") held in June 1992 at Rio de Janeiro, Brazil. The conservation of biodiversity had never been more hotly debated or discussed than during these times. For several years prior to UNCED, there was an outburst of activities that led to the development of environmentally directed strategies, many focused on conserving biodiversity (IUCN/UNEP/WWF 1991; Keystone Center 1991; Lubchenco et al. 1991; United Nations Environment Programme 1992a, b; WRI/IUCN/UNEP 1992). The genesis of these efforts arose from concerns about accelerating losses of species and continued destruction of the world's ecosystems particularly in the tropics.

UNCED had a tremendously ambitious goal: to make environmental concerns a central issue in international relations (Raeburn 1992). At UNCED and during negotiations for a global "Convention for the Conservation of Biological Diversity," there was almost universal agreement that conservation of biodiversity is a serious global concern. However, there was and continues to remain intense disagreement on how to balance preservation and conservation with social, economic, and sustainable use factors.

Biodiversity is an issue that often brings questions and confusion in any discussion. Intuitively we all have a basic understanding of the meaning of biodiversity. We may not individually be able to provide a textbook definition but there is no real mystery about it. Perhaps the simplest and at the same time most complete definition of biodiversity is that "Biodiversity is the variety of life and its processes" (Keystone Center 1991). When we have concerns for biodiversity we are identifying we have a concern for all life and its relationships. As arguably the most intelligent species on earth we have a responsibility to try as much as possible to preserve all forms of life. Biodiversity means we must expand our view to encompass not just forests but riparian systems, ponds, alpine meadows, grasslands, and deserts as well. This includes more than vascular plants and traditional vertebrate species such as pocket mice, hummingbirds, spiny lizards, and native trout but also bees, butterflies, and fungi.

Because biodiversity is so complex, it is useful to consider the different levels of biodiversity, from the smallest to the largest perspective (Szaro and Shapiro 1990; Szaro and Salwasser 1991).

Genetic diversity—the variety of genetic building blocks found among individual representatives of a species. Although less obvious than species diversity, genetic diversity is crucial to a species survival. A varied gene pool provides for resilience in the face of environmental stresses, a hedge against an unknown future that allows a species to adapt to changing conditions.

Species diversity—the variety of living organisms found in a particular place, for example, the hundreds of different species found in a ponderosa pine forest stand, including plants, birds, mammals, and a host of less visible organisms. This is the level of biodiversity that usually receives the most attention. To address species diversity properly, one must consider how it changes in place and time.

Ecosystem diversity—the variety of species and ecological processes—both their kind and their number—that occur in different physical settings. Examples of ecosystems include an old-growth forest, a riparian area, or Sonoran desert.

Landscape diversity—the geography of different ecosystems across a large area and the connections among them. For example, a landscape interspersed with grasslands, shrublands, meadows, ponds, streams, wetlands, and forests has more diversity than one with a broad expanse of mostly grassland.

Biodiversity may be defined simply but little else about it is simple. Biodiversity includes all life forms, from one-celled fungi, protozoa, and bacteria to complex organisms such as plants, insects, fishes, and mammals. And it includes the millions of processes, pathways, and cycles that link living organisms into populations, ecosystems, and ultimately the entire biosphere that is the planet Earth.

This variety of life is dynamic and constantly changing. Some changes are subtle, such as alterations in the genetic composition of populations; others are dramatic, such as plant succession after fires or floods. Change, which is neither good nor bad in itself, is the reason that biodiversity cannot be assessed simply at a single point in time.

The conservation of biodiversity has profound implications for managing natural resources. At the root of this issue are concerns about the accelerating losses of species because of the cumulative impacts of human activities. While past as well as current rates of species extinctions are difficult to determine, biologists generally agree that we are currently experiencing losses of diversity that are unprecedented in human history and that these losses are a significant problem relating to human welfare. Increasingly, land managers will be asked to assess the impacts of their management practices on biodiversity. This is no small task—almost every land management practice affects biodiversity in some way.

For a long time, concerns about biodiversity have focused on threatened and endangered species of plants and animals, but these represent only one aspect of a larger issue: conservation of the full variety of life, from genetic variation in species populations to the full richness of ecosystems on Earth.

A unique challenge and opportunity is presented by this new perspective. We now are being asked to manage not just for traditional products such as timber, water, game animals and fish, recreation, and minerals but also for complex ecological values, aesthetics, and socio-economic issues of global concern. These include both the maintenance and sustainability of natural ecosystems and also their enrichment in terms of species, size, and relative numbers. But with more intensive use of biological resources there is increasing concern about the influence of management practices. This is especially apparent for those that result in massive habitat alterations and impact both the capability of the land to yield commodities on a sustained basis and on the quality of the environment. Unfortunately, the long-term effects of land management practices on ecosystem productivity and biodiversity have received relatively little systematic scientific attention. The problems are now system-oriented and the thrust of research and management must now deal with a wide range of issues and at a variety of spatial scales.

The need for scientific information as a foundation for resource management decisions continues to increase dramatically. Even a decision to not manage is essentially a management decision. There is a great need to assemble and format new and existing research results into packages that can be used by managers. As such, efforts to synthesize science, and in so doing identify information voids and set priorities, should be increased. Scientists and resource managers need to develop new knowledge and technologies about biodiversity concepts such as population viability, ecosystem restoration, landscape ecology, habitat isolation and fragmentation, analyses of cumulative effects, and genetic variation. Extensive cooperation between researchers, managers and policy-makers is needed to determine research priorities and develop adaptive management strategies.

Biodiversity crosses all traditional resource boundaries and requires a true interdisciplinary approach to develop potential solutions for its conservation. Many of the concepts and theories on biodiversity are still in the formulative stages but this does not prevent our conducting a variety of projects designed for its conservation. Many of the basic building blocks needed for conservation have been under study for years and provide an adequate base for developing a comprehensive conservation program. Threatened, endangered, and sensitive species (TES) programs have long been the focal point of biodiversity concerns on the species level but they represent only a small part of world's biodiversity. Trying to develop a conservation strategy that deals with all species and processes makes it clear that we do not have the resources for intense management programs for all species if only because of the enormity of the problem. For example, there are more than 500 TES species in the United States alone with at least another 1,000 proposed for listing. There is even less hope in tropical environments where the potential numbers of species, particularly insect species, number in the millions (Erwin 1988). Perhaps because biological diversity is so much greater than previously imagined, it has proved difficult to express it as a coherent subject of scientific inquiry (Wilson 1985).

The greatest challenge for managing biodiversity lies in preparing for the disturbances and changes that are a fundamental feature of natural ecosystems. Many of these disturbances are the direct result of human activities. Expanses of pristine forest with their enormous diversity of wildlife, plants, and richness of human cultures are being rapidly converted into vast wastelands that support a few tough, fire-resistant weeds and perhaps some cattle, while people scrounge for food and fuelwood from the newly-degraded soils and sparse vegetation (Gradwohl and Greenberg 1988). Yet, we cannot conserve biodiversity simply by preserving areas and trying to prevent all changes, whether naturally occurring or human-caused. Nor can we conserve biodiversity by trying to maximize diversity on any particular site.

The responsibility for conserving biodiversity belongs to all people and institutions, both public and private (Salwasser 1990). The repeated association of diversity with preservationist approaches leads to the perception that diversity requires wilderness and preserves and cannot be sustained where human activities are prevalent. This view has disastrous consequences. The World Resources Institute in 1986 reported that only 3% of the surface of the earth had major protection status as a national park, nature preserve, wilderness, or sanctuary. It is clear that the major accomplishments on behalf of biological diversity must occur in conjunction with human activities.

To maintain biodiversity, we must ensure that a sufficient amount of each ecosystem is conserved and managed through a variety of actions that address different yet related concerns. And because these actions must occur on lands under a variety of ownerships, goals, and uses, considerations for biodiversity must be blended into a myriad of management approaches (Salwasser 1990). We must strive to understand the functions and processes of natural ecosystems, and make the wise, tough decisions that are necessary to maintain and enhance the productivity of those systems for all purposes and uses. This means that biodiversity, and an understanding of ecosystems, should be the underlying basis for the management of all natural systems.

Our hopes for conserving the world's biodiversity hinge on our being able to instill a sense of the value of biodiversity in present and future generations. This can only be accomplished through well conceived and implemented environmental education programs. Biodiversity provides a new focus for conservation programs. It builds on the rich legacy of forestry, wildlife, range, fisheries, park, reserve, and watershed management. But even the best-conceived conservation program could not have anticipated the enormous changes of the 20th century, with their resultant pressures on natural ecosystems. Current global concerns about losses of ecosystems and their associated species provide an opportunity to build on past accomplishments and chart new courses in conservation for multiple uses and values.

Policy decisions need to balance the needs of society with the best scientific knowledge. To facilitate this, the interface between social, physical-biological, and ecological models must be improved (Szaro 1994). An essential task is to improve the ability to quantify social demands for both consumptive and non-consumptive goods. These demands must then be weighed against the need to maintain ecosystems and their attributes. To provide the scientific basis for ecosystem management requires a combination of (1) basic long-term scientific inquiry; (2) issue- focused, problem solving interdisciplinary research and technology development; (3) scientific support to the ongoing management of lands and resources, such as through ecological classifications, inventories, ecological analyses, and monitoring and evaluation; and (4) integrating science into the policy-making process (Szaro et al. 1994). Only through better cooperation among scientists, managers, and a concerned society can we achieve the goal of a rich and varied natural world for the future. However this issue is viewed, an inescapable fact is that biodiversity is the foundation for all human resources. Future generations are entitled to expect, and will need, an environment as biologically rich as the one inherited by today's generations.

The intent of this book is to look beyond the theory of biodiversity to the principles, practices, and policies needed for its conservation. Its objectives are to provide the scientific basis for understanding biodiversity, document case examples of theory and concepts applied at differing scales, and examine policies that affect its conservation. Conserving biodiversity in parks and reserves alone is not enough. Successful biodiversity conservation is dependent upon policies and strategies that concentrate on patterns and processes in managed landscapes. This book weaves together biological, economic, and social concerns by focusing on actual accomplishments and strategies for conserving biodiversity that address: genetic diversity, species diversity, community diversity, landscape diversity, setting objectives and priorities, inventory, monitoring, and assessment, management strategies, and policy and social considerations.

September, 1994 Robert C. Szaro

LITERATURE CITED

Erwin, T. L. 1988. The tropical forest canopy: the heart of biotic diversity. In *Biodiversity*, eds. E. O. Wilson and F. M. Peter, pp. 123-129. National Academy Press, Washington,

D.C. Gradwohl, J., and R. Greenberg. 1988. *Saving the Tropical Forests.* Island Press, Washington, D.C.

IUCN/UNEP/WWF. 1991. *Caring For The Earth: A Strategy for Sustainability.* IUCN, UNEP and WWF.

Keystone Center. 1991. *Final Consensus Report of the Keystone Policy Dialogue on Biological Diversity on Federal Lands.* The Keystone Center, Keystone, Colo.

Lubchenco, J., A. M. Olsen, L. B. Brubaker, S. R. Carpenter, M. M. Holland, S. P. Hubbell, S. A. Levin, J. A. MacMahon, P. A. Matson, J. M. Melillo, H. A. Mooney, C. H. Peterson, H. R. Pulliam, L. A. Real, P. J. Regal, and P. G. Risser. 1991. The sustainable biosphere initiative: an ecological research agenda. Ecology 72: 371-412.

Raeburn, P. 1992. The Convention on Biological Diversity: Landmark Earth Summit pact opens uncertain new era for use and exchange of genetic resources. *Diversity* 8:4-7.

Salwasser, H. 1990. Conserving biological diversity: A perspective on scope and approaches. *For. Ecol. Manage.* 35:79-90.

Szaro, R. C. 1994. Biodiversity Maintenance. In *Encyclopedia of Energy Technology and the Environment,* eds. A. Bisio and S. G. Boots. Wiley-Interscience, John Wiley & Sons, Inc., New York (In press).

Szaro, R. C., G. A. Lessard, and W. T. Sexton. 1994. Ecosystem Management: An Approach for Conserving Biodiversity. In *Biodiversity: Science and Development—Towards a New Partnership: Panel N°.5—Biodiversity and Agricultural/Pastures/Forestry,* [International Union of Biological Sciences' International Forum, September 5-9, 1994, Paris, France] Biology International (Special Issue) (In press).

Szaro, R. C., and H. Salwasser. 1991. The management context for conserving biological diversity. 10th World Forestry Congress [Paris, France, September 1991]. *Revue Forestiere Francaise, Actes Proceedings* 2:530-535.

Szaro, R., and B. Shapiro. 1990. *Conserving Our Heritage: America's Biodiversity.* The Nature Conservancy, Arlington, Va.

United Nations Environmental Programme (UNEP). 1992a. *Convention on Biological Diversity.* UNEP document No. 92-5509. Nairobi, Kenya.

United Nations Environment Programme. 1992b. *Conference for the adoption of the agreed text of the Convention on Biological Diversity.* No. 92-7825. Nairobi, Kenya.

Wilson, E. O. 1985. The biological diversity crisis. *BioScience* 35:700-706.

World Resources Institute (WRI). 1986. *World Resources 1986: An Assessment of the Resource Base that Supports the Global Economy.* World Resources Institute, New York.

WRI/IUCN/UNEP. 1992. *Global Biodiversity Strategy: Guidelines to Save, Study and Use the Earth's Biotic Wealth Sustainably and Equitably.* WRI, IUCN, UNEP, Washington, D.C., Gland, Switzerland, and Nairobi, Kenya.

I

PRINCIPLES OF BIODIVERSITY

1

THE GEOLOGIC HISTORY OF DIVERSITY

Douglas H. Erwin

Department of Paleobiology
National Museum of Natural History
Washington, D. C.
U.S.A.

Our current biodiversity crisis is frequently compared to the mass extinction events documented by the fossil record, but these events provide both good and bad news. The good news is that life has prospered following mass extinctions far more extensive than that occurring today. The bad news is that such recoveries occur over millions of years. The geologic record provides the only historical information on the response of populations, species, communities, and ecosystems to large-scale environmental perturbations, information with important implications for our response to the current biodiversity crisis. The effects of mass extinction events on biotic diversity have received considerable attention lately, largely because of suggestions that the Cretaceous/Tertiary mass extinction was caused by an extra-terrestrial impact. But geologic history contains abundant examples of biotic responses to other environmental perturbations, among them regional climatic change, oceanic anoxia, drastic reductions in habitat, catastrophic volcanic eruptions, and long-term shifts from icehouse to greenhouse climates. Studies of these events can provide important information on their ecological effects and long-term evolutionary consequences. Investigations of mass extinctions reveal that the current diversity of life is a consequence of the creative effects of past mass extinctions. We must not lose sight of the fact that in an evolutionary context (over tens of millions of years), the species extinction and community reorganization caused by environmental perturbations are a source of innumerable evolutionary opportunities.

Studying the history of life is not unlike dealing with the federal budget—in both cases one deals with numbers far too large for easy comprehension. Just as no one in Washington seems to be quite sure how much a billion dollars is, so paleontologists are not entirely sure how long a million years was. Nonetheless, paleontologists are accustomed to operating on two time scales: deep time, involving processes occurring over millions and tens of millions of years, and briefer interludes of tens to hundreds of thousands of years. There is a tradeoff in resolution between these two scales. In deep time we can study processes and events on a continental or global scale with fairly good resolution over millions of years, but it is difficult to examine events or processes that occur over shorter time scales, or local events over long spans of time. In other words, usually the temporal resolution of the fossil record does not allow us to examine ecological processes on ecological time scales. Nor can we follow the record of species diversity over long time scales. The fossil record is simply too spotty. However, within local areas we can often achieve a high degree of temporal resolution, particularly in the terrestrial realm. The tradeoff here is that these intervals of increased temporal acuity are limited to brief periods of geologic time.

The emphasis of this paper will be on deep time, on the large scale patterns of biotic diversity over millions and tens of millions of years. This paper presents an overview of the history of life, with particular emphasis on the controls on diversity, past biodiversity crises and their aftereffects. The fossil record has several important lessons for understanding the current biodiversity crisis, including the constant nature of extinction, the lengthy period of depauperate ecosystems for millions of years after mass extinction and the fact that mass extinctions actually enhance diversity over even longer times (tens of millions of years).

THE FOSSIL RECORD

Paleontologists have long recognized that the taxonomic diversity of families, genera, and species all appear to be increasing through the Phanerozoic, with a particularly sharp increase since the Jurassic and Cretaceous. However, Raup (1972, 1976 a,b) showed that the same trend exists in the area of exposed rock of different ages and the volume of rock available for sampling (Figure 1.1), suggesting that the more rock available, the more species paleontologists will discover. In addition, paleontologists appear to be drawn to those periods of geologic time with more exposed outcrop, largely the Cretaceous and Cenozoic systems (Signor 1982). How severely do these problems bias the geologic history of diversity? Although it appeared initially that diversity patterns largely reflected the biases of the fossil record, eliminating any possibility of uncovering the true record of diversity, this no longer appears to be true. Studies which correct for biases in the fossil record suggest it is a relatively faithful reflection of the actual pattern of family or generic diversity (Sepkoski et al. 1981; Signor 1990). A more important issue is how well the family or generic diversity pattern reflects species diversity.

Paleontologists would prefer to use species-level data, but have recognized that vagaries of preservation and other problems make this impractical. Consequently, we have compromised on the generic and family level of the taxonomic hierarchy. Although this results in some loss of information, considerable research indicates that generic and family diversity trends are a useful proxy for species diversity from the Cambrian (600 million years ago (Ma)) at least to the Miocene (20 Ma). The most detailed data come from the marine fossil record and J. John Sepkoski, Jr. of the University of Chicago has developed a comprehensive family and generic database for marine taxa through the Phanerozoic (Sepkoski 1992). The family and generic diversity data used in this paper were compiled by identifying the first and last occurrence of a species assigned to a genus or family species. Higher-level clade diversity plots are constructed by charting family diversity through time.

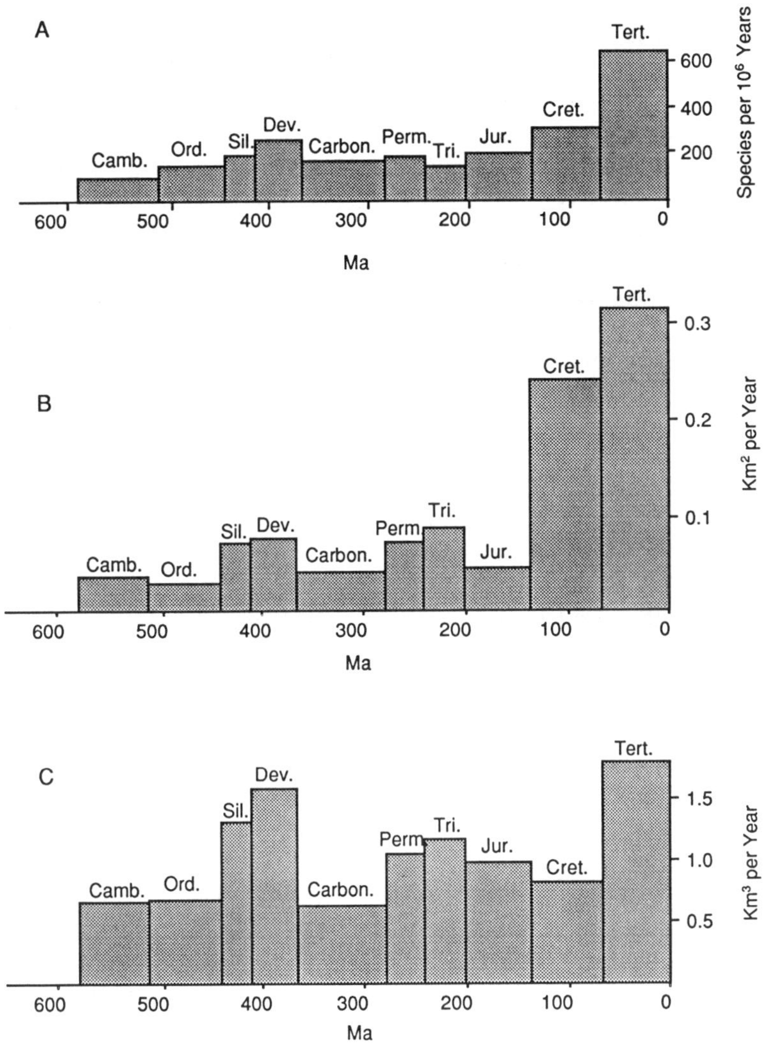

Figure 1.1. Estimated species diversity (A), area of rock exposed (B) and estimated volume of rock preserved (C) during geologic periods of the Phanerozoic. After Raup 1976 a,b.

A BRIEF HISTORY OF PHANEROZOIC DIVERSITY

Life evolved by at least 3.7 billion years ago, but well-skeletonized animals do not appear until about 600 Ma. Molecular evidence suggests that during the intervening three billion years, unicellular prokaryotes and eukaryotes diversified extensively. Unfortunately the fossil record of this time is relatively spotty. Beginning about 700 Ma, a variety of unskeletonized marine animals appeared; these might be related to modern sponges and cnidarians. About 545 Ma, a very rapid diversification of metazoans gave rise, as near as paleontologists can tell, to all animal phyla as well as a large number of bizarre creatures whose taxonomic affinities remain uncertain. This Cambrian Metazoan Radiation is marked by the appearance of skeletonized marine animals, but the record of plankton, soft-bodied fossil assemblages like the Middle Cambrian Burgess Shale Fauna in British Columbia, and an increase in trace fossils demonstrates that all animals, both soft-bodied and skeletonized, underwent an explosive increase in diversity during this interval. The causes of this rapid diversification are the subject of active research. A variety of triggering mechanisms are receiving renewed attention, from changes in the environment, perhaps an increase in atmospheric oxygen, to the acquisition of important developmental innovations.

Following the Cambrian explosion, diversity increased rapidly through the Cambrian and Ordovician before reaching a plateau in the Silurian. In the marine realm, family (and presumably species) diversity remained relatively constant for several hundred million years until the great end-Permian mass extinction. Following the extinction, family diversity quickly rebounded and has continued to increase up to the present. These broad diversity patterns are depicted on a plot of Phanerozoic family diversity (Figure 1.2). As noted above, debate continues over whether this tremendous increase in Cenozoic diversity is real, or simply reflects preservational bias.

Paleontologists have long recognized that the history of marine life was far more complicated than this simple picture might suggest. As early as 1840, John Phillips recognized that major changes occurred in the composition of communities through the Phanerozoic, and he used these changes to divide the Phanerozoic into three eras: the Paleozoic, Mesozoic and Cenozoic. More recently, analysis of marine family diversity has identified three Evolutionary Faunas, sets of classes with similar diversity histories during the Phanerozoic. The earliest is the Cambrian Evolutionary Fauna, dominated by trilobites, which accounts for the bulk of Cambrian marine diversity. The Cambrian Evolutionary Fauna was replaced in the Ordovician by the Paleozoic Evolutionary Fauna. During the reign of this fauna, marine communities were dominated by sessile, epifaunal filter-feeders, particularly crinoid and blastoid echinoderms, articulate brachiopods, and stenolaemate bryozoans. The classes of the Modern Evolutionary Fauna also arose during the Ordovician. This fauna is dominated by bivalves, gastropods, malacostracan echinoderms, echinoids and bony fish, all groups of generally minor importance until after the end-Permian mass extinction. During the post-Paleozoic, these groups all expanded tremendously as a new complex of communities, dominated by mobile, infaunal forms with a greater number of predators, came into their own. Several evolutionary innovations during the Mesozoic and Cenozoic apparently aided the expansion of the Modern Fauna. For example, the development of advanced chemosensory organs led to the appearance of many predatory gastropod clades, greatly expanding the trophic repertoire of the class.

Terrestrial diversity patterns are less accurately known than marine diversity patterns simply because the record is poorer (Figure 1.3). Marine algae date to at least 1.4 billion years ago, but vascular plants first appeared during the Late Ordovician or Silurian. The appearance of the first seed plants during the Devonian was part of an extensive radiation of plant morphologies during this period. Land plant diversity (Figure 1.3A) climbed

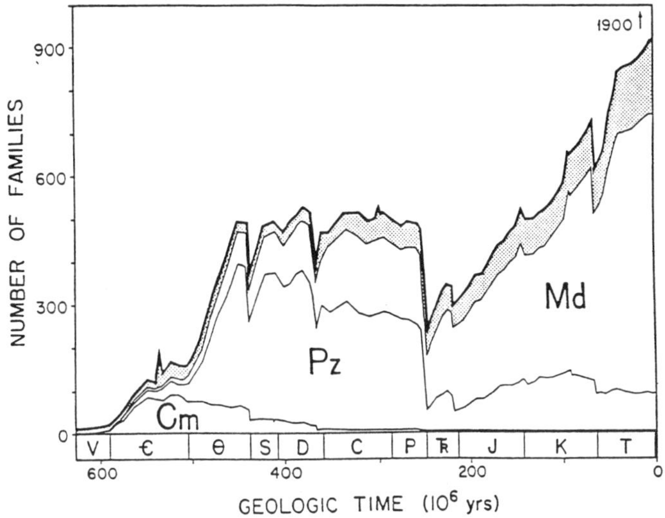

Figure 1.2. Taxonomic diversity of marine families during the Phanerozoic based on Sepkoski's compilation of stratigraphic ranges. The upper curve shows total family diversity in each stratigraphic stage. The number 1900 in the upper right corner is the estimated number of marine families today, including soft-bodied and other rarely preserved forms. The stippled region shows the diversity of poorly-preserved families during the Phanerozoic, whereas the lower two lines indicate the changing diversity patterns of durably skeletonized families assigned to the three Evolutionary Faunas: the Cambrian Evolutionary Fauna (Cm), the Paleozoic Evolutionary Fauna (Pz) and the Modern Evolutionary Fauna (Md). Symbols along the bottom indicate the geologic systems: Vendian (V); Cambrian (€); Ordovician (Θ); Silurian (S); Devonian (D); Carboniferous (C); Permian (P); Triassic (Tr); Jurassic (J); Cretaceous (K); and Tertiary (T). Time scale is from Harland et al. (1982). Figure from Sepkoski (1984).

slowly but steadily until the mid-Cretaceous, when the appearance of the angiosperms unleashed a rapid diversification during the Cenozoic (Knoll et al. 1979; Niklas et al. 1980, 1985). This steady increase conceals considerable turnover in the composition of plant floras between the Carboniferous and Cretaceous. Most of the groups which dominated Carboniferous and Early Permian floras became extinct or were substantially reduced in diversity by the Triassic. They were replaced by groups better adapted to the drier climates of the Mesozoic, which were in their turn displaced by the expansion of flowering plants beginning in the mid-Cretaceous (Lidgard and Crane 1988).

The record of non-marine tetrapods has been analyzed in some detail, revealing a pattern not unlike that of terrestrial plants. The tetrapod record begins in the Devonian and diversity increases gradually through the remainder of the Paleozoic. A series of extinction events during the Permian and Triassic had the cumulative effect of eliminating the earlier amphibian groups and permitting the diversification of therapsids (the incorrectly named "mammal-like reptiles"). In the late Triassic, dinosaurs begin their expansion and successive clades dominate Mesozoic terrestrial ecosystems. The elimination of the dinosaurs at the close of the Cretaceous is one of the few times during

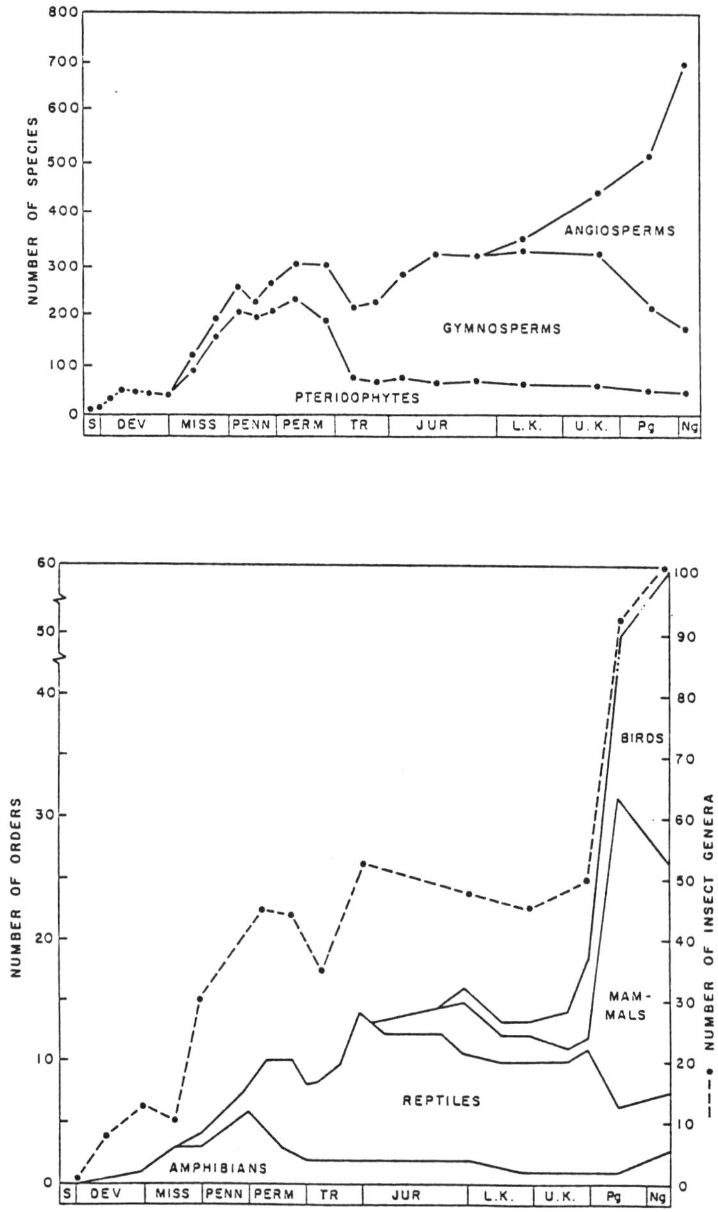

(A)

(B)

Figure 1.3. Taxonomic diversity of terrestrial groups. (A) Species diversity of vascular plants from the Silurian to the present, showing the importance of the various "pteridophyte" groups during the Upper Paleozoic, gymnosperms during the Mesozoic, and the explosive diversification of angiosperms since the Lower Cretaceous. Symbols as in Fig. 2, except Lower Cretaceous (L. K.); Upper Cretaceous (U. K.); Paleogene (Pg); Neogene (Ng). (B) Taxonomic diversity of vertebrate orders (solid lines) and number of insect genera (dotted line). Note the rapid diversification of mammals and birds following the end-Cretaceous mass extinction. Both figures from Niklas (1986).

the Phanerozoic when a vertebrate mass extinction was demonstrably coincident with a mass extinction in the marine realm (Table 1.1), suggesting that for much of the time the two environments are relatively separate. During the Cenozoic, mammals, which first appeared by the Triassic, and birds, expanded greatly in diversity.

The climatic disturbances produced by Late Cenozoic glaciation and de-glaciation severely strained global ecosystems. High latitude faunas and floras generally responded by migration, when possible. When migration routes are cut off, regional extinction may result. For example, Stanley (1986) documented high extinction along the Atlantic margin between the Early and Late Pleistocene when Florida Straits prevented warm-water species from migrating farther south (but see Allmon et al. 1993). Perhaps the most interesting aspect of these climatic perturbations is the differential response of taxa, destroying previous communities and producing radically altered ecosystems (Overpeck et al. 1993). This suggests that communities do not respond as integrated units to climatic change, but rather each species responds differently. This discovery has important implications both for community ecology and understanding the response of communities to future global warming. Several common trends emerge from the examples cited above. First, on a global scale, Cenozoic diversity is far greater than at any other time in the Phanerozoic, and this pattern appears to be real, rather than a result of preservational bias. Second, marine invertebrates, insects, plants, and tetrapods all exhibit greater morphologic innovation early in the history of their clades than later. Indeed, there appears to be a general trend toward declining morphologic breadth as species diversity increases. In other words, the number of variations may increase dramatically, but the number of morphologic themes decreases with time. Finally, and unremarkably, new clades are constantly replacing older clades. What processes control these patterns of diversity?

Table 1.1. Marine generic extinction percentages, estimated species extinction percentages and non-marine tetrapod familial extinction percentages for some of the major Phanerozoic mass extinctions.

Extinction[1]	Age (Ma)	Generic Extinction	Species Extinction	Non-marine Families
Early Oligocene	30	—	—	10%
Late Eocene	35	15%	35±8%	—
End-Cretaceous *	65	47	76±5	14
Late Cenomanian	90	26	53±7	—
End-Jurassic	146	21	45±8	—
Early Jurassic	187	26	53±7	—
Late Triassic *	208	47	76±5	22
End-Permian *	250	84	96±2	58
Late Devonian	367	55	82±4	—
Late Ordovician	439	61	85±3	—

[1] The starred marine and non-marine extinction episodes appear to be correlative. Marine generic and species data from Jablonski (1991) and Sepkoski (1989); non-marine tetrapod extinction data from Benton (1988). Ages in millions of years (Ma).

CONTROLS ON THE GEOLOGIC HISTORY OF DIVERSITY

Paleobiologists and ecologists have divided global diversity into three components: species richness within a single community (alpha diversity), differentiation between communities (beta diversity) and differentiation between regions (gamma diversity) (Sepkoski 1988). Deciphering the relative importance of changes in each of these components is a complex process; indeed paleontologists have chiefly concentrated on the first and last.

Bambach (1977) demonstrated that mean within-habitat species diversity in open marine environments more than doubled from the Lower Paleozoic to the Cenozoic, while at the same time species diversity in high-stress environments remained relatively constant. Diversity in variable nearshore environments increased, but less than in the open marine settings. The most interesting conclusion Bambach reached was that these increases were episodic, although the driving force behind the changes is unclear. The data behind these apparent trends remain open to question and several potential biases have yet to be ruled out, but if the pattern is real (and most paleontologists agree that it is), what caused these changes?

One explanation involves increasing occupation of previously unfilled modes of life (or "guilds") through the Phanerozoic (Bambach 1983). For example, during the Cambrian the only infaunal, shallow-burrowing, active deposit feeders were a few trilobites and some worms assigned to the polychaetes. Today, this guild includes many taxa, including bivalves, many polychaetes, echinoids, holothuroidians, and others. Another possibility is an increase in species packing within communities. Because the "communities" sampled by paleontologists are generally a long-term average of a variety of ecological communities in a local area, it is difficult to test this hypothesis, although Bambach (1983) argues against it. A final possibility involves the increase in the spatial structure of marine communities through the Phanerozoic. Paleozoic marine communities were dominated by epifaunal suspension feeders, and through the Paleozoic the number of tiers (analogous to the number of layers within a forest) and the maximum height of the suspension feeders both increased. Similarly, the number of infaunal tiers increased (Ausich and Bottjer 1982, 1985; Bottjer and Ausich 1986). Whatever the factors involved in alpha diversity, paleontologists have long recognized that these changes are unable to explain more than about 50% of the increase in total diversity between the Cambrian and the present.

Global (gamma) diversity tracks the number of marine provinces quite well, leading to suggestions that an increase in the number of provinces might be sufficient to account for the missing diversity (Valentine and Moores 1970; Valentine 1973; Flessa and Imbrie 1973; Valentine et al. 1978; Schopf 1979). The number of marine provinces was low throughout most of the Paleozoic, rising to about 12-16 during the mid-Permian. Following the formation of the supercontinent of Pangea and the end-Permian mass extinction, the number of provinces dropped to perhaps 3-5. The dispersion of the continents over the past 100 Ma, coupled with the development of a strong north-south temperature gradient during the Cenozoic has generated about 35 provinces in modern seas. However, increasing provinciality apparently cannot account for all of the Cenozoic increase. (Paleontologists have been slow to capitalize on this problem of "missing diversity"; look at all the physicists have accomplished with "missing matter"!)

Climate and sea level (more accurately, the extent of epicontinental seas) are two other factors frequently invoked as controls on global diversity. Depending on topography, an increase in sea-level might appreciably increase available shelf area and convert two-dimensional, linear coastlines into three-dimensional epicontinental seas, with attendant consequences for diversity. However, the impact of sea-level change might have been oversold. Although apparent species diversity tracks sea-level in a variety of clades, for example in Mesozoic ammonites (Kennedy 1977), sea-level changes during the Pleistocene glaciations and de-glaciations had relatively little effect on global diversity (Erwin 1993). Climate is a more powerful control on gamma diversity than sea-level, at least over time scales ranging from hundreds of thousands of years to tens of millions of years. Stanley (1986, 1988) has argued that occasional global cooling events have been a major cause of mass extinctions during the Phanerozoic. Clearly this is not true for some events, such as the end-Permian mass extinction (Erwin 1993), but cooling may have reduced diversity during other intervals.

Beta (between community) diversity has been the subject of far less investigation than either alpha or gamma diversity. In an analysis of changing patterns of beta diversity during the Paleozoic, Sepkoski (1988) concluded that increased habitat selection and overall specialization increased significantly during the Ordovician, and accounted for about 50% of the global diversity increase. He argued, however, that the remaining Ordovician missing diversity stems from the development of new community types (reefs, hardgrounds and other organic buildups) rather than an increase in the number of provinces; this remains a hypothesis, however.

Clearly a number of questions concerning the controls on diversity during the Phanerozoic remain to be resolved, for neither increased within-habitat diversity, proliferation of community types nor increased provinciality seems to account for the tremendous increase in diversity toward the Recent. The fact that both terrestrial plants and vertebrates show similar patterns suggests that similar forces might be acting on them as well. Benton (1990) argues that the history of tetrapod diversity suggests that expansion of adaptive space, specialization and increased provinciality, in that order, are the primary causes of expanding diversity.

MASS EXTINCTIONS

Mass extinctions punctuate the history of life, episodically rearranging the dominance relationships within both marine and terrestrial communities. Some mass extinctions have had only minor effects, others have far more severe effects. Their role in affecting the geologic history of biotic diversity has received considerable attention over the past decade, but in this paper I have emphasized long-term trends in diversification to make the point that mass extinction events are at best a temporary phenomenon. A mass extinction often sets the history of life on a new course, as did the end-Permian mass extinction, but there is little evidence that they reduce diversity over the long-term. Indeed, a growing boody of evidence suggests that the reorganization of communities which accompanies mass extinctions might trigger further diversification.

Some mass extinction episodes might have been short-lived, others spanned at least several million years, but all caused global declines in diversity in a wide variety of groups (Table 1.1). Within the marine realm, paleontologists generally recognize major extinction events: at the close of the Ordovician, Devonian, and Permian, in the Late Triassic, and at the end of the Cretaceous. Some other, smaller events have been recognized as well, particularly during the Mesozoic and Cenozoic, although debate continues over whether these events are truly mass extinctions, and indeed over whether mass extinctions are distinct from the normal processes of background extinction.

The causes of these five major events appear to be quite varied. The end-Ordovician event is associated with global cooling and a pronounced regression of the oceans (Brenchley 1989). Global cooling and oceanic anoxia occurred during the Devonian mass extinction (one of the more poorly understood events), but the actual causes of the extinction remain unclear (McGhee 1988). The end-Permian mass extinction lasted several million years and brought the Paleozoic to a close. In the ecological vacuum after the extinction, a variety of new groups appeared, and groups which had been only minor components of marine ecosystems took more prominent places. Until recently the cause of this extinction has been enigmatic, but appears to be more complex than many other extinctions, involving loss of marine habitat through marine regression, climatic change, and perhaps some level of anoxia (Erwin 1993, 1994). The causes of the end-Triassic extinction are also enigmatic, with various authors suggesting an extra-terrestrial impact, global cooling, marine anoxia, and other agents as the cause. None of these explanations are satisfying, and considerable work remains to be done to understand this event. The end-Cretaceous mass extinction eliminated the dinosaurs, of course, and a growing body of evidence suggests the cause was one or more extra-terrestrial impacts. The causal connection between the impact(s) and the extinction is somewhat unclear, but global cooling and widespread wildfires have recently received the most attention.

The discovery of an periodic cycle of mass extinctions over the past 250 Ma with an approximate cycle of 26 Ma (Figure 1.4) (Raup and Sepkoski 1984, 1986; Sepkoski 1989) suggests a common cause, perhaps extra-terrestrial, to many of these extinctions. Although considerable doubt remains within the paleontological community about the existence of the periodic extinction cycle, most criticisms have been ably rejected by Raup and Sepkoski (Raup 1987; Sepkoski 1989). My own qualms stem from the fact that the end-Permian mass extinction, the event with which I am the most familiar, and the first event in the sequence, has little in common with the end-Cretaceous extinction, the best understood event in the series. No evidence supports an extra-terrestrial impact at the close of the Permian.

The periodicity hypothesis raises several interesting questions for the history of biodiversity, however. How quickly do ecosystems recover from these occasional events? On a human time-scale all recoveries from mass extinctions are extremely slow. If the extinction events are frequent enough, and the periods of recovery long enough, many ecosystems might be essentially in a recovery phase much of the time. Second, as discussed in the following section, if mass extinctions are blind to adaptation, as some studies suggest, much of the history of life might be driven not by adaptation, but by contingent artifacts of history. For example, the probably fortuitous survival of the Permian echinoid *Miocidaris* insured the survival of the clade, a group which otherwise might seem as odd as the many other Paleozoic echinoderm groups which perished during the extinction. This is a very active area of research for paleobiologists and one with profound implications for how we view the history of life.

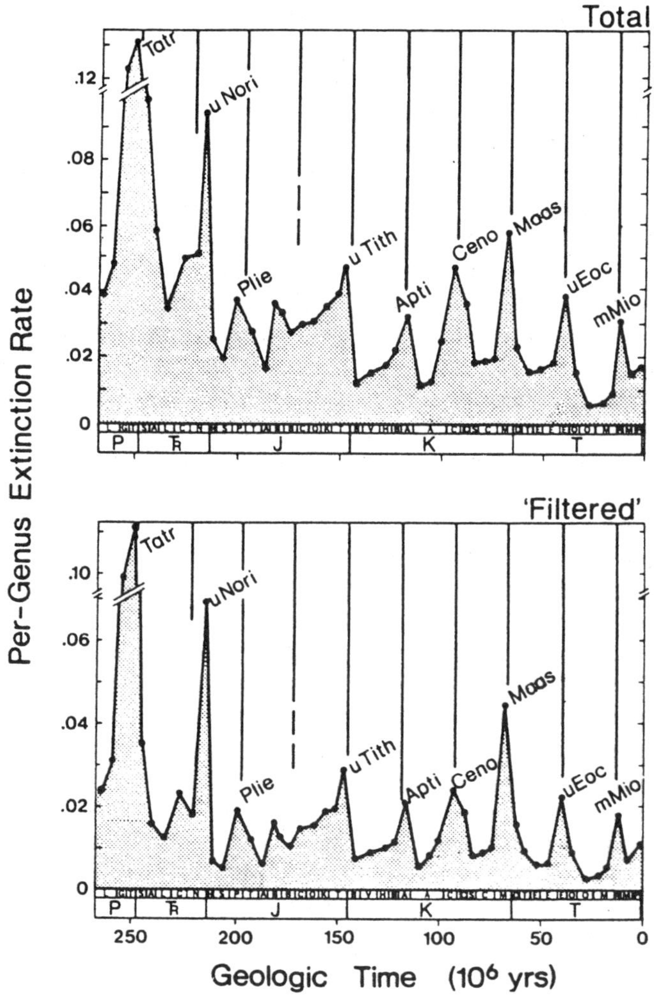

Figure 1.4. Variations in the per-genus extinction rates (extinctions/genus/Ma) for durably skeletonized marine genera from the Permian to the Recent. The vertical lines are best-fit positions of a 26 Ma periodicity. The upper graph includes the entire data set of 17,500 genera. The lower figure, labled "filtered", is based on a subset of 11,000 genera which excludes all genera known only from a single stratigraphic interval. From Sepkoski (1989).

CONCLUSIONS: INSIGHTS FROM THE FOSSIL RECORD

What lessons can we draw from the geologic history of diversity? First, both continuing loss of species ("background extinctions") and mass extinctions are a fact of life. Past mass extinctions appear to be caused by a buildup of environmental stresses which push ecosystems beyond their limits (Jablonski 1991). Considerable fossil evidence demonstrates that these stresses may be local, regional, or global in scope. Second, mass extinctions often result in the replacement of once dominant groups by others that are new, or were previously minor elements of the fauna. It is very difficult, if not impossible, to predict which clades will rise to prominence during post-extinction recoveries. Jablonski (1986, 1989) showed that attributes which favor survival during the period before a mass extinction might be of no help in surviving a mass extinction. In his study of Upper Cretaceous bivalves and gastropods, he found that the synergistic effect of broad geographic range of a species and species-richness within clades, which enhanced survival prior to the end-Cretaceous mass extinction, conferred no advantage during the extinction. This appears not to be true during the end-Permian mass extinction, when the synergistic effect is maintained (Erwin 1989), but during the latter event gastropod clades which survived the extinction relatively unscathed were unable to diversify, while other clades, which suffered greater extinction, became the leading components of the Triassic radiation. Thus the fossil record provides little support for claims that one can identify, before or during an extinction, the likely long-term winners and losers of mass extinction events.

A more general lesson might emerge as well. Despite our uncertainties about the sources of the increase in global diversity during the Phanerozoic, each of the factors discussed above is associated with reduced environmental stress and increased environmental heterogeneity. In contrast, reduced biotic diversity, whether on a local, regional or global scale, is associated with increased environmental stress and reduced environmental heterogeneity.

LITERATURE CITED

Allmon, W. D., G. Rosenberg, R. W. Portell, and K. S. Schindler. 1993. Diversity of Atlantic coastal plain mollusks since the Pleistocene. *Science* 260:1626-1629.

Ausich, W. I., and D. J. Bottjer. 1982. Tiering in suspension-feeding communities on soft substrata throughout the Phanerozoic. *Science* 216:173-174.

Ausich, W. I., and D. J. Bottjer. 1985. Phanerozoic tiering in suspension-feeding communities on soft-substrata. In *Phanerozoic Diversity Patterns: Profiles in Macroevolution*, ed. J. W. Valentine, pp. 255-274. Princeton University Press, Princeton, N.J.

Bambach, R. K. 1977. Species richness in marine benthic habitats through the Phanerozoic. *Paleobiol.* 3:152-167.

Bambach, R. K. 1983. Ecospace utilization and guilds in marine communities through the Phanerozoic. In *Biotic Interactions in Recent and Fossil Benthic Communities*, eds. M. J. S. Tevesz and P. L. McCall, pp. 719-746. Plenum Press, New York.

Benton, M. J. 1988. Mass extinctions in the fossil record of reptiles: paraphyly, patchiness, and periodicity (?). In *Extinction and Survival in the Fossil Record*, ed. G. P. Larwood, pp. 269-294. Clarendon Press, Oxford.

Benton, M. J. 1990. The causes of the diversification of life. In *Major Evolutionary Radiations*, eds. P. D. Taylor and G. P. Larwood, pp. 409-430. Clarendon Press, Oxford.

Bottjer, D. J., and W. I. Ausich. 1986. Phanerozoic development of tiering in soft substrata suspension-feeding communities. *Paleobiol.* 12:400-420.

Brenchley, P. J. 1989. The late Ordovician extinction. In *Mass Extinctions: Processes and Evidence*, ed. S. K. Donovan, pp. 104-132. Belhaven Press, London.

Erwin, D. H. 1989. Regional paleoecology of Permian gastropod genera, southwestern United States and the end-Permian mass extinction. *Palaios* 4:424-438.

Erwin, D. H. 1993. *The Great Paleozoic Crisis: Life and Death in the Permian.* Columbia University Press, New York.

Erwin, D. H. 1994. The Permo-Triassic extinction. *Nature* 367:231-235.

Flessa, K. W., and J. Imbrie. 1973. Evolutionary pulsations: evidence from Phanerozoic diversity patterns. In *Implications of Continental Drift to the Earth Sciences*, eds. D. H. Tarling and S. K. Runcorn, pp. 247-294. Academic Press, London.

Jablonski, D. 1986. Background and mass extinctions: the alternation of macroevolutionary regimes. *Science* 231:129-133.

Jablonski, D. 1989. The biology of mass extinction: a paleontological view. *Phil. Trans. Royal Soc., London*. Ser. B. 325:357-368.

Jablonski, D. 1991. Extinctions: a paleontological perspective. *Science* 253:754-757.

Kennedy, W. J. 1977. Ammonite evolution. In *Patterns of Evolution*, ed. A. Hallam, pp. 251-304. Elsevier, Amsterdam.

Knoll, A. H., K. J. Niklas, and B. H. Tiffney. 1979. Phanerozoic land-plant diversification. *Science* 206:1400-1402.

Lidgard, S., and P. R. Crane. 1988. Quantitative analysis of the early angiosperm radiation. *Nature* 331:344-346.

McGhee, G. R., Jr. 1988. The Frasnian-Famennian extinction event. In *Mass Extinctions: Processes and Evidence*, ed. S. K. Donovan, pp. 133-151. Belhaven Press, London.

Niklas, K. J. 1986. Large-scale changes in animal and plant terrestrial communities. In *Patterns and Processes in the History of Life*, ed. D. M. Raup and D. Jablonski, pp. 383-405. Springer-Verlag, Berlin.

Niklas, K. J., B. H. Tiffney, and A. H. Knoll. 1980. Apparent changes in the diversity of fossil plants. *Evol. Biol.* 12:1-89.

Niklas, K. J., B. H. Tiffney, and A. H. Knoll. 1985. Patterns in vascular land plant diversification: an analysis at the species level. In *Phanerozoic Diversity Patterns: Profiles in Macroevolution*, ed. J. W. Valentine, pp. 97-128. Princeton University Press, Princeton, N. J.

Overpeck, J. T., R. S. Webb, and T. Webb, III. 1993. Mapping eastern North American vegetation change of the past 18 ka: No-analogs and the future. *Geology* 20:1071-1074.

Raup, D. M. 1972. Taxonomic diversity during the Phanerozoic. *Science* 177:1065-1071.

Raup, D. M. 1976a. Species diversity in the Phanerozoic: a tabulation. *Paleobiol.* 2:279-288.

Raup. D. M. 1976b. Species diversity in the Phanerozoic: an interpretation. *Paleobiol.* 2:289-297.

Raup, D. M. 1987. Mass extinction: a commentary. *Palaeontology* 30:1-13.

Raup, D. M., and J. J., Sepkoski, Jr. 1984. Periodicity of extinctions in the geologic past. *Proc. Nat. Acad. Sci. (USA)* 81:801-805.

Raup, D. M., and J. J. Sepkoski, Jr. 1986. Periodic extinction of families and genera. *Science* 231:833-836.

Schopf, T. J. M. 1979. The role of biogeographic provinces in regulating marine faunal diversity through geologic time. In *Historical Biogeography, Plate Tectonics and the Changing Environment*, eds. J. Gray and A. J. Boucot, pp. 449-457. Oregon State University Press, Corvallis, Ore.

Sepkoski, J. J., Jr. 1984. A kinetic model of Phanerozoic taxonomic diversity: III. Post-Paleozoic families and mass extinctions. *Paleobiol.* 10:246-267.

Sepkoski, J. J., Jr. 1988. Alpha, beta, or gamma: where does all the diversity go? *Paleobiol.* 14:221-234.

Sepkoski, J. J., Jr. 1989. Periodicity in extinction and the problem of catastrophism in the history of life. *J. Geol. Soc. London* 146:7-19.

Sepkoski, J. J., Jr. 1992. A compendium of fossil marine animal families. 2nd ed. Milwaukee Public Museum Contributions in Biology and Geology. No. 83. 1-175.

Sepkoski, J. J., Jr., R. K. Bambach, D. M. Raup, and J. W. Valentine. 1981. Phanerozoic marine diversity and the fossil record. *Nature* 293:435-437.

Signor, P. W., III. 1982. Species richness in the Phanerozoic: compensating for sampling bias. *Geology* 10:625-628.

Signor, P. W. 1990. The geologic history of diversity. *Ann. Rev. Ecol. Syst.* 21:509-539.

Stanley, S. M. 1986. Anatomy of a regional mass extinction: Plio-Pleistocene decimation of the Western Atlantic bivalve fauna. *Palaios* 1:17-36.

Stanley, S. M. 1988. Paleozoic mass extinctions: shared patterns suggest global cooling as a common cause. *Am. J. Sci.* 288:334-352.

Valentine, J. W. 1973. *Evolutionary Paleoecology of the Marine Biosphere.* Prentice Hall, Englewood Cliffs, N.J.

Valentine, J. W., T. C. Foin, and D. Peart. 1978. A provincial model of Phanerozoic marine diversity. *Paleobiol.* 4:55-66.

Valentine, J. W., and E. M. Moores. 1970. Plate-tectonic regulation of faunal diversity and sea-level: a model. *Nature* 228:657-659.

2

SPECIES EXPIRATIONS AND CURRENT EXTINCTION RATES: A REVIEW OF THE EVIDENCE

Alwyn H. Gentry[1]

Missouri Botanical Garden
St. Louis, Missouri
U.S.A.

Today we hear much about the "biodiversity crisis," by which is meant the anticipated imminent loss of a significant portion of the earth's species because of human activities. In both the popular and scientific literature one can read statements like this one from Norman Myers (1981): [The devastation of moist tropical forests] "will represent a biological debacle to surpass all others that have occurred since life first emerged 3.6 billion years ago." At a biodiversity conference in Washington, D.C. in September 1986, Wilson concluded that the one sin for which future generations will least forgive us is the extinction of a massive portion of the earth's living creatures (Wilson and Peter 1988).

On the other hand, our highest elected officials recently claimed at the "Earth Summit" in Rio de Janeiro, implicitly if not explicitly, that there is no "biodiversity crisis," or at least none so serious that we should concern ourselves with it (Leggett 1992; Bawa and Wilkes 1992). In addition to Republican politicians, those who claim that the "biodiversity crisis" is grossly exaggerated tend to be disproportionately economists (e.g., Simon 1986). Critics point out that, through time, extinction has been approximately as frequent as speciation, that many species survive deforestation, and that little direct evidence exists for massive human-induced extinction (Lugo 1988; Simon 1986).

Which of these very different opinions is correct? Has a massive and unprecedented spasm of extinctions actually begun or are ecologists "crying wolf" (Mann 1991)? This paper will attempt to review and evaluate the available evidence on the extent of human-caused extinction to date. I will focus mostly on plants, not only because botany is my field of expertise, but also because plants have been somewhat neglected in the "biodiversity crisis" debate and because some of the conservation conclusions suggested by botanical data are rather different from those based on a more zoological viewpoint. Moreover, flowering plants, second only to insects in number of described species (220,000-240,000 species worldwide according to May 1988 and Raven 1976), are of utmost importance to mankind as the ultimate source of our food and the basis of our agriculture and commerce. I will also focus mostly on the tropics, an appropriate prejudice given that most of the earth's plant and animal species are concentrated in this small portion of the planet's land surface.

[1] Deceased

EXTINCTION IN EVOLUTIONARY PERSPECTIVE

It sometimes tends to be overlooked that extinction has been a normal part of the evolutionary process since life first arose. Indeed well over 99% of all species that have evolved have become extinct, some in massive extinction events like the now-famous end-Cretaceous meteor impact (Alvarez et al. 1980; Florentin et al. 1991; Kerr 1991) or the Permian-Triassic boundary when an estimated 96% of the earth's species might have perished (Raup 1979; D. Erwin this volume). However, the great majority of species extirpations through geological history have occurred as part of the "background extinction" associated with the evolutionary process (Erwin this volume), a rate estimated from the fossil record as from 10^{-6} to 10^{-7} per species year, depending on the taxon (Raup 1978, 1988). According to Wilson (1988), estimated current extinction rates are 1,000 to 10,000 times as great as this background extinction.

The putatively impending man-induced extinction episode has often been likened in scale to major extinction episodes of the geologic record like the end-Cretaceous one wherein the dinosaurs disappeared (Simberloff 1986). Is such a comparison justified?

RECORDED HUMAN-CAUSED EXTINCTIONS

Compiled figures by Reid and Miller (1989) and Reid (1992) indicate only 724 recorded species extinctions since 1600 worldwide. This is hardly a biological catastrophe, but the number is likely a gross underestimate. Only two major groups of organisms, birds and mammals are well enough known taxonomically to evaluate directly the current worldwide extinction trends. Extinctions of birds and mammals have clearly increased in recent years and the numbers are accelerating. According to the data collated by Reid and Miller (1989), 42 bird species and 27 mammal species have become extinct since 1900, as compared to 32 birds and 16 mammals in the preceding half century, and 12 and 2 species, respectively, in the half century before that.

Human-caused extinctions have long been with us. Most authors believe that the elimination of 73% of the late Pleistocene genera of North American large mammals and 80% of the South American mammalian genera 11,000 years ago was due to overkill by paleo-Indian hunters, newly arrived across the Bering Strait (Martin 1986; Diamond 1989; Wilson 1992). Extinctions have been especially prevalent on islands. For example, two-thirds of the 50 native Hawaiian land birds at the time of European discovery have since become extinct, and another 50 species are known from fossils to have been extirpated by the Polynesians prior to European arrival (Olson and James 1991; Wilson 1992). Nevertheless, the number of recent bird and mammal extinctions recorded to date does not approach the predicted biological holocaust.

Although it might be argued that the recorded loss of a tenth of the planet's bird species represents a major biological debacle, the documented human-caused extinctions to date do not yet rank with the major extinction events of the geological record. Two major problems arise with this optimistic conclusion. The first is that our knowledge of the earth's biota, especially in the extremely species-rich and poorly known tropical forests, is inadequate to evaluate the size of the planetary biota, let alone how much of it might have already disappeared. The second is that projections based on the theory of island biogeography (MacArthur and Wilson 1967) and the rates of tropical deforestation predict that a major extinction episode looms on the horizon if current patterns continue.

Unfortunately, most organisms are not birds or mammals (or even vertebrates), but are taxonomically less known groups like plants and insects. Only 98 extinct invertebrates and 384 plants are included in Reid and Miller's (1989) list of extinct taxa, fewer than two-thirds of the total documented extinctions, although these groups constitute at least 97% (and probably well over 99%) of the earth's species. By extrapolation from the better known taxa, there seems little doubt that many additional, unrecorded, extinctions of plants and invertebrates must have occurred in recent times.

In this light, it is instructive to consider the current roster of extinct plant species in the United States—118 from Hawaii, 51 from the continental United States, and 3 from Puerto Rico (J. Cablish, Center for Plant Conservation [CPC], pers. comm.), for a total of 172 or 45% of the world's total as reported by Reid and Miller (1989: based on data from the World Conservation Monitoring Centre). An earlier CPC count of 213-228 extinct plant species in the United States (cited by Wilson 1992) included 39 additional taxonomically questionable species (plus a couple that have been rediscovered), and would represent 55% of the world's total. That half of the recorded extinct plants of the entire world should be from the United States, which contains only about 7% of the world's flora, is far more likely to reflect profound ignorance of plant extinctions than exceptional levels of extinction in the United States. Clearly we cannot effectively evaluate extinction levels by simply counting historically recorded extinctions; some kind of estimation must be used.

ESTIMATIONS OF SPECIES DIVERSITY

One of the major problems with estimating future extinctions is that such estimates depend on size of the extant biota. About 1.4 million living species have been described (Ehrlich and Wilson 1991). A decade or two ago, an informed concensus among biologists would have suggested that this represented about one-half of the earth's species. With the advent of new techniques for studying biodiversity, the best estimate for the number of living species has increased dramatically in recent years. For example, far more species of insects live in the canopy of tropical rainforests than had been previously supposed. Based on data from fumigation of rainforest canopies, Erwin (1982) suggested that there might be closer to 30 million than 3 million species, most of them insects. Although the majority of presumed undescribed species might be insects, similarly high estimates are known for other taxonomic groups. For example, the retiring president of the International Mycological Society recently estimated 1.5 million species of fungi, or perhaps even 13.5 million, as compared to the ca. 69,000 described species (Hawkworth 1991). Using new techniques, several thousand undescribed new species of bacteria can be found in a single gram of soil or a single drilling, about as many as the total 4,000 bacterial species described to date from the entire world (Fliermans and Balkwill 1989; Torsvik et al. 1990; Wilson 1992).

Although such data suggest to some that there might be as many as 100 million living species (Ehrlich and Wilson 1991), dissenters consider some of these estimates exaggerated. For example, Gaston (1991) estimated no more than 5-10 million insects, a tenth of these already described. The only reasonably firm conclusion as to the size of the earth's biota is that it is much larger than the current catalog of known taxa and that we know incredibly little about the biota. This basic taxonomic uncertainty leads to major problems in attempting to estimate anticipated extinctions.

In the case of flowering plants to be emphasized here, an intermediate level of taxonomic inventory has been attained. Unlike birds, where only one or two new species are described each year in the entire world (ca. 55 in the past 25 years, T. Parker, pers. comm.), numerous new plants continue to be discovered, especially in the tropics. However, the vast majority of flowering plants have surely been discovered and described already. Current estimates suggest that perhaps 10,000 additional species of flowering plants will eventually be added to the current roster of ca. 240,000 species (Gentry 1982, 1994a). According to Raven's (1976) tabulation, approximately two-thirds of these (ca. 150,000) species are found in the tropics.

Although plants are better known than many other kinds of organisms, their basic inventory and cataloguing are incomplete. This problem is especially acute in parts of the Neotropics where large numbers of new species are being discovered every year. It is estimated (Prance 1977; Mori pers. comm.) that at current rates the Flora Neotropica monograph series will be completed in 300-400 years, long after much of the flora has been lost if the extinction projections are anywhere near accurate. Worse, even new monographs quickly become out of date as new collections continue to be made. For example, Prance (1989) has described 67 additional species of Chrysobalanaceae from collections made during the 15 years subsequent to his original Flora Neotropica monograph of that family in 1972.

A few examples from my own work serve as anecdotal evidence that this problem is of sufficient magnitude to affect significantly the floristic estimations based on compilations from published data. The commonest climbing species in the seasonally inundated forests near well-collected Iquitos, Peru proved to be a new species of *Memora* (Gentry 1981). Only recently has a *Styloceras* species been described, even though it is so common in parts of Manu National Park that it had long been infamous among local primatologists for impeding their progress through the forest (Gentry and Foster 1981). In the same area the commonest Lauraceae (and one of the commonest tree species) was only described in 1986; it turned out to belong to the Asian genus *Caryodaphnopsis*, only recently reported for the Western Hemisphere (Werff and Richter 1985; Werff 1986). This "Asian" genus is now known to include nine neotropical species, at least one of which is an important timber tree (Gentry 1992a). The most important timber tree of the Iquitos, Peru, area, producing 72-74% of the round wood used in construction locally, turns out to be a *Caraipa* new to science (Vasquez 1991). The main construction timber on the lower Ucayali of Amazonian Peru turns out to be a new species of *Haploclathra* so distinctive that we originally thought it might represent a new genus (Vasquez 1993).

In the cloud forests along the base of the Andes, the situation is even worse. At the Rio Palenque Science Center in western Ecuador, 92 of the ca. 1,100 plant species proved new to science (Dodson and Gentry 1978, 1991). On the nearby Centinela Ridge ca. 90 species were undescribed (Gentry 1986; Dodson and Gentry 1991). From such patterns I have estimated that ca. 20% of the species in ecologically unexplored isolated patches of Andean foothill cloud forest might typically be undescribed (Gentry 1986). Overall, the world's estimated 10,000 undescribed flowering plant species seem to be disproportionately concentrated in the Neotropics, especially in such poorly known cloud forest regions as the Choco (Gentry 1982, 1992b).

Despite such glaring inadequacies in the available data base, we might nevertheless conclude that the floristic, if not the faunistic, inventory is near enough to completion to provide a useful baseline against which to evaluate projected extinctions.

DEFORESTATION

The second element in the extinction equation is tropical deforestation. It is now widely accepted that tropical forests are being destroyed very rapidly. With the advent of new satellite imagery and monitoring techniques, estimates of rates of deforestation in the tropics now have a much firmer basis. Although minor differences remain over how altered a forest should be to be counted as eliminated and over whether edge-effect should be considered in evaluating deforestation, the general grim conservation state of the world's tropical forests is no longer seriously disputed (Myers 1991; Whitmore and Sayer 1992). As summarized by Raven (1988) based on FAO statistics, about one-half (46%) of the world's tropical forests had been destroyed by the late 1970s, with 1.1% of the remainder being clear-cut per year. In more graphic terms, an area the size of Ireland is being clear-cut from a total remaining tropical forest area two-thirds the size of Europe. Using a 1990 data set, Reid (1992) calculated tropical deforestation during the 1980s as 10.5 million ha./year, about half in the Neotropics, a third in Australasia, and a sixth in Africa.

From the perspective of extinction, it is significant that tropical deforestation is not uniformly distributed across continents. Take the Neotropics as an example. Tropical America retains the majority of the world's remaining tropical forest (Lanly 1982) but also has the most rapid rate of deforestation (Myers 1980). As recently reported in *Science* (Aldhous 1993), South America lost 6.2 million ha./year of tropical forest during the 1980s, an annual deforestation of 0.6%; the 1.5%/year rate of deforestation in Central America during the 1980s was even higher. What is obscured in such statistics is the regional and local picture. Because Amazonia constitutes the largest block of neotropical rainforest, it contributes disproportionately to the continental figures. Although Amazonian deforestation continues unabated, with estimates ranging from 1.4 million ha./year to 8 million ha./year (Myers 1989, 1991), only about 8% of Brazil's legal Amazonia had been deforested by 1989 (Fearnside 1990). Thus, at the continental scale, the Neotropical forests seem relatively unthreatened.

In contrast to Amazonia, however, some neotropical regions of high species endemism like coastal Brazil and western Ecuador are nearly completely deforested. The Mata Atlantica forest of coastal Brazil has plant species endemism estimated at 73% of a rich flora of 13,000 angiosperms (Gentry 1992a) and has been reduced to highly fragmented remnants, variously estimated as constituting 2-5% of its original area (IUCN/WWF 1982) to 12% (Brown and Brown 1992). Western Ecuador below 900 m, with a vascular plant flora estimated at 6,300 species, 20% of them endemic, retains only 4% of the original forest cover (Dodson and Gentry 1991).

Certain ecological regions are also in especially critical shape. Dry forests, which are floristically distinctive and often more suitable for human habitation than are wetter forests, are a good example. Of the dry seasonal forest of Central America and Mexico only 2%, mostly highly disturbed, remains (Raven 1988). In coastal Ecuador, only 1% of the original dry forest area remains, as compared to the 4% for other forest types (Dodson and Gentry 1991). This dry forest area has a flora of ca. 1,000 species with an estimated 19% of them being endemic, and shows generally little floristic overlap with adjacent moist and wet forest areas (Dodson and Gentry 1991). Even the Brazilian cerrado, which a decade ago was largely intact, has been ravaged by large-scale cattle-ranching and mechanized agriculture. By 1990 it was estimated that 56% of the original 2 million km^2 was in managed use, 37% under cultivation, and only 7% in pristine condition (Dias 1990). Perhaps only in southeastern Bolivia do substantial tracts of largely intact dry forest remain, and these are currently under increasing pressure (Gentry 1994b).

Northern Andean cloud forests are also largely gone. Estimates of extant Andean forest in Colombia suggest that <10% remain (Henderson et al. 1991), perhaps even <5% of the high-altitude upper montane forest (Hernandez 1990); only 4% of the Andean forests of Ecuador's western cordillera remain (Dodson and Gentry 1991). Yet this region is floristically one of the richest in the world with an especially large complement of locally endemic species and a regional endemism rate estimated at 56% (Gentry 1992a). For example, 39-40% of the montane species of Ecuador are endemic to that country alone, as compared to 16% of the lowland flora (Balslev 1988). Circumstantial evidence suggests that evolution might be somehow accelerated in these Andean cloud forests, with speciation in nature occurring in as little as 15 years (Gentry 1989; Gentry and Dodson 1987). The result is exceedingly high local endemism.

We can conclude that large areas of the tropics have been deforested, that deforestation continues unabated, and that many areas of high endemism are approaching total deforestation. What conclusions about extinction can be drawn from this disturbing picture?

PROJECTIONS OF EXTINCTIONS

Suggestions of massive extinction in tropical forests are based on the relationship between species number and area. The projection of extinction rates is largely based on the theory of island biogeography (MacArthur and Wilson 1967) and the observation that large islands have predictably more species than do small islands. According to this theory the diversity of an island's biota depends on an equilibrium between immigration and extinction. Other factors being equal, immigration is higher and extinction lower on large islands, and thus the equilibrium number of species will be higher. This theory has been tested not only by counting species on islands but also directly by defaunation of a series of mangrove islets of different sizes and observing their recolonization (Simberloff and Wilson 1970). Both observation and experiment accord closely with the theoretical expectations.

Habitat islands on the mainland behave similarly to islands, reaching an equilibrium number of species depending on their size and the difficulty of immigration. As tropical forests are reduced in area and fragmented into smaller and smaller habitat islands, they might be expected to contain fewer species at equilibrium, just as would real islands. The details of the relationship between an area of habitat and the number of species expected therein at equilibrium are described by the "species-area curve", usually expressed by the equation $S=cA^z$, where S is the equilibrium number of species, c is a constant, A is area, and z is a parameter that varies from taxon to taxon. The shape of this curve is dependent on the exponent z which has been determined empirically to vary between .15 and .35 in natural systems (Connor and McCoy 1979; Wilson 1992). As a general rule and using a z-value of .30, a reduction of island size by 90% results in a 50% loss of species at equilibrium (Wilson 1992).

Given initial species numbers and rate of habitat loss, an estimate of extinction rate can be calculated for tropical forests as a whole or for given habitat patches. Using this process, based on my (1982) estimate of 90,000 neotropical flowering plant species and assuming a "z" of .25, Simberloff (1986) calculated that 15% of neotropical angiosperms are likely to become extinct by the end of the century if an additional 1 million km^2 is deforested, with 66% extinction anticipated by the middle of the next century if only the ca. 100,000 km^2 of forest contained in reserves remains. For Amazonian birds the comparable projections are 12% and 69% (Simberloff 1986). Similarly, Reid (1992) calculated that for z values between .15 and .35 over the next 25 years an estimated 4-8% of tropical forest species are likely to be committed to extinction, with 17-35% of all tropical forest species committed to eventual extinction by the middle of the next century if deforestation continues at its current rate.

For birds, at least, some evidence exists that reduction in size of a forest patch does lead to the predicted loss of species. For example, 66 bird species have been lost from Barro Colorado Island (15 km^2) subsequent to its isolation by formation of Gatun Lake during construction of the Panama Canal. Many of these were second-growth species lost as a result of loss of second-growth habitat when regrowth of formerly disturbed forest occurred. Thirteen forest species also disappeared, presumably because of reduction of population size on such a small island. The number of species lost is remarkably similar to the 17 species expected from the theory of island biogeography (Terborgh 1974; Wilson 1992).

Similarly, areas like western Ecuador or coastal Brazil which have high endemism and have lost more than 90% of their forests, might be expected to have lost half or more of their species when they reach the MacArthur/Wilson equilibrium. Interestingly, few plant species yet seem to have become extinct in either region (Dodson and Gentry 1991; Gentry 1992). In western Ecuador this might relate to the fact that most of the deforestation has been very recent with insufficient time to reach an equilibrium. Deforestation in coastal Brazil has a longer history, suggesting that small plant populations might be able to persist under non-equilibrium conditions for a longer time than do birds or other animals.

As emphasized by Raven (1988) and by Dodson and Gentry (1991) this gives botanists and conservationists a ray of hope. Even though a major spasm of extinction from tropical deforestation might be imminent, it has not yet happened. Very few plant species have yet been lost. Thus, there might still be time to attempt to save what is left. Moreover, even the very small habitat fragments that are left in such regions would appear capable of playing a significant role in maintaining most of a region's plant species. These last small patches, often written off by zoologists who are more concerned with habitat requirements of top predators, might be of utmost conservation significance from the botanical viewpoint.

Of course, the persistence, for an as yet undetermined time, of most plant species in tiny forest remnants will be irrelevant to conservation if these patches are not conserved. When the last habitat patch of a given forest type is cut, that region's endemics become extinct, no matter how capable they might be of potential survival in reduced populations. There is a documented example of a massive extinction of endemic plants on the Centinela ridge in western Ecuador when the last remnant of the forest was cut in 1986 (Gentry 1986; Dodson and Gentry 1991). We estimated that about 90 angiosperm species, many of them undescribed, became extinct when deforestation of Centinela was completed. Although a few of the supposedly endemic Centinela species still exist on a nearby spur ridge (itself rapidly being deforested) and several more have subsequently been discovered in a surviving patch of cloud forest farther north near Lita, several additional new species have been described from the extant collections. Overall, the number of presumed extinct species remains near our original estimate (Dodson and Gentry 1991). Presumably similar, but undocumented, extinctions of mostly undescribed plant species have occurred as other similarly isolated cloud forest patches have been cleared.

There seems little doubt that the earth is on the verge of major extinctions of plants as well as animals. However, to date, gratifyingly few extinctions of plant species are known to have occurred. We might still have time to save nearly all the plant species of the Neotropics, if we act now to save other forest remnants from sharing the fate that befell Centinela.

ACKNOWLEDGMENTS

I thank D. Falk and J. Cablish of the Center for Plant Conservation for data on North American plant extinctions and the Pew Scholars Program in Conservation and the Environment for supporting preparation of this manuscript.

LITERATURE CITED

Aldhous, P. 1993. Tropical deforestation: not just a problem in Amazonia. *Science* 259: 1390.

Alvarez, L. W., W. Alvarez, F. Asaro, and H. V. Michel. 1980. Extraterrestrial cause for the Cretaceous-Tertiary extinction. *Science* 108:1095-1108.

Balslev, H. 1988. Distributional patterns of Ecuadorean plant species. *Taxon* 37:567-577.

Bawa, K., and H. G. Wilkes. 1992. Who shall speak for biodiversity? *Conserv. Biol.* 6:473-474.

Brown, K. S., and G. G. Brown. 1992. Habitat alteration and species loss in Brazilian forests. In *Tropical Deforestation and Species Extinction*, eds. T. Whitmore and J. Sayer, pp. 119-142, Chapman and Hall, London.

Connor, E. F., and E. D. McCoy. 1979. The statistics and biology of the species-area relationship. *Am. Nat.* 113:791-833.

Diamond, J. 1989. Quaternary megafaunal extinctions: variations on a theme by Paganini. *J. Archaeol. Sci.* 16:167-175.

Dias, B. F. 1990. Conservação da natureza no cerrado Brasileiro. In *Cerrado: Caracterização, Ocupação, e Perspectivas*, ed. M. Pinto, pp. 583-640, Universidade de Brasilia, Brasilia.

Dodson, C., and A. Gentry. 1978. Flora of the Rio Palenque Science Center, Los Rios, Ecuador. *Selbyana* 4:1-628.

Dodson, C., and A. Gentry. 1991. Biological extinction in western Ecuador. *Ann. Missouri Bot. Gard.* 78:1-24.

Ehrlich, P., and E. O. Wilson. 1991. Biodiversity studies: Science and policy. *Science* 253:758-762.

Erwin, T. 1982. Tropical forests: their richness in Coleoptera and other arthropod species. *Coleopterists Bull.* 36:74-75.

Fearnside, P. 1990. The rate and extent of deforestation in Brazilian Amazonia. *Environ. Conserv.* 17:213-226.

Fliermans, C. B., and D. L. Balkwill. 1989. Microbial life in deep terrestrial subsurfaces. *BioScience* 39:370-377.

Florentin, J.-M., R. Maurrasse, and G. Sen. 1991. Impacts, tsunamis, and the Haitian Cretaceous-Tertiary boundary layer. *Science* 252:1690-1693.

Gaston, K. J. 1991. The magnitude of global insect species richness. *Conserv. Biol.* 5:283-296.

Gentry, A. H. 1981. New species and a combination in Palmae, Theaceae, Araliaceae, Apocynaceae, and Bignoniaceae from the Choco and Amazonian Peru. *Ann. Missouri Bot. Gard.* 68:112-121.

Gentry, A. H. 1982. Neotropical floristic diversity: phytogeographical connections between Central and South America, Pleistocene climatic fluctuations, or an accident of the Andean orogeny. *Ann. Missouri Bot. Gard.* 69:557-593.

Gentry, A. H. 1986. Endemism in tropical vs. temperate plant communities. In *Conservation Biology*, ed. M. Soulé, pp. 153-181. Sinauer Press, Sunderland, Mass.

Gentry, A. H. 1989. Speciation in tropical forests. In *Tropical Forests: Botanical Dynamics, Speciation, and Diversity*, eds. L. Holm-Nielsen, I. Nielsen, and H. Balslev, pp. 113-134, Academic Press, London.

Gentry, A. H. 1992a. Tropical forest biodiversity: distributional patterns and their conservational significance. *Oikos* 63:19-28.

Gentry, A. H. 1992b. New nontimber forest products from Western South America. In *Sustainable Harvest and Marketing of Rain Forest Products*, eds. M. Plotkin and L. Famalore, pp. 125-136, Island Press, Washington, D. C.

Gentry, A. H. 1994a. Diversidad floristica de los Neotropicos: tamano y estado de conservacion de la flora. In *Conservacion y Manejo de Recursos Naturales en America Latina*, eds. R. Dirzo, M. Pineros and M. K. Arroyo (in press).

Gentry, A. H. 1994b. A new South American vegetation type: the conservational significance of the dry forest of Santa Cruz, Bolivia. *Conserv. Biol.* (in press).

Gentry, A. H., and C. Dodson. 1987. Diversity and biogeography of neotropical vascular epiphytes. *Ann. Missouri Bot. Gard.* 74:205-233.

Gentry, A. H., and R. Foster. 1981. A new Peruvian *Styloceras* (Buxaceae): discovery of a phytogeographical missing link. *Ann. Missouri Bot. Gard.* 68:122-124.

Hawkworth, D. L. 1991. The fungal dimension of biodiversity: magnitude, significance, and conservation. *Mycol. Res.* 95:641-655.

Henderson, A., S. Churchill, and J. Luteyn. 1991. Neotropical plant diversity. *Nature* 351:21-22.

Hernandez, J. 1990. Las Selvas Andinas de Colombia. In *Selva y Futuro*, eds. J. Carrizosa and J. Hernandez, INDERENA, Bogota, Colombia.

Heywood, V. H., and S. N. Stuart. 1992. Species extinctions in tropical forests. In *Tropical Deforestation and Species Extinction*, eds. T. Whitmore and J. Sayer, pp. 91-118, Chapman and Hall, London.

Kerr, R. A. 1991. Dinosaurs and friends snuffed out? *Science* 251:160-162.

Lanly, J. 1982. Tropical Forest Resources. *FAO Forestry Paper* 30, Rome.

Leggett, J. 1992. Running down to Rio. *New Scientist,* May 1992:38-42.

Lugo, A. 1988. Estimating reductions in the diversity of tropical forest species. In *Biodiversity*, ed. E. O. Wilson, pp. 58-70, National Academy Press, Washington, D.C.

MacArthur, R. H., and E. O. Wilson. 1967. *The Theory of Island Biogeography*. Princeton University Press, Princeton, N.J.

Mann, C. 1991. Extinction: Are ecologists crying wolf? *Science* 253:736-738.

Martin, P. S. 1986. Refuting late Pleistocene extinction models. In *Dynamics of Extinction*, ed. D. Elliott, pp. 107-130. John Wiley and Sons, New York.

May, R. 1988. How many species are there on earth? *Science* 241:1441-1449.

Myers, N. 1980. *Conversion of Moist Tropical Forests*. National Academy of Sciences, Washington, D.C.

Myers, N. 1981. Conservation needs and opportunities in tropical moist forests. In *The Biological Aspects of Rare Plant Conservation*, ed. H. Synge, pp. 141-154, John Wiley and Sons, New York.

Myers, N. 1989. *Deforestation Rates in Tropical Forests and their Climatic Implications*. Friends of the Earth, London.

Myers, N. 1991. Tropical deforestation: the latest situation. *BioScience* 41:282.

Olson, S., and H. James. 1991. Descriptions of thirty-two new species of birds from the Hawaiian Islands, Part 1: Non-Passeriformes; Part 2: Passeriformes. *Ornithol. Monogr.* 45:1-88.

Prance, G. 1972. Chrysobalanaceae. *Flora Neotropica Monogr*. 9:1-406.

Prance, G. 1977. Floristic inventory of the tropics: where do we stand? *Ann. Missouri Bot. Gard.* 64:659-684.

Prance, G. 1989. Chrysobalanaceae (Supplement). *Flora Neotropica Monogr.* 9S: 1-267.

Raup, D. M. 1978. Cohort analysis of generic survivorship. *Palaeobiol.* 4:1-15.

Raup, D. M. 1979. Size of the Permo-Triassic bottleneck and its evolutionary implications. *Science* 206:217-218.

Raven, P. H. 1976. Ethics and attitudes. In *Conservation of Threatened Plants*, eds. J. Simmons et al., pp. 155-179, Plenum Press, New York.

Raven, P. H. 1988. Tropical floristics tomorrow. *Taxon* 37:549-560.

Reid, W. V. 1992. How many species will there be? In *Tropical Deforestation and Species Extinction*, eds. T. Whitmore and J. Sayer, pp. 55-74, Chapman and Hall, London.

Reid, W. V., and K. R. Miller. 1989. *Keeping Options Alive: the Scientific Basis for Conserving Biodiversity*. World Resources Institute, Washington, D. C.

Simberloff, D. 1986. Are we on the verge of a mass extinction in tropical rain forests? In *Dynamics of Extinction*, ed. D. K. Elliott, pp. 165-180. John Wiley and Sons, New York.

Simberloff, D., and E. O. Wilson. 1970. Experimental zoogeography of islands: a two-year record of recolonization. *Ecology* 51:934-937.

Simon, J. 1986. Disappearing species, deforestation and data. *New Scientist,* May 15:60-63.

Terborgh, J. 1974. Preservation of natural diversity: the problem of extinction prone species. *BioScience* 24:715-722.

Torsvik, V., J. Goksoyr, and F. L. Daae. 1990. *Appl. Environ. Microbiol.* 56:782-787.

Vasquez, R. 1991. *Caraipa* (Guttiferae) del Peru. *Ann. Missouri Bot. Gard.* 78:1002-1008.

Vasquez, R. 1993. Una nueva *Haploclathra* (Guttiferae) de la Amazonia Peruana. *Novon 3* (in press).

Werff, H. van der. 1986. A new species of *Caryodaphnopsis* from Peru. *Syst. Bot.* 11:415-418.

Werff, H. van der, and H. G. Richter. 1985. *Caryodaphnopsis* Airy-Sahwe, a genus new to the Neotropics. *Syst. Bot.* 10:166-173.

Whitmore, T., and J. A. Sayer. 1992. Deforestation and species extinction in tropical moist forests. In *Tropical Deforestation and Species Extinction*, eds. T. Whitmore and J. Sayer, pp. 1-14. Chapman and Hall, London.

Wignall, P. 1992. The day the world nearly died. *New Scientist* 133(1805):51-55.

Wilson, E. O. 1988. The current state of biological diversity. In *Biodiversity*, ed. E. O. Wilson, pp. 3-18. National Academy Press, Washington, D. C.

Wilson, E. O. 1992. *The Diversity of Life*. Belknap Press, Harvard University, Cambridge, Mass.

3

THE MEANING OF QUANTITATIVE GENETIC VARIATION IN EVOLUTION AND CONSERVATION

Russell Lande

Department of Biology
University of Oregon
Eugene, Oregon
U.S.A.

Many factors affect the persistence of a species. In the short-run, genetic variability may usually be less important than other factors, but in the long-run, it can play a decisive role, allowing a population to adapt and persist in a changing environment. The genetic constitution of a population and the state of the environment determine its demographic vital rates, i.e., age-specific birth and death rates and dispersal or migration rates, which, in turn, determine the population abundance and spatial distribution through time. However, the overwhelming causes of population reductions and extinctions in many areas of the world are habitat destruction and alteration, and introductions of exotic species of parasites, predators, and competitors.

Much recent conservation effort has focused on genetic events in small populations, such as threatened and endangered species on the verge of extinction. The restoration and maintenance of healthy habitats and ecosystems should be of equal or greater concern to a mature science of conservation biology. Long-term preservation of biodiversity requires understanding not only the demography and genetics of small populations, but also the ecology and evolution of abundant species.

In this paper, I review some genetic and demographic factors affecting the persistence of single populations or species. I outline the genetic basis of inbreeding depression, and the role of single-locus and quantitative (polygenic) variation in adaptive evolution, considering their implications for management of captive and wild populations. An overview of the risks of population extinction from deterministic and stochastic demographic factors follows.

GENETIC VARIABILITY

Inbreeding Depression

It is well known that mating between close relatives tends to result in decreased fitness of the offspring. Decrease in the mean of a character upon inbreeding is known as inbreeding depression. A typical human or *Drosophila* fly carries one or two nearly recessive lethal or sublethal mutations in heterozygous form, whereas in some species of coniferous trees, individuals may be heterozygous for several recessive lethal equivalents (Sorensen 1969; Dobzhansky 1970). In historically large outbreeding populations, rapid inbreeding decreases the mean of characters like body size and fitness components such as seed yield in corn or litter size in mice and pigs by a few to several percent for every 10% increase in the inbreeding coefficient (Ralls and Ballou 1983; Falconer 1989). In most species, close inbreeding as by brother-sister mating results in rapid extinction of the great majority of lineages (Soulé 1980). Habitual inbreeding decreases this hidden (largely recessive) "genetic load" by eliminating recessive lethal and sublethal mutations from the population when they are expressed in homozygous form. Predominantly self-fertilizing plant species, therefore, manifest reduced levels of inbreeding depression (Lande and Schemske 1985).

The genetic basis of inbreeding depression in *Drosophila* has been studied more carefully than in other species. Using special genetic techniques available in *Drosophila*, chromosomes have been sampled intact, without recombination, from wild populations and tested in homozygous and in heterozygous form. The distribution of egg-to-adult viability in chromosomal heterozygotes is approximately normal with a coefficient of variation of several percent of the mean. In chromosomal homozygotes, egg-to-adult viability has a strongly bimodal distribution, with a lethal and sublethal mode and a quasinormal mode that has a smaller mean and larger variance than that of chromosomal heterozygotes (Dobzhansky 1970). Analysis of these and other experiments on recombination and mutation shows that about one-half of the inbreeding depression is caused by rare nearly recessive lethal and sublethal mutations occurring at roughly 5,000 genetic loci, with the remainder attributable to mutations of small, more nearly additive effect (Simmons and Crow 1977).

An important, but often neglected, point is that the amount of inbreeding depression manifested by a population depends on the rate of inbreeding and the opportunity for selection to purge recessive lethal and semilethal mutations. Gradual inbreeding by incremental reductions in population size over several or many generations allows selection to eliminate at least the lethal and sublethal mutations when they become homozygous (Falconer 1989). The component of inbreeding depression from more nearly additive mutations of small effect may be difficult or impossible to purge by inbreeding (Lande and Schemske 1985; Charlesworth and Charlesworth 1987).

Adaptive Evolution

A range of mutational effects is available for the evolution of most characters. Mutations of large effect on a particular trait tend to be deleterious because of their extreme effect or because of side-effects on other traits, and arise much less frequently than mutations of small effect (Muller 1949; Fisher 1958). These observations imply that adaptive evolution by major mutations is most likely to occur under strong sustained selection, powerful enough to overcome their deleterious effects, e.g., following a sudden, extreme environmental change (Wright 1977; Lande 1983). Because major mutations are generally kept at very low frequencies in populations by selection, only large populations are likely to contain them. For example, the evolution of insecticide resistance in large natural populations of insect pests usually is based on single gene mutations that confer a high level of resistance, whereas in small laboratory populations the evolution of the same resistance generally has a polygenic basis (Roush and McKenzie 1987).

To maintain deleterious single-locus mutations, or selectively neutral heterozygosity, with potential adaptive value with respect to future environmental changes, populations may have to be quite large, on the order of 100,000 individuals or more (Lande and Barrowclough 1987). This is well beyond the capacity of most captive breeding programs involving large organisms, and emphasizes the value of conserving large populations in intact ecosystems. Nevertheless, large amounts of genetic variability can be maintained temporarily in small populations, and long-term adaptive evolution can occur based on spontaneous mutation in polygenic characters (see below).

Quantitative genetic variance and rate of evolution.— Quantitative characters are of great importance in adaptive evolution, e.g., in adaptive radiations of morphological traits revealed in the fossil record. Most of the characters of interest to ecologists and evolutionary biologists are quantitative, continuously varying characters influenced by multiple genes and environmental factors. Meristic (countable) traits and discontinuous variation (threshold characters) also can often be analyzed using quantitative genetic methods (Wright 1968; Falconer 1989). Although adaptive evolution can occur by mutations of large effect as described above, the evolution of quantitative characters usually has a polygenic basis between divergent populations within a species or between closely related species (Wright 1968; Lande 1981; Coyne 1985).

The rate of evolution in the mean phenotype in response to selection on a single quantitative character is equal to the additive genetic variance in the character, multiplied by the intensity of directional selection (Lande 1976; Falconer 1989). However, genetic variability is thought not to be a rate-limiting factor in long-term evolution. Long-term rates of evolution, and adaptive radiations, are instead limited by ecological opportunity (Simpson 1953; Wright 1978). The consistency of the short- and long-term views can be seen in a model of the common situation where natural selection acting on a quantitative character (other than fitness itself or age-specific vital rates) favors an intermediate optimum phenotype (Fisher 1958; Wright 1968), or a fluctuating optimum (Charlesworth 1984; Lande 1986). The following model also illustrates that genetic variability can be either detrimental or beneficial, depending on the pattern of environmental change.

Evolution in a changing environment.— Consider the evolution of a single trait, where the expected Malthusian fitness per generation of individuals with phenotype z at time t in a population of size N, denoted as r(z, N, t), can be expressed as their intrinsic rate of increase at low density, $r_0(z, t)$, minus a density-dependent term, f(N), a smooth positive function that approaches 0 as N approaches 0,

$$\text{(1a)} \quad r(z, N, t) = r_0(z, t) - f(N)$$

The intrinsic rate of increase is assumed to be a quadratic function of the deviation of the phenotype from the optimum,

$$\text{(1b)} \quad r_0(z, t) = r_{max} - (s/2)[z - \theta(t)]^2 .$$

rmax is the maximum rate of increase achieved by the optimum phenotype at low population density. θ(t) is the optimum phenotype at time t, and s represents the strength of stabilizing selection toward the optimum phenotype. The intrinsic rate of increase of a population at time t is determined by both the mean phenotype, $Z \equiv Z(t)$, and by the state of the environment, $\theta(t)$. Averaging equation (1a) over the phenotype distribution in the population gives the population growth rate, R, (Wright 1935)

$$\text{(2)}\quad R(Z, N, t) = R_0(Z, t) - f(N) = r_{max} - (s/2)\{[Z(t) - \theta(t)]^2 + V_z\} - f(N)\,.$$

V_z is the total phenotypic variance in the population, which is assumed to be nearly constant, with the genetic component of the phenotypic variance in the population maintained by mutation, recombination, and (in the spatial model below) migration.

The population size obeys the dynamics

$$\text{(3)}\quad dN/dt = R(Z, N, t)\, N\,.$$

N cannot increase from low values if $R_0(Z, t)$ is negative. Thus, a sufficient condition for extinction is that the equilibrium, or average value through time, of the intrinsic rate of increase of the population is negative, $E[R_0] < 0$.

Assuming a normal phenotype distribution, the mean phenotype evolves at a per generation rate equal to the additive genetic variance multiplied by the selection gradient (Lande 1976),

$$\text{(4)}\quad dZ/dt = V_g \partial R/\partial Z = - V_g s[Z(t) - \theta(t)]$$

where V_g is the additive genetic variance in the character, the component of genetic variance contributing to the resemblance between parents and offspring. Evolution is always toward the optimum phenotype at any time.

The influence of various genetic and environmental factors on the persistence of the population can be described in terms of different types of "loads" that decrease the populations's intrinsic rate of increase below its maximum possible value of rmax , as shown in Table 3.1. Increased load from any of the following factors decreases $E[R_0]$ and hence increases the likelihood of population extinction. Each of the loads arises from variance that it creates in the deviations of individual or mean phenotypes away from the optimum. In the quadratic optimum model of selection the different types of loads are additive when occurring simultaneously. Results below are from Lynch and Lande (1993) and Lande (unpubl. data).

In a constant environment, the optimum phenotype is a constant, $\theta(t) = \theta$, and at equilibrium $Z = \theta$. As a component of the total phenotypic variance, V_z, the genetic variance increases the load from stabilizing selection toward an intermediate optimum phenotype. Clearly, when the mean phenotype is at the optimum, any source of phenotypic variance is detrimental because it increases the average deviation of individual phenotypes from the optimum.

Finite population size causes random genetic drift of the mean phenotype away from the optimum phenotype, described by adding a white noise term with mean 0 and variance V_g/N_e to equation (4). The load due to random genetic drift in the mean phenotype, $-1/N_e$ is independent of the strength of stabilizing selection or the amount of genetic variance (Lande 1976), and depends only on N_e, the effective size of the

Table 3.1. Loads caused by different factors in a population under stabilizing selection toward a moving optimum phenotype in a changing environment. Each of the loads decrements the intrinsic rate of increase in the population below its maximum value of r_{max}, increasing the likelihood of extinction.

Factor[1]	Load
Stabilizing selection	$-(s/2)V_z$
Random genetic drift in mean phenotype	$-1/N_e$
Random fluctuation in optimum phenotype	$-(s/2)V_\theta[1 + (s/2)V_g]$
Cyclical change in optimum phenotype	$-(s/4)A^2b^2/[b^2 + s^2V_g^2]$
Sustained directional change in optimum	$-k^2/(2sV_g^2)$

[1] V_z = total phenotypic variance in the population;
s = strength of stabilizing selection toward the optimum phenotype;
V_g = additive genetic component of phenotypic variance;
N_e = effective population size;
V_θ = temporal variance in optimum phenotype, $\theta(t)$;
A = amplitude and $2\pi/b$ = period of cyclical fluctuations in optimum;
k = rate of sustained directional change in optimum.

population. In wild populations, N_e is generally considerably smaller than the actual population size, N, because of uneven sex ratios, variance in reproductive success exceeding the mean, and temporal fluctuations in population size. In artificial populations, N_e can be increased up to 2N by enforcing a 1:1 sex ratio and equalizing reproductive success per individual, keeping the population size constant (Lande and Barrowclough 1987). Random genetic drift also alters other components of the total load by decreasing the amount of additive genetic variance maintained in the population.

Random fluctuation in the optimum phenotype, without autocorrelation, can be described by a white noise process, $\theta(t)$, with mean 0 and variance V_θ per generation. The load due to a fluctuating optimum increases with increasing genetic variance, because any evolutionary response to completely unpredictable fluctuations only increases the average deviation of the mean phenotype from the optimum. The influence of additive genetic variance is small unless there is strong stabilizing selection toward the optimum phenotype ($sV_g > 1$). Positive autocorrelation of environmental fluctuations in the optimum phenotype decreases the load, and negative autocorrelation increases the load. In highly autocorrelated (predictable) environments, additive genetic variance can greatly reduce the load by allowing the mean phenotype to track closely a gradually moving optimum.

Cyclical environments that cause oscillations in the optimum phenotype with amplitude A and period $2\pi/b$, can be described by $\theta(t) = \theta + A\sin(bt)$. With long-period fluctuations (much greater than a generation), $b << 1$, and strong stabilizing selection, additive genetic variance greatly reduces the load by allowing the mean phenotype to track closely the optimum. Under sustained directional change in optimum phenotype at a constant rate, $\theta(t) = kt$, the mean phenotype follows the optimum with an equilibrium lag of $-k/(sV_g)$. Genetic variance is crucial in tracking the optimum since the load caused by this lag is inversely proportional to the square of the additive genetic variance. The importance of additive genetic variance can be seen in the critical rate of environmental change, k_c , which is the maximum value of k that the population can track without going extinct. For a character under weak stabilizing selection, this is nearly proportional to the additive genetic variance,

$$(5)\quad k_c = V_g(2sR_{max})^{1/2}$$

where $R_{max} = r_{max} - (s/2)V_z$. It is also interesting to note that the equilibrium lag and the lag load are inversely proportional to the strength of stabilizing selection, so that the critical rate of environmental change increases with the strength of stabilizing selection.

Comparison of the different loads shows that additive genetic variance is crucial in allowing a population to track an optimum phenotype that is changing in a predictable manner, as in a highly autocorrelated environment, or one with long-period cycles or sustained directional change. However, the population also pays a price for its genetic variance by incurring an increased load from stabilizing selection within generations and from unpredictable environments with no (or negative) autocorrelation.

Evolution in a changing environmental gradient.—The preceding model describes a single population, such as on an island or in a spatially restricted preserve, where the population must either adapt to sustained environmental changes in situ or become extinct. In natural populations without definite spatial boundaries, there is another mechanism of coping with environmental change: individuals can disperse or migrate to follow suitable habitat as it moves in space, e.g., populations can migrate toward higher latitudes or elevations during global warming, tracking the optimum temperature to which they are preadapted.

In a spatial model of this situation developed by Pease et al. (1989), there is a one-dimensional environmental gradient in space, such as a latitudinal or elevational change in temperature, and the entire gradient shifts at a constant rate, e.g., from global climate change. There is a position on the gradient at any time where the intrinsic rate of increase is maximized, and neighboring environments are suboptimal, but in each environment there is a different optimum phenotype. At any moment there is a geographic cline in the optimum phenotype, that moves through space with the changing environment but the slope of the cline does not depend on the rate of environmental change. The population density of the species is concentrated in an area that moves through space to track the changing position of the optimal environment. The mean phenotype at the mode of population density approaches a constant value, despite the fact that in any given locality there is a great deal of evolution as the local population adapts toward a changing locally optimum phenotype, and the local population density waxes and wanes as the optimum environment approaches and passes its location.

In the spatial model, genetic variance actually *decreases* the ability of the population to track the moving environment, by increasing the spatial lag, the distance between the mode of the population density and the position of the optimum environment. Genetic variance, nevertheless, does augment the mean intrinsic rate of increase in the population, by creating a spatial cline in the mean phenotype adapted to the environmental gradient at any time, thereby broadening the geographic range of the population. However, the influence of genetic variance on the intrinsic rate of increase of the population is independent of the rate of environmental change.

Genetic variance is not necessary for the population to track a moving environment, which can occur through an active mechanism of behavioral habitat selection by individual organisms seeking out the habitats to which they are adapted, and through a passive mechanism of habitat fidelity, by local populations multiplying most rapidly in environments to which they are preadapted and declining in relative abundance in suboptimal environments. These mechanisms can both maintain relative evolutionary stasis over long periods of time in changing environments (Lande 1986).

Thus, the meaning of genetic variability in adaptation to changing environments depends on whether the population is restricted to a particular geographic range to which it must adapt to persist, or whether it can disperse to follow the spatial movements of environments to which it is preadapted, maintaining a nearly constant phenotype.

Rates of Restoration of Genetic Variability by Mutation

Following intense inbreeding for several generations at a reduced population size all types of genetic variability will be severely depleted. If the population regains a large size, different types of genetic variability will be replenished at different rates, because they have different rates of mutation and because of the different forms of natural selection on them. Selectively neutral heterozygosity occurring at the rate μ per locus per gamete per generation has an equilibrium value of $H = 4N_e\mu/(4N_e\mu + 1)$ which is approached on a time scale equal to the shorter of $2N_e$ or $1/(2\mu)$ generations (Kimura and Crow 1964). In a large population with typical mutation rates at single loci on the order of $\mu = 10^{-6}$, the time to approach typical levels of heterozygosity in the range of a few to several percent at electrophoretically detectable loci could easily be on the order of tens of thousands or hundreds of thousands of generations. In contrast, inbreeding depression and quantitative genetic variability are regenerated much more rapidly.

In *Drosophila melanogaster* the 5,000 lethal-producing loci mutate at an average rate of 2×10^{-6} per locus generation, giving a gametic mutation rate of 10^{-2} per generation. Lethal and sublethal mutations are not completely recessive, decreasing the fitness of heterozygotes by about 2% on average (Simmons and Crow 1977). In large populations, selection keeps such strongly deleterious mutations at low equilibrium frequencies ($\approx 10^{-4}$), and they approach equilibria on time scales on the order of 100 to 1,000 generations. The expected frequency of partially dominant lethals and sublethal mutations in a population of constant size is approximately independent of population size, provided that N_e is large enough for selection to keep them at low frequency ($N_e >> 2$) (Crow and Kimura 1970). Hence, the equilibrium genetic load and inbreeding depression due to such slightly dominant lethals and sublethals are approximately independent of population size unless N_e is very small.

For a variety of quantitative characters in plants and animals, the additive genetic variance created each generation by spontaneous mutation is roughly 10^{-3} times the environmental (nongenetic) variance in the character, V_e (Lande 1975; Lynch 1988). For any typical quantitative trait, in excess of one in a hundred gametes contains a new mutation with a small effect on the character (Russell et al. 1963; Hoi-Sen 1972). Recent experiments indicate that many of these mutations, especially those with large phenotypic effects, are highly detrimental, and that the rate of production of quasineutral mutational variance per generation, denoted as V_m (which is likely responsible for much of the standing variation in quantitative traits of natural populations) may be an order of magnitude lower, $V_m \approx 10^{-4}V_e$ (Lopez and Lopez-Fanjul 1993a,b).

This rate of production of quasineutral genetic variance is sufficient for mutation to replenish typical levels of additive genetic variance in a large population within a few thousand generations. The expected additive genetic variance maintained in the absence of selection is $2N_eV_m$. Even characters under a typical strength of stabilizing selection (a stabilizing selection load of a few to several percent) with effective population sizes on the order of several hundred can maintain enough additive genetic variance for substantial adaptive evolution (Lande and Barrowclough 1987). Therefore, lack of electrophoretic variability in rare species (e.g., Barrowclough and Gutiérrez 1990) does not preclude inbreeding depression or long-term adaptive evolution in quantitative characters if N_e is in the hundreds or in the thousands.

The N_e = 50/500 Rule

Franklin (1980) and Soulé (1980) proposed that an effective population size $N_e = 50$ is sufficient to avoid the short-term deleterious effects of inbreeding depression. This is supported by extensive experience in applied animal and plant breeding. For populations at or above this effective size, the average inbreeding coefficient for selectively neutral genes increases at a rate of at most $1/(2N_e) = 1\%$ per generation. They also proposed that $N_e = 500$ is sufficient for long-term maintenance of genetic variability in quantitative characters, and this number was quickly adopted as the basis of several management plans for captive as well as wild populations, including the northern spotted owl (USDA Forest Service 1984).

The Franklin-Soulé number is based on a theoretical argument using measured rates of production of additive genetic variance by mutation in quantitative characters, assuming that all of the spontaneous mutational variance is selectively neutral, and that for a typical quantitative character half of the phenotypic variance is heritable (cf. Lande and Barrowclough 1987). If we incorporate the finding that only 10% of the spontaneous mutational variance is quasineutral, the Franklin-Soulé number would have to be increased by a factor of 10, up to $N_e = 5{,}000$. Analysis of the risk of population extinction from fixation of new mildly deleterious mutations also suggests that effective population sizes on the order of a few thousand may be necessary to avoid significant risks of extinction (Lande 1994).

Such numbers derived from simple genetic considerations may be appropriate for captive populations in which demographic factors are largely under control, but should not be applied to wild populations without careful consideration of demographic risks of extinction. For example, the original spotted owl management plan based on $N_e = 500$ was shown to be inadequate on demographic grounds (Lande 1988).

The 50/500 rule has also been abused to suggest abandonment of conservation efforts for populations that do not meet these sizes. This too is naive, in that mutation will eventually replenish genetic variability lost during temporary periods of small population size, and even populations with N_e on the order of 100s can maintain substantial quantitative genetic variation for adaptive long-term evolution, if the population can survive demographic threats of extinction. For instance, small stock populations of

Drosophila fixed for major visible mutations have been observed to evolve gradually over a period of years toward the wild phenotype through the cumulative action of selection on polygenic modifiers (Lewontin 1974)

The 90%-200 Year Goal for Captive Breeding

Soulé et al. (1986) suggested a different goal for captive breeding plans: to preserve in the captive population 90% of the genetic variability originally present in the (wild) population for a period of 200 years. They considered the number of individuals founding the captive population, the population growth rate in captivity, and the generation time of the species to calculate the final or equilibrium population size necessary to meet the goal. In many cases, particularly for species with long generation times, the management goal can be met with moderate or low population size, thereby allowing more efficient utilization of limited resources for captive breeding. However, for populations with short generation times (undergoing many generations of random genetic drift in 200 years), their model required enormous population sizes. This occurred because their analysis did not allow for mutation.

By accounting for polygenic mutation in quantitative characters, a substantial reduction in population size can be justified for species with short generations (Lande unpubl. data). To meet the goal of preserving 90% of the original genetic variance, the effective population size, if constant through time, would never have to be larger than 4,500, i.e., 90% of the modified Franklin-Soulé number ($N_e = 5{,}000$), based on mutation parameters and genetic variability in a typical quantitative character.

Further reductions in population can be accomplished by immigration from the wild into the captive breeding stock, or by subdivision of the population into noninterbreeding subpopulations if the increased inbreeding within subpopulations can be tolerated. For characters involved in adaptation to the captive environment, it may not be possible to meet the management goal even with an indefinitely large population, although a population with a modest effective size may maintain almost as much additive genetic variance as an infinite population.

Selective Breeding and Harvesting of Wild Populations

Genetic problems in forestry and fishery management involve evolution in intensively managed populations that live largely in the wild but are subject to selective breeding and harvesting.

Intense harvesting can have a strong selective effect on life history parameters such as size and age at maturity, and characters that affect the chance (if any) of escaping harvest. Ricker (1981) documented such changes in harvested populations of Pacific salmonids, and similar changes have been observed in sperm whales that were formerly overharvested (Berzin 1972). Recent declines and extinctions of salmonid runs in the Pacific northwest are likely to be caused in part by past inbreeding and use of nonlocal germplasm in hatcheries, and adaptation to early growth and reproduction in hatchery environments, but the relative importance of genetic factors in comparison to demographic factors, including habitat deterioration caused by dams and logging, overfishing, and parasites introduced with hatchery populations, is not known (Ryman and Utter 1987; Hindar et al. 1991).

In forestry, the long generation time of trees reduces problems of evolutionary responses to harvesting, but exacerbates the problem of selective breeding to anticipate long-term environmental changes such as global warming. One method of accelerating responses to artificial selection that is being investigated in various crop plants, including trees, involves the use of molecular genetic polymorphisms linked with genetic loci affecting characters of economic importance (Lande 1992). Past restrictions of the breeding population, use of nonlocal germ plasm, and planting of exotic or nonlocally dominant species, contributed to current declines in forest health (Ledig 1988). However, past errors of genetic management of forest crop species may have less overall ecological significance for the entire forest community than clearcutting and overharvesting.

DEMOGRAPHIC FACTORS

Demographic factors are often of overwhelming importance among the various risks of population extinction, as when extreme habitat destruction or fragmentation causes the extinction of a population regardless of its genetic composition. Following is an outline of deterministic and stochastic factors in demography that constitute some of the main threats of extinction.

Deterministic Factors

Allee effects arise from cooperative effects among individuals in a population that increase individual fitness, e.g., in social groups, feeding flocks, chemical and physical conditioning of environments, and mate finding. These can determine a minimum size or density of population necessary for persistence (Allee et al. 1949; Dennis 1989).

Edge effects usually refer to ecological deterioration of suitable habitat for a species near a boundary with less suitable or unsuitable habitat. Another meaning of edge effects is demographic. Dispersal of individuals across a boundary between suitable and unsuitable habitat determines a critical size of a patch of suitable habitat necessary for population persistence. For populations with random dispersal and a low mean intrinsic rate of increase per generation, the critical patch size may be several or many times larger than the typical individual dispersal distance per generation (Okubo 1980; Pease et al. 1989).

Local extinction and colonization allows regional persistence of a population when local populations or their habitats are unstable or ephemeral, if the rate of colonization at least equals the rate of local extinction (Levins 1969). Many species naturally have this type of metapopulation structure, and many others are acquiring it through artificial habitat fragmentation. The population of a territorial, habitat specialist can be analyzed as a metapopulation by identifying the spatial unit as the individual territory and local extinction with the death of a resident individual. This allows age-structured demography and searching behavior during dispersal to be incorporated into a metapopulation model (Lande 1987, 1988). Two conclusions from this analysis are of importance in conservation:

(1) With further habitat destruction or fragmentation, the equilibrium population size for a given amount of suitable habitat declines faster than the amount of suitable habitat. (2) An extinction threshold occurs in the form of a minimum amount of suitable territory in a region that is necessary for population persistence. Vacant patches of suitable habitat may be as important as occupied ones for regional persistence of a population undergoing local extinction and recolonization: continued destruction of temporarily unoccupied habitat will lead to regional decline and extinction of the population.

Stochastic Factors

Relative risks of population extinction from different stochastic demographic factors can be assessed by comparing asymptotic scaling relationships describing how the average time to extinction, T, scales with the carrying capacity of a population, K (Leigh 1981; Shaffer 1987). Lande (1993) added single stochastic factors to a simple deterministic model of exponential population growth up to K. A critical parameter affecting the extinction dynamics is the expected long-run growth rate of a population below K, including stochastic factors. For populations in which the expected long-run growth rate is positive, with initial population size at or near K, the results are as follows.

Demographic stochasticity is caused by chance realizations of probabilities of survival and reproduction among individuals in a finite population. The demographic variances of vital rates and of population growth rate are inversely proportional to population size. Hence, demographic stochasticity is most important in small populations. With demographic stochasticity alone, T scales almost exponentially with carrying capacity, in proportion to e^{aK}/K (Leigh 1981; Goodman 1987b). The constant a depends on the average population growth rate below K and on the variance in Malthusian fitness among individuals.

Environmental stochasticity arises from nearly continuous perturbations of small or moderate magnitude that similarly affect the survival and reproductive rates of all individuals. It is, therefore, of potential importance in populations of all sizes. With environmental stochasticity, T scales as a power function of carrying capacity, proportional to K^c. For T to increase faster than linearly with K (for $c > 1$) requires only that the mean population growth be larger than its environmental variance. This result is in agreement with formulas in Leigh (1981), but contradicts the claim by Goodman (1987a,b) that under environmental stochasticity the mean persistence time of a population increases at most linearly with K.

Random catastrophes are large environmental perturbations that produce sudden major reductions in population size at random times. With random catastrophes, T again scales as a power function of carrying capacity, proportional to K^c. Here the constant c depends on the frequency and magnitude of catastrophes, and for infrequent catastrophes it is easily possible for T to increase faster than linearly with K. This contradicts the assertion of Shaffer (1987), based on a random catastrophe model of Ewens et al. (1987), that under random catastrophes the mean persistence time of a population increases only logarithmically with carrying capacity. Logarithmic scaling of T with K is characteristic of declining populations in which the expected long-run growth rate is negative, as in the model of Ewens et al. (1987).

Thus, prevailing views (Goodman 1987a,b; Shaffer 1987) regarding demographic extinction risks for populations in fluctuating environments are too pessimistic. A small population subject to environmental stochasticity or random catastrophes alone can persist for a very long time if its expected long-run growth rate is positive. Unfortunately, for populations of many species, deterministic or stochastic demographic and genetic factors act or interact to reduce the expected long-run growth rate to near zero or negative, putting them critically at risk of extinction.

CONCLUSIONS

Genetic considerations are often of critical importance in management of harvested populations and in the conservation of biodiversity, particularly for endangered species in small populations. Risks of population extinction from genetic factors, including inbreeding depression, fixation of new deleterious mutations, and lack of genetic adaptability in a changing environment, are still not completely understood, especially in terms of their interactions with each other and with demographic factors. Effective management of living resources and endangered species should be conservative in allowing ample margins for our uncertainties. Long-term management goals should involve the restoration and maintenance of healthy ecosystems, rather than maximum exploitation of resources and maintenance of endangered species at minimum viable population sizes.

LITERATURE CITED

Allee, W. C., A. E. Emerson, O. Park, T. Park. and K. P. Schmidt. 1949. *Principles of Animal Ecology.* Saunders, Philadelphia, Pa.

Barrowclough, G. F., and R. J. Gutiérrez. 1990. Genetic variation and differentiation in the spotted owl (Strix occidentalis). *Auk* 107:737-744.

Berzin, A. A. 1972. *The Sperm Whale.* Keter Press, Jerusalem.

Charlesworth, B. 1984. Some quantitative methods for studying evolutionary patterns in single characters. *Paleobiol.* 10:308-318.

Charlesworth, D., and B. Charlesworth. 1987. Inbreeding depression and its evolutionary consequences. *Ann. Rev. Ecol. Syst.* 18:237-268.

Coyne, J. A. 1985. Genetic studies of three sibling species of *Drosophila* with relationship to theories of speciation. *Genetical Res.* 46:169-192.

Crow, J. F., and M. Kimura. 1970. *An Introduction to Population Genetics Theory.* Harper and Row, New York.

Dennis, B. 1989. Allee effects: population growth, critical density, and the chance of extinction. *Natural Resource Modeling* 3:481-538.

Dobzhansky, Th. 1970. *Genetics of the Evolutionary Process.* Columbia University Press, New York.

Ewens, W. J., P. J. Brockwell, J. M. Gani, and S. I. Resnick. 1987. Minimum viable population sizes in the presence of catastrophes. In *Viable Populations for Conservation*, ed. M. E. Soulé, pp. 59-68. Cambridge University Press, New York.

Falconer, D. S. 1989. *Introduction to Quantitative Genetics.* 3rd ed. Longman, New York.

Fisher, R. A. 1958. *The Genetical Theory of Natural Selection.* 2nd ed. Dover, New York.

Franklin, I. R. 1980. Evolutionary changes in small populations. In *Conservation Biology. An Evolutionary-Ecological Approach*, eds. M. E. Soulé and B. A. Wilcox, pp. 135-149. Sinauer Associates, Sunderland, Mass.

Goodman, D. 1987a. The demography of chance extinction. In *Viable Populations for Conservation*, ed. M. E. Soulé, pp, 11-43. Cambridge University Press, New York.

Goodman, D. 1987b. Consideration of stochastic demography in the design and management of biological reserves. *Natural Resource Modeling* 1:205-234.

Hindar, H., N. Ryman, and F. Utter. 1991. Genetic effects of cultured fish on natural fish populations. *Can. J. Fish. Aquat. Sci.* 48:945-957.

Hoi-Sen, Y. 1972. Is subline differentiation a continuing process in inbred strains of mice? *Genetical Res.* 19:53-59.

Kimura, M., and J. F. Crow. 1964. The number of alleles that can be maintained in a finite population. *Genetics* 49:725-738.

Lande, R. 1975. The maintenance of genetic variability by mutation in a polygenic character with linked loci. *Genetical Res.* 26:221-235.

Lande, R. 1976. Natural selection and random genetic drift in phenotypic evolution. *Evolution* 30:314-334.

Lande, R. 1981. The minimum number of genes contributing to quantitative variation between and within populations. *Genetics* 99:541-553.

Lande. R. 1983. The response to selection on major and minor mutations affecting a metrical trait. *Heredity* 50:47-65.

Lande, R. 1986. The dynamics of peak shifts and the pattern of morphological evolution. *Paleobiol.* 12:343-354.

Lande R. 1987. Extinction thresholds in demographic models of territorial populations. *Am. Nat.* 130:624-635.

Lande, R. 1988. Demographic models of the northern spotted owl. *Oecologia* 75:601-607.

Lande, R. 1992. Marker-assisted selection in comparison with traditional methods of plant breeding. In *Plant Breeding in the 1990s*, eds. H. T. Stalker and J. P. Murphy, pp. 437-451. CAB International, Wallingford, UK.

Lande, R. 1993. Risks of population extinction from demographic and environmental stochasticity, and random catastrophes. *Am. Nat.* (in press).

Lande, R. 1994. Risk of population extinction from fixation of new mutations. *Evolution* (in press).

Lande, R., and G. F. Barrowclough. 1987. Effective population size, genetic variation, and their use in population management. In *Viable Populations for Conservation*, ed. M. Soulé, pp. 87-123. Cambridge University Press, New York.

Lande, R., and D. W. Schemske. 1985. The evolution of self-fertilization and inbreeding depression in plants. I. Genetic models. *Evolution* 39:24-40.

Ledig, F. T. 1988. The conservation of diversity in forest trees. *BioScience* 38:471-479.

Leigh, E. G. 1981. The average lifetime of a population in a varying environment. *J. Theoret. Biol.* 90:213-239.

Levins, R. 1969. Some demographic and genetic consequences of environmental heterogeneity for biological control. *Bull. Entomol. Soc. Am.* 15:237-240.

Lewontin, R. C. 1974. *The Genetic Basis of Evolutionary Change.* Columbia University Press, New York.

Lopez, M. A., and C. Lopez-Fanjul. 1993a. Spontaneous mutation for a quantitative trait in *Drosophila melanogaster.* I. Response to artificial selection. *Genetical Res.* 61:107-116.

Lopez, M. A., and C. Lopez-Fanjul. 1993b. Spontaneous mutation for a quantitative trait in *Drosophila melanogaster.* II. Distribution of mutant effects on the trait and fitness. *Genetical Res.* 61:117-126.

Lynch, M. 1988. The rate of polygenic mutation. *Genetical Res.* 51:137-148.

Lynch, M., and R. Lande. 1993. Evolution and extinction in response to environmental change. In *Evolutionary, Population, and Community Responses to Global Change*, eds. P. Karieva, R. Huey, and J. Kingsolver, pp. 243-250. Sinauer Associates, Sunderland Mass.

Muller, H. J. 1949. Redintegration of the symposium on genetics, paleontology, and evolution. In *Genetics, Paleontology and Evolution*, eds. G. S. Jepsen, G. G. Simpson, and E. Mayr, pp. 421-445. Princeton University Press, Princeton, N.J.

Okubo, A. 1980. *Diffusion and Ecological Problems: Mathematical Models.* Springer-Verlag, New York.

Pease, C. M., R. Lande, and J. J. Bull. 1989. A model of population growth, dispersal and evolution in a changing environment. *Ecology* 70:1657-1664.

Ralls, K., and J. Ballou. 1983. Extinction: lessons from zoos. In *Genetics and Conservation: A Reference for Managing Wild Animal and Plant Populations*, eds. C. M. Schonewald-Cox, S. M. Chambers, B. MacBryde and L. Thomas, pp. 164-184. Benjamin-Cummings, New York.

Ricker, W.E. 1981. Changes in the average size and average age of Pacific salmon. *Can. J. Fish. Aquat. Sci.* 38:1636-1656.

Roush, R. T., and J. A. McKenzie. 1987. Ecological genetics of insecticide and acaricide resistance. *Ann. Rev. Entomol.* 32:361-380.

Russell, W. A., G. F. Sprague, and L. H. Penny. 1963. Mutations affecting quantitative characters in long-time inbred lines of maize. *Crop Sci.* 3: 175-178.

Ryman, N., and F. Utter. 1987. *Population Genetics & Fishery Management.* University of Washington Press, Seattle, Wash.

Shaffer, M. L. 1987. Minimum viable populations: coping with uncertainty. In *Viable Populations for Conservation*, ed. M. E. Soulé, pp. 69-86. Cambridge University Press, New York.

Simmons, M. J., and J. F. Crow. 1977. Mutations affecting fitness in *Drosophila* populations. *Ann. Rev. Genetics* 11:49-78.

Simpson, G. G. 1953. *The Major Features of Evolution.* Columbia University Press, New York.

Sorensen, F. 1969. Embryonic genetic load in coastal Douglas fir, Pseudotsuga menziesii var. menziesii. *Am. Nat.* 103:389-398.

Soulé, M. 1980. Thresholds for survival: maintaining fitness and evolutionary potential. In *Conservation Biology. An Evolutionary-Ecological Approach*, eds. M. E. Soulé and B. A. Wilcox, pp. 151-169. Sinauer Associates, Sunderland, Mass.

Soulé, M., M. Gilpin, W. Conway, and T. Foose. 1986. The millennium ark: how long a voyage, how many staterooms, how many passengers? *Zoo Biol.* 5: 101-114.

USDA Forest Service. 1984. *Final Regional Guide and Final Environmental Impact Statement for the Pacific Northwest Region.* USDA Forest Service, Portland, Ore.

Wright, S. 1935. Evolution in populations in approximate equilibrium. *J. Genetics* 30:257-266.

Wright, S. 1968. *Evolution and the Genetics of Populations.* Vol. 1. Biometric Foundations. University of Chicago Press, Chicago, Ill.

Wright, S. 1977. *Evolution and the Genetics of Populations.* Vol. 3. Experimental Results and Evolutionary Deductions. University of Chicago Press, Chicago, Ill.

Wright, S. 1978. *Evolution and the Genetics of Populations.* Vol. 4. Variability within and among Natural Populations. University of Chicago Press, Chicago, Ill.

4

CONSERVING GENETIC DIVERSITY IN NATURAL SYSTEMS

Gary K. Meffe

University of Georgia's Savannah River Ecology Laboratory
Aiken, South Carolina
U.S.A.

INTRODUCTION: EVOLUTION AS THE CENTRAL CONCEPT

The great population geneticist Theodosius Dobzhansky once stated that "Nothing in biology makes sense except in the light of evolution," a theme which I fervently urge conservation practitioners to adopt. Evolution is the single, unifying concept that integrates all levels of biology and through which a deeper understanding of biological systems may be gained. Adaptation, morphological constraints, life history traits, and social behavior, for example, can best be understood, and often *only* understood, in an evolutionary context. Conservation conducted outside of this context is done apart from biological reality.

Because conservation is ultimately a biological problem, the various issues cannot be effectively understood and confronted without an evolutionary perspective as the central mode of thought. There remains a need to cultivate "evolutionary thinking" among conservation biologists and resource managers. To do otherwise is to run counter to the ways that ecological systems operate.

Evolutionary thinking with respect to conservation involves adopting a mind-set that at a minimum accepts the following principles: (1) Think in terms of a long time scale. A human reference scale is largely irrelevant to conservation *in perpetuity*. Humans generally have about a five generation view of their world: grandparents, parents, self, children, and grandchildren. This is sorely inadequate for biological systems based on evolutionary events with time scales of tens of thousands to hundreds of millions of years. Conservation efforts and goals must ultimately be placed on an evolutionary time scale; anything less is short-sighted and often for immediate human return and gratification at the cost of long-term environmental degradation. Most declines in biodiversity or changes in ecological systems have occurred instantaneously on an evolutionary time scale, and their recoveries must also be oriented toward that longer-range view. This is, of course, in direct contrast to typical economic or political goals, with time frames of several months to several years, often tied to election cycles. (2) Evolution is not directed or goal-oriented; it is without foresight. Thus, management should not direct species or systems toward any particular long-term goal other than an ability to continue evolutionary processes. Our management practices should not lock species or systems into a configuration from which they cannot change, even if that configuration seems appropriate in our present world. Ability for continued evolutionary change is critical, further emphasized by the next point. (3) Adaptation can lead to a dead end if the environment changes. Today's well-adapted population may suffer if the environment changes at a later date and if it cannot appropriately track that new environment. (4) Evolution is more than natural selection. Drift, founder effect, and mutation also play a role in local population structure, and their influences should not be ignored. This means that the present genetic structure of a population may not be due entirely to selection and might therefore not be the best "fit" to local conditions; random processes could have influenced the population and it may not be particularly well adapted. Do not assume adaptedness. (5) Evolution is dynamic; by definition it means change. Our programs should not adopt a *preservationist* mentality of keeping populations as they now are (even if they seem in especially good condition), but a *conservationist* mentality of allowing populations to continually change and adapt. A *status quo* philosophy is inappropriate for long-term conservation.

Evolutionary thinking forces conservationists to consider their actions relative to long-term interests of the species or system of concern, rather than the perhaps well-intended but ultimately misguided goals of human-oriented interests. Evolution is a pervasive element of all biological systems and should be *the* central paradigm of management concern. Ignoring evolution will only compound existing conservation problems. If evolution is the basis of conservation, then genetics is the basis of evolution, which brings us to the next question.

WHY CONSERVE GENETIC DIVERSITY?

Genetic diversity should be conserved in natural systems for at least three biological reasons. First, the Fundamental Theorem of Natural Selection states that "the rate of increase in fitness of any organism at any time is equal to its genetic variance in fitness at that time" (Fisher 1930). This means that a loss of genetic diversity in a population could result in slower adaptive change or an inability to change adaptively in response to environmental challenges. Lower genetic diversity might result in lower adaptability.

Second, high individual genetic variation may be related to individual fitness. A consensus opinion (although not unanimous) among population geneticists is that individual fitness increases with genetic variation (Mitton and Grant 1984), often measured as heterozygosity level. High heterozygosity has been related to survivorship, disease resistance, growth rate, and developmental stability (Allendorf and Leary 1986, and references therein). Consequently, loss of heterozygosity can lead to a decline in these fitness-related characters. The mechanism of apparent heterozygote advantage is not clearly understood, but may be attributable to lack of inbreeding depression, or might involve superior physiological performance of heterozygote genotypes (Crow 1948; Frankel 1983).

Third, the genetic variation contained in an organisms' DNA may be likened to a library of information. The information content of the DNA of a single house mouse, for example, is nearly equivalent to the amount of information in all the editions of the

Encyclopedia Brittanica published since 1768 (Wilson 1985). This represents a tremendous global biotic resource that is eroded as genetic variation disappears. This genetic diversity is the "blueprint" for an individual's development and the basis for much of its subsequent biological characteristics. The global pool of genetic diversity contains all the information for all biological processes on the planet. Loss of genetic diversity at the individual or population level erodes the material upon which natural selection acts and will likely decrease the ability of organisms to respond to environmental change in a biologically meaningful way.

THE UNITS OF CONSERVATION

Accepting the inherent importance of genetic variation to the biological world, a difficult and practical question then confronts the genetic resource manager in the field: what should we conserve? What are the units of genetic conservation? Obviously we cannot save every genotype, every variant, every rare allele in nature. We must concentrate limited fiscal and energetic resources in areas that will provide tangible benefit and will protect the greatest amount of genetic diversity.

In theory, we should identify and protect something like evolutionarily significant units (ESUs), defined by Waples (1991) as populations that (1) are reproductively isolated from other conspecific units, and (2) represent an important component in the evolutionary legacy of the species. The first part of this definition is essentially the definition of a "deme," which is the local unit of free genetic exchange, and is often difficult to identify in the field. The second component is a comparative value judgment that requires knowledge of the population within the context of other populations of the species in general. Consequently, this definition, appealing in a theoretical sense, is limited as a functional working definition.

As an extension of the ESU definition, I recommend a hierarchical gene diversity analysis as the basic methodology to identify and protect important conservation units. Such an analysis defines the geographic distribution of genetic diversity (see below), permits estimation of the relative times since separation (and thus relative evolutionary significance) of different genetic groups, and allows concentration of efforts and resources on those populations or groups that contain the highest diversity or are most divergent. In this spirit, K. P. Currens and colleagues (unpubl. data) defined a genetic conservation unit as "the most-inclusive genetic, geographic, and temporally persistent group of individuals of a species for which management action will not result in loss of genetic diversity of less-inclusive, less-widely distributed, and less-persistent groups that compose it." That is, a genetic conservation unit is the lowest unit in an identifiable genetic hierarchy.

Currens et al. discuss genetic conservation units in terms of salmonid fishes of the Pacific Northwest, and state the following: "The key concept is that the long-term productivity and evolutionary potential of more- inclusive groups of salmon depend on the continued existence of multiple, less-inclusive groups, which maintain their evolutionary independence for shorter periods. The challenge of identifying conservation units is to characterize and protect the smallest, least-inclusive groups of genetically isolated or differentiated fish that have persisted or may persist over multiple generations with only limited or intermittent gene flow. Once the smallest groups are identified, conservation units may be conveniently defined at more-inclusive levels according to the potential impacts of management plans, but *less-inclusive conservation units should not be combined for convenience if such management would reduce their number or evolutionary independence*" (italics in original). Thus, a viable genetic structure consists of smaller, independent or connected genetic units (demes) that are the building blocks of a larger, hierarchically structured genetic framework. What is the basis for such a hierarchical genetic structure?

HIERARCHICAL GENE DIVERSITY APPROACH

A genetic hierarchy derives from the following. Most species do not consist of single, panmictic units, and genetic diversity is rarely if ever evenly distributed among individuals or populations. Typically, geographically based structure to genetic diversity exists in most species. Selander and Whittam (1983) stated that "studies of protein polymorphisms indicate that a great variety of organisms, ranging from bacteria to humans..., are strongly structured genetically and that their evolution cannot be understood without reference to this structure."

Consider a hypothetical species consisting of three populations (Figure 4.1). Genetic diversity of that species consists of two components, within-population diversity (individual heterozygosity) and among-population divergence (genetic differences among locations). A simple genetic model can then be written as $H_t = H_c + D_{ct}$, where H_t = total genetic diversity (or heterozygosity) in the species, H_c = diversity within colonies, or average local heterozygosity, and D_{ct} = average divergence among colonies across the total species range. Conservation of the species must consider both intra- and inter-populational genetic variation in order to retain the highest amount of variation and to maintain a natural population genetic structure.

The divergence component of diversity can be further subdivided based on any geographically reasonable hierarchy. For example, consider an insect that is parasitic on tree leaves and that has high site fidelity. Genetic divergence in this case might conceivably

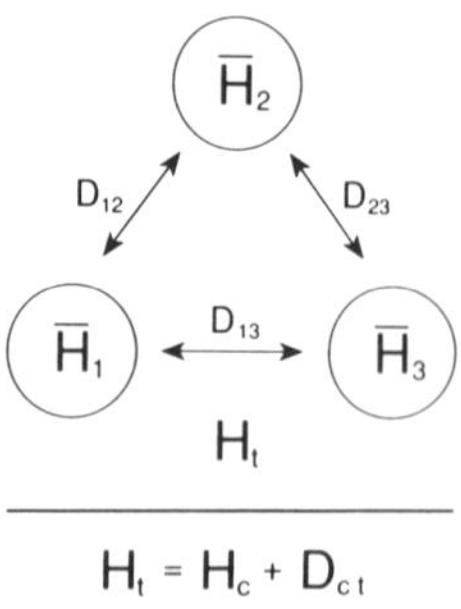

Figure 4.1. A simple model of population genetic structure. Total genetic diversity (H_t) consists of the average heterozygosities of the three populations, or colonies (H_c), plus the mean genetic divergence among populations (D_{ct}).

occur among branches with a tree (D_{bt}), among trees within a forest patch (D_{tp}), among forest patches within a watershed (Dpw), and among different watersheds across the total range of the species (D_{wr}). The degree of divergence at any level in the hierarchy is a function of the amount of gene flow across units at that level. High gene flow will reduce among-population divergence and increase within-population heterozygosity. In the above example, high gene flow among branches in a tree would result in little to no genetic divergence at that level in the hierarchy. Alternatively, strong isolation through site fidelity would tend to reduce local heterozygosity (especially if populations are small) and increase differences among populations. Divergence occurs through both random (founder effect, genetic drift, mutation) and directional (selection) forces.

A genetic hierarchy is then based on the geographic distribution of genetic diversity. The simplest (and least interesting) hierarchy would be one interbreeding population that constitutes the whole species; all of the genetic diversity is represented in this one population, and there is no hierarchical structure. A more realistic hierarchy is represented by the three populations in Figure 1, where some of the genetic diversity of the species consists of within-population heterozygosity (H_c), and the rest is manifested in divergence among the populations (D_{ct}). A more complicated hierarchy would obtain if the situation in Figure 1 were replicated several more times, with further identifiable divergence among the various major groups. The hierarchy in this case would consist of the heterozygosity component (H_c), divergence among colonies within the local range (D_{cr}), and divergence among the major groups (D_{rt}).

All levels in the hierarchy are potential genetic resources and can contribute to individual fitness and local population viability. The value of within-population heterozygosity was discussed above relative to fitness, and need not be further pursued. Among-population divergence at any level in the hierarchy may prove critical through local adaptations that typically develop over long periods of isolation with low gene flow. Many examples exist of strong local adaptation, including different growth forms of plants at different elevations (Clausen et al. 1940), local tolerance of plants to mine tailings (Antonovics et al. 1971), and life history differences in different populations of fish (Reznick et al. 1990).

Divergence might also be important because certain levels of the hierarchy may contain unique genetic information absent at other levels or in other geographic regions. Such information is an important genetic resource to the species at large. For example, Echelle et al. (1989) studied genetic diversity in the endangered Pecos gambusia (*Gambusia nobilis*), a livebearing fish, in Texas and New Mexico. They found that one of the four geographic areas supporting extant populations had a much higher allelic diversity than the other areas. These divergent populations contain unique allelic information that would be lost to the species if not identified and maintained. Such information could be important in a fitness context. Consequently, in addition to determining a geographic hierarchical structure, this approach can also identify and protect genetic uniqueness.

How are within- and among-population genetic diversities lost? Within-population heterozygosity is lost as an inverse function of genetically effective population size (Ne), through genetic drift or founder effect. Small populations lose variation faster than large populations, and loss of rare alleles is a greater problem than is loss of overall quantitative variation. Among- population diversity is lost as a function of gene flow that homogenizes populations, eroding divergence. Details of these processes are summarized elsewhere (Allendorf 1983; Meffe 1986), and will not be pursued further here.

Sound management of species in nature should incorporate genetic information at least to the extent of understanding the hierarchical geographic distribution of variation, and attempting to maintain or restore that structure as much as possible. Knowledge of the relative amounts of variation at different levels in the hierarchy is basic information necessary to approach genetic management of a species. The degree to which genetic diversity is a function of within-population heterozygosity and among-population divergence will dictate to some degree the management approach for the particular case.

The approach can be demonstrated with a data set compiled by Echelle (1991) regarding endangered and threatened fishes of western United States. The proportion of total genetic diversity consisting of within-population heterozygosity versus among-population divergence varies widely among species (Table 4.1). In some, such as *Xyrauchen texanus* (razorback sucker) or *Cyprinodon bovinus* (Leon Springs pupfish), the vast bulk of genetic variation is contained within populations, with little to no divergence among populations. In others, such as *Gambusia nobilis* (Pecos gambusia) or *Catostomus discobolus yarrowi* (Owens sucker), much genetic diversity is represented as among-population divergence.

These two extremes require different management approaches. In the first, it is more critical to maintain large local populations, because that is where the bulk of genetic diversity lies. Movements of individuals among populations could be included in this approach if that helps to maintain local populations, because little among-population

Table 4.1. Distribution of genetic diversity within and among populations in selected threatened and endangered fishes of western United States.

Family/*species*	H_t[1]	H_c	D_{ct}
Salmonidae			
Oncorynchus nerka	.046	94.4	5.6
O. clarki henshawi	.065	55.5	44.5
O. clarki lewisi	.029	67.6	22.4
O. gilae	.059	86.4	13.6
Catostomidae			
Catostomus discobolus yarrowi	.030	54.8	45.2
Xyrauchen texanus	.029	98.9	1.1
Cyprinodontidae			
Cyprinodon bovinus	.046	98.6	1.4
C. macularius	.049	70.1	29.9
C. tularosa	.028	81.0	19.0
Poeciliidae			
Gambusia nobilis	.063	48.4	51.6
Poeciliopsis o. occidentalis	.033	59.3	60.7
Cottidae			
Cottus confusus	.028	53.9	46.1

[1] H_t is total diversity as measured by percent heterozygosity; H_c and D_{ct} is the percentage of total genetic diversity within versus among populations, respectively (modified from Echelle 1991).

genetic divergence can be lost. In the latter examples, it is critical to maintain population distinctness, and not to move individuals among sites, because that would erode the uniqueness developed in each population. The exception to this rule is the case where human intervention has prevented former gene flow among populations. In this situation, restoration of past gene flow is warranted and recommended to mimic the natural pattern that man has interrupted (Meffe and Vrijenhoek 1988).

A hierarchical gene diversity analysis was implemented by Vrijenhoek et al. (1985) for the endangered Sonoran topminnow (*Poeciliopsis occidentalis*), a livebearing fish native to desert regions of Arizona and New Mexico. The range of the topminnow has declined precipitously in the last 50 years because of habitat loss and introduction of exotic fishes (Minckley and Deacon 1968; Meffe et al. 1983). The species now occurs in a few remnant populations in Arizona, but remains abundant in Mexico.

Vrijenhoek et al. (1985) surveyed 21 populations of *P. occidentalis* from Arizona and Sonora, Mexico, for allozyme diversity at 25 gene loci, and analyzed the data in a hierarchical fashion. The model they developed partitioned total genetic diversity (H_t) into heterozygosity within colonies (H_c), divergence among colonies within rivers (H_{cr}), divergence among rivers within major subdivisions (D_{rs}), and divergence among subdivisions across the total range of the species (D_{st}). The percentage of total genetic diversity found at each level was, respectively, 21.3%, 25.5%, 0.4%, and 52.8%.

Over one-half of the genetic diversity in this species was found to occur among three major geographic subdivisions, with the remaining significant portions at the individual population level and among populations within drainages. The estimated time necessary for the three genetic divergences observed among the three subgroups ranged from 1.7 to 4.3 million years, obviously not trivial evolutionary events. Management scenarios for the Sonoran topminnow should strive to keep separate the three major subdivisions, because combination across these geographic barriers would erode the largest component of genetic diversity and destroy a long evolutionary separation.

LEVELS OF EVOLUTIONARY DIVERGENCE

It is evident that the hierarchical approach not only outlines an overall genetic hierarchy, but also can distinguish "deep" (older) evolutionary divisions from more "shallow" (recent) divergences that have lesser evolutionary significance. In the Sonoran topminnow, the three major subgroups have divergence times of perhaps a few million years, evolutionarily important entities. Because genetic divergence among drainages or populations within each group is far less, these separations are more recent in evolutionary time, and less significant (although not necessarily *in*significant) in a conservation scenario. In fact, some regular gene flow probably occurred in recent times among many populations within drainages (Meffe and Vrijenhoek 1988).

The theme of evolutionary depth also has been emphasized by Avise and his coworkers (Avise et al. 1987; Avise 1989; Avise and Ball 1990). They argue that molecular genetic information (especially mitochondrial DNA, but also traditional allozyme data) should be used to determine phylogeographic relationships of populations and to identify deep evolutionary separations. Once the population structure is identified, it is easier to make decisions regarding conservation efforts. Avise and Ball (1990) stated that "...populations isolated from one another for increasing lengths of time should be genealogically differentiable by increasing numbers of loci. Thus, if the phylogenetic histories of many independent genes in an empirical survey were to separate individuals into concordant arrays, such a finding would be consistent with long-term reproductive separation of those arrays... Therefore, we suggest that such population subdivisions concordantly identified by multiple independent genetic traits should constitute the population units worthy of recognition as phylogenetic taxa", or, I would add, as conservation units. The message once again is to identify the deeper (older) evolutionary separations and concentrate conservation efforts at that level in the hierarchy.

Sometimes the likely hierarchical structure and depth of evolutionary separation among populations of a species is suggested by the known geographic distribution. Meffe and Vrijenhoek (1988) proposed two zoogeographic models of isolation and gene flow in desert fishes based on the likely connectance of habitats and, therefore, probability of gene flow over recent evolutionary time. Their "Death Valley Model" described the case of populations of fish in highly isolated desert springs or streams that had no connections to other habitats since the last pluvial period, 10,000-12,000 years ago. In this situation, it is unlikely that gene flow has occurred among these populations over the last 10-12 millennia because of their extreme isolation, and, therefore, likely that among-population genetic divergence is high. In fact, morphological, physiological, behavioral, and genetic differences among populations support this claim (Turner 1974; Naiman and Soltz 1981). In this zoogeographic model, the divergences among populations appear to be deep and of evolutionary significance, and they should be managed to retain their distinctness; no artificial gene flow among habitats should be allowed.

The "Stream Hierarchy Model," in contrast, applies to desert fishes in dendritic river systems that have various degrees of connectance among populations. Some level of gene flow has historically occurred among populations in this model, the magnitude depending on geographic proximity and the nature of local hydrography. In this model, geographically proximate populations probably have not been isolated historically, and thus do not have deep evolutionary divergences. Stronger divergence, and, therefore, of greater significance to conservation, is more likely to occur across drainages or geographic regions. This was the case in the Sonoran topminnow, discussed above, where more than one-half of genetic diversity consisted of divergence among geographic subgroups.

Remnant populations of this formerly abundant species generally occur in the more remote and peripheral areas of its former range. Because of man's influences in the region, many are now geographically isolated from other populations with which they undoubtedly exchanged genetic information in the past. It might, therefore, be necessary for the resource manager to restore natural levels of gene flow among such populations by periodically moving small numbers of fish among selected localities. This would mimic the natural situation in the Stream Hierarchy Model and maintain high levels of heterozygosity of these now small and isolated populations. Such artificial gene flow is necessary to restore former connections that have been severed by human development and concomitant water loss. Otherwise, these populations, newly isolated in evolutionary time, will ultimately lose the within-population genetic diversity that was formerly maintained by gene flow from other sites. Any such artificial gene flow, however, must be conducted only within levels of the hierarchy that would not jeopardize among-population divergence. There should absolutely be no movement, for example, among geographic subgroups.

A PROPOSED USE FOR THE HIERARCHICAL GENETIC APPROACH IN MANAGEMENT

Salmonid fishes of the Pacific Northwest are in decline because of overfishing and habitat degradation and loss through damming, water diversion, mining, pollution, and siltation from deforestation and road construction (NPPC 1987; NMFS 1991). Many stocks have been lost and many others are in serious jeopardy of following suit (Nehlsen et al. 1991). A great deal of effort and money is being devoted to management and recovery of five species of salmon in the region, with a major emphasis on hatchery production. An important aspect of management is conservation of genetic diversity of remaining natural populations, and an important question is the identification of conservation units (CBFWA 1991). Several workshops of geneticists and managers have been held to deal with this and similar genetic issues.

One approach is to adopt the hierarchical genetic structure methodology to define conservation units, as follows. A clear and objective hierarchical geographic structure exists in salmonid habitat in the form of stream orders. Every point in a watershed can clearly be identified in a stream order classification system (Horton 1945). First order streams are the smallest, headwater units. Second order streams consist of two or more first order streams; third order streams are formed by two or more second order streams, and so forth. A likely hierarchical genetic structure exists at the level of stream order (Figure 4.2). The following hierarchical genetic model can be applied to salmonids: total genetic diversity (H_t) consists of average within-population heterozygosity (H_c) plus divergence among populations within first order streams ($D_{c\cdot 1}$) plus divergence among first order streams within second order ($D_{1\cdot 2}$), plus divergence among second order streams within third order ($D_{2\cdot 3}$), and so on through the highest order. This approach would indicate the partitioning of genetic diversity at each level of stream order and help to define the smallest reasonable conservation unit.

A hypothetical example demonstrates this approach (Table 4.2). For simplicity, I have made the heterozygosity component constant in all cases. In the first instance, 14% of genetic diversity occurs as the average divergence among populations within first order streams. This is a large value and indicates that individual demes within close proximity in first order streams are strongly divergent and are the basic conservation units. In the second case, there is essentially no divergence at that level, but high divergence occurs among populations inhabiting different first order streams, indicating that is the significant level of conservation unit. Finally, the third example illustrates the case where the unit of conservation would be second order streams.

Such an approach would allow the resource manager to reveal the geographic structure of genetic variation and to identify a reasonable conservation unit for concentration of efforts. It would also reveal any especially divergent populations or

$$H_t = H_c + D_{c\cdot 1} + D_{1\cdot 2} + D_{2\cdot 3} + D_{3\cdot 4} + D_{4\cdot 5}$$

Figure 4.2. A hypothetical example of a hierarchical structure in salmonid fishes based on stream order. Hierarchical population structure could be examined based on an objective system of stream order.

unique alleles that might need protection, and would allow maintenance of a natural hierarchical genetic structure. The latter goal would be attained through establishment of genetic resource reserves.

GENETIC RESOURCE RESERVES

An important management tool for conservation of genetic diversity in natural systems is the establishment of genetic resource reserves, which would allow normal evolutionary

Table 4.2. A hypothetical example of the results of three genetic hierarchy studies in a Pacific salmon species. For simplicity, in all cases mean heterozygosity within colonies is the same.

	Case		
	1	**2**	**3**
H_c	63	63	63
$D_{c\cdot 1}$	14	0.1	0.1
$D_{1\cdot 2}$	10	15	2
$D_{2\cdot 3}$	6	7	19
$D_{3\cdot 4}$	4	9	11
$D_{4\cdot 5}$	3	6	5

processes to largely continue within and among populations, in a hierarchical fashion. Such reserves must be designed with knowledge of the geographical genetic structure of the species in hand.

Historically, wildland reserves have been designated based on aesthetic value or built around a charismatic species or set of species. Examples include Yellowstone, Glacier, Yosemite, and Grand Canyon National Parks. Although this has preserved some outstanding natural areas, it has ignored ecological and evolutionary processes, which are retained (or not) as a by-product of other design criteria. More recently, reserve design has adopted size and shape criteria as central features (Terborgh 1975; Diamond and May 1981); this is a step toward a more sound biological basis for conservation.

A more significant step, however, would be reserve design based on dynamic ecological and evolutionary processes, regardless of political or aesthetic boundaries. Such an approach would strike at the heart of the needs of biological systems and would be more likely to protect and maintain biological processes, and thus biodiversity, at all levels. One important process, of course, is maintenance of genetic diversity for continued evolutionary change.

A genetic resource reserve would consist of a network (*sensu* Noss and Harris 1986) of reserves and linkages that mimics the natural genetic hierarchical structure of a species. It would rely on knowledge of geographical genetic structure, identification of critical conservation units, and preservation of areas that would maintain those units and the natural connections (or isolation) among them.

Currens et al. listed several principles to be followed in developing genetic resource reserves for Pacific salmon. I have adapted their list for more general application of the concept of genetic resource reserves.

(1) *Protection of genetic structure requires a network of reserves that parallels the hierarchy of genetic organization of the species of interest.* For example, if the species has a wide geographic range with much genetic substructure, then reserves must include representatives of all of those units, as well as migration corridors among them, if gene flow was a historical feature in the species. Also, multiple representation of conservation units is desirable to (a) maintain more overall diversity, (b) reduce the chances of genetic loss through local catastrophic events, and (c) serve as sources of gene flow among similar units to maintain or restore diversity within a hierarchical level.

(2) *Natural evolutionary rates of gene flow will be maintained or restored by protecting or restoring historical patterns of movement among reserves.* The natural level of gene flow among habitats must be estimated through genetic data, and that level should be maintained in the reserve system. This was discussed above with respect to the two models of gene flow in desert fishes. Such an approach is necessary to determine whether opportunities for gene flow, or even forced movement of individuals among habitats is desirable, or natural gene flow was minimal or absent, and therefore undesirable in a management scenario.

(3) *Maintenance of demographic stability decreases the likelihood of loss of genetic variation in reserves.* Very small or unsuitable reserves could result in undesirably small populations of the target species. This could result in loss of genetic diversity through drift, as well as population loss through demographic declines (see below).

(4) *Identification and protection of critical habitats corresponding to the life history of the species will help maintain demographic stability and evolutionary processes.*

Knowledge of the natural history requirements of target species is obviously necessary for design of genetic resource reserves. High habitat quality is more likely to maintain larger populations of the species, and avoid demographic problems.

(5) *Harvest management (if applicable) must work within the constraints of genetic guidelines.* If the species is a game animal or harvested plant, then harvest regulations must work within guidelines set by genetic principles. Differential mortality through selective capture is known to alter genetic structure and phenotypic characteristics of harvested populations (Spangler et al. 1977; Goedde and Coble 1981). Regulations must be developed at the level of the entire genetic resource reserve, incorporating knowledge of the hierarchical genetic structure of the species.

(6) *Genetic reserves will best succeed if they address diverse regional and local human concerns.* The greatest opportunity for management success is when reserves are designed with potentially conflicting human interests taken into account. This demands planning across the entire range of the resource reserve and incorporation of social, aesthetic, religious, economic, and political concerns into the design. It is also important to consider a spectrum from "pristine" to heavily modified habitats in the reserve design. Reliance solely on natural, unimpacted habitats will greatly restrict the options and opportunities for reserve development. Modified habitats are typically more abundant, can provide important sources of genetic material, and offer essential educational opportunities during heavy use by the public.

These points would all be strengthened and coalesced under an integrated landscape approach to the problem. A landscape-level approach such as GAP analysis (Scott et al. 1991) could be used to identify ecosystems that would provide the most species benefits, to identify major spatial or temporal gaps in conservation genetic units, and to ensure that necessary and sufficient levels in the genetic hierarchy and appropriate life history stages are identified and included. It would also focus on appropriate areas for restoration efforts, and would indicate where multi-species conservation efforts should be pursued.

OTHER CONSIDERATIONS RELATED TO GENETIC CONSERVATION

Taxonomic Uncertainties

Traditional taxonomies are typically based upon morphological data. Such data can accurately reflect evolutionary relationships, or might be heavily influenced by phenotypic plasticity and adaptation to local conditions, and therefore be less relevant to determining evolutionary relationships. Genetic data can contribute to clarification of species or subspecies designations, and make conservation decisions more objective because it more closely reflects evolutionary events. Avise (1989) discussed two cases of "taxonomy-based mismanagement" of endangered species that could have been clarified by genetic information. These were the colonial pocket gopher (*Geomys colonus*) from Georgia, and the dusky seaside sparrow (*Ammodramus nigrescens*) from Brevard County, Florida, both described as distinct species (the sparrow later reduced to a subspecies, *A. maritimus nigrescens*) based on morphological or color characters.

In both cases, molecular genetic information revealed that the endangered taxon was not what it appeared. The pocket gopher was not demonstrably different from adjacent populations of *G. pinetus*. The dusky seaside sparrow was genetically indistinguishable from other Atlantic coast populations of *A. maritimus*, but these populations were quite distinct from Gulf coast populations of the species, with which it was hybridized in a last-ditch effort at saving a remnant of the gene pool. In both cases, conservation efforts were directed toward taxa that were not evolutionarily distinct, nor showed any genetic reason for special attention.

Demography and Genetics

Lande (1988) has argued that loss of genetic diversity is likely to be less of a problem in population or species extinction than are demographic concerns. Demography, he stated, "...may usually be of more immediate importance than population genetics in determining the minimum viable sizes of wild populations." This is because demographic influences are likely to be more prevalent in the short-term, making long-term genetic concerns irrelevant if extinction due to demography occurs first. I agree with his view in general, and his further observation that demographics has been neglected in conservation planning, although I maintain that genetic considerations may also be critical under many short-term and certainly all long-term circumstances.

Part of Lande's (1988) criticism of a genetically based approach to management of small populations is directed toward those who rely on the "50/500 rule." This guideline states that a genetically effective population size of at least 50 is necessary for short-term conservation of genetic diversity (i.e., a few generations) and at least 500 is needed for long-term maintenance of diversity. I agree completely with Lande that these numbers are misleading and potentially dangerous, for at least two reasons. First, demographic, ecological, or behavioral concerns could require that substantially more than 50/500 individuals are needed for population survival in the short-/long-term, irrespective of genetic needs. This is most evident in species that breed or feed in large colonies that are dependent on numbers for success; five hundred might be far less than is needed for successful behavioral repertoires. Second, adherence to such a rule runs the danger of "writing off" smaller populations that do not have a reasonable chance of attaining those higher numbers, even though they may be flourishing at smaller sizes. Many populations of desert fishes, for example, have existed at small population sizes for many generations with no apparent ill effects (Deacon and Deacon 1979). Some have gone through gradual inbreeding, have purged deleterious alleles, and could go on indefinitely at small population sizes. The 500 rule would dictate that these are not worthy of conservation efforts because they are doomed to genetic loss and extinction.

Rather than adhere to quantitative "rules of thumb," management efforts need to accumulate as much ecological and genetic information as possible for each particular species or scenario, and make decisions based on that information. Case-specific data collection and analysis by competent people are needed, rather than reliance on cookbook prescriptions; the latter either do not exist or are dangerously simplistic. Demography must certainly be considered, along with genetics, habitat quality and quantity, and the other myriad factors, biological and otherwise, that constitute contemporary conservation. Genetics is certainly not the final word in this process, but should be considered along with all other appropriate data, all evaluated in an evolutionary context.

Managerial Problems in Conserving Genetic Diversity

In the United States, there are a number of deeply entrenched attitudes and historical approaches among some resource agencies that make genetic conservation difficult. First, the philosophical orientation toward the utilitarian philosophy of Gifford Pinchot (Callicott 1990) is often inconsistent with conservation of genetic resources. The multiple-use approach toward natural areas adopted by many U.S. Federal Agencies automatically places genetic conservation in conflict with other goals such as harvest, grazing, or recreational use, typically to the detriment of conservation. This partly stems from a poor understanding of evolution within these agencies, which in turn is a product of university curricula that have not adequately incorporated evolutionary ecology into resource management programs.

Entrenched multiple-use attitudes are perpetuated by political and economic pressures from lawmakers who are even further removed from the biological world, and even less capable of making informed decisions that consider realistic biological scenarios. Often, such individuals do not care about long-term biological interests relative to short-term political gains. They do not understand that all economic growth and security ultimately depends on ecological capital and ecosystem services, which in turn depends on conserving biodiversity. There are signs, however, that trends in the political arena are beginning to shift toward more enlightened thinking (Gore 1992; Scheuer 1993).

In other parts of the world, managerial problems can run much deeper. Many developing countries do not have the luxury of even thinking about conservation of genetic diversity. Their problems are much more basic, including how to feed rapidly expanding human populations, how to service huge international debts, or how to stabilize shaky and sometimes corrupt governments. A common trend in such developing countries is to erode natural, ecological capital now for short-term gain, while contributing to long-term misery by having sold off their non-renewable resources, and in the process contributing to physical disasters such as flooding and earth movement. Policies of groups such as the U.S. Agency for International Development and the World Bank have contributed to and stimulated such trends, but now are beginning to incorporate ecological concerns into their lending policies (Goodland 1988).

The solution, of course, is education at all levels. Our resource managers need an evolutionary perspective that balances the utilitarian approach that has dominated federal agencies for more than a century (Nash 1989). This is as much a philosophical readjustment as a biological education. Likewise, the political arena and the public are sorely lacking in a basic foundation with which to evaluate adequately environmental decline on the planet. This is largely due to failure of academic scientists to communicate effectively their findings and philosophies to the public at large. However, if political and economic concerns continue to constrain and direct conservation, and fly in the face of evolutionary forces, then we as a society are not truly serious about conservation of biodiversity or the future of life on the planet. Evolution is the final arbiter in conservation and will ultimately determine the degree to which our efforts on behalf of biodiversity succeed or fail.

CONCLUSIONS

Because long-term biodiversity protection requires a long-term perspective, evolutionary concepts must form the core of conservation management at all levels, including genetics. I suggest a hierarchical gene diversity analysis as a practical, evolutionary approach to guide gene conservation. This analysis defines the geographic distribution of genetic diversity, permits estimation of relative evolutionary significance of genetic groups, and directs efforts and resources toward priority groups. The example of the hierarchical approach, using a desert fish species, demonstrated its utility in real management scenarios. A hypothetical management scenario for Pacific salmonid fishes, which included establishment of genetic resource reserves, further demonstrates the potential for a genetically based management system. Such approaches, although foreign to some resource managers, offer the best hope for long-term, sustainable resource protection.

ACKNOWLEDGMENTS

Writing was supported by contract DE-AC09-76SROO-819 between the U.S. Department of Energy and the University of Georgia. Many of these ideas were developed in discussion with other members of a salmonid genetics workshop sponsored by the Northwest Power Planning Council, October 1991, especially Ken Currens, Graham Gall, David Philipp, Brian Riddell, and Nils Ryman. Fred Allendorf, Robert Naiman, and Willa Nehlsen commented on earlier versions of the manuscript.

LITERATURE CITED

Allendorf, F. W. 1983. Isolation, gene flow, and genetic differentiation among populations. In *Genetics and Conservation. A Reference for Managing Wild Animal and Plant Populations*, eds. C. M. Schonewald-Cox, S. M. Chambers, B. MacBryde and L. Thomas, pp. 51-65. Benjamin/Cummings Publ. Co., Menlo Park, Calif.

Allendorf, F. W., and R. F. Leary. 1986. Heterozygosity and fitness in natural populations of animals. In *Conservation Biology. The Science of Scarcity and Diversity*, ed. M. E. Soulé, pp.57-76. Sinauer Associates, Sunderland, Mass.

Antonovics, J., A. D. Bradshaw, and R. G. Turner. 1971. Heavy metal tolerance in plants. *Adv. Ecol. Res.* 7:1-85.

Avise, J. C. 1989. A role for molecular genetics in the recognition and conservation of endangered species. *Trends Ecol. Evol.* 4:279-281.

Avise, J. C., and R. M. Ball, Jr. 1990. Principles of genealogical concordance in species concepts and biological taxonomy. *Oxford. Surv. Evol. Biol.* 7:45-67.

Avise, J. C., J. Arnold, R. M. Ball, E. Bermingham, T. Lamb, J. E. Neigel, C. A. Reeb, and N. C. Saunders. 1987. Intraspecific phylogeography: the mitochondrial DNA bridge between population genetics and systematics. *Ann. Rev. Ecol. Syst.* 18:489-522.

Callicott, J. B. 1990. Whither conservation ethics? *Conserv. Biol.* 4:15-20.

Clausen, J., D. D. Keck, and W. M. Heisey. 1940. Experimental studies on the nature of species. I. Effects of varied environments on western North American plants. *Carnegie Inst. Washington, Publ.* 520:1-452.

Columbia Basin Fish and Wildlife Authority (CBFWA). 1991. Integrated system plan for salmon and steelhead production in the Columbia River basin. CBFWA, Portland, Ore.

Crow, J. F. 1948. Alternative hypotheses of hybrid vigor. *Genetics* 33:477-487.

Deacon, J. E., and M. S. Deacon. 1979. Research on endangered fishes in national parks with special emphasis on the Devils Hole pupfish. In *Proceedings on the First Conference on Scientific Research in the National Parks*, ed. R. M. Linn, pp. 9-19. U.S. National Park Service, Transactions and Proceedings, Series 5, Washington, D.C.

Diamond, J. M., and R. M. May. 1981. Island biogeography and the design of nature reserves. In *Theoretical Ecology. Principles and Applications*, ed. R. M. May, pp. 228-252. Sinauer Associates, Sunderland, Mass.

Echelle, A. A. 1991. Conservation genetics and genic diversity in freshwater fishes of western North America. In *Battle Against Extinction. Native Fish Management in the American West*, eds. W. L. Minckley and J. E. Deacon, pp. 141-153. University Arizona Press, Tucson, Ariz.

Echelle, A. F., A. A. Echelle, and D. R. Edds. 1989. Conservation genetics of a spring-dwelling desert fish, the Pecos gambusia, *Gambusia nobilis* (Poeciliidae). *Conserv. Biol.* 3:159-169.

Fisher, R. A. 1930. *The Genetical Theory of Natural Selection*. Clarendon Press, Oxford.

Frankel, O. H., ed. 1983. *Heterosis: Reappraisal of Theory and Practice*. Springer Verlag, Berlin.

Goedde, L. E., and D. W. Coble. 1981. Effects of angling on a previously fished and an unfished warmwater fish community in two Wisconsin lakes. *Trans. Am. Fish. Soc.* 110:594-603.

Goodland, R. J. A. 1988. A major new opportunity to finance the preservation of biodiversity. In *Biodiversity*, ed. E. O. Wilson, pp. 437-445. National Academy Press, Washington, D.C.

Gore, A., Jr. 1992. *Earth in the Balance. Ecology and the Human Spirit*. Penguin Books, New York.

Horton, R. E. 1945. Erosional development of streams and their drainage basins: hydrophysical approach to quantitative morphology. *Bull. Geol. Soc. Am.* 56:275-370.

Lande, R. 1988. Genetics and demography in biological conservation. *Science* 241:1455-1460.

Meffe, G. K. 1986. Conservation genetics and the management of endangered fishes. *Fisheries* 11(1):14-23.

Meffe, G. K., and R. C. Vrijenhoek. 1988. Conservation genetics in the management of desert fishes. *Conserv. Biol.* 2:157-169.

Meffe, G. K., D. A. Hendrickson, W. L. Minckley, and J. N. Rinne. 1983. Factors resulting in decline of the endangered Sonoran topminnow Poeciliopsis occidentalis (Atheriniformes: Poeciliidae) in the United States. *Biol. Conserv.* 25:135-159.

Minckley, W. L., and J. E. Deacon. 1968. Southwestern fishes and the enigma of "endangered species." *Science* 159:1424-1433.

Mitton, J. B., and M. C. Grant. 1984. Associations among protein heterozygosity, growth rate, and developmental homeostasis. *Ann. Rev. Ecol. Syst.* 15:479-499.

Naiman, R. J., and D. L. Soltz, eds. 1981. *Fishes in North American Deserts*. John Wiley, New York.

Nash, R. F. 1989. *The Rights of Nature. A History of Environmental Ethics*. University of Wisconsin Press, Madison, Wisc.

National Marine Fisheries Service (NMFS). 1991. Factors for decline. A supplement to the notice of determination for Snake River spring/summer Chinook salmon under the Endangered Species Act. NMFS, Environ. & Tech. Serv. Div., Portland, Ore.

Nehlsen, W., J. E. Williams, and J. A. Lichatowich. 1991. Pacific salmon at the crossroads: stocks at risk from California, Oregon, Idaho and Washington. *Fisheries* 16(2):4-21.

Northwest Power Planning Council (NPPC). 1987. Columbia River Basin and Wildlife Program. Portland, Ore.

Noss, R. F., and L. D. Harris. 1986. Nodes, networks, and MUMs: preserving diversity at all scales. *Environ. Manage.* 10:299-309.

Reznick, D. A., H. Bryga, and J. A. Endler. 1990. Experimentally induced life-history evolution in a natural population. *Nature* 346:357-359.

Scheuer, J. H. 1993. Biodiversity: Beyond Noah's ark. *Conserv. Biol.* 7:206-207.

Scott, J. M., B. Csuti, and S. Caicco. 1991. Gap analysis: assessing protection needs. In *Landscape Linkages and Biodiversity*, ed. W. E. Hudson, pp. 15-26. Island Press, Washington, D.C.

Selander, R. K., and T. S. Whittam. 1983. Protein polymorphism and the genetic structure of populations. In *Evolution of Genes and Proteins*, eds. M. Nei and R. K. Koehn, pp. 89-114. Sinauer Associates, Sunderland, Mass.

Spangler, G. R., N. R. Payne, J. E. Thorpe, J. M. Byrne, H. A. Regier, and W. J. Christie. 1977. Responses of percids to exploitation. *J. Fish. Res. Bd. Can.* 34:1983-1988.

Terborgh, J. 1975. Faunal equilibria and the design of wildlife preserves. In *Tropical Ecological Systems: Trends in Terrestrial and Aquatic Research*, eds.F. Golley and E. Medina, pp. 369-380. Springer-Verlag, New York.

Turner, B. J. 1974. Genetic divergence of Death Valley pupfish species: biochemical versus morphological evidence. *Evolution* 28:281-294.

Vrijenhoek, R. C., M. E. Douglas, and G. K. Meffe. 1985. Conservation genetics of endangered fish populations in Arizona. *Science* 229:400-402.

Waples, R. S. 1991. Definition of "species" under the Endangered Species Act: application to Pacific salmon. NOAA Tech. Mem. NMFS F/NWC-194.

Wilson, E. O. 1985. The biological diversity crisis. *BioScience* 35:700-706.

5

ON POPULATION VIABILITY ANALYSIS AND MANAGEMENT

Bruce G. Marcot

USDA Forest Service
Pacific Northwest Research Station
Portland, Oregon
U.S.A.

Dennis D. Murphy

Center for Conservation Biology
Stanford University
Stanford, California
U.S.A.

Reflecting a dramatic increase in public awareness of wildlife management issues and a proliferation of programs that involve the conservation of threatened and endangered organisms and ecosystems, a science that focuses on species viability has become central to public and private lands planning. Numerous examples of populations extirpated, species in decline, and ecological communities disrupted exist that involve nearly all taxonomic groups. Birds provide but one telling example: nearly a thousand species, more than 11% of all bird species, are at some risk of global extinction. And, particularly sensitive species groups are faring even worse; nearly one-third of all parrots can be considered to be threatened or otherwise warrant special concern (Collar and Andrew 1988).

The major contributing factor to this global crisis is the loss of natural habitats, particularly those in tropical biomes. Only a small percentage of the world's remaining natural habitats have been protected as reserves or conserved for other purposes. Losses experienced outside of those areas are stultifying. In the Afrotropical realm, for example, by 1986, just 42% of an original 8.22 million km^2 of upland montane and other dry forest types remained, and but 15% of that is protected. In the Indomalayan realm, the situation is even more grim; also as of 1986, only 28% of an original 3.41 million km^2 of dry forest remained, and just 11% of that remaining was protected (Harmon 1990). The status of other habitats, particularly wetlands, mangroves, and ancient forests, is substantially worse.

Additional concerns for species viability have arisen from more direct over-exploitation of wildlife by humans for subsistence or sport hunting, and from the secondary effects associated with increasing urbanization and accompanying environmental pollution. The toll is a rapidly expanding list of wildlife species at risk of local extirpation or global extinction. The USDI Fish and Wildlife Service has officially listed as endangered or threatened more than 1,200 taxa of plants, invertebrates, and vertebrates (including species, subspecies, or distinct vertebrate populations). Of those, however, only 311 taxa enjoy legally mandated plans for their recovery (Anonymous 1992). Hardly encouraging is the fact that nearly five times as many candidate species warrant listing, but are not currently afforded protection (Bean, this volume).

Clearly, there is a pressing need for procedures that will allow land managers and decision-makers to identify, analyze, and plan to ensure the persistence of viable populations of wildlife. This chapter draws from existing treatises that have discussed at length species extinctions (e.g., Hoage 1985; Pimm et al. 1988) and population viability (e.g., Soulé 1987), and presents a framework for analysis and management of viable wildlife populations. The literature now includes numerous case studies that highlight a wide array of animal groups; however, an unambiguous application of biological data and conservation procedures seems to evade land managers. The purpose of this chapter is to introduce basic concepts and to outline a procedure for population viability analysis (PVA) and its subsequent management application. First, we briefly define population viability and describe its analysis. We then review some of the legal charges pertinent to the management of viable populations on public lands in the United States and outline the major ecological factors that affect the continued existence of populations. Next, we provide a procedure to apply population viability information in a risk management framework. Finally, we consider the need for multiple-species and whole ecosystem approaches to conservation of viable populations.

DEFINITIONS OF POPULATION VIABILITY AND POPULATION VIABILITY ANALYSIS

Population viability addresses the *probability* of the continued existence of geographically *well-distributed populations* over *specified time periods*. Such a probabilistic description demands clear definitions of several key terms (similarly, see Murphy and Noon 1991).

The *probability of continued existence* can be estimated by various qualitative and quantitative means and can refer to: the likelihood of global or local extinction of target taxa; the probability of population sizes dropping below low, but non-zero, thresholds (that is, quasiextinction, see Ginzburg et al. 1982); the length of time to extinction; the likely duration that populations can exist above quasiextinction levels; the likelihood that populations will remain spatially well-distributed; as well as numerous other measures that consider demographic and genetic bottlenecks, population recovery, and habitat occupancy rates (see Table 5.1). No single measure alone is sufficient to assess population viability; different suites of data and analyses are necessary to meet each conservation challenge where each involve distinct sets of species and habitats.

Central to all planning efforts that involve population viability analysis is provision of interconnected and *well-distributed populations* and the habitats that support them. Although some species are naturally patchily distributed, in general the maintenance of

Table 5.1. Some expressions of the numerical viability of a population.

- *extinction likelihood* — the probability of a specific population becoming extinct (zero individuals remaining) within a specified geographic area over a specific duration of time (or at several points in time)
- *pseudoextinction likelihood* — the probability of a specific population declining to or below a specific population density or overall population size (typically non-zero) within a specified geographic area over a specific duration of time
- *time to extinction* — the duration of time from the present when the population is most likely to reach extinction or pseudoextinction levels (densities or population sizes)
- *duration of persistence of well-distributed population* — the duration of time that the population will likely continue to exist well distributed throughout at least its current range (well-distributed would be defined in terms of specific geographic locations or a range of ecological conditions); note that a population could fail to be well distributed but could be highly likely to continue to persist above extinction or pseudoextinction numbers or densities
- *probability of persistence of well-distributed population* — the likelihood that the population will continue to remain at or above a specific size or density; similar to pseudoextinction likelihood except that the level is not describing threatened status but rather one describing a minimally well-distributed pattern
- *degree of population bottleneck* — the expected size or density of populations passing through bottleneck periods, and associated anticipated effects
- *duration of population bottlenecks* — the duration of time through which the population would likely remain at small size or low density, secondarily affecting local or global extinction likelihoods or long-term adverse genetic effects
- *likelihood of recovery* — related to likelihoods of population bottlenecks are likelihoods that the population would recover from low densities or sizes over a specified geographic area and over a specified time period; such likelihoods might include consideration of recovery growth rates and qualities of habitats, and colonization rates of new habitats by dispersing organisms (founder rates)
- *rate of recovery* — the expected duration of time over which a population would expand in geographic range, density, or population size to specified levels
- *occupancy rate of habitats* — the average proportion of potentially occupied sites that are predicted to be occupied over specific time intervals, or the proportion predicted to be occupied at a particular point in time; expressed also as proportion of habitat patches or reserves occupied by at least one individual or a breeding unit, the proportion of the landscape occupied by individuals or breeding units, the proportion of the landscape consisting of potentially occupied habitats or reserves.

distributions and connectivity is key to sustaining genetically diverse and demographically healthy populations. We describe a population or habitat as well-distributed if it is maintained over the long term across at least its existing range of geographic and ecological conditions. In some cases, the maintenance of well-distributed populations or habitats requires restoration of suitable conditions in portions of historic geographic ranges, particularly where needed to reconnect populations isolated by human activities.

Operational definitions of the term "well-distributed" have varied. For example, planning for viable wildlife populations on National Forests commonly pits contending interests in competition for scarce resources, such as with the commercial timber-use of long-leaf pine forests (*Pinus palustris*) that provide optimal habitat for the endangered red-cockaded woodpecker (*Picoides borealis*) in the Southeast. Typically in such cases, much less than the full range of existing or historic conditions is targeted by conservation efforts due to economic concerns. The appropriate definition of "well-distributed," then, can be a policy decision that considers social welfare and other factors. The role of the biologist in such a multifactorial decision-making process is to estimate, quantitatively if possible, and to articulate clearly the effects on species viability of a range of possible decisions.

The *specified period of time* over which a population is to be maintained has been treated equivocally in past planning activities on public lands in the United States. Planning time-frames for management of public lands typically are one to ten years. In contrast, to ensure long-term genetic adaptability, phenotypic diversity, and high probabilities of population persistence, an appropriate planning time-frame should be up to a hundred times that of a single generation of a target species. Recommendations from the conservation biology literature for the appropriate planning time-frames for population viability have ranged from a century to a millennium (Soulé and Wilcox 1980, Shaffer 1981, Frankel and Soulé 1981). Some have suggested a standard time-frame for all PVAs, such as the 200-year length of a full successional cycle in many complex vegetation communities.

Such time-frames are biologically highly desirable, but may not prove to be practical either in analysis or in setting management guidelines, particularly on public lands where revisions of management plans can occur every several years. Planning for long-term viability of species of ancient forests of the Midwest, including restoration of those forests already lost to logging and urbanization, would likely entail rigorous management planning for time horizons of even more than two centuries—longer than the United States has been a nation. Thus, we make a practical recommendation of setting a time-frame of at least 10 generations of a species targeted for planning for analyzing viability effects of proposed management plans; longer time periods should be considered if species have particularly short generation times relative to either planning horizons or frequencies of disturbance events. Longer planning time-frames allow biologists to detect population dynamic responses that may be masked over the short term, although viability analyses for conservation plans might apply for shorter time periods reflecting competing, non-biological constraints. Indeed, there is professional disagreement on whether one time-frame should be set for all PVAs for all purposes, or, as we could suggest here, time-frames should be allowed to vary to account for differences in planning objectives, species ecologies, and disturbance dynamics.

Population viability analysis is the formal process by which one can estimate the likelihood of persistence of populations, metapopulations, or species, over specified time periods, given one or more management alternatives. PVA is neither a monolithic concept nor a cookbook procedure. No single method for conducting a PVA exists. The host of factors that potentially influence the viability of a species, and how those factors are analyzed and influenced by management activities, thus, are case-specific, although a general assessment procedure can be and is described below.

The types of data brought to bear in PVA likewise vary widely between species. Viability analysis for cheetahs (*Acinonyx jubatus*), for example, has focused on the genetic risks associated with population bottlenecks. O'Brien et al. (1985) have shown that cheetahs are monomorphic at essentially all gene loci investigated. Combined with low overall numbers and restricted geographic range, a high risk of inbreeding depression appears to exist; hence, conservation planners must consider gene flow among isolated populations for their management planning. In contrast, even though Barrowclough and Gutiérrez (1990) have shown a similar lack of genetic variation in the northern spotted owl (*Strix occidentalis caurina*), its comparatively wide range and greater numbers have focused conservation planning on demographic factors that put the species at risk of local extirpation. Current conservation strategies, among other considerations, call for the provision of ample suitable habitat to sustain stable demographic units of the owl. Differently, the threatened Bay checkerspot butterfly (*Euphydryas editha bayensis*) can be found in high population numbers, but is at risk to drought that can lead to dramatic population declines and disruption of metapopulations. Conservation planning for that species thus has focused on environmental perturbations, habitat heterogeneity, and adult butterfly dispersal dynamics within and among habitats (Murphy et al. 1990).

Unfortunately for most species, reliable empirical data on historic population dynamic trends, the role of environmental conditions in regulating populations, and other crucial information simply do not exist. In such circumstances, only qualitative predictions can be made regarding potential likelihoods of persistence and effects of different management alternatives. In a very few cases, as with some especially well-studied threatened or endangered species, more is known, and quantitative estimates of extinction probabilities or other factors can be made through use of spatially explicit simulation models of dynamic habitats and metapopulations (e.g., Harrison et al. 1988; Murphy et al. 1990; Harrison 1991; Murphy and Noon 1992) and use of demographic and genetic analyses of populations *in situ* (e.g., Eberhardt 1990; Haig and Norstrom 1991; Hedrick and Miller 1992). Yet, when such data are lacking, proceeding with at least interim, qualitative evaluations of population viability status can be useful (Marcot and Salwasser 1991).

LEGAL CHARGES TO MANAGE FOR VIABLE POPULATIONS

Although much has been written on population viability and its analysis, no nation has laws or formal conservation policies that directly mandate both analysis *and* management of viable populations. Most regulations that deal with conservation of sensitive, rare, threatened, or endangered species or their habitats only indirectly encourage population viability analysis and management. For example, India's 1988 National Forest Policy calls for widespread protection and restoration of forest habitats throughout the country, but does not directly mention definition, analysis, and management of viable fish and wildlife populations. Similarly, Great Britain's 1967 Forestry Act (as amended in 1985) mandates a balance between afforestation and "conservation and enhancement of natural beauty and ... flora [and] fauna," and their conservation guidelines emphasize habitat protection, but not population viability explicitly (Forestry Commission 1990:2).

In the United States, the USDA Forest Service has a mandate to address maintenance of viable populations of fish and wildlife on public lands, although no explicit policy calling for application of PVA exists. The National Forest Management Act (NFMA) of 1976 specified that the National Forest System be managed to provide for diversity of plant and animal communities to meet multiple-use objectives. Subsequent regulations for planning land and resource management (36 CFR 219), adopted in 1979 (refined in 1982 and undergoing current revision), augmented that diversity policy by requiring that management of habitats serve to maintain viable populations of vertebrates. This requirement has been interpreted by the Secretary of Agriculture, upon the advice of a committee of scientists appointed to help implement the National Forest Management Act, to include the maintenance of viable populations of vertebrates (Committee of Scientists 1979; 36 CFR 219.19; see Salwasser et al. 1986). The regulations now stipulate that "fish and wildlife habitat shall be managed to maintain viable populations of existing native and desired non-native vertebrate species in the planning area" (36 CFR 219.19). The regulations also mandate that "all management prescriptions shall...provide for adequate fish and wildlife habitat to maintain viable populations of existing native vertebrate species" (36 CFR 219.27(a)(6)). The NFMA regulations go further, defining a viable population as "... one which has the estimated numbers and distribution of reproductive individuals to insure its continued existence is well distributed in the planning area."

Other Federal resource laws in the United States provide impetus for managing for viable wildlife populations on public lands. These other laws include the National Wilderness Preservation Act of 1964, the National Environmental Policy Act of 1969, and the Endangered Species Act of 1973 (as amended). Together, these laws reflect society's desire to maintain healthy and productive ecosystems and the species they support, and implicitly call for population viability analyses in scientifically credible conservation plans.

FACTORS AFFECTING VIABILITY

Conditions that influence sizes and trends of wild populations can be categorized as demographic, genetic, and environmental (Shaffer 1981; see Table 5.2 for a generally applicable list of factors that should be explored in all PVA's). A thorough viability assessment should account for potential influences from each of these factors.

Demographic Factors

Demographic factors that must be considered in PVAs focus especially on variation in birth and death rates, which can cause small and isolated populations to become extinct or to decline to levels where other risk factors take effect, the so-called quasiextinction levels (Pimm et al. 1988; Iwasa and Mochizuki 1988; McGraw and Garbutt 1990). For example, Smith and McDougal (1991) found that variation in birth rates and first year survival in tigers (*Panthera tigris*) in Royal Chitwan National Park in Nepal are as important to population stability as is quality of a female's territory. Demographic variation, therefore, can reflect individual variation endogenous to the population, or effects of environmental noise on individual performance from exogenous factors.

Source and sink dynamics of local populations constitute a species' metapopulation structure, that is, the interactions among individual demographic units, which in turn can influence local and global likelihoods of persistence (Harrison 1991, Pulliam 1988). The aforementioned conservation strategy for the northern spotted owl in the Pacific Northwest was founded on metapopulation dynamic considerations; the size and spacing of habitat conservation areas were designed to ensure local demographic persistence and successful dispersal of owls among habitats and populations (Thomas et al. 1990).

Table 5.2. Factors potentially affecting population viability. (After Marcot et al. 1986)

1. Demographic
   - (a) Variation in birth and death rates
   - (b) Variation in size of reproductive population
   - (c) Dynamics of source and sink populations
2. Genetic
   - (a) Inbreeding depression
   - (b) Genetic drift
   - (c) Loss of genic diversity through range subtraction
3. Environmental
   - (a) Loss of habitat
     - (i) Chronic effects
       - (1) Succession
       - (2) Reduction of habitat quality or quantity by human use
       - (3) Shift of geographic locations of habitat
       - (4) Climatic cycles
     - (ii) Catastrophic effects
       - (1) Storms
       - (2) Fires
       - (3) Floods
       - (4) Volcanoes
       - (5) Insect or disease epidemics
       - (6) Droughts
       - (7) Human-caused acute alteration of habitat conditions
   - (b) Biological interactions
     - (i) Predators
     - (ii) Competitors
     - (iii) Disease
     - (iv) Parasites
     - (v) Reduction in prey abundance or availability
   - (c) Pollutants, toxicants

Genetic Factors

Population viability additionally can be compromised by losses of genetic and allelic variation associated with population declines that are caused by habitat loss and geographic range reduction. Such losses then can lead to inbreeding depression and genetic drift (Barrowclough 1980). The likelihood of deleterious recessive alleles becoming fixed in a population is greater in species that are less vagile and that exist in small, isolated populations. Those likelihoods can be estimated with analytic models like those of Emigh and Pollak (1979).

Variation in the size of a reproductive population can depress the genetically effective population (or breeding) size (N_e), resulting in a population given to the characteristics of much smaller populations (or, more specifically, smaller genetic pools). For example, the study of tigers by Smith and McDougal (1991) also showed that substantial lifetime variation in reproduction typical of the species results in lower effective population sizes (N_e:N = 0.41 and N_e = 26) and higher risks of inbreeding, than if reproductive rates were more constant through time. Variation in the reproductive output of individuals generally is much greater in smaller populations than in larger ones, because fewer breeders are active each season.

Treatments of effective population size have varied. Hill (1979) and Lande and Barrowclough (1987) provided formulae for estimating N_e from life history parameters. In contrast, the method of Pollack (1983) was based on changes in allele frequencies. Nei and Tajima (1981) linked genetic drift to declines in N_e. Whichever the method, the calculation of N_e, a central step in estimating potential effects of drift and inbreeding depression, is at best an inexact science. Harris and Allendorf (1989) underscored this observation in a report on the wide deviation in estimates of N_e from applying various published formulae on identical data on grizzly bears (*Ursus arctos horribilis*).

Environmental Factors

Environmental factors virtually always serve as the mediating forces that reduce population sizes to levels at which demographic variation and genetic threats put populations at risk. Environmental factors that affect population viability include systematic (or deterministic) sources of threats including anthropogenically caused losses of habitat, interactions with alien species, over-hunting, and impacts from pollutants and toxicants. Stochastic sources include normal environmental fluctuations (such as drought and deluge) and catastrophic events (such as wildfire). Studies of North American songbirds provide numerous examples of factors that contribute to population and species declines, including losses of wintering habitats relative to breeding habitats (Wilcove and Terborgh 1984), fragmentation of forest breeding habitats (Lynch and Whigham 1984; Rosenberg and Raphael 1986; Lehmkuhl and Ruggiero 1991), invasion by competitors and nest parasites, particularly the brown-headed cowbird (*Molothrus ater*) (Brittingham and Temple 1983), and other factors.

Few generic computer programs are available commercially (for example, Ferson's 1988 RAMAS) or in the public domain (for example, POPDYN from USDA Forest Service, Pacific NW Region). Such programs can aid analyses of demographic, genetic, and environmental effects on population viability.

POPULATION VIABILITY RISK MANAGEMENT

In Situ Versus *Ex Situ* Management

The purpose of viability risk management is to implement planning guidelines to ensure long-term, *in situ* (in the field) conservation, productivity, and persistence of wildlife species. We emphasize *in situ* management for viable populations for several reasons. First, the most common cause of concerns over population viability is habitat loss. Successful *in situ* management demands conservation of an adequate amount and distribution of habitats to support target species. Second, *ex situ* (not in the field) conservation methods, such as captive propagation or zoo stock management, simply cannot replicate the diversity of natural conditions essential for long-term adaptation and evolution of new genetic lineages. Third, most methods of translocation, reintroduction, and captive propagation of small populations remain untested, and, therefore, have dubious value as long-term conservation strategies.

For example, all reintroductions of bighorn sheep (*Ovis canadensis*) carried out with fewer than 50 individuals have resulted in populations that have gone extinct within 50 years (Berger 1990). Likewise, captive breeding programs should be viewed as last-resort, high-cost, and high-risk manipulations of critically endangered species, such as the program for the Puerto Rican parrot (*Amazona vittata*) (Wiley et al. 1992); moreover, alternatives to captive breeding are usually much less expensive. Finally, waiting until populations in decline reach levels at which they become a legal or litigative concern, as with federal threatened species listings or court actions, hardly promotes conservation leadership. Population viability management is best used to ensure the continued security of species and taxa currently not threatened in their natural habitats, as well as to help plan the recovery of those species that are threatened or endangered.

A 9-Step Process for Viability Risk Management

Specific guidelines and policies used to manage for viable populations vary by species and with landowner goals. A general procedure to plan the management of population viability risk, however, can be described as a 9-step process (after Marcot et al. 1986; Table 5.3).

Step 1. Screen for species at risk. In a first step, species are screened to determine the immediacy of viability concerns. Species that should automatically be prioritized as

Table 5.3. Components and steps in a population viability risk management (decision-making) process. (After Marcot et al. 1986)

1. *Describe the species.* Identify and screen for species at risk.
2. *Describe pertinent regulations and laws.* Define the range of acceptable conditions that meet policy regulations or law. Identify responsible agencies and institutions and coordinate their planning activities.
3. *Describe species conditions and ecology.* Describe the species environmental relationships and reasons for viability concern or population decline.
4. *Develop planning alternatives.* Develop a range of planning or management strategies. Describe risk attitudes of publics and responsible agencies and institutions. Describe perceived and expected utilities of planning alternatives. Identify factors and value ranges in decision criteria to be used to select among the alternatives.
5. *Evaluate viability effects of the alternatives.* Conduct a viability risk analysis. Analyze how the alternatives affect population size, distribution, and persistence, resource tradeoffs, and social concerns.
6. *Array the alternatives according to anticipated effects.* Assess the benefits to population viability, indirect costs, opportunity costs, and foregone options for each alternative. Identify and array the range of acceptable alternatives that meet decision criteria.
7. *Select an alternative.* Apply decision criteria to results and select a course of action.
8. *Implement the alternative.* Institutionalize management guidelines of the selected alternative. Develop budgets and track implementation.
9. *Monitor results.* Compile and interpret monitoring information; revisit steps 1-8 as suggested by monitoring information and changing conditions.

targets of conservation concern include those legally identified as threatened or endangered by the USDI Fish and Wildlife Service or equivalent agencies in other countries (particularly if a species recovery plan has not been developed and implemented). Useful screens also include "alert lists" such as *The ICBP World Checklist of Threatened Birds* (Collar and Andrew 1988), CITES (Council on International Trade of Endangered Species) appendices (although CITES addresses only international trade species), The Nature Conservancy's Natural Heritage Inventories, and the *Red Data Books* of International Union for Conservation of Nature and Natural Resources (IUCN). IUCN in particular has categorized species as threatened, endangered, vulnerable, rare, indeterminate, and insufficiently known (e.g., King 1978-1979; Collar and Stuart 1985).

Schonewald-Cox (1983) presented a nine-class scale that depicts the degree of population security as a simple function of approximate effective population size. This approach proved helpful to identify the current level of viability concern for the northern spotted owl in the Pacific Northwest (USDA 1986). Other simple and useful screening tools follow rarity ratings like that of Rabinowitz et al. (1986), as modified by B. Marcot and colleagues (unpubl. data) to depict management priorities (Table 5.4). For example, a species that has restricted habitat specificity, narrow geographic distribution, and small population size throughout its range, such as Honey Creek cave blind salamander (*Eurycea tridentifera*) in Texas, would rank as having high viability concern and management priority. A species that similarly has restricted habitat specificity, but a wider geographic range and larger population size, such as the Oregon silverspot butterfly (*Saxeria zerene hypolitta*), would have a lower ranking. Other viability ranking criteria were developed by Lehmkuhl and Ruggiero (1991) for old-growth forest wildlife in the Pacific Northwest. Rankings are based on general life history traits, such as the vagility and habitat specificity of each species, and anticipated levels of vulnerability to likely management actions.

The depiction of current viability concerns can be aided by mapping habitats and species distributions. For example, Burke et al. (1991) depicted areas of viability concern by mapping habitat distribution for Stephen's kangaroo rat (*Dipodomys stephensi*).

A mapping procedure, used to assess the viability status of spotted owls and other terrestrial species on National Forests lands, can be useful with a wide variety of other species and habitats:

(1) map the current distribution of the target species.

Table 5.4. A method for prioritizing species for viability concern. Table entries depict priorities for viability concern. (Source: B. Marcot and colleagues (unpubl. data), after Rabinowitz et al. 1986).

Geographic distribution:	Wide		Narrow	
Habitat specificity:	Broad	Restricted	Broad	Restricted
Local population size:				
Somewhere large	very low	low	low	moderate
Everywhere small	low	moderate	moderate	high

(2) map the geographic range of suitable habitat. Use habitat suitability models to depict grades of habitat quality.

(3) delineate on map overlays the distributional considerations of particular concern, including population isolates, dispersal barriers, areas of low population density, and areas of special concern, such as where populations have bottlenecked or where landscapes may serve as corridors linking habitats.

(4) overlay locations of proposed activities that could compromise habitat quality, for each planning alternative under consideration. From this overlay, the areas of conservation concern are prioritized for planning.

Mapping can also be used to identify areas of high endemic species richness for conservation action. Examples include the range map by Collar and Juniper (1992) for neotropical parrot species at risk of extinction, and maps by Scott et al. (1988) for the distribution of Hawaii's endangered native avifauna. The goal of any map-based approach is the identification of target species that warrant viability analysis.

Step 2. Describe pertinent regulations and laws. The second step in the population viability management process is to describe regulations and laws pertinent to the conservation of wildlife species and habitats. This step includes a definition of the range of acceptable viability conditions to meet policy regulations or legal mandates, and identification of the key agencies or institutions responsible for carrying out and coordinating viability management activities.

Step 3. Describe conditions and ecology of target species. The third step is an in-depth status review of the species, including descriptions of species life histories, ecologies, environmental relationships, trends, and reasons for viability concern or population decline. This step uses basic field inventories of species and their habitats, as well as species-habitat relationship models (see Morrison et al. 1992).

Also useful (some would argue essential) in this step are demographic models and analyses. For example, Crouse et al. (1987) used a stage-based population model to assess population status, thereby aiding in the conservation of loggerhead sea turtles (*Caretta caretta*). The use of demographic models by Kinnaird and O'Brien (1991) suggested a high probability of extinction of a primate, the Tana River crested mangabey (*Cercocebus galeritus galeritus*), over the next 50-100 years, thus helped to identify population sizes that would better ensure long-term viability. Similarly, Knight and Eberhardt (1984) and Shaffer (1983) assessed future population sizes of Yellowstone grizzly bears using stochastic demographic life table models. Many other examples are available from the literature. Such models are best used for species with adequate field data and to generate hypotheses for monitoring population trends and dynamics.

Simulation models are useful but not absolutely necessary for conducting viability assessments. And, it is useful to differentiate quantitative PVAs, that should make use of such models, from qualitative population viability assessments, that use a mix of professional judgment and empirical evidence to pose working hypotheses concerning population responses to existing or proposed management actions. Examples of qualitative population viability assessments include the evaluations of the persistence of fungi, lichen, bryophyte, nonvascular and vascular plants, invertebrates, fish, and wildlife species and species groups in recent region-wide conservation planning exercises (Thomas et al. 1993; FEMAT 1993). These evaluations relied on the judgment of panels of species and taxonomic group experts to rank qualitatively each species or species group according to their likelihoods of persisting well-distributed for a century or longer on National Forests (Thomas et al. 1993) and other federal lands (FEMAT 1993) in the Pacific Northwest. Although distributions and amounts of late-successional forest habitats were projected through time, no population simulation models per se were used; thus, results should be treated as tentative working hypotheses on population responses until additional monitoring and research studies can validate them.

Steps 4 and 5. Develop planning alternatives and evaluate viability effects. The next steps entail the development and evaluation of one or more management alternatives that display a spectrum of potential effects on population viability and the status of other resources. Each alternative should be evaluated for its effect on population size, distribution, and persistence, as well as on resource and economic tradeoffs and other non-biological considerations. Perceived and expected utilities of each planning alternative should be described in the context of agency mandates and public expectations (Nash 1991).

Murphy and Noon (1992) formalized a mapping process that reflected biological and non-biological constraints toward development of a habitat conservation strategy for northern spotted owls. The process included the mapping of habitats and tentative conservation areas in an iterative process of proposing and testing hypotheses in the context of existing, real-world conditions, including the explicit expectation of continued timber harvest.

Step 6. Array the alternatives by effect. The sixth step arrays alternatives according to anticipated effects, such as on benefits to population viability and availability of other resources. Indirect costs, opportunity costs, and foregone options should be clearly delineated for each alternative. The range of acceptable alternatives that meet the decision criteria set by policy-makers can then be identified and arrayed.

Steps 7, 8, and 9. Selection, implementation, and monitoring. The final steps entail selection of an alternative, implementation of the alternative, and monitoring the results. Monitoring should be used to revisit all previous steps to adapt management as warranted by the data.

POPULATION VIABILITY RISK ANALYSIS

Objectives of Viability Risk Analysis

Population viability risk analysis evaluates potential effects on viability of populations from alternative courses of action. It is used to help craft conservation plans, inform the public, and aid decision-makers in differentiating among management alternatives. Specific objectives of a population viability risk analysis are:

- *To estimate the overall likelihoods of success or failure of implementing possible management actions.* What are possible outcomes? What is the overall range of possible outcomes, the set of more likely outcomes, and the most likely outcome? What are the probabilities of each outcome under distinct management actions?
- *To identify the key factors contributing to likelihoods of success or failure of management actions.* What are the key biological parameters that most affect predictions of outcomes and their associated likelihoods? How do variations in

the set of potential management guidelines influence success or failure of the strategies? Sensitivity tests of simulation models are useful for developing hypotheses of key factors to test through implementation and monitoring.

- *To reveal the degree with which likelihoods of success or failure are accurately and precisely portrayed, and the sources and magnitudes of uncertainty in such predictions.* How well estimated are the parameters that influence success or failure? Is the biological system understood adequately to make predictions of system responses? How well are potential future short-term and long-term conditions predicted? How do various factors compound to affect success or failure?
- *To evaluate the need for new information.* How much will it cost to add to our knowledge to reduce uncertainties associated with key parameters or to gather further basic biological information? How can new knowledge help in refining the likelihoods of predicted outcomes, or in devising new management strategies with lower social costs and equal (or better) likelihoods of maintaining viable populations?

Use of Decision Theory for Analyzing Viability

In decision theory, risk analysis serves to evaluate qualitatively or quantitatively the likelihoods of different outcomes of potential management actions. In the ecological literature, population risk analyses have typically included evaluations of the likelihoods of extinction within specified time periods, which are estimated from predictions of stochastic models of population demography, although estimating likelihoods of other conditions and parameters is also tenable (see Table 5.1 which lists the variety of parameters idiosyncratically estimated by various PVAs).

Population viability risk analysis and management can be conducted in the context of decision analysis (Maguire 1986b, 1991). This approach requires the identification of key management decision sequences that affect environments, habitats, and populations, as well as estimation of various responses of species and habitats to each management decision. Unfortunately, such data normally are not available; however, sometimes they can be approximated by extrapolating from available data or from using best professional guesses. More formal methods to derive professional judgments of potential outcomes include variations on the Delphi technique (Richey et al. 1985a,b). Also, various forms of Bayesian statistics, such as use of sequential and empirical Bayes algorithms (Gazey and Staley 1986; Milne et al. 1989), can be used initially to select outcome probabilities and correct them over time from results of research and monitoring. Each of these approaches, however, has shortcomings that can add uncertainties to predictions, in ways that are often unknown or not even quantifiable. Many decision analysis methods, such as use of decision trees, are more useful for the risk management stage of conservation planning than for the initial and more clinical phases of risk analysis. In using decision analysis for viability risk management, clear decision criteria should be articulated along with the factors and value ranges used to rank and select alternatives.

Decision theory has been used for prioritizing conservation of fish and wildlife species in Florida (Millsap et al. 1990). It also has been used for viability assessments of Concho water snake, *Natrix harteri paucimaculata* (Soulé 1989), grizzly bear (Maguire

1986a), black-footed ferret, *Mustella nigripes* (Maguire et al. 1988), tiger (Maguire and Lacy 1990), and Sumatran rhinoceros, *Dicerorhinus sumatrensis* (Maguire et al. 1987).

Definition of Risk

The term "risk" has been used in literature on decision theory and in conservation biology in a wide variety of contexts and (mostly unarticulated) meanings (Marcot and Salwasser 1991). As examples, the term has been used to depict means and variances of projected future population sizes in probabilistic and deterministic modeling, to describe scientific uncertainties in available studies and data, and to depict the degree of conservatism in decision-making attitudes. In a conservation planning strategy designed to provide for viable populations, risk is the likelihood that the implementation of a particular strategy will not result in viable populations that can be sustained though time. Technically, viability risk can be estimated as the frequency of replicate runs of a stochastic demographic simulation that fail to remain above a particular population level over a specific simulation time-frame.

Describing Uncertainties

Depicting uncertainties in scientific data and knowledge is one important aspect of evaluating risk. Risk analysis demands clearly displayed sources and degrees of uncertainty associated with the basic scientific information used to predict effects of various potential courses of action. Such uncertainties can include quantitative statistical estimates of the precision with which key parameters are known, such as estimation of confidence intervals or standard errors with predictions of finite rates of change of population size. Also pertinent is the description of both confidence and power of statistical hypothesis tests, such as tests of the null hypothesis that lambda for a given population is greater than or equal to 1 (for example, see Murphy and Noon 1992).

Uncertainties can also include qualitative statements concerning the degree of applicability of particular ecological concepts, theories, data sets, and models to cases at hand. Concepts, theories, data, and models might be less directly applicable where environmental conditions shift or where they are applied in unknown settings or different ecological settings than those for which the model was originally constructed.

In summary, predictions of potential effects of management alternatives on maintaining viable populations are influenced by: (1) the degree of certainty associated with parameters and basic data used in projecting outcomes; (2) the degree of applicability of concepts, theories, models, and data to the area of interest, especially in the context of the certainty of future conditions; (3) the propagation of errors caused by compounding parameters in multi-factor models; (4) the sampling error inherent in estimates of central tendencies and spread values (e.g., means and standard errors) of parameters; and (5) the degree of completeness and realism of the specific models used. Also, statistical properties of specific estimators, including bias, accuracy, and consistency, affect predictions of potential effects. Each of these influences should be discussed in a risk analysis of population viability.

TOWARD A MULTIPLE-SPECIES APPROACH

Conservation of individual species is a central facet of nearly all efforts to protect biotic diversity (Kelly and Harwell 1990). Species-specific analyses and management guidelines are most prudently applied in the context of multiple species and habitat conservation plans. Such biodiversity, or ecosystem-centered approaches, more economically and efficiently provide for multiple species needs than do species-by-species approaches. It is critical, however, that attention be paid to threatened, endangered, sensitive, and rare species on a species-specific level, as well.

The viability of populations depends on the sustenance and vitality of whole ecological communities and ecosystems, ultimately including humans. This concept was highlighted by Clark et al. (1991) for the management of resources in the Greater Yellowstone Ecosystem. Population viability analysis and management, therefore, are best integrated into more expansive resource management plans designed to provide for conservation of entire habitats and ecosystems.

One example of the need to integrate species-specific evaluations into a broader, coarse-filter approach to habitat management is the qualitative evaluation of species viability for planning management activities in late-successional and old-growth forests on National Forests of the Pacific Northwest (Thomas et al. 1993). Among other things, this evaluation was a test of how well management planning for old forests in the Northwest that focused on the maintenance of population viability of northern spotted owls would serve to ensure long-term viability of 667 other species of concern that are also closely associated with late-successional forests. Results suggested that habitats provided for spotted owls would ensure viability for only about one-third of those species. The other two-thirds required more site- or substrate-specific management guidelines than the spotted owl management plan provided, or those species were so poorly known scientifically that a judgment simply could not be made. Thomas et al. (1993) recommended additional management mitigations to provide for these other species, or suggested inventory and monitoring activities designed to provide basic empirical information on their habitat requirements. Thus, in this case, providing for the needs of northern spotted owls would not ensure viability of all other associated species.

Such approaches to multiple-species viability planning are just being developed elsewhere as well. For example, on Tongass National Forest in southeast Alaska, USDA Forest Service uses steps for screening species for viability concern, describing the current viability status of each species, and developing conservation strategies to meet multiple-species needs (L. Suring, pers. comm.). This approach uses data on life history characteristics, notably home range area and dispersal capabilities, to craft size and spacing of old-forest habitat conservation areas. A different approach taken by Raphael (this volume) uses linear programming to describe the optimal combination of various vegetation successional stages needed to meet multiple species needs. Both approaches remain untested and more are sure to be developed. However, it is absolutely essential that species-specific analysis, planning, and management is retained in more holistic approaches to ecosystem management and biodiversity conservation.

LITERATURE CITED

Anonymous. 1992. Final listings. *Endangered Species* 2(2):23.

Barrowclough, G. F. 1980. Gene flow, effective population sizes, and genetic variance components in birds. *Evolution* 34:789-798.

Barrowclough, G. F., and R. J. Gutiérrez. 1990. Genetic variation and differentiation in the spotted owl (*Strix occidentalis*). *Auk* 107:737-744.

Berger, J. 1990. Persistence of different-sized populations: an empirical assessment of rapid extinctions in bighorn sheep. *Conserv. Biol.* 4:91-98.

Brittingham, M. C., and S. A. Temple. 1983. Have cowbirds caused forest songbirds to decline? *BioScience* 33:31-35.

Burke, R. L., J. Tasse, C. Badgley, S. R. Jones, N. Fishbein, S. Phillips, and M. E. Soulé. 1991. Conservation of the Stephen's kangaroo rat (*Dipodomys stephensi*): planning for persistence. *Bull. Southern Calif. Acad. Sci.* 90(1):10-40.

Clark, T. W., E. D. Amato, D. G. Whittemore, and A. H. Harvey. 1991. Policy and programs for ecosystem management in the Greater Yellowstone Ecosystem: an analysis. *Conserv. Biol.* 5:412-422.

Collar, N. J., and P. Andrew. 1988. Birds to watch: the ICBP world checklist of threatened birds. ICBP Technical Publication No. 8, Smithsonian Institution Press, Washington, D.C.

Collar, N. J., and A. T. Juniper. 1992. Dimensions and causes of the parrot conservation crisis. In *New World Parrots in Crisis: Solutions from Conservation Biology*, eds. S. R. Beissinger and N. F. R. Snyder, pp. 1-24. Smithsonian Institution Press, Washington, D.C.

Collar, N. J., and S. N. Stuart. 1985. Threatened birds of Africa and related islands: the ICBP/IUCN red data book, part 1. 3rd ed. International Council for Bird Preservation and International Union for Conservation of Nature and Natural Resources, Cambridge, England.

Committee of Scientists. 1979. Final report. *Federal Register* 44(181):53967-53976.

Crouse, D. T., L. B. Crowder, and H. Caswell. 1987. A stage-based population model for loggerhead sea turtles and implications for conservation. *Ecology* 68:1412-1423.

Eberhardt, L. L. 1990. Survival rates required to sustain bear populations. *J. Wildl. Manage.* 54:587-590.

Emigh, T., and Pollak. 1979. Fixation probabilities and effective population numbers in diploid populations with overlapping generations. *Theoret. Pop. Biol.* 15:86-107.

FEMAT. 1993. Forest ecosystem management: an ecological, economic, and social assessment. Report of the Forest Ecosystem Management Assessment Team. Government Printing Office, Washington, D.C.

Ferson, S. 1988. Microcomputer software for stochastic demography and ecological risk analysis. *Am. Statistician* 42:273.

Forestry Commission. 1990. Forest nature conservation guidelines. Forestry Commission Publications, HMSO, London, England.

Frankel, O. H., and M. E. Soulé. 1981. Conservation and evolution. Cambridge University Press, Cambridge.

Gazey, W. J., and M. J. Staley. 1986. Population estimation from mark-recapture experiments using a sequential Bayes algorithm. *Ecology* 67:941-951.

Ginzburg, L. R., L. B. Slobodkin, K. Johnson, and A. G. Bindman. 1982. Quasiextinction probabilities as a measure of impact on population growth. *Risk Analysis* 2:171-181.

Haig, S. M., and L. H. Nortsrom. 1991. Genetic management of small populations. In *Challenges in the Conservation of Biological Resources. A Practitioner's Guide*, eds. D. J. Decker, M. E. Krasny, G. R. Goff, C. R. Smith, and D. W. Gross, pp. 119-138. Westview Press, Boulder, Colo.

Harmon, D. 1990. Wildlife and habitat. In *World Resources 1990-91*. The World Resources Institute. Pp. 121-140. Oxford University Press, New York.

Harris, R. B., and F. W. Allendorf. 1989. Genetically effective population size of large mammals: an assessment of estimators. *Conserv. Biol.* 3(2):181-191.

Harrison, S. 1991. Local extinction in a metapopulation context: an empirical evaluation. *Biol. J. Linn. Soc.* 42:73-88.

Harrison, S., D. Murphy, and P. Ehrlich. 1988. Distribution of the Bay checkerspot butterfly, *Euphydryas editha bayensis*: evidence for a metapopulation model. *Am. Nat.* 132:360-382.

Hedrick, P. W., and P. S. Miller. 1992. Conservation genetics: techniques and fundamentals. *Ecol. Appl.* 2:30-46.

Hill, W. G. 1979. A note on effective population size with overlapping generations. *Genetics* 92:317-322.

Hoage, R. J., ed. 1985. *Animal Extinctions*. Smithsonian Institution Press, Washington, D.C.

Iwasa, Y., and H. Mochizuki. 1988. Probability of population extinction accompanying a temporary decrease of population size. *Res. Pop. Ecol.* 30:145-164.

Kelly, R. J., and M. A. Harwell. 1990. Indicators of ecosystem recovery. *Environ. Manage.* 14(5):527-545.

King, W. B. 1978-1979. Red data book, 2: Aves. 2nd ed. International Union for Conservation of Nature and Natural Resources, Morges, Switzerland.

Kinnaird, M. F., and T. G. O'Brien. 1991. Viable populations for an endangered forest primate, the Tana River crested mangabey (*Cercocebus galeritus galeritus*). *Conserv. Biol.* 5(2):203-214.

Knight, R. R., and L. L. Eberhardt. 1984. Projected future abundance of the Yellowstone grizzly bear. *J. Wildl. Manage.* 48:1435-1438.

Lande, R., and G. F. Barrowclough. 1987. Effective population size, genetic variation, and their use in population management. In *Viable Populations for Conservation*, ed. M. E. Soulé, pp. 87-123. Cambridge University Press, New York.

Lehmkuhl, J. F., and L. F. Ruggiero. 1991. Forest fragmentation in the Pacific Northwest and its potential effects on wildlife. In *Wildlife and Vegetation of Unmanaged Douglas-fir Forests*, eds. L. F. Ruggiero, K. B. Aubry, A. B. Carey, and M. H. Huff, pp. 35-46. General Technical Report PNW-GTR-285. USDA Forest Service, Portland Ore.

Lynch, J. F., and D. F. Whigham. 1984. Effects of forest fragmentation on breeding bird communities in Maryland, USA. *Biol. Conserv.* 28:287-324.

Maguire, L. A. 1986a. An analysis of augmentation strategies for grizzly populations: the Cabinet-Yak Ecosystem as an example. Report to USDA Forest Service, Missoula, Montana. Contract No. 40-3187-4-1748.

Maguire, L. A. 1986b. Using decision analysis to manage endangered species populations. *J. Environ. Manage.* 22:345-360.

Maguire, L. A. 1991. Risk analysis for conservation biologists. *Conserv. Biol.* 5(1):123-125.

Maguire, L. A., T. W. Clark, R. Crete, J. Cada, C. Groves, M. L. Shaffer, and U. S. Seal. 1988. Black-footed ferret recovery in Montana: a decision analysis. *Wildl. Soc. Bull.* 16:111-120.

Maguire, L. A., and R. C. Lacy. 1990. Allocating scarce resources for conservation of endangered subspecies: partitioning zoo space for tigers. *Conserv. Biol.* 4:157-166.

Maguire, L. A., U. S. Seal, and P. F. Brussard. 1987. Managing critically endangered species: the Sumatran Rhino as a case study. In *Viable Populations for Conservation*, ed. M. E. Soulé, pp. 141-158. Cambridge University Press, Cambridge.

Marcot, B. G., R. Holthausen, and H. Salwasser. 1986. Viable population planning. In *The Management of Viable Populations: Theory, Applications, and Case Studies*, eds. B. A. Wilcox, P. F. Brussard, and B. G. Marcot. Center for Conservation Biology, Stanford, Calif.

Marcot, B. G., and H. Salwasser. 1991. Views on risk analysis for wildlife planning and management in USDA Forest Service. Presented at Society for Risk Analysis Annual Meeting, 8-11 Dec. 1991, Baltimore Md.

McGraw, J. B., and K. Garbutt. 1990. Demographic growth analysis. *Ecology* 71:1199-1203.

Millsap, B. A., J. A. Gore, D. E. Runde, and S. I. Cerulean. 1990. Setting priorities for the conservation of fish and wildlife species in Florida. *Wildl. Monogr.* 111:1-57.

Milne, B. T., K. M. Johnston, and R. T. T. Forman. 1989. Scale-dependent proximity of wildlife habitat in a spatially-neutral Bayesian model. *Landscape Ecol.* 2:101-110.

Morrison, M. L., B. G. Marcot, and R. W. Mannan. 1992. *Wildlife Habitat Relationships: Concepts and Applications,* University of Wisconsin Press, Madison, Wisc.

Murphy, D. D., K. E. Freas, and S. B. Weiss. 1990. An environment-metapopulation approach to population viability analysis for a threatened invertebrate. *Conserv. Biol.* 4:41-51.

Murphy, D. D., and B. D. Noon. 1991. Coping with uncertainty in wildlife biology. *J. Wildl. Manage.* 55:773-782.

Murphy, D. D., and B. R. Noon. 1992. Integrating scientific methods with habitat conservation planning: reserve design for northern spotted owls. *Ecol. Appl.* 2(1):3-17.

Nash, S. 1991. What price nature? Future ecological risk assessments may chart the values, and the odds. *BioScience* 41(6):677-680.

Nei, M., and F. Tajima. 1981. Genetic drift and estimation of effective population size. *Genetics* 98:625-640.

O'Brien, S., M. Roelke, L. Marker, A. Newman, C. Winkler, D. Meltzer, L. Colly, J. Evermann, M. Bush, and D. Wildt. 1985. Genetic basis for species vulnerability in the cheetah. *Science* 227:1428-1434.

Pimm, S. L., H. L. Jones, and J. Diamond. 1988. On the risk of extinction. *Am. Nat.* 132:757-785.

Pollak, E. 1983. A new method for estimating the effective population size from allele frequency changes. *Genetics* 104:531-548.

Pulliam, H. R. 1988. Sources, sinks, and population regulation. *Am. Nat.* 132:652-661.

Rabinowitz, D., S. Cairns, and T. Dillon. 1986. Seven forms of rarity and their frequency in the flora of the British Isles. In *Conservation Biology: The Science of Scarcity and Diversity,* ed. M. E. Soulé, pp. 182-204. Sinauer Associates, Sunderland, Mass.

Richey, J. S., R. R. Horner, and B. W. Mar. 1985a. The Delphi technique in environmental assessment. II. Consensus on critical issues in environmental monitoring program design. *J. Environ. Manage.* 21:147-159.

Richey, J. S., B. W. Mar, and R. R. Horner. 1985b. The Delphi technique in environmental assessment. I. Implementation and effectiveness. *J. Environ. Manage.* 21:135-146.

Rosenberg, K. V., and M. G. Raphael. 1986. Effects of forest fragmentation on vertebrates in Douglas-fir forests. In *Wildlife 2000: Modeling Habitat Relationships of Terrestrial Vertebrates,* eds. J. Verner, M. L. Morrison, and C. J. Ralph, pp. 263-272. University of Wisconsin Press, Madison Wisc.

Salwasser, H., R. S. Holthausen, and B. G. Marcot. 1986. Implications of laws and regulations for viable population planning in the National Forest system. In *The Management of Viable Populations: Theory, Applications, and Case Studies,* eds. B. A. Wilcox, P. F. Brussard, and B. G. Marcot. Center for Conservation Biology, Stanford, Calif.

Schonewald-Cox, C. M. 1983. Conclusions: guidelines to management: a beginning attempt. In *Genetics and Conservation: A Reference for Managing Wild Plant and Animal Populations,* eds. C. M. Schonewald-Cox, S. M. Chambers, B. MacBryde, and W. L. Thomas, pp. 414-445. Benjamin/Cummings, Menlo Park, Calif.

Scott, J. M., C. B. Kepler, C. van Riper, and S. I. Fefer. 1988. Conservation of Hawaii's vanishing avifauna. *BioScience* 38(4):238-253.

Shaffer, M. L. 1981. Minimum population sizes for species conservation. *BioScience* 31:131-134.

Shaffer, M. L. 1983. Determining minimum viable population sizes for the grizzly bear. *Internat. Conf. Bear Res. Manage.* 5:133-139.

Smith, J. L. D., and C. McDougal. 1991. The contribution of variance in lifetime reproduction to effective population size in tigers. *Conserv. Biol.* 5(4):484-490.

Soulé, M. E., ed. 1987. *Viable Populations for Conservation.* Cambridge University Press, Cambridge.

Soulé, M. E. 1989. Risk analysis for the concho water snake. *Endangered Species Update* 6(10):19,22-25.

Soulé, M. E., and B. A. Wilcox, eds. 1980. *Conservation Biology: An Evolutionary-ecological Perspective.* Sinauer Associates, Sunderland Mass.

Thomas, J. W., E. D. Forsman, J. B. Lint, E. C. Meslow, B. R. Noon, and J. Verner. 1990. A conservation strategy for the northern spotted owl. Interagency Scientific Committee to Address the Conservation of the Northern Spotted Owl. USDA Forest Service, Portland, Ore.

Thomas, J. W., M. G. Raphael, R. G. Anthony, E. D. Forsman, A. G. Gunderson, R. S. Holthausen, B. G. Marcot, G. H. Reeves, J. R. Sedell, and D. M. Solis. 1993. Viability assessments and management considerations for species associated with late-successional and old-growth forests of the Pacific Northwest. USDA Forest Service National Forest System and Forest Service Research, U.S. Government Printing Office, Washington, D.C.

USDA. 1986. Draft supplement to the environmental impact statement for an amendment to the Pacific Northwest Regional Guide. 2 vols. USDA Forest Service, U.S. Government Printing Office, Washington, D.C.

Wilcove, D. S., and J. W. Terborgh. 1984. Patterns of population decline in birds. *Am. Birds* 38:10-13.

Wiley, J. W., N. F. R. Snyder, and R. S. Gnam. 1992. Reintroduction as a conservation strategy for parrots. In *New World Parrots in Crisis: Solutions from Conservation Biology*, eds. S. R. Beissinger and N. F. R. Snyder, pp. 165-200. Smithsonian Institution Press, Washington, D.C.

6

AN ORGANISM-BASED PERSPECTIVE OF HABITAT FRAGMENTATION

Scott M. Pearson,[1] Monica G. Turner,[2] Robert H. Gardner,[3] and Robert V. O'Neill

Environmental Sciences Division
Oak Ridge National Laboratory
Oak Ridge, Tennessee
U.S.A.

Environmental heterogeneity in both space and time has been important in the evolution and maintenance of biodiversity. Moreover, this heterogeneity is hierarchical (O'Neill et al. 1986). Differences occur between biomes, between landscapes (within biomes), and in nested levels of organization within landscapes. For example, climatic factors, such as temperature and rainfall, determine the spatial distribution of life zones on a continent (Raunkaier 1934; Whittaker 1956). Landscapes in the southern Appalachian Mountains are dominated by deciduous forest. Although deciduous forest covers most of these landscapes, the community composition of dominant tree species varies with soil type and topography. In a given area of the same soil type and topographic exposure, the density, age structure, and seral status of trees can vary depending on the history of disturbance. These local patterns of structural variation influence levels of light levels reaching the forest floor and, therefore, the herb community. Thus, hierarchical patterns of heterogeneity are a consequence of the complexity within ecological communities (Dayton 1984). The maintenance of biodiversity depends upon the preservation of complex ecological patterns.

Natural landscapes are dynamic systems with temporal and spatial heterogeneity. However, human exploitation tends to simplify landscapes (Krummel et al. 1987). For example, agriculture and forestry tend to reduce spatial heterogeneity in biota. These land uses often change diverse communities to monocultures. Attempts to control disturbances in ecosystems, such as fire suppression and flood control, interrupt natural disturbance/recovery regimes. The challenge of preserving biodiversity in managed landscapes is to incorporate natural levels of spatial and temporal heterogeneity into management schemes.

The concept of scale has emerged as an important topic among ecologists who recognize the role of heterogeneity in natural ecosystems. Subjects related to scale, such as grain (level of detail) and extent (size of area or duration of time), are frequently used

[1] Present address: Biology Department, Mars Hill College, Mars Hill, North Carolina.
[2] Present address: Department of Zoology, University of Wisconsin, Madison, Wisconsin.
[3] Present address: Appalachian Environmental Laboratory, Frostburg, Maryland.

to determine the appropriate interpretation of ecological data (Turner et al. 1989; Wiens 1989). Likewise, considering scale is important for the application of ecological principles to the protection of biodiversity and conservation. The scale of a particular conservation effort will affect its ability to fulfill its goals. For example, the spatial extent of a reserve system must be large enough to sustain the ecological process it is intended to protect. However, the geographic boundaries are often determined by the distribution of surviving populations, political boundaries, or fiscal constraints rather than on an accurate understanding of the ecological process. Temporal scales can also be included in conservation strategies. Estimating the probability of a single population surviving for a specified period of time based on population size and environmental variability (e.g., Shaffer 1981) is an example of incorporating temporal scale. Therefore, scaling issues should be considered when designing and evaluating conservation strategies. Techniques for identifying areas deserving protection have been developed (e.g., Scott et al. 1987). However, preservation of these systems depends on protecting the broader landscapes in which they are embedded.

This chapter addresses issues related to biodiversity protection at landscape scales, areas ranging from tens to thousands of hectares. However, the working definition of landscape scale should depend on the relevant organisms and ecological processes (Wiens 1989; Wiens and Milne 1989). For example, beetles and eagles operate at different spatial scales. In this chapter, we will emphasize the importance of natural patterns of spatial and temporal heterogeneity, encourage a broader-scale perspective for conservation, and illustrate the interaction between landscape-level heterogeneity and organism-based scales of resource utilization with a simulation experiment.

RETHINKING CONCEPTS OF HETEROGENEITY

Strategies for the protection of biodiversity depend on an understanding of the ecological function of the relevant species, communities, or ecosystems. The perception of these functions might be affected by scale. For example, conclusions about a species' persistence might depend on the spatial extent over which it is studied. In source-sink populations (sensu Pulliam 1988), some populations exhibit birth rates lower than mortality rates. Although some reproduction occurs in these sink populations, they are dependent on demographic subsidies from source populations, where birth rates exceed mortality rates. A study of only one population might conclude that the population is healthy and growing, or that it is declining. Only a study of the larger metapopulation could identify the critical source populations.

The process-oriented concept of the ecological neighborhood is useful for management of biodiversity because it encourages ecologists and managers to evaluate spatial patterning of resources in the context of ecological functions. An ecological neighborhood is defined by an ecological process, the time and space scale appropriate for that process, and an organism's relationship to the process during a specific time period (Addicott et al. 1987). This concept can be applied to ecological activities as diverse as foraging by individual animals, gene flow, predator-prey interactions, or population growth and persistence. For an individual breeding bird, the ecological neighborhood might be its territory. For a flowering plant, the neighborhood would include the activity ranges of its pollinators and the distribution of the plants they visit. Neighborhoods for community-level processes are more complex because they involve the population dynamics of all the component species, which can operate at different scales.

Addicott et al. (1987) used ecological neighborhoods to describe the relative scale of heterogeneity in the environment. Heterogeneity is coarse if the neighborhood includes only a portion of one patch, but fine-scaled if the neighborhood includes several patches. Thus, the scale of heterogeneity depends on both spatial pattern and the ecological neighborhoods of populations and communities. Moreover, in fine-scaled landscapes the spatial arrangement of resources can affect persistence of a population. For example, the network structure of a metapopulation will be disrupted if the distance between subpopulations exceeds the organism's dispersal capabilities. Changes in the composition and spatial arrangement of habitats will affect the community composition of individual patches (Pearson 1993).

Role of Spatial Heterogeneity

Landscapes have natural levels of temporal and spatial heterogeneity. These spatial and temporal patterns are driven by disturbance/recovery regimes (White 1979; Romme 1982; Baker 1989; Turner et al. 1993) in addition to variation in physical forces like geology and climate. For example, the tree communities of deciduous forests of the eastern United States change in composition with elevation, topography, and soil characteristics (Whittaker 1956). They are also affected by a large number of small disturbances. Tree falls affect about 1% of the landscape annually (Runkle 1981), but recovery from these mild disturbances is rapid. Thus, natural landscapes in eastern deciduous forests are dominated by mature forests with isolated, ephemeral patches of early successional habitat. In contrast, lodgepole pine forests of the northern Rocky Mountains regularly experience forest fires (Romme and Despain 1989). Variation in the spatial extent of fires coupled with the long recovery times because of harsh climate results in a landscape with a mosaic of forests of different ages (Romme 1982).

Changing the patterns of resources and habitats at the landscape level can affect the species composition. Increasing spatial heterogeneity can increase species (gamma) diversity by increasing the number of habitat types available in the landscape. This heterogeneity provides habitat for species adapted to different habitats and seral stages. However, excessive levels of heterogeneity can, depending on the actual spatial pattern, result in the loss of species sensitive to habitat fragmentation. Thus, landscape-level changes can have desirable or undesirable effects depending upon a particular conservation goal and the specific changes that occur. For example, for the survival of a population of black bears, heterogeneity, as measured by variation in type and phenology of food items, could provide a varied, nutritious diet. Such heterogeneity should have a positive effect on population persistence by insuring that suitable food items are consistently available. Conversely, increasing spatial heterogeneity by adding unsuitable habitats, such as roads (McLellan and Shackleton 1988), would have a negative effect. Therefore, heterogeneity per se is not good or bad; its value depends on the effect of changing specific spatial and temporal patterns on specific ecological processes. All species are adapted to particular spatial and temporal patterns in the abiotic and biotic characteristics of their environment. In general, the risk of losing biodiversity and disrupting ecological functions is greatly increased when natural patterns of heterogeneity are altered.

Characterizing Spatial Heterogeneity

The terms patch, matrix, and corridor are boldly written in the lexicon of landscape ecology (Forman and Godron 1986). The definition and use of these terms have been useful in describing the spatial structure of habitats at landscape scales. The widespread acceptance and popularity of these terms will provide a basis for the next step in understanding landscape-level processes.

These terms are especially useful in landscapes where the contrast between the patches and the matrix is sharp and has ecological meaning. The concepts of patches and corridors evolved in the context of highly manipulated landscapes as in western Europe and the northeastern United States and often reflect an anthropocentric perspective. But, in natural landscapes, patches and corridors are seldom discrete (Kotliar and Wiens 1990). If the variation between patch and matrix is greater than the variation within either patch or matrix, then patches can be easily delineated. However, discernable, nonarbitrary patches might not be apparent in landscapes showing continual gradual variation, but spatial structure might still be present (Musick and Grover 1991). Therefore, patch, matrix, and corridor can be of limited use in describing natural pattern in many landscapes.

It is difficult to define specific patches in a landscape without being arbitrary. Patches have been defined in an ecological context as a discontinuity in an ecological variable affecting an organism (Wiens 1976) or as a place where the abundance of resources or organisms is low or high relative to its surroundings (Roughgarden 1977). Both of these definitions use an organism-based perspective. Analyzing a landscape from an organism's perspective, particularly of a non-vertebrate, can reveal a very different set of environmental patterns and gradients (Wiens and Milne 1989; Buechner 1989) than the patterns apparent to humans. Even so, natural landscapes tend to have hierarchical levels of variation (O'Neill et al. 1986; Kotliar and Wiens 1990). Nested levels of heterogeneity might exist from the perspective of a single organism. If a landscape could be evaluated from the viewpoint of many species, any single map of the landscape would be arbitrary because that map would reflect the perspective of one species or a small group of similar species (Risser et al. 1984).

Community-level interactions can complicate the optimal design of patches and corridors for preserving biodiversity. For example, several pairs of antbirds existed in 1-10 ha fragments of tropical forest until corridors connecting forests were cut (Bierregaard 1990). Antbirds are obligate commensals with foraging army ants; they disappeared after army ant activity was lost in the forest fragment. The ant colonies have home ranges of several hundred hectares (Willis and Oniki 1978), and colonies did not visit the forest fragment when the corridor was cut. Bierregaard (1990) also reported that bird species dependent on foraging flocks disappeared from forest fragments when the nuclear species (bird species critical to flock organization) left the fragment.

Conservation strategies using patch networks to preserve populations rely heavily on dispersal between patches to maintain genetic variation and to avoid stochastic extinctions (Diamond 1975). Corridors are intended to facilitate movement between patches. However, satisfying the habitat preferences and dispersal capabilities of different species requires corridors of various habitat types. Although satisfactory corridors permit animals to move between patches, they must be able to sustain plant populations if the distance between patch requires more than one generation to traverse. Corridors that satisfy the dispersal needs of the entire community must be short, wide, and free of environmental gradients acting as a barrier to any species. This perfect corridor might not exist. However, multiple corridors to each patch could solve this problem for some patches. Moreover, a small archipelago of patches isolated from the mainland of biodiversity can exhibit island-like dynamics like high extinction probabilities because of small population sizes. In addition to being arbitrary, a management strategy that concentrates on patches and corridors might be too narrowly focused for conservation.

The value of patches and corridors for conservation of biodiversity depends on the larger landscape in which they are embedded. The patch/matrix/corridor concepts imply that landscape biodiversity can be preserved in a selected number of refugia (patches) while ignoring the matrix. However, this implication ignores the vital role of the matrix in maintaining the biodiversity of a single site. For example, Glenn and Collins (1992) found that individual sites (9-50 ha in extent) in a Kansas prairie exhibited a high turnover of species between years. High rates of extinction were balanced by frequent immigration, resulting in a relative stability in species richness. An open system like this cannot be preserved by a small number of patches because maintenance of diversity at any one site depends on processes such as immigration which operate at the scale of the larger system. In an Amazonian forest, a 70-m wide strip of clearcut eliminated movement by forest understory birds from the unbroken forest to a forest island. However, as second growth began to fill the former clearcut, movement by these birds between the island and the larger forest became more common (Bierregaard 1990). Thus, movement between patches increased as the contrast between the patches and matrix decreased.

In a landscape dominated by a matrix of unsuitable habitats for native species, ecological processes in the matrix might overwhelm patches and corridors, depending on the size of the patches and the ability of the matrix processes to penetrate the patch interiors (i.e. edge effects; Wilcove 1985; Lovejoy et al. 1986; Janzen 1988). Therefore, it might be necessary to manage not only the patches and corridors, but also the matrix.

Whereas concepts and terms such as patches are useful for describing and discussing landscape-level patterns, it is time to take the next step in recognizing broad-scale heterogeneity and its importance to biodiversity. This step entails viewing the landscape as a whole from the perspective of different species instead of subdividing the landscape into arbitrary patch and matrix categories. This approach is more useful for formulating preservation strategies because it promotes the creation of management strategies that preserve patterns of heterogeneity critical for the persistence of biodiversity. This broad-scale perspective has already been recognized. In a recent book, Hudson (1991) noted the importance of linkages between protected regions. This book described regional-scale corridors (Noss 1991) that encompass a great deal of environmental heterogeneity and are wide enough to be considered landscapes themselves because populations of many species can complete their life cycle within the corridor itself.

The following portion of this chapter will describe an example of how landscape-level characteristics such as the abundance of suitable habitat and its spatial arrangement affect habitat fragmentation, an important, broad-scale characteristic that can affect population persistence. Moreover, this example shows how evaluating habitat fragmentation (or connectivity) from the perspective of different organisms affects conclusions about the suitability of different landscapes. This example involves the use of a computer simulation model. Although the model simulates a simple landscape that consists of only two habitat types, it illustrates the complex interactions between habitat abundance, spatial pattern, and resource utilization of different species. The technical details of the simulations have been documented for interested readers.

LANDSCAPE CONNECTIVITY: THE INTERACTION BETWEEN LANDSCAPE PATTERN AND ORGANISMAL SCALES

Habitat fragmentation is one of the greatest threats to biodiversity (Whitcomb et al. 1981) and, therefore, conservation strategies should maintain habitat connectivity within and among landscapes. Connectivity depends on both the abundance and spatial arrangement of habitats in a landscape. In raster maps where cells of a given habitat are arranged randomly and connectivity is defined by nearest neighbor adjacencies, the habitat spans the map if it occupies about 60% or more of the landscape (Gardner et al. 1987). Below this critical percentage, connectivity is severely reduced on random maps, and the habitat occurs as numerous small isolated clusters. This trend in connectivity has also been observed in forested habitat in real landscapes (Gardner et al. 1991).

Thresholds of connectivity can exert strong influences on the dynamics of several ecological processes, such as spread of disturbance (Turner et al. 1991; O'Neill et al. 1992), utilization of resources (O'Neill et al. 1988), and the movement and dispersal of organisms (Gardner et al. 1989; Gardner et al. 1991). However, connectivity also depends on the ability of the organism or process to move across the landscape. For example, a plant with wind-dispersed seeds is more likely to colonize a small, apparently disconnected cluster of habitat than a heavy-seeded plant lacking a mechanism for long-range dispersal. Similarly, a river or a highway can be a barrier to movement for a mouse, but a bird or deer might regularly cross such obstacles. Critical thresholds in habitat abundance and connectivity can be identified for a variety of organisms, but these thresholds will depend on both landscape pattern and the scale at which the organism can use the landscape.

A series of simulation experiments was designed (1) to explore the implications of landscape pattern for organisms that perceive the landscape at different scales, and (2) to illustrate how a particular organism might respond to alternative patterns of suitable habitat across the landscape. A set of landscape maps was developed in which the abundance and spatial distribution of habitat and the degree of scale-dependent structure were varied. In addition, a set of rules was defined to describe alternative resource utilization scales (sensu O'Neill et al. 1988) for a variety of organisms. Resource utilization scale, as discussed here, represents an organism's ability to use disjunct habitat fragments. We were particularly interested in how thresholds of connectivity might differ with the scale at which an organism can utilize resources and how the scale-dependent structure of the landscape (spatial arrangement of habitat) might affect the ability of an organism to use different landscapes.

Landscape Patterns

The simulated landscape was represented as a 512 by 512 grid in which the abundance and spatial distribution of suitable habitat was varied. Suitable habitat was distributed at random across the landscape with probability p (Gardner et al. 1987; O'Neill et al. 1988; Turner et al. 1989) or by using a hierarchical map-generating algorithm (O'Neill et al. 1992; Gardner et al. in press). In random maps, the size, shape, and arrangement of suitable habitat is controlled by p (Gardner et al. 1987) and no spatial scale is imposed on the pattern. In hierarchical maps (O'Neill et al. 1992), scale-dependent structure is introduced with curdling algorithms from fractal geometry (Mandelbrot 1983).

To create a hierarchical map, consider that a landscape contains three successively finer (nested) scales at which pattern will be generated. At the broadest scale, the landscape is considered a m_1 x m_1 grid in which potentially suitable habitat is distributed randomly within $(m_1)^2$ cells by selecting cells with probability p_1. Each of these selected cells then contains an m_2 x m_2 grid in which potentially suitable habitat is distributed randomly in p_2 of the cells that were selected at the first level. This process is then repeated a third time by subdividing each of these sites into a m_3 x m_3 grid in which suitable habitat occurs in p_3 of the cells selected at the previous level. Cells that are selected at this third level are considered suitable habitat. The total number of cells in the landscape is $(m_1$ x m_2 x $m_3)^2$, and the overall proportion, p, of the landscape occupied by suitable habitat is p_1 x p_2 x p_3.

An illustration of the curdling algorithm for a map with three scales in which $m_1 = m_2 = m_3 = 3$ (i.e., a 27 x 27 grid) is shown to the left in Figure 6.1. Black cells represent cells at each level l that were selected at a probability of p_l. The overall p of the curdled map at the finest level is 0.44 x 0.66 x 0.55 = 0.16.

Curdled maps represent an extreme case of scaled pattern. More subtle patterns can be introduced by adding "wheys" to the "curds." A wheyed map can be described by the curdled map discussed above, with the following additions. At the second level, cells that were designated as unsuitable at the first level are now considered to contain a proportion q_2 of suitable habitat. Similarly, at the third level, cells designated as unsuitable at the second level are now considered to contain a proportion q_3 of suitable habitat. In the curdled and wheyed map (right side, Fig. 1), the overall proportion p of the landscape occupied by suitable habitat is now:

$$p = [p_1*p_2*p_3] + [p_1*(1-p_2)*q_3] + [(1-p_1)*q_2*p_3] + [(1-p_1)*(1-q_2)*q_3].$$

If $p_1 = p_2 = q_2 = p_3 = q_3$, the map is quite similar to a random map.

Resource Utilization Scales

A set of three rules for how an organism might perceive and move in a landscape was defined to generate a range of scales (Figure 6.2). Rule 1 represents an organism that can move to suitable habitat only within the four cardinal directions from the organism's present site. Rule 2 represents an organism that can move to suitable habitat within the four cardinal and four subcardinal directions. Under Rule 2, an organism can move to a

Figure 6.1. Hierarchically structured maps created by using curdling and wheying algorithms. The three maps in the left column were created by curdling at three nested levels (p_1, p_2, and p_3). The maps on the right were created using both curdling and wheying. Wheys are introduced at the second level by turning on some cells (at q_2=0.22) which were turned off at the coarser first level. See text for a further explanation of these techniques.

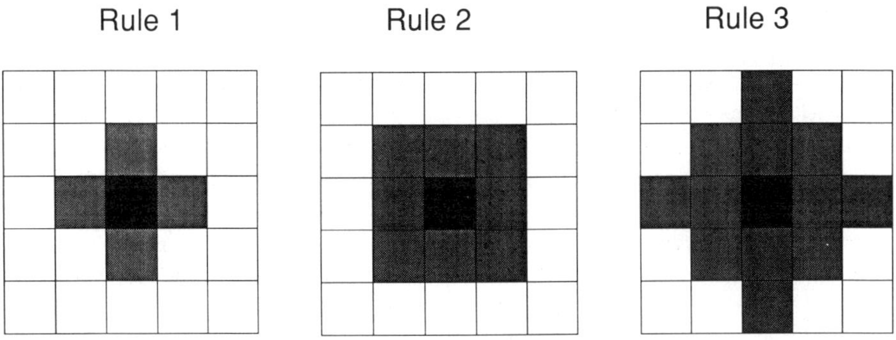

Figure 6.2. Three resource utilization rules used to identify clusters of connected habitat. The solid square indicates the cell containing suitable habitat. Other cells of suitable habitat covered by the shaded cells are considered connected to solid cell. The rules represent three successively broader scales of resource use.

suitable habitat on adjacent or diagonal cells and exploit more suitable habitat than under Rule 1. Rule 3 expands the scale of movement from Rule 2 to include not only the eight adjacent and diagonal neighbors, but also cells that are two cells away from the organism's current position. Thus, Rule 3 allows an organism to move to adjacent or diagonal sites or to jump over a pixel of unsuitable habitat in the cardinal directions. We used these rules in two sets of simulations to explore the interaction between the scale at which organisms perceive the landscape the scale of pattern on the landscape.

Experiment I: Organisms Operating at Different Scales in Random Landscapes.

First, we address the question of how organisms that perceive the landscape at different scales respond to the abundance of suitable randomly distributed habitat. A series of random maps was generated with the proportion, p, of the landscape occupied by suitable habitats ranging from 0.1 to 0.9. Three replicate maps were created for each p level. To compare how organisms operating at different scales can use these landscapes, the pattern of suitable habitat on each map was quantified with each of the three rules described above. Resource clusters were identified by the movement rule. A single cluster is defined as a set of resource cells are accessible by an organism moving within the cluster. Resource cells not accessible to an organism within one cluster belong to separate clusters. Resource clusters were identified using each resource utilization rule, and the weighted mean size of resource clusters (S_{avg}) was determined. Because the size of the largest contiguous area that an organism can use is often important for conservation, we then identified the largest resource cluster on the map. The relative size of the largest cluster (LC_{rel}) was determined by dividing its size (LC) by the maximum cluster size possible (LC_{max}). The maximum cluster size is given by pm^2, when all suitable habitat occurs in a single cluster. Thus, suitable habitat across the landscape becomes contiguous as LC_{rel} approaches 1. LC_{rel} is inversely related to the degree of fragmentation of suitable habitat.

The weighted mean size of resource clusters (S_{avg}) varied with both the rule and the proportion, *p*, of the landscape occupied by suitable resources (Figure 6.3) on the random maps in which *p* was varied. Values of S_{avg} increased rapidly as *p* increased, but the value of *p* at which this increase occurred varied among the rules. For Rule 1, which represents the most restricted movement, mean cluster size increased rapidly near $p = 0.6$, the critical threshold for connectivity reported from percolation theory (Stauffer 1985). Rapid increases in S_{avg} occurred near $p = 0.4$ for Rule 2 and near $p = 0.3$ for Rule 3. As *p* increased to approximately 0.7, there was little or no difference in the change in S_{avg} among the different rules. Similarly, when *p* was very low on these random maps, e.g., < 0.25, S_{avg} was very low, and no differences among the rules were found.

The relative size of the largest resource cluster, LC_{rel}, also varied with *p* and among the rules (Figure 6.4). The value of LC_{rel} approached zero for all four rules at low *p* and reached a value of 1 at some higher value of *p*. The transition between 0 and 1 was abrupt, but the value of *p* at which this abrupt transition occurred varied among the rules. For Rule 1, the rapid increase again occurred near $p = 0.6$. As with the rapid increases in S_{avg}, the value of LC_{rel} increased rapidly near $p = 0.4$ for Rule 2 and near $p = 0.3$ for Rule 3. The *p* values at which these rapid changes occur represent the critical threshold, p_{crit}, at which randomly distributed sites of suitable habitat become connected across the landscape for organisms that operate at the scales defined by each rule.

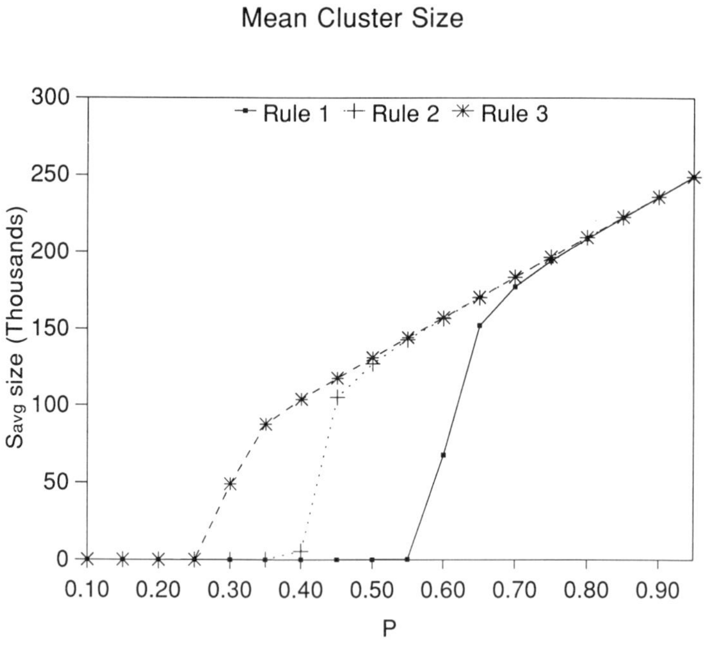

Figure 6.3. Relationship between average cluster size and frequency of suitable habitat. The average number of cells in habitat clusters (S_{avg}) increases as the proportion of the landscape covered by suitable habitat, *p*, increases. Habitat clusters were measured by three resource utilization rules (see Figure 6.2).

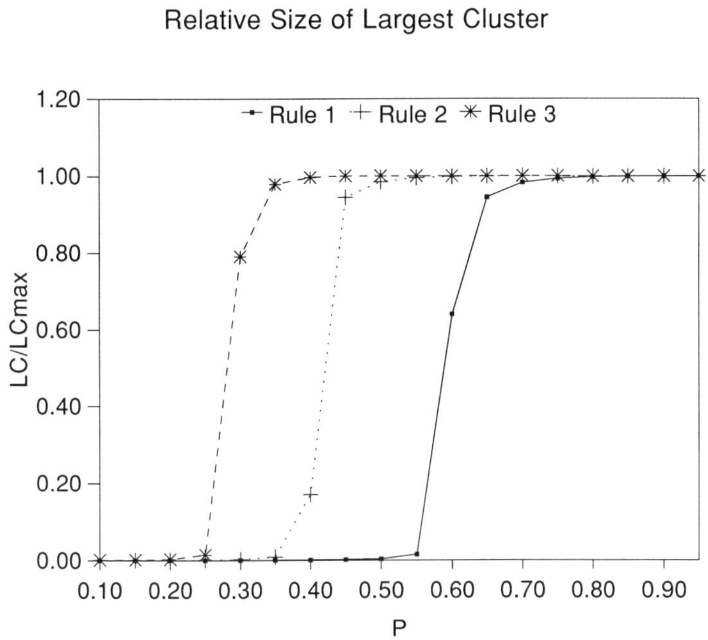

Figure 6.4. Relationship between habitat connectivity and frequency of suitable habitat. The relative size of the largest cluster (LC/LC_{max}) indicates the level of connectivity among the suitable habitat on the landscape. Habitat is highly connected when LC/LC_{max} is near 1 and highly fragmented when LC/LC_{max} is near 0. LC/LC_{max} for a random landscape is shown here for clusters measured by three resource utilization rules (See Fig. 2). LC/LC_{max} increases as the proportion of the landscape covered by suitable habitat, p, increases.

Experiment II: Scaled Landscape Patterns

Next, we address how organisms with the same resource utilization rule respond to landscapes which have the same abundance of suitable sites, but the spatial patterns of habitat is different. We held p constant at three levels (p = 0.328, 0.422, 0.656) and varied the pattern in the landscape by generating curdled maps, curdled and wheyed maps, and random maps. Suitable habitat was generated at three levels in which $m_1 = m_2 = m3 = 8$ (i.e., a 512 x 512 grid). For both curdled and curd-and-whey maps, we generated two alternative patterns to represent a range of potential patterns. One pattern had the higher value of p at the finest scale, i.e., the third level, and we refer to these maps as curdled-bottom or wheyed-bottom. The other pattern had the higher value of p at the coarsest level, and we refer to these maps as curdled-top or wheyed-top. An illustration of a set of these maps with the same p is shown in Figure 6.5, and the values of p_1 and q_1 used to generate the maps are shown in Table 6.1. The landscape pattern on each map was again quantified with each of the three rules and determining S_{avg} and LC_{rel} as described above.

Consider first a landscape in which suitable habitat is not very abundant, e.g., p = 0.328. If an organism uses resources as defined by Rule 1 (the most restrictive rule

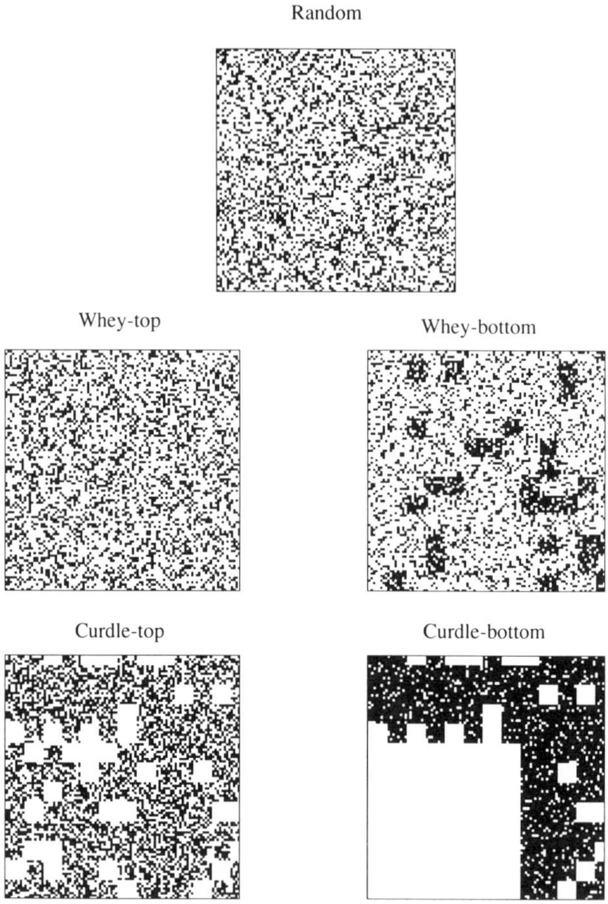

Figure 6.5. Hierarchical landscape patterns. Landscape-level pattern is strongly affected by hierarchical structure. These 100 x 100 cell maps are covered by the same amount of suitable habitat (32.8%) but were produced by random, curdling, and curdling-and-wheying algorithms. Shaded cells represent suitable habitat, and open cells represent unsuitable habitat.

allowing movement only to the adjacent neighbors), the weighted mean cluster size (S_{avg}), as perceived by this organism, is very low on the random and wheyed-top maps, slightly higher on curdled-top and wheyed-bottom, but dramatically larger on the curdled-bottom map (Table 6.2). Similarly, the relative size of the largest cluster (LC_{rel}) is approximately 0.5 on the curdled-bottom map but is near zero for the other patterns. Thus, the suitable habitat of this organism is more connected on the curdled-bottom map than on the other maps.

An organism using resources as defined by Rule 2 perceives small clusters on the random, wheyed-bottom, and wheyed-top maps, but perceives large, more connected

Table 6.1. Values of the proportion (p_l) of the landscape occupied by suitable habitat that were used at each of three nested levels (l) to generate hierarchical maps. See text for description of the methods for generating hierarchical landscape maps.

Map pattern[1]	p_l			q_l		
	p_1	p_2	p_3	q_1	q_2	q_3
			$p = 0.328$			
Curdled-bottom	0.500	0.750	0.875	0	0	0
Wheyed-bottom	0.125	0.375	0.750	0	0.125	0.250
Curdled-top	0.875	0.750	0.500	0	0	0
Wheyed-top	0.875	0.625	0.375	0	0.625	0.250
			$p = 0.422$			
Curdled[2]	0.75	0.75	0.75	0	0	0
Wheyed-bottom	0.250	0.375	0.625	0	0.125	0.375
Wheyed-top	0.875	0.625	0.375	0	0.625	0.500
			$p = 0.656$			
Curdled-bottom	0.750	0.875	1.000	0	0	0
Wheyed-bottom	0.125	0.250	0.750	0	0.250	0.625
Curdled-top	1.000	0.875	0.750	0	0	0
Wheyed-top	1.000	0.750	0.625	0	0.875	0.750

[1] The terms 'bottom' and 'top' refer to whether the highest values of p and q were used at the most coarse level (l = 1) or at the finest level (l = 3).

[2] There is a finite set of values that can be used to generate a hierarchical map with a given p, and for p = 0.422 it was impossible to generate a curdled-top and curdled-bottom.

clusters on both the curdled-bottom and curdled-top maps, as indicated by S_{avg} and LC_{rel} (Table 6.2). Thus, increasing the resource utilization scale to include diagonal neighbors permits an organism under Rule 2 to perceive a relatively connected landscape with two scale-dependent patterns. Further increasing the resource utilization scale to Rule 3 again results in an increase in S_{avg} and LC_{rel} on most of the maps (Table 6.2). However, S_{avg} and LC_{rel} remain relatively low on the wheyed-bottom map, even with Rule 3.

When the abundance of suitable habitat is increased to p = 0.422, the trend for a Rule 1 organism is similar to that observed when p = 0.328. The Rule 1 organism still perceives relatively small disconnected resource clusters except on the curdled landscape, when the largest cluster comprises more than 90% of the suitable habitat in the landscape (Table 6.2). The Rule 2 organism perceives a more connected landscape, but resource clusters are still relatively small and discontinuous on the wheyed-bottom map. The Rule 3 organism, in contrast, perceives connected suitable habitat in all the landscapes. The values of S_{avg} are all high and consistent, and LC_{rel} approaches 1.0 (Table 6.2), indicating that almost all suitable habitat occurs in a single resource cluster.

When the abundance of suitable habitat increases again to p = 0.656, there is little effect of scale-dependent landscape pattern among the different resource utilization rules. That is, suitable habitat is sufficiently abundant for habitat connectivity to be high for all the organisms. Even for the Rule 1 organism, more than 95% of the suitable habitat is

Table 6.2. Values of the weighted mean cluster size (S_{avg}) and the relative size of the largest cluster (LC_{rel}) as determined using three rules that represent different resource utilization scales. Within each value of *p*, the proportion of the landscape occupied by suitable habitat, five different spatial arrangements of habitat were simulated by generating scale-dependent landscape pattern. See text for definition of the rules and the scale-dependent pattern.

	Weighted mean cluster size (S_{avg})			Relative size of largest cluster (LC_{rel})		
Map pattern	Rule 1	Rule 2	Rule 3	Rule 1	Rule 2	Rule 3
			$p = 0.328$			
Random	6	55	77,671	0.001	0.004	0.951
Curdled-bottom	31,581	54,398	54,399	0.529	0.742	0.742
Wheyed-bottom	68	176	3,550	0.009	0.020	0.141
Curdled-top	26	62,448	84,695	0.002	0.844	0.993
Wheyed-top	6	57	77,469	0.000	0.005	0.949
			$p = 0.422$			
Random	16	69,174	110,819	0.001	0.790	0.999
Curdled	95,622	106,776	106,824	0.926	0.982	0.982
Wheyed-bottom	37	12,800	109,705	0.004	0.304	0.996
Wheyed-top	15	71,341	109,949	0.001	0.802	0.997
			$p = 0.656$			
Random	156,085	171,914	172,064	0.953	1.000	1.000
Curdled-bottom	166,510	166,685	166,685	0.983	0.983	0.983
Wheyed-bottom	156,888	171,913	172,030	0.955	1.000	1.000
Curdled-top	169,519	172,002	172,032	0.993	1.000	1.000
Wheyed-top	156,830	171,921	172,031	0.955	1.000	1.000

located in a single resource cluster (Table 6.2), and Rule 2 and Rule 3 organisms essentially perceive one resource cluster.

Discussion of Simulation Results

Connectivity in the landscape is a direct consequence of the abundance of suitable habitat, spatial patterning of the habitat in the landscape, and the organism's scale of resource utilization. Results of the simulations illustrate that when suitable habitat is abundant, landscape-level patterning (i.e., random or hierarchical patterns) and resource utilization scale are not important because the landscape is sufficiently connected when suitable habitat occupies more than 60% of the landscape.

When suitable habitat is less abundant, patterning and resource utilization scales became increasingly important. Whereas the patterning in the random and wheyed-top landscapes appears fine, the curdled-bottom landscape appears more coarse. That is, large contiguous blocks of habitat occur more frequently in the curdled-bottom maps than in random and wheyed-top maps (Figure 6.5).

Landscape-level patterning interacts with the resource utilization scales to affect connectivity at intermediate levels of resource abundance. At p=0.422, organisms with broad scales of resource utilization (Rule 3, Table 6.2) perceive all landscapes as connected. However, organisms with more restrictive scales of resource use (Rule 1) perceive only the coarser curdled-bottom map (Figure 6.3) as connected.

At intermediate levels of habitat abundance, small amounts of change in the abundance of habitat can have profound effects on the connectivity of the landscape. The abundance level at which these changes in connectivity occur (p_{crit}) depends on spatial patterning in the landscape and the resource utilization scale of the organism. For example, p_{crit} is around 0.60 for Rule 1 organisms operating on random maps (Figure 6.2). P_{crit} on random maps decreases to around 0.35 for Rule 2 and 0.25 for Rule 3. However, landscape-level spatial patterning can change this relationship. For Rule 2, some hierarchical maps, like wheyed-bottom maps with the finest scale of heterogeneity (Figure 6.5), disconnect the habitat even when the abundance of habitat is above p_{crit} for the random map (see p=0.422, Table 6.2). Similarly, connectivity can occur at habitat abundances less than p_{crit} for random maps, if heterogeneity in the landscape is coarser. For example, Rule 2 organisms find the curdled landscapes connected at p=0.328 but perceive the other landscapes as disconnected.

These results suggest that fine-scale fragmentation poses a greater risk to landscape connectivity than the same percentage reduction of habitat arranged in a coarser pattern. Moreover, fine-scale fragmentation increases the amount of edge in the landscape. Therefore, edge-avoiding species find less suitable habitat in a landscape with fine-scale fragmentation. This result agrees with the findings of Franklin and Forman (1987) whose recommendations for forest management include reducing the number of small, dispersed clearcuts and preserving large patches of primeval forest.

The benefits of coarse-scale structure can be observed in real landscapes. Gardner et al. (1991) compared landscapes from the ridge-and-valley and piedmont regions of Georgia. Forest species could easily traverse the ridge-and-valley landscapes with 43% forest. In this region, forest dominated the ridges whereas non-forest lands were confined to the valleys, imposing structure on the landscape. Some landscapes in the piedmont region were relatively disconnected when forest covered 68% of the landscape because areas of non-forest land use, not restricted by topography, were dispersed in a fine-scaled manner throughout the landscape.

The results also suggest that the greatest opportunity for intervention with conservation is at low and intermediate levels of habitat abundance. It is difficult to predict the consequences of habitat fragmentation at intermediate values of habitat abundance because of complex interactions between pattern and resource utilization scale. Biodiversity, by definition, refers to a variety of organisms which differ in their mobility and dispersal capabilities. Such organisms are differentially affected by the amount and spatial pattern of habitat reduction. Imposing coarse scale structure on the landscape, like protecting large blocks of habitat or creating regional-scale corridors, can preserve connectivity in the face of some acceptable amount of habitat destruction. The same amount of habitat loss in a fine-scale pattern is more likely to fragment the landscape further.

INCORPORATING SCALING ISSUES INTO CONSERVATION

The following suggestions should be useful for protecting biodiversity in managed landscapes. The suggestions are phrased negatively to emphasize management choices which might endanger biodiversity.

(1) *Do not focus on single, simple concepts like patches and corridors* (regional scale corridor excepted). Expand the management perspective to include viewing the landscape as a whole. Use landscape-level indices to measure pattern at multiple scales. The spatial extent of protection must be equal to or broader than the spatial extent of the ecological system needing protection.

(2) *Do not destroy the natural levels of heterogeneity* (spatial and temporal). Instead, match exploitative/disruptive activities to natural patterns of heterogeneity. Natural processes (such as fire and flooding) create and maintain heterogeneity and should not be disrupted. Consider the question, "Under the proposed management scheme, will the appearance of (levels of heterogeneity in) future landscape be similar to those of a natural landscape?"

(3) *Do not fragment the landscape.* Maintain connectivity by keeping the amount of native habitats above potential thresholds for connectivity and/or imposing coarse-scale structure on the landscape. Limiting the amount and spatial extent of exploitative activities that result in habitats which starkly contrast with native habitats (e.g., agriculture versus forest) promotes connectivity.

(4) *Do not ignore the effects of critical thresholds.* The methods and results presented here show that small changes in habitat abundance and pattern can suddenly fragment an otherwise well-connected landscape at some (but not all) resource utilization scales.

CONCLUSIONS

Natural patterns of heterogeneity are related to biodiversity and affect ecological processes. However, the role of heterogeneity differs between species and between ecological processes. Therefore, protection strategies using a single species perspective can result in an array of complex management strategies likely to contain contradictory prescriptions. Taking a broader scale perspective and implementing protection at the landscape level is a more robust protection strategy. An intact landscape will preserve the natural processes and patterns of heterogeneity that its ecosystems and species require. A landscape-level characteristic such as habitat connectivity can be preserved by controlling the spatial extent and pattern of habitat loss. The challenge to conservation is to recognize critical landscape-level patterns and to minimize human effects on natural patterns and processes.

ACKNOWLEDGMENTS

V. H. Dale, D. DeAngelis, G. Orians, R. C. Szaro, D. W. Johnston, and an anonymous reviewer contributed useful suggestions on earlier drafts of this chapter. Funding for this research was provided by the Office of Health and Environmental Research, Department of Energy under contract No. DE-AC05-84OR21400 with Martin Marietta Energy Systems, Inc. S. M. Pearson received support from an Alexander Hollaender Postdoctoral Fellowship from Oak Ridge Associated Universities. This is publication number 4060 of the Environmental Sciences Division of Oak Ridge National Laboratory.

LITERATURE CITED

Addicott, J. F., J. M. Aho, M. R. Antolin, D. K. Padilla, J. S Richardson, and D. A. Soluk. 1987. Ecological neighborhoods: scaling environmental patterns. *Oikos* 49:340-346.

Baker, W. L. 1989. Landscape ecology and nature reserve design in the Boundary Waters Canoe Area, Minnesota. *Ecology* 70:23-35.

Buechner, M. 1989. Are small-scale landscape features important factors for field studies of small mammal dispersal sinks? *Landscape Ecol.* 2:191-199.

Bierregaard, R. O., Jr. 1990. Avian communities in the understory of Amazonian forest fragments. In *Biogeography and Ecology of Forest Bird Communities*, ed. A. Keast, pp. 333-343. SPB Academic Publishing, The Hague.

Dayton, P. 1984. Processes structuring some marine communities: are they general? In *Ecological Communities: Conceptual Issues and the Evidence*, eds. D. R. Strong, Jr., D. Simberloff, L. G. Abele, and A. B. Thistle, pp. 181-195. Princeton University Press, Princeton, N. J.

Diamond, J. M. 1975. The island dilemma: lessons of modern biogeographic studies for the design of natural reserves. *Biol. Conserv.* 7:129-146.

Franklin, J. R., and R. T. T. Forman. 1987. Creating landscape patterns by forest cutting: ecological consequences and principles. *Landscape Ecol.* 1:5-18.

Forman, R. T. T., and M. Godron. 1986. *Landscape Ecology*. John Wiley and Sons, New York.

Gardner, R. H., B. T. Milne, M. G. Turner, and R. V. O'Neill. 1987. Neutral models for the analysis of broad-scale landscape pattern. *Landscape Ecol.* 1:19-28.

Gardner, R. H., R. V. O'Neill, M. G. Turner, and V. H. Dale. 1989. Quantifying scale-dependent effects of animal movement with simple percolation models. *Landscape Ecol.* 3:217-227.

Gardner, R. H., M. G. Turner, R. V. O'Neill, and S. Lavorel. 1991. Simulation of the scale-dependent effects of landscape boundaries on species persistence and dispersal. In *Ecotones: The Role of Landscape Boundaries in the Management and Restoration of Changing Environments*, eds. M. M. Holland, P. G. Risser, and R. J. Naiman, pp. 76-89. Chapman and Hall, New York.

Glenn, S. M., and S. L. Collins. 1992. Effects of scale and disturbance on rates of immigration and extinction of species in prairies. *Oikos* 63:273-280.

Hudson, W. 1991. *Landscape Linkages and Biodiversity*. Island Press, Washington, D.C.

Janzen, D. H. 1988. Tropical dry forests: the most endangered tropical ecosystem. In *Biodiversity*, ed. E. O. Wilson, pp. 130-137. National Academy Press, Washington, D.C.

Kotliar, N. B., and J. A. Wiens. 1990. Multiple scales of patchiness and patch structure: a hierarchical framework for the study of heterogeneity. *Oikos* 59:253-260.

Krummel, J. R., R. H. Gardner, G. Sugihara, R. V. O'Neill, and P. R. Coleman. 1987. Landscape patterns in a disturbed environment. *Oikos* 48:321-324.

Lovejoy, T. E., R. O. Bierregaard, A. B. Rylands, J. R. Malcom, C. E. Quintela, L. H. Harper, S. Brown, Jr., A. H. Powell, G. V. N. Powell, H. O. R. Schubart, and M. Hays. 1986. Edge and other effects of isolation on Amazon forest fragments. In *Conservation Biology*, ed. M. Soulé, pp. 257-285. Sinauer Associates, Inc. Sunderland, Mass.

McLellan, B. N., and D. M. Shackleton. 1988. Grizzly bears and resource-extraction industries: effects of roads on behavior, habitat use and demography. *J. Appl. Ecol.* 25:451-460.

Mandelbrot, B. B. 1983. *The Fractal Geometry of Nature*. W. H. Freeman and Co., San Francisco, Calif.

Musick, H. B., and H. D. Grover. 1991. Image textural measures as indices of landscape pattern. In *Quantitative Methods in Landscape Ecology: the Analysis and Interpretation of Landscape Heterogeneity*, eds. M. G. Turner and R. H. Gardner, pp. 77-103. Ecological Studies, vol. 82. Springer-Verlag Inc., New York.

Noss, R. F. 1991. Landscape connectivity: different functions at different scales. In *Landscape Linkages and Biodiversity*, ed. W. Hudson, pp. 27-39. Island Press, Washington, D.C.

O'Neill,, R. V., D. L. DeAngelis, J. B. Waide, and T. F. H. Allen. 1986. *A Hierarchical Concept of Ecosystems*. Princeton University Press, Princeton, N.J.

O'Neill, R. V., B. T. Milne, M. G. Turner, and R. H. Gardner. 1988. Resource utilization scales and landscape pattern. *Landscape Ecol.* 2:63-69.

O'Neill, R. V., R. H. Gardner, M. G. Turner, and W. H. Romme. 1992. Epidemiology theory and disturbance spread on landscapes. *Landscape Ecol.* 7:19-26.

Pearson, S. M. 1993. The spatial extent and relative influence of landscape-level factors on wintering bird populations. *Landscape Ecol.* 8:3-18.

Pulliam, H. R. 1988. Sources, sinks, and population regulation. *Am. Nat.* 132:652-661.

Raunkaier, C. 1934. *The Life Form of Plants and Statistical Plant Geography*. Clarendon, Oxford.

Risser, P. G., J. R. Karr, and R. T. T. Forman. 1984. *Landscape Ecology, Directions and Approaches.* Special Publication No. 2, Illinois Natural History Survey, Champaign, Ill.

Romme, W. H. 1982. Fire and landscape diversity in subalpine forests of Yellowstone National Park. *Ecol. Monogr.* 52:199-221.

Romme, W. H., and D. G. Despain. 1989. Historical perspective on the Yellowstone fires of 1988. *BioScience* 39:695-699.

Roughgarden, J. D. 1977. Patchiness in the spatial distribution of a population caused by stochastic fluctuations in resources. *Oikos* 29:52-59.

Runkle, J. R. 1981. Gap regeneration in some old-growth forests of the eastern United States. *Ecology* 62:1041-1051.

Scott, J. M., B. Csuti, J. D. Jacobi, and J. E. Estes. 1987. Species richness: a geographic approach to protecting future biological diversity. *BioScience* 37:782-788.

Shaffer, M. L. 1981. Minimum population sizes for species conservation. *BioScience* 31:131-134.

Shaffer, M. 1987. Minimum viable populations: coping with uncertainty. In *Viable Populations for Conservation*, ed. M. E. Soulé, pp. 69-86. Cambridge University Press, Cambridge.

Stauffer, D. 1985. *Introduction to Percolation Theory*. Taylor and Francis, London.

Turner, M. G., V. H. Dale, and R. H. Gardner. 1989. Predicting across scales: theory development and testing. *Landscape Ecol.* 3:245-252.

Turner, M. G., R. H. Gardner, and R. V. O'Neill. 1991. Potential responses of landscape boundaries to global climate change. In *Ecotones: The Role of Landscape Boundaries in the Management and Restoration of Changing Environments*, eds. M. M. Holland, P. G. Risser and R. J. Naiman, pp. 52-75. Chapman and Hall, New York.

Turner, M. G., W. H. Romme, R. H. Gardner, R. V. O'Neill, and T. K. Kratz. 1993. A revised concept of landscape equilibrium: disturbance and stability on scaled landscapes. *Landscape Ecol.* 8:213-227.

Whitcomb, R. F., C. S. Robbins, J. F. Lynch, B. L. Whitcomb, M. K. Klimkiewicz, and D. Bystrak. 1981. Effects of forest fragmentation on avifauna of eastern deciduous forest. In *Forest Island Dynamics in Man-dominated Landscapes*, eds. R. L. Burgess and D. M. Sharpe. Springer-Verlag, New York.

White, P. S. 1979. Pattern, process, and natural disturbance in vegetation. *Bot. Rev.* 45:229-299.

Whittaker, R. H. 1956. Vegetation of the Great Smoky Mountains. *Ecol. Monogr.* 26:1-80.

Wiens, J. A. 1976. Population responses to patchy environments. *Ann. Rev. Ecol. Syst.* 7:81-120.

Wiens, J. A. 1989. Spatial scaling in ecology. *Functional Ecol.* 3:385-397.

Wiens, J. A., and B. T. Milne. 1989. Scaling of 'landscapes' in landscape ecology, or, landscape ecology from a beetle's perspective. *Landscape Ecol.* 3:87-96.

Wilcove, D. S. 1985. Nest predation in forest tracts and the decline of migratory songbirds. *Ecology* 66:1211-1214.

Willis, E. O., and Y. Oniki. 1978. Birds and army ants. *Ann. Rev. Ecol. Syst.* 9:243-263.

7

BIODIVERSITY AND ECOLOGICAL PROCESSES

Mary F. Willson

Forestry Sciences Laboratory
Juneau, Alaska
U.S.A.

"Biodiversity" is a buzzword for what we used to call diversity. These two terms can be usefully applied at various scales or levels, from landscapes to within-population genetics. I will focus chiefly on species diversity within a community or ecosystem.

The goal of biodiversity management is not to achieve the highest possible level of species diversity per unit area. Humans are capable of increasing diversity by very intensive management, but the end product is a zoo (for animals) or an arboretum (for plants), useful enough for certain purposes, but not as a measure of success in maintaining biodiversity of their respective taxa. The animals in a zoo, or the trees in an arboretum, retain few of their natural ecological functions or relationships. Less extreme increases in species diversity might be obtained (in North America) by adding dandelions to a grassland, or house sparrows to a parkland, thereby increasing the number of species found on these sites, but, again, this is not successful management of biodiversity. Not only are the added species exotic, but also their introduction may both disrupt natural processes and eventually reduce the diversity of native species. The goal of biodiversity management is not to maximize diversity but rather to maintain natural levels of diversity despite human impacts. For this reason, I focus here on species in their native (not necessarily static, however) ranges and habitats and on the ecological processes and interactions that affect these species.

The choices regarding what to DO about biodiversity problems are seldom based on biological or even scientific considerations. There exists a spectrum of decisions that reflect the goals of policy-making bodies and determine how biodiversity issues are incorporated into policy. At one end of the spectrum, the decision might be to preserve/maintain natural diversity (at whatever level). Here, the real goal is conserving diversity, and success is measured by the duration of the desired diversity. At the next position on the spectrum, the decision could be to mitigate the effects of some human activity on natural diversity. In this case, the real goal is to salvage what can be, given that some disruption has or will have happened, and success is measured by the amount of diversity salvaged. The following position is represented by a decision to make the policy LOOK like it will either mitigate disruptive effects or maintain diversity. Here, the real goal has nothing to do with diversity itself, and success is measured by the appointment of a committee or the hiring of a person under a title that mentions biodiversity. And the far end of the spectrum is the assertion that diversity considerations are irrelevant or trivial; the goal does not include any consideration of diversity, and success is measured in terms of whatever human activity engendered the changes in diversity. The choice among these positions on the spectrum is entirely political, and occasionally determined by history. My concerns here are with the kind of biological science we bring to bear, and, thus, I will deal with the first two steps in the spectrum just described, in which diversity is accorded some real value.

This paper is organized in three principal sections: some problems with past studies of biodiversity, some emergent patterns of biodiversity, and some potentials for future studies of biodiversity. I will emphasize the central importance of focusing more on processes than on taxonomic units, more on interactions than on species. Good land-use planning, in my opinion, is based on an understanding of ecological and evolutionary processes, interactions between organisms, and variation in the progress and outcome of these processes.

SOME PROBLEMS

To design land-use plans that incorporate genuine concern for diversity, we must have information that is presently lacking. An obvious and oft-mentioned need is for more and better inventories and surveys of existing diversity (of species, genotypes), not only in virtually unstudied regions in the Third World, but in many areas of North America as well. But surveys alone are clearly insufficient. Ecological and evolutionary processes need to be considered explicitly and thoroughly. In conservation projects for particular species, certain processes are often included. But in the design of land-use plans, in assessing the consequences of certain land-use practices, in environmental impact statements, ecological and evolutionary processes commonly get little or no attention. The elementary concept of food chains frequently gets a mention. "Habitat use" is often a central focus, but too often the latter term is simplified to unreality or reduced to measuring home ranges. There has been considerable interest in nutrient cycling and productivity, but their relationship to the abundance and diversity of terrestrial organisms remains unclear. Predation and brood parasitism have only recently been actively brought to bear on problems of diversity in terrestrial communities.

Species are too often reduced to head counts, with little regard for the composition and distribution of their populations, their reproductive success, their interactions with each other, or for variation in their responses and requirements. To give a simplistic example: blueberry shrubs (of several species) are very common in the understory of the rainforests of Southeast Alaska, and frequently resprout from rootstocks after clearcutting. A wildlife biologist once told me that, therefore, it was unimportant to study fruit consumption by animals and subsequent seed dispersal for blueberries because recruitment was obviously no problem for these species. Such a statement makes many assumptions—that differences among genetic lineages, gene flow, and genetic structure of the population do not really matter, that the use of these plants by bears, birds, deer, and insects is of no relevance, and that variation in any of these things is without ecological significance. In my opinion, this is a cripplingly limited view.

In general, past studies have often been limited in many ways:

(1) They have often assumed an equilibrium situation—that, on some scale at least, the many processes within a community (or landscape or species) are in some kind of balance. Yet, there is abundant evidence that this may seldom be so, both within particular mini-systems and on geographic scales. Examples range from time lags in vegetation responses to climatic change (Cwynar and Spear 1991) to successional changes (Lertzman 1992) to dramatic fluctuations in population size in response to weather and resource variation (Stiles 1992). Indeed, many systems may not be in equilibrium on an ecological time-scale (of a few years to a few generations), so that many theoretical models and many assumptions about dynamics may not be applicable.

(2) Past studies have often been far too unifactorial—considering a single kind of effect rather than alternate and/or multiple effects. Community ecology suffered a major setback from the bandwagon that touted competition as the governing regulator of community structure, to the neglect of other important ecological forces (Wiens 1989). Both competition and predation have received intense study, theoretical and empirical, and warrant many pages in ecology texts, while less antagonistic and mutualistic interactions are typically reduced to "just-so" stories (May 1982; Bronstein 1991). Yet mutualisms are extremely common in many kinds of ecosystems, involve vast numbers of organisms, and are critical to the survival and reproduction of many species (see below).

(3) Previous studies commonly have been too typological. The emphasis is on the mean (or mode, or median), as if there were no important variation in time or space. Yet most processes vary in both dimensions, and variation in the nature and outcome of interactions is essential to understanding the systems we study and try to manage sensibly (Thompson 1988). As Thompson points out, the proper data point is a variance, not a mean.

(4) Past studies have often been too limited in temporal and spatial scope, restricted to a single study site or a short time span. This limitation may often reflect financial constraints, or simply (and reasonably) represent a starting point, but in other cases the problem reflects a serious intellectual constraint. Studies of natural biodiversity need to recognize not only the necessity of including multiple sites and times (as in no. 3), but also comparisons among categories. It is all very well to study, for instance, the bird communities of aspen forests across the continent, but eventually someone should compare aspen forests with oak or pine forests, in an effort to extract general results. A second instance: if evapotranspiration is commonly a good predictor of plant or animal productivity, diversity, or abundance, but exceptions are seen in certain systems, those exceptions are highly valuable in helping to identify additional factors that explain more of the observed pattern (M. L. Rosenzweig pers. comm.). A major long-term goal should be to address such comparative questions, to extract the most general principles possible, encompassing extant variation.

(5) Too often, they neglect evolutionary aspects of ecology. Problems of inbreeding (or of genetic drift) related to small populations are more or less routinely addressed. The problem of resistance to disease, as a function of population structure and isolation, has received some attention in certain cases, and serves as a reminder of the need to consider the process of natural selection in populations. But the general question of patterns and intensity of natural selection, their effects on genetic structure of the focal population, and the importance of genetic variation (in addition to inbreeding effects) on

the fitness of organisms and on population viability in a varying environment seem seldom to impinge on our consciousness, despite a rich theoretical literature. It is hardly possible to study all the aspects of selection in any population, but it is important to recognize that alteration of coexisting populations must often alter patterns of selection. An important goal of good land management is the maintenance of ecological systems in such a way that selection and other evolutionary processes could exert their effects in their natural ways, in ecological if not geological time.

(6) Many studies substitute correlation for explanation. Correlation can be useful, if all that is needed is description of a trend, and the trend can then be used to substitute an easily measured variable for one that is less tractable (and provided that the correlation is sufficiently tight). But a correlation seldom explains a pattern: the renowned correlation of foliage height diversity and bird species diversity eventually lost its fascination, in part because the correlation turned out to be so variable among regions and so often erratic even in a single locale, and in part because it did not explain much. We still do not really know what it is about multi-layered vegetation that supports more avian species or whether the multiple layers are directly involved at all. There are indications that foliage volume is related to avian density, perhaps by affecting food resources or nest sites (Holmes et al. 1979; Mills et al. 1991, Martin 1988a,b), but the actual mechanisms of interaction remain to be documented.

SOME PATTERNS

Despite all these carpings, cavilings, and complaints, some more or less robust biological patterns have emerged, although in no case have we arrived at a completely sound ability to predict effects or explain patterns in general. We know that very local, episodic, physical disturbances of forests (e.g., by wind-throw, forming a gap in the canopy) are often critical to regeneration (e.g., Schupp et al. 1989) and to patterns of animal activity and abundance (e.g., Willson et al. 1982). But we generally lack comparative information for different kinds of forests or for parallel kinds of disturbances in other vegetation types, such as badger mounds in prairies.

We know that species richness commonly increases with size of land area, even when habitat variety is constant. But much less is known about the relative contributions of area per se and habitat variety in different ecosystems, or about differences in the slope or position of the asymptote of the species/area curve for different taxa or habitat types. Ignorance about these things is one factor that has contributed to the debates on the requisite size and shape of preserves.

We also know that, at least in some systems, decreasing forest area and increasing edge effects diminish not only the proportion of the community composed of forest-interior bird species but may drastically reduce the breeding success of most of the remaining breeders because of intense brood parasitism and predation near habitat edges. But how general is the effect of predation on forest-edge breeders, and whether the magnitude of the effect is similar in many different kinds of forests, remains to be ascertained. Where brood-parasitic cowbirds are absent, how much does avian nesting success decline in response to predation alone? More generally, forest fragmentation may exert its effects on organisms in several ways (Terborgh 1992), which may differ among kinds of organisms (Robinson et al. 1992), but comparative information on the relative impact of these several factors in different kinds of forests is still lacking.

On a grander scale, there is evidence that evapotranspiration is often a good predictor of primary productivity and that resource availability (to some extent related to primary productivity) bears a regular relationship with animal diversity (e.g., Brown 1981; Owen 1988). But the emerging relationships still require comparisons that include greater ranges of evapotranspiration and productivity, and more taxonomic groups.

At the genetic level, recent work has shown that genes can "jump" from host to parasite to new host (references in Travis 1992). Thus, to some extent, genetic variation may depend directly on ecological interactions. This may be a relatively rare event, but its discovery is so recent that its implications for biodiversity remain to be explored.

As another example, the study of snags for cavity-nesting birds has been a fad, but (important as snags may be) the matter is seldom generalized to a process of habitat and nest-site selection and the factors that govern nesting success. Snags may be merely a kind of nest-site that is relatively easy to identify (by researchers) and that is known to limit some breeding densities; there has been much less study of birds that do not require cavities for nests (Martin 1988a,b). Conversely, not all snags are equivalent, saving a few snags does not constitute a general concern for biodiversity, and adding nest boxes potentially alters mating systems, clutch sizes, nesting success, foraging patterns, and competitive relationships both intra- and interspecific. I do not mean to denigrate the efforts that have led to this much understanding; they form the basis for future detailed studies.

SOME POTENTIALS

I DO insist that our recognition of the importance of interactions between organisms be made both broader and deeper at all scales. This plea is centered principally on ecological processes that involve species interactions within communities and their relationship to understanding and conserving biodiversity (although other levels may also be affected). In particular, I argue for serious consideration of the multiplicity of interactions, mutualistic kinds of interactions, and variation in interactions and processes.

Multiplicity of Interactions

The ecological literature is replete with examples of one-on-one interactions between species; in more recent years, many examples of multiple interactions have emerged. For example, we now know that the outcome of competition between A and B may be mediated by C (a predator/parasite), or that the outcome of predation of X on Y may be mediated by Z (Y's food supply, or abiotic factors), for example. We also know that plants can derive some protection from their herbivores, not merely by their own evolved defenses, but through association with other plants (Atsatt and O'Dowd 1976), and that food choices by small herbivores can be determined in part by the protective refuges provided by host plants (Hay 1991). It has become clear that we must now start with the assumption that interactions are *networks* of critical importance to biodiversity (Price et al., 1980; Pianka 1987; Price 1992).

Mutualistic Interactions

Mutualisms have been the neglected step-child in the family of interactions (May 1982; Bronstein 1991); their inclusion as full members of the family is long overdue. At issue here are *direct* mutualisms, in which the participants interact directly with each other.

I provide a reminder of the wealth of direct mutualisms known in terrestrial systems (Table 7.1), merely to emphasize their ubiquity and variety. Many of these mutualisms are extremely common and widespread, and by no means limited to the tropics. For example, animals provide the sole method of pollen transfer (and hence of seed set, and

Table 7.1. A catalog of some direct mutualisms in terrestrial systems.[1]

1. Plants and animals
   - (a) flowering plants and animal pollinators: all communities, all latitudes, many orders of plants, chiefly insects, birds and mammals;
   - (b) seed plants and animal dispersal agents: most communities, most latitudes, many orders of plants, chiefly vertebrates and ants;
   - (c) woody plants and protective arthropods: most woody-plant communities, most latitudes, many orders of plants, chiefly ants and mites;
   - (d) vascular plants with extrafloral nectaries and animal protectors: many latitudes and communities, many plant orders, chiefly hymenopterans.
2. Plants and fungi
   - (a) mycorrhizal associations: many communities, many latitudes, many taxonomic orders of plants, a variety of fungi;
   - (b) endophytic fungi that decrease attacks on host by natural enemies: a variety of plants and fungi, in several kinds of communities, at many latitudes;
   - (c) lichens: a restricted array of taxa, many communities at many latitudes.
3. Plants and microorganisms
   - (a) nitrogen-fixing microbes and angiosperms: many communities, many latitudes, many taxonomic groups.
4. Animals and other animals
   - (a) plant-feeding insects and ants: several insect taxa, a variety of communities and latitudes;
   - (b) interspecific flocks and herds: best known in birds and mammals, all latitudes, many communities;
   - (c) an assortment of specific cases of more local distribution, including hornbills and mongooses, honeyguides and mammals, tickbirds and mammals in Africa; cowbirds and caciques in Latin America; burying beetles and mites.
5. Animals and fungi
   - (a) spore dispersal of mycorrhizal fungi by small mammals: many communities, many latitudes, many taxonomic groups of plants, several genera of mammals;
   - (b) leaf-cutter ants and fungal gardens: many taxa of tropical plants.
   - (c) wood-boring ambrosia beetles and fungal gardens, many communities, many latitudes.
6. Animals and microorganisms
   - (a) digestive aids: best developed in termites and plant-feeding vertebrates, many communities, all latitudes;
   - (b) surface-dwelling microbes that help protect vs. pathogens: probably widespread and common.

[1] In addition, there is now abundant indication that eukaryotes constitute a permanent symbiosis between the "host" cells and in-dwelling particles; even this interaction is subject to modification by human interference via genetic engineering.

sexual/outcrossing reproduction) for prodigious numbers of angiosperms. The availability of floral resources for animals and of biotic pollen vectors for plants are essential to the maintenance of both groups. Similarly, animals are the principal vectors of seeds for huge numbers of seed-plants, ranging up to almost 100% of the species in some communities (Willson et al., 1989; Willson et al., 1990). Some participants in certain of these mutualistic systems are seen as "keystone" species that maintain their interacting partners through times of food scarcity and stress (e.g., Gilbert 1980; Terborgh 1986). Gilbert emphasizes that many mutualists serve as critical "mobile links" between segments of the community or even between communities.

Mutualisms between plants that provide housing and/or food and arthropods that protect the plant from natural enemies, once thought to be centered in tropical zones, are now known to range to the treeline at high latitudes (O'Dowd and Willson 1989, 1991; Willson 1991 and unpublished data for Alaska). Mycorrhizal associations are likewise common, widespread, and known to be biologically important. Interactions so pervasive are clearly worthy of inclusion in studies related to biodiversity; preserving a population of plants without the requisite pollinators, dispersal agents, protective and nutritive mutualists is more akin to collecting and preserving pressed specimens than to conservation of living systems (see also Redford 1992).

In the next paragraphs, I briefly discuss three of these mutualisms to indicate the potential magnitude of changes associated with disruption of mutualistic relationships and the desirability of learning more about these interactions.

(1) There is evidence that we have already conducted a vast, inadvertent, and uncontrolled experiment on mycorrhizal mutualisms (Cherfas 1991). Precipitous declines in mushroom diversity, abundance, and size across Europe and possibly North America, perhaps induced by air pollution, are likely to have major repercussions for trees in both natural and silvicultural systems. The growth and survivorship of trees can be directly affected and, in turn, are likely to influence plant community organization. Because many of the fungal species in these associations are also food resources for small mammals, which disperse the fungal spores (Maser and Maser 1988), there could be negative effects on small mammal populations and further, cascading, difficulties for the fungi and their seed-plant associates.

(2) Seed-dispersal mutualisms are often quite unspecific or "diffuse", in that numerous species of both plant and animals in any region may participate in the same generalized interaction. Evidence of the biological importance of dispersal of offspring away from the parent is partly indirect (e.g., the existence of many dispersal mechanisms), partly theoretical (e.g., Hamilton and May 1977; Comins et al. 1980), and partly experimental (references in Howe and Smallwood 1982). Loss of mutualistic seed vectors should be expected to alter the seed shadow, rates of colonization of new sites, patterns of gene flow, and genetic structure of the population. Two recent papers have emphasized the ecological importance of flying foxes, major dispersal agents for many Old-World tropical trees (Fujita and Tuttle 1991; Cox et al. 1991), thus indicating increased recognition of the importance of these mutualistic interactions.

Furthermore, habitat fragmentation commonly results in avifaunal shifts and changing composition of the frugivore fauna. Concomitant changes in seed dispersal should be expected. If the migratory forest thrushes are lost from the northern forests, a major component of the disperser assemblage will be missing; even some of the wood-warblers eat fruit on fall migration and the wintering grounds. Thus, our contemporary concern for neotropical migrants is linked to the maintenance of seed-dispersal mutualisms and of integrated forest ecological systems. Birds and mammals are not the only important agents of seed dispersal—tortoises and ants, even lizards and fishes in some places, are also critical interactors in certain systems. However, the quantitative effect of losing one or more dispersal agents from these diffuse interactions has not been measured. This lacuna is partly the result of the difficulty of making the requisite measurements, but it is also partly a result of the general failure to consider mutualistic interactions as major parts of community structure and species biology (see also Redford 1992).

Mutualisms between fruits and frugivores also affect other aspects of community structure in a broad sense. For example, bears are important agents of seed dispersal for many North American plant species, and fruits are critical dietary components for bears in many areas. Bears also prey on salmon during spawning runs. Some reports suggest that berry crops may draw bears away from salmon runs, which may affect salmon recruitment, but a few reports suggest that salmon draw bears away from fruit crops, which may affect seed dispersal rates and patterns (Willson 1993). This represents a web of potential interaction that spans terrestrial and aquatic systems, with possible ramifications for plant population biology and colonization patterns, for avian frugivore mutualists that share fruit resources with bears, for fish recruitment, and for other predators and scavengers of the salmonids. Yet no concrete data can be found to describe the most fundamental aspects of these unifying interactions.

For another example, treefall gaps (and the altered microclimate therein) in a variety of forests are known to affect pollinator activity, local distributions of birds and sometimes mammals, fruit and seed production, fruit and seed consumption, as well as infestation levels of insect pests of fruits and seeds, tannin production in the leaves of several plants (which may affect their herbivores), and the recruitment of many plants. Yet we lack any comprehensive, integrated picture of the entire network, the relative strengths of the interactions, or of variations in the patterns of interaction (see next section) as a function of vegetation structure.

(3) The ecological ramifications of another mutualism, between predatory/fungus-eating mites and plants with foliar domatia, are even less well known, because this interaction has been so recently rediscovered, a hundred years after its initial description (O'Dowd and Willson 1989; Pemberton and Turner 1989). The consequences of impairing the interaction cannot yet be predicted, but could involve both domestic and wild plants. The history of this long-neglected but now well-documented interaction suggests that still other mutualisms will probably be revealed by further research.

Variation in Interactions

Two points can be made here. The nature and outcome of interactions often vary in both space and time. The outcome of a competitive interaction varies with conditions; a mutualistic interaction can shift toward parasitism in some circumstances. The status of critical interactions commonly changes among seasons and years, in relation to shifts in weather, population fluctuations (which may result from events elsewhere; e.g., on the wintering grounds of migrants, or in the center versus the edge of a species range), and habitat changes induced by weather or geological forces. Seldom is the full range of such variation known for any system, but a successful program directed to maintaining biodiversity must allow for it, by avoiding the simple use of averages and incorporating explicit use of variances.

Furthermore, much variation in interacting taxa probably remains to be discovered, even in well-studied taxa such as birds. For example, Red Crossbills (*Loxia curvirostra*) are known to be quite nomadic, moving to regions with good crops of the conifer seeds on which they depend. Recent work, however, has shown that there is probably not just one species, but several closely-related species (or at least well-defined varieties). The bills of the birds of each variety in western North America have characteristics that make them most efficient at opening the cones and handling the seeds of certain conifers—for instance, one variety specializes on Douglas fir, one on hemlock, one on lodgepole pine, one on ponderosa pine, and one on Sitka spruce (C. Benkman 1993 and pers. comm.). Thus, it is no longer appropriate to treat red crossbills as a single entity interacting with conifers. Instead, there are several distinct interactions, each with its own problems of variation in conifer seed production and its own population responses.

Red crossbills offer us another object lesson related to biodiversity, by illustrating the consequences of management tactics employed in ignorance of critical ecological (and evolutionary) interactions. The Newfoundland subspecies of red crossbill has declined drastically in the past several decades, since the introduction of red squirrels to the island (Benkman 1989; Pimm 1990). The squirrels, now at high densities, are thought to out-compete the crossbills for the seeds of black spruce. Furthermore, squirrels may cause strong selection for cone traits that reduce the access to spruce seeds by the crossbills and, thus, impede possible recovery by the crossbill population. This chain of events was begun by the overtrapping of marten. To build up the populations of marten, squirrels were introduced as marten prey, but they apparently have had the additional effect of near-extinction of the Newfoundland crossbill. Red squirrels and red crossbills coexist on the mainland, in part because of some differences in habitat and food preferences, in part because of the ability of mainland crossbills to travel widely in search of good cone crops, and in part perhaps because of a history of coexistence; none of these conditions obtain on the island. Differences among the populations of crossbills and among their habitats, and a network of interactions, seem to be critical in determining their fates in the face of man-made changes.

DISCUSSION

So far I have emphasized the complexity of the web of interactions and the lacunae in our knowledge. Confrontation by complexity and ignorance sometimes engenders merely distress or denial. Where emergency action is required, there may not be time to deal with these complexities. But more reasoned responses to ecological complexity are both appropriate and more constructive in the longer run. The first positive response that is required is a different mind-set, one that focuses on multiple and variable interactions. This mind-set would be useful and constructive for land-managers charged with maintaining biodiversity or conservationists desirous of doing so.

Some tools exist for establishing or estimating the magnitude and direction of interactions. For example, Paine (1992) offers a procedure for quantifying a consumer's effect on its community; his method requires experimentation, which is certainly more feasible in some communities than in others. Another approach, less definitive but still potentially useful, is the analysis of connectance, as exemplified by Jordano's (1987) application of this method to pollination and seed-dispersal mutualisms.

The spread of human activities is often so fast, so extensive, and/or so destructive to natural communities that conservation measures commonly require either great foresight or remedial action. Several tactics have been used for this situation. One is the establishment of preserves large enough to encompass all the activities of a substantial number of some large, mobile, and often popular and attractive beasts; conservation of a variety of smaller organisms is a side effect. However, large preserves are often infeasible for a variety of reasons. Sometimes we use "indicator species" to identify relevant habitat patches or types, but this tactic often ignores variation in the ecology of the indicator and always neglects potentially relevant features not indicated by the chosen species. Small enclaves stet aside for rare and patchily distributed species may temporarily preserve a particular population but can only provide a delay in the slide to oblivion. We try to declare certain species or populations to be threatened or endangered, and eventually accord them a certain degree of legal and only partially effective protection. All such tactics can be useful and often may be the only thing we can do. However, even these measures often take an unconscionable amount of time (Gibbons 1992). In those cases in which emergency measures are not required and in which we have a chance to use our much-vaunted sapience, it may be possible to do some real planning about conservation of biodiversity, something more than an emergency reaction. Given the length of time and the resources it takes to provide protection to particular species even on an emergency basis, the time and resources required to make the studies needed for real planning need not be prohibitive.

A focus on process, function, and interaction provides a different basis for making the difficult choices among species to which salvage efforts are to be allocated. None of us wishes to be in the situation in which triage is necessary, but since we are, we should use the best biological rationales available. It may be more effective to focus our efforts on species that have strong interactions with several others, because changes in those species will have a greater impact on total community function than those interacting only weakly or with only one other. To go a further step: I suggest that the unit of preservation/conservation often should be the interactions themselves, rather than single species. If we focus more on ecological networks, sets of interacting populations in a variable environment, instead of single species, our chances of maintaining biodiversity are likely to improve.

CONCLUSIONS

Maintenance of natural levels and patterns of biodiversity requires serious consideration of many kinds of ecological interactions and of variation in their intensities and outcomes. Preserves must be designed to maintain functioning interactions. Land-management plans should incorporate explicit goals directed at maintaining interactions. And basic ecological research on the diversity and variation of interactions is needed to provide the scientific background for well-founded strategies of maintaining biodiversity.

ACKNOWLEDGMENTS

I am grateful to T. C. Comet, M. L. Dilger, K. C. Halupka, S. Scher, J. N. Thompson, C. J. Whelan, and J. Zasada for helpful comments on the manuscript.

LITERATURE CITED

Atsatt, P. R., and D. J. O'Dowd. 1976. Plant defense guilds. *Science* 193:24-29.

Benkman, C. W. 1989. On the evolution and ecology of island populations of crossbills. *Evolution* 43:1324-1330.

Benkman, C. W. 1993. Adaptation to single resources and the evolution of crossbill (*Loxia*) diversity. *Ecol. Monogr.* 63: 305-325.

Bronstein, J. L. 1991. Mutualism studies and the study of mutualism. *Bull. Ecol. Soc. Am.* 72:6-8.

Brown, J. H. 1981. Two decades of homage to Santa Rosalia: toward a general theory of diversity. *Am. Zool.* 21:877-888.

Cherfas, J. 1991. Disappearing mushrooms: another mass extinction. *Science* 254:1458.

Comins, H. N., W. D. Hamilton, and R. M. May. 1980. Evolutionarily stable dispersal strategies. *J. Theor. Biol.* 82:205-230.

Cox, P. A., T. Elmqvist, E. D. Pierson, and W. E. Rainey. 1991. Flying foxes as strong interactors in South Pacific island ecosystems: A conservation hypothesis. *Conserv. Biol.* 5:448-454.

Cwynar, L. C., and R. W. Spear. 1991. Reversion of forest to tundra in the central Yukon. *Ecology* 72:202-212.

Fujita, M. S., and M. D. Tuttle. 1991. Flying foxes (Chiroptera: Pteropodidae): threatened animals of key ecological and economic importance. *Conserv. Biol.* 5:455-463.

Gibbons, A. 1992. Mission impossible: Saving all endangered species. *Science* 256:1386.

Gilbert, L. E. 1980. Food web organization and the conservation of neotropical diversity. In *Conservation Biology: An Evolutionary - Ecological Perspective*, eds. M. E. Soulé and B. A. Wilcox, pp. 11-33. Sinauer Assoc, Sunderland, Mass.

Hamilton, W. D., and R. M. May. 1977. Dispersal in stable habitats. *Nature* 269:578-581.

Hay, M. E. 1991. Marine-terrestrial contrasts in the ecology of plant chemical defenses against herbivores. *Trends Ecol. Evol.* 6:362-365.

Holmes, R. T., R. E. Bonney, and S. W. Pacala. 1979. Guild structure of the Hubbard Brook bird community: A multivariate approach. *Ecology* 60:512-520.

Howe, H. F., and J. Smallwood. 1982. Ecology of seed dispersal. *Ann. Rev. Ecol. Syst.* 13:210-228.

Jordano, P. 1987. Patterns of mutualistic interactions in pollination and seed dispersal: connectance, dependence asymmetries, and coevolution. *Am. Nat.* 129:657-677.

Lertzman, K. P. 1992. Patterns of gap-phase replacement in a subalpine, old-growth forest. *Ecology* 73:657-669.

Martin, T. E. 1988a. Processes organizing open-nesting bird assemblages: competition or nest predation? *Evol. Ecol.* 2:37-50.

Martin, T. E. 1988b. Habitat and area effects on forest bird assemblages: is nest predation an influence? *Ecology* 69:74-84.

Maser, C., and Z. Maser. 1988. Interactions among squirrels, mycorrhizal fungi, and coniferous trees in Oregon. *Great Basin Nat.* 48:358-368.

May, R. M. 1982. Mutualistic interactions among species. *Nature* 296:803-804.

Mills, G. S., J. B. Dunning, and J. M. Bates. 1991. The relationship between breeding bird density and vegetation volume. *Wilson Bull.* 103:468-479.

O'Dowd, D. J., and M. F. Willson. 1989. Leaf domatia and mites on Australasian plants; ecological and evolutionary implications. *Biol. J. Linn. Soc.* 37:191-236.

O'Dowd, D. J., and M. F. Willson. 1991. Associations between mites and leaf domatia. *Trends Ecol. Evol.* 6:179-182.

Owen, J. G. 1988. On productivity as a predictor of rodent and carnivore diversity. *Ecology* 69:1161-1165.

Paine, R. T. 1992. Food-web analysis through field measurement of per capita interaction strength. *Nature* 355:73-75.

Pemberton, R. W., and C. E. Turner. 1989. Occurrence of predatory and fungivorous mites in leaf domatia. *Am. J. Bot.* 76:105-112.

Pianka, E. R. 1987. The subtlety, complexity and importance of population interactions when more than two species are involved. *Rev. Chilena Hist. Nat.* 60:351-361.

Pimm, S. L. 1990. The decline of the Newfoundland crossbill. *Trends Ecol. Evol.* 5:350-351.

Price, P. W. 1992. The resource-based organization of communities. *Biotropica* 24(2b):273-282.

Price, P. W., C. E. Bouton, P. Gross, B. A. McPheron, J. N. Thompson, and A. E. Weis. 1980. Interactions among three trophic levels: Influence of plants on interactions between insect herbivores and natural enemies. *Ann. Rev. Ecol. Syst.* 11:41-65.

Redford, K. H. 1992. The empty forest. *BioScience* 42:412-422.

Robinson, G. R., R. D. Holt, M. S. Gaines, S. P. Hamburg, M. L. Johnson, H. S. Fitch, and E. A. Martinko. 1992. Diverse and contrasting effects of habitat fragmentation. *Science* 257:524-526.

Schupp, E. W., H. F. Howe, C. K. Augspurger, and D. J. Levey. 1989. Arrival and survival in tropical treefall gaps. *Ecology* 70:562-564.

Stiles, F. G. 1992. Effects of a severe drought on the population biology of a tropical hummingbird. *Ecology* 73:1375-1390.

Terborgh, J. 1986. Community aspects of frugivory in tropical forests. In *Frugivores and Seed Dispersal*, eds. A. Estrada and T. H. Fleming, pp. 371-384. Junk, Dordrecht.

Terborgh, J. 1992. Maintenance of diversity in tropical forests. *Biotropica* 24:283-292.

Thompson, J. N. 1988. Variation in interspecific interactions. *Ann. Rev. Ecol. Syst.* 19:65-87.

Travis, J. 1992. Possible evolutionary role explored for "jumping genes." *Science* 257:884-885.

Wiens, J. A. 1989. *The Ecology of Bird Communities*, 2 vols. Cambridge University Press, Cambridge.

Willson, M. F. 1991. Foliar shelters for mites in the eastern deciduous forest. *Am. Midl. Nat.* 126:111-117.

Willson, M. F., E. A. Porter, and R. S. Condit. 1982. Avian frugivore activity in relation to forest light gaps. *Carib. J. Sci.* 18:1-4.

Willson, M. F., A. K. Irvine, and N. G. Walsh. 1989. Vertebrate dispersal syndromes in some Australian and New Zealand plant communities, with geographic comparisons. *Biotropica* 21:133-147.

Willson, M. F., B. L. Rice, and M. Westoby. 1990. Seed dispersal spectra: A comparison of temperate plant communities. *J. Veg. Sci.* 1:547-562.

Willson, M. F. 1993. Mammals as seed-dispersal mutualists in North America. *Oikos* 67:159-176.

8

RELATING PHYLOGEOGRAPHIC POPULATION STRUCTURE TO ISSUES IN CONSERVATION BIOLOGY

Brett R. Riddle

University of Nevada Las Vegas
Las Vegas, Nevada
U.S.A.

K. Bruce Jones

Environmental Protection Agency
Las Vegas, Nevada
U.S.A.

One of the notable aspects of recent literature in conservation biology is the recognition that biodiversity can be understood within a hierarchical framework combining genealogical and ecological features of taxa and biotas (e.g., Noss 1990; this volume). However, only rarely do analyses of ecological and genetic questions in conservation biology include an explicit recognition of the historical responses of individual taxa as well as entire biotas to alterations in habitat configurations (e.g., connectivity or geographic distribution). The kinds of responses that might be expected in such an extended historical framework include speciation, extinction, and dispersal (Vrba 1985). Many fundamental issues in conservation biology rely on an accurate assessment of the geographic distribution of genetic variation, either within a single widespread species, or among several species or higher taxa. This, in turn, leads to several important questions. What forces are responsible for the patterns observed? How long have different populations been isolated from one another, and how much genetic divergence has occurred since isolation? What do patterns of isolation and divergence tell us about the responses of species to episodes of climatic change? In what ways would such information be relevant to the issue of endangered and threatened species? At a more inclusive level, one may wish to understand the history of species associations (i.e., the historical integrity) in regional biotas for a variety of reasons. Such an understanding would allow one to address the degree to which coexisting species within a biota share a single history of responses (e.g., cospeciation or dispersal) to habitat alterations (Vrba 1985; Brooks and McLennan 1991) across different spatial and temporal scales (Delcourt and Delcourt 1991). What insights could such information provide for predicting biotic responses to future global climatic changes? Are certain topographic features of the earth predictably associated with patterns of speciation, extinction, and dispersal (Vrba 1992)? Do patterns of species richness predict locations of evolutionary "hot spots," or alternatively, ecotonal areas between two or more historically separated biotas? Which of these attributes of biodiversity should be prioritized in conservation decisions (i.e., species richness [Scott et al. 1987]; or areas of endemism [Vane-Wright et al. 1991])?

A recent approach to addressing the geographic structure of populations in an extended historical context incorporates phylogenetic systematics, biogeography, population genetics, and molecular evolution, and has been referred to as the study of phylogeographic population structure (Avise et al. 1987). In this chapter, we summarize the implications of applying a phylogeographic approach to conservation biology.

APPROACHES TO UNDERSTANDING POPULATION STRUCTURE

The most common approaches to investigating genetic variation within and among populations rely on concepts and methods derived from the field of population genetics (Vrijenhoek 1989; Hedrick and Miller 1992). One example utilizes the hierarchical gene diversity method (see Meffe, this volume). Commonly addressed questions include the estimation of amount of genetic diversity (heterozygosity and polymorphism) within and among populations and levels of gene flow among populations. Often, the aim of such investigations is to ascertain the current and future viability of an endangered species through estimates of genetic diversity relative to inbreeding or outbreeding depression (e.g., several papers in Soulé 1986). Genetic information is also used to assess the integrity of metapopulation structure relative to landscape attributes (Gilpin 1991).

Genetic differentiation among populations might be initiated by geographic isolation related to physical or ecological barriers. Much of the genetic variation present within a widespread species might, therefore, be a consequence of vicariant isolation and subsequent divergence resulting from large-scale climatic (e.g., Pleistocene glacial cycles) or geological events. As a consequence of long-term isolation, populations in a particular geographic region might have accumulated distinct and diagnostic characters (genetic and morphological) through evolutionary change. Subsequently, geographically separated populations might be recognizable as distinct lineages with their own evolutionary histories and trajectories into the future (see evolutionary species concept, Wiley 1981). The distribution of newly acquired characters can be used to reconstruct evolutionary relationships among geographically isolated populations through methods of phylogenetic analysis (see reviews by Swofford and Olsen 1990; Brooks and McLennan 1991). The specific components of a reconstructed phylogeny of interest to conservation biologists include: times of divergence (and hence, degree of persistence) of evolutionary lineages, relative levels of phylogenetic diversity across evolutionary lineages, amount of evolutionary divergence along different lineages, and the geographic distribution of different lineages in relation to postulated historical events (e.g., climatic cycles, geological activity). Additionally, if physical or general habitat factors are responsible for initiating phylogenetic divergence within a single geographically widespread lineage, then the same geographic pattern might be replicated across multiple independent taxa, to the extent that different taxa have responded to a given "barrier" in a similar way (e.g., either isolation and subsequent divergence, or dispersal because of shifts in habitat distributions).

WHAT IS PHYLOGEOGRAPHY?

Understanding the geographic distribution of phylogenetic structure among species and higher taxa is known as historical biogeography or historical ecology (Brooks and McLennan 1991). A similar approach applied to populations within a single species has been called phylogeography (Avise et al. 1987). Biologists have not arrived at a consensus as to what entities or criteria constitute a species (i.e., Otte and Endler 1989); thus, any actual delineation between phylogeography, historical biogeography, and historical ecology is somewhat arbitrary. However, many of the questions of interest to conservation biologists pertain to the geography of population structure among closely related species, or among populations within a single geographically widespread species (Avise 1989, 1992; Dizon et al. 1992). We, therefore, follow Avise and his co-workers in using phylogeography to define the discipline of conservation biology concerned with: identifying phylogenetic and biogeographic structure among populations, species, and biotas, understanding processes involved in the origin and maintenance of that pattern, and predicting the ecological and evolutionary consequences of disrupting such a pattern.

WHAT DATA ARE USED IN PHYLOGEOGRAPHY?

What kinds of data are appropriate for addressing population genetic structure in a geographic and phylogenetic context? We speculate that many phylogeographic patterns of interest to conservation biologists have developed within a Quaternary (i.e., the past 2 million years) or somewhat longer time frame, because this is the temporal window most directly related to the composition of modern biotas and ecological communities (Delcourt and Delcourt 1991). To emphasize a point made previously, within this time frame, most patterns of interest will generally occur within single widespread species (Bennett 1990). In short, one must identify a genetic system that is evolving at a sufficiently rapid rate to provide a signal of phylogenetic and geographic structure associated with recent episodes of isolation and divergence. If populations have been physically isolated from one another over an appreciable period of time, they are likely to represent distinct and diagnosable lineages with their own evolutionary histories and trajectories into the future. Two processes that can accelerate the attainment of recognizable phylogeographic structure are the random extinction rate of genes or genomes within a population, and the rate of mutation and fixation of new genotypes (Figure 8.1). Both of these processes are theoretically more rapid for the haploid vertebrate mitochondrial DNA (mtDNA) genome than for single-copy nuclear DNA (Brown et al. 1982; Avise et al. 1984; Vawter and Brown 1986). In addition, the phylogenetic signal is enhanced for mtDNA over short time frames because of the lack of recombination among

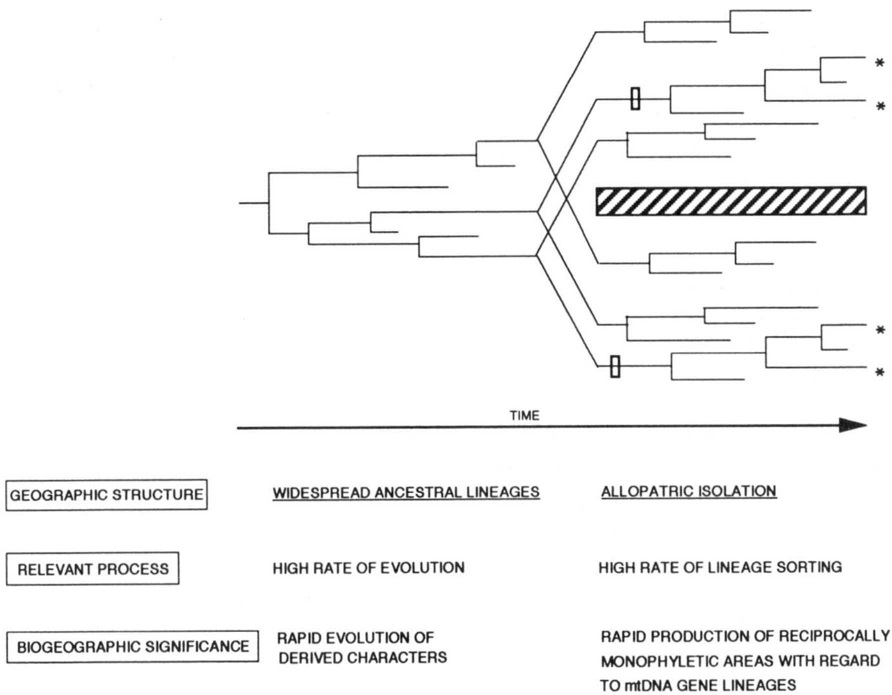

Figure 8.1. An illustration of the relevance of several attributes of mitochondrial DNA evolution (high rate of evolution, high rate of lineage sorting) to phylogeography in an idealized biogeographic scenario. The diagram shows the isolation of formerly connected populations by an external physical barrier (vicariance event), and the subsequent trajectories of phylogenetic branching events. Asterisks denote extant mtDNA lineages derived from a common ancestor (lineages marked by open bars) following geographic isolation. All other lineages go extinct at some point during evolutionary history. Note that the common ancestor to extant lineages in different geographic areas does not necessarily coincide with, and theoretically might precede in time, the origination of geographically isolated populations.

mtDNA molecules, thus allowing one to treat genomes, rather than populations, as fundamental units for phylogenetic analysis. For these reasons, as well as ease of laboratory manipulation, phylogeography has to date been defined largely in terms of mtDNA geography and phylogeny (Avise et al. 1987; Avise 1992). One should be aware, however, of the possibility that organelle genomes will reflect a different population structure than nuclear genomes (Harrison 1989). Reasons for possible discordance among genomes include random lineage sorting of organelle genomes, different rates of evolution, differential effects of historical population sizes, differential introgression across hybrid zones, and different population demographics for males and females. Recent advances have been made, however, in addressing phylogeographic population structure from the perspective of nuclear DNA with levels of resolution that are comparable to mtDNA (e.g., Hadrys et al. 1992; Karl and Avise 1992; Ostrander et al. 1992). Therefore, future studies will likely involve comparative analyses of patterns derived from nuclear and mitochondrial genomes. Nevertheless, mtDNA will reflect the history of population isolation and divergence among geographic regions if: (1) time between successive vicariance events is sufficiently long that extant mtDNA lineages in each region derive from a common ancestor unique to that region (Figure 8.1), and (2) dispersal of mtDNA lineages among geographic regions following deterioration of a barrier has not been extensive.

PHYLOGEOGRAPHIC PATTERNS AND CONSERVATION IMPLICATIONS

Single-taxon Patterns

One can summarize the phylogeographic partitioning of variation within a single taxon in a flow diagram of patterns and postulated processes (Figure 8.2). The structure of Figure 8.2 is designed to suggest the interplay between historical and current levels of association between populations in a geographic context, and the general approach that one can take to infer the relative contributions of history and current population dynamics to the development of phylogeographic population structure. The degree of phylogeographic subdivision of a species can often be associated with a specific attribute of an organism that might influence either its historical response to habitat alterations or

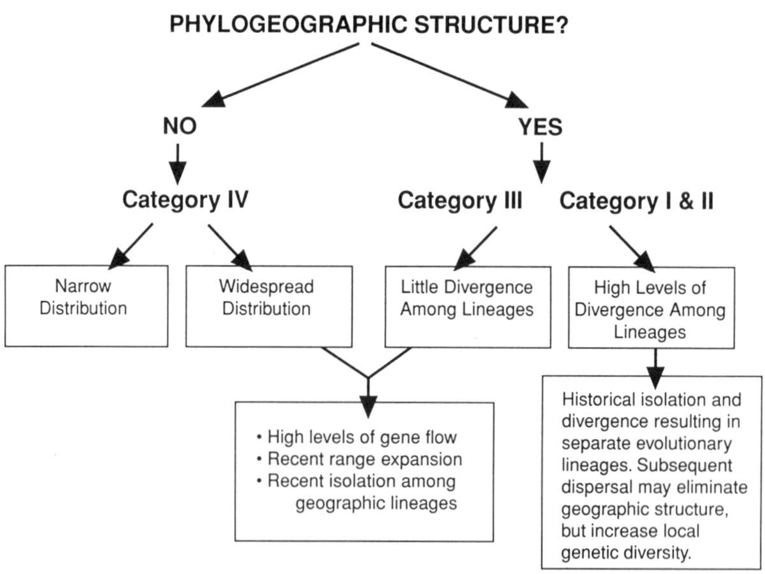

Figure 8.2. Flow diagram of patterns and postulated historical processes resulting in observed phylogeographic structure at the scale of a single taxon (see text for discussion). Roman numerals I - IV relate phylogeographic categories identified here with those of Avise (1989).

current patterns of gene flow among populations. Such attributes could include dispersal capability (e.g., migrant vs. non-migrant species, or species with pelagic larvae), ecological requirements (e.g., degree of habitat or resource specialization), or ecophysiological tolerances (e.g., responses to temperature or precipitation extremes). Organismal features can, therefore, establish a basis for understanding differences in the degree of phylogeographic structure among different species.

Many examples are now available in the literature documenting high levels of mtDNA divergence among geographically separated populations within a single widespread species or among several closely related species. Often, this diversity can be summarized into phylogenetic hypotheses among mtDNA genomes that demonstrate a strong underlying geographic component ("phylogeographic category I" of Avise 1989). If the phylogeographic structure is concordant with inferred barriers to population connectivity, then historical biogeography might provide the most plausible explanations for development of the pattern (Figure 8.1). For example, Lamb et al. (1989) attributed deep phylogenetic divergence among populations of desert tortoises (*Xerobates agassizii*) to the long-term isolation effects of the Colorado River system. Avise and Nelson (1989) determined that seaside sparrow (*Ammodramus maritimus*) populations could be partitioned phylogenetically into Atlantic Coast and Gulf Coast evolutionary lineages, attributable to vicariance caused by Pleistocene episodes of sea level change. Dizon et al. (1992) reported strong phylogeographic subdivisions between Northern and Southern Hemisphere Minke whale (*Balaenoptera acutorostrata*) populations, as well as a nested phylogeographic structure among Northern Hemisphere populations.

Several studies have identified significant phylogenetic subdivisions without associated geographic structure ("phylogeographic category II" of Avise 1989; Figure 8.2). The most likely explanation for this pattern invokes original geographic isolation and divergence of populations, followed by the deterioration of the original isolating feature, with subsequent gene flow among previously separated populations. Such an explanation has been advanced to account for mtDNA phylogeography in black-backed jackals (*Canis mesomelas*; Wayne et al. 1990), blue tits (*Parus caeruleus*; Taberlet et al. 1992), and northern grasshopper mice (*Onychomys leucogaster*; Riddle et al. 1993).

Phylogeographic structure might not be discernable within some species (or even among two or more "taxonomic" species), or might be present but with low levels of divergence among populations ("phylogeographic categories IV and III", respectively, of Avise 1989; Figure 8.2). In each case, one might infer that long-term isolating processes have not played a significant part in determining current levels of divergence and geographic structure. For example, Neigel et al. (1991) suggested that weak phylogeographic structure can often be attributable to recent mutation and fixation of new mtDNA genotypes within local populations followed by dispersal of animals carrying the new genotypes away from those populations. Avise (1992) related variation in the extent of shallow phylogeographic structure among several species of freshwater and marine fishes and invertebrates to species-level differences in gene flow regimes. Ball et al. (1988) found surprisingly little geographic variation in mtdna among red-winged blackbird (*Agelaius phoeniceus*) populations across the North American continent, and attributed this pattern to a high level of gene flow among populations, possibly in the form of a recent range expansion from Pleistocene refugia. In these and many more examples in the literature, the causal explanations for geographic variation in mtDNA genotypes usually include high levels of gene flow among populations, recent increases in overall geographic distribution of the species, or relatively recent isolation among populations.

Single-species phylogeographic patterns have several implications for conservation biology. For example, Dizon et al. (1992) developed a phylogeographically-based protocol to identify "evolutionarily significant units" (ESUs), and, thus, facilitate the management of marine mammal "stocks" as mandated by the Marine Mammal Protection Act of 1972. Their approach is likely to have utility relative to management decisions required by the Endangered Species Act of 1973 as well. The identification of ESUs informs management decisions in at least two ways: (1) by suggesting a priority ranking of populations relative to their evolutionary distinctness as well as amount of genetic information present, and (2) by understanding natural phylogeographic breaks among populations, and, therefore, when not to mix populations from different geographic regions in captive or wild breeding programs (see Ryder, this volume). The latter point has recently become controversial in light of developing molecular approaches to characterizing populations and species relative to one another, especially within the context of so-called "hybrid" populations and species (O'Brien and Mayr 1991; Wayne and Jenks 1991; Dowling et al. 1992). Part of the issue involves a debate over which ESUs to emphasize: the traditional "species," which this paper and others (e.g., Crozier 1992) argue is not the appropriate unit, or selected populations within a species. If the population is considered the appropriate ESU for conservation purposes, then a phylogeographic approach must be taken to identify and prioritize populations.

Multiple-taxa Patterns

Once the phylogeographic patterns for each of several unrelated taxa with similar geographic distributions are analyzed, one can ask whether a signal of concordant geographic and phylogenetic structure emerges (e.g., Avise 1992). Assemblages of different evolutionary lineages can demonstrate a concordant phylogeographic structure if each has responded in a similar fashion to geological or climatic events (e.g., through vicariance) during evolutionary history (Figure 8.3). Spatially, this concordance could range from landscapes (10^2 km^2) to larger biogeographic regions (10^6 km^2). In an ecological context, this information is valuable for addressing the degree of "openness" of species assemblages (Brown 1987; Ricklefs 1987) at community, landscape, and regional scales through evolutionary time. Ecologists have recently emphasized the need to examine the effects of global and regional climatic change on biodiversity (Lubchenco et al. 1991). One of their identified research goals includes an investigation of patterns of genetic and species diversity and underlying processes that account for those patterns. Is it possible to identify the influence of past environmental changes on the current structure of community and regional biotic assemblages, and can this insight be used to predict future ecological and evolutionary responses? The common approach to such a question involves reconstruction of distributions and ecological associations through analysis of paleoecological data (e.g., pollen and woodrat midden records; Betancourt et al. 1991; Delcourt and Delcourt 1991). However, these kinds of data have several

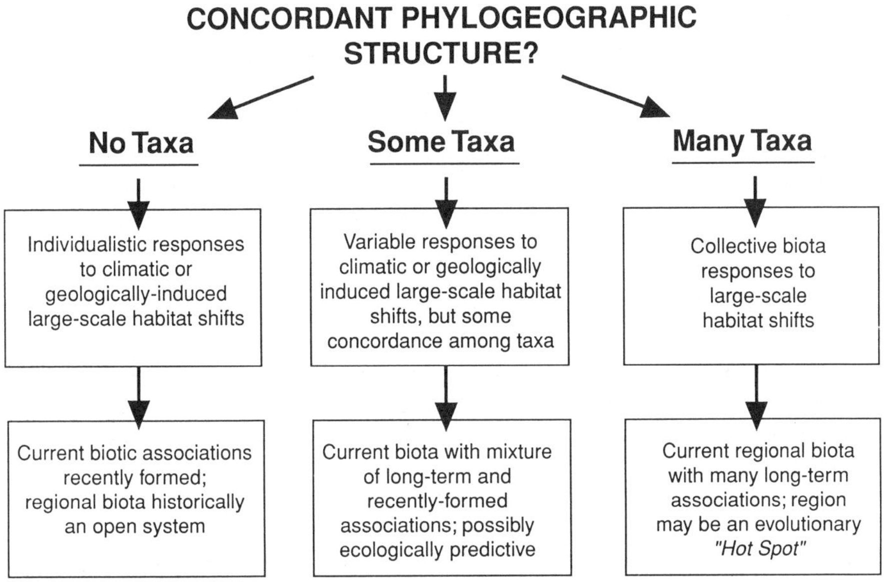

Figure 8.3. Flow diagram of hypothetical levels of concordance in phylogeographic structure across a set of independent taxa at the scale of regional biotas, and postulated extent of historical associations among taxa (see text for discussion).

limitations, including unequal sampling of local floras and faunas both in space and time. It should be possible to use phylogeographic data on the distribution of genetic variation in extant species as a complementary test of hypotheses constructed from fossil data. The reason for suggesting that phylogeographic and paleoecological data are complementary to one another is that each uses a different approach, with different underlying assumptions, to estimate historical changes in biotas. Thus, even though paleoecological data might provide accurate and detailed "spot samples" of temporal changes in biotas, phylogeographic data would tend to provide insight into the evolutionary significance of locally-recorded biotic fluctuations.

If some, but not all taxa in a biota demonstrate a concordant phylogeographic structure (Figure 8.3), can one relate this finding to common ecological traits (e.g., resource use, life-history, ecophysiology, or trophic level), and perhaps even tighter functional associations (see e.g., McDade 1992; Stein 1992)? It might be possible to partition a biota into long-term associates versus "fugitive" species. It is of interest to conservation biologists to identify levels of functional redundancy within an ecosystem (Walker 1992), and decisions about the "value" of particular species to local ecosystems can be enhanced by knowing the degree of persistence of functional equivalents within a local area through time.

Biotas that consist of a number of taxa separated from sister-taxa in other geographic areas by deep phylogenetic branches might reflect the location of evolutionary "hot spots" (Erwin 1991). These are of interest to conservation biology for several reasons, including: (1) prioritizing areas of the earth according to their contribution to preserving maximum phylogenetic diversity within networks of reserves (Vane-Wright et al. 1991; Crozier 1992; Faith 1992), and (2) evaluating the tempo and mode of speciation throughout evolutionary history in relation to climatic and environmental changes (Vrba 1992), thus being able to predict future ecological and evolutionary responses to environmental change. A pattern of widespread concordance in phylogeographic structure has been documented for fish, birds, and invertebrates ranging from the Gulf of Mexico northward along the Atlantic coast of North America (Avise 1992). This strong signal of phylogeographic concordance was attributed to climatically-induced Pleistocene vicariance among intraspecific populations, and provided the basis for recommendations concerning the locations of biotic reserves as well as a suggestion to limit intraspecific transplantations across phylogeographic boundaries (for a detailed case study, see Avise and Nelson 1989). Not all studies have produced the same level of regional biotic concordance. Lamb et al. (1992) studied mtDNA phylogeography in the desert tortoise and two species of lizards (*Dipsosaurus dorsalis* and *Sauromalus obesus*) with broadly overlapping geographic distributions in western North American deserts. Although the desert tortoise demonstrated a deep phylogenetic split associated with the Lower Colorado River drainage, the two species of lizards did not: each species has responded differently to historical changes in habitats along this river. This is not to say that the southwestern herpetofauna does not demonstrate a concordant phylogeographic structure (currently being investigated in our lab), but that if such a pattern exists, it might be present at a larger spatial scale (e.g., among the six major desert regions of western North America).

WHERE SHOULD BIOLOGICAL RESERVES BE LOCATED?

Scott et al. (1987) advocated location of biological reserves in areas of greatest species richness. Others have indicated that maximal species richness might be associated with current ecotonal areas between two habitat types (Delcourt and Delcourt 1991), or with areas of secondary contact between areas of endemism (e.g., evolutionary "hot spots"; Erwin 1991; Vane-Wright et al. 1991). In either case, species composition in such areas might be inherently dynamic during times of rapid climatic change, thereby decreasing the long-term value of preserving such areas if the potential for future responses of populations, species, and biotas are not also taken into consideration.

Vane-Wright et al. (1991) advocated a two-step process in reserve design that takes into account both evolutionary and ecological concerns. The first step is to prioritize areas based on their contributions to the phylogenetic diversity contained within a given species of concern or several species that would comprise a biota of concern (i.e., locate *areas of endemism* or *critical areas*). Crozier (1992) and Faith (1992) have extended the approach of Vane-Wright et al. (1991) to include consideration of the amount of diversification along each branch of a phylogeny in addition to the branching sequence itself in determining the relative contribution of an evolutionary lineage to overall phylogenetic diversity. In a phylogeographic context, this first step amounts to determining whether a "phylogeographic category I" pattern exists within a taxon (Figure 8.2), and whether multiple taxa within a biota share a similar pattern (Figure 8.3). If a "phylogeographic category II" pattern emerges, the task is to decide whether it is a product of secondary contact between two historically separated areas of endemism (in which case a record of evolutionarily significant critical areas might still be recoverable). The second step of Vane-Wright et al. (1991) is to use ecological and population demographic data to optimize reserve location and connectivity *within* each of the critical areas prioritized during the first step (e.g., Margules et al. 1988). For this purpose, "phylogeographic category III or IV" patterns (Figure 8.2) can be viewed in terms of understanding the ecological genetics of population connectivity (e.g., metapopulation structure) through methods summarized elsewhere (Meffe, this volume; Vrijenhoek 1989; Hedrick and Miller 1992).

PHYLOGEOGRAPHIC SURVEYS IN CONSERVATION BIOLOGY

Several biologists recently advocated the implementation of broad-based biological inventories for various regions of the earth (e.g., Soulé and Kohm 1989; Lubchenko et al. 1991). To the extent that such action is taken in the future, all funding agencies should require the collection and preservation of biological material that is suitable for extraction of DNA (see protocols in Hillis and Moritz 1990). We emphasize here the need to obtain samples of taxa from areas which are currently undergoing rapid development resulting in greatly accelerated extinction rates. Populations already extinct might be represented in museum collections, thus providing the opportunity to obtain adequate DNA samples for phylogeographic investigations (Thomas et al. 1989; Thomas et al. 1990; Patton and Smith 1992). Whether one is interested in issues related to endangered species, or in prioritizing areas of the earth relative to preserving maximum biotic diversity, we suggest that phylogeographic information can provide an important basis for making informed decisions.

CONCLUSIONS

A number of questions of interest to conservation biology require knowledge about the geographic distribution of population genetic diversity. In particular, we believe that one can obtain important insights into current patterns of biodiversity as well as predicting consequences of disrupting those patterns through analysis of the historical responses of populations, species, and biotas to habitat alterations. Recent attempts to incorporate population genetics, systematics, molecular evolution, ecology, and biogeography into the analysis of phylogeographic population structure offer a promising approach to understanding the causal factors associated with current patterns of biodiversity. As such, we argue that a number of issues in conservation biology should include phylogeographic information, including delineation of evolutionarily significant units (ESUs) and the location of evolutionary hot spots. Therefore, we strongly urge agencies to require collection of material suitable for extraction of DNA in any biological survey research, and encourage development of molecular phylogeographic protocols as a necessary component to understanding the distribution of biodiversity.

ACKNOWLEDGMENTS

We thank D. I. Orange and D. Nickle for helpful comments on this manuscript. Writing was conducted while B. R. R. was supported by NSF Grant BSR9107520.

LITERATURE CITED

Avise, J. C. 1989. Gene trees and organismal histories: a phylogenetic approach to population biology. *Evolution* 43:1192-1208.

Avise, J. C. 1992. Molecular population structure and the biogeographic history of a regional fauna: a case history with lessons for conservation biology. *Oikos* 63:62-76.

Avise, J. C., J. E. Neigel, and J. Arnold. 1984. Demographic influences on mitochondrial DNA lineage survivorship in animal populations. *J. Molecular Evol.* 20:99-105.

Avise, J. C., J. Arnold, R. M. Ball, E. Bermingham, T. Lamb, J. E. Neigel, C. A. Reeb, and N. C. Saunders. 1987. Intraspecific phylogeography: the mitochondrial DNA bridge between population genetics and systematics. *Ann. Rev. Ecol. Syst.* 18:489-522.

Avise, J. C., and W. S. Nelson. 1989. Molecular genetic relationships of the extinct Dusky Seaside Sparrow. *Science* 243: 646-648.

Ball, R. M., Jr., S. Freeman, F. C. James, E. Bermingham, and J. C. Avise. 1988. Phylogeographic population structure of Red-winged Blackbirds assessed by mitochondrial DNA. *Proc. Nat. Acad. Sci. USA* 85:1558-1562.

Bennett, K. D. 1990. Milankovitch cycles and their effects on species in ecological and evolutionary time. *Paleobiol.* 16:11-21.

Betancourt, J. L., T. R. Van Devender, and P. S. Martin, eds. 1991. *Packrat Middens: The Last 40,000 Years of Biotic Change*. University of Arizona Press, Tucson, Ariz.

Brooks, D. R., and D. A. McLennan. 1991. *Phylogeny, Ecology, and Behavior*. University of Chicago Press, Chicago, Ill.

Brown, J. H. 1987. Variation in desert rodent guilds: patterns, processes, and scales. In *Organization of Communities: Past and Present*, eds. J. H. R. Gee and P. S. Giller, pp. 185-203. Blackwell Scientific, Oxford.

Brown, W. M., E. M. Prager, A. Wang, and A. C. Wilson. 1982. Mitochondrial DNA sequences of primates: tempo and mode of evolution. *J. Molecular Evol.* 18:225-239.

Crozier, R. H. 1992. Genetic diversity and the agony of choice. *Biol. Conserv.* 61:11-15.

Delcourt, H. R., and P. A. Delcourt. 1991. *Quaternary Ecology: A Paleoecological Perspective*. Chapman and Hall, London.

Dizon, A. E., C. Lockyer, W. F. Perrin, D. P. Demaster, and J. Sisson. 1992. Rethinking the stock concept: a phylogeographic approach. *Conserv. Biol.* 6:24-36.

Dowling, T. E., B. D. DeMarais, W. L. Minckley, M. E. Douglas, and P. C. Marsh. 1992. Use of genetic characters in conservation biology. *Conserv. Biol.* 6:7-8.

Erwin, T. L. 1991. An evolutionary basis for conservation strategies. *Science* 253:750-752.

Faith, D. P. 1992. Conservation evaluation and phylogenetic diversity. *Biol. Conserv.* 61:1-10.

Gilpin, M. E. 1991. The genetic effective size of a metapopulation. *Biol. J. Linn. Soc.* 42:165-175.

Hadrys, H., M. Balick, and B. Schierwater. 1992. Applications of random amplified polymorphic DNA (RAPD) in molecular ecology. *Molecular Ecol.* 1:55-63.

Harrison, R. G. 1989. Animal mitochondrial DNA as a genetic marker in population and evolutionary biology. *Trends Ecol. Evol.* 4:6-11.

Hedrick, P. W., and P. S. Miller. 1992. Conservation genetics: techniques and fundamentals. *Ecol. Appl.* 2:30-46.

Hillis, D. M., and C. Moritz, eds. 1990. *Molecular Systematics*. Sinauer and Associates, Sunderland, Mass.

Karl, S. A., and J. C. Avise. 1992. Balancing selection at allozyme loci in oysters: implications from nuclear RFLPs. *Science* 256:100-102.

Lamb, T, J. C. Avise, and J. W. Gibbons. 1989. Phylogeographic patterns in mitochondrial DNA of the Desert Tortoise (*Xerobates agassizi*), and evolutionary relationships among the North American Gopher Tortoises. *Evolution* 43:76-87.

Lamb, T., T. R. Jones, and J. C. Avise. 1992. Phylogeographic histories of representative herpetofauna of the southwestern United States: mitochondrial DNA variation in the Desert Iguana (*Dipsosaurus dorsalis*) and the Chuckwalla (*Sauromalus obesus*). *J. Evol. Biol.* 5:465-480.

Lubchenco, J., A. M. Olsen, L. B. Brubaker, S. R. Carpenter, M. M. Holland, S. P. Hubbell, S. A. Levin, J. A. MacMahon, P. A. Matson, J. M. Melillo, H. A. Mooney, C. H. Peterson, H. R. Pulliam, L. A. Real, P. J. Regal, and P. G. Risser. 1991. The sustainable biosphere initiative: an ecological research agenda. *Ecology* 72: 371-412.

Margules, C. R., A. O. Nicholls, and R. L. Pressey. 1988. Selecting networks of reserves to maximise biological diversity. *Biol. Conserv.* 43:63-76.

McDade, L. A. 1992. Pollinator relationships, biogeography, and phylogenetics. *BioScience* 42:21-26.

Neigel, J. E., R. M. Ball, Jr., and J. C. Avise. 1991. Estimation of single generation migration distances from geographic variation in animal mitochondrial DNA. *Evolution* 45:423-432.

Noss, R. F. 1990. Indicators for monitoring biodiversity: a hierarchical approach. *Conserv. Biol.* 4:355-364.

O'Brien, S. J., and E. Mayr. 1991. Bureaucratic mischief: recognizing endangered species and subspecies. *Science* 251:1187-1188.

Ostrander, E. A., P. M. Jong, J. Rine, and G. Duyk. 1992. Construction of small-insert genomic DNA libraries highly enriched for microsatellite repeat sequences. *Proc. Natl. Acad. Sci. USA* 89:3419-3423.

Otte, D., and J. A. Endler, eds. 1989. *Speciation and its Consequences*. Sinauer and Associates, Sunderland, Mass.

Patton, J. L., and M. F. Smith. 1992. mtDNA phylogeny of andean mice: a test of diversification across ecological gradients. *Evolution* 46:174-183.

Ricklefs, R. E. 1987. Community diversity: relative roles of local and regional processes. *Science* 235:167-171.

Riddle, B. R., R. L. Honeycutt, and P. L. Lee. 1993. Mitochondrial DNA phylogeography in northern grasshopper mice (*Onychomys leucogaster*)—the influence of Quaternary climatic oscillations on population dispersion and divergence. *Molecular Ecol.* 2:183-193.

Scott, J. M., B. Csuti, J. D. Jacobi, and J. E. Estes. 1987. Species richness: a geographic approach to protecting future biological diversity. *BioScience* 37:782-788.

Soulé, M. E., ed. 1986. *Conservation Biology: The Science of Scarcity and Diversity*. Sinauer and Associates, Sunderland, Mass.

Soulé, M. E., and K. A. Kohm. 1989. *Research Priorities for Conservation Biology*. Island Press, Washington, D.C.

Stein, B. A. 1992. Sicklebill hummingbirds, ants, and flowers. *BioScience* 42:27-33.

Swofford, D. L., and G. J. Olsen. 1990. Phylogeny reconstruction. In *Molecular Systematics*, eds. D. M. Hillis and C. Moritz, pp. 411-501. Sinauer and Associates, Sunderland, Mass.

Taberlet, P., A. Meyer, and J. Bouvet. 1992. Unusual mitochondrial DNA polymorphism in two local populations of blue tit *Parus caeruleus*. *Molecular Ecol.* 1:27-36.

Thomas, R. H., W. Schaffner, A. C. Wilson, and S. Pääbo. 1989. DNA phylogeny of the extinct marsupial wolf. *Nature* 340:465-467.

Thomas, W. K., S. Pääbo, F. X. Villablanca, and A. C. Wilson. 1990. Spatial and temporal continuity of kangaroo rat populations shown by sequencing mitochondrial DNA from museum specimens. *J. Molecular Evol.* 31:101-112.

Vane-Wright, R. I., C. J. Humphries, and P. H. Williams. 1991. What to protect—systematics and the agony of choice. *Biol. Conserv.* 55:235-254.

Vawter, L., and W. M. Brown. 1986. Nuclear and mitochondrial DNA comparisons reveal extreme rate variation in the molecular clock. *Science* 234:194-196.

Vrba, E. S. 1985. Environment and evolution: alternative causes of the temporal distribution of evolutionary events. *South African J. Sci.* 81:229-236.

Vrba, E. S. 1992. Mammals as a key to evolutionary theory. *J. Mammalogy* 73:1-28.

Vrijenhoek, R. C. 1989. Population genetics and conservation. In *Conservation for the Twenty-first Century*, eds. D. Western and M. C. Pearl, pp. 89-98. Oxford University Press, New York.

Walker, B. H. 1992. Biodiversity and ecological redundancy. *Conserv. Biol.* 6:18-23.

Wayne, R. K., and S. M. Jenks. 1991. Mitochondrial DNA analysis implying extensive hybridization of the endangered red wolf *Canis rufus*. *Science* 351:565-568.

Wayne, R. K., A. Meyer, N. Lehman, B. Van Valkenburgh, P. W. Kat, T. K. Fuller, D. Girman, and S. J. O'Brien. 1990. Large sequence divergence among mitochondrial DNA genotypes within populations of eastern African black-backed jackals. *Proc. Nat. Acad. Sci. USA* 87:1772-1776.

Wiley, E. O. 1981. *Phylogenetics: The Theory and Practice of Phylogenetic Systematics*. John Wiley and Sons, New York.

9

THE RELATIONSHIP BETWEEN SOCIO-ECONOMIC FACTORS AND THE LOSS OF BIODIVERSITY: FIRST EFFORTS AT THEORETICAL AND QUANTITATIVE MODELS

Gary E. Machlis and Deborah J. Forester

Department of Forest Resources
University of Idaho
Moscow, Idaho
U.S.A.

The loss of biodiversity is a significant ecological problem and an important component of global environmental change (Ehrenfeld 1972; Soulé 1986; Wilson 1988a; Lovejoy 1988). The major cause of biodiversity loss is human action, primarily land-use that alters habitat (Pimm and Gilpin 1989; Freedman 1989). Yet, the ability to forecast the effect of specific human actions on biodiversity is not developed, and practical techniques for conducting such analyses do not exist (OTA 1987; Soulé and Kohm 1989).

The purposes of this chapter are to (1) describe the relation between socio-economic factors and biodiversity loss, (2) argue for an interdisciplinary approach to developing models of this relation, and (3) describe first efforts at theoretical and quantitative models. First, the central problem is presented; biodiversity loss is largely caused by human actions but the relation is poorly understood. There is significant scientific and policy-making value in attempting to model the relation. Second, a general introduction to model-building is presented, with a special emphasis on the unique demands of interdisciplinary models. The methodological steps in constructing models and criteria for evaluating alternative models are briefly discussed.

Next, first efforts at interdisciplinary models of biodiversity loss are reviewed, beginning with general models of interactions between society and Nature, and culminating in specific models of biodiversity loss. These models, first emerging in the 1970s and taken from a disparate literature, represent an intellectual history of attempts to understand relations between society and Nature and are the precursors to predictive, quantitative, and useful models of biodiversity loss. An example of each kind is illustrated. The chapter concludes with a modest research agenda, emphasizing the need for careful data collection, rigorous model-testing, and strengthened interdisciplinary cooperation.

THE CENTRAL PROBLEM

Definitions and Background

Biodiversity has numerous definitions (Huntley 1988); most treat diversity as a qualitative state at genetic, species, ecosystem, or landscape levels (Wilson 1988a). In the sciences of population biology and ecology, diversity is a multi-dimensional concept (MacArthur 1985). Wilcox (1984:640) defines biodiversity as "the variety of life forms, the ecological roles they perform, and the genetic diversity they contain." A definition by the U.S. Office of Technology Assessment (OTA 1987, p. 1) is similar:

> Biological diversity refers to the variety and variability among living organisms and the ecological complexes in which they occur. Diversity can be defined as the number of different items and their relative frequency. For biological diversity, these items range from complete ecosystems to the chemical structures that are the molecular basis of heredity. Thus, the term encompasses different ecosystems, species, and genes, and their relative abundance.

There is disagreement over how to measure biodiversity (Hurlbert 1971; Norton 1986); current measures select different components of ecosystems for emphasis. Potential indicators include number of species, or "richness" (Scott et al. 1987; Magurran 1988), abundance and distribution of populations (Krebs 1972; Westman 1990), number of endangered species, centers of species-richness with high endemism (Myers 1988), and degree of genetic variability (Allen 1963; Ruffie 1982). Other approaches treat ecosystem functions (Ray 1988), interactions (Janzen 1988), natural communities (TNC 1975; Western et al. 1989), successional stages (Franklin 1988), or ecological redundancy (Walker 1992) as key measures of diversity.

There is widespread agreement that global biodiversity is being reduced at an accelerated rate (Myers 1980; Wilson 1988a). There is less agreement about the actual level of biodiversity loss (Harwood 1982; Lovejoy 1986), compounded by the wide range of operational measures, variation between biomes, and lack of baseline knowledge about the number of species and taxa (Freedman 1989). Even when there is a common measure (e.g., species-richness) and biome (tropical forests), differences in estimated rates are large. Lugo (1988) compared several estimates of species extinction in the tropics; they ranged from 15% to 50% of all species by the year 2000.

Using any of these estimates, the loss of biodiversity is an important social and biological concern. Lost species may have commodity values; an example is the rosy periwinkle, a single species of plant yielding pharmaceuticals worth $100 million annually (Farnsworth 1988). Species may have intrinsic values in a socio-cultural system (Ehrenfeld 1976; Norton 1987; Shiva et al. 1991), and landscapes that provide habitat may have symbolic meanings (Burch 1971; Kellert 1986), provide an economic resource base (Brown 1985), or serve legally mandated functions such as wilderness (Hendee 1990). Biologically, loss of species can lead to synergistic effects on other species, altered energy flows and nutrient cycles (Freedman 1989), reduced ecosystem services such as oxygen production and climate modification (Mueller-Dombois et al. 1983; Norton 1986), and a decline in ecosystem resilience (Margalef 1968; Wilson 1988a). Hence, biodiversity loss is a global concern (Soulé 1986) and in some locations, such as the Amazonian tropical forests, a critical problem (Myers 1988).

Homo sapiens plays a dominant role in contemporary biodiversity loss; Wilson (1988b) suggests that for tropical rain forests, the current per-species rate and absolute loss is 1,000 to 10,000 times higher than before human intervention. Contemporary extinctions are almost entirely caused by human action, although natural forces such as demographic stochasticity, natural catastrophes, social dysfunction, and genetic deterioration contribute a "background" extinction rate (Soulé and Simberloff 1986). Diamond (1984), echoed by Pimm and Gilpin (1989), suggests four major causes of biodiversity loss: 1) destruction and degradation of habitat, 2) anthropogenic over-exploitation of plants and animals, 3) introduction of alien species, and 4) secondary effects of extinctions. Three of the four are directly related to the activities of *Homo sapiens*. Soulé (1991) suggests six classes of "human interference": loss of habitat; fragmentation of remaining habitat; over-exploitation; spread of exotic species; air, water and soil pollution; and climate change.

Habitat destruction is considered the most widespread anthropogenic cause of biodiversity loss (Brown 1985; Myers 1986; Wilson 1988a). IUCN/UNEP (1986a,b) estimate that 65% of the original ecosystems south of the Sahara have been subject to major ecological disturbance; 67% of natural habitat has been lost in southeast Asia, and in western Europe the percentage of habitat loss is even higher (Schücking and Anderson 1991). Habitat fragmentation may differentially increase vulnerabilities (Lovejoy 1986; Vermeij 1986); extinction selectivity suggests that certain habitats and species are more at risk (Terborgh and Winter 1980; Slobodkin 1986). Particular attention has been paid to the tropical rain forests because of their high biodiversity and accelerated rate of deforestation (Myers 1988; WRI 1986). Other affected tropical habitats include coral reefs (Salm 1984) and mangrove forests (Johannes and Hatcher 1986). Case studies of species decline due to anthropogenic habitat change also exist in the temperate zone, for example, the ivory-billed woodpecker, dusky seaside sparrow, and black-footed ferret (Wilson 1988a; Freedman 1989).

A decline in biodiversity has been associated with specific land-use changes such as powerline construction (Nickerson et al. 1989), urbanization (Leidy and Fiedler 1985), colonization adjacent to protected areas (Neumann and Machlis 1989), and fragmentation of forests (Harris 1984; Hanson et al. 1990). Currently, the literature is anecdotal and descriptive rather than systematic and analytical. Comparative studies are rare, and predictive ability is especially meager. Hence, there is a need to better understand the relation between human action and biodiversity loss.

The Need for Interdisciplinary Models

Because of the limited state of knowledge, the relationship between human action and biodiversity is part of most biodiversity and global change research agendas (OTA 1987; Soulé and Kohm 1989; Reid and Miller 1989; NRC 1990; NRC 1992; WRI 1992). For example, Soulé and Kohm (1989) argue for research on "the effects of social and economic change on biodiversity" and "the relationship between land-use patterns and biodiversity." Scott et al. (1989) ask, "How will changes in land-use affect the number of species not found in protected areas?"

To answer these questions, there has been a call for interdisciplinary and theory-grounded research (see Machlis 1990 for a review), with a "common framework of biological and social science research" (Soulé and Kohm 1989). The OTA report (1987, p. 127) stated:

> Social and political processes influencing how biological diversity is perceived and valued are probably the least well-understood and, in the long run, the most important factors affecting success of on-site diversity maintenance. Geographers, sociologists, anthropologists, historians, and biologists who have ventured outside their field of technical expertise have developed important descriptions of social factors affecting diversity maintenance at specific sites. But the analysis needed to develop a broader understanding and theories from which to generalize has yet to be undertaken.

A necessary step is the construction of specific, testable models, because the advancement of scientific knowledge relies heavily on the process of modeling (Pease and Bull 1992). Such models have been called for as part of several formal research programs, such as the Human Dimensions of Global Change Program, a component of the IGBP (International Geosphere-Biosphere Programme [Robinson 1991]).

If human action is treated as a variable influencing habitat change and biodiversity loss, the scope of the modeling problem is clearly interdisciplinary. Purely biological models are limited to investigating intervening variables and proximate causes; the explanatory variables are likely to be socio-economic and political. Describing the problem in the neotropics, Machlis (1992, p. 164) stated:

> The causes of habitat destruction are ultimately linked to unique demographic patterns, national histories, land tenure rules, distribution of wealth, and the role of agricultural monocultures (coffee, sugar, and now cocaine). The worldwide trend toward industrialization, increased per capita energy consumption, and economic interdependence are also critical factors. Understanding these socio-economic trends is prerequisite for predicting the rate, extent, and consequences of biodiversity decline.

Thus, several of the social sciences (such as sociology, economics and geography) can potentially contribute to the development and testing of models that attempt to organize our understanding about the causes of biodiversity loss.

MODELS AND MODEL BUILDING

Systems and Their Models

A model is the simplification of what we know and understand about a system (Hall and Day 1977b). A system is a group of interrelated or interacting components that can best be understood by viewing them as a whole. Systems form a nested hierarchy; simple systems serve as subsystems or components of more complex systems (Miller 1978; Higashi and Burns 1991). Systems at higher and lower levels impose constraints and bounds on the system of interest (Overton 1977; O'Neill 1989). Models are useful for conceptualizing, organizing, and communicating complex phenomena and for understanding, assessing, and optimizing estimates of system interactions (Hall and Day 1977b; Starfield and Bleloch 1986).

Models attempt to portray systems generally, realistically, and precisely. Yet, it is difficult, if not impossible, to optimize all three characteristics simultaneously (Levins 1966; Kingsland 1985). The modeler must determine which characteristics are most important to understanding the system of interest. As Peters (1991) notes, models must be "tuned" to specific problems; often a system must be represented by a suite of different models.

Kinds of Models

There are numerous topologies of models. The varying terminologies often confuse and require clarification, as many models can be placed in several categories (Robertson et al. 1991). Hall and Day (1977b) describe four kinds: conceptual, diagrammatic, mathematical, and computer. A *conceptual* model is an idea of how the components of a system fit together. It can be more formally represented as a *diagrammatic* model; system components or variables are represented in boxes with arrows representing flows between components. Diagrammatic models portray the relations between variables, identify exogenous and endogenous variables, and illustrate the driving forces in a system.

Mathematical models can be an outgrowth of diagrammatic models. They are a series of mathematical statements that represent the relations between variables in a model. Statistical techniques such as regression and correlation analysis may be used to specify such models. Computers can be used to iteratively process a series of more complicated equations. *Computer* models can give the researcher wide latitude for testing and revising a model. These quantitative models help to objectively determine the validity of conceptual models and can be used to simulate changes in the modeled system (Robertson et al. 1991).

An alternative topology is suggested by Turner (1989). He describes three categories of models: "wiring diagrams," flow diagrams, and predictive models. A *wiring diagram* demonstrates the structure of cooperation and information flows among different disciplines and aids in conceptualizing how different theoretical approaches can be combined to promote interdisciplinary studies (Meyer and Turner 1992). The relations represented in the wiring diagram begin to suggest the structure and components of a conceptual model.

Flow diagrams are analogous to diagrammatic models; model variables are in boxes, and arrows represent flows between variables. The variables are based on the various theoretical approaches of the different disciplines that lend insight into the system. The flows may be one or several measures (energy, dollars, information), or causal relations.

Finally, all or part of the flow diagram can be transformed into a *predictive* or quantitative model that represents hypotheses about how model variables influence one another. The model is then operationalized to determine how well it reflects or predicts the real world; appropriate indicators are selected and statistical tests to confirm or deny the model are carried out. Once a model is confirmed (which may take several revisions), it can be used to predict future outcomes. Predictive models can be mathematical (analytic) or numeric (simulation). Analytic models are relatively simple and provide an approximate solution to the model. Simulation techniques are more complex and can be used to explore systems through iterative tests (Hall and Day 1977b; Shoemaker 1977).

A synthesis of these topologies suggests three distinctive types of models: general frameworks, conceptual models and predictive models. Table 9.1 illustrates the basic

Table 9.1. A taxonomy of models.

	Hall and Day (1977)	Turner (1989; 1991)
General frameworks		wiring diagrams
Conceptual models	conceptual models diagrammatic models	flow diagrams
Predictive models	mathematical models computer models	predictive models (analytic or simulation)

taxonomy. General frameworks are the foundation for conceptual models, which in turn provide the basis for predictive models.

The Process of Modeling

There is a significant methodological literature on the development of models, both in the social sciences (Duncan 1969; Johnston 1970; Klein 1986) and in ecology (Odum 1970; Hall and Day 1977a; Starfield and Bleloch 1986). Overton (1977) outlines the general idealized strategy: specify the objectives of the model; reduce the model to sub-models; construct and validate sub-models; reassemble and validate complete model; answer objective questions; and continue analyzing and validating the model.

Hall and Day (1977b) provide a pragmatic model of the model-building process (Figure 9.1). It includes a description of what may actually happen as opposed to the ideal, and emphasizes the iterative nature of modeling. Of special note is their recognition of interdisciplinary issues ("natural and/or social system"), the importance of assumptions about the system, and the relation between modeling and decision-making.

Criteria for Evaluating Models

Modeling is a dynamic process in which successive advances in understanding a system are used to revise and improve the model. Hence, evaluation criteria are critical to model building (Peters 1991). An important criterion is concise and clear scope conditions; Overton (1977:54) listed the key elements:

1. The objective question (question to be answered by the model);
2. Perturbations and stimuli to be accommodated;
3. Exact target system to which the question applies;
4. Exact higher level system of which the target system is a part;
5. Exact environment to which the question applies;
6. Region of prescribed behavior;
7. Region of extrapolation and prediction;
8. Factual and hypothetical base (data sets, assumptions, source of knowledge);
9. Criteria of empirical validation (objective functions for curve fitting and criteria for prediction testing);
10. Criteria of theoretical validation (realism, agreement with accepted knowledge and theory).

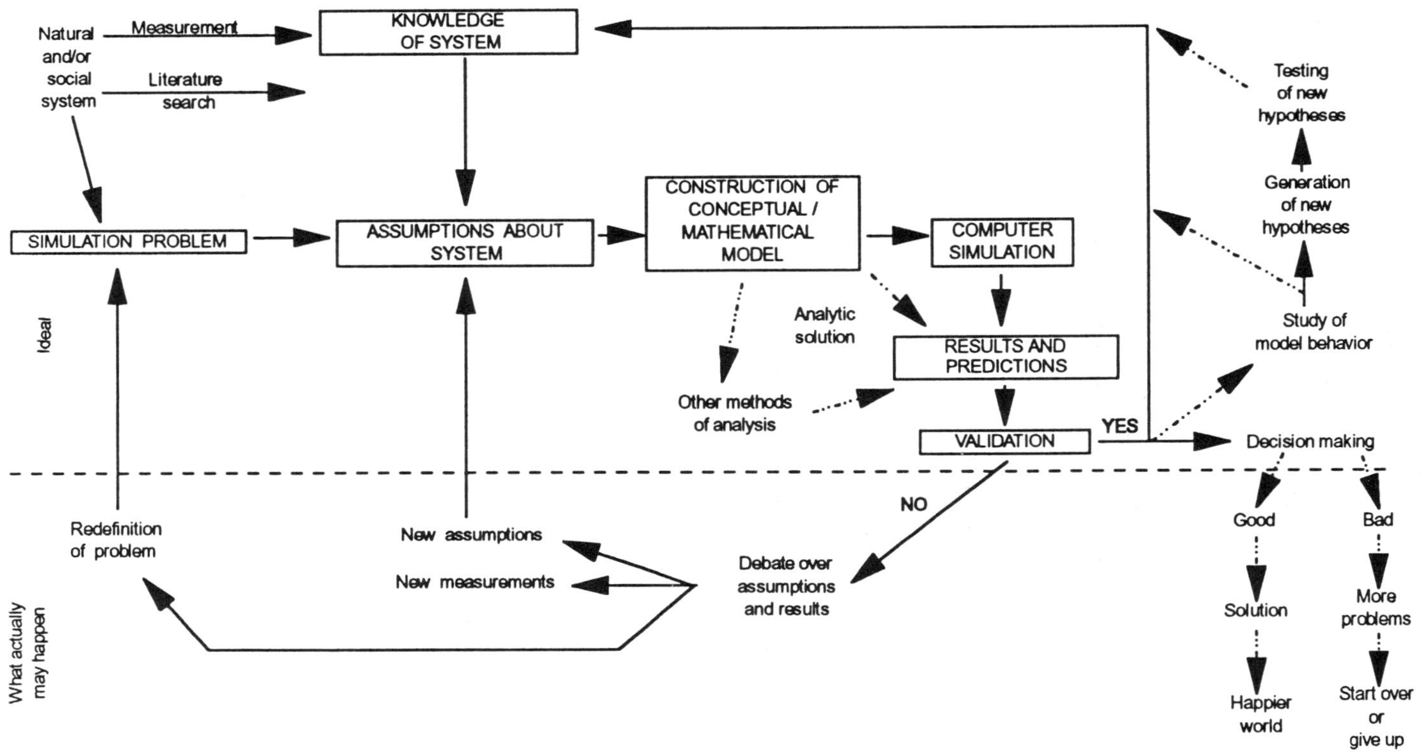

Figure 9.1. A diagrammatic model of the model-building process. (From Hall and Day 1977b).

Not all models (particularly general frameworks and conceptual models) can meet such strict scope conditions. However, all models can be evaluated by more general criteria, described below:

Specification: The model should be adequately specified, and include most, if not all, elements listed above. The objectives, target system, and scope of the model should be explicitly stated.

Explanatory and Predictive Power. The model and its variables should adequately explain the system of interest and prove acceptably accurate in predicting future observations (Friedland 1977).

Parsimony. The model should be as simple as possible (or only as complex as necessary) and logically represent the system. Extraneous variables or relations should be avoided.

Testability. The model should be testable, that is, there must be some objective means of validating the model and subjecting the model variables to sensitivity analysis.

Insight Power. The model should help to illuminate system processes and demonstrate an understanding of the nature of the system.

These general criteria can be applied to modeling the relation between socio-economic factors and biodiversity loss.

TOWARD INTERDISCIPLINARY MODELS OF BIODIVERSITY LOSS

General Frameworks

The development of a forecasting model for biodiversity loss that includes biological and socio-economic variables requires an interdisciplinary framework. General frameworks exist; an example is Hawley's (1950; 1986) classic "POET" model in which the "ecological complex" includes four key variables: population, organization, environment and technology. Bailey's (1990) "PISTOL" model is similar: the key variables are population, information, space, technology, organization and level of living, and all are interrelated. Gallopin et al. (1989) describe the societal/eco-system nexus: society and ecological systems are influenced by each other and the external environment. Clark (1989) illustrates the interactions between human and environmental systems (Figure 9.2). In Clark's model, external and internal sources of change (which are not defined) are stressed; human and environmental systems influence each other, and neither is dominant.

Such general frameworks are broad, largely useful as heuristic devices and difficult to operationalize. Often they are poorly specified. Some are parsimonious (Hawley's POET model has four variables), others relatively complex with full interactions. Few are testable, because the relations are rarely described in sufficient detail. Insight varies, of course, and most reflect the specific assumptions and scope conditions of the modeler.

Conceptual Models

Conceptual models of human/nature interactions have been widely attempted, and these efforts form the basis for developing specific models of biodiversity loss. Most reflect

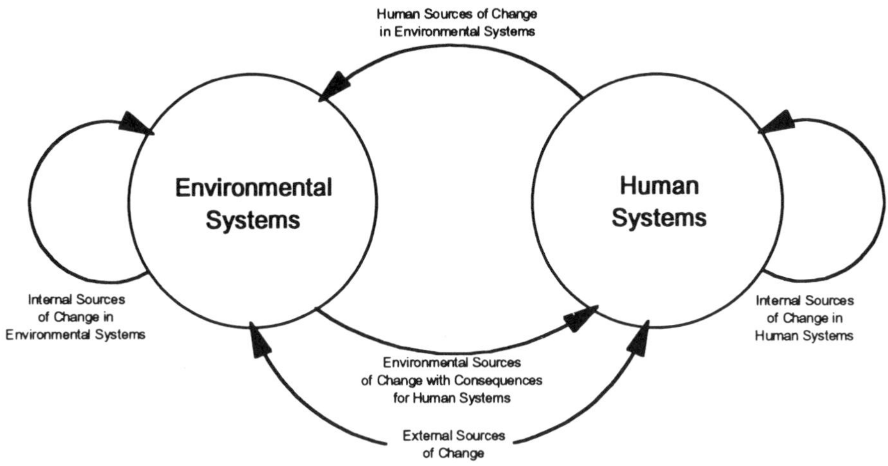

Figure 9.2. Interactions between human and environmental systems. (From Clark 1989).

the disciplinary focus of the modeler. For example, Bennett (1976) provides a conceptual model from a cultural anthropologist's perspective. The physical environment is transformed by culture into natural resources, and socio-cultural variables such as social organization and formal controls are stressed. Costanza and Daly (1992) provide a conceptual model from an economic perspective; it treats natural capital as a key variable and models both flows and relationships.

Figure 9.3 shows an example of a conceptual model focusing on the human causes of global change. Broad variables (such as human agency) are included, and driving forces are shown to influence proximate sources of change that ultimately have environmental impacts.

At a smaller scale, Machlis and colleagues have developed a series of models for protected areas, treating them as human ecosystems and using human ecology as a general framework (Machlis and Wright 1984; Machlis and Tichnell 1985; Machlis 1989). The models are built from the interactions of subsystems (air, water, fauna, and vegetation), and each relation is determined by flows of energy, materials, information, money, and individuals.

Finally, conservation biologists have provided conceptual models linking habitat loss and specific ecological effects (Miller and Botkin 1974; Roughgarden et al. 1989). For example, Gilpin and Soulé (1986) model several extinction vortices, which represent theoretical pathways for ecological disturbances to have species-diversity effects. "Human impacts" are the driving force behind change (Soulé 1986).

These conceptual models share several strengths and weaknesses. Most are poorly specified, with general variables such as ideology or social organization loosely used and ill-defined. The model objectives and scope conditions might not be stated, and issues of scale are often ignored. Explanatory power is often potentially high because the modelers are targeting large systems. Yet, because they are by definition conceptual, most are unvalidated. Parsimony varies, and the exclusion or inclusion of variables is sometimes neither consistent nor logical. The models are testable (given better specification) although the reality of collecting the necessary data gives cause for concern.

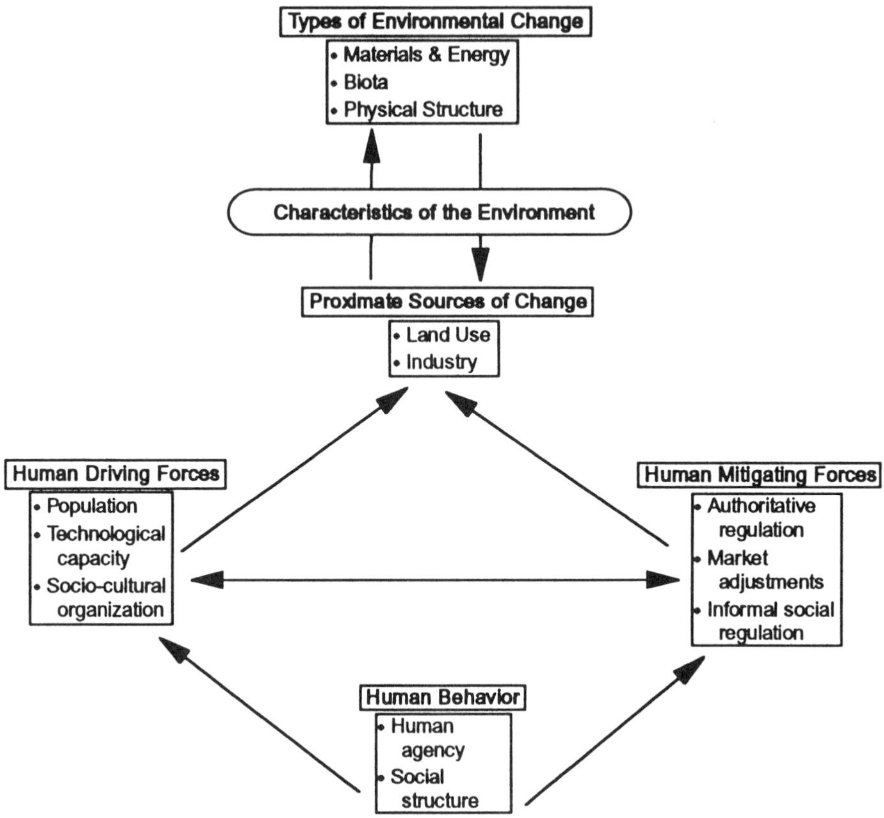

Figure 9.3. Components and linkages of the human cause of global environmental change. (From Turner 1989).

Insight is often quite high, because these conceptual models attempt to integrate biological and social systems previously not considered as explicit components of larger human/Nature systems.

Predictive Models

Although a large number of disciplinary models that predict ecosystem change are available (for example, models of wildlife populations by ecologists), the number of interdisciplinary, predictive models is actually quite sparse. The model in Figure 9.4 is clearly interdisciplinary and predictive; its scope is limited to deforestation in the Brazilian Amazon. Road construction is modeled as a driving force leading to deforestation, and the relations are implicitly treated as linear, explicitly as positive.

Yoshino (1991) models how increases in human activity lead to a variety of specific environmental changes that impact agriculture, forestry, and fisheries. A feedback loop emphasizes reciprocal relations between natural resource use and environmental change.

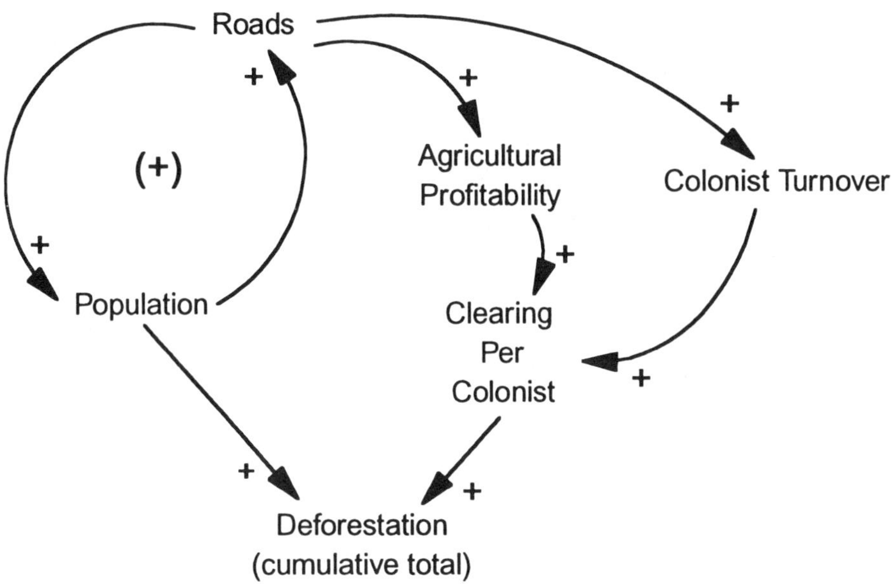

Figure 9.4. Causal loop diagram of the relation between road building and deforestation. (From Fearnside 1987).

Evaluation of available predictive models is difficult, because they vary profoundly. Specification is much improved in most of these models, and, provided scope conditions are clearly stated, many are testable. Explanatory power is largely dependent on the modeler's willingness to consider intervening effects and focus on specific aspects of ecosystem change. Few directly attempt to predict biodiversity loss. Parsimony is often lost because of wide or ill-defined scope conditions, and insight varies by the model's ability to predict counter-intuitive consequences (such as human *em*igration leading to habitat loss because of conversion from subsistence to plantation agriculture).

A SPECIFIC MODEL OF BIODIVERSITY LOSS

A specific interdisciplinary model for biodiversity loss requires that (1) socio-economic indicators serve as measures for key social variables, (2) the social variables have specific impacts upon environmental variables, (3) the variables and relationships be derived from biological and social science theory, and (4) biodiversity loss be operationalized for measurement over time. To be specific, the model should be constructed for a particular locale and spatial scale; to be generalizable it should contain variables likely to be operating at other locales and scales (Turner and Meyer 1991). In Levins' (1966) terms, this exploratory model-building trades off *precision* for *generality* and *realism*.

To understand better the specific processes and causative pathways that lead to biodiversity loss, Machlis and colleagues are attempting to develop a series of testable models (McKendry and Machlis 1991; Machlis and Forester 1992; Machlis 1992). The modeling is explicitly interdisciplinary, and assumes that changes in human social systems are the driving forces behind most contemporary biodiversity loss. A conceptual model is shown in Figure 9.5. Social, economic, and political forces are considered as independent variables driving changes in the human uses of resources. These uses have specific ecosystem impacts, and these ecological fluxes (such as heightened ecosystem stress) lead to biodiversity loss. Feedback loops explicitly complicate the conceptual model.

From the conceptual model, a preliminary version of a predictive model was developed (Figure 9.6). The various subsystems or component variables of the system are represented in boxes. The arrows represent causal relations, and the hypothesized nature of the relationship is identified as increasing (+), decreasing (-), or causing an unspecified change (Δ). Model variables are aggregated by the general class of activities to which they belong, such as economic exchange, politics, resource use, and industry. Definitions for each of the variables are outlined in Table 9.2.

The model illustrates that a variety of variables and relations are important for understanding human effects on biodiversity. However, different variables may play a larger or smaller role in explaining biodiversity loss, depending on the spatial and temporal scales and on the geographic locations to which the model is applied. For example, demographic change might be an important driving force behind changes in human use of resources at the local level, whereas government policies may play a bigger role at the state or national level. Industrial activity may play a large role in ecosystem stress in industrialized countries, whereas the over-harvesting of selected species (natural resource production activity) might be more important in agricultural or primary production regimes.

A literature review provided the basis for the predictive model structure. While some of the component interactions are based on empirical evidence, others are based on descriptive case studies. For example, several studies empirically establish a positive link between population growth (an aspect of demographic change) and deforestation (an aspect of land use conversion) in developing countries (Allen and Barnes 1985; Rudel 1989). Reh and Seitz (1990) demonstrate a link between land-use conversion (roads) that fragments habitat and reduced genetic diversity in animal populations (frogs) residing in smaller habitat patches. Norton (1991) explains how one of the factors that harmed a mistletoe in New Zealand was habitat loss because of settlement (land-use conversion) and mining (industrial activity). This combined with other factors such as over-collecting, habitat fragmentation, introduced species, and the natural rarity of the mistletoe led to the extinction of the mistletoe. The eclectic and unsystematic attention by the literature is obvious.

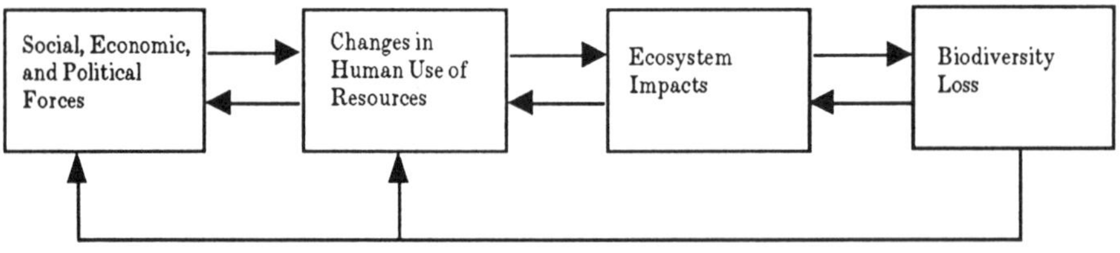

Figure 9.5. Conceptual model of human-caused biodiversity loss. (From Machlis 1992).

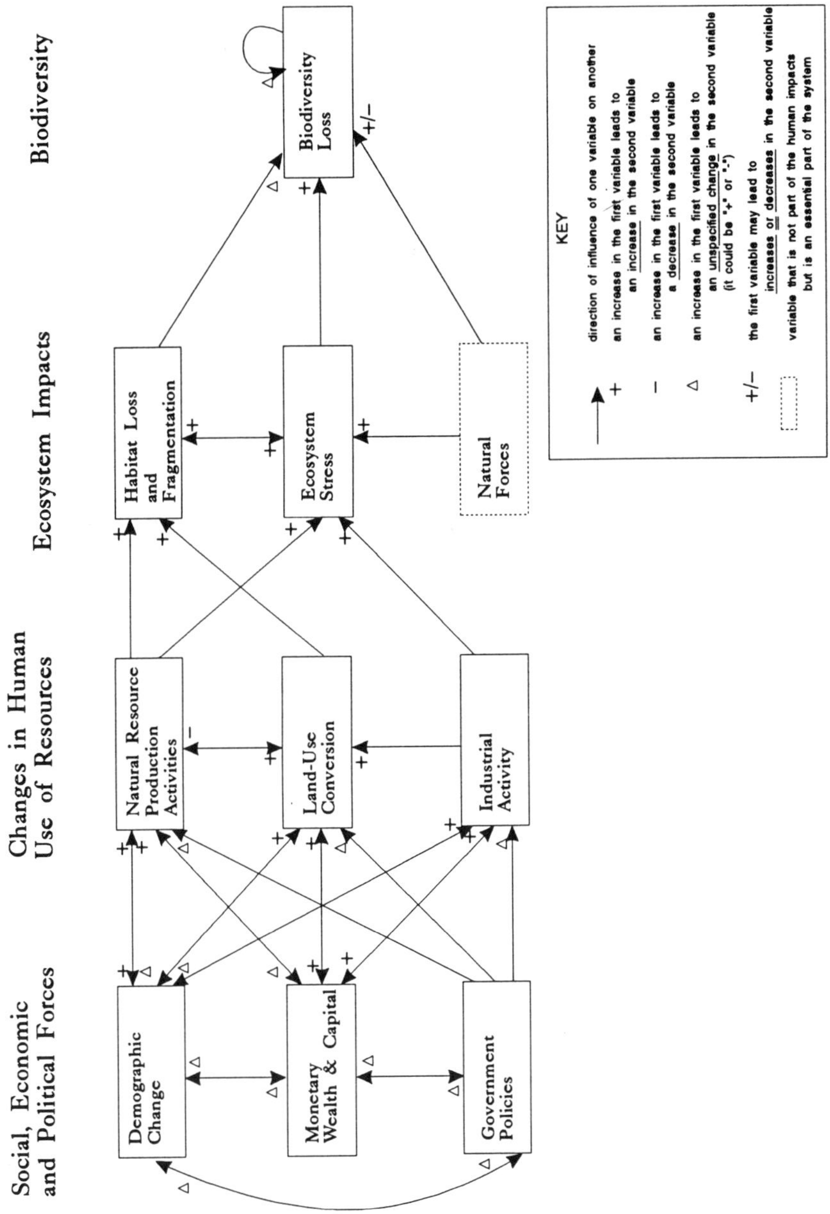

Figure 9.6. Predictive model of anthropogenic effects on biodiversity.

Table 9.2. Definition of model components for Figure 9.6.

1. Social, Economic and Political Forces
   (a) *Demographic change* refers to changes in the size, distribution and density of human populations in a given area. Examples include population growth and migration.
   (b) *Monetary wealth and capital* refers to the presence of money and money available for use in producing more money. Note that wealth and capital do not refer to natural resources, labor or other human and biological resources. Examples include income and savings.
   (c) *Government policies* refers to laws, regulations and practices adopted or enforced by government at any of its several levels (national, regional, local). Inaction by a government also constitutes a government policy. Examples include taxes, hunting regulations and presence of (or lack of) zoning laws.
2. Changes in Human Use of Resources
   (a) *Natural resources production activities* refers to the use and exploitation of natural resources to produce goods and services which may be used directly or in industrial activity. Examples include forestry, agriculture, hunting, mining, and fishing activities.
   (b) *Land-use conversion* refers to the transformation of land from one use to another, usually from a less to a more developed state. Examples include the building of roads and railways where they did not exist previously, and the conversion of land from one use to another, e.g. agriculture to housing.
   (c) *Industrial activity* refers to practices and processes involved in the manufacture of goods by industry (which is usually the transformation of natural resources). Examples include gas emissions, waste disposal, transport, and energy use.
3. Ecosystem Impacts
   (a) *Habitat fragmentation* refers to the physical division of a habitat into smaller units. An example would be the division of one area into several due to road construction or timber harvesting.
   (b) *Habitat loss* refers to changes in a habitat such that it can no longer support a keystone species. Examples include the outright destruction of a habitat by clearing an area or a significant change in a habitat due to cattle grazing.
   (c) *Ecosystem stress* refers to changes in natural systems beyond a dynamic equilibrium state to compensate for a disturbance. Examples include harvesting a species beyond some threshold level, loss of a keystone species, or forest die-back due to acid rain.
   (d) *Natural forces* refers to biological, geological, climatological, and meteorological activities that occur without a dominant influence from humans. This would include natural (background) extinctions, genetic adaptation and speciation, and natural climate change.
4. Biodiversity
   (a) *Biodiversity loss* refers to a significant reduction (e.g. beyond dynamic fluctuation) in the variety and variability of genes, species or habitats. This reduction may change or impair ecological processes. Species extinctions and loss of heterozygosity are examples.

Limitations to the Model

The general predictive model has several limitations, all of which must be considered in applying the model. The aggregation level may be too coarse to adequately capture all of the important processes that ultimately influence biodiversity loss. Reciprocal relations or feedback loops may be difficult to conclusively test and validate. There is no standard way to measure biodiversity or its loss. The lack of a consistent and precise measure of the dependent variable may limit the usefulness of analyses. The model fails to take into account the potential time lag between the main system components. Various indicators could be used for each of the model variables. Although this flexibility of application was one of the considerations behind building the model, it may be difficult to compare the usefulness of different analyses. Its usefulness to resource managers, planners, policy makers, and others remains theoretical until it is validated.

Using the evaluation criteria described earlier, the model has potential value. It can be properly specified, though some variables (wealth and capital) may require refinement. Its explanatory power is directed specifically at biodiversity loss and, if validated, could provide predictions of vulnerability to loss for specific locales. It is reasonably parsimonious, and intervening social and biological variables are explicitly considered. The model is testable given clear scope conditions, because it is constructed of a series of individual hypotheses about human/Nature interactions. Evaluating its insight awaits its application.

First Quantitative Test

This predictive model was operationalized for testing with data from a sample of 60 developing countries (Forester 1992). To do so, the predictive model was abbreviated (only portions have data available) and the reciprocal (feedback) relations ignored. Two to four indicators for each variable were identified. Most data were taken from *World Resources 1990-91* (WRI 1990), which contains statistics on a variety of socio-economic and environmental topics. The indicators used for each variable are listed in Table 9.3.

Structural equation modeling techniques were used to determine the appropriateness of indicators and the validity of the model. SYSTAT 5.01 (SYSTAT 1990) and EzPATH (Steiger 1989), a SYSTAT add-on module, were used for all statistical analyses. Structural

Table 9.3. Model variables and their indicators.

1. Demographic Change
   (a) Percent average population change, 1985-1990
   (b) Rural population density, 1990 (people/hectare)
   (c) Percent average annual growth of labor force, 1980-1990
2. Monetary Wealth and Capital
   (a) Gross national product per capita, 1987 ($)
   (b) Official development assistance per capita, 1987 ($)
   (c) Percent average annual change in real gross national product, 1977-1987
3. Government Policies
   (a) Percent of national land area protected, 1989
   (b) Number of protected areas, 1989
   (c) State environmental scale (on a scale of 1 to 9) (Dietz and Kalof 1992)
4. Natural Resources Production Activities
   (a) Percent change in Index of Agricultural Production, 1976, 1978—1986, 1988
   (b) Percent change in average annual fertilizer use per hectare of cropland, 1975, 1977—1985, 1987
   (c) Percent change in roundwood production: 1975, 1977—1985, 1987
5. Land-Use Conversion
   (a) Percent change in cropland, 1975, 1977—1985, 1987
   (b) Percent change in permanent pasture, 1975, 1977—1985, 1987
   (c) Percent change in livestock, 1976, 1978—1986, 1988
   (d) Percent change in forest and woodland, 1975, 1977—1985, 1987
6. Biodiversity Loss
   (a) Percent threatened mammals, 1989
   (b) Percent threatened swallowtail butterflies (UNEP 1991)

equation analysis involves solving a series of simultaneous equations. A covariance matrix of indicators is used as the basis for the analysis. Taking into account the specified relations the computer program attempts to reproduce the matrix.

Confirmatory factor analysis was used to determine the relationship between indicators and their model variables. Indicators that play little role can be identified and removed. Path analysis was used to evaluate the relation between independent and intervening model variables and between them and the dependent variable (biodiversity loss). Path coefficients (or partial regression coefficients) were calculated. They provide an evaluation of the structure of the model, testing whether or not the model is consistent with the data (see Bollen 1989 for a discussion of structural equation modeling).

Initial results are displayed in Figure 9.7. The sign of the path coefficients ("+" or "-") does not always agree with what was hypothesized. Indicators for some variables, notably land-use conversion, proved useful. Indicators for other variables (such as demographic change) did not adequately represent the variable.

Various goodness-of-fit indices were calculated. The chi-square value and its probability level were calculated (chi-square = 268.43, d.f. = 126, $\underline{P}$ = <.001). This would indicate poor model fit, though the chi-square value is not always a good estimate (Steiger 1989). The low ratio of the chi-square value to degrees of freedom (2.13) suggests a reasonably good fit (Tanaka 1987; Bollen 1989). Two other estimates of model fit were evaluated. The Steiger-Lind adjusted root mean square (RMS) index is the root mean square standardized residual. The adjusted population gamma index (PGI) is similar to a coefficient of model determination. Both are adjusted for model complexity. Values of the RMS below .10 indicate reasonably good fit, while values of the PGI above .90 indicate good fit (Steiger 1989). The Steiger-Lind adjusted RMS index was .10, while the adjusted PGI was .82 for the tested model.

These results suggest that the operational model will need revision, though it does demonstrate some value. Other tests are needed to better assess the usefulness of the model.

Further testing of the model at the cross-national level is anticipated. Attention is focusing on collecting the best available data for indicators of each variable. Multidimensional variables may call for more indicators than were used in the first test. Of particular concern are the ecosystem impact variables, which were left out of the first test due to a lack of data, and biodiversity loss, data for which are scant and incomplete. Further testing may suggest ways to revise the model's structure or choose better indicators to improve its predictive power.

KEY CHALLENGES FOR AN EMERGING RESEARCH AGENDA

The root causes of biodiversity loss are primarily human actions, imbedded in social systems that are inextricably intertwined with natural systems. Hence, understanding the relation between socio-economic factors and biodiversity loss is a prerequisite for wise and effective management for protection of biodiversity. Such understanding will largely rest on sound and careful science. The nature of the problem demands interdisciplinary efforts. The process of modeling and first efforts at constructing models are necessary steps.

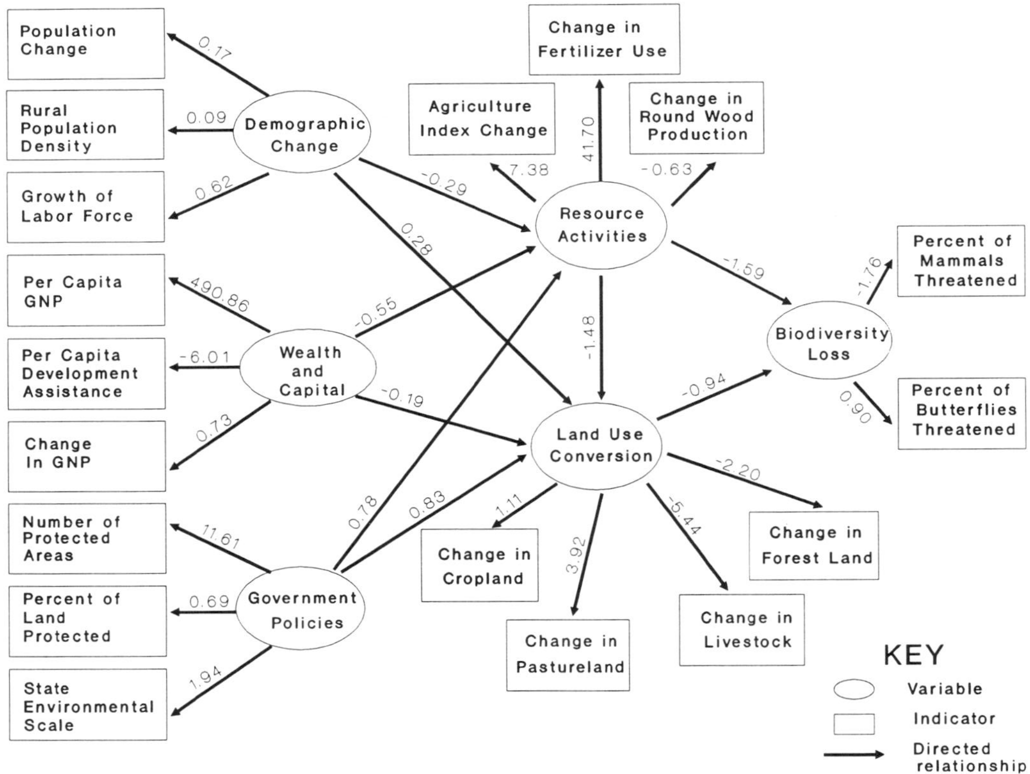

Figure 9.7. Results of model analysis. Note: structural relations are presented in standardized form. Residuals were calculated, but are not illustrated here. (From Forester 1992).

A modest but important research agenda emerges. Ecology and conservation biology will be, of course, crucial partners with the social sciences (Machlis 1992; NRC 1992). Among the social sciences, no single discipline (be it sociology, geography or economics) will or should dominate. Nor will a particular paradigm in the social sciences (such as functionalism or Marxist analysis) have any special claim; all must be evaluated as to their ability to provide usable knowledge about biodiversity loss. A wide range of research methods (from case studies to cross-sectional comparisons) might prove fruitful. Interdisciplinary teams of scientists should be encouraged, but individual efforts within a narrower, disciplinary focus may also be useful. Hence, a contemporary research agenda for modeling biodiversity loss is relatively general. Several key challenges are described below.

The Construction of Explicit Biodiversity Models

Although a large number of conceptual and predictive models of human/Nature interactions exist, explicit models of biodiversity loss are sparse and incomplete. Although many of the generic models treat environmental change or ecosystem alteration as the dependent variable, it is not at all clear that biodiversity loss can simply be substituted for these more general factors. Biodiversity loss is a special case of environmental change, and the socio-economic factors that influence it may not have generic effects. For example, biodiversity loss measured as species-richness may be so contextual (dependent upon natural biodiversity levels at a specific locale or the scale used) that certain generic models fail to explain even the most profound levels of biodiversity loss. The importance of habitat in the preservation of biodiversity may suggest that spatial relations at the local scale play a more significant role in biodiversity models than, say, models of climate change.

Hence, the task is to construct conceptual, and eventually predictive, models that explicitly treat biodiversity loss as the dependent variable. Well-considered conceptual models should provide the interdisciplinary information discussed above, and predictive models should provide testable hypotheses to guide research efforts in fruitful directions. Flow models, if they go beyond single-flow explanations (such as energy) will be particularly useful, because they have the potential to trace causal effects of human action. These models, even at the conceptual stage, should be subject to wide and interdisciplinary review.

Sustained Efforts at Model Testing and Revision

As first efforts at modeling biodiversity loss produce conceptual and predictive models, it should become possible to test these emerging models for validity, realism and accuracy. Success will depend on patience. First efforts are likely to ignore important variables, mis-specify driving forces or give them inappropriate weights, and be predictive at one scale and wildly irrelevant at another. A *sustained* effort to construct these models, test them, revise, them and test them again (and perhaps again) is necessary. First failures are the precursors of more successful (that is predictive and useful) models. They will need to be shared among interested scientists, and critically evaluated. The criteria described in this chapter may serve as guidelines and might profitably be applied in the review and evaluation of biodiversity models.

A Diversity of Model-Testing Strategies

Because biodiversity loss is so clearly a multi-scale and multi-dimensional problem, no single approach to testing models will produce definitive results. A variety of testing strategies should be encouraged. Biodiversity loss is clearly a time-dependent variable and, hence, several time scales may be useful. For example, models of biodiversity loss could be tested with historical records for specific locales, cross-sectionally across locales, or by predicting loss for a specific locale and measuring actual loss in the future. Retrospective models of certain megafauna populations (such as great whales, elephants and rhinos) have potential, because data on population size, age structure, harvest levels and even market forces are reasonably well documented (see, for example, MacNab 1991). Also important is the contemporary establishment of baseline data for long-term research; variables to be monitored might be drawn from conceptual models currently available, with flexibility and addition of new variables built into the monitoring process. A critical challenge is to separate out, for a variety of time scales, those socio-economic variables that are invariant (constants) and those undergoing significant change (NRC 1992).

The issue of scale is crucial, because biodiversity loss is probably embedded in a complex human/Nature system that operates at several hierarchical levels. Socio-economic factors important at one scale may be less significant at another. At each different scale, new variables and relations may emerge as critical driving forces. To be useful for decision-makers, model resolution will need to vary from scale to scale; a national park superintendent requires a significantly different model than those responsible for managing the Global Biodiversity Strategy (for a description of research needs associated with the Strategy, see WRI et al. 1992).

Biodiversity loss is also site-specific; the high species richness of the neotropical forest may react very differently than the low-level richness of the English heathlands. Hence, models of biodiversity loss must be tested in a variety of ecosystem types and biomes. In addition, the human social systems and economies that overlay such ecosystems vary profoundly. Models with narrow scope conditions can be tested for specific socio-economic systems, but those with broader scope will require a more varied testing regime. For example, the predictive model described above requires testing in several biomes and economic systems (from underdeveloped mixed economies to developed capitalist ones) before its true scope conditions and validity can be described.

Improved Data Sources and Analysis Techniques

Rigorous model-testing is dependent upon good data. Such data will have several qualities: (1) they will be available at appropriate levels of aggregation, (2) they will have known and acceptable levels of accuracy, consistent throughout the database, (3) they will correspond to key social and biological variables, i.e., act as useful indicators, (4) sources and methods of data collection will be clearly understood, and (5) the data will be widely accessible to the scientific community. These are demanding requirements, and currently, such interdisciplinary databases are not available.

In the past few years, several statistical abstracts dealing with environmental issues (and occasionally including biodiversity data) have been published; the World Resources Institute (WRI 1990, 1992) and the World Conservation Monitoring Centre (WCMC 1992) have made admirable efforts. However, these compilations are primarily aimed at policy-makers and public interest groups; the scientific community has not yet been well-served. The development and distribution of scientific data is necessary if the modeling of biodiversity loss is to proceed.

In addition, practical and accessible analysis techniques are required. The testing of predictive models requires statistical techniques such as structural equation modeling and path analysis; easily accessible software is required if more than a few statisticians are to be involved. Geographic information systems (GIS) offer important opportunities to test models that emphasize spatial relations. Yet, protocols and techniques to merge biological and socio-economic data are not available, and the ability to quickly and efficiently move from GIS output to cartographically correct (i.e., faithful to cartographic design principles) and high quality maps (a requirement for using the models in decision-making) is close, but not currently at hand. Solving these technological problems will greatly aid the progress of modeling biodiversity loss.

Improved Communication Among the Scientific Community

The construction of models of biodiversity loss is only one part of a larger interdisciplinary research agenda dealing with biodiversity (see Machlis 1992, for a discussion). This agenda has the potential to bring together scientists from a variety of disciplines, perspectives, institutions, and cultures. Yet currently, the relevant fields of study are held apart by the orthodoxies of academic organization.

Significant advances in interdisciplinary problem-solving require expanding and deepening of the communication between the social and biological sciences, individual disciplines, schools of thought, and what sociologists call invisible colleges. Creative approaches, such as interdisciplinary symposia, electronic networks, organized data-sharing, and funding programs that support exploratory and interdisciplinary research have to be explored. Specific to the modeling exercise, it may be useful to convene a workshop dedicated to developing a series of models of biodiversity loss, and bring together biologists, social scientists, and modeling specialists for that purpose.

Traditional lines of communication should be exploited as well. The scientific journals that deal with issues of ecosystem functioning, social behavior, and other concerns related to biodiversity should broaden their treatments and take the initiative to publish interdisciplinary research that meets all criteria for scientific merit and importance. Lists of editorial staff and reviewers should be expanded to include a variety of disciplines, so that informed and critical reviews of relevant yet non-traditional (for that particular journal) manuscripts can be objectively evaluated. Special issues dedicated to biodiversity research or an annual review may be acceptable alternatives. The objectives should be to disseminate the emerging research to as wide a range of scientists as possible, subject the research to varied and thorough critique, and share advances in theory, methods, and empirical findings.

CONCLUSIONS

Because biodiversity loss is so crucial a problem in human affairs, its treatment will, like most environmental issues, be primarily as a political, economic, and management problem. Decision-makers are most likely to not be scientists, yet, rely heavily on the scientific community for facts, predictions, and advice. Models of biodiversity loss are likely to be utilized by such decision-makers and, if the pattern is similar to other environmental issues (such as risk assessment or global climate change), with mixed results.

Beyond first efforts at modeling the relations between socio-economic factors and biodiversity loss, there is a realistic possibility of truly usable models, perhaps within the next ten years. The conversion of such scientific research into instruments of policy has its own challenges. All models carry the weight of assumptions, biases, deeply held values,and ideology. Models can become political tools and, like all political tools, be used to oppress or liberate. We need to be prepared.

ACKNOWLEDGMENTS

The authors would like to thank Doug Houston, Jean McKendry, Tim Tear, J. Michael Scott and Gerry Wright for their comments on earlier drafts of this chapter. This work was supported by the Pacific Northwest Region, National Park Service, and the Idaho State Board of Education, Competitive Grants Program.

LITERATURE CITED

Allen, J. M., ed. 1963. *The Nature of Biological Diversity.* McGraw Hill, Inc, New York.

Allen, J. C., and D. F. Barnes. 1985. The causes of deforestation in developing countries. *Ann. Assoc. Am. Geographers* 75(2):163-184.

Bailey, K. D. 1990. From POET to PISTOL: Reflections on the ecological complex. *Sociological Inquiry* 60(4):386-394.

Bennett, J. W. 1976. *The Ecological Transition.* Pergamon Press, New York.

Bollen, K. A. 1989. *Structural Equations with Latent Variables.* John Wiley and Sons, New York.

Brown, R. L., ed. 1985. *State of the World 1985: A Worldwatch Institute report on Progress Toward a Sustainable Society.* W. W. Norton, New York.

Burch, W. R., Jr. 1971. *Daydreams and Nightmares: A Sociological Essay on the American Environment.* Harper and Row, New York.

Clark, W. C. 1989. The human ecology of global change. *Internat. Social Sci. J.* 41(3):315-345.

Costanza, R., and H. E. Daly. 1992. Natural capital and sustainable development. *Conserv. Biol.* 6(1):37-46.

Diamond, J. M. 1984. 'Normal' extinctions of isolated populations. In *Extinctions,* ed. M. H. Nitecki, pp. 191-246. University of Chicago Press, Chicago, Ill.

Dietz, T., and L. Kalof. 1992. Environmentalism among nation-states. *Social Indicators Research* 26(4):353-366.

Duncan, O. D. 1969. Social forecasting—the state of the art. *The Public Interest* 17:88-118.

Ehrenfeld, D. 1972. *Conserving Life on Earth.* Oxford University Press, New York.

Ehrenfeld, D. 1976. The conservation of non-resources. *Am. Sci.* 64:654.

Farnsworth, N. R. 1988. Screening plants for new medicines. In *Biodiversity,* ed. E. O. Wilson, pp. 83-97. National Academy Press, Washington, D.C.

Fearnside, P. M. 1987. Causes of deforestation in the Brazilian Amazon. In *The Geophysiology of Amazonia*, ed. R. E. Dickinson, pp. 37-57. John Wiley and Sons, New York.

Forester, D. 1992. Human impacts on biodiversity loss in developing countries: A preliminary analysis. M.S. thesis, University of Idaho, Moscow, Ida.

Franklin, J. F. 1988. Structural and functional diversity in temperate forests. In *Biodiversity*, ed. E. O. Wilson, pp. 166-175. National Academy Press, Washington, D.C.

Freedman, B. 1989. *Environmental Ecology: The Impacts of Pollution and other Stresses on Ecosystem Structure and Function.* Academic Press, Inc., San Diego, Calif.

Friedland, E. I. 1977. Values and environmental modeling. In *Ecosystem Modeling in Theory and Practice: an Introduction with Case Histories*, eds. C. A. Hall and J. W. Day, pp. 115-131. University Press of Colorado, Niwot, Colo.

Gallopin, G., P. Gutman, and H. Maletta. 1989. Global impoverishment, sustainable development and the environment: A conceptual approach. *Internat. Social Sci. J.* 41(3):375-397.

Gilpin, M. E., and M. E. Soulé. 1986. Minimum viable populations: Processes of species extinction. In *Conservation Biology: The Science of Scarcity and Diversity*, ed. M. E. Soulé, pp. 19-34. Sinauer Associates, Inc., Sunderland, Mass.

Hall, C. A. S., and J. W. Day, eds. 1977a. *Ecosystem Modelling in Theory and Practice: An Introduction with Case Histories.* University Press of Colorado, Niwot, Colo.

Hall, C. A. S., and J. W. Day. 1977b. Systems and models: terms and basic principles. In *Ecosystem Modelling in Theory and Practice: An Introduction with Case Histories*, ed. C. A. S. Hall and J. W. Day, pp. 5-36. University Press of Colorado, Niwot, Colo.

Hanson, J. S., G. P. Malanson, and M. P. Armstrong. 1990. Landscape fragmentation and dispersal in a model of riparian forest dynamics. *Ecol. Modelling* 49:277-296.

Harris, L. D. 1984. *The Fragmented Forest.* The University of Chicago Press, Chicago, Ill.

Harwood, M. 1982. Math of extinction. *Audubon Mag.* 84:18-21.

Hawley, A. H. 1950. *Human Ecology: A Theory of Community Structure*. Ronald Press, New York.

Hawley, A. H. 1986. *Human Ecology*. The University of Chicago Press, Chicago, Ill.

Hendee, J. C. 1990. *Wilderness Management*. Fulcrum Press, Fort Collins, Colo.

Higashi, M., and T. P. Burns, eds. 1991. *Theoretical Studies of Ecosystems*. Cambridge University Press, New York.

Huntley, B. J. 1988. Conserving and monitoring biotic diversity—some African examples. In *Biodiversity*, ed. E.O. Wilson, pp. 284-260. National Academy Press, Washington, D.C.

Hurlbert, S. H. 1971. The nonconcept of species diversity: A critique and alternative parameters. *Ecology* 52:577-586.

International Union for the Conservation of Nature and Natural Resources and Natural Resources and United Nations Environmental Programme (IUCN/UNEP). 1986a. *Review of the Protected Areas System in the Indo-Malayan Realm.* IUCN, UNEP, Gland, Switzerland.

International Union for the Conservation of Nature and United Nations Environmental Programme (IUCN/UNEP). 1986b. *Review of the Protected Areas System in the Afro-tropical Realm.* IUCN, UNEP, Gland, Switzerland.

Janzen, D. H. 1988. Tropical dry forests—the most endangered major tropical ecosystem. In *Biodiversity*, ed. E.O. Wilson, pp. 130-137. National Academy Press, Washington, D.C.

Johannes, R. E., and B. G. Hatcher. 1986. Shallow tropical marine environments. In *Conservation Biology*, ed. M. E. Soulé, pp. 371-382. Sinauer Associates, Inc., Sunderland, Mass.

Johnston, D. F. 1970. Forecasting methods in the social sciences. *Technological Forecasting and Social Change* 2:173-187.

Kellert, S. R. 1986. Social and perceptual factors in the preservation of animal species. In *The Preservation of Species*, ed. B. G. Norton, pp. 50-73. Princeton University Press, Princeton, N.J.

Kingsland, S. E. 1985. *Modeling Nature—Episodes in the History of Population Ecology.* University of Chicago Press, Chicago, Ill.

Klein, L. R. 1986. Macro-economic modeling and forecasting. In *Behavioral and Social Science: Fifty Years of Discovery*, ed. N. J. Smelser and D. R. Gerstein, pp. 95-110. National Academy Press, Washington, D.C.

Krebs, C. J. 1972. *Ecology*. Harper & Row Publishers, Inc., New York.

Leidy, R. A., and P. L. Fiedler. 1985. Human disturbance and patterns of fish species diversity in the San Francisco Bay drainage, California. *Biol. Conserv.* 33:247-267.

Levins, R. 1966. The strategy of model building in population biology. *Am. Sci.* 54:421-431.

Lovejoy, T. E. 1986. Species leave the ark one by one. In *The Preservation of Species,* ed. B. G. Norton, pp. 13-27. Princeton University Press, Princeton, N.J.

Lovejoy, T. E. 1988. Diverse considerations. In *Biodiversity,* ed. E. O. Wilson, pp. 421-427. National Academy Press, Washington, D.C.

Lugo, A. E. 1988. Estimating reductions in the diversity of tropical forest species. In *Biodiversity,* ed. E. O. Wilson, pp. 58-70. National Academy Press, Washington, D.C..

MacArthur, H. 1985. Patterns of species diversity. *Biol. Rev.* 40:510-533.

Machlis, G. E. 1989. Managing parks as human ecosystems. In *Public Places and Spaces,* Vol. 10 (HBE series), ed. I. Altman and E. H. Zube, pp. 255-273. Plenum Publishing Corporation, New York.

Machlis, G. E. 1990. "The application of sociology to biodiversity problems." Paper presented at the Applied Sociology Annual meeting, October 20-22 1989, Denver, Colo.

Machlis, G. E. 1992. The contribution of sociology to biodiversity research and management. *Biol. Conserv.* 62(3):161-170.

Machlis, G. E., and D. J. Forester. 1992. "Extended gap analysis: A technique for biodiversity management of protected area systems." Paper presented at the World Parks Congress, February 10-21, 1992, Caracas, Venezuela.

Machlis, G. E., and D. L. Tichnell. 1985. *The State of the World's Parks: An International Assessment for Resource Management, Policy, and Research.* Westview Press, Boulder, Colo.

Machlis, G. E., and R. G. Wright. 1984. "Potential indicators for monitoring biosphere reserves." Paper presented at the Miami International Symposium on the Biosphere, April 16-18, 1984, Miami Beach, Fla.

MacNab, J. 1991. Does game cropping serve conservation? A reexamination of the African data. *Reviews/Synthèse* 69:2283-2290.

Magurran, A. E. 1988. *Ecological Diversity and its Measurement*. Princeton University Press, Princeton, N.J.

Margalef, R. 1968. *Perspectives in Ecological Theory.* University of Chicago Press, Chicago, Ill.

McKendry, J. E., and G. E. Machlis. 1991. "The role of geography in extending biodiversity gap analysis." Paper presented at the American Association of Geographers Annual Meeting, April 13, 1991, Miami, Fla.

Meyer, W. B., and B. Turner. 1992. Human population growth and global land-use/land cover change. *Ann. Rev. Ecol. Syst.* 23:39-62.

Miller, J. G. 1978. *Living Systems.* McGraw-Hill, New York.

Miller, R. S., and D. B. Botkin. 1974. Endangered species: Models and predictions. *Am. Sci.* 62:172-180.

Mueller-Dombois, D., K. Kartawinata, and L. L. Handley. 1983. Conservation of species and habitats: A major responsibility in development planning. In *Natural Systems for Development: What Planners Need to Know,* ed. R. A. Carpenter, pp. 1-51. MacMillan Publishing Company, New York.

Myers, N. 1980. The problem of disappearing species: What can be done? *Ambio* 9(5):229-235.

Myers, N. 1986. "Tackling mass extinction of species: A great creative challenge." Albright Lecture, University of California, Berkeley, Calif.

Myers, N. 1988. Tropical forests and their species—going, going,...?. In *Biodiversity,* ed. E. O. Wilson, pp. 218-235. National Academy Press, Washington, D.C.

National Research Council (NRC). 1990. *Forestry Research: A Mandate for Change.* National Academy Press, Washington, D.C.

National Research Council (NRC). 1992. *Global Environmental Change.* National Academy Press, Washington, D.C.

Neumann, R. P., and G. E. Machlis. 1989. Land-use and threats to parks in the neotropics. *Environ. Conserv.* 16(1):13-18.

Nickerson, N. H., R. A. Dobberteen, and N. M. Jarman. 1989. Effects of power-line construction on wetland vegetation in Massachusetts, USA. *Environ. Manage.* 13(4):477-483.

Norton, B. G. 1986. On the inherent danger of undervaluing species. In *The Preservation of Species,* ed. B. G. Norton, pp. 110-137. Princeton University Press, Princeton, N.J.

Norton, B. G. 1987. *Why Preserve Natural Variety?* Princeton University Press, Princeton, N.J.

Norton, D. A. 1991. *Trilepidia adamsii:* an obituary for a species. *Conserv. Biol.* 5(1):52-57.

Odum, H. T. 1970. *Environment, Power and Society.* University of North Carolina Press, Chapel Hill, N.C.

Office of Technology Assessment (OTA). 1987. *Technologies to Maintain Biological Diversity.* OTA-F-330, Government Printing Office, Washington, D.C.

O'Neill, R. 1989. Perspectives in hierarchy and scale. In *Perspectives in Ecological Theory,* ed. J. Roughgarden, R. May, and S. Levin, pp. 140-156. Princeton University Press, Princeton, N.J.

Overton, W. S. 1977. A strategy of model construction. In *Ecosystem Modelling in Theory and Practice: An Introduction with Case Histories,* ed. C. A. S. Hall, and J. W. Day, pp. 49-73. University Press of Colorado, Niwot, Colo.

Pease, C. M., and J. Bull. 1992. Is science logical? *BioScience* 42(4): 293-298.

Peters, R. H. 1991. *A Critique for Ecology.* Cambridge University Press, Cambridge.

Pimm, S. L., and M. E. Gilpin. 1989. Theoretical issues in conservation biology. In *Perspectives in Ecological Theory,* ed. J. Roughgarden, R. M. May, and S. A. Levin, pp. 287-305. Princeton University Press, Princeton, N.J.

Ray, G. C. 1988. Ecological diversity in coastal zones and oceans. In *Biodiversity,* ed. E. O. Wilson, pp. 36-50. National Academy Press, Washington, D.C.

Reh, W., and A. Seitz. 1990. The influence of land use on the genetic structure of populations of the common frog *Rana temporaria. Biol. Conserv.* 54(3):239-249.

Reid, W. V., and K. R. Miller. 1989. *Keeping Options Alive: The Scientific Basis for Conserving Biodiversity.* World Resources Institute, Washington, D.C.

Robertson, D., A. Bundy, R. Muetzelfeldt, M. Haggith, and M. Uschold. 1991. *Eco-logic: Logic-based Approaches to Ecological Modelling.* The MIT Press, Cambridge, Mass.

Robinson, J. B. 1991. Modelling the interactions between human and natural systems. *Internat. Social Sci. J.* 43(4):629-648.

Roughgarden, J., R. M. May, and S. A. Levin, eds. 1989. *Perspectives in Ecological Theory.* Princeton University Press, Princeton, N.J.

Rudel, T. K. 1989. Population, development, and tropical deforestation: A cross-national study. *Rural Sociology* 54(3):327-338.

Ruffie, J. 1982. *The Population Alternative: A New Look at Competition and the Species.* Pantheon Books, New York.

Salm, R. V. 1984. *Marine and Coastal Protected Areas: A Guide for Planners and Managers.* International Union for the Conservation of Nature, Gland, Switzerland.

Schücking, H., and P. Anderson. 1991. Voices unheard and unheeded. In *Biodiversity: Social & Ecological Perspectives,* ed. V. Shiva, P. Anderson, H. Schücking, A. Gray, L. Lohmann, D. Cooper, pp. 13-41. Zed Books, Lt., London.

Scott, J. M., B. Csuti, J. E. Estes, and H. Anderson. 1989. Status assessment of biodiversity protection. *Conserv. Biol.* 3(1):85-87.

Scott, J. M., B. Csuti, J.D . Jacobi, and J. E. Estes. 1987. Species richness: A geographic approach to protecting future biological diversity. *BioScience* 37(11):782-788.

Shiva, V., P. Anderson, H. Schücking, A. Gray, L. Lohmann, and D. Cooper. 1991. *Biodiversity: Social & Ecological Perspectives.* Zed Books, N.J.

Shoemaker, C. A. 1977. Mathematical construction of ecological models. In *Ecosystem Modeling in Theory and Practice: An Introduction with Case Histories,* eds. C. A. Hall and J. W. Day, pp. 75-114. University Press of Colorado, Niwot, Colo.

Slobodkin, L. B. 1986. On the susceptibility of different species to extinction: Elementary instructions for owners of a world. In *The Preservation of Species,* ed. B. G. Norton, pp. 226-242. Princeton University Press, Princeton, N.J.

Soulé, M. E. 1986. *Conservation Biology: The Science of Scarcity and Diversity.* Sinauer Associates, Inc., Sunderland, Mass.

Soulé, M. E. 1991. Conservation: tactics for a constant crisis. *Science* 253(5021): 744-750.

Soulé, M. E., and K. A. Kohm, eds. 1989. *Research Priorities for Conservation Biology.* Island Press, Washington, D.C.

Soulé, M. E., and D. Simberloff. 1986. What do genetics and ecology tell us about the design of nature reserves? *Biol. Conserv.* 35:19-40.

Starfield, A. M., and A. L. Bleloch. 1986. *Building Models for Conservation and Wildlife Management.* Macmillan Publishing Co., New York.

Steiger, J. H. 1989. *EzPATH.* SYSTAT, Inc., Evanston, Ill.

SYSTAT. 1990. *SYSTAT.* SYSTAT, Inc., Evanston, Ill.

Tanaka, J. S. 1987. How big is big enough?: Sample size and goodness of fit in structural equation models with latent variables. *Child Development* 58:134-146.

Terborgh, J., and B. Winter. 1980. Some causes of extinction. In *Conservation Biology: An Evolutionary-Ecological Perspective,* ed. M. E. Soulé and B. A. Wilcox, pp. 119-133. Sinauer Associates, Sunderland, Mass.

The Nature Conservancy (TNC). 1975. *The Preservation of Natural Diversity: A Survey and Recommendations.* The Nature Conservancy, Washington, D.C.

Turner, B. L. 1989. The human cause of global environmental change. In *Global Change and our Common Future: Papers from the Forum,* eds. R. S. DeFries and T. S. Malone, pp. 90-99. National Academy Press, Washington, D.C.

Turner, B. L., and W. B. Meyer. 1991. Land use and land cover in global environmental change: Considerations for study. *Internat. Social Sci. J.* 43(4):669-679.

United Nations Environment Programme (UNEP). 1991. *Environmental Data Report,* 3rd ed. Basil Blackwell, Oxford.

Vermeij, G. J. 1986. The biology of human-caused extinction. In *The Preservation of Species,* ed. B. G. Norton, pp. 28-49. Princeton University Press, Princeton, N.J.

Walker, B. H. 1992. Biodiversity and ecological redundancy. *Conserv. Biol.* 6(1):18-22.

Western, D., M. C. Pearl, S. L. Pimm, B. Walker, I. Atkinson, and D. S. Woodruff. 1989. An agenda for conservation action. In *Conservation for the Twenty-First Century,* eds. D. Western and M. C. Pearl, pp. 304-323. Oxford University Press, New York.

Westman, W. 1990. Managing for biodiversity. *BioScience* 40(1):26-33.

Wilcox, B. A. 1984. Concepts in conservation biology: Applications to the management of biological diversity. In *National Parks, Conservation, and Development: the Role of Protected Areas in Sustaining Society,* eds. J. A. McNeely and K. R. Miller, pp. 639-647. Smithsonian Institution Press, Washington, D.C.

Wilson, E. O. 1988a. *Biodiversity.* National Academy Press, Washington, D.C.

Wilson, E. O., ed. 1988b. The current state of biological diversity. In *Biodiversity,* ed. E. O. Wilson, pp. 3-18. National Academy Press, Washington, D.C.

World Conservation Monitoring Centre (WCMC). 1992. *Global Diversity: Status of the Earth's Living Resources.* Chapman and Hall, New York.

World Resources Institute (WRI). 1986. *World Resources 1986.* World Resources Institute, New York.

World Resources Institute (WRI). 1990. *World Resources 1990-91.* Oxford University Press, New York.

World Resources Institute (WRI). 1992. *World Resources 1992-93.* Oxford University Press, New York.

World Resources Institute, The World Conservation Union, United Nations Environment Programme (WRI/WCU/UNEP). 1992. *Global Biodiversity Strategy.* World Resources Institute, Washington, D.C.

Yoshino, M. 1991. The impact of climatic change on agriculture from the viewpoint of east Asia. In *The Global Environment,* eds. K. Takeuchi and M. Yoshino, pp. 16-41. Springer-Verlag, Berlin.

II

INVENTORY, MONITORING AND ASSESSMENT

10

BIODIVERSITY INVENTORY: BUILDING AN INVENTORY AT SCALES FROM LOCAL TO GLOBAL

John G. Dennis and Michael A. Ruggiero[1]

Wildlife and Vegetation Division
National Park Service
Washington, D.C.
U.S.A.

People from all walks of life conduct inventories of biodiversity all the time. Whether it is the urbanite who catalogs the birds at the backyard feeder, the sanitarian who assesses the kinds and population sizes of kitchen insects, the field biologist who assembles species lists and community relationships for all major groups of plants and animals found within the boundaries of a land management unit, or the international planner who tries to assess the status of biodiversity in countries throughout the world, the goal is the same—to know what elements of biodiversity are found where. As long as each of these efforts remains individual, and no effort is made to use them for larger scale syntheses, there is little need to coordinate them or to require compatibility of timing, methods, or taxonomy.

The growing awareness of the interconnectedness of elements of biodiversity across land management units, regions, continents, and the globe, as well as through time, brings with it a concomitant awareness of a need to foster coordination and cooperation. We realize this need every time we try to compare or aggregate data from different land management units and discover that we cannot because of incompatibilities of methods, terms, or data handling systems. We believe it is important to rectify these incompatibilities at the planning stage, before data collection starts. Therefore, drawing on concepts presented in the National Park Service Natural Resources Inventory and Monitoring Guideline (NPS 1992), in this paper we offer ideas that we believe will be useful in coordinating inventories at one scale, as well as making them available to other scales. We believe the process to ensure compatibility across scales requires that thought be given to what an inventory is; its purpose, orientation, completeness, and desired intensity; and the influences of operational capabilities, scale, and biodiversity focus.

DEFINITION OF INVENTORY

In our view, a biodiversity inventory is a formal cataloging of the occurrences at a point in time of defined elements of biodiversity that are contained in a defined geographic

[1] Present address: National Biological Survey, Washington, D.C.

unit. To be conducted formally, this cataloging must be done in response to a clearly stated purpose, according to well established field procedures, within statistically valid sampling designs, and using rigorous quality control and data administration practices. This cataloging might involve data only about the biodiversity element, itself, or might also include data on biological or non-biological factors found in association with the biodiversity elements being cataloged. Because of lack of funds, time, and trained personnel, this cataloging almost invariably will represent a sampling, rather than a complete listing, of the biodiversity of an area (di Castri et al. 1992).

PURPOSE OF CONDUCTING AN INVENTORY

We see three possible purposes for conducting an inventory: an end in itself, developing the baseline needed for follow-up monitoring, and developing a capability to predict future biological events. As an end in itself, a biodiversity inventory will provide only spatial information about the distributional characteristics of the elements of biodiversity being inventoried. In providing baseline information for a monitoring program, a biodiversity inventory provides both spatial and temporal information about the subject elements. In developing a capability to predict future biological events from a one-time inventory, such as vegetational succession in the absence of future disturbance, or community response to perturbation by spillage of hazardous materials, a biodiversity inventory provides only temporal information about the biodiversity elements and the corrolaries that are being inventoried.

ORIENTATION OF AN INVENTORY

We suggest four possible approaches for orienting an inventory: survey of major elements, identification of indicator elements, identification of targeted elements, and comprehensive assessment of all elements. An inventory oriented toward the major elements would focus on those elements that are common and are considered general descriptors of the system being inventoried, and would consciously not be concerned with rare or indicator elements. More specifically oriented inventories, such as inventories of either indicator or targeted elements (e.g., snags as an indicator of old growth forest conditions, or wetland communities or listed threatened or endangered species as targets of an inventory) would focus on finding the indicator or targeted elements and would not be concerned with providing a general description of the system. A comprehensive assessment would combine the orientations of survey, indicator, and targeted inventories and would seek to provide information on the location and abundance of all elements in proportion to their occurrence in the system.

COMPLETENESS OF AN INVENTORY

As Solbrig (1991) points out and Stohlgren and Quinn (1991) discuss, the purpose and orientation of an inventory will determine choice of methods which in turn will influence the completeness of the inventory in terms of taxonomy, community/ecosystem representation, geographic space, and seasonal/temporal representation. An inventory which searches out all occurrences of chosen taxonomic groups will be complete taxonomically, but not necessarily either ecologically or spatially. One which uses stratified sampling designs to ensure inventorying each type of major ecological element thought to be in the study area will be complete in terms of ecological systems, but not necessarily either taxonomically or spatially. On the other hand, an inventory that conducts samples at each point of an overlaid grid will be complete geographically, but not necessarily either taxonomically or ecologically.

INTENSITY OF INVENTORY

An inventory can have three general levels of intensity: qualitative, quantitative, and relational (integrated or holistic in the terminology of Conant et al. 1983). A qualitative inventory merely provides presence or absence information about the desired elements of biodiversity, whereas a quantitative inventory identifies population sizes, distributional frequencies, or areal extents of coverage of the desired elements of biodiversity. A relational inventory combines either or both qualitative and quantitative approaches with an inventory of other biotic or abiotic elements to permit their being correlated with the distributions and abundances of the desired biodiversity elements.

OPERATIONAL INFLUENCES

Six general classes of operational considerations influence inventorying of biodiversity. A fundamental influence is the existing knowledge base from which an inventory is to be constructed—the more pre-existing resource knowledge, the more definitive, detailed, and comprehensive the inventory can be made. A second influence is the level of personnel and technical capabilities available to be assigned to the inventory. Scientifically trained personnel working with the latest technological advances in remote sensing, optics, geographical positioning, GIS and data base computing, statistics, and laboratory analyses can provide a much more intensive, precise, and comprehensive inventory than can less well trained personnel provided limited or no technological support. A third influence is level of funding—the larger the funding base, the greater the number of trained personnel, amount of specialized equipment, and availability of complex logistical support that can be employed in the inventory. A fourth operational influence are the designed purpose and intensity of the inventory—the more complex the purpose and greater the intensity, the greater the level of expert personnel, technological, and logistical resources needed to produce a successful result. A fifth influence comes to bear when inventories are designed to have multiple performers contributing to a common network of data administration and analysis. In such cases, the success of the networking will depend on the ability of all partners to bridge any language, equipment, or scientific terminology differences that exist among them. A sixth influence is exerted by the level of enthusiasm, dedication, and commitment of the people and institutions involved in conducting the inventory.

GEOGRAPHIC SCALE OF INVENTORY

For the purposes of our discussion here, we identify four levels of geographic scale of an inventory: within the boundaries of an individual land management unit, within a biogeographic region involving two or more land management units and definable in terms of biological commonalities, within a continent and generally encompassing more than one biogeographic region, and global. An inventory conducted within a single land management unit generally involves only one decision-maker, one type of technological system, a more limited number of elements of biodiversity, and generally can be intensively conducted to meet clearly definable purposes. In contrast, inventories conducted at biogeographic region or larger scales usually involve multiple decision-makers and technological systems, potentially many elements of biodiversity, and a multiplicity of possibly conflicting purposes which, taken together, can make it quite difficult to articulate a clearly defined purpose.

BIODIVERSITY FOCUS OF AN INVENTORY

As summarized by Noss (1990) and in the report of the Keystone Policy Dialogue (Keystone Center 1991), an inventory of biodiversity can be focused on one or more of five general levels of inquiry: genetic, species, community, landscape, and process. The most detailed levels (genetic, species, and community) can be conducted either within a single land management unit or among two or more units. The broader levels (landscape and process) can be conducted within a single land management unit but often are more productively conducted at biogeographic or larger scales.

INTERACTION OF FACTORS—EXAMPLES FROM U.S. NATIONAL PARKS

Inventory data exist for lands contained within units of the U.S. National Park System starting from prior to the time of establishment of most of the units. The earliest data generally were qualitative, species-oriented, not well-referenced geographically, and of limited correlative value. As the levels of personnel training, technology, logistic support, and scientific knowledge have increased, so have the breadth and intensity of park inventories increased. We estimate that, today, 40 to 60 parks have fairly comprehensive species lists for vertebrates and vascular plants, terrestrial vegetation descriptions and maps, distribution and abundances of selected aquatic biota and systems, soils and geology maps, and correlative process information. Some parks have functioning geographic information systems and are using geopositioning devices to reduce the size of georeferencing errors. Some parks are developing distributional information about non-vascular plants and invertebrates, and a few have begun to develop information about genetic diversity of either indicator or disjunct species. Many parks are beginning to develop intensive data sets about biodiversity process elements, especially related to fire and fire suppression, deposition of air-borne pollutants, fragmentation of landscapes and loss of migration corridors, influence of introduced species, and roles of irregular events such as forest gap formation and hurricanes. In recent years, parks have begun to build networks, especially within biogeographic regions, but also on continental and global scales.

In developing this inventory data base and in using it to support monitoring, research, interpretation, and management efforts, the Park Service has had to accept a large number of trade-offs among choices of inventory purpose, orientation, completeness, intensity, operations, and scale. In most cases, a tacit assumption has been made that parks already have achieved surveys of major elements and that, from a system view, their inventories are acceptable. It is clear, however, that from taxonomic, geographic, comprehensive, targeted element, baseline, or correlative points of view, all these inventories are incomplete (Stohlgren et al. 1991; Ruggiero et al. 1992). From the data for the best inventoried vertebrate group, the birds, we currently estimate completeness levels of 80% or better in only 43% of the parks on a taxonomic basis, 31% of the parks on an ecological basis, and 32% of the parks on a geographical basis. Given this awareness, parks have begun to identify the monetary, trained personnel, and logistics support needed to develop comprehensive, quantitative, and correlative inventories of park genetic, species, community, and process elements of biodiversity. In addition, the National Park Service in several areas is developing networking opportunities at biogeographic, continental, and global scales and is coming face to face with problems of equipment, terminology, and language differences which impede the ability to construct functional networks.

BUILDING A REGIONAL TO GLOBAL SCALE INVENTORY

In our view, probably the hardest task in building an inventory at a bioregional or larger scale is establishing a common purpose and orientation. In the context of biodiversity in those managed landscapes where one management purpose is to contribute to the goal of perpetuating representative examples of all elements of biodiversity of the larger bioregion, however, we believe the task becomes much easier. In such a landscape, the purpose of the inventory should be to create baseline information to support a cooperative monitoring program, and the orientation of the inventory should include both a survey of major elements of each landscape and a focus on those elements cooperatively agreed to be targeted elements. To achieve this focus and orientation, we believe the landscape cooperators in the inventory should design the inventory to achieve geographic completeness at the bioregional scale and should conduct the inventory to obtain quantitative, rather than qualitative, information. Because different partners in any regional to global scale inventory effort will be starting from different levels of pre-existing knowledge, trained personnel, available technology and logistics, it is important, in the words of Conant et al. (1983, p.28), "to anticipate integration by using a common data base system." Hence, the cooperators at the outset should give much thought to what information will constitute the smallest functional unit of value to both the individual partners and to the network as a whole. Once they have established this smallest functional unit, the cooperators in the network should adopt that unit of value as the base program, while recognizing that different partners might choose to build hierarchically (Quinn and Van Riper 1990; Stohlgren and Quinn 1991) on the base in compatible, but not necessarily identical ways. Once the basic unit of value is established, the partners also should establish clearly how each operating site in the network will be connected around that basic unit of value and how potential barriers of language, equipment, and scientific terminology will be avoided.

Another question that partners might find important to discuss and resolve is whether or not to accept data from volunteers, some of whom will be skilled and others of whom will be enthusiastic but unskilled. Volunteers can provide a much needed source of intensive field labor, and we suggest that modern computer support can permit handling the large volumes of data they generate in ways that can manage and isolate the increased error terms introduced as a result of the differing levels of human skills involved. Examples of volunteer-supported inventories and monitoring programs from the United States include the existing Breeding Bird Survey, Christmas bird counts, and annual survey of butterflies, and potential National Park Service inventories being tested through partnerships with Earthwatch and Partners in Parks.

Recent National Park Service experience in building inventories can offer some insight to the choices that have to be made in the development process. At Sequoia and Kings Canyon National Parks, where much inventory information already exists, for example, the inventory manager has been able to construct a grid-based sampling design to focus on developing unbiased and geographically distributed coverage of the two parks. Because of logistic difficulties, however, he is unable to visit all grid points with equal ease and so has designed a system to obtain multi-disciplinary data from single trips to remote sites. Although this approach yields valuable cost savings, it does result in some compromising of methodological requirements (D. Graber, pers. comm.). In contrast, for the Service as a whole, a nearly completed inventory of existing park inventory data has been designed as a presence/absence survey of major taxonomic and system elements. This survey draws on existing, but highly variable, knowledge bases by requiring that the individual park survey reports must conform to common standards. These standards require use of defined numerical scales for reporting the status of biological inventories for vascular plants and vertebrates for each park in terms of taxonomic, geographic, ecological, and seasonal completeness. This effort since has been combined in a pilot project with The Nature Conservancy's Heritage Methodology in parks of the eastern United States to provide what looks like a robust inventory for the park or reserve manager with common elements that can be synthesized to hierarchically broader scales. Drawing in part from this National Park Service experience in building an inventory of inventories, the EuroMAB Committee of the Man and the Biosphere Programme has initiated an inventory of EuroMAB biosphere reserve activities to form the basis for a developing Biosphere Reserves Integrated Monitoring program. The development of this inventory has been facilitated by agreement on a common language, use of a paper-based survey instrument to increase the commonality of terminology, and use of an already existing, but nascent, international network created by the biosphere reserve program. Similarly, the World Conservation Monitoring Center recently completed an inventory of references to inventory data for protected areas within 131 tropical countries (Murray et al. 1992). Among other recommendations, Murray et al. urged the use of standard formats.

Looking ahead to any global biodiversity inventory which might be initiated as a result of discussions arising out of the biodiversity convention, we propose that the same basic network-building considerations be brought to bear. From a global perspective, we suggest that networking and operational considerations of language, equipment, commonality of terminology, existing knowledge, and availability of trained personnel and technological capabilities will be paramount for building a global inventory. As a means of getting started, we suggest that, because of the integrative strength of biogeographic relationships, the global network be secondary, and that primary networks first be organized around biogeographic region or continental scale relationships. Given this level of organization, we expect some networks would focus on survey level, spatially focused, presence or absence inventories of targeted elements, whereas other networks would focus on creating comprehensive baselines for quantitatively monitoring the relationships of selected elements of biodiversity to other elements of the environment. Given proper design, underlying this diversity of networks would be a lowest functional value of providing inventory data on a global scale to help land managers in participating countries jointly develop the monitoring programs needed for assessing long term trends in the sustainability of the earth's genetic, species, community, landscape, and process elements of biodiversity.

CONCLUSIONS

The simple elements derived from an inventory, such as verified species occurrences at a particular time and at a particular place, or descriptions of biological communities occurring within a stated biogeographic area, might by themselves seem inconsequential. When done with quality and consistency, and when repeated over a broad geographic range, however, they can become an extremely important contribution to understanding biological diversity at landscape or larger scales. We suggest that the model for inventories developed from the National Park Service inventory of inventories and now being extended to the EuroMAB scale be tested and further developed by being applied across managed landscapes within a particular biogeographic region, such as, within the United States, the Southern Appalachians, Sierras, Colorado Plateau, Sonoran Desert, or Chihuahuan Desert.

LITERATURE CITED

Conant, F., P. Rogers, M. Baumgardner, C. McKell, R. Dasmann, and P. Reining, eds. 1983. Resource inventory & baseline study methods for developing countries. AAAS Publ. No. 83-3. Am. Assoc. Adv. Sci., Washington, D.C.

di Castri, F., J. R. Vernhes, and T. Younes, eds. 1992. Inventorying and monitoring biodiversity a proposal for an international network. *Biol. Internat.*, Special Issue No. 27.

Keystone Center. 1991. *Final Consensus Report of the Keystone Policy Dialogue on Biological Diversity on Federal Lands*. The Keystone Center, Keystone, Colo.

Murray, M. G., M. J. B. Green, and K. S. Walter. 1992. Status of plant and animal inventories for protected areas in the tropics. World Conservation Monitoring Centre, Cambridge.

National Park Service (NPS). 1992. Natural Resources Inventory and Monitoring Guideline NPS-75. National Park Service, Washington, D.C.

Noss, R. F. 1990. Indicators for monitoring biodiversity: A hierarchical approach. *Conserv. Biol.* 4:355-364.

Quinn, J. F., and C. van Riper III. 1990. Design considerations for national park inventory databases. In *Examples of Resource Inventory and Monitoring in National Parks of California*, eds. C. van Riper III, T. J. Stohlgren, S. D. Veirs, Jr., and S. C. Hillyer, pp. 5-13. Trans. and Proc. Series No. 8. National Park Service, Washington, D.C.

Ruggiero, M., T. Stohlgren, and G. Waggoner. 1992. Towards a biological survey of the U.S. National Park System. In *Science and the Management of Protected Areas. Developments in Landscape Management and Urban Planning*, 7, eds. J. H. M. Willison, S. Bondrup-Nielsen, C. Drysdale, T. B. Herman, N. W. P. Munro, and T. L. Pollock, pp. 31-35. Elsevier, Amsterdam, Netherlands.

Solbrig, O. T., ed. 1991. *From Genes to Ecosystems: A Research Agenda for Biodiversity.* International Union of Biological Sciences, Paris, France.

Stohlgren, T. J., and J. F. Quinn. 1991. Status of natural resources databases in national parks: western region. National Park Service, Cooperative Park Studies Unit, Technical Report 44, University of California, Davis, Calif.

Stohlgren, T., M. Ruggiero, J. Quinn, and G. Waggoner. 1991. National park biotic inventories assessed. *Park Sci.* 11(4):16-17.

11

DIVERSITY MEASUREMENT AND COMPARISON WITH EXAMPLES

Jeffrey H. Gove

USDA Forest Service
Northeastern Forest Experiment Station
Durham, New Hampshire
U.S.A.

Ganapati P. Patil and Charles Taillie

Center for Statistical Ecology and Environmental Statistics
Department of Statistics
The Pennsylvania State University
University Park, Pennsylvania
U.S.A.

The concept of community diversity and its measurement has been popular in community ecology studies for more than four decades. Diversity is the condition of being different; a useful synonym for diversity is variety. It can be thought of as playing a similar role to qualitative observations (i.e., tree species) as the variance does to quantitative measures (i.e., tree diameter) (Pielou 1975). Diversity is considered here a property of the community and, therefore, the trophic, taxonomic or structural levels that we perceive diversity in terms of, depend upon how the community containing these assemblages is defined. Most often, it is unrealistic to identify the components of a community as all of the living organisms within a specific region. Therefore, the nature of the organisms for which diversity assessment is desired needs to be clearly determined; this is termed the *taxocene* by Pielou (1974), and can be applied with equal validity to taxonomic groups, structural components, or even communities within communities (i.e., forest stands within a landscape mosaic).

This paper is concerned with elucidating a recent study dealing with the issues of *diversity* in specifically defined *communities*. Its intent is to facilitate "technology-transfer" and encourage future research in the subject area of ecological diversity measurement and comparison. To this end, the complete data set for several communities discussed is reproduced. The overwhelming issue of how to integrate diversity considerations into management decisions and policy analysis is not discussed here. Gove et al. (1994) provide an initial treatment of this issue involving forest management decisions and policy analysis.

MEASURING DIVERSITY—AN OVERVIEW

This section presents the background material needed to understand the different diversity measures used in the applications throughout the rest of the paper. The literature on diversity measurement is immense with numerous indices and other methods available for use in measuring diversity. For this reason, only those measures encountered in the rest of the chapter are reviewed here; no attempt has been made at being encyclopedic. In addition, problems with misapplication of diversity measures are evidently commonplace in the literature. Other sources, such as Grassle et al. (1979), Hurlbert (1971), Magurran (1988), and Pielou (1974, 1975, 1977), discuss many of these other measures and their correct use; so accessible is the treatment of this subject by these authors that it need not be repeated here.

The semantics of a few terms used throughout this paper need some attention. The term "species" as used when referring to diversity is simply a convenient label for a countable set of distinct categories comprising the community; therefore, "species" could mean a set of vegetative height strata (e.g., 0-20 ft, 21-40 ft, 41+ ft), forest stands, or true biological species. The "community," therefore, is used in an equally general sense to mean a recognizable collection of species in space-time continuum.

Average Species Rarity

Traditionally, the diversity of a community is composed of two components: the number of species, or species richness and the evenness, or equitability of species as given by their relative abundance. Let s be the species richness for a community, and define the relative abundance vector for this community as $\pi=(\pi_1,\ldots,\pi_s)$, such that

$$\sum_{i=1}^{s} \pi_i = 1.$$

The relative abundance vector can be computed from any one of a number of different measures, such as number of individuals, percent cover, or biomass per unit area. The quantity used will depend on the definition of species (i.e., the composition of the s categories of interest) and there is as yet no consensus about which measure to use in a given situation (Pielou 1974; Magurran 1988). A conceptual community is written as $C(\mu)$, where μ can be some function of s and π, or μ can have some space-time connotation. The *completely even community* is a special case where

$$\pi_i = 1/s = \pi_E$$

for all i, so that s alone determines the relative abundance vector. The completely even community is denoted as $C_E(s)$.

In accordance with Patil and Taillie (1979a, 1982), community diversity is defined here as the average species rarity within the community. If a numerical measure of rarity can be associated with each species in the community, then the community diversity can be determined as its expected value. Thus, if the rarity of the ith species in community $C(\pi)$ is given as $R(i;\pi)$, then

$$\Delta = \Delta(\pi) = \sum_{i=1}^{s} R(i;\pi)\pi_i \tag{1}$$

is the community diversity. The choice of how to assign a rarity value to each individual species will determine the diversity measure.

Patil and Taillie (1979a, 1982) discussed two types of rarity measures: *dichotomous-type* and *rank-type*. With the dichotomous-type rarity measures, the rarity of the ith species depends only on its own relative abundance π_i. In this case rarity is denoted as $R(\pi_i)$ and the measure of community diversity is defined as

$$\Delta = \sum_{i=1}^{s} R(\pi_i)\pi_i .$$

For rank-type measures, however, species rarity depends only on its rank. Consider the *ranked* relative abundance vector, $\pi^{\#}$ which is determined by ranking the components of π from greatest to least so that

$$\pi^{\#} = \left(\pi_1^{\#}, \ldots, \pi_s^{\#}\right),$$

where

$$\pi_1^{\#} \geq \ldots \geq \pi_s^{\#} .$$

Then, if the rarity of the ith-ranked species is denoted as $R(i)$, community diversity is given as

$$\Delta = \sum_{i=1}^{s} R(i)\pi_i^{\#} .$$

Examples of both types of rarity measures and their associated diversity functions are given in the following sections.

Diversity Indices

Several of the most frequently used diversity indices can be conveniently expressed under the umbrella of average species rarity through judicious choice of rarity functions. Species richness, species count, Shannon's, and Simpson's indices can all be derived from this theory as follows

$$\Delta_{SR} = \sum_{i=1}^{s} \left\{\frac{1}{\pi_i}\right\}\pi_i = s \qquad \text{Species Richness, (2)}$$

$$\Delta_{SC} = \sum_{i=1}^{s} \left\{\frac{1}{\pi_i} - 1\right\}\pi_i = s - 1 \qquad \text{Species Count, (3)}$$

$$\Delta_{Sh} = \sum_{i=1}^{s} \{-\log \pi_i\}\pi_i = -\sum_{i=1}^{s} \pi_i \log \pi_i \qquad \text{Shannon, (4)}$$

$$\Delta_{Si} = \sum_{i=1}^{s} \{1 - \pi_i\}\pi_i = 1 - \sum_{i=1}^{s} \pi_i^2 \qquad \text{Simpson, (5)}$$

where the term in brackets denotes the species rarity function used in each case.

A hypothetical example of three forest stands composed of just five or fewer species of trees is presented in Table 11.1. The example clearly shows how the different indices are inconsistent in their ranking of these three communities. For example,

$$\Delta_{SC}(\text{Stand 1}) > \Delta_{SC}(\text{Stand 2}),$$

$$\text{but } \Delta_{Sh}(\text{Stand 1}) < \Delta_{Sh}(\text{Stand 2}) \text{ and } \Delta_{Si}(\text{Stand 1}) < \Delta_{Si}(\text{Stand 2}).$$

This is an interesting comparison because it illustrates how one might be lead to the conclusion that a community with fewer species (Stand 2) can be more diverse than one with more species (Stand 1) using either Shannon's or Simpson's index. Similar inconsistencies among the indices are found by comparing Stands 1 and 3. The only comparison that is consistently ordered with all indices is

$$\Delta(\text{Stand 2}) > \Delta(\text{Stand 3}).$$

This inconsistency of different diversity indices is evidently quite common when making comparisons between communities (Hurlbert 1971) and arises from a lack of intrinsic diversity ordering between the communities being compared.

Diversity Profiles

Diversity profiles allow the graphical comparison of diversity between communities. One set of profiles which incorporates the species count, Shannon and Simpson indices as point estimates along the curve, are the so-called Δ_β profiles of Patil and Taillie (1979a, 1982). Because the Δ_β profile incorporates indices developed from dichotomous-type rarity measures, it too can be developed in the same manner

$$\Delta_\beta = \sum_{i=1}^{s} \left\{ \frac{\left(1 - \pi_i^\beta\right)}{\beta} \right\} \pi_i = \frac{1 - \sum_{i=1}^{s} \pi_i^{\beta+1}}{\beta}, \ \beta \geq -1. \tag{6}$$

The restriction that $\beta \geq -1$ assures that Δ_β has certain desirable properties (Patil and Taillie 1979a, 1982). The species count, Shannon and Simpson indices, are related to Δ_β by $\Delta_{SC} = \Delta_{-1}, \Delta_{Sh} = \Delta_0$ and $\Delta_{Si} = \Delta_1$.

The Δ_β diversity profiles for the three stands in Table 11.1 are presented in Figure 11.1. The profile for Stand 1 crosses both profiles for Stands 2 and 3. The profile for Stand 1 crosses that of Stand 2 at $\beta = -0.45$, which explains why both Δ_{Sh} and Δ_{Si} rank diversity of these two communities differently than Δ_{SC}. On the other hand, the profiles for Stands 1 and 3 cross at $\beta = 0.62$ showing how Δ_{SC} and Δ_{Sh} rank these two communities differently than Δ_{Si}. In general, it is also possible for two Δ_β profiles to cross at $\beta > 1$ or for them to cross more than once. In either case, even calculating all three indices (Δ_{SC}, Δ_{Sh} and Δ_{Si}) alone might not be enough to show the inconsistent ranking of communities at larger β. Calculating and plotting Δ_β profiles for $\beta > 1$ might not be helpful either because the profiles tend to converge quickly beyond this point and intersections do not resolve. An algorithm for numerically finding the intersections of any two Δ_β profiles is required in this case.

Table 11.1. Three hypothetical forest stand communities composed of five or fewer species of trees.

	Stand		
Species	1	2	3
	Relative Abundance		
Pinus strobus	0.50	0.25	0.35
Quercus rubra	0.30	0.25	0.35
Tsuga canadensis	0.10	0.25	0.30
Acer rubrum	0.05	0.25	0.00
Betula papyrifera	0.05	0.00	0.00
Total:	1.00	1.00	1.00
	Diversity Index		
Δ_{SR}	5	4	3
Δ_{SC}	4	3	2
Δ_{Sh}	1.24	1.39	1.10
Δ_{Si}	0.65	0.75	0.67

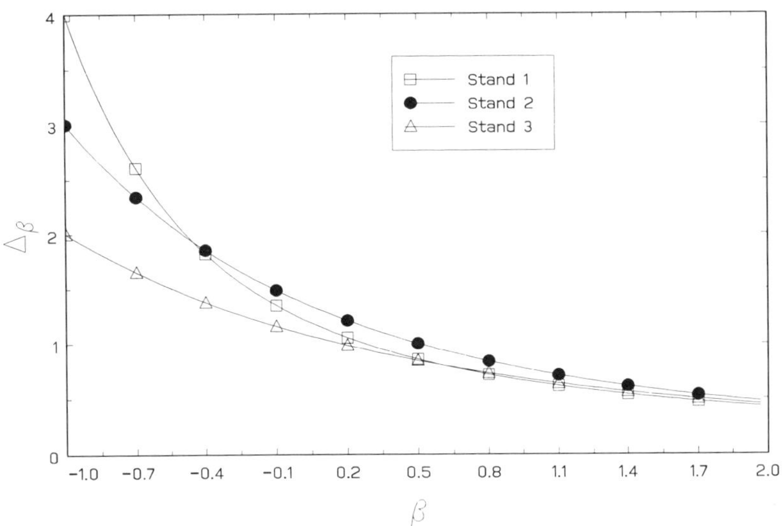

Figure 11.1. Δ_β profiles for the three hypothetical forest stand communities in Table 11.1.

Perhaps the most useful way to compare diversity between communities is by the concept of *intrinsic diversity ordering*. This concept is defined as follows (Patil and Taillie 1979a, 1982):

> Community C_1 is intrinsically more diverse than community C_2 (written C_1 IMD C_2) provided C_2 leads to C_1 by a finite sequence of (1) introducing a species, (2) transferring abundance from more to less abundant species without reversing the rank-order of the species, and (3) relabeling species (i.e., permuting the components of the abundance vector).

Patil and Taillie noted that this ordering is only partial and, like Solomon (1979), they stressed that two given communities need not be intrinsically comparable.

A diversity profile approach has been developed by Patil and Taillie (1979a,b, 1982) using a rank-type rarity measure on $\pi^{\#}$ that incorporates the concepts of intrinsic diversity ordering defined above. Let

$$R(i) = \begin{cases} 1 \text{ if } i > j; \\ 0 \text{ if } i \le j, \end{cases}$$

for $1 \le j \le s$. Then average species rarity is given as

$$T_j = \sum_{i=j+1}^{s} \pi_i^{\#}, \; j = 1, \ldots, s-1, \tag{7}$$

where T_s=0 and T_0=1. The quantity in (7) is termed the right tail-sum of the ranked relative abundance vector $\pi^{\#}$, and when a plot of the (j,T_j) pairs is constructed for each community, the resulting profiles are termed *intrinsic diversity profiles* (Patil and Taillie 1979b). Any intrinsic orderings of the communities, if they exist, can be determined with the intrinsic diversity T_j profiles.

The right tail-sum profiles for the three stands in Table 11.1 are plotted in Figure 11.2. The profile for Stand 1 crosses both those for Stands 2 and 3, but the profile for Stand 2 is everywhere above that for Stand 3. It follows that the only intrinsic diversity ordering for these stands is C(Stand 1) IMD C(Stand 3). This is consistent with the previous findings of the indices and the Δ_β profiles. The Δ_β profiles are isotonic to intrinsic diversity ordering in that, if an intrinsic diversity ordering exists, they will preserve it. However, the Δ_β profiles might not cross even if the T_j profiles do; therefore, the Δ_β profiles do not necessarily reflect intrinsic diversity ordering. Because the diversity indices discussed have the same properties as the Δ_β profiles, it should be emphasized that the T_j profiles are the most reliable measure of intrinsic diversity ordering between communities of the methods presented thus far.

COMPARATIVE STUDIES ON NORTHERN HARDWOODS, HUBBARD BROOK, NEW HAMPSHIRE

Gove et al. (1992) presented a study of plant species diversity using the concept of intrinsic diversity ordering (Patil and Taillie 1979a, 1982) to compare pre- and post-treatment plant communities on two watersheds within the northern hardwood ecosystem. The treatments on these watersheds were initiated in 1970, and the plant data used had

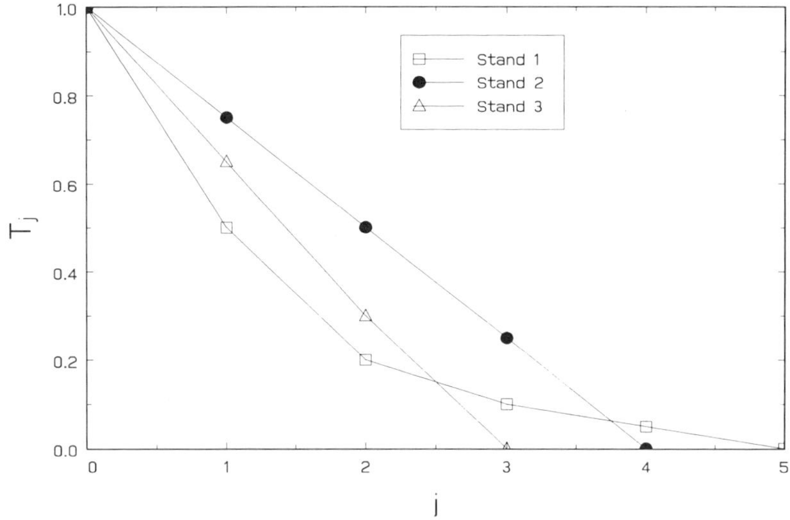

Figure 11.2. Right tail-sum (T_j) profiles for the three hypothetical forest stand communities in Table 11.1.

already been collected to meet other objectives (Martin and Hornbeck 1989). These initial objectives included determining the impacts of even-aged harvests on hydrologic and nutrient cycles and the composition of tree species in the regenerating forest. The diversity comparisons on the watersheds presented by Gove et al. (1992) laid the foundation for long-term monitoring of diversity in response to temporally changing environmental and ecological conditions on the treated watersheds. This objective is consistent with the overall goal of the long-term ecological research at the Hubbard Brook Experimental Forest, that is, to develop a better understanding of the response of northern hardwood-ecosystems to large-scale disturbance.

Here the details of the analytical methods used to make diversity comparisons in the Hubbard Brook study are presented. These comparisons, although similar to the diversity profile methods previously discussed, differ in that a nonparametric jackknifing technique is used here to correct for possible biases which can occur when sample-based community abundance vectors are simply substituted in place of the unknown population vectors. The jackknifing procedure also allows simple hypotheses to be tested among communities so that inferences can be made about intrinsic diversity orderings.

Study Site

Two watersheds located at Hubbard Brook Experimental Forest within the White Mountain National Forest in central New Hampshire were used for the study. Watershed 101 was a 12-ha block clearcut that ranged in elevation from 470 to 595 m. Watershed 4 was a 36-ha progressive strip cut that ranged in elevation from 440 to 730 m. Both watersheds had southerly aspects and slopes of 20 to 30%. Other details including pre-harvest forest conditions and soil characteristics are given by Martin and Hornbeck (1989).

In preparation for harvest, Watershed 4 was surveyed into 49 strips, each 25 m wide, oriented from east to west roughly parallel to the watershed contour. Harvest began in autumn of 1970 by cutting every third 25-m strip. After a 2-year period for establishment of regeneration, a second series of 25-m strips was cut on the southernmost side of the original cuts. The remaining strips were then cut after a second establishment period of 2 years, completing the harvest over a 4-year interval. After the first strip was cut, it received shade from the trees on its southerly border, providing favorable conditions for germination and early survival. When the second strip was cut, it received shade for initial establishment, but also allowed the first strip to receive direct sunlight for growth. The block clearcut on Watershed 101 was carried out in a single operation immediately upon completion of the first strip cut in late autumn of 1970.

On both the strip cut and block clearcut, all trees and snags whose diameter at breast height (dbh) was > 2.5 cm were felled and commercial portions removed. Scarification was encouraged during logging to provide exposed mineral soil for seedling establishment. Regeneration on the block clearcut and strip cut has been surveyed on permanent plots at 1- to 4-year intervals since harvest. On the strip cut, 57 25×25-m plots were selected. These plots were cross-classified by year of cut and elevation and were randomly allocated. The plot allocation produced 19 plots for each year of cut, and 19 plots at low elevation (440-550 m), 18 at mid-elevation (550-650 m), and 20 at high elevation (650-730 m). On the block clearcut, 48 25×25-m plots were chosen at random with no stratification since the harvest was completed in a single operation, and the elevation range of the entire harvest area was about the same as for the mid-elevation plots on the strip cut. Vegetation surveys were conducted before cutting on each of the strips, but not on the block clearcut (Table 11.2).

On each watershed the entire 25×25-m plot was used to sample trees whose dbh was ≥10 cm. Trees between 2 and 10 cm dbh were tallied on a 1×25-m strip plot running along either the east or west border (decision by coin toss) of each plot. Saplings measured on the 1×25-m nested strip plots before cutting were considered to be advanced regeneration. There were a total of 57 1×25-m strips on Watershed 4, and 48 on Watershed 101. Lesser vegetation was surveyed on four 1-m^2 subplots spaced evenly along each strip plot, for a total of 228 subplots on Watershed 4 and 192 on Watershed 101. In addition, stem counts were made by species for trees, shrubs, and herbs in two height classes, < 0.5 m and > 0.5 m tall. Data from the subplots were averaged to give stems/m^2 for each 1×25-m strip plot. Individual stems of all species of trees, shrubs, and herbs were counted on each subplot except for sedges, grasses, liverworts, rushes, clubmosses, mosses, oxalis, and spikenard, which were inventoried as percent cover and were not included in this analysis.

Table 11.2. Pre- and post-treatment community measurement dates in the two watersheds.

Treatment	Watershed	Precut	1 year post	10 year post
Strip 1	WS 4	1970	1971	1980
Strip 2		1972	1973	1982
Strip 3		1974	1975	1984
All	WS 101	—[1]	1971	1980

[1] No data were available in this period.

Analysis Methods

Diversity indices such as Shannon's index and diversity profiles such as T_j and Δ_β were developed for use when the π have been calculated from complete enumeration of the population. Patil and Taillie (1979b) noted that "most diversity indices, when evaluated on a sample, consistently underestimate community diversity by an amount varying with sample size." They applied a jackknifing method (Zahl 1977) that might help to correct for this bias and, in addition, provides estimates of precision which can be used for hypothesis testing. In particular, the jackknifing method was used to test whether intersections which occur in the T_j profiles are real or simply a sampling effect. The jackknife approach of Patil and Taillie (1979b) is used here on all Δ_β profiles for bias correction and intersection testing. The procedure is summarized as follows:

(1) Calculate and graph the T_j profiles with all plots included for the communities of interest and identify any intersections.

(2) Calculate and graph the Δ_β profiles with all plots included.

(3) Jackknife the Δ_β profiles for each community as follows

(i) Calculate $\Delta_\beta^{(-i)}$ the Δ_β profile with the *i*th plot data omitted.

(ii) Calculate a pseudo value as given in (8) at each β

$$\Psi_\beta^{(-i)} = n\Delta_\beta - (n-1)\Delta_\beta^{(-i)}. \tag{8}$$

(iii) Repeat steps (i) and (ii) n times, where n is the number of plots.

(iv) Calculate the average of the pseudo values and the estimated variance at each β

$$\hat{\Delta}_\beta = \frac{1}{n}\sum_{i=1}^{n}\Psi_\beta^{(-i)}$$
$$V\hat{a}r\left(\hat{\Delta}_\beta\right) = \frac{1}{n(n-1)}\sum_{i=1}^{n}\left(\Psi_\beta^{(-i)} - \hat{\Delta}_\beta\right)^2 \tag{9}$$

(4) Graph the jackknifed $\hat{\Delta}_\beta$ profiles for each community.

In implementing the above procedure, m=100 evenly spaced β values in the range $-1 \le \beta \le 2$ were generated in steps (3) and (4) for each profile. The parameter β was allowed to range to 2 because some intersections were identified in the range $+1 \le \beta \le 2$. The actual intersections for the Δ_β profiles (non-jackknifed) were determined with a root-finding algorithm. The approximate intersections for the $\hat{\Delta}_\beta$ profiles were found by subtraction of pairs of profiles. The reason for using subtraction in the case of jackknifing is that the jackknifing procedure above does not provide an estimate $\hat{\pi}$ to evaluate equation (6) analytically; it only provides m discrete points on the $\hat{\Delta}_\beta$ curve.

The estimation phase entailed applying the jackknifing algorithm, matching up the intersections on the T_j profiles with those on the $\hat{\Delta}_\beta$ profiles, and subsequently testing the significance of these intersections. In so doing, it is helpful to keep in mind that small j in the T_j profiles corresponds to large β in the $\hat{\Delta}_\beta$ or Δ_β profiles. The inference procedure involves forming the composite null hypothesis given by Patil and Taillie (1979b). To illustrate, Figure 11.3 shows $\hat{\Delta}_\beta$ profiles for two communities (C_1 and C_2) with an intersection found at the point β^*. In this case, the procedure would be to choose β' and β'' so that $\beta' < \beta^* < \beta''$ and form the pair of null hypotheses

$$H_0' : \hat{\Delta}_{\beta'}(C_1) \geq \hat{\Delta}_{\beta'}(C_2)$$
$$H_0'' : \hat{\Delta}_{\beta''}(C_1) \geq \hat{\Delta}_{\beta''}(C_2)$$

Simultaneous rejection of these null hypotheses at some specified level of significance would establish that $\Delta_{\beta'}(C_1) < \Delta_{\beta'}(C_2)$ and $\Delta_{\beta''}(C_1) < \Delta_{\beta''}(C_2)$; that is, the two community profiles $\hat{\Delta}_\beta(C_1)$ and $\hat{\Delta}_\beta(C_2)$ *do* intersect and, therefore, the two communities are *not* intrinsically comparable.

To test the composite null hypothesis calculate the two-sample t statistics

$$t_\beta = \frac{\hat{\Delta}_\beta(C_1) - \hat{\Delta}_\beta(C_2)}{\sqrt{V\hat{a}r\left(\hat{\Delta}_\beta(C_1)\right) + V\hat{a}r\left(\hat{\Delta}_\beta(C_2)\right)}}, \ \beta = -1, -.97, -.94, \ldots, 2. \tag{10}$$

If at least one t_β (viz., the maximum t_β) on each side of the intersection is significant when compared to the one-tailed Bonferonni-adjusted $t(1\text{-}\alpha/g;n\text{-}1)$ value (where g=2 and α=0.05 here), then the implication is that the profiles intersect, and the communities C_1 and C_2 are *not* intrinsically comparable. If none of the t_β values calculated are significant, then it can be inferred that no difference has been found in diversity between the two communities being compared.

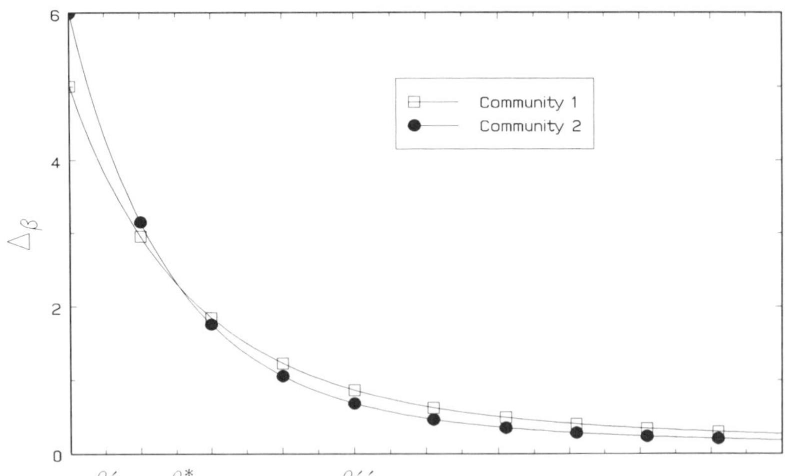

Figure 11.3. Δ_β profiles illustrating an intersection and hypothetical points $(\beta', \beta^*, \beta'')$ for the composite null hypothesis described in the text.

This procedure must be viewed as only an approximate test because it involves "difficult and unresolved questions of simultaneous inference" (Patil and Taillie 1979b). However, even with this qualification, it remains a valuable procedure over simply treating the π as population values. Also, by no means does it diminish the usefulness of the bias correction aspect of the jackknife.

Community Diversity Comparisons

Both temporal and spatial community comparisons were of interest. The temporal comparisons tracked individual communities within a treatment over time, whereas the spatial comparisons contrasted treatments at certain relative (i.e., since time of treatment) points in time. For example, comparisons within the first set of cut strips fall into the within-treatment category, whereas comparisons between strips, 1 year after cut, fall into the between-treatment category. Four within-treatment comparisons were made corresponding to the rows in Table 11.2. Between-treatment comparisons were made at 1- and 10-years postharvest. Right tail-sum, Δ_β, and jackknifed $\hat{\Delta}_\beta$ profiles were generated for each comparison. The following presents a detailed analysis of the temporal comparisons on strip 2, Watershed 4; the results of the other temporal and spatial comparisons are summarized at the end of this section.

Table 11.3 presents the relative abundances (π) for the precut, 1- and 10-year postcut plant communities in strip 2. These relative abundances can be ranked from greatest to

Table 11.3. Relative abundances of plant species on strip 2 of Watershed 4 on the Hubbard Brook Experimental Forest based on frequency of individuals.[1]

Species	1972	1973	1982
Herbs			
Aralia nudicaulis	0.001462150	0.001691946	0.005539921
Arisaema triphyllum	0.003216729	0.003759879	0.007755889
Aster acuminatus	0.102935342	0.097944850	0.135174064
Clintonia borealis	0.002047010	0.008083740	0.024375651
Coptis Trifolia var. *groenlandica*	0.000292430		
Cornus canadensis	0.003509159	0.001879940	0.070356992
Dennstaedtia punctilobula	0.006725889	0.009211704	0.025483635
Dryopteris spinulosa	0.033921874	0.031770978	0.094732643
Erythronium umbilicatum	0.000292430		
Galium triflorum	0.002924299	0.005827813	0.001107984
Maianthemum canadense	0.070475617	0.072377672	0.093070667
Medeola virginiana	0.000584860	0.009775686	0.002769960
Mitchella repens	0.000292430	0.001127964	0.000553992
Polygonatum biflorum			0.003323952
Prenanthes spp	0.000292430	0.001315958	
Smilacina racemosa	0.000877290	0.000563982	
Solidago flexicaulis	0.007310749	0.013347571	0.011079841
Solidago macrophylla	0.002631870		
Streptopus amplexifolius	0.000292430	0.007707752	

continued

Table 11.3.—*continued*

Species	1972	1973	1982
Thelypteris noveboracensis		0.001315958	0.019389722
Thelypteris phegopteris	0.001754580	0.004135867	0.013295810
Trientalis borealis	0.007895609		0.006647905
Trillium erectum	0.002631870	0.003759879	0.004431937
Trillium undulatum	0.001462150	0.000939970	0.003877944
Uvularia sessilifolia	0.071937767	0.029515051	0.077004897
Viola incognita	0.052052531	0.139115526	0.046535334
Viola rotundifolia	0.007895609	0.010527661	0.001107984
Shrubs			
Cornus alternifolia	0.002339440	0.002443921	0.002393246
Prunus virginiana	0.003509159		0.004431937
Ribes glandulosum	0.000877290	0.000939970	0.002215968
Rubus allegheniensis	0.004678879	0.003383891	0.057615175
Rubus idaeus	0.017253367	0.051134356	0.043211381
Rubus spp		0.000187994	
Sambucus canadensis	0.006433459	0.019739365	
Sambucus racemosa var. *pubens*			0.010614488
Sorbus americana	0.000584860	0.000939970	0.000908547
Viburnum lantanoides	0.032167294	0.017483438	0.027699603
Trees			
Abies balsamea	0.000301788	0.000088639	
Acer pensylvanicum	0.037226566	0.028575081	0.016176568
Acer rubrum	0.000001872	0.001127964	0.002215968
Acer saccharum	0.253125023	0.096628892	0.098721386
Acer spicatum	0.021536413	0.009211704	0.009440025
Betula alleghaniensis	0.154730534	0.209425265	0.027322889
Betula papyrifera	0.000061761	0.001127964	0.002149489
Fagus grandifolia	0.037263062	0.068865945	0.024065415
Fraxinus americana	0.037170420	0.018047420	0.007467813
Picea rubens	0.000041174		0.000088639
Prunus pensylvanica	0.004397679	0.013535565	0.015556097
Prunus serotina	0.000584860	0.001503952	

[1] Nomenclature follows USDA Soil Conservation Service (1982), and Little (1979).

least for each year to form the ($\pi^{\#}$) vectors and the T_j profiles (Figure 11.4). Examination of the T_j profiles without the benefit of the inset graph tend to suggest that the profiles do not cross because of the resolution at large j. However, the inset clearly shows that each pair of profiles intersect once. The points of intersection can be easily found by examining the inset graph or by taking the differences between the right tail-sums for each pair of communities and noting where the change in sign occurs (Table 11.4). Note that the T_j profiles all cross in the "tail"—where j is most sensitive to rare species.

The population Δ_β profiles (i.e., those constructed by treating π as the population values), and the jackknifed $\hat{\Delta}_\beta$ profiles for strip 2 are shown in Figure 11.5. Comparison of these two sets of profiles shows that they are very similar in shape, but that the

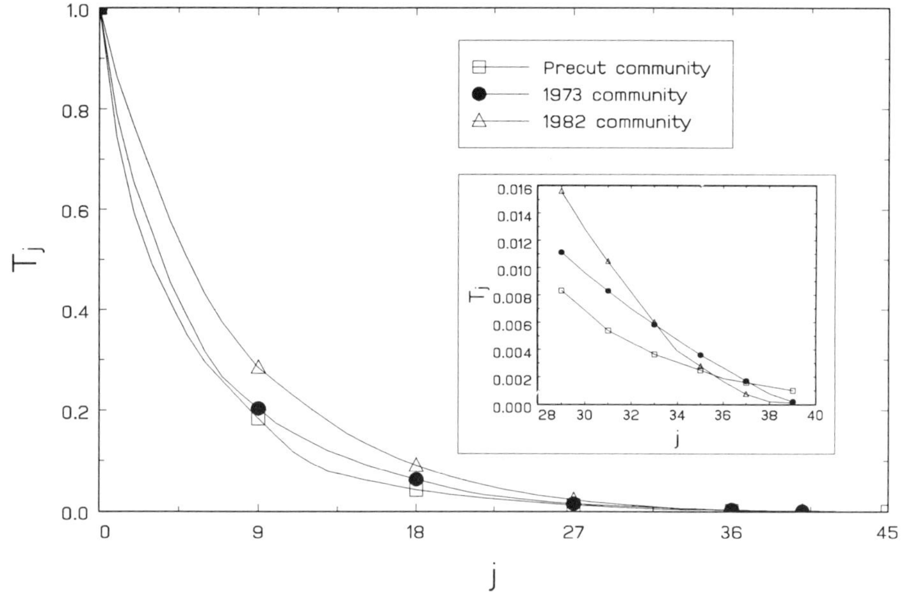

Figure 11.4. Right tail-sum (T_j) diversity profiles for the antecedent, 1- and 10-year postharvest communities on strip 2, Watershed 4 (inset shows details in the tail of the profiles where intersections occur).

jackknifed profiles have changed considerably for β near -1, illustrating the effect of the jackknife bias correction. The crossing points for the two sets of profiles, however, have not been greatly affected by jackknifing as the following shows:

Δ_β Profile	Precut vs. 1-year posttreatment	Precut vs. 10-year posttreatment	1-year vs. 10-year posttreatment
Population	-0.768	-0.852	——
Jackknife	-0.655	-0.775	——

Even though the two estimates of the species count index (Δ_{-1}) are quite different, jackknifing seems to preserve the approximate crossing points in β—where β is sensitive to rare species. Therefore, both the Δ_β and $\hat{\Delta}_\beta$ profiles are consistent with the T_j profiles in Figure 11.4 in that they all cross where their respective parameters are sensitive to rare species.

To evaluate the significance of the intersections on the $\hat{\Delta}_\beta$ profiles, t_β values were calculated and are plotted in Figure 11.6. Comparing the precut community with the 10-year postharvest community, the precut profile is above the 1982 profile for $-1 \le \beta < -0.775$ and the reverse is true for $\beta > -0.775$. The largest t_β statistics found to the left and right of the intersection were $t_{-1}=2.00$ and $t_{0.08}=-3.09$, respectively. Comparison with the Bonferroni-adjusted $t(0.025;18)=2.101$ leads to a significant difference only for $\beta > -0.775$ (Figure 11.6). Therefore, given the current data, there is not enough evidence to reject the hypothesis that $C(1982)$ IMD $C(\text{precut})$.

Table 11.4. Right tail-sum differences for strip 2 on Watershed 4 of the Hubbard Brook Experimental Forest[1].

j	$T_j(1972) - T_j(1973)$	$T_j(1972) - T_j(1982)$	$T_j(1973) - T_j(1982)$
0	0.00000000	0.00000000	0.00000000
1	-0.04369470	-0.11794593	-0.07425123
2	-0.05931226	-0.17395762	-0.11464536
3	-0.06430045	-0.18215802	-0.11785757
4	-0.03960770	-0.16102352	-0.12141582
5	-0.03770408	-0.15449268	-0.11678861
6	-0.02088950	-0.13618703	-0.11529754
⋮	⋮	⋮	⋮
32	-0.00244935	-0.00371484	-0.00126549
33	-0.00219867	-0.00237613	-0.00017746
34	-0.00165556	-0.00081152	0.00084404
35	-0.00111245	-0.00028838	0.00082408
36	-0.00075734	0.00023476	0.00099211
37	-0.00012092	0.00083978	0.00096069
38	0.00052662	0.00110135	0.00057473
39	0.00079818	0.00089755	0.00009937
40	0.00069375	0.00069375	0.00000000
41	0.00040132	0.00040132	0.00000000
42	0.00010890	0.00010890	0.00000000
43	0.00004445	0.00004445	0.00000000
44	0.00000222	0.00000222	0.00000000
45	0.00000000	0.00000000	0.00000000

[1] Changes in sign indicate points of intersection for the respective T_j profiles.

The intersection between the precut and 1-year postharvest communities finds a significant difference only at the species count index (t_{-1}=2.16). This would seem to suggest a second apparent ordering: C(precut) IMD C(1973). However, because only one extreme test statistic t_{-1} is significant out of all 100 calculated, it seems instructive to scrutinize the graphs before settling on this apparent ordering. The T_j, Δ_β, and $\hat{\Delta}_\beta$ profiles all show that C(1973) plots above C(precut) over most of their lengths. This would seem to contradict the intrinsic diversity ordering above. In addition, there really seems to be very little difference between the two community profiles (see especially Figure 11.4). Therefore, a more reasonable interpretation would be that the apportionment of individuals within these two communities is probably quite similar for the major (most dominant) species. However, the significance difference in species richness between these two communities suggests that C(precut) is significantly more diverse in the "tail" than C(1973). That is, the rarer species are more diverse in C(precut) than in C(1973), but there seems to be no difference in diversity between the two communities overall, that is, C(precut)=C(1973).

The t_β curves (Figure 11.6) illustrate that the intersections for the $\hat{\Delta}_\beta$ profiles can be found by noting where these curves cross zero, indicating the point where the numerator in equation (10) becomes zero, viz., the point where the profiles are the same. Neither the Δ_β nor the $\hat{\Delta}_\beta$ profiles shows an intersection for C(1973) and C(1982). Yet the T_j

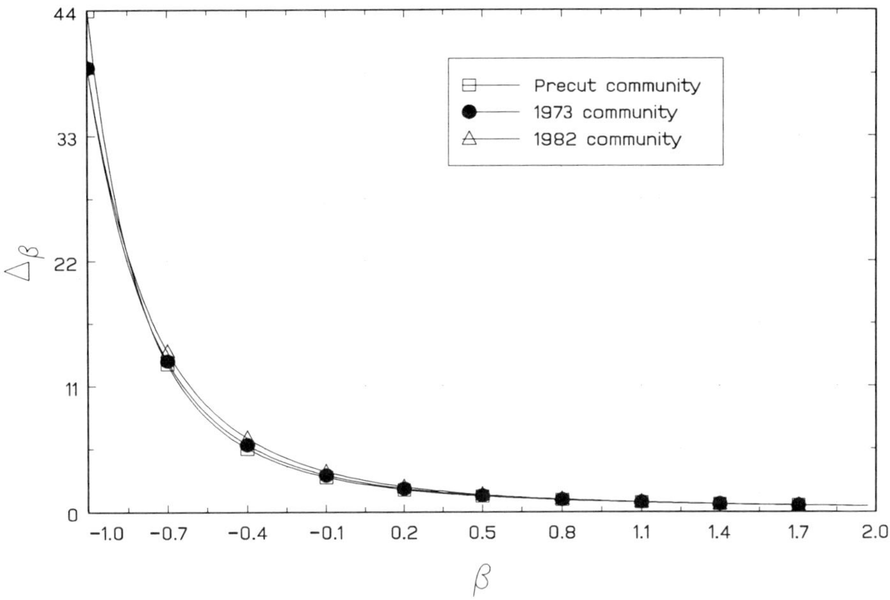

(a)

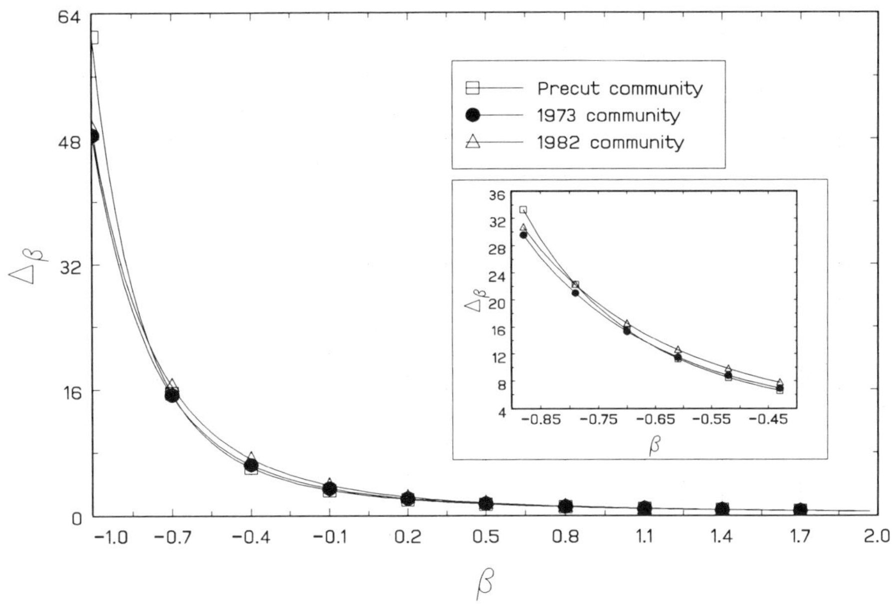

(b)

Figure 11.5. Δ_β profiles for the antecedent, 1- and 10-year postharvest communities on strip 2, Watershed 4: (a) population profiles; (b) jackknifed profiles (inset shows details at points of intersection).

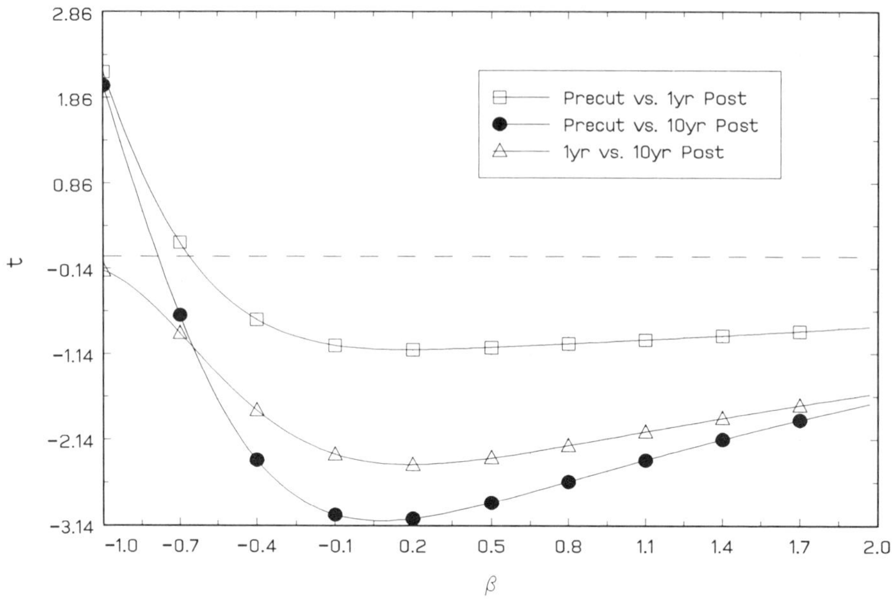

Figure 11.6. Two-sample t_β statistics for the jackknifed community comparisons showing intersection points where the profiles cross zero; significant and nonsignificant differences in the profiles are also shown (see text for details).

profiles *do* record such an intersection. This anomaly points out a potential problem with this technique of intersection testing: the Δ_β and $\hat{\Delta}_\beta$ profiles might not cross even though the T_j profiles do, or the Δ_β or $\hat{\Delta}_\beta$ profiles might cross more than once (this problem was alluded earlier). Patil and Taillie (1979b) recognized this problem and had to use other profiles to provide test statistics in such cases to draw inferences about the T_j intersections. In this case, that is an option too, but an alternative approach is also possible. It was noted above that *C*(1982) IMD *C*(precut), and that *C*(precut)=*C*(1973) through the testing procedure. In addition, Figure 11.6 shows numerous significant T_β when comparing *C*(1973) and *C*(1982); that is, there seems to be some difference between the two community profiles. These observations would strongly support the final intrinsic ordering for strip 2: *C*(1982) IMD *C*(precut)=*C*(1973).

It is interesting to note that the species count and Shannon indices rank communities *C*(precut) and *C*(1982) in the opposite sense. This can be seen from the $\hat{\Delta}_\beta$ profile because $\hat{\Delta}_{-1}(\text{precut}) > \hat{\Delta}_{-1}(1982)$, whereas $\hat{\Delta}_0(\text{precut}) < \hat{\Delta}_0(1982)$. Had only these two point estimates of Δ_β been calculated, without the benefit of a large range of β values (and hence diversity indices) or hypothesis testing through the jackknife, contradictory conclusions would have been reached concerning the ordering of these two communities. A subsequent check of Simpson's index at $\hat{\Delta}_{-1}$ might have been enough to sway the investigator toward the same conclusions drawn here with jackknifing, however, because $\hat{\Delta}_{-1}(\text{precut}) < \hat{\Delta}_1(1982)$; but it would have been based on subjective judgment. Similar results can be found for the *C*(precut) and *C*(1973) profiles.

Applying this same procedure to the rest of Watershed 4 and to Watershed 101 yielded similar results. On strip 1, C(1980) IMD C(1970) IMD C(1971); on strip 3 C(1984) IMD C(1974)=C(1975). Unfortunately, no inventory data were taken for the precut community before the block clearcut. Therefore, all that could be done was to study the trend in diversity following clearcutting on Watershed 101; this trend showed C(1980) IMD C(1971). As in the strip cut, diversity seems lowest right after the cut and then increases.

Between-treatment (spatial) comparisons were made within Watershed 4 (i.e., between strips) at 1- and 10-years postharvest. The results showed a change in the spatial diversity ranking over time. In the 1-year communities the intrinsic ordering C(strip 2) IMD C(strip 3) IMD C(strip 1). In the 10-year communities, however, there has been a change in ranking so that C(strip 2)=C(strip 3) IMD C(strip 1).

Comparisons were also made between strip 1 and the block clearcut for both the 1- and 10-year postharvest communities. After intersection testing, no significant difference was found between the 1-year communities, but that after 10 years, C(strip 1) IMD C(block clearcut). Because strips 2 and 3 were both intrinsically more diverse than strip 1 for the 1- and 10-year comparisons mentioned, it is easily argued that all three strips were intrinsically more diverse than the block clearcut at both points in time; indeed, graphical analysis (not shown) supported this argument.

Bias Reduction and the Jackknife

The jackknifing procedure of Patil and Taillie (1979b) provides a useful, approximate method for testing intrinsic diversity ordering when relative abundance data are based on samples rather than complete enumeration. Other assumptions and uses of the jackknife have been discussed in detail in relation to general forestry usage by Gregoire (1984), and in relation to diversity indices by Zahl (1977) and Palmer (1991). Quenouille (1956) originally introduced the jackknife as a bias reduction technique. The effect of bias reduction on diversity indices can be seen using the easily comprehensible species count index $(s-1)$ as an example. The sample-based $(s-1)$ and jackknifed $(\hat{s}-1)$ estimates are presented in Table 11.5 along with the variances estimated through jackknifing. Comparison of $(s-1)$ and $(\hat{s}-1)$ supports the *a priori* expectation that the simple sample-based estimates will underestimate the jackknifed estimates of the species count index. These comparisons give strong evidence that the species count index is a negatively biased estimate of the true species richness (minus 1) when sampling variability is not accounted for in the communities on these watersheds. Unfortunately, the true population species richness remains unknown; therefore, it is not possible to judge how closely either of these estimates are to the true population values in this analysis.

The jackknife's usefulness in correcting for bias is also apparent in the between-treatment comparisons. For example, Table 11.5 shows that it would be difficult to compare the two watersheds on the basis of the uncorrected species counts alone. The main reason for this is the disparity in sample size between watersheds: n=19 on the strip cut and n=48 on the block clearcut. Although the differences look real, especially for strips 2 and 3 when compared to block clearcutting, the differences in sample size leave doubt without the jackknife correction.

Table 11.5. Sample-based $(s-1)$ and jackknifed $(\hat{s}-1)$ estimates of the species count index in the two watersheds[1].

Treatment	Watershed	Precut	1 Year Post	10 Years Post
			$(s-1)$	
Strip 1	WS 4	35	33	40
Strip 2		44	39	39
Strip 3		41	39	46
All	WS 101	——[2]	36	35
			$(\hat{s}-1)$	
Strip 1	WS 4	43.5 (6.382)	38.7 (7.679)	49.5 (15.856)
Strip 2		61.1 (17.950)	48.5 (15.856)	49.4 (15.756)
Strip 3		50.5 (10.172)	48.5 (12.066)	59.3 (60.332)
All	WS 101	——[2]	45.8 (15.585)	40.9 (7.099)

[1] Jackknife estimates of the variance for the species count index are given in parentheses.
[2] No data were available in these periods.

The profiles and comparisons presented in this section clearly show the dynamic nature of plant diversity in managed northern hardwood forests even over time spans as short as 10 years. To what extent the different regeneration methods influenced the diversity orderings encountered here, and to what extent natural factors were responsible, cannot be tested with present techniques.

CONCLUSIONS

Monitoring biodiversity in managed landscapes requires quantitative methods for measuring and comparing diversity. This paper has illustrated the use of diversity profiles based on the concept of intrinsic diversity ordering as one means to help meet this monitoring requirement. The Δ_β and T_j diversity profiles are both useful methods for determining intrinsic orderings of communities if they exist. Unfortunately, diversity indices and profiles all assume that the population has been enumerated completely. In cases where this is not possible and sampling is the only alternative, jackknifing can be applied for bias correction and hypothesis testing. The detailed example of the Hubbard Brook Experimental watersheds shows how jackknifed diversity profiles can be applied to spatial and temporal monitoring problems in managed landscapes to determine intrinsic diversity orderings from sampled data.

LITERATURE CITED

Gove, J. H., C. W. Martin, G. P. Patil, D. S. Solomon, and J. W. Hornbeck. 1992. Plant species diversity on even-aged harvests at the Hubbard Brook experimental forest: 10-year results. *Can. J. For. Res* 22(11):1800-1806.

Gove, J. H., G. P. Patil, and C. Taillie. 1994. A mathematical programming model for maintaining structural diversity in uneven-aged forest stands with implications to other formulations. *Ecol. Modelling* (in press).

Grassle, J. F., G. P. Patil, W. K. Smith, and C. Taillie, eds. 1979. *Ecological Diversity in Theory and Practice.* Statistical Ecology: Vol. 6. International Co-operative Publishing House, Burtonsville, Md.

Gregoire, T. G. 1984. The jackknife: an introduction with applications in forestry data analysis. *Can. J. For. Res.* 14:493-497.

Hurlbert, S. H. 1971. The nonconcept of species diversity: A critique and alternative parameters. *Ecology* 52(4):577-586.

Little, E. L., Jr. 1979. *Checklist of United States Trees (Native and Naturalized).* USDA Forest Service, Agricultural Handbook No. 541, Washington, D.C.

Magurran, A. E. 1988. *Ecological Diversity and its Measurement.* Princeton University Press, Princeton, N.J.

Martin, C. W., and J. W. Hornbeck. 1989. Revegetation after strip cutting and block clearcutting in northern hardwoods: a 10-year history. USDA Forest Service Research Paper NE-625, Broomall, Pa.

Palmer, M. W. 1991. Estimating species richness: the second-order jackknife reconsidered. *Ecology* 72(4):1512-1513.

Patil, G. P., and C. Taillie. 1979a. An overview of diversity. In *Ecological Diversity in Theory and Practice.* Statistical Ecology: Vol. 6, eds. J. F. Grassle, G. P. Patil, W. K. Smith, and C. Taillie, pp. 3-27. International Co-operative Publishing House, Burtonsville, Md.

Patil, G. P., and C. Taillie. 1979b. A study of diversity profiles and orderings for a bird community in the vicinity of Colstrip, Montana. In *Contemporary Quantitative Ecology and Related Ecometrics.* Statistical Ecology: Vol. 12, eds. G. P. Patil and M. L. Rosenzweig, pp. 23-48. International Co-operative Publishing House, Burtonsville, Md.

Patil, G. P., and C. Taillie. 1982. Diversity as a concept and its measurement. *J. Am. Statistical Assn.* 77:548-567.

Pielou, E. C. 1974. *Population and Community Ecology.* Gordon and Breach, New York.

Pielou, E. C. 1975. *Ecological Diversity.* John Wiley and Sons, New York.

Pielou, E. C. 1977. *Mathematical Ecology.* John Wiley and Sons, New York.

Quenouille, M. H. 1956. Notes on bias in estimation. *Biometrika* 43:353-360.

Solomon, D. L. 1979. A comparative approach to species diversity. In *Ecological Diversity in Theory and Practice.* Statistical Ecology: Vol. 6, eds. J. F. Grassle, G. P. Patil, W. K. Smith, and C. Taillie, pp. 29-35. International Co-operative Publishing House, Burtonsville, Md.

USDA Soil Conservation Service. 1982. *National List of Scientific Plant Names, Volume 1: List of Plant Names.* SCS-TP-159, Washington, D.C.

Zahl, S. 1977. Jackknifing an index of diversity. *Ecology* 58:907-913.

12

NATURAL HERITAGE DATA CENTER NETWORK: MANAGING INFORMATION FOR MANAGING BIODIVERSITY

Robert E. Jenkins, Jr.[1]

The Nature Conservancy
Arlington, Virginia
U.S.A.

With the natural landscape continuously shrinking and pervasively subjected to massive ecological degradation, more and more species and natural communities will be unable to survive without our intelligent intervention. As the term "managed landscapes" implies, we must intervene in increasingly sophisticated ways. To know where and how to intervene, we will require more and better knowledge about biota, ecosystems, and factors affecting them— all organized in ways that make it accessible and usable by the managers. A great many conservation experts have been urging appropriate programs to gather and organize the relevant information for such purposes (TNC 1975; Hoose 1981; OTA 1987; Miller 1989; Reed and Miller 1989; Botkin 1990; McNeely et al. 1990; Salwasser 1990; McAllister 1991).

We are surprisingly unprepared to do this. Our major device for gathering, recording, and conveying information has been the individual research project with results published as papers. Larger works are usually distilled from many papers. Although much admired, especially by those engaged in it, this process has many shortcomings for the purpose at hand. It is too disorganized and too much factual information is lost along the way. Conclusions are published while the original observations or data are discarded, thus impeding additive investigation. Most research topics are not chosen to address management needs. Rarely do multiple researchers coordinate their work or combine to address common problems. Results are often jealously guarded, sometimes indefinitely, pending future publication. Innumerable publication outlets scatter results entropically. Peer review is overestimated as a quality control device. Errors are infrequently corrected and never erased from the original record, so sorting fact from fiction is impractical except for narrowly focused specialists. The means to combine factual knowledge from disparate fields, geographic areas, or subjects areabsent or primitive. The manager does not have time to conduct literature reviews, it is expensive to contract for them, and even then, the results are of limited quality and utility.

[1] Present address: 2200 Wilson Blvd. #102-218, Arlington, Virginia

Something different is needed: something directed, organized, cumulative, continuously improved, and religiously maintained. The general idea is summarized by the word "database," but this inadequately conveys the magnitude, complexity, scope, scale, or character of what is meant. The computer revolution has enabled nearly anyone to create databases. A few individually-maintained databases have become quite useful to a wider usership in the biodiversity field (e.g., John Kartesz's continuously evolving checklist of North American plants). Unfortunately, such efforts have obvious limitations in scope, scale, and most regrettably, continuity. To organize the knowledge required to conserve and manage biodiversity demands a much larger effort and strong institutional support for the long term. Ideally such an effort will engage many institutions and innumerable individuals. It will not only derive data from the usual sources but will identify data needs and fill them. It will make information accessible to the managers and organize it in ways that facilitate its use.

We can call such an effort a "biodiversity data bank." There are many impediments to creating and operating such a data bank including many purely technical difficulties in information management. More importantly, people and institutions are not naturally inclined to cooperate to this degree. Given the enormous scope of biodiversity information and the difficulty of organizing and managing it, I believe that we cannot meet the information demands of the biodiversity conservation crisis except through a greater concentration of resources, greater cooperation between institutions, and persistent continuity of effort.

So far I know of only two thoroughgoing efforts at biodiversity databank creation of the sort that I mean—the Environmental Resources Information Network (ERIN) in Australia and the network of Natural Heritage Data Centers in the United States and Western Hemisphere (Jenkins and Moyseenko 1976; Jenkins 1975, 1988). The latter is older, larger, and much better known to me. Over a hundred independent land management agencies and institutions are currently cooperating in the Heritage network. Their collective intent is to provide a universal source of information on biodiversity resources. By networking in this way, they hope to avoid duplication of effort and to provide consistently the best available information to all decision-makers. The network currently covers the entire United States, and has spread to Canada, Latin America, and the Caribbean.

The remainder of this paper describes the Heritage network and makes some observations about information management from two decades of Heritage experience in attempting it. Ten of the most important questions for conservation, development, and management decisions are enumerated, with an explanation of how the Heritage network attempts to answer them.

NETWORK OF NATURAL HERITAGE DATA CENTERS

In 1974, the first of what are generally called State Natural Heritage Data Centers was launched as a cooperative venture of the South Carolina Department of Wildlife and Marine Resources and the Science Division of The Nature Conservancy. The purpose of the program was to compile information about native species, biotic communities, and the protected and unprotected landscape of the state to plan systematically for biodiversity conservation and resource management. We used the term "natural diversity" to highlight the fact that the program was to include a concern for community, ecosystem, and landscape diversity as well as species and genetic diversity. Both the Conservancy and many state governments had been concerned with "biological inventory" in the context of biodiversity and natural area conservation for many years, but had not accomplished much of lasting value because their efforts were small and discontinuous to develop the necessary machinery.

The South Carolina experiment showed that state governments and the Conservancy were well suited to mutually supportive roles in pursuit of this mission, so that additional cooperative data centers were subsequently started in other states. We started three or four new ones a year until by 1987 they had been established in all 50 states. As early as the late 1970s, such centers were also begun outside the United States and by 1992, there were 3 provincial programs in Canada and 14 mostly national data centers in Latin America. Various other supplementary centers have been established at institutions like TVA and the Navajo Nation, large federal and Conservancy reserves, and various government or private operating units offices that fall within the geographic matrix of the tessellating state and national programs. One measure of the extent of the network is that the Biological and Conservation Database System (BCD) developed for these data centers is now used to manage data in about 150 different office locations.

Most of the real information gathering and management is done in the data centers by the cooperating agency or partner. The Nature Conservancy's Science Division's role is to develop, document, and support the computer systems, operating methods, and related technical aspects of the programs. It also trains new staff, maintains an important set of central databases, and mediates data exchanges throughout the network. The central databases were created primarily to facilitate the work of the individual data centers by reconciling taxonomy, efficiently storing commonly useful data for retrieval by multiple programs, and acting as a transfer point for network data exchanges, but they have now become important in themselves as well.

A vast amount of information has now been compiled within the network centers, from secondary sources such as publications and museum collections and from original field investigations. A huge amount of field verification of secondary source information has been carried out by Heritage staff and cooperators which, along with some rigorous quality control procedures of checking and rechecking, has produced a very high quality data set. Information is compiled on a number of "major logical entities," but the special focus has been on the "Elements of Biodiversity," mainly species, natural communities, and their "Element Occurrences" (EOs) in the landscape. The combined databases for the United States contain information on over 35,000 species and about 7,000 subspecific taxa. Each is designated by the currently most acceptable scientific name, and many thousands of additional names are tracked in synonym databases to accommodate regional usages and to avoid confusion. By 1992, there were between 300,000 and 400,000 EOs that have been identified and deemed significant enough to be tracked in detail by the programs. These data sets are used extensively by managers and decision makers, more than for any other databases we know of. The information contained in the individual State Heritage Data Centers is consulted an estimated 200,000 times a year for endangered species planning, environmental impact review, development permit applications, land-use decisions, natural resource management, and scientific research.

The information in the data centers is managed in a balanced information system consisting of manual files, manual maps, and the computerized BCD database management systems. Most of the centers are now using computerized GIS systems to take over the earlier manual mapping functions and to facilitate analysis. Nearly all of the programs interact extensively with state multi-resource GIS programs where these exist. The BCD is the sixth generation of database management systems developed and supported by the TNC Science Division for use by the network, each evolving from the one before as computer technology and Heritage methods have advanced. The BCD is a super-relational database management system utilizing Advanced Revelation software running on DOS and Banyan VINES network operating systems on PC microcomputers. By 1992, it consisted of over 40 data files or tables organized around 11 major logical entities, with over 2,300 data fields, and tens of thousands of controlled values defined for many of these fields. Every one of these entities, files, fields, and values has had to be identified, distinguished as sufficiently important for inclusion, and defined, frequently in tightly parameterized ways. This has not happened in a day and everything continues to evolve. A third upgrade to the BCD system was released in 1992; it contained changes and additions to over 1,000 data fields.

At the present time the main data centers employ several hundred scientific staff whose primary responsibilities focus on biodiversity data collection and management. An unknown number, but certainly hundreds, of other agency staff with multiple responsibilities interact with the database part time. Thousands of others—cooperating scientists, natural historians, agency personnel—supply information and/or obtain information from the network on an episodic basis. The total investment going into the network of data centers, not counting the efforts of volunteers and unpaid cooperators, is currently estimated to be between $50 and $80 million per year with a cumulative investment of perhaps $400 or $500 million.

BIODIVERSITY INFORMATION AND SYSTEMS FROM HERITAGE EXPERIENCE

The task of accumulating, organizing, storing, retrieving, manipulating, and analyzing data or information is referred to as information management. To manage information casually is relatively easy and constitutes a large part of ordinary human activity. To do it with rigor, however, especially for such a vast subject as biodiversity at the scale and complexity being discussed here, is much more difficult than it appears. As a result, many people and institutions enter upon the business with less thought and preparation than they should, underestimate the expertise and commitment required, waste a lot of resources that could be put to better use, and often wind up hindering the enterprise they intended to further. Competently managing biodiversity information involves much more than computerizing data and simply turning the job over to someone with a degree in computer science, which will not produce the right result.

As implied above, although modern computers are undeniably important tools in the management of biodiversity information, one must begin with a deep understanding of the biodiversity crisis, of what actions could help alleviate it, and a vision of how information can be organized to facilitate and guide the efforts of those taking the actions. As the main designer of the largest and most elaborate computerized biodiversity database system in the world, I can say with confidence that for successfully managing a biodiversity information system, computers, per se, are really secondary to the more important business of conceptualizing the problem, determining what information is important, identifying data needs, defining terms, determining relationships among major factors, designing basic system architecture, and detailing data elements. No successful information management occurs until someone discovers how to file documents in a drawer, what should be put on a map with what notation, and how to apply the information to the problem at hand. The operation parts of the system—the people, institutions, the physical plants, the budgets, the financing, the logistics, and the administrative details—are more crucial than the computers. Successful management of biodiversity information requires a high order of general management—of personnel, budgets, operating procedures, institutional arrangements, and cooperative relationships. All these disparate aspects must be brought together into a single enterprise. It is helpful to think of the effort as a business venture in which a problem in finance, marketing, or public relations is just as apt to bring about general failure as anything that can go wrong in electronic data processing.

Certainly, computers are important too, but there is more to be said on the subject than can be fitted into this paper. One certain thing is that with the rapid pace of advancement in the computer industry, whatever works best now will no longer be the best option in three to five years. As stated above, the Natural Heritage network is now using its sixth generation of database management systems on its fourth or fifth generation of hardware platforms—and this does not include the parallel use of Geographic Information Systems of a variety of sorts. Each of the newer computer configurations has brought increasing power and ease of use, but the basic system architecture has remained remarkably stable, expanding upon the core logical entities and relationships that reflect the real world the system attempts to model.

Bringing the disparate threads together is harder work than would-be system designers and promoters usually imagine. No year passes that some group does not announce that it is just beginning to "develop a database on all the natural areas of the United States" or on "all neotropical migrant songbirds" or on "U.S. endangered species," and it turns out they plan to do this in two years with a $100,000 grant. These are perfectly sincere initiatives and everyone has at some point begun in the same way. But for such ventures to be successful the annual budget estimates must be increased by a factor or ten and the time estimates from any fixed period to permanent operation. Continuity of effort is essential.

The vast majority of such undertakings simply fail. Those who are interested in accomplishment of the particular objectives would be better advised to see what systems are already operating and how their interests could either be incorporated or made into complementary parts of the greater whole, led by the experiences of those who have gone before. Unfortunately, this rarely happens—everyone underestimates the difficulties and imagines that they can do better than those already operating in the field. This is partly because all operating systems have detectable defects, whereas the unimplemented systems are expected to be flawless—but they will most certainly have defects of their own. On average they will have more flaws than the operating system they are intended to supplant, simply because they are younger. The inescapable truth is that it takes time to develop, debug, and implement any system. Doing this, and further, actually putting the system to effective use in problem solving is, in fact, a permanent process and operating systems gradually tend to get better.

For an example, a federal procurement unit was recently choosing between two computer systems. One was operational (well over a million dollars had been invested in an application of a widely used basic system) and the other existed only as a set of preliminary specifications for a system to be created from scratch. The operational system was still being debugged and had a number of imperfections. The users complained about many problems. Of course, no one had encountered flaws yet in the nonexistent system, and whatever anybody did not like about it was seemingly fixed with a few facile adjustments in the design diagrams. The procurement specialists chose the vaporware. The operational program was soon debugged and is currently producing outstanding and continuously improving results. Meanwhile, many other would-be users are standing by while millions of dollars are being spent attempting to duplicate the same functionality *de novo*. My prediction is that something else will supercede before this is achieved. Even if the new system is developed, it will then have to undergo the same test by fire that the other has already gone through and by that time, the already operating system may well have advanced to a yet higher plane.

Another reason that everyone tends to underestimate the difficulties in biodiversity information management is that beginning is deceptively easy. Every programming job begins with a single key stroke. Anyone can easily create that first flat file in the memomaster that came bundled with their PC. With modern relational database software, even creating the first cluster of a half dozen files is pretty easy. Ironically, it is often easy to obtain support also, because funding sources like to start new things. Everyone wants to support a "pilot project."

But a day of reckoning is at hand. Systems will grow. Other boxes from the diagram must be added to the first cluster. The diagram will change. Things will not work out quite as envisioned. More than one programmer will be needed, and their work will be hard to integrate. Actually, the system will have to be installed somewhere, possibly in more than one place. Actual data will have to be obtained and input. Data another source was to supply will prove to be seriously flawed. Someone will actually try to employ the system, and their needs and preference will have to be taken into account. Users will demand to know definitions; documentation will prove inadequate; the users will begin misusing the data fields and redefining the files. Bugs will appear. Operational breakdowns will intrude. Operations will take longer than expected and be more expensive. The computer industry will release new and incompatible hardware and operating software. The system will collide with those of others attempting to do the same things. Some will want to cooperate by cross-walking their systems, and this will tremendously complicate everyone's efforts. Another three specialists will be needed to mediate the process. It will be time to get the second and the third and the fourth round of funding, and the funding sources that were drawn by the novelty will be disappointed with prosaic reality and recoil from continuing commitments. Half of the energy will now go into proposal writing and seeking new funding sources. The principal system architects will establish reputations with the result that much energy will be drawn off into dozens of committee assignments to ponder big things and much networking.

The above paragraph does not describe system development gone wrong. It describes system development under even the best of conditions, as everyone who has ever done much of this work knows well.

Some years ago, many problems encountered by system developers were brought together in a pithy book by John Gall called *Systemantics* (Gall 1975). The book applies equally to computer systems and to the encompassing systems of coordinated human enterprise. The book's style is droll but its purpose is serious; it should be required reading. Among the many important rules and admonitions the book advances are several worth repeating here for anyone contemplating biodiversity information system development:

> A complex system that works is invariably found to have evolved from a simple system that worked.
>
> A complex system designed from scratch never works and cannot be made to work. You have to start over, beginning with a working simple system.
>
> In setting up a new system, tread softly. You may be disturbing another system that is actually working.
>
> A system can fail in an infinite number of ways.
>
> In complex systems, malfunction and even total nonfunction may not be detectable for long periods, if ever.

This last point bears further examination. A database represents a model of reality. But because the model is separated in space and time from that of reality, it is difficult to verify the model, and it is easy to mistake the model for the real thing. Even a completely erroneous model can look good, but even the most assiduously-derived prescriptions from it will wreak harmful results on the real world.

Instead of speaking of biological diversity we could as well have chosen the term biological complexity, and this complexity makes information management in the biodiversity field especially challenging. Paraphrasing J. B. S. Haldane, Frank Egler used to say that "ecosystems are not only more complex than we think, they are more complex than we *can* think." When species diversity, gene pools, and total landscapes are included and further combined with the various aspects of human society that are causing biological endangerment or conversely, those that could help alleviate it, one is dealing with as complex a body of information as exists.

Seemingly simple questions, such as, "What species occur in Yellowstone National Park?," turn out to be tremendously hard to answer definitively. People underestimate the difficulty because it is relatively easy to answer the related question, "What common plants and animals are regularly encountered in Yellowstone NP?" But to provide a complete list, even just of higher plants and vertebrates, requires an enormously greater amount of field work, prior knowledge of systematic biology, ecology, and field survey methods, well-developed information systems (in the sense indicated above), and coordinated effort by many investigators. Any federal land manager can readily answer the second question, but almost none can answer the first.

In fact, the seemingly simple question posed above is practicably unanswerable, comprehensively, for even a single area and yet, as soon as it is posed, the more difficult question leaps to mind of comparing the element diversity list of a given area to that of other areas, the region, the continent, or the whole earth. Ambitious proposals are currently being advanced to discover the complete species composition of just a few selected areas. This would be interesting and might tell us some useful things about biodiversity and ecological relationships. However, I am not sure how it will help us with biodiversity conservation. As any scientific researcher knows, asking the right questions is nine-tenths of dealing with any problem. We should carefully ask ourselves what are the important questions that need to be answered for biodiversity conservation and management and what data is needed to answer them.

Because nature is infinitely varied and rich, it would be easy to overwhelm a system that was too open-ended. Of millions of observations made in the field in Yellowstone, for example, some are of great importance, some of none at all. The design of a workable system will depend largely on distinguishing between the two extremes. In the next section I deal with what we have found in the Heritage network to be among the ten most important questions for biodiversity conservation.

Even when addressing the correct questions, there are further requirements for biodiversity information to be effectively useful. It must be accurate, and, with some qualifications, it must be complete enough to fully answer the question. It must represent current reality, or, for many applications, such as restoration, change detection, or monitoring, it must be specifically dated as to when it was current. It must be accessible. It must be comprehensible to the user, although in fairness, the user might have to accept some training. Ideally, it should be in a form the user likes to work with.

The unavailability of the information we need is partly due to the fact that we have not treated the steady accumulation of accurate information as an important end in itself. We have been careless with information, treating it, as one of my scientist friends used to say, "as a one-time use, throw-away commodity." Thus, of the useful observations made over the years in Yellowstone or any other area subject to management we can readily retrieve only a tiny fraction. Frequently the manager must make his decisions based on superficial knowledge and general impressions. Even when such general ideas are correct (often they are not), they are insufficiently precise to deal with many management decisions. For example, the welfare of a target species may hang on whether the manager chooses to manage for condition A or B, one beneficial to the species, the other not.

If we are to improve our knowledge base to meet the needs of management we must develop better practices to capture, organize, and accumulate important information. These practices need to engage nearly everyone who can contribute to the knowledge base so generated. An obvious idea is for as many contributors as possible to utilize standard systems to concentrate information. At Yellowstone, such a system ideally would capture important information from even the tourists. Certainly every actual researcher should be required to take the time to incorporate what he or she has learned there. A good, well-maintained system would become a resource for the researcher as well as the manager and, thus, generate enthusiasm for its use.

The Natural Heritage system and network is offered as the best current basis for such applications. It has been under continuous development for over 20 years, and it is already used by a wide variety of institutions with more joining all the time. It already has an enormous amount of carefully assembled and verified biodiversity data available, and readily accepts additions to these data sets. It has had the time and continuing focus to meet and solve many problems. The reader is referred to discussions (Jenkins 1985, 1987) important attributes that the network has developed over the years.

TEN IMPORTANT QUESTIONS

Biodiversity is being conserved in several different ways—by protecting native habitats in nature preserves, by avoiding areas important to biodiversity in development projects, by preserving species or their propagules outside their native habitats, by accommodating biodiversity through sustainable resource management schemes, and by appropriately modifying public policy. The Natural Heritage network has identified 10 important questions that must be answered for success in any of these types of actions:

What are the elements of biodiversity?
Which of these elements needs help?
Where can these elements be found?
Where should our efforts be concentrated?
How are these elements related to one another?
How should or can these elements be managed?
What changes are occurring?
What are the threats?
Who are the actors?
Where can this information be obtained?

The Heritage data centers make it their chief responsibility to answer these ten questions. Confining this discussion to the United States and North America, I will attempt to show how they are answered and to what effect. Data centers in Latin America and the Caribbean are attempting to answer these questions also, but there are many more gaps in geographic coverage, greater biodiversity complexity to be contended with, and less time has transpired for developing powerful databases. Efforts are continuing to solve these problems.

What Are the Elements of Biodiversity?

The entire Biological and Conservation Data (BCD) system used by Heritage programs, with all its files, fields, and controlled values is an attempt to answer the question of what exists and what vocabulary should we use to define it. Heritage programs attempt to define the universe of what biodiversity entities exist by creating checklists of species and infrataxa (a "fine filter"), and classifications of natural community types (a "coarse filter"). Also, in existence are lists of "other types" for entities that are neither species nor communities but which seem to merit attention, such as outstanding specimens, unique landscape features, or key migratory concentration points for various birds. These species, communities, and other types are collectively known as the "Elements of Biodiversity."

The idea of the "coarse filter" and "fine filter" was invented for Heritage biodiversity information and planning purposes. The idea is that there are more species than we can track or even know about. Therefore, makes sense first to identify the kinds of major communities found in a region, identify the best location of each of these, and, by protecting sites containing selected examples of each, simultaneously protect populations of most species. One can increase the capture rate by subdividing the community classification to a finer level and by certain other means elucidated in the Heritage operations manual, but eventually diminishing returns will occur. The species most likely to fall through the "coarse filter" are the rare ones, so these must be identified individually. They are referred to as a "fine filter" because their populations are mingled with others that are also protected when reserves are established for the target species. It is further assumed that many rare species are habitat specialists found together in their own microcommunities with unknown species which are probably rare for similar reasons.

Heritage has had an easier time dealing with species than with natural communities, which are a much more difficult objects for classification. The network has managed to compile what we think is a comprehensive checklist of all vertebrate and vascular plant species for North America with some basic information about each of them. Comprehensive checklists from the United States have also been compiled for certain invertebrate groups, but fewer for other groups. In the latter, only those species believed by experts to be imperiled have been listed, along with certain species included because they are management pests or for other reasons. Some units of the network have been working through the nonvascular plants for the last several years with the help of some cooperating museum scientists, who have compiled apparently complete lists of both bryophytes and lichens. As noted above, only those subspecies or varieties singled out for conservation concern have generally been included.

These are not perfect checklists but they must be as good or better than the best available for any other country. Hundreds of Heritage staff and volunteers have combed the literature, the systematic biology collections, and the landscape in the various states for many years. Much documentary data and expert judgment lie behind these checklists.

Heritage Data Centers have all developed community classifications for their states. Both terrestrial and aquatic communities have been treated, although aquatic community classification is even more challenging and less developed. To the extent possible, preexisting classifications have been built upon or cross-walked to facilitate secondary information utilization and communication. Several kinds of efforts are being undertaken to bring all of the state classifications together into a comprehensive national system, but this has yet to be achieved. Some progress is being made by adjacent states to synthesize shared community types. Conservancy regional ecologists are also drawing together and reconciling the classifications for all the states in their regions. Reconciliation of ground-based classifications with classified satellite imagery is becoming a strong tool, especially in cooperation with the U.S. Fish and Wildlife Service's Gap Analysis Program. The U.S. Forest Service has also been cooperating and providing support to community classification efforts across the network.

Beyond the elements of biodiversity are a number of other major logical entities (many of them geographic entities) which we have to know about to effect conservation planning: existing reserves, sites of known conservation significance, specific land ownership parcels, key individual and institutions, and sources of secondary information. Heritage systems incorporate data on all of these.

Which of These Elements Needs Help?

Fortunately, many species and natural community types are currently faring well, and every established reserve or site of conservation significance is not necessarily currently under threat. Although any species could be exterminated by a sudden unpredictable event like the chestnut blight, many species are apparently safe for the foreseeable future. Some species are virtually ineradicable; others have been reduced to a few individuals. To use scarce conservation resources efficiently and make the most of our opportunities, we must know where species fall along this security continuum. We must be able to allocate our own resources and to show others objectively that these priorities are worthy objects of attention. At the other end of the spectrum, we must be able to identify apparently secure species to avoid squandering scarce resources on their management.

One of the more useful things that Heritage has done is to develop a formal process for tracking and documenting Element status, known as Element Ranking. For each community and species, a standard series of questions are asked and the answers recorded in specific databases. Some of the standard questions are: number of remaining Occurrences for the Element, number of total individuals or aerial extent (for communities); number of protected Occurrences, degree of threat, and innate vulnerability or fragility. On the basis of the answers, each element is given an Imperilment Rank ranging from 1 to 5, critically imperiled to demonstrably secure. The scale is roughly logarithmic and for most groups, only about 10% to 20% of the Elements have ranks of 1 to 3. Elements are ranked on a global or range-wide basis (G ranks), a national basis (N-ranks), and a state or subnational basis (S-ranks). In addition to 1 to 5 ranks, there are also X ranks (extinct or extirpated) and H ranks ("historical," meaning that no occurrences can be found but the search for them continues). Various other special ranks and postscripts are used to denote migratory particularities, questionable taxonomy, and other matters. Special notation is also used for infra-specific taxa, which are also ranked where needed (Master 1991).

This process is well advanced, with the number of "unknown" or questionable ranks for Elements in better known groups rapidly trending toward zero. Because of incompatibilities of state community classifications, most communities have been ranked only at the state level.

Element ranking has proved an incredibly useful device. The ranks are used to set priorities for both data collection and conservation (resources are concentrated on the highly ranked Elements). The ranks allow Element lists and databases to be sorted by imperilment classes and the generation of Element lists in sequential imperilment order. The terse annotation can be handily attached to the Element name to convey a great amount of information. The ranking forms in the databases are revised as new information comes to light and ranks revised accordingly.

Where Can These Elements Be Found?

This question logically follows the question of what elements need help because it is impossible to discover the detailed distributions of all biodiversity elements, and it is very important, for resource allocation reasons, to avoid the attempt. Some people have argued that counting and mapping every white oak tree would be tremendously valuable because it would provide such a marvelous device for change detection. No doubt, but these people have not tried to do it and given limited resources this could only be done at the expense of finding out about more important things. By rigorously concentrating information efforts on the Elements that need help, the detailed data needed to manage them can be obtained— beginning with their exact localities, and beyond that, the aerial extent of these, their current condition, and a host of other factors that affect their survival and future.

At another scale are efficient ways to use general distribution data on less imperiled Elements for various conservation purposes. Remote sensing data to depict entire landscapes is an example of an efficient device which Heritage is increasingly using to effect conservation planning at the community and landscape levels. The growing use of various statistical sampling techniques promises to provide probabilistic ways of informing ourselves about the locations and distribution of certain biodiversity variables across wide areas.

Using the above element ranks to prioritize their work, the Heritage Data Centers concentrate much of their activity on finding, verifying, and describing individual Element Occurrences (populations of species or outstanding community examples at exact localities). As noted above, hundreds of thousands of such localities have been found, pin-pointed, mapped, and surveyed in the field. This is the indispensable nugget of information for most managers because they cannot manage any part of biodiversity without knowing where it is located.

In addition to EOs, the Heritage Data Centers also pinpoint the locations, map, and describe existing reserves or "Managed Areas," Sites of conservation significance (landscape units which might contain one or many EOs), and associated land ownership Tracts. Managers need to know the location, extent, and features of these entities as well. Remote sensing and GIS are increasingly augmenting traditional manual field survey and mapping processes, which among other things will facilitate the accessibility and sharing of this information. The use of procedures, such as county aerial surveys, "gradsect" transects, Gap Analysis, and Rapid Ecological Assessment, are continuously expanding the efficiency and comprehensiveness of Heritage landscape inventory.

Where Should Our Efforts Be Concentrated?

This is not simply the last question repeated. Even when the most important Elements and other entities have been located or a coarse view of the conservation status has been determined for entire landscapes, choices remain. A given endangered species might still have several extant populations, and one will have to be saved before another. Information is needed to identify which one should come first. Because landscapes contain complex interactions among species and communities, decisions involve some compound calculations— for example, three rare species and two outstanding community remnants in an easily defensible topographic setting versus two rarer species in a badly damaged ecosystem but with exceptional funding opportunities available versus an average landscape with no endangered species but with good representation of half the major communities of the region.

This question of where to concentrate effort must be made at many different scales. Do we concentrate on Brazil or Indonesia, California or Georgia? What are the most significant areas within a given county? Should we restore this 40-acre old field or run prescribed burns in the adjacent pine woods? Where do we build the visitors' comfort station? And although we must concentrate our information efforts on the upper end of the significance scale, the continuum always extends to the entire landscape. As long as destructive human impacts are increasing, more and more of the landscape is going to be altered or destroyed, and the places that we decide are not conservation priorities are being consigned to development. To the extent possible, we must be prepared to give guidance to developers and consumptive resource users about where we believe their activities will do the least harm.

One of the most important applications of Heritage data is in preserve selection and design. For many years each Heritage program has generated Biodiversity Scorecards and Priority Site Tracking lists to establish priorities for land conservation by its host agency, the Conservancy, and other conservation institutions. In some states there are Interagency Protection Planning Committees that meet periodically to coordinate conservation activities of the various institutions at the priority sites. At the priority sites, the associated land ownership tracts are also prioritized on the basis of their relative importance to the viability of the reserve for the target elements.

Larger and smaller scale priorities can also be addressed. Heritage data can illuminate important variables affecting a regional policy decision, such as the allocation of endangered species efforts or help the individual preserve manager decide where to place his interpretive trail. For a smaller scale example, at one federal land unit with a data center, the Heritage maps are the first thing they turn to when deciding how to fight wildfires.

How Are These Elements Related to One Another?

Most of what we need to know for biodiversity conservation relates to geography and the importance of co-location comes up repeatedly— not only in regard to combinatorial calculus for efficient priority setting but in defining ecological relationships important to the viability of species, communities, and landscape complexes. Mapped information is one of the main needs of the biodiversity manager. Other important questions regard species-habitat relationships, mutualisms, predator-prey relationships, and species dependence on specific ecosystem functions.

Much of what Heritage programs do involves mapping; this shows how things are geographically related. It is carried to considerable lengths in the EO and site databases and field forms, in which the community ecologists, for example, set out permanent plots and collect great quantities of phytosociological data. Species and community databases also contain much information about habitat relationships and interspecific interactions.

How Should or Can These Elements Be Managed?

Or to put it another way, what are the ecological requirements of the various target species, communities, and landscapes complexes. The complexity of these questions makes everything else look simple by comparison, and once again, we have to focus our efforts efficiently on the most important knowledge gaps. The same schemes for setting priorities alluded to above must be employed with special rigor here because of frequent contention between scientists and managers. Academic scientists have traditionally chosen their research subjects on unapplied grounds, either to extend some line of inquiry within their particular discipline or simply on the basis of personal curiosity. Meanwhile, managers are desperate for information on particular species, communities, ecosystems, and geographical localities that are uninvestigated. Except for a few journals, such as *Conservation Biology* and *Range Science*, little in the scientific literature is useful to managers. One hopes this can be changed.

Heritage databases and manual files are full of information that can help the manager understand the needs, opportunities, and management priorities of given elements, sites, and preserves. Element Ranking forms contain data fields specifically for rating Element management needs and citing specific prescriptions where known. Site Basic Records rank sites on several criteria, for example, Management Urgency and known management needs. Managed Area records obviously contain much information on management and administration. Actions databases are used to record and track specific management actions planned for specific sites, elements, and managed areas.

Management information is perhaps most intensively developed in the Element Stewardship Abstracts, which pull together as much information as possible about the management needs and proven management practices on particularly endangered species or ecosystems and lately, on efficacious forms of control for intrusive exotic species. Hundreds of ESAs have now been developed, and these are added to and revised as cooperating managers learn from their field experiences what works and what does not.

Heritage staff have more recently been cooperating with Conservancy field staff and others in inventories and planning of major reserve complexes called Bioreserves. This work is beginning to involve major ecosystem modeling efforts on hydrology, fire phenology, and related matters in the management of entire landscapes.

What Changes Are Occurring?

With the accumulating evidence of impending or current global climate change, "monitoring" is suddenly the hot topic, and we are almost astonished at how unprepared we are to deal with change detection. This goes to the very heart of our biodiversity information management dilemma. We have done such a shockingly poor job of defining, creating, and retaining tight data sets that we look around in bewilderment about how we came to know so little with enough specificity to see whether change has occurred. In general, the most certain way of detecting change is by knowing that at an exact time and place certain things were found and certain conditions existed, and then rechecking at a later time. But data collection and recording have historically been inexact. Locational data on older museum specimens may not be as precise as county-of-occurrence. Synecological data have been taken without much reference to exact localities. Monumented permanent plots and transects have been few and far between, and many of these have been neglected and lost. We have a fair amount of older aerial photography, but almost none of it has adequate ground truth information to be sure what it shows. Systems of soils benchmarks and vegetation exclosures have been established and later abandoned. One can read entire journals without ever encountering a paper that says exactly where the work was done. With the advent of remote sensing, GIS, and global positioning systems, one hopes that precise locational information will become the norm.

The beginning point for change detection is to have a baseline, and the geographic exactness of so much of the Heritage data and the careful dating of the records make them exceptional resources for monitoring purposes. Heritage biologists and cooperators, such as Conservancy state land stewards, are carrying out specific monitoring projects on individual rare plant and animal populations and community stands at mapped EOs. The BCD system has recently been expanded to facilitate this by the addition of specific monitoring modules for optimal handling of the time-sequenced resampling data that results.

In addition, Heritage Data Centers have been increasingly using aerial photographs and satellite imagery in carrying out county inventories and Rapid Ecological Assessments, documenting ground truth conditions with standard field surveys and plots, and accumulating the results in databases, maps, and digital libraries. With the proliferation of GIS systems, GPS use, and orthoimagery, more and more researchers from many institutions are generating classified and annotated digital data sets, and the Heritage network offers a permanent, organized repository for such data for reliable future retrieval and resampling.

What Are the Threats?

Also, what unnatural (and natural) forces are exerting an impact or have in the past? It is obviously important to recognize threats to important sites and elements, but attempts to estimate threats are often overdone. Significant threats and impacts are usually self-evident and easily recorded, but speculations about possible threats often substitute for action or other more important data activities. The uncertainties surrounding threat estimates also offer endless opportunities to introduce bias into the evaluation of priorities, where threat estimates are used to justify preferred courses of action.

Heritage programs attempt to make threat estimation as precise as they can. Threats to sites, elements, and EOs are recorded in many places in the Heritage data systems (e.g., on the Element Ranking forms). As protection and management planning become more elaborate, specific threat reduction strategies are being developed for more and more sites with the Action databases and other Heritage modules being used for this.

Who Are the Actors?

This may be too peripheral to treat as biodiversity information but it opens a window on subject matter that biodiversity conservationists must know to be effective—who can help, who is causing the problem, how are they all connected, what is going on? It is also worth mentioning here because information on individuals and institutions is very open ended (it could be extended to public opinion monitoring) and surprisingly difficult to master or keep track of. This is partly because of the frequency of change. The average federal official, for example, seems to change jobs about once a year and phone numbers once a month. All institutions are undergoing continual reorganization. The National Wildlife Federation publishes a Conservation Directory which many of us find indispensable. They have to publish a new edition annually to catch up momentarily with these changes. It must be a huge effort, and very few such directories of any scope are kept up with equal success. This is another example of the unexpected difficulty of the data management business and the need for combined efforts.

Information about people and institutions is recorded by Heritage data centers in both Contact databases and Source Abstracts. Because of frequent feedback resulting from the high degree of use of Heritage data sets, the associated databases are probably kept more current on certain information about important conservation actors like preserve managers (over 20,000 existing Managed Areas are tracked in the United States) and subject matter experts (tens of thousands scattered through more than 140,000 Source Abstracts) than nearly any other source.

Where Can This Information Be Obtained?

And how? And what must the user do personally? What data gaps exist and can the additional needed information be obtained from others or from others in combination with the user's own efforts? No single institution gathers information by itself in this field, even from secondary sources in the tradition of literature review. If a user group has enough time, money, managerial skill, and dogged perseverance, it can contribute much to the greater whole. If not, it is probably important not to waste the effort but to see instead how the problem can be attacked with existing knowledge, native intelligence, and intuition.

The Heritage network is a very good place to start. Most of the primary data centers in the network are found conveniently in state agencies in the state capitals. Heritage information has been amassed to be made available to any "legitimate" user (some information, like certain endangered species localities, must be distributed carefully to protect the resource), usually for only the cost of reproduction. Much of the information is contained in the computerized BCD system so that it can be sorted and reported in a variety of formats, briefly or at length, to facilitate specific applications. Not only does the Heritage network itself constitute by far the largest compendium of accurate and current biodiversity-related information, but its Source Abstracts and manual files offer a window on additional information, including other databases and repositories.

CONCLUSIONS

We have an enormous job to do in obtaining and managing biodiversity information. We are not accomplishing nearly enough now because of an insufficiency of coordination. Amassing and organizing the amount of information we need must be a cumulative process, so continuity of effort is crucial. The amount and diversity of information exceeds that which any single institution can be expected to master, so multi-institutional cooperation is indispensible.

This paper focuses on the existing network of Heritage Data Centers which have developed into the kind of multi-institutional cooperative venture required to master collectively biodiversity information for conservation and management purposes. For the United States, at least, and for the western hemisphere generally, it is by far the leading compendium of biodiversity knowledge and the one of most use for conservation decisions. The description of the Heritage network offered herein is intended to provoke not emulation but cooperation. The existing network is more than willing to provide, through appropriate reciprocal arrangements, its data, technology, operations manuals, and software, to anyone who is in a position to fill geographic or subject matter data gaps in the existing network. Of course the data and information produces are available to managers on a first come-first serve basis.

LITERATURE CITED

Botkin, D. B. 1990. *Discordant Harmonies: A New Ecology for the 21st Century*. Oxford University Press, New York.

Gall, J. 1975. *Systemantics: How Systems Work and Especially How They Fail*. Quandrangle/The New York Times Book Co., New York.

Hoose, P. M. 1981. *Building an Ark: Tools for the Preservation of Natural Diversity Through Land Protection*. Island Press, Covelo, Calif.

Jenkins, R. E. 1975. Heritage programs: inventory progress report. *TNC News* 25(2):26-27

Jenkins, R. E. 1985. Information methods: Why the Heritage programs work. *TNC News* 35(6):21-23.

Jenkins, R. E. 1987. Natural Heritage Inventory Overview. In *Natural Heritage Program Operations Manual*. The Nature Conservancy, Arlington, Va.

Jenkins, R. E. 1988. Information management for biodiversity conservation. In *Biodiversity*, ed. E. O. Wilson, pp. 231-239. National Academy Press, Washington, D.C.

Jenkins, R. E., and H. Moyseenko. 1976. Progress in developing a national data base on ecological diversity. *TNC News* 26(1):25-26.

Master, L. L. 1991. Assessing threats and setting priorities for conservation. *Conserv. Biol.* 5(4): 559-563.

McAllister, D. E. 1991. What is biodiversity? Unpublished paper prepared for Keystone Center for dialogue on biodiversity conservation (available from The Nature Conservancy).

McNeely, J. A., K. R. Miller, W. V. Reid, R. A. Mittermeier, and T. B. Werner. 1990. *Conserving the World's Biological Diversity.* International Union for Conservation of Nature and Natural Resources, World Resources Institute, Conservation International, World Wildlife Fund-U.S., and the World Bank, Gland, Switzerland.

Miller, K. R. 1989. *Global Strategy for Conserving Biodiversity.* World Resources Institute, Washington, D.C.

Office of Technology Assessment (OTA). 1987. *Technologies to Maintain Biological Diversity.* OTA-F-330. Government Printing Office, Washington, D.C.

Reid, W. V., and K. R. Miller. 1989. *Keeping Options Alive: The Scientific Basis for Conserving Biodiversity*. World Resources Institute, Washington, D.C.

Salwasser, H. 1990. Conserving biological diversity: A perspective on scope and approaches. *For. Ecol. Manage.* 35:79-90.

The Nature Conservancy (TNC). 1975. *Final Report. The Preservation of Natural Diversity: A Survey and Recommendations*, eds. J. Humke et al. U.S. Department of the Interior, Washington, D.C.

13

REGIONAL SCALE MONITORING OF BIODIVERSITY

K. Bruce Jones

U.S. Environmental Protection Agency
Las Vegas, Nevada
U.S.A.

Brett R. Riddle

Department of Biology
University of Nevada, Las Vegas
Las Vegas, Nevada
U.S.A.

Estimates of global decline in species richness and decline in processes that lead or contribute to stability and/or resiliency of ecosystems have led to concern among scientists and managers about the future of biotic potential on the earth (Schlesinger et al. 1990; Lubchenco et al. 1991; UNEP 1992). These concerns range from local to global scales and from present to future generations.

Even though scientists and environmental managers generally agree that we are experiencing significant declines in species richness, several opinions exist on the type and magnitude of data needed to understand and address the issue. For example, Wilson (1985) and Ehrlich (1992) expressed a need to act immediately in surveying the earth's diversity in an effort to set priorities and prevent loss of species, whereas others (Hubbell and Foster 1992; Franklin 1993), urged greater emphasis on studies of ecosystems. Numerous variations in both of these approaches have been implemented or are being proposed. For example, the All Taxa Biodiversity Inventory is a refinement of the global survey, but it proposes an initial exhaustive survey of all organisms in a single area (Yoon 1993). The idea is to conduct an exhaustive survey in areas where species richness is known or expected to be high (e.g., "hot spots" of biodiversity).

Fragmentation of natural habitats is believed to affect biodiversity significantly at regional and continental scales (Saunders et al. 1991). Additionally, ecosystem fragmentation and simplification (e.g., regular as opposed to complex shapes) are thought to change dramatically disturbance regimes which affect successional trajectories and resiliency of landscapes (Turner et al. 1989, 1991). Despite this knowledge, little is known about the status, magnitude, rate, and nature of change (e.g., relationship to land-use) in habitats and landscapes as they relate to biodiversity.

Broadening the scale of research and monitoring of biodiversity is needed, both temporally and spatially (Brown and Mauer 1989; Ricklefs 1989; Baron and Galvin 1990; Brown and Roughgarden 1990; Bailey 1991). Throughout the world and especially in the United States there has been a focus on local and community level processes that maintain species richness (Brown and Roughgarden 1990; Fox 1992). Regional and continental composition (e.g., source pool) and processes (e.g., colonization) of biodiversity are now thought to play important roles in determining the composition and structure of local diversity and ecosystem processes (Wiens 1986; Addicott et al. 1987; Ricklefs 1987; Delcourt and Delcourt 1988; Brown and Mauer 1989; Brown and Roughgarden 1990; Hunsaker et al. 1990a; Graham et al. 1991; Avise 1992; Walker 1992; Franklin 1993), and these coarse-scale processes, such as dispersal, are being affected by mankind on a global scale (Romme and Turner 1991).

The United States has no comprehensive regional or national scale biodiversity monitoring program (OTA 1987), although a few programs, such as the U.S. Breeding Bird Survey (BBS), provide regional and national monitoring of certain elements of biodiversity (Droege and Sauer 1989). Additionally, status and trends of ecosystems and habitats in the United States are virtually unknown (Brown and Roughgarden 1990), although at least one new initiative, the Environmental Monitoring and Assessment Program (EMAP 1993; Messer et al. 1991), holds promise for producing such assessments. Comprehensive and coordinated monitoring programs are needed to address status and trends in elements of biodiversity at regional and national scales, as well as concurrent status and trends in habitats, ecosystems and landscapes. This paper highlights a three-component process to monitor biodiversity and discusses issues, concepts, and methods relevant to monitoring biodiversity and habitats on regional and national scales.

RELATIONSHIP OF MONITORING TO INVENTORIES AND SURVEYS

Confusion exists over exactly what is meant by monitoring and how it differs from an inventory or survey. An inventory differs from monitoring in that it attempts to characterize or determine components of a community, ecosystem, or area at any given point in time, but generally is designed without regard to future, independent samples to determine change (e.g., difference between two time periods) or trends (trajectory based on three or more time periods). An inventory is generally synoptic (e.g., everything is measured), whereas a survey is generally based on samples. The GAP Analysis Program (Scott et al. 1993) combines the attributes of an inventory and survey; habitat is measured synoptically via remote sensing, whereas species are surveyed from existing data (e.g., Heritage Databases) and related to synoptic habitat data via models. Monitoring, on the other hand, attempts to determine changes and trends in resources over time; samples are independent (e.g, samples taken at T_1 are independent of those taken at T_2 through T_n). Monitoring establishes a baseline upon which to assess changes or trends; this is often referred to as the "status" part of the program. Some surveys, such as the BBS and the U.S. Census, have strong monitoring components.

Monitoring can be either synoptic (e.g., land-cover changes or trends over time as derived from satellite imagery) or sample-based. Sample-based approaches are founded on either non-statistical (e.g., qualitative selection of ecological representativeness) or statistical (e.g., probability-based design) extrapolations (see later discussion). The Long-term Ecological Research (LTER) Site program (Franklin et al. 1990; Magnuson 1990; Swanson and Sparks 1990) consists of a series of ecological research sites which were selected as qualitatively representative of biomes or major ecosystems (e.g., Chihuahuan Desert scrub).

Monitoring can provide a mechanism to evaluate changes and trends in biodiversity and habitats over a variety of spatial and temporal scales. Monitoring data also help identify the levels of spatial and temporal hierarchy into which patterns and functions of biodiversity are organized, especially when monitoring is designed to collect data over long time-frames and over a range of spatial scales (e.g., the LTER program, Franklin et al. 1990; Magnuson 1990; Swanson and Sparks 1990).

A MULTI-SCALED APPROACH FOR MONITORING BIODIVERSITY

Three general monitoring approaches have been used to address status and trends in biological and ecological resources: (1) those that evaluate broad patterns of diversity to identify geographic areas undergoing rapid change, (2) those that evaluate changes in biodiversity and/or habitats within specific areas (e.g., a watershed), and (3) those that evaluate the effectiveness of a specific management prescription in counteracting decline or enhancing diversity. Monitoring of broad-scale status and trends in biodiversity is generally based on few hypotheses (e.g., vertebrate species richness is declining across region X). This approach attempts to address a few stressors that potentially could be associated with patterns of biodiversity, including climate and land-use. Monitoring programs that address trends in specific areas generally deal with specific species or habitats. The third approach is often referred to as "cyclic-incremental" monitoring, and data obtained from these studies help verify that a policy or management prescription has had the desired effect (Jones 1986). This approach tends to be more hypo-deductive, that is, a specific management prescription is hypothesized to have a positive effect on diversity and the monitoring project is designed to evaluate or test the hypothesis. This approach is consistent with the "ecosystem capabilities" approach described by Rapport (1993).

Rarely have these three approaches been applied concordantly in environmental monitoring programs. However, Seidl and Murray (1991) outlined a conceptual framework for the design of an ecological monitoring program for the Great Lakes, one that incorporates multiple monitoring levels. This framework involves monitoring at three different levels: a regional level to evaluate broad patterns in condition of ecosystems (hereafter referred to as regional detection level monitoring), a more diagnostic level of monitoring to evaluate associations between stressors and ecosystems in areas of concern (identified by detection monitoring), and cyclic-incremental monitoring which specifically addresses the relative success of management or policy.

We propose a national scale monitoring program for biodiversity and habitats that involves three distinct levels: (1) a broad-scale, regional and national detection component; (2) a finer-scale diagnostic component, and (3) a component that addresses effectiveness of management and policy (Figure 13.1). Although we describe each level, greater attention is given to the regional detection component because it is the central theme of this paper.

The regional detection monitoring component would involve measurement of indicators to identify broad scale status and trends in biodiversity and habitats, and to help identify geographic areas where more diagnostic monitoring would take place. Status and trends data on a regional scale are then analyzed to determine which areas should be evaluated in greater detail. In this sense, the status and trends analysis is a filter for more diagnostic monitoring within specific areas. Criteria are needed to determine characteristics of areas to be analyzed in greater detail. These criteria could involve a certain level or magnitude or change in indicators, including changes in configuration and amounts of habitat.

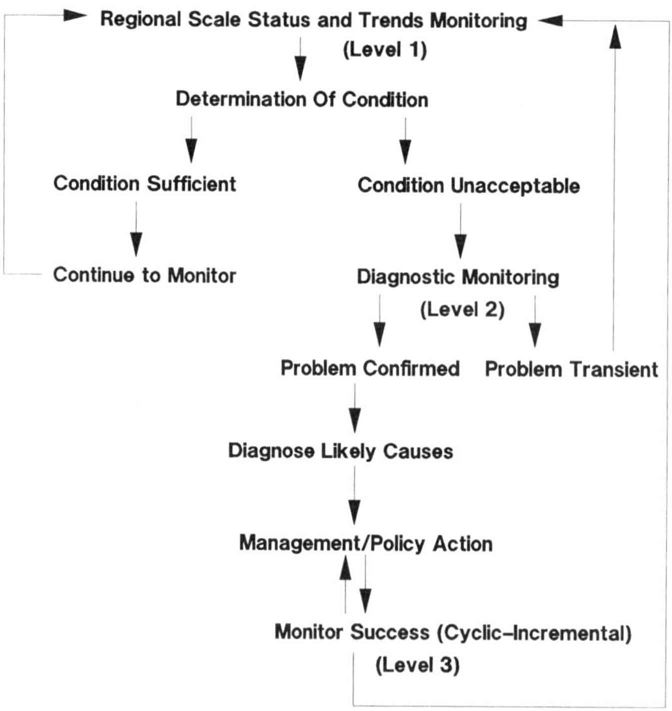

Figure 13.1. A three-level monitoring process for biodiversity (adapted from Seidl and Murray 1991).

The diagnostic level component would involve more detailed measurement of indicators of biodiversity in areas that meet criteria established for change and status in indicators at the regional scale (see above), as well as more detailed investigations of relationships between observed condition and stressors. However, it is possible that other areas not meeting these criteria might be monitored at a diagnostic level. For example, an area might be selected for diagnostic monitoring because of public concern or other reasons (see later discussion).

Diagnostic level monitoring might involve: (1) determining within-area status and trends of biodiversity, (2) expanding coverage to include other components of biodiversity and habitat, or narrowing the focus to specific components, and (3) expanding analysis of relationships between specific stressors and status, and trends in components of biodiversity. Greater focus on specific components of biodiversity and relationship to stressors should improve our ability to diagnose problems related to biodiversity within specific areas. However, diagnostic monitoring does not imply determination of cause-and-effect, but rather a weight or convergence of evidence through association. Indicators of biodiversity and habitats at this level could address ecosystem disabilities (e.g., indicators of declining conditions) and capabilities (e.g., indicators of potential to improve conditions)(Rapport 1993).

The cyclic-incremental component would involve measurement of indicators of biodiversity and habitats to evaluate effectiveness of management and policy in specific areas. Management and policy adopted for a specific area could result from data obtained from diagnostic monitoring, or from detailed studies on cause-and-effect. Indicators at this level would address special components of or issues related to biodiversity, such as habitat of a federally Endangered Species, or more general components, such as species richness within major taxonomic groups (e.g., animals and plants).

Traditionally, cyclic-incremental monitoring has involved local scale assessment of management and policy effectiveness (Jones 1986). However, with the advent of watershed and regional scale planning and management, it might be possible to design and implement cyclic-incremental monitoring on coarser scales. For example, local scale management prescriptions are being implemented within the Murray-Darling Basin of Australia to improve regional water quality. The magnitude and distribution of water quality improvement measures on a local scale (e.g., tree planting) is constrained or dictated by regional scale models of water quality and movement. It should be possible to design a regional scale water quality monitoring program within the Murray-Darling Basin which tests the hypothesis that regionally-constrained, but locally applied, land management prescriptions improve regional scale water quality. However, other variables, including climate, must be considered in this analysis.

Each of the three monitoring levels should address a set of different assessment questions (Table 13. 1); these help define the scope of monitoring at each level. Regional detection level assessment questions should address broad scale patterns in biodiversity; diagnostic level assessment questions should be specific to certain areas, components of biodiversity, and stressors; cyclic-incremental assessment questions should focus on responses of components of biodiversity to specific management prescriptions and policy.

Table 13.1. Examples of Assessment Questions at Three Levels of Monitoring

1. Regional Detection Monitoring
   (a) What proportion of watersheds in Region X have natural cover types with decreasing connectivity?
   (b) What proportion of watersheds in Region X have natural cover types with decreasing patch sizes?
   (c) What proportion of watersheds in Region X have lost more than 25% of at least one natural cover type in the last 10 years?
2. Diagnostic Level Monitoring
   (a) Of those watersheds in region X showing a 25% loss of at least one cover type, what percentage is associated with increases in urban encroachment? with increases in agriculture? other land uses?
   (b) Of those watersheds in region X showing a 25% loss of at least one cover type, has there been a decline in the abundance and diversity of breeding birds?
3. Cyclic-Incremental Level
   (a) On those areas with landscape level habitat improvements (e.g., corridor restoration), has there been an increase in species abundance and richness (this assumes that increased species abundance and richness is the desired outcome)? If so, is this increase correlated with other variables, such as variation in climate?

REGIONAL DETECTION MONITORING FOR BIODIVERSITY

The primary aim of the regional detection component is to identify regional and national scale status and trends in indicators of biodiversity, as well as habitats and landscapes. The broad-scale component also identifies areas undergoing rapid change (e.g., decline in species richness and certain habitats); these areas are then monitored in greater detail to look at within-area status and trends, and associations with anthropogenic and natural stressors (e.g., the diagnostic component).

Four general analyses offer potential for identifying areas for diagnostic level monitoring, but only two are based on change or trend detection. The two change and trend analyses are: (1) monitoring indicators of biodiversity (e.g., species richness) to detect areas undergoing rapid change and (2) monitoring indicators of habitats and landscapes to detect areas undergoing rapid change.

Indicators at this level should integrate a number of important elements of biodiversity and habitats because status and trends in these will be used to select areas for diagnostic monitoring.

Two other broad scale analyses, taxic hot-spot and phylogenetic analyses, can be used as filters to select areas for diagnostic monitoring, but neither is based on change or trends. The GAP Analysis Program (Scott et al. 1993) is an example of a taxic hot-spot approach. The principal aim of the program is to identify areas of the United States with high species richness (e.g., hot spots) and little or no protection (e.g., gaps in protective coverage). Although the GAP Analysis Program uses primarily vertebrate data to establish hot spots, the approach allows for addition of other taxa. Data obtained via the GAP program could be used to identify areas where more intensive monitoring would take place.

Phylogenetic analysis determines the relative evolutionary uniqueness of biota within an area (Avise 1992), and such an approach could be used to prioritize areas to be monitored in greater detail (see Riddle and Jones this volume). Vane-Wright et al. (1991) have developed a conservation prioritization strategy that incorporates both the number of species and phylogenetic uniqueness of each individual species within an area or habitat.

INDICATORS FOR REGIONAL DETECTION LEVEL MONITORING

Process for Selecting Indicators

Selection, evaluation, and implementation of indicators are critical in establishing a regional scale monitoring program for biodiversity. Indicators are defined as measurable surrogates of assessment endpoints or societal values (Hunsaker et al. 1990b). These can be individual measurements or variables or a combination of measurements or variables, such as indices. The Environmental Monitoring and Assessment Program (EMAP 1993) is attempting to describe status and trends in ecological resources through measurement of condition indicators, and correlation between ecological condition and indicators of stress. The latter allows EMAP to generate hypotheses about potential relationships between certain stressors and ecological resource condition. EMAP has two general types of indicators, stressor and condition. Stressor indicators include both natural and anthropogenic sources of stress, such as climate and land use, whereas condition indicators are those that indicate the condition of ecological resources, such as populations and communities (see Hunsaker et al. 1990b for lists examples of indicators being considered by EMAP). The ecological indicator approach of EMAP (Hunsaker et al. 1990b; Knapp et al. 1991) provides a good framework upon which to characterize and select indicators of biodiversity. This indicator approach is being applied in regional ecological monitoring programs, such as EMAP (Hunsaker et al. 1990b) and in ecological risk assessment (Suter 1990; Hunsaker et al. 1990a; Graham et al. 1991). The approach involves selection of societal values, formulation of a series of assessment questions relative to biodiversity (see earlier discussion), development of conceptual models, and selection, evaluation and implementation of indicators and associated sample designs (Figure 13.2). The general societal value of biodiversity must be refined into more detailed statements or assessment endpoints (Suter 1990). This is necessary because the broad values, including biodiversity, are not typically amenable to direct measurement (Riitters et al. 1992). Examples of assessment endpoints include "desired levels" of faunal richness, breeding bird species richness, fish species richness, and trophic level diversity. "Desired levels" can refer to goals set for specific areas or habitats based on site-specific trends, regional average or representative conditions, or a best attainable status (Riitters et al. 1992). Identification of assessment endpoints of biodiversity allows for formulation of explicit assessment questions and conceptual models, as well as indicators (see later discussion).

Assessment questions should be derived for assessment endpoint within each region. Examples of assessment questions are listed in Table 13.1.

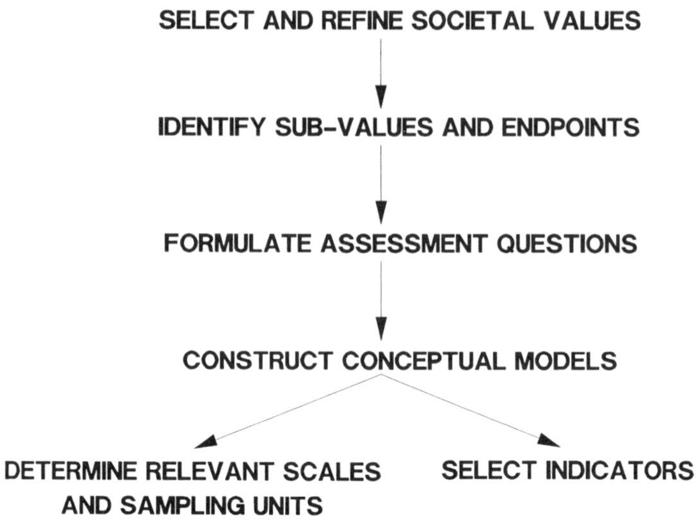

Figure 13.2. A top-down approach for designing regional level monitoring.

Once assessment endpoints and questions are identified, it remains to develop conceptual models around each assessment endpoint; these models qualitatively define how components of biodiversity function, including interactions among components and critical pathways and scales (see Figure 13.3 for an example). Some of these models are based on published accounts (e.g., MacArthur-Wilson equilibrium model, MacArthur and Wilson 1967), derivations of published accounts, combinations of models (e.g., hydrological-ecological), and in some cases, new models. Indicators and measurements are then selected from key pathways or components within conceptual models; generally, they are sensitive to fundamental changes in function, structure, or composition (Hunsaker et al. 1990b). The scales and ecological units upon which indicators are measured are also derived from critical pathways and components identified by conceptual models (Figure 13.3). Because different assessment endpoints of biodiversity likely have different critical pathways and elements (as defined by conceptual models for each), regional detection level monitoring will probably involve multiple scales and ecological classifications (e.g., watersheds and biogeographic regions).

A series of tests and evaluations must be conducted to evaluate each indicator's potential to determine status and trends in components of biodiversity. To do so requires development of a list of selection criteria by which indicators will be evaluated. Table 13.2 lists examples of criteria from Schaeffer et al. (1988) and Seidl and Murray (1991).

Types of Indicators

Of primary importance is selection of a small suite of indicators that integrates many aspects of condition related to biodiversity. Generally, a suite of indicators representing

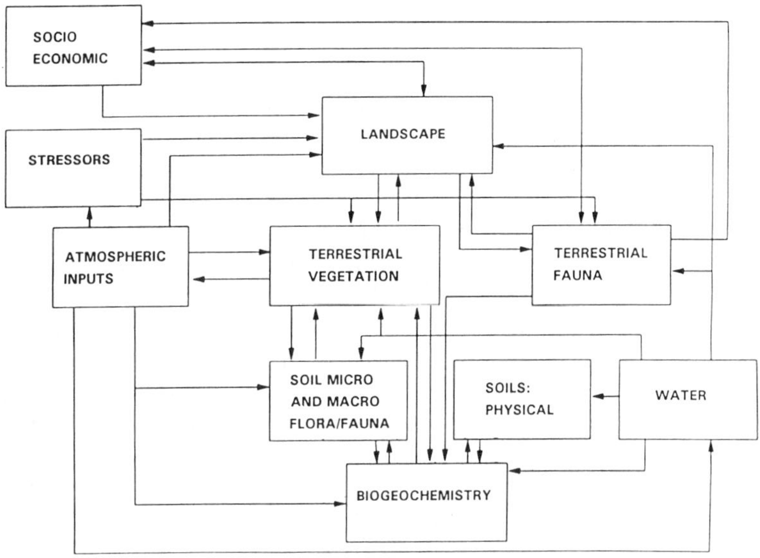

Figure 13.3. An example of a general conceptual model.

Table 13.2. Potential criteria for selection of indicators (adapted from Schaeffer et al. (1988) and Seidl and Murray (1991)).

Attribute	Description
Biologically relevant	Related to structure, composition, function of biological diversity and ecosystems
Sensitive	Responsive to stressors of concern
Geographic coverage	Applicable over large geographic areas
Diagnostic	Helps uncover potential source of the problem
Low measurement error	Measurement error contributes little to overall variation (e.g., signal to noise)
Interpretable	Unambiguous with respect to distinguishing between acceptable and unacceptable conditions
Cost-effective	Relatively inexpensive to measure with the maximum amount of data per unit effort. Is not dependent upon a census or inventory of many species
Integrative	Represent a few to several important processes of biological diversity and ecosystems
Historical database	Historical data that allows for establishing nominal variability, trends, and possibly acceptable and unacceptable condition
Anticipatory	Provides an early warning of declining conditions before serious harm has occurred
Nondestructive	Data collection methods do not impair biological diversity and ecosystems such that the methods do not change the trajectory of trends
Capable of scaling	Provides an ability to aggregate through temporal and spatial scales
Synergistic	Adds value or interpretability to other indicators being measured (suite of indicators concept) yet is sufficiently different to provide some unique information. Is not dependent upon presence, absence, or condition of a single species

a series of critical structural, compositional, and functional attributes is desirable (Rapport 1985; Noss 1990; Seidl and Murray 1991; Walker 1992). This results in a coarse monitor of biodiversity on regional scales but with minimal ability to explain the nature or cause of status and trends. Ability to determine the nature of change and potential association to stressors is achieved through the diagnostic component of our three-component approach. We discuss three basic types of indicators and indices, although all three are interrelated: (1) species metrics, (2) metrics that integrate composition, structure and functions of biodiversity, and (3) ecological indicators applied at landscape scales. Similar to Franklin (1993), we argue that status and trends in indicators of habitats, ecosystems, and landscapes provide the best monitors of biodiversity on a regional scale.

Species Metrics. Status and trends in metrics related to species, including the number of species or species richness, species diversity indices (e.g., Shannon-Weaver). The number or percentage of federally-threatened or endangered species have been used widely to assess the condition of biodiversity (Noss 1990). Unfortunately, emphasis has been limited mostly to plants and vertebrates (Franklin 1993), with the latter the best represented. Invertebrates, fungi, and bacteria have typically been omitted from such assessments, yet they comprise approximately 90% of the species (Franklin 1993). Adding these additional taxa to improve our estimates of species richness and diversity would require considerable additional resources.

Additionally, species richness and diversity estimated from a "biological species" classification might not identify those areas undergoing the most significant change in diversity. It might be more important to know losses of taxa that share few close relatives. A phylogeographic approach (Avise 1992) has been used to determine the relative uniqueness of taxa within an area. Phylogenetic relationships are derived through a comparative approach using molecular (e.g., mtDNA) or morphological data (see Riddle and Jones this volume).

Integrated Metrics of Biodiversity. The guild concept is a method of evaluating composition, structure, and function of biodiversity. Numerous definitions of guilds exist (Hawkins and MacMahon 1989), and these have lead to inconsistent results in their application. Generally, a guild is defined as "a group of species that exploit the same class of environmental resources in a similar way" (Root 1967). Most recent application of the definition involves defining guilds based on breeding and feeding habitat and/or behavior (Landres and MacMahon 1980; Short and Burham 1982; Short and Williamson 1987; Brown and Kurzius 1989; Hawkins and MacMahon 1989). The idea is to determine a guild structure for the taxa of interest and then sample representative species of each guild (this was also known as the "Indicator Species" approach). This allows one to reduce the number of species sampled. This application of the guild concept has some significant drawbacks, including the need to have extremely detailed information on the life-histories of all taxa being considered, and community dynamics (e.g., competition). Furthermore, spatial and temporal variability in community dynamics (and, hence, guild composition) makes it difficult to use individual species to represent entire guilds.

The other application of the guild approach hypothesizes that losses of guilds represent significant changes in ecological function and condition. The primary issue with this approach is whether it would be more efficient to measure ecological function directly. However, Walker (1992) suggests that guild analysis might provide a measurement of redundancy in ecosystems; this condition might not be detected through other ecological measures. Therefore, change in guild composition and structure might be a useful measure of biological function and structure at the regional scale.

The Index of Biotic Integrity (IBI) combines a series of structural, compositional, and functional attributes of aquatic biota into one value (Karr 1981). The IBI is receiving greater attention in aquatic monitoring programs and, hopefully, a similar index can be developed for terrestrial ecosystems.

Ecological Indicators at Landscape Scales. Ecological indicators might be the most important monitors of biodiversity on a regional scale. Franklin (1993) states that "larger-scale approaches -at the level of ecosystems and landscapes- are the only way to conserve the overwhelming mass- the millions of species- of existing biodiversity."

Indicators of configuration of habitats and ecosystems on a regional scale can provide an important measure of status and trends in biodiversity. As mentioned earlier, ecosystem and habitat fragmentation, losses of or vast reductions in certain ecosystems, and simplification of landscapes are implicated in loss of biodiversity. Landscape indicators capture fundamental changes in habitat and ecosystem configuration and can be calculated for large areas at multiple scales from remote sensing imagery. O'Neill et al. (1993) developed a multi-scaled approach which would evaluate landscape status and trends on a regional scale using a variety of remote sensing and existing data. Data on landscape configuration from this program could be evaluated against changes in indicators of biodiversity at the regional scale. This approach would allow us to correlate status and trends in indicators of biodiversity with indicators of landscape pattern, such as fractal dimensions (a measure of shape complexity), contagion (a measure of clumping or clustering of habitats), percolation (a measure of habitat connectivity), and land-use patterns and dominance (see O'Neill et al. 1988). The ability to evaluate landscape pattern at multiple scales is critical because species diversity patterns result from ecological and biological processes operating within and among a variety of scales (Addicott et al. 1987; Ricklefs 1987, 1989; Holling 1992).

A landscape approach also provides a mechanism to evaluate species diversity and ecological indicators within and among traditional ecological classifications. Not only is the pattern and organization of ecosystems considered in a landscape approach, but also the flow of nutrients, energy, and species among landscape components (Costanza et al. 1990). This departs from the typical evaluation of indicators within ecosystem types (e.g., a forest, an agroecosystem, a wetland to name a few). Status and trends of physical and functional relationships, such as water movement, among individual ecosystems (e.g., wetlands) within an area (e.g., watershed) are now being used to evaluate individual ecosystem health and vulnerability (Preston and Beford 1988; Kesner and Meentemeyer 1989; Knight and Wallace 1989; Costanza et al. 1990; Gosselink et al. 1990; Schlosser 1991; Skopek et al. 1991; Fox 1992).

Finally, it might be possible to use landscape data to assess relative risks to biodiversity (EPA 1992), and to develop regional strategies to conserve biodiversity and enhance ecosystem sustainability.

It might be necessary to supplement landscape data with status and trends data on vertical structure and species composition. Although satellite data might detect changes in the spatial configuration of certain ecological resources (e.g., forest patches), generally, they are insufficient to detect changes in plant species composition and structure. Changes in plant species composition and structure can have dramatic effects on animal community structure and ecological processes. Therefore, it might be important to include measures of vertical vegetation and composition in regional detection level monitoring (Short 1992).

A combination of remote sensing and field data measurements is being applied congruently to evaluate condition of ecological function over large regions (Mouat et al. 1992). For example, surface reflectance values derived from satellite imagery (e.g., Landsat) can be used as an indicator of photosynthetic activity in plants (Mouat et al. 1992). Shifts in spectral absorption curves can indicate reduced energy assimilation by plants and declining plant condition. The satellite integrates the response of several plants over the minimal area detected by the satellite (e.g., 30 m × 30 m area using Landsat-Thematic Mapper). The key to this approach is understanding the information content of the minimum area of resolution (e.g., the satellite pixel)—how does the spectral range observed relate to light absorbed by plants? This is accomplished by relating spectral values for pixels with values of plant absorption that are derived simultaneously from field spectral measurements. Other regional scale estimates of ecological function, such as net primary productivity (Burke et al. 1991), have been derived from a combination of field measurements and satellite imagery. The primary benefit of this approach is the widespread coverage of the United States with satellite imagery and its relatively low cost (as compared to field sampling). These types of measurements can add significantly to our ability to perform detection level monitoring of biodiversity on a regional scale.

SAMPLING DESIGN CONSIDERATIONS FOR REGIONAL SCALE MONITORING OF BIODIVERSITY

Once decisions have been made about which indicators of biodiversity will be monitored, it remains to decide on an appropriate monitoring design to address the series of assessment questions. Several important considerations are related to sampling designs. One of the fundamental decisions to be made is whether to attempt to monitor all entities or whether to attempt to use a sample-based approach. Rarely on a regional scale will it be possible to measure all populations or communities or attributes about populations or communities; therefore, a sampling strategy must be devised. Exceptions to this include monitoring of species' populations occurring over a large area but only on a few sites, and ecological measurements obtained by remote sensing (Burke et al. 1991; Scott et al. 1993).

A statistically-based sampling design, such as a probability sample, increases the chances that an unbiased sample of known quality is achieved (provided standard sampling protocols and quality assurance procedures are followed). EMAP (Messer et al. 1991) and several resource survey programs in the United States (e.g., the Forest Inventory and Analysis, Ohmann 1990) utilize a statistically-based probability sample to produce estimates of status and trends. EMAP, as far as we are aware, is the only monitoring program to apply such a design to estimate status and trends of ecological indicators on a regional scale. One of the key issues of this approach is defining the types of statistical populations to be sampled. These can be populations of organisms, communities or ecological resources (e.g., wetlands) distributed over a region. If an ecological resource or community type is sampled, one must assume that these are discrete ecological or biological resources.

A non-statistically-based sampling procedure usually involves the selection of ecologically-representative sites which are then used to represent the remainder of similar sites within the region. This is the fundamental assumption applied to the LTER program (Franklin et al. 1990). Fundamental to selecting a design is whether or not to use a stratified random sample or a systematic random sample. The former requires a decision on a classification system a priori, whereas the latter only requires identification of statistical populations. Few ecological classification systems meet a broad number of needs. For example, watersheds can be used to describe water movement through a landscape, whereas a biogeographical region might be more appropriate for describing the distribution of and association among species. Therefore, regional-scale monitoring programs attempting to address a series of issues might find it difficult to stratify samples based on a single classification system. EMAP is attempting to address a series of regional-scale issues, and it employs a systematic random grid consisting of evenly spaced points projected onto the earth's surface (Messer et al. 1991). A random systematic grid eliminates the requirement of deciding, a priori, on a classification scheme, yet allows the monitoring program to address a number of issues at the regional scale. However, a hierarchical classification strategy based on multi-scaled ecological factors (climate, topography, soils, etc.), such as that of Bailey (1988, 1991), might provide a stratification approach suitable to most ecological and biological monitoring. Bailey's classification hierarchy is based on physical and ecological factors that produce general ecosystem patterns at multiple scales over the entire globe. Until we learn more about relationships between ecosystem processes and diversity at and among a variety of scales, monitoring should occur concomitantly at a variety of spatial and temporal scales.

Another major question with regard to sampling design at regional scales is the number of samples necessary to represent ecological or biological resources of interest (e.g., via indicators). This question can be addressed by conducting simulations and field tests of indicator variability and sensitivity (see Larsen and Christie 1993 for examples).

CONCLUSIONS

We recommend a multiple scale monitoring approach for addressing components of biodiversity. This includes: (1) a broad, regional scale monitoring component, (2) a diagnostic component to evaluate, in greater detail, areas meeting specific status and change criteria, and (3) a cyclic-incremental component to evaluate effectiveness of specific management prescriptions in achieving desired conditions in components of biodiversity. The regional component should include measurement of a small suite of indicators that integrates several key components of biodiversity.

We suggest a rigorous process for selecting and evaluating indicators of biodiversity and habitats, such as that developed for the Environmental Monitoring and Assessment and the Great Lakes programs.

Disclaimer

The information in this document has been funded in part by the United States Environmental Protection Agency. It has been subjected to Agency review and approved for publication. Mention of trade names or commercial products does not constitute endorsement or recommendation for use.

LITERATURE CITED

Addicott, J. F., J. M. Ahoy, M. F. Antolin, D. K. Padilla, J. S. Richardson, and D. A. Soluk. 1987. Ecological neighborhoods: scaling environmental patterns. *Oikos* 49:340-346.

Avise, J. C. 1992. Molecular population structure and the biogeographic history of a regional fauna: a case history with lessons for conservation biology. *Oikos* 63:62-76.

Bailey, R. G. 1988. Ecogeographic analysis: a guide to the ecological division of land for resource management. U.S. Forest Serv., Misc. Publ. No. 1465., Washington, D.C.

Bailey, R. G. 1991. Design of ecological networks for monitoring global change. *Environ. Conserv.* 18:173-175.

Baron, J., and K. A. Galvin. 1990. Future directions of ecosystem science: toward an understanding of the global biological environment. *BioScience* 40:640-642.

Brown, J. H., and B. A. Maurer. 1989. Macroecology: the division of food and space among species on continents. *Science* 243:1145-1150.

Brown, J. H., and M. Kurzius. 1989. Spatial and temporal variation in guilds of North American granivorous desert rodents. In *Patterns in the Structure of Mammalian Communities*, eds. D. W. Morris, Z. Abramsky, B. J. Fox, and M. R. Willig, pp. 71-90. Spec. Publ. Mus., Texas Tech. Univ., No. 28, Lubbock, Tex.

Brown, J. H., and J. Roughgarden. 1990. Ecology of a changing earth. In *In Our Hands: Survival on Earth*, eds. J. Roughgarden, J. Brown, E. Lehman, B. Mendelsohn, and J. Unruth, pp. 173-186. Stanford University Press, Palo Alto, Calif.

Burke, I. C., T. G. F. Kittel, W. K. Lauenroth, P. Snook, C. M. Yonker, and W. J. Parton. 1991. Regional analysis of the Central Great Plains. *BioScience* 41:685-692.

Costanza, R., F. H. Sklar, and M. L. White. 1990. Modeling coastal landscape dynamics. *BioScience* 40:91-107.

Delcourt, H. R., and P. A. Delcourt. 1988. Quaternary landscape ecology: relevant scales in space and time. *Landscape Ecol.* 2(1):23-44.

Droege, S., and J. R. Sauer. 1989. Annual summary of North American breeding bird survey. U.S. Fish and Wildlife Service, Office of Migratory Bird Management, Laurel, Md.

Ehrlich, P. R. 1992. Population biology of checkerspot butterflies and the preservation of global biodiversity. *Oikos* 63:6-12.

Environmental Monitoring and Assessment Program (EMAP). 1993. Environmental Monitoring and Assessment Program Guide. In Press.

Fox, J. 1992. The problem of scale in community resource management. *Environ. Manage.* 16:289-297.

Franklin, J. F. 1993. Preserving biodiversity: species, ecosystems, or landscapes. *Ecol. Appl.* 3:202-205.

Franklin, J. F., C. S. Bledsoe, and J. T. Callahan. 1990. Contributions of the Long-term Ecological Research Program. *BioScience* 40:509-523.

Gosselink, J. G., G. P. Shaffer, L. C. Lee, D. M. Burdick, D. L. Childers, N. C. Leibowitz, S. C. Hamilton, R. Boumans, D. Cushman, S. Fields, M. Koch, and J. M. Visser. 1990. Landscape conservation in a forested wetland watershed. *BioScience* 40:588-600.

Graham, R. L., C. T. Hunsaker, R. V. O'Neill, and B. L. Jackson. 1991. Ecological risk assessment at the regional scale. *Ecol. Appl.* 1:196-206.

Hawkins, C. P., and J. A. MacMahon. 1989. Guilds: the multiple meanings of a concept. *Ann. Rev. Entomol.* 34:423-451.

Holling, C. S. 1992. Cross-scale morphology, geometry, and dynamics of ecosystems. *Ecol. Monogr.* 62:447-502.

Hubbell, S. P., and R. B. Foster. 1992. Short-term dynamics of a neotropical forest: why ecological research matters to tropical conservation and management. *Oikos* 63:48-61.

Hunsaker, C. T., R. L. Graham, G. W. Suter, Jr., R. V. O'Neill, L. W. Barnthouse, and R. H. Gardner. 1990a. Assessing ecological risk on a regional scale. *Environ. Manage.* 14:325-332.

Hunsaker, C. T., D. Carpenter, and J. Messer. 1990b. Ecological indicators for regional monitoring. *Bull. Ecol. Soc. Am.* 71:165-172.

Jones, K. B. 1986. Data types. In *Inventory and Monitoring of Wildlife Habitat*, ed. A. Cooperrider, pp. 11-28. U.S. Bureau of Land Management, Denver, Colo.

Karr, J. R. 1981. Assessment of biotic integrity using fish communities. *Fisheries* 6(6):21-26.

Kesner, B. T., and V. Meentemeyer. 1989. A regional analysis of total nitrogen in an agricultural landscape. *Landscape Ecol.* 2:151-163.

Knapp, C. M., D. R. Marmorek, J. P. Baker, K. W. Thorton, J. M. Klopatek, and D. P. Charles. 1991. The indicator and development strategy for the Environmental Monitoring and Assessment Program. EPA/600/3-91/023. U.S. EPA, Environ. Res. Lab, Corvallis, Ore.

Knight, D. H., and L. L. Wallace. 1989. The Yellowstone fires: issues in landscape ecology. *BioScience* 39:700-706.

Landres, P. B., and J. A. MacMahon. 1980. Guilds and community organization: analysis of an oak woodland avifauna in Sonora, Mexico. *Auk* 97:351-365.

Larsen, D. P., and S. J. Christie, eds. 1993. EMAP-Surface 1991 Pilot Report. EPA/620/R-93/003, Washington, D.C.

Lubchenco, J., A. M. Olson, L. B. Brubaker, S. R. Carpenter, M. M. Holland, S. P. Hubbell, S. A. Levin, J. A. MacMahon, P. A. Matson, J. M. Melillo, H. A. Mooney, C. H. Peterson, H. R. Pulliam, L. A. Real, P. J. Regal, and P. G. Risser. 1991. The sustainable biosphere initiative: an ecological research agenda. *Ecology* 72:371-412.

MacArthur, R. H., and E. O. Wilson. 1967. *The Theory of Island Biogeography*. Princeton University Press, Princeton, N. J.

Magnuson, J. J. 1990. Long-term ecological research and the invisible present. *BioScience* 40:495-501.

Messer, J. J., R. A. Linthurst, and W. S. Overton. 1991. An EPA program for monitoring ecological status and trends. *Environ. Mon. Assess.* 17:67-78.

Mooney, H. A. 1991. Emergence of the study of global ecology: is terrestrial ecology an impediment to progress? *Ecol. Appl.* 1:2-5.

Mouat, D., C. A. Fox, and M. R. Rose. 1992. Ecological indicator strategy for monitoring arid ecosystems. In *Ecological Indicators*, eds. D. H. McKenzie, D. E. Hyatt, and V. J. McDonald, pp. 717-738. Elsevier Applied Science, London.

Noss, R. F. 1990. Indicators for monitoring biodiversity: a hierarchical approach. *Conserv. Biol.* 4:355-364.

Ohmann, J. L. 1990. Multiresource inventories in the Pacific Coast states—progress and future direction. In *Proceedings for the Symposium on the State-of-the-art Methodology for Forest Inventory*, eds. V. J. LaBau and T. Cunia, pp. 529-537. USDA Forest Service Gen. Tech. Rpt. PNW-GTR-263, Portland, Ore.

O'Neill, R. V., J. R. Krummel, R. H. Gardner, G. Sugihara, B. Jackson, D. L. DeAngelis, B. T. Milne, M. G. Turner, B. Zygmunt, S. W. Christensen, V. H. Dale, and R. L. Graham. 1988. Indices of landscape pattern. *Landscape Ecol.* 1:153-162.

O'Neill, R. V., C. T. Hunsaker, and D. A. Levine. 1992. Monitoring challenges and innovative ideas. In *Ecological Indicators*, eds. D. H. McKenzie, D. E. Hyatt, and V. J. McDonald, pp. 1443-1460. Elsevier Applied Science, London.

O'Neill, R. V., S. J. Turner, V. I. Cullinan, D. P. Coffin, T. Cook, W. Conley, J. Brunt, J. M. Thomas, M. R. Conley, and J. Gosz. 1991. Multiple landscape scales: an intersite comparison. *Landscape Ecol.* 5:137-144.

O'Neill, R. V., J. Wickham, K. H. Riitters, I. Goodman, and K. B. Jones. 1993. A research strategy to implement a national landscape monitoring program. U.S. EPA, Environ. Mon. Sys. Lab, Las Vegas, Nev.

Office of Technology Assessment (OTA). 1987. *Technologies to Maintain Biological Diversity.* OTA-F-330. Government Printing Office, Washington, D.C.

Preston, E. M., and B. L. Bedford. 1988. Evaluating cumulative effects on wetland functions: a conceptual overview and generic framework. *Environ. Manage.* 12:565-583.

Rapport, D. J. 1985. Ecosystem behavior under stress. *Am. Nat.* 125:617-640.

Rapport, D. J. 1993. Evaluating ecosystem health. *J. Aquatic Ecosystem Health* in press.

Ricklefs, R. E. 1987. Community diversity: relative roles of local and regional processes. *Science* 235:167-171.

Ricklefs, R. E. 1989. Speciation and diversity: the integration of local and regional processes. In *Speciation and its Consequences*, eds. D. Otte and J. A. Endler, pp. 599-622. Sinauer Associates, Sunderland, Mass.

Riitters, K. H., B. E. Law, R. C. Kucera, A. L. Gallant, R. L. DeVelice, and C. J. Palmer. 1992. A selection of forest condition indicators for monitoring. *Environ. Mon. Assess.* 20:21-33.

Romme, W. H., and M. G. Turner. 1991. Implications of global climate change for biogeographic patterns in the greater Yellowstone ecosystem. *Conserv. Biol.* 5:373-386.

Root, R. B. 1967. The niche exploitation pattern of the blue-gray gnatcatcher. *Ecol. Monogr.* 37:317-350.

Saunders, D. A., R. J. Hobbs, and C. R. Margules. 1991. Biological consequences of ecosystem fragmentation: a review. *Conserv. Biol.* 5(1):18-32.

Schaeffer, D. J., E. E. Herricks, and H. W. Kerster. 1988. Ecosystem health: I. Measuring ecosystem health. *Environ. Manage.* 12:445-455.

Schlesinger, W. H., J. F. Reynolds, G. L. Cunningham, L. F. Huenneke, W. M. Jarrell, R. A. Virginia, and W. G. Whitford. 1990. Biological feedbacks in global desertification. *Science* 247:1043-1048.

Schlosser, I. J. 1991. Stream fish ecology: a landscape perspective. *BioScience* 41:704-711.

Scott, J. M., F. Davis, B. Csuti, R. Noss, B. Butterfield, G. Groves, H. Anderson, S. Caicco, F. D'Erchia, T. C. Edward, Jr., J. Ulliman, and R. G. Wright. 1993. GAP Analysis: a geographic approach to protection of biological diversity. *Wildl. Monogr.* 123:1-41.

Seidl, P., and P. Murray, eds. 1991. A proposed framework for developing indicators of ecosystem health for the Great Lakes region. Council of Great Lakes Res. Man, International Joint Commission, United States and Canada.

Short, H. L. 1992. Use of the habitat linear appraisal system to inventory and monitor the structure of habitats. In *Ecological Indicators*, eds. D. H. McKenzie, D. E. Hyatt, and V. J. McDonald, pp. 961-974. Elsevier Applied Science, London.

Short, H. L., and K. P. Burnham. 1982. Technique for structuring wildlife guilds to evaluate impacts on wildlife communities. U.S. Fish and Wildlife Service, Spec. Sci. Rep. Wildl., No. 244.

Short, H. L., and S. C. Williamson. 1987. Evaluating the structure of habitat for wildlife. Part I: Development, testing and application of wildlife-habitat models, U.S. Fish and Wildlife Service, Fort Collins, Colo.

Skopek, V., J. Vachal, and Z. Sterbacek. 1991. A method of approach to landscape stability. Part 1: fundamentals and methodology. *Environ. Manage.* 15:205-214.

Soulé, M. E., A. C. Alberts, and D. T. Bolger. 1992. The effects of habitat fragmentation on chaparral plants and vertebrates. *Oikos* 63:39-47.

Suter, G. W., Jr. 1990. Endpoints for regional ecological risk assessment. *Environ. Manage.* 14:9-23.

Swanson, F. J., and R. E. Sparks. 1990. Long-term ecological research and the invisible place. *BioScience* 40:502-508.

Turner, M. G., R. H. Gardner, V. H. Dale, and R. V. O'Neill. 1989. Predicting the spread of disturbance across heterogeneous landscapes. *Oikos* 55:121-129.

Turner, M. G., R. H. Gardner, and R. V. O'Neill. 1991. Potential responses of landscape structure to global environmental change. In *Ecotones: The Role of Landscape Boundaries in the Management and Restoration of Changing Environments*, eds., M. M. Holland, P. G. Risser, and R. J. Naiman, pp. 52-75. Chapman and Hall, New York.

Ulfstrand, S. 1992. Biodiversity—how to reduce its decline. *Oikos* 63:3-4.

United Nations Environmental Programme (UNEP). 1992. *World Atlas of Desertification*. Edward Arnold, London.

U.S. Environmental Protection Agency (EPA). 1992. Synoptic national assessment of comparative risks to biological diversity and landscape types. U.S. EPA, Environ. Res. Lab., Corvallis, Ore.

Vane-Wright, R. I., C. J. Humphries, and P. H. Williams. 1991. What to protect?—Systematics and the agony of choice. *Biol. Conserv.* 55:235-254.

Walker, B. H. 1992. Biodiversity and ecological redundancy. *Conserv. Biol.* 6:18-23.

Wiens, J. A. 1986. Spatial scale and temporal variation in studies of shrubsteppe birds. In *Community Ecology*, eds. J. Diamond and T. Case, pp. 154-172. Harper and Row, New York.

Wilson, E. O. 1985. The biological diversity crisis: a challenge to science. *Iss. Sci. Technol.* 2(1):20-29.

Yoon, C. K. 1993. Counting creatures great and small. *Science* 260:620-622.

14

COMPARATIVE UTILITY OF VEGETATION MAPS OF DIFFERENT RESOLUTIONS FOR CONSERVATION PLANNING

Peter A. Stine and Frank W. Davis

Department of Geography
University of California
Santa Barbara, California
U.S.A.

Blair Csuti

Idaho Cooperative Fish and Wildlife Research Unit
College of Forestry, Wildlife and Range Sciences
University of Idaho
Moscow, Idaho
U.S.A.

J. Michael Scott

U.S. Fish and Wildlife Service
College of Forestry, Wildlife and Range Sciences
University of Idaho
Moscow, Idaho
U.S.A.

Most conservation efforts have addressed immediate, local issues related to impacts of development on selected native species and habitats. Unfortunately, these efforts are being conducted without a regional perspective on the distribution and conservation status of the biological resources. This often results in incoherent or poorly coordinated conservation planning and misallocation of financial resources.

Gap analysis seeks to provide a regional conservation planning perspective by using relatively coarse distribution maps to assess the protection status of plant communities and vertebrate species (Scott et al. 1991). "Gaps" are those natural communities or species not adequately represented in existing reserves (or, more broadly, in areas managed for the long-term maintenance of biodiversity). One goal of gap analysis is to help identify and prioritize areas, communities, and species in need of more detailed conservation planning. The data are also used to locate areas where these systems or species are best represented (both in terms of quantity and quality).

Gap analysis uses a relatively large, 100-ha minimum mapping unit (mmu) for mapping vegetation, and it is important to understand how well this coarse resolution representation serves for conservation planning. Our approach in California has been to compare our map to local, larger-scale vegetation maps in terms of the distribution, amount, and modeled conservation status of plant communities.

In this paper we present preliminary results of such a comparison between the gap analysis map of coastal sage scrub in southwestern California and a much larger scale map (1 ha mmu) prepared concurrently by an independent group, Regional Environmental Consultants Inc. (RECON). First, we summarize the status of biodiversity planning in the study region of southern California. Next we describe the methods used to prepare each map and quantify the differences between these two representations of the resource.

BIODIVERSITY PLANNING AND COASTAL SAGE SCRUB IN SOUTHERN CALIFORNIA

Increasing human population growth and further land conversion in California have triggered regulatory requirements to mitigate impacts and some voluntary long-range planning. In many areas of the state, efforts have ranged from protecting local species or communities in areas of less than one hectare to multi-agency efforts which address multiple species and communities covering regions of tens to thousands of square kilometers.

The Gap Analysis Program in California is attempting to provide information on the current status of vegetation communities and vertebrate species within the geographical context of ecoregions (Davis and Stoms 1992). Ecoregions provide logical geographic boundaries for conservation planning at the landscape level, encompassing the habitat needs of most component plant and animal species or populations as well as community and ecosystem processes. We have adopted the ecoregion system described in the revised Jepson Flora of California (Hickman, 1993). This is a hierarchical system that describes 10 ecological regions for the state based on climate, geology, physiography, and general vegetation cover.

Our focus here is on the southwestern ecoregion, delineated by the Santa Ynez mountains at the north, the Transverse Ranges to the north and east, and the Peninsular Ranges to the east, covering some 34,000 km^2 (Figure 14.1). It is an area of considerable and immediate interest for several reasons. Almost 20 million people live within the bounds of this ecoregion which includes some of the most rapidly urbanizing areas in the United States. Ongoing major habitat destruction and fragmentation, compounded with significant historical losses because of widespread agricultural development, have placed many native habitats and species in jeopardy.

One of the critical conservation issues within the southwestern ecoregion is loss of the coastal sage scrub plant community. Coastal sage scrub is characterized by a series of structurally similar, facultatively drought-deciduous shrublands found along the coast

Figure 14.1. Ecoregions of California (based upon the Jepson Manual: Higher Plants of California).

of central and southern California, and south to El Rosario in Baja California (Westman 1981). Significant attention has been focused on this community recently with the federal listing (under the Endangered Species Act) of the California gnatcatcher (*Polioptila californica californica)* (Federal Register 1993). This species, as well as several others considered "candidates" for either federal or state listing, depends in large measure on the availability of certain floristic associations of undisturbed coastal sage scrub habitat (Atwood 1990).

Coastal sage scrub in the southwestern ecoregion is generally restricted to elevations below 800 m. Although the current limits of coastal sage scrub are fairly well established, the original distribution of this community is not well known. Kuchler (1964) estimated that some 12,000 km^2 of coastal scrub existed or potentially existed, based on his map of natural vegetation of the United States. Estimates of the losses over the last 100 or more years range from 36% (Klopatek et al. 1979) to 85% (Westman 1981) of the projected original extent of this community. Losses have accelerated in the last decade, and despite recent, well organized efforts to plan for the conservation of coastal sage scrub, significant acreages of this habitat continue to be converted to human uses.

Five floristic associations are recognized as coastal sage scrub along the California coast, as well as two additional associations in Baja California that are described as coastal succulent scrub (O'Leary 1990). The Venturan, Riversidian and Diegan associations occur exclusively within the southwestern ecoregion. The Venturan association is a coastal form, extending from Santa Barbara County south through coastal Los Angeles County. The Riversidian association is a more interior form, extending some 150 km inland in Los Angeles County through Riverside and interior San Diego Counties to the lower slopes of the mountains adjoining the desert. The Diegan association extends from Orange County along the coast to just south of Ensenada (Westman 1983). Conservation efforts to date have focused on the Diegan association, and to a lesser extent on the Riversidian association. This is largely because of the concern for the regulatory requirements associated with the conservation of the gnatcatcher and the conflict with land development activity in these areas.

METHODS AND RESULTS

Mapping Coastal Sage Scrub—Gap Analysis Technique

Vegetation maps for the gap analysis program are being prepared using a minimum mapping unit of 100 ha and a Series-level classification of vegetation based on vegetation structure and dominant or co-dominant overstory species. In California, our map units or "polygons" are conceptualized as landscapes containing vegetation mosaics. A landscape is defined as a kilometers-wide area of uniform climate, geology and physiography containing repeating vegetation, soil, and land use complexes (Forman and Godron 1986). For each map polygon, we identify a primary (most widespread) upland vegetation series, a secondary upland series, and the most prominent wetland series. Each series is defined by up to three co-dominant species, and data are included on crown closure and areal extent (proportion of polygon).

LANDSAT Thematic Mapper imagery is visually interpreted to locate landscape boundaries. Ancillary data, such as existing large-scale vegetation maps, aerial photography, and ground surveys, are used to label vegetation units according to floristic composition and structure. We should note that the RECON map was not used in developing the gap analysis map of the study area.

Conservation planning for coastal sage scrub has focused on southern Orange County and the western portions of Riverside and San Diego Counties. In the vegetation data base developed for this three-county area, we have identified approximately 800 vegetation polygons (i.e., landscape units). Of these 800 polygons, 130 contain coastal sage scrub as the primary (i.e., dominant cover of polygon) vegetation type, with one of four plant species dominating the association (*Artemesia californica*, *Eriogonum fasciculatum*, *Encelia californica*, and *Salvia apiana*). Coastal sage scrub was the secondary vegetation type (i.e., secondary cover, 40% or less of any polygon) in 143 polygons. A fifth species, *Salvia mellifera*, occurs as a dominant in one additional secondary vegetation type. Coastal scrub was the primary and/or secondary vegetation in a total of 262 polygons (Table 14.1). The distribution of coastal sage scrub polygons in the study area is shown in Figure 14.2.

Table 14.1. Summary of Coastal Sage Scrub Habitat as Detected by the Gap Analysis Technique (in hectares)

	Artemisia	*Eriogonum*	*Encelia*	*Salvia a.*	*Salvia m.*
Primary[1]	100,258	61,705	681	9,133	----
Number	77	46	1	6	----
Area Occupied[2]	63,045	37,182	681	5,433	----
Mean	823	808	681	905	
Secondary[1]	92,564	88,969	1,647	34,292	1,697
Number	65	58	1	17	2
Area Occupied[2]	24,289	25,743	494	7,610	512
Mean	373	443	494	447	256
Total Area (by type)	87,694	52,925	903	13,043	512
Grand Total	155,077				

[1] Primary and Secondary refer to relative proportion of the polygon dominated by the stated vegetation type. See text for further detail.

[2] Area Occupied refers to the estimated actual coverage of a given polygon by the stated vegetation type (i.e. the size of the polygon multiplied by the percentage of the polygon covered by the type).

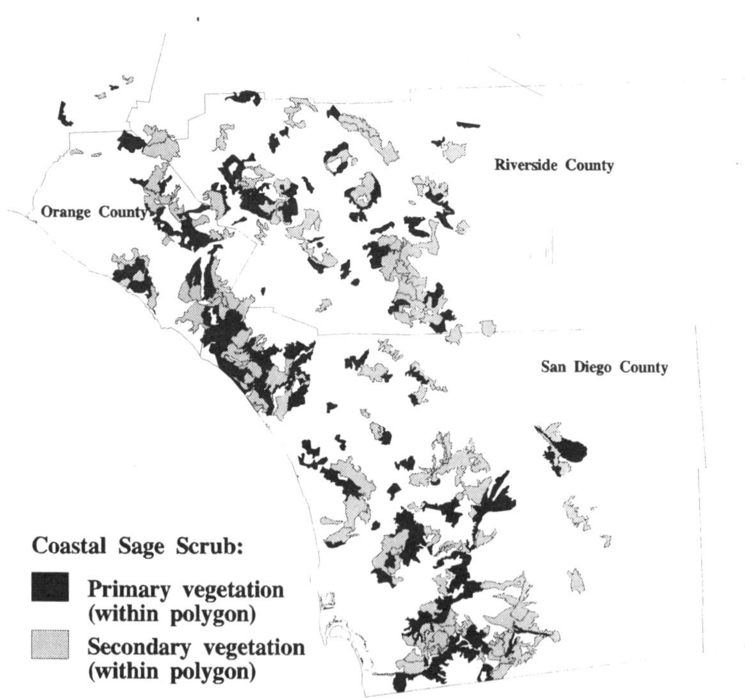

Figure 14.2. Landscape units containing extant coastal sage scrub; southwestern ecoregion as mapped by Gap Analysis Project.

The data in Table 14.1 represent all forms of coastal sage scrub we encountered, including stands that had some mixture of other vegetation, such as non-native annual grassland. These landscape units also include many areas that are disturbed and subject to continuing human impacts such as off-road vehicle use, fire, encroaching urbanization, and grazing by livestock. These data also include areas of coastal sage scrub that have a strong desert influence and have substantially different climatic conditions and floristic composition, particularly many *Eriogonum* dominated regions.

Although on the surface it appears that coastal sage scrub is still widespread and relatively common, it has been highly fragmented, and this habitat has declined substantially in the three-county area. Moreover, the landscape fragmentation complicates options for effective conservation and management of large, self-sustaining reserves of the coastal sage scrub community.

Local Scale Investigation

RECON's mapping effort was restricted to the urbanized or rapidly urbanizing area of western San Diego county. Color aerial photography (acquired in 1989, 1:48,000 scale) was used in conjunction with field surveys of approximately 75% of the areas mapped as coastal sage scrub (P. Fromer pers. comm.). The outlines of stands of coastal sage scrub defined on these aerial photos were then manually transferred to USGS 7 1/2 minute topographic maps. The criteria for mapping were similar to the gap program; however, in this case they only mapped stands of coastal sage scrub, including those stands disturbed or partly intermixed with grassland or chaparral. However, the spatial resolution is much greater; the minimum mapping unit (mmu) is approximately 2.5 acres (1 ha). This spatial resolution is approximately 100 times greater than that used for the gap analysis mapping. A summary of the data from the RECON survey appears in Table 14.2.

The contrast in level of detail provided in the mapping is shown in Table 14.2. The mean area covered in the polygons we have mapped range between 284 and 712 ha for the different primary types. The mean area of the RECON polygons (essentially homogeneous polygons) was 25.6 ha. There were almost 15 times the number of polygons as well, 1,625 RECON polygons compared with 105 (some individual polygons contain coastal sage scrub vegetation associations as both the primary and the secondary type) included within the subset of the gap coverage.

We have overlaid the RECON map on a version of our map, which combines all coastal sage scrub polygons (Figure 14.3). Results are summarized in Table 14.3.

As perhaps expected, the larger stands of coastal sage scrub are captured by both mapping efforts. The mean size of the RECON polygons that were also classified as coastal sage scrub in the Gap Analysis map is over 3 1/2 times the size of those mapped only by RECON.

Of the 1,383 polygons found outside the gap analysis polygons, 1,368 (98.9%) are less than 100 ha, the minimum mapping unit of the gap analysis effort. In fact, 1,031 (74.5%) of the coastal sage scrub polygons mapped only by the RECON effort are actually less than 10 ha and 189 (13.7%) are less than 1 ha (Figure 14.4).

Table 14.2. Comparison of Recon Map Data with Gap Analysis Data

Map of Coastal Sage Scrub for Same Three County Area (gap analysis data)

	Artemisia	*Erigonium*	*S. apiana*	*S. melli*
Primary				
Number	43	12	4	
Area	30,613	7,419	1,135	
Mean	712	618	284	
Secondary				
Number	35	16	5	1
Area	11,664	4,441	993	167
Mean	333	277	199	—
Total Area[1]	42,277	11,860	2,128	167

Recon Mapping of Coastal Sage Scrub Habitat in Western San Diego County

Polygons	Area (ha)	Mean Size	Max Size	Min Size
1,625	42,948	25.6	1,207	0.36

[1] Total area of Coastal Sage Scrub within this study region is 56,432 hectares

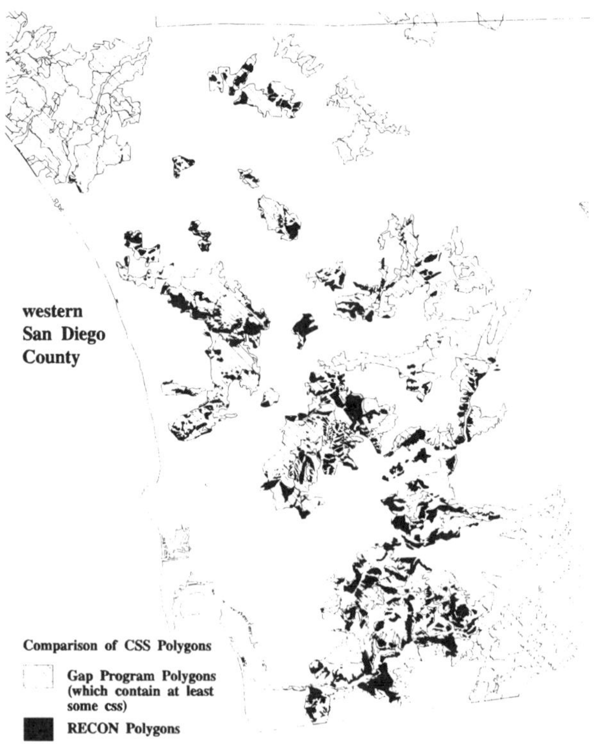

Figure 14.3. Recon polygons of CSS within Gap CSS polygons; western San Diego County (RECON map excludes Camp Pendleton—NW Corner).

Table 14.3. Comparison of Recon and Gap Polygons.

Number	Area (ha)	Mean (ha)	Max (ha)	Min (ha)
RECON Polygons *within* Gap Polygons[1]				
797	28,734	36.0	1,150	0.059
RECON Polygons *outside* Gap Polygons[1]				
1,383	14,213	10.2	426	0.01

[1] Again, the GIS operation of "clipping", where two separate maps are overlaid and the features of one map are extracted from within the outlines of the other map, can create additional artificial "sliver" polygons at the edges of the map. Although this slightly influences the figures, the comparison still illustrates the differences.

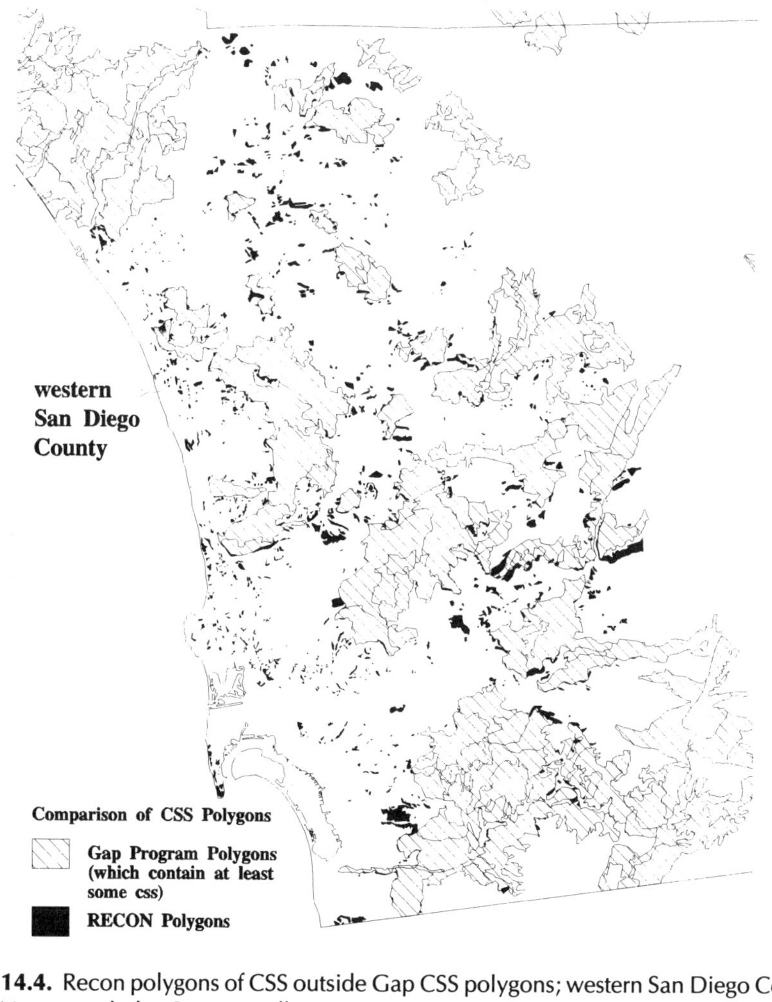

Figure 14.4. Recon polygons of CSS outside Gap CSS polygons; western San Diego County (RECON map excludes Camp Pendleton—NW Corner).

Fifteen polygons mapped by RECON and larger than 100 ha were not mapped by the Gap Analysis project. These polygons range in size from 104 to 426 ha. Possible causes for these discrepancies include mapping error, actual changes in land cover between the dates of the source data(April 1989 for RECON and June and 1990 for the Gap Analysis), and slight differences in classification rules. An examination of the 15 polygons revealed that two areas had, in fact, been cleared in the intervening 14-month period between dates of the source data. Most remaining areas were islands of coastal scrub surrounded by urban or agricultural use which were overlooked or misclassified in producing the Gap Analysis map. This suggests a bias in our map towards over-generalization of urban and agricultural areas.

DISCUSSION

We found the comparison of the Gap Analysis vegetation map to larger scale "insets" to be extremely useful in depicting plant community patterns. Overlay of the two representations shows patterns of disagreement that arise from many different sources ranging from positional and attribute errors to inconsistencies arising from the map generalization process (Chrisman 1991). The insights obtained by such an overlay cannot be obtained through conventional point sampling procedures generally used in statistical map accuracy assessments (e.g., Congalton 1991).

Gap Analysis is using maps and GIS overlay to estimate the amount and protection status of biological resources. It is well established that estimates can vary widely depending on the minimum mapping unit (O'Neill et al. 1991). We have tried to reduce this scale-dependence in the Gap Analysis maps by including information on both primary and secondary vegetation types in mosaics that are recognizable only at resolutions too detailed to map on a state-wide basis. Nevertheless, we estimate 56,432 ha of coastal scrub in the RECON study area compared to their estimate of 42,948 ha, a 31% difference.

Perhaps more interesting is the difference in distribution of coastal sage scrub as represented at these different map scales. Nearly all of the low elevation, coastal occurrences of this type are small remnants in canyons surrounded by development. These remnants are too small to be represented even as a secondary type in the Gap Analysis map, and are certainly not likely candidates for core conservation areas given their size and isolation. However, we quickly point out that these areas, as an archipelago, can be significant in providing connections to larger patches, and possibly as refugia for rare and localized endemics, particularly plant species. The important lesson here is that the finer-grain representation of the resource does not necessarily nest inside the coarse-grain representation. If the coarse-grain map is used to guide more detailed local studies, potentially significant sub-regions might never be considered.

We believe, as suggested by Wiens (1981), that the objectives of a spatial analysis should define the scale of investigation. The objectives of Gap Analysis are to map the extensive areas of occurrence of plant communities and wildlife habitats that we consider to be of greatest significance in the long-term maintenance of biodiversity, and to assess the current ownership and management status of those areas. Although data on these physical and biological phenomena must be reliable, a high degree of spatial precision might not be necessary for purposes of regional planning. The high level of agreement between the Gap Analysis map and the RECON map for extensive areas of coastal sage scrub suggests that our current mapping method is imperfect, but adequate for providing a regional overview of the distribution of core areas of plant communities and habitats. However, the comparison also highlights the danger of applying Gap Analysis maps locally.

Our analysis suggests the need to develop ecoregional conservation strategies with at least two levels of detail. The broader concerns, such as distribution of the vegetation communities throughout the ecoregion and the entire distribution of species (e.g., the California gnatcatcher), can be identified over larger domains with relatively coarse resolution. More local conservation concerns, on the other hand, must be addressed using more detailed maps and ground survey data.

CONCLUSIONS

Conservation efforts are inescapably constrained by financial resources, including those available to assess and plan. The logical trend towards addressing entire ecosystems as a whole and inseparable entity will require even more cost effective methods. The Gap Analysis program attempts to make a meaningful contribution towards a first stage assessment of the conservation status of natural communities within an ecoregion. Although the distribution and abundance of some communities are such that they are only detectable at detailed levels of resolution (e.g., most wetland types), we believe this approach will be an important first step for the major terrestrial community types.

Coverage of an entire ecoregion requires some compromises in mapping methods. The primary constraint of a 100-ha mmu will dictate what features will be detectable and thus addressed by this first stage assessment. Therefore, it is important that significant features of the those communities we assume to be adequately covered across the landscape are not missed. To test this assumption we examined more carefully an important terrestrial community type in southern California, the coastal sage scrub community. Comparison of our data base with another prepared at roughly 100 times the level of resolution confirms that with virtually no unexplained exception we are able to capture all landscape units with meaningful stands of coastal sage scrub.

We do not recommend that Gap Analysis be used as the only tool for assessing the landscape in the process of designing ecosystem reserves. Many factors must be considered in this complex science of preserve design. However, we believe that this approach does identify those remaining landscapes that will provide the possible foundation for a regional reserve system. We hope that this information will make an important first contribution towards the construction of a system of reserves that will maintain all of the component parts and the physical processes that will allow the system to maintain itself indefinitely.

ACKNOWLEDGMENTS

We would like to thank Paul Fromer of RECON for providing the map of coastal sage scrub for San Diego County, and Mike Bueno, Violet Gray, and Mark Borchert for assistance in preparing the Gap Analysis map. Nick Chrisman suggested the use of large scale map insets for map accuracy assessment. We received helpful comments from several reviewers including Ross Kiester, John Wiens, Robert Szaro, and Reed Noss. Funding for the Gap Analysis project was provided by the National Fish and Wildlife Foundation and the Southern California Edison Company.

LITERATURE CITED

Atwood, J. L. 1990. Status review of the California gnatcatcher (*Polioptila californica*). Unpubl. Tech. Rep., Manomet Bird Observatory, Manomet, Mass.

Chrisman, N. R. 1991. The error component in spatial data. In *Geographical Information Systems: Principles and Applications*, eds. D. M. Maguire, M. F. Goodchild, and D. W. Rhind, pp. 165-174. Longman Scientific and Technical, New York.

Congalton, R. G. 1991. A review of assessing the accuracy of classifications of remotely sensed data. *Remote Sensing Environ.* 37(1):35-46.

Davis, F. W., and D. M. Stoms. 1992. Gap Analysis of biodiversity in California. In *Proceedings of the Symposium on Biodiversity of Northwestern California*, ed. H. M. Kerner, pp. 23-29. Wildland Resources Center Report 29, University of California, Berkeley, Calif.

Federal Register. 1993. Final rule to list the coastal California gnatcatcher as threatened. Vol. 58, No. 59, 16742-16757.

Forman, R. T. T., and M. Godron. 1986. *Landscape Ecology*. John Wiley and Sons, New York.

Hickman, J. C., ed. 1993. *The Jepson Manual: Higher Plants of California*. University of California Press, Berkeley, Calif.

Klopatek, J. M., R. J. Olson, C. J. Emerson, and J. L. Joness. 1979. Land-use conflicts with natural vegetation in the United States. *Environ. Conserv.* 6(3):191-199.

Kuchler, A. W. 1964. Manual to accompany the map-potential natural vegetation of the conterminous United States. *Am. Geogr. Soc. Spec. Publ.* No 36.

O'Leary, J. F. 1990. Californian coastal sage scrub: general characteristics and considerations for biological conservation. In *Endangered Plant Communities of Southern California*, ed. A. A. Schoenherr, pp. 24-41. Proceedings of the 15th Annual Symposium of Southern California Botanists No. 3.

O'Neill, R. V., S. J. Turner, V. I. Cullinan, D. P. Coffin, T. Cook, W. Conley, J. Brunt, J. M. Thomas, M. R. Conley, and J. Gosz. 1991. Multiple landscape scales: An intersite comparison. *Landscape Ecol.* 5:137-144.

Scott, J. M., B. Csuti, K. Smith, J. E. Estes, and S. Caicco. 1991. Gap analysis of species richness and vegetation cover: an integrated biodiversity conservation strategy. In *Balancing on the Brink of Extinction: the Endangered Species Act and Lessons for the Future*, ed. K. A. Kohm, pp. 282-297. Island Press, Washington, D.C.

Westman, W. E. 1981. Diversity relations and succession in Californian coastal sage scrub. *Ecology* 62:170-184.

Westman, W. E. 1983. Xeric Mediterranean-type shrubland associations of Alta and Baja California and the community-continuum debate. *Vegetatio* 52:3-19.

Wiens, J. A. 1981. Scale problems in avian censusing. In *Estimating Numbers of Terrestrial Birds*, eds. C. J. Ralph and J. M. Scott, pp. 513-521. *Studies Avian Biol.* 3:385-397.

15

BIODIVERSITY INVENTORIES AND MONITORING: ESSENTIAL ELEMENTS FOR INTEGRATING CONSERVATION PRINCIPLES WITH RESOURCE DEVELOPMENT PROJECTS

Francisco Dallmeier

Smithsonian/MAB Biodiversity Program
Smithsonian Institution
Washington, D.C.
U.S.A.

Problems associated with managing terrestrial ecosystems are rapidly accelerating throughout the world. Conflicts arise when proposals for extracting natural resources threaten efforts to conserve biodiversity and promote sustainable use of resources. In many developing nations where a large number ofidentified and unidentified species exist, the situation is often exacerbated by burgeoning human populations that put additional pressure on the resource base. Complicating matters is our incomplete understanding of the exact role that elements of biodiversity play in creating a healthy environment, especially in developing nations where infrastructures for biodiversity research are relatively weak. Even developed countries fail to allocate adequate personnel and funding to the task.

One result is a skeptical and sometimes misinformed public that views researchers and land managers as obstacles to resource development and an improved standard of living. Those people often find themselves in the center of controversial issues such as accelerated commercial logging of primary forests, conversion of forests to agricultural uses, or zoning regulations that do not accommodate imperiled forest species.

Fortunately, the mounting body of evidence concerning the value that biodiversity holds for us appears to be creating a climate for more rational use of natural resources. Documentation of the tremendous rate of loss of species and habitats that otherwise might benefit humans points to the need to integrate conservation and sustainable use principles with resource development projects (Dallmeier and Devlin 1993). But successful integration urgently requires additional data about how ecosystems function, particularly in relation to maintaining biodiversity, so that the impacts of changes in ecosystem functions caused by development can be mitigated.

In line with this reasoning, Smithsonian Institution and UNESCO Man and the Biosphere Program established an entity called the Smithsonian Institution/Man and the Biosphere Biodiversity Program (SI/MAB) (SI/MAB Biodiversity News 1991, 1992,

1994). The goal for the program was the establishment and coordination of a global network of long-term biodiversity inventory and monitoring plots, focussing on temperate, subtropical, and tropical forest ecosystems. Work was to revolve around what Davis (1993) describes as the driving purpose of monitoring—to suggest potential agents of change and then, through research, document patterns of cause and effect. Originators of SI/MAB emphasized that the effort must entail cooperation with host-country officials, resource managers, and educational institutions as well as training to enhance host-country capabilities for continuing the projects into the future.

It was also expected that SI/MAB would play a unique and complementary role in conserving global biodiversity (Dallmeier 1993). Therefore, SI/MAB coordinates all phases of its work with international sponsors and researchers and with counterparts in other research programs. It appears this approach is eliminating unnecessary duplication while adding important information to the body of knowledge needed to reach the goal of ameliorating conflicts between resource conservation and development.

This paper describes the SI/MAB program, from project design and implementation to use of results (Figure 15.1). It discusses the methodology of SI/MAB for establishing appropriate project frameworks, managing and reporting data from inventory and monitoring plots, and conducting training. Examples from SI/MAB projects illustrate how the program works.

SETTING THE PARAMETERS

The involvement of SI/MAB in designing and implementing biodiversity inventory and monitoring projects usually begins with the request of a host country seeking help in carrying out biodiversity conservation programs. This sometimes comes about through international channels such as the UNESCO Man and the Biosphere offices around the world. Other times, we may be contacted by host-country institutions such as the University of Guyana or by managers of protected areas.

Siting the Project

The first task is to select the research site. SI/MAB works with host countries to determine the appropriate location, mindful of each country's conservation priorities. Usually, biosphere reserves or conservation units are chosen. We have found that protected areas, whether basically undisturbed or in the process of recovery, offer the best opportunities to gather baseline data and monitor changes in ecological processes. In relation to goals of SI/MAB, protected areas have an additional advantage. They are usually dedicated to conserving critically threatened or representative ecosystems to maintain as much diversity as possible, paying particular attention to plant and animal species that are habitat specialists, that occur in low densities, or that may be found nowhere else (Probst and Crow 1990). Furthermore, at nearly all such sites, broad conservation mandates are already in place. Often they are based on sound scientific principles for conserving biodiversity that are applicable worldwide. But at each research site, specific circumstances can dictate additional steps or modifications of tested methodologies.

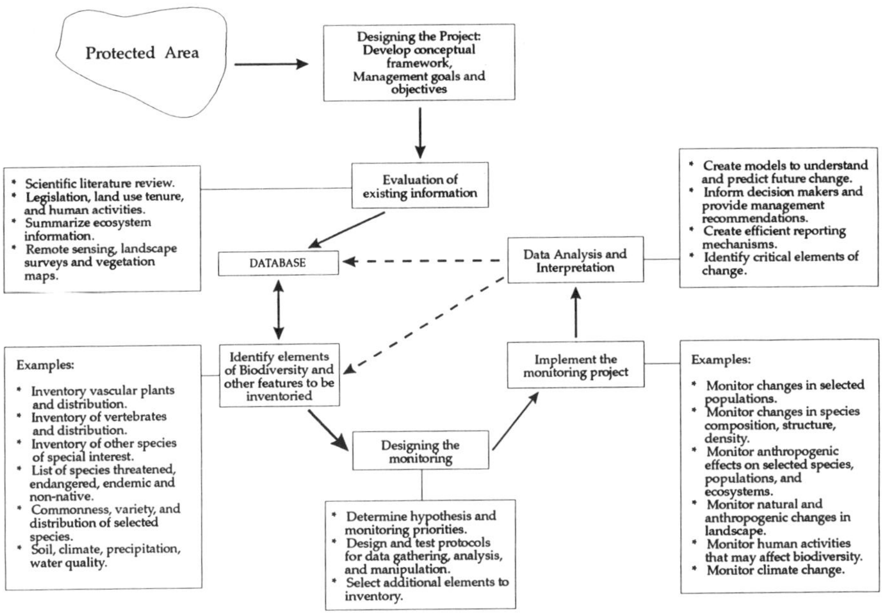

Figure 15.1. Steps in the Smithsonian Institution/UNESCO Man and the Biosphere Biodiversity Program inventory and monitoring program.

In selecting the site, two critical factors come into play: (1) the protected area or conservation unit must be off limits to resource development for at least ten years, and (2) representatives of cooperating entities and the host country need to be involved in siting the project. It is also necessary to conduct site visits before the final decision is reached and project design can continue.

Conceptual Framework, Management Goals, and Project Objectives

The cornerstones of successful biodiversity inventory and monitoring projects are carefully delineated conceptual frameworks, management goals, and project objectives. To determine these factors, SI/MAB forms a project team consisting of representatives from relevant host-country institutions and organizations and the management staffs that will be carrying out the work. The team meets in an initial "brainstorming" session, at which time all participants must agree on the framework, goals, and objectives. Decisions must be clearly stated and understood by all parties.

The process proceeds fairly smoothly at research sites where managers have already defined goals and objectives and local and national governments or research institutions are seeking assistance in carrying out biodiversity inventories and monitoring. Where those conditions are not met or where international political obstacles exist, it may take longer to complete this essential phase of project design.

On a cautionary note, SI/MAB has found that whereas enthusiasm and commitment are high, in many cases the human and financial resources needed to maintain the projects over the long run may be uncertain. The brainstorming session should always include an assessment of the potential for future adequate support; otherwise, unrealistic planning will lead to disappointment. An important consideration in determining likely support is the need to maintain close contact with relevant government institutions. Governments are ultimately responsible for most of the areas where the inventory and monitoring projects are located. No matter how solid the research is, governments, which may change over the life of the project, and their bureaucracies will have the final word concerning the areas and possibly the continuance of the projects. It is, therefore, advisable to keep appropriate individuals in government agencies up to date through letters, other written documents, and face-to-face meetings.

The meeting also provides the opportunity to discuss the responsibilities of each collaborating entity. Once decided, they should be recorded in letters of intent to avoid future difficulties. For example, the collection and final disposition of voucher specimens is of special concern in biodiversity inventories and monitoring. Voucher collections are needed to validate the biological information gathered and interpreted at the research sites. Each country has different laws and regulations for collecting and exporting specimens. Past misunderstandings, especially over specimen export, have made it difficult to conduct biodiversity research in certain countries. Letters of agreement laying out responsibilities and conditions that adhere to laws for the handling of specimens are essential.

Example: Dinghushan Biosphere Reserve

The experience of SI/MAB at Dinghushan Biosphere Reserve illustrates the steps just described.

Dinghushan is found in South China's Guangdong Province near the Pacific Ocean to the east and the South China Sea to the south. It was established in 1956 as a national natural reserve and was accepted into the UNESCO Man and the Biosphere program in 1979. Dinghushan Arboretum, an affiliate of the Chinese Institute of Botany (itself an organization of the Chinese Academy of Sciences), is responsible for managing Dinghushan. Located close to the large urban center of Guangzhou, the reserve is an island of primary forest in the midst of land that has been subjected to intense human disturbances. Buddhist monks protected the area for centuries, and its old temples and scenic beauty attract hundreds of thousands of tourists annually. In the buffer zone, local people have long gathered large quantities of forest litter and understory plants for their fuel needs.

The reserve is recognized as ideal for studying transitions between tropical and subtropical forest. Mean annual temperature (21° C), rainfall (1,956 mm.), and relative humidity (81.5%), combined with the reserve's location near major water bodies, have created a landscape very different from many other areas of the world that are also near the Tropic of Cancer, two-thirds of which are desert or semi-desert. Nearly 79% of Dinghushan's 1,155 ha is covered primarily by monsoon evergreen broad-leafed forest, pine and broad-leafed mixed forest, and pine forest.

Biodiversity is high at Dinghushan. To date, 1,843 wild plant and 349 cultivated plant species have been recorded, 23 of which are rare or endangered and have been accorded national protection. Of the 799 insect, 176 bird, and 38 mammal species identified so far, eight are receiving national protection.

Management plans for the reserve have as their basic tenet the creation of a biodiversity conservation program that will protect species diversity while allowing moderate human use. More recently, managers have added the goal of studying the reserve to gain knowledge of changes in the monsoon subtropical forest ecosystem as a result of changes in global climate. Managers and researchers are seeking ways to improve their on-going program to meet these goals.

Through the UNESCO Man and the Biosphere Program and the South China Institute of Botany, SI/MAB was asked to join a team discussion about establishing a long-term inventory and monitoring project at the reserve, in conjunction with training in inventory and monitoring methods. Over a period of three years that included a site visit, we worked with representatives of the South China Institute of Botany, the Chinese Man and the Biosphere Program, UNESCO Man and the Biosphere Program, and managers of Dinghushan Biosphere Reserve to design a project. The eventual outcome was a set of objectives to meet the management goals for Dinghushan. The team agreed that the project and training should contribute to: (1) an increased understanding of the structure, function, and dynamics of Dinghushan's monsoon subtropical forest ecosystem, (2) devising sound methods and models for restoring and sustaining the zonal forest in Dinghushan and nearby areas so that it might continue to serve local residents, (3) conserving and regenerating rare and endangered plant and animal species in South China, and (4) studying the impacts of global climate change and the response of forest species and ecosystems to that change.

The team also devised a plan for future participation of SI/MAB in the biodiversity conservation program. In late 1993, SI/MAB staff traveled to Dinghushan to conduct the first training session and begin implementation of the inventory and monitoring project.

EVALUATING EXISTING INFORMATION

After the framework, goals, and objectives are determined, an evaluation of existing information relevant to the site takes place. This procedure usually involves extensive bibliographic searches for published and unpublished reports of historical and scientific material; perusal of records of legislation, land-use patterns, and rare natural events (hurricanes, long droughts, extreme temperatures); and, where materials are available, the study of vegetation maps, collections of specimens, remote sensing documents, and landscape surveys.

Relevant data should be collected, summarized, and stored in databases at national depositories such as the Conservation Data Center in Bolivia or the Biodiversity Center of the University of Georgetown in Guyana. At sites where little research has been done and information is lacking from other sources, it may be necessary to conduct a rapid assessment of biodiversity in the area to gain enough data to proceed with the project.

Returning to the example of Dinghushan, much information related to the reserve had been collected and organized in a database, making it much simpler to continue with the next steps in designing the inventory and monitoring project.

SELECTING ELEMENTS OF BIODIVERSITY AND OTHER SITE FEATURES FOR INVENTORY

Following the data evaluation, the project team plans a biodiversity inventory to collect baseline data documenting the current on-the-ground situation and for use in selecting the elements of biodiversity, as well as other features, that will be monitored. The results can confirm the occurrence of species reported in the past and the presence of new species. In some cases, the inventory can provide information on species that require special attention such as endemic, threatened, endangered, or alien species. Where site vegetation maps are available, the inventory can validate existing information; where they are not, the inventory may be useful in generating them.

Again, in selecting elements for monitoring, it is necessary to keep funding and personnel capabilities in mind. Each element that is selected will require a complete review of existing relevant information or possibly new research efforts. Depending on available funding and the extent of supporting infrastructures, research may necessarily be limited to monitoring a few selected elements such as indicator species, endangered species, or species threatened by development proposals.

Figure 15.1 lists some of the desirable elements to be inventoried. In relation to individual species, it is extremely difficult and expensive to inventory all taxa. Potential limitations should be discussed by the project team and priority placed on monitoring those elements that best fulfill the management goals and project objectives. Because the focus of the SI/MAB program is forest ecosystems, we recommend that vascular plants receive top priority, followed by vertebrates. Vascular plants are the dominant biota in most terrestrial ecosystems, and vertebrates are generally well known because many are the objects of special management or regulations; therefore, data about both groups are likely to be more readily available. If funding is not a limiting factor, inventories of invertebrates and non-vascular plants should be planned. SI/MAB also encourages the inclusion of soils, climate, precipitation, and water quality in the inventory. Soil and precipitation data are especially important in reaching an understanding of distribution patterns for species.

DESIGNING THE MONITORING PROGRAM

Once the inventory plan is in place, the project team establishes the hypothesis underlying the monitoring and sets monitoring priorities for testing the hypothesis. The team devises a model of ecosystems, communities, populations, and species that will be monitored. It also chooses a protocol for data gathering, analysis, and interpretation, and decides on procedures for determining the effectiveness of the model and protocol. During this step, team members may agree that additional elements need to be monitored to meet the project objectives.

One of the most important aspects to be determined is the sampling method. It must be cost efficient, reliable, and ensure that the data needs are fulfilled. It must also be flexible, possibly adding other variables should they be necessary. Usually SI/MAB uses standardized sampling methods that provide numerous and precise details for inventory and monitoring purposes in our network of plots because we are interested in comparing data among the research sites. Such methods are not always necessary. Some monitoring programs are designed to gather and interpret data to help solve problems at a single site. In such cases, rapid sampling techniques designed to determine dominant species and general patterns, instead of providing detailed data, may suffice.

The process of designing the monitoring aspect of the project should clearly define assignments for conducting monitoring. SI/MAB encourages the formation of a monitoring working group consisting of professionals experienced in inventory and monitoring methods to act as primary researchers, trained workers (usually hired from local communities) to handle day-to-day tasks such as tree tagging and measuring, and students or young people in resource conservation programs to enter and verify data under supervision of technical experts in data management. Publications should be prepared by the group members who are actively involved in analyzing and interpreting the data.

The group must be informed in advance of schedules for monitoring and estimated time-frames for completing each task. In setting schedules, it is advisable to take into account climate and logistical difficulties such as seasonal flooding, extremes in humidity and temperatures that can damage expensive equipment, and the need to transport large amounts of equipment and human necessities to remote sites.

IMPLEMENTING THE MONITORING PROJECT

Implementing the inventory and monitoring project is the most complicated step in the overall process. It entails the completion of a number of tasks (Figure 15.1), all of which require long-term funding and sufficient numbers of trained personnel.

Over the past seven years, SI/MAB has tested and refined a systematic methodology for conducting monitoring, managing the resulting data, and preparing detailed reports and other publications (see below, Data Analysis and Interpretation). The methodology calls for the establishment of a network of small inventory and monitoring plots (one hectare in size) that provide a snapshot of the larger forest landscape. The size of the plots means that virtually all important elements of biodiversity can be inventoried precisely and relatively quickly and that other important features, such as topography, rainfall, soil, and climate, can also be accurately described.

The general objectives at the plots are to (1) inventory woody plants and their distribution, (2) record forest structure, composition, and diversity and monitor natural and human-induced changes in intervals of three to five years, (3) detect patterns of recruitment, mortality, and regeneration of forest trees in a wide range of natural and managed environments, (4) determine how abiotic factors (e.g., soil and precipitation) affect all of the functions just noted, and (5) contribute data for the preparation of vegetation maps.

At the time of plot establishment, the project team carries out the initial inventory, following the guidelines set forth earlier by the team members when they identified the elements of biodiversity and other site features for inventory.

SI/MAB has established permanent inventory and monitoring plots at eight long-term research sites, primarily in Latin America and the Caribbean. We intend to expand our global monitoring network to 300 plots in Africa, Asia, Latin America, the Caribbean, and North America by the year 2000, thus creating the world's largest grouping of biodiversity inventory and monitoring plots in a diverse range of forest habitats. The network will be linked by the methodology of SI/MAB.

Specific techniques for laying out the plots and conducting monitoring are described in Dallmeier (1992), and Dallmeier and Comiskey (in press). For purposes of this paper, the following discussion centers on work at several of the research sites to demonstrate how the overall program is implemented and how the results are used in efforts to resolve controversies between resource conservation and development.

Beni Biosphere Reserve, Bolivia

Bolivia's 135,000-ha Beni Biosphere Reserve, consisting primarily of flat land with poorly drained areas, is covered by a mosaic of forest and savanna. Approximately 90,000 ha are blanketed by closed forest on both flooded and non-flooded soils. About 70% of the area is seasonally inundated. The Beni reserve is typical of many forests in the interior of South America in that indigenous peoples have long used its plant and animal species for essential commodities such as food, medicines, and construction materials. Also, Beni is threatened with larger-scale nontraditional development of forest resources.

Permanent monitoring plots of SI/MAB at Beni were established in 1987 as a cooperative project of the Beni Biological Station, the Bolivian National Academy of Sciences, the National Museum of Natural History at La Paz, and SI/MAB (Dallmeier et al. 1991c,d). They are located in four primary vegetation zones: low drainage inundation forest and non-flooded forest at the core of the reserve, forest island surrounded by open savanna, and savanna with three dominant species of fire resistant trees in an intensively managed buffer zone. Censuses were conducted of the plots in 1990, 1991, and 1992; species identification was completed in 1991. Six additional plots will be established in 1994 and 1995.

Monitoring objectives focus on documenting the composition, structure, diversity, and dynamics of the forest to validate vegetation maps of the area and to devise management policies for the regulation of nontraditional resource development, conservation of natural areas, and restoration of damaged sites. The policies are being formulated in a management plan for the reserve (Miranda 1991). Additional data are being gathered from local populations of Chimanes and "mestizos" to estimate the value of sustainable use of forest resources for traditional purposes. More than 70% of all species present in the four plots is utilized in some way by Chimane and mestizo inhabitants of the area.

Manu Biosphere Reserve, Peru

Located in the departments of Madre de Dios and Cuzco in the upper Amazon Basin of Peru, Manu Biosphere Reserve stretches across nearly 2 million ha ranging from 240 to 4,000 m in elevation. Nearly all ecological zones of eastern Peru are found in the reserve, which also boasts one of the highest levels of biodiversity found in a protected area anywhere in the world (Gentry and Terborgh 1992). SI/MAB has established four plots in Manu since 1987 in a cooperative venture with the Ministry of Agriculture of Peru and the University of San Marcos.

Manu's substantially undisturbed condition presents an excellent opportunity to gain an understanding of natural processes in tropical forests for application in restoring degraded tropical forests and tempering the impacts of resource development. That is the primary objective of the SI/MAB monitoring project, and it is shared by researchers in other parts of the reserve. Combined, the monitoring projects are providing the basic ecological information needed by Peruvians and other Latin American biologists to influence management decisions in Manu.

SI/MAB plots are located near the Manu River on terrace with highly dissected old-growth forest, low alluvial terrace with inundation forest, well-drained terrace old-growth forest, and old river bed forest (Dallmeier et al. 1992a,b, Dallmeier et al. in press). We have completed two censuses over four years. The research shows a high level of diversity—up to 172 tree species on a single hectare at the plots. Species diversity in the surrounding forest may be even higher. We selected our plots to illustrate diversity among vegetation zones, and they may understate species diversity in the forest as a whole.

We found much variation in abundance of species at the plots. One of the plots may be a source area for *Rinorea* and *Iriartea* species, and another may be a source area for *Laetia corymbulosa*, *Allophylus scrobiculatus*, and *Inga dumosa*. The large number of species represented by a single individual suggest the plots are also potential sinks for many species with source areas located elsewhere.

Tree mortality rates ranged from 2-5% annually over the study period, a rate substantially higher than that recorded from same-sized plots at nearby Cocha Cashu (Dallmeier et al. in press). However, the Cocha Cashu inventories (recording annual mortality at 1-2%) were done ten years apart, with no intermediate tally. It is possible that mortality rates at the SI/MAB plots will drop to a comparable level when the tenth year census is taken. Other research shows that tree mortality varies greatly across sites and over time at the same site. The advantage of conducting several censuses in the same plots over short time intervals (three to five years) is that differences in mortality can be tracked with a higher degree of accuracy.

Guatopo National Park, Venezuela

Guatopo National Park, located in north central Venezuela near Caracas, was established in 1958 to help conserve natural resources. The rainforest in the area is regenerating following intense human use. Until the late 19th century, it remained virtually untouched by human influences because of the impenetrability of its vegetation. Colonization by Spanish, French, and Italian immigrants began in 1880. With their arrival and increasingly mechanized means of production, large portions of the area were subjected to agriculture, cattle ranching, and some commercial ventures. Several development sites were abandoned as people sought work elsewhere when the petroleum boom began. By 1950, mostly small, non-technical agriculture production was dispersed throughout the area.

In a cooperative project with INPARQUES (Venezuela's park service) and EcoNatura (a nongovernmental organization), SI/MAB has established two monitoring plots in the La Raya section of the park between 658 and 799 m above sea level. One of the plots has experienced moderate to intense human-caused disturbance and is now characterized by secondary recuperation growths averaging 16 m in height. The other plot is located on relatively undisturbed ground. The first census of the plots has been completed. The results are being documented to understand forest regeneration in one of the most important watersheds in northern Venezuela, to devise proposals for restoring similarly degraded areas in and out of protected land units, and to document forest vegetation types in Venezuela for mapping through a program of INPARQUES.

Kwakwani, Guyana

In the Kwakwani region of Guyana, the monitoring project of SI/MAB is designed to document composition, structure, mortality, and regeneration in low lying rainforest on sandy soils for use in restoring areas mined for bauxite and to serve as the basis for a restoration model in other similar environments. The project is an effort of SI/MAB and the University of Guyana and is sponsored by Aroaima Mining Company (Comiskey et al. 1993, Comiskey et al. 1994). The Kwakwani region consists of undulating low elevation terrain with swampy areas in the depressions and near rivers and mixed forest on the higher ground. Two plots of SI/MAB, located in spots that are representative of the larger area, are within near reach of each other. Soils are a mix of sand and clay; porosity and drainage are high. Soil depth is shallow, averaging about 1.5 m on top of 9 m of bauxite over 60 m of volcanic rock.

Five primary forest types are generally found in the Kwakwani region: Mora forest, Morabukea forest, mixed forest, Greenheart forest, and Wallaba forest. These forest types are not the result of different climatic conditions, and there is no successional relationship between one type and the next. Soils and drainage directly affect each, however, thus making the distributions of forest types predictable.

In relation to the objectives of the monitoring project, a significant observation is the high level of mammalian biodiversity coupled with the complexity of the community trophic structure, suggesting intricate relationships between plant and animal species. The evidence points to a need for wildlife conservation and management to avoid the local extinction of mammals and minimize the impact of forest disturbances.

Luquillo Biosphere Reserve, Puerto Rico

The Luquillo Biosphere Reserve is located in the Luquillo Mountains of eastern Puerto Rico, approximately 35 km from San Juan. Four life zones are generally recognized in the area: subtropical wet forest, subtropical rainforest, subtropical lower montane wet forest, and subtropical lower montane rainforest (Ewel and Whitmore 1973).

The Bisley biodiversity monitoring project of SI/MAB in Luquillo is carried out in conjunction with the University of Puerto Rico and the Institute of Tropical Forestry (Dallmeier et al. 1991a,b). The primary objective is to increase understanding of forest dynamics and regeneration after natural cataclysmic events, in this case, Hurricane Hugo which hit the island with devastating force in 1989. Researchers recorded data for all live standing trees in the plot during its establishment in 1988 (Dallmeier et al. 1991a,b). Subsequent censuses in 1990, 1991, and 1992 accounted for all trees present before Hugo; they were then classified into several categories of damage. New growth is also being documented. The data recorded prior to the hurricane are serving as a baseline against which to measure the natural regeneration that is taking place. The results of the monitoring project at Bisley might also serve as the basis for a model restoration program in the human-impacted zones of other Caribbean areas that once contained abundant rainforest.

Virgin Islands Biosphere Reserve, St. John

The Caribbean island of St. John is the home of the Virgin Islands Biosphere Reserve, a 6,127-ha area that varies in elevation from sea level to about 400 m. Subtropical dry forest and brush cover 70-85% of the island, with subtropical moist forest located in higher elevations, ravines, and protected valleys. The dry forest of St. John has been regenerating for more than a century, since extensive sugar cane cultivation was terminated in the mid-1800s. Thus, the reserve is an excellent area for accomplishing the major objective of the monitoring project: providing data for use in predicting dry forest regeneration and for planning restoration and management in similar ecosystems throughout the Caribbean region.

The project team established Plot 1 at St. John in 1991. Results of the first census were published in the SI/MAB user's guide to the plot (Dallmeier et al. 1993, Dallmeier and Ray 1992). In addition, independent research at the plot on the highly diverse beetle populations of the Virgin Islands will soon be published.

We are setting up a second plot under a three-year grant from the U.S. National Park Service, which manages the reserve. The plot will be located in dry evergreen scrubland, one of the major formations of the Caribbean dry forest that occurs in wind-exposed coastal areas of low rainfall (< 1,000 mm annually) at low elevation (< 200 m). Traditionally, the scrublands have been used to feed livestock (sheep and goats) and have suffered in many cases from overgrazing. The site selected for the new plot has not been grazed for several decades. It is now a compositionally diverse area covered with scrub thickets and containing federally and locally listed endangered plant species. The low canopy height and nearly impenetrable, dense thorny thickets indicate that plot establishment methods of SI/MAB will require modification because they were designed for the forest floors and higher canopies of rainforests.

DATA ANALYSIS AND INTERPRETATION

For long-term biodiversity monitoring projects to succeed, they must be able to identify elements of change in ecosystems, create models for predicting future changes, report new information from the monitoring program to all key individuals, and inform decision-makers and managers of practical management alternatives based on the research. A data management system that promotes efficient data collection, verification, interpretation, and dissemination is essential to accomplish these tasks. Where such systems are lacking, research usually remains separate from the decision-making process and less effective than it could be.

SI/MAB has developed and refined a user-friendly biodiversity monitoring data management system (BioMon) with integrated, computerized components to ensure consistent data collection, interpretation, and dissemination by researchers who have little prior training in its use (Comiskey et al. in press). We incorporated off-the-shelf software and created portable links that, along with documentation, are made available to other researchers. Whereas SI/MAB primarily records vegetation data, BioMon is designed so that additional links can be added to handle the results from different types of research. Possibilities include faunal, hydrologic, and soil surveys.

Field and Office Modules

Two modules of BioMon for the field and the office are menu-driven for ease in manipulating data. On site at each plot, researchers use the field module to record and store standard observations (tree types, measurements, locations, diameters, stem conditions, crown conditions, degree of coppicing, among others) along with notations of other pertinent features or occurrences. The field module displays an array of basic calculations for each site, including the number of species and their relative density, dominance, and frequency. The program can also generate distribution maps for any or all species in the plot. Thus, further investigations can be done while researchers are in the field to help in immediate data interpretation. Built-in controls help catch errors and maintain the quality of the data. For example, data can be entered into the columns available for recording information only within specified ranges. Another feature is the error message that appears if users enter nonexistent codes.

The office module is a more complex version of the field module, providing access to data from all sites where SI/MAB maintains biodiversity inventory and monitoring plots. To make the data more manageable, information from each site is stored within its own directory in a series of tables. The structure of the tables is identical for each site, enabling data comparisons among all sites. Data collected and entered on portable computers in the field can be downloaded to the office module through a network connection or directly from backup diskettes. The office module allows movement from database management into statistical analysis, map creation, and use of data for publications. Its capacity for storing and processing large amounts of information is especially important given the millions of individual data being collected at existing monitoring plots and the SI/MAB plans for expanding its plot network over the next decade.

Publications

Our methodology for conducting biodiversity inventories and monitoring addresses methods, formats, and appropriate review standards for publishing the data that are processed in BioMon (Comiskey et al. in press). User and field guides from each monitoring site are made available soon after data are interpreted. User guides contain up-to-date, detailed biological information about the plots; the companion field guides are condensed versions of the user guides for on-site application. The data also form the basis for numerous articles and reports destined for a wide audience through journals and other means.

BioMon is improving the content of SI/MAB publications, saving time and labor, and allowing new staff and our international counterparts to carry out more effective monitoring and reporting.

TRAINING

Ultimately, inventory and monitoring projects such as those described above depend on cadres of trained professionals in host countries with adequate infrastructures to support their efforts (Ulloa-Vaca et al. 1990, Wemmer et al. 1993). SI/MAB has neither the resources nor the mandate to manage long-term monitoring and research efforts from our offices in Washington, D.C. We have found, however, that the number of people needing training often overwhelms the capacity of well-prepared professionals in host countries. Many need assistance in building up their expertise and support systems.

That is the goal of the SI/MAB training courses. Over the past six years, we have conducted 28 biodiversity inventory and monitoring workshops in the Caribbean, Latin America, and China. More than 600 participants have been trained in the methods of long-term monitoring, in data verification and interpretation, and in publication of results. As with the inventory and monitoring component of our program, SI/MAB training sessions are carried out with the cooperation and support of host-country agencies, institutions, and individuals and through international entities such as UNESCO's Man and the Biosphere Program, the U.S. Agency for International Development, the World Wildlife Fund, Conservation International, and the Nature Conservancy.

Our training sessions are usually divided into three parts. First are classroom lectures delivered by local, national, and international experts. Participants usually attend at least 20 hours of lectures and engage in one-on-one discussions with the instructors. Next, trips to selected sites allow participants a first-hand look at professionals who are applying conservation management methods to real situations. Finally, field exercises at a research station challenge participants to apply the knowledge, skills, and techniques that they gain in the other phases of the training to assigned field problems.

The courses are designed for different levels of expertise, from on-site technical personnel and professionals responsible for inventories and monitoring to decision-makers and managers from protected areas. Whenever possible and appropriate, we encourage participation from students of biological, ecological, and land management disciplines and from young professionals working in conservation and natural resource programs.

In the summer of 1993, we held our first annual International Biodiversity Measuring and Monitoring Training Course at the Smithsonian Conservation and Research Center in Front Royal, Virginia. The five-week training session brought together 19 people from several countries in Africa and Latin America and from Malaysia, Canada, the United States, and China. Participants spent more than 300 hours in the classroom and field and heard from nearly 70 instructors and guest lecturers on the complete spectrum of approaches for conducting monitoring programs. The training proved very successful. Participants completed first-year inventories and monitoring at the Front Royal plots, documented the results in a preliminary report, and wrote a long-term plan for future monitoring and research at the site. In addition, all participants made presentations about important aspects of biodiversity in their countries. Plans are underway for the second annual course in 1994 at Front Royal.

The practical results of our training courses include increased commitment to sound resource management on the part of the participants, who are able to share their concerns, observations, and experiences with individuals from different backgrounds and countries. Through the field work at the training courses, inventory and monitoring plots are established and baseline data are gathered and documented for use in future monitoring efforts. The sessions also provide cadres of knowledgeable individuals who can continue monitoring at the sites and train other people in the monitoring and data management methodology.

SUMMARY

All phases of the SI/MAB program revolve around the goal of promoting data collection and increased knowledge of ecological functions for practical application in reaching conservation-based decisions about land management. We believe that sustainable use of resources is possible if reliable data about changes in ecosystems and their impacts on biodiversity can be brought to bear in considering the consequences of human-caused disturbances. Through the implementation of well designed monitoring projects, the use of consistent and proven methods for data interpretation and dissemination, and the training of professional teams in the proper methods of biodiversity inventories and monitoring, we increase the opportunities for integrating resource conservation principles with development projects at appropriate sites.

ACKNOWLEDGMENTS

The work described in this paper is possible because of many dedicated people associated with the organizations and agencies that have made the SI/MAB program a reality and because of the extraordinary individuals who, often under adverse circumstances, continue to provide the strong leadership that is essential to conserving global biodiversity. There are too many names to list here, but I would be remiss in not recognizing the following for their special contributions: Mario Baudoin, Carmen Miranda, Don Wilson, Abelardo

Sandoval, Caroline Rogers, Gary Ray, Ariel Lugo, Fred Scatena, Gerardo Aymard, and many others. Special thanks go to Margo Kabel, Richard Rice, James Comiskey, Deanne Kloepfer, and the anonymous reviewers for their comments.

LITERATURE CITED

Comiskey, J., F. Dallmeier, G. Aymard, and A. Hanson. 1993. Biodiversity survey of Kwakwani, Guyana. Smithsonian Institution, SI/MAB Program, Washington, D.C.

Comiskey, J., G. Aymard, and F. Dallmeier. 1994. Floristic composition, structure and diversity of the Kwakwani forest in Guyana. *Diversitas*: in press.

Comiskey, J., G. Ayzanoa, and F. Dallmeier. In press. A data management system for monitoring vegetation plots: BioMon. *J. Trop. For. Sci.*

Dallmeier, F. 1992. Long-term monitoring of biological diversity in tropical forest areas: methods for establishment and inventory of permanent plots. MAB Digest 11. UNESCO, Paris.

Dallmeier, F. 1993. Tracking biodiversity for practical applications. Canadian Museum of Nature. *Global Biodiversity* 3(1):24-27.

Dallmeier, F., and J. Comiskey. In press. From the forest to the user: a methodology up-date. In *The Biodiversity of Pakistan and Its Environments*, eds. D. Wilson and A. Sandoval. Proceedings of the Biodiversity Programs, Smithsonian Institution, Washington, D.C.

Dallmeier, F., J. Comiskey, and G. Ray. 1993. User's guide to the Virgin Islands Biosphere Reserve biodiversity plot 01, U.S. Virgin Islands. Smithsonian Institution, SI/MAB Program, Washington, D.C.

Dallmeier, F. and F. Devlin. 1993. Forest biodiversity in Latin America; reversing the losses? *J. Trop. For. Sci.* 5(2):232-270.

Dallmeier, F., R. Foster, and J. Comiskey. 1992a. User's guide to the Manu Biosphere Reserve biodiversity plots, Peru. Smithsonian Institution, SI/MAB Program, Washington, D.C.

Dallmeier, F., R. Foster, and J. Comiskey. 1992b. Field guide to the Manu Biosphere Reserve biodiversity plots, Peru. Smithsonian Institution, SI/MAB Program, Washington, D.C.

Dallmeier F., R. Foster, C. Romano, R. Rice, and M. Kabel. 1991c. User's guide to the Beni Biosphere Reserve Biodiversity Plots, Vol. I and II. Smithsonian Institution, Washington, D.C.

Dallmeier F., R. Foster, C. Romano, R. Rice, and M. Kabel. 1991d. Field guide to the Beni Biosphere Reserve Biodiversity Plots, Vol. I, II, III, IV. Smithsonian Institution, Washington, D.C.

Dallmeier, F., M. Kabel, R. Foster, and J. Comiskey. In press. Measuring floristic composition, diversity, mortality and recruitment on different substrates: lowland tropical forest, Pakitza, Rio Manu, Peru. In *The Biodiversity of Pakistan and Its Environments*, eds. D. Wilson and A. Sandoval. Proceedings of the Biodiversity Programs, Smithsonian Institution, Washington, D.C.

Dallmeier, F., and G. Ray. 1992. Smithsonian Institution's permanent monitoring plot: research opportunities in a rare, semi-evergreen dry woodland. *Park Sci.* 12(2):5.

Dallmeier F., M. C. Taylor, C. Romano, and R. Rice. 1991a. User's guide to the Bisley Biodiversity Plots, Luquillo Biosphere Reserve, Puerto Rico. Smithsonian Institution, Washington, D.C.

Dallmeier F., M. C. Taylor, C. Romano, and R. Rice. 1991b. Field guide to the Bisley Biodiversity Plots, Luquillo Biosphere Reserve, Puerto Rico. Smithsonian Institution, Washington, D.C.

Davis, G. E. 1993. Design elements of monitoring programs: the necessary ingredients for success. *Environ. Mon. Assess.* 26:99-105.

Ewel, J., and J. Whitmore. 1973. The ecological life zones of Puerto Rico and the U.S. Virgin Islands. Forest Service Research Paper ITF-18. Institute of Tropical Forestry. USDA Forest Service, Southern Forest Experimental Station, Rio Piedras, Puerto Rico.

Gentry, A. L., and J. Terborgh. 1992. Composition and dynamics of the Cocha Cashu "mature" floodplain forest. In *Four Neotropical Rainforests*, ed. A. L. Gentry. Yale University Press, New Haven and London.

Miranda, C. L., ed. 1991. Plan de Manejo de la Reserva de la Biofera, Estacion Biologica del Beni, Bolivia. Academia Nacional de Ciencias Bolivia, La Paz, Bolivia.

Probst, J. R., and T. R. Crow. 1990. Integrating biological diversity and resource management, an essential approach to productive, sustainable ecosystems. *J. For.* 89(2):12-17.

SI/MAB Biodiversity News. 1991. Smithsonian Institution/Man and the Biosphere Biological Diversity Program. Smithsonian Institution, Washington, D.C.

SI/MAB Biodiversity News. 1992. Smithsonian Institution/Man and the Biosphere Biological Diversity Program. Smithsonian Institution, Washington, D.C.

SI/MAB Biodiversity News. 1994. Smithsonian Institution/Man and the Biosphere Biological Diversity Program. Smithsonian Institution, Washington, D.C.

Ulloa-Vaca, R., G. Mullins, and F. Dallmeier. 1990. Biological diversity conservation program workshops for Latin America—A Case Study. Society for Conservation Biology 1990 Annual Meeting.

Wemmer, C., R. Rudran, F. Dallmeier, and D. Wilson. 1993. Training developing-country nationals in the critical ingredient to conserving global biodiversity. *BioScience* 43(11): 762-67.

III

STRATEGIES FOR MAINTAINING BIODIVERSITY

16

BIODIVERSITY, FISHES, AND THE INTRODUCTION PARADIGM

Walter R. Courtenay, Jr.

Department of Biological Sciences
Florida Atlantic University
Boca Raton, Florida
U.S.A.

Peter B. Moyle

Department of Wildlife and Fisheries Biology
University of California, Davis
Davis, California
U.S.A.

Transfer of fishes by humans to waters outside their native regions has a long history, dating back at least to the Roman Empire (Welcomme 1988) and perhaps earlier. Most fishes were introduced intentionally to enhance native faunas that were judged to be inadequate for providing fishes for food, sport, biological control, ecosystem enhancement (e.g., forage species), and aesthetics (aquarium fishes). Other intentional introductions have been made by individuals dumping unwanted aquarium or bait fishes into lakes or streams. Increasingly common have been accidental introductions—species carried in ballast water of transoceanic ships, transported through canals, or escaped from aquaculture operations. Introduction of fishes by all these means is a worldwide phenomenon, spurred by development of modern transportation systems and by exploitative western attitudes towards Nature (Courtenay and Robins 1975; Moyle 1976a,b, 1986; Arratia 1978; McDowall 1984; McKay 1984; Courtenay et al. 1986; Pollard and Burchmore 1986; Carlton 1987, 1992; Arthington 1989; de Moor and Bruton 1989; Courtenay 1990; Pollard and Hutchings 1990; Courtenay and Williams 1992). More recently, some endangered fishes have been transferred to new waters to protect them from extinction (Johnson and Rinne 1982; Johnson and Hubbs 1989; Wikramanayake and Moyle 1989).

Regardless of how or why a species was introduced, it has the effect of increasing local species richness of fishes, at least temporarily. From a global perspective, however, introductions result in a net loss of species because local endemic forms can be driven extinct by introduced species. The purpose of this paper is to provide evidence for these statements and to show why fisheries managers need to re-assess seriously their use of introduced species as a management tool.

SCOPE OF SPECIES TRANSFERS

Welcomme (1988) listed 237 species of inland fishes that have been introduced outside their historical ranges, with 1,354 introductions made into 140 countries. This listing represents only a small fraction of the introductions, however, as it does not include marine species, expansions of species ranges within a country by introductions, and introductions that have not been officially recorded. The fresh and marine waters of the United States now contain at least 70 established fishes of foreign origin (exotic species). Of the 50 states, only Alaska lacks established exotic species (Courtenay 1993; Courtenay and Moyle 1992). In addition, more than 150 species of fish have become established beyond their native ranges in the United States, the result of interdrainage transfers (Courtenay and Taylor 1984; Courtenay 1991). The percentage of the fauna of each state that is introduced (including exotics) ranges from 2 to 59% (Moyle 1986).

Three basic eras of fish introductions are recognized (Welcomme 1988)—a premodern period, a period of unbridled introductions (ca. 1850-1950), and the modern period (1950-present). During the premodern period, fishes were moved mainly short distances overland, for pond culture or to improve fisheries. The most frequently moved fish was the common carp (*Cyprinus carpio*), which became established throughout Eurasia as a result. During the period of unbridled introductions, fishes were moved about at an ever-accelerating rate, primarily by fisheries managers hoping to improve upon nature by introducing sport and commercial species. In fact, some modern fisheries agencies had their origins in 19th century fish acclimatization societies. In the modern era, the rate of fish introductions has continued to be high as the result of not only approved introductions, but of introductions by amateurs following the lead of the agencies, as well as accidental introductions, especially from bait buckets and aquaculture operations. In the modern era, however, many introductions have been heavily criticized and are officially regulated by strict guidelines in many countries, although guidelines and laws are widely ignored.

Trends in fish introductions can be seen in the history of releases into California (Figure 16.1). The first introduction was American shad (*Alosa sapidissima*) brought from the Hudson River in New York, followed by common carp from Germany and a wide variety of species from the east coast of North America. Eastern species were carried across the continent in special railroad cars on the newly completed transcontinental railroad. Virtually all introductions were sanctioned as a way of improving local fisheries, even though at least seven kinds of salmonids were introduced into a region with the most species-rich salmonid fauna in the world, and common carp and other cyprinids were introduced into a region with a diverse cyprinid fauna (Moyle 1976a,b). After 1950, official introductions declined but were replaced by unauthorized and accidental introductions. Today, the California Department of Fish and Game spends substantial time and money trying to eradicate such introductions, especially predatory

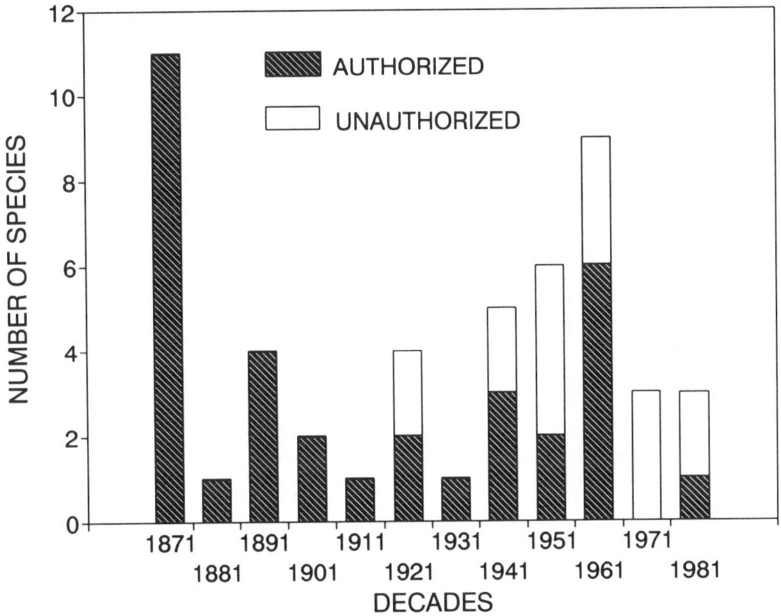

Figure 16.1. Patterns of fish introductions into California, 1871 to present.

fishes like white bass (*Morone chrysops*) and northern pike (*Esox lucius*). Such introductions are increasingly being recognized as major causes of decreasing biodiversity in California waters (Moyle and Williams 1990).

EFFECTS OF INTRODUCTIONS ON BIODIVERSITY

Introductions have increased local species richness in many of the world's fresh waters. Nevertheless, in the short time in which fishes have been widely introduced, in all too many instances they have become a form of biological pollution—living organisms that expand their ranges and populations rapidly, and disrupt, decrease, or destroy the native biota (Miller 1961; Minckley and Deacon 1968, 1991; Moyle 1976a, 1986; Courtenay and Robins 1975, 1989; Taylor et al. 1984; Barel et al. 1985; Moyle et al. 1986; Deacon 1988; Pierce 1990; Moyle and Leidy 1992).

The degree to which an introduced fish species alters the recipient fish fauna varies greatly. Species with the least effect are generally small herbivores, insectivores, or omnivores introduced into nearby drainages; those with the greatest effect are generally large predators, the type favored as game fishes. Thus, the fish faunas of coastal drainages of the eastern United States have been greatly enriched through bait-bucket introductions of small fishes, but no extinctions have been attributed to such introductions, except through hybridization (e.g., Horwitz 1982; Hocutt et al. 1986). Likewise, the introduction of five species of endangered omnivorous cyprinids into a nearby drainage in Sri Lanka does not seem to have affected the established fish communities (Wirkamanayake and Moyle 1989), nor did the introduction of the omnivorous California roach (*Lavinia symmetricus*) into the Eel River of California seem to have had much effect on native fishes (L. Brown and P. Moyle unpubl. data). In contrast, the introduction of predatory green sunfish (*Lepomis cyanellus*) into North Carolina streams caused major changes in fish communities (Lemly 1985), and the release of piscivorous Sacramento squawfish (*Ptychocheilus grandis*) into the Eel River induced major shifts in the resident fish community, which can eventually result in elimination of one species from the system (Brown and Moyle 1991). The most depressingly spectacular result of a predator introduction has been the apparent extinction of 100-200 cichlid species in Lake Victoria, Africa, by the Nile perch (*Lates nilotica*).

Additionally, even seemingly harmless species can alter aquatic communities through a variety of means, including competition, interruption of reproductive behavior of native fishes, introduction of new parasites and diseases to which native fishes are not resistant, and through hybridization (Deacon 1979, 1988; Hoffman and Schubert 1984; Courtenay and Stauffer 1984; Moyle et al. 1986). Effects like these tend to be gradual and hard to detect. For example, if closely related species hybridize, genetic fitness that evolved in the native species might be reduced (Phillip et al. 1983; Murphy and Terre 1984; Waples et al. 1990; Behnke 1991). A situation in which effects were extremely rapid occurred in the early 1980s when the sheepshead minnow (*Cyprinodon variegatus*) was released, probably from bait buckets, into the Pecos River in western Texas; within five years, introgressive hybridization with the endemic Pecos pupfish (*C. pecosensis*) was found over 430 river-km (Echelle and Connor 1989). With one introduction, the genetic spectrum of a species that took perhaps more time to evolve than the entire history of *Homo sapiens* can be dramatically and forever altered.

Effects of introductions on biodiversity also vary from region to region and with how much the waters into which introductions are made have been altered by human activity. In general, regions with a combination of low natural diversity and highly modified waterways are most impacted by introduced species. This can be seen by examining the effects of introduced species on Clear Lake, a large natural lake in California, and on reservoirs in the southeastern and southwestern United States.

Clear Lake is one of the oldest lakes in North America. Its original fish fauna consisted of 11 species, at least three of them endemic to the lake (Moyle 1976a). Over the years, 16 species of fish were introduced successfully into the lake, and now 21 fish species are there (Figure 16.2). Whereas total diversity has nearly doubled, six native fishes were extirpated. Two of them are species that are now globally extinct. It is likely that extinction was caused by the introduction of predatory centrarchids, combined with diversion of spawning streams of native lake fishes.

Southeastern reservoirs were constructed in a region with high rainfall and high fish diversity. Most introductions made into these reservoirs were of predatory fishes to enhance sport fishing and to replace riverine species that declined in abundance as reservoirs filled and aged. These introductions have impacted few native species, and positive effects (fisheries) have outweighed negative ones (Courtenay 1990). Nevertheless, Sheldon (1988) noted that reservoirs serve to isolate fish faunas of tributary streams from one another because they are huge areas of habitat unfavorable to small stream fishes, in part because they are filled with large predators. Consequently, when a

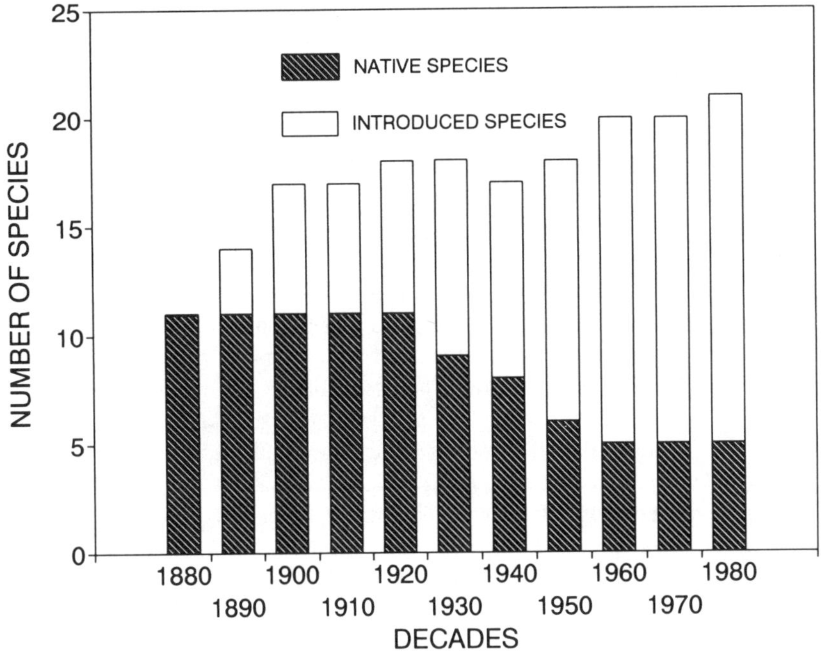

Figure 16.2. History of native and introduced fishes in Clear Lake, California, 1880 to present.

local extinction occurs from natural causes, streams cannot be recolonized naturally from populations in other streams.

Southwestern reservoirs, in contrast to those in the southeastern United States, are in a region with low rainfall and low diversity of fishes. These differences resulted in introduced sport fishes being very detrimental to native fishes. Dam construction on the Colorado River, for example, severely disrupted life histories of all native fishes, but probably most would have persisted if predatory fishes had not been introduced (Minckley 1991). These fishes prey on the young of the natives and effectively prevent reproduction from occurring both in reservoirs and in the river itself. Predation on stocked fish also greatly reduces the success of hatchery programs designed to restore native fishes (Marsh and Brooks 1989). As a result, nearly all native fishes in the Colorado River basin are listed as threatened or endangered and on the verge of extinction.

INTRODUCTIONS, BIODIVERSITY, AND FISHERIES MANAGEMENT

Only in the last decade or so has preservation of biodiversity become an issue of worldwide concern, leading to the emergence of Conservation Biology as a discipline. Protection of aquatic biodiversity is gradually becoming a concern of fisheries management agencies, but remains secondary to their primary goal of improving fisheries. Therefore, fish introductions continue to be a major management tool, and this results in serious proposals to introduce species that are known for their ability to alter ecosystems. Thus, European zander (*Stizostedion lucioperca*) have been proposed for introduction into the Missouri River basin and rainbow smelt (*Osmerus mordax*) for introduction into the Colorado River in North America (Courtenay and Robins 1989).

Introductions continue to be proposed because of the undoubted past success of many and the perceived success of others. Thus, in western states, inland sport fisheries, with the exception of those for anadromous species, are largely for introduced species. The value of these fisheries is so large that the loss of a few native fish species would be considered by most managers today to have been a small price to pay, even though loss of endemic species was not even considered when most introductions were made. Intentional introductions are increasingly hard to justify, however, now that it is possible to make better predictions of the ecological cost/benefit ratios and now that protection of biodiversity is (or should be) part of the calculation of these ratios. Also, intentional introductions, when properly evaluated, in fact rarely meet their stated objectives (Courtenay 1990). They usually create additional problems in waters already burdened by other human impacts, such as canalization, channelization, dams and diversions, and a wide spectrum of pollutants (Moyle and Leidy 1992).

Although there seems to be growing doubt among fisheries managers in North America and Europe about the effectiveness of further introductions in management, much of their previous enthusiasm for the seemingly instant successes wrought by introduced species persists among anglers in the West and in developing countries. For example, introductions of white bass and northern pike into central California reservoirs were presumably made by anglers who thought they could improve the notoriously variable fishing success typical of such reservoirs (the result of managing them for water-users, not fishes). The California Department of Fish and Game is now forced to spend millions of dollars trying to eradicate these predators because of potential harm to populations of salmon and other fishes in the main rivers. Their efforts to carry out eradication programs have been continuously delayed by lawsuits of anglers who want to maintain the fisheries that have developed, and by people afraid of the effects of a pesticide (rotenone) in the water supply. So far, little concern has been expressed by anglers or the public for the potential effects of these predators on biodiversity.

In developing countries, the use of introductions to improve fisheries has been enthusiastically adopted, and introductions are frequently made on the advice of western fisheries experts. Nobody can deny the good intentions of the experts to provide protein for humans in protein-short areas. Uncertainties arise, however, from two aspects—will these supposed resources be used by the local peoples and (the typically unasked question) what might be the effects on native aquatic resources? Developing nations will argue that it is more important to feed their increasing populace in the short term rather than to worry about other long-term biological effects. This attitude is similar to that long practiced in developed countries but for different reasons—how can we please the recreational fishing public or make the greatest profits from utilizing non-native species? Both societies fail to recognize that humans are only one part of the intricate biological communities involved and that our manipulations, made in ignorance of how the systems function, are bound to change them, often in unexpected ways. The long-term results of an introduction may be less protein available to local people, not more. Nevertheless, introductions will continue to be made, for better or worse. This can be illustrated by two examples—the introduction of Nile perch into Lake Victoria, Africa, and the program of introductions now underway in Papua New Guinea.

Lake Victoria

Lake Victoria is one of the Great Lakes of East Africa, each with a fauna made up predominantly of small haplochromine cichlids. Lake Victoria alone had 200-300 species of cichlids, most of them endemic and many of them undescribed. The verb in the previous sentence is past tense because predatory Nile perch were introduced into the lake about 1960 and seem to have eliminated most of the endemic haplochromines (Hughes 1986; Miller 1989). This centropomid was introduced as the result of advice of western scientists who thought it would "control" the haplochromines by making them forage for more useful, larger fish (Gee 1965). It was thought that Nile perch would not only improve the indigenous fishery but would attract tourists for sport fishing. An initial introduction, in the 1950s, was done as an experiment in a small lake nearby, where the perch also eliminated the cichlids (Ribbink 1987). The perch subsequently and surreptitiously appeared in the main lake. Although the introduction was justified on the basis that local people did not adequately utilize haplochromines, at the same time the haplochromines showed signs of overfishing (Acere 1988)! Following the explosion of the Nile perch population, there was a 75% drop in the total fish biomass, presumably because the incredibly diverse native fauna was better able to use the total food resources available than the perch was by itself. Although a prosperous fishery for Nile perch has developed on the lake, with a surplus for export, it is likely to collapse in the near future because of overexploitation. Moreover, the perch is so large that it cannot be sun-dried for preservation like the smaller fishes once were. Instead, it must be cooked or smoked, and large tracts of land are currently being deforested to provide the fuel necessary for cooking or smoking (Miller 1989). Thus, the long-term economic well-being of riparian peoples of the Lake Victoria region and 200+ species of fishes have probably been sacrificed because of a single misguided introduction.

Papua New Guinea

The disaster of Lake Victoria has had one positive side effect; it was used as a justification for an evaluation of a program of introductions in the Sepik River of Papua New Guinea (PNG) (Coates 1987). Introductions were felt necessary because PNG has a problem with protein deficiency, and fish yields were much lower from the Sepik River than from comparable systems elsewhere, even in New Guinea (Coates 1985). Also, already much of the catch in the fisheries was the introduced Mozambique tilapia, *Oreochromis mossambicus*, and common carp. Both had escaped from aquaculture operations, and it was likely that more introductions would be made, legally or illegally, whether or not a study was done. To ensure that introductions would have maximum benefit with minimum harm, an extensive study of the Sepik River and its biota was undertaken (including life histories of selected native species), sponsored by the Food and Agriculture Organization of the United Nations. Once studies were completed, a list of species was compiled for possible introduction, according to the "fit" they would have into the local fish community. Information on the Sepik River and its fishes, together with the list of recommended introductions, was then submitted to an international panel of experts for evaluation. At the end of this process, three species were recommended for introduction: *Tilapia rendalli*, *Trichogaster pectoralis*, and *Osphronemus gouramy* (Coates 1990). The latter two species already occur in New Guinea, but not in the Sepik drainage. The three species are all omnivores, but feed primarily on aquatic plants, algae, and detritus, so their potential for preying on native species or competing with them for food might be minimal. The following excerpt from a letter of P. Moyle to D. Coates (3 January 1990) in regard to the introductions reflects the opinions of the expert panel:

> I agree that some introductions are necessary [to the Sepik River], especially *Tilapia rendalli*. I must admit that I am not wildly enthusiastic about [the introductions] as I think alterations, and perhaps extinctions, of the local biota are an inevitable consequence. However, planned introductions, with international blessings, are at least likely to be fewer and less harmful than the unplanned ones that have been perpetrated on so many aquatic systems. I just hope that the government of PNG will limit itself to your recommendations and that follow-up studies will be conducted.

It is unfortunate that unplanned introductions are so inevitable, as it limits the role of informed fisheries management to damage control. Most people making introductions, including fisheries managers, are so ignorant of the possible ecological consequences of an introduction that they see it only in the short-term, as the least costly method of solving a perceived problem. What they usually fail to realize is that introductions often create new problems for the future, especially problems associated with the disruption of ecosystems and the loss of native species (Paxton 1990). These problems, such as those created by the introduction of Nile perch, cannot be corrected by any expenditure of funds and effort. Personnel overturn within agencies is such that replacements inherit problems created by their predecessors, predicaments that they did not create but are expected to correct, often with more introductions (Courtenay 1990).

ETHICAL CONSIDERATIONS AND THE INTRODUCTION PARADIGM

Species introductions have been based on an agricultural paradigm that humans can improve on nature. Much of world agriculture in both developed and emerging nations is built on introduced species. Paradigms, however, require review to test their value. Nearly all agricultural plants and many farm animals are now so far removed genetically from their ancestral stocks that they require culture or husbandry for continued production. Rarely do such altered stocks create environmental problems, although less intensely selected stocks of pigs, goats, sheep, and other herbivores are major ecological pests in some systems.

This introduction paradigm also encouraged movement of plants and animals not destined for culture and husbandry, but rather for use in narrower purposes. In all cases, wild stock (or wild stock then in culture) was chosen mostly because one or a few persons determined it might be useful. Rarely have such introductions proven beneficial for some perceived target reason or for financial gain (except, perhaps, to the promoters). In fact, most that have been examined on a biological basis have been shown to be costly mistakes (Elton 1958; Laycock 1966, 1992; Mooney and Drake 1986).

Biodiversity is the current end product of evolution of which we are only one part. We are also the only species that can decide the fate of thousands of other species. No other living organism ever held the potential to conserve or destroy the biota of this planet. As human populations increased, an attitude of superiority and dominance developed in most societies and, as history has shown, biodiversity has suffered severely, particularly within this century. The one species that has the ability to be the steward for biodiversity is systematically destroying it. The attitude that biodiversity exists only for our use is the root of the problem.

By "playing God" we are severely altering the course of natural evolution. Most biologists recognize this, yet their pleas to the public to assist and support conservation efforts still center around our potential use of other living organisms. Introductions of non-native species provide evidence that resource management is not supporting conservation efforts. Management goals are often centered on perceptions of how they can best serve human interests, no matter how broad or narrow. Nevertheless, most resource management agencies consider themselves as conservation agencies and many have the word "conservation" in their title.

Rarely are introductions based on sound biological data. They are often made before all other management options have been examined and exhausted. Reasons for this are simple—management agency directors and political entities are looking for a quick fix of a perceived problem (often of past management debacles) and want the cheapest way out. They would rather spend large sums to conduct introductions than to investigate possible short- and long-term impacts on other biota or habitat. In the following quotation from Pierce (1990), one could substitute any nation or state for the word "Australian":

> The message for Australian fisheries managers is clear; destroy the fishery you are meant to be saving and you will be rewarded with more work, a larger budget, and ever increasing dependence upon your hatcheries to augment depleted stocks. Enhance the ability of wild fish populations to support themselves through careful habitat and harvest management and you do yourself out of a job!

Although this might be an overstatement, there are indications that many so-called conservation agencies are drifting in this direction.

Introductions can be a good and powerful management tool if evaluated on the basis of long-term ecological principles, rather than primarily on immediate anthropocentric interests. Toward this goal, the American Fisheries Society approved a position on introductions of exotic aquatic animals in 1972. This policy was modified in 1986 to encompass all introductions of non-native aquatic biota, but an important procedure for assuring compliance was inadvertently omitted (Courtenay and Robins 1989). The International Council for Exploration of the Seas (Sindermann 1986) and

European Inland Fisheries Advisory Commission (Welcomme 1986) have established policies with regard to introductions. What needs to happen to assure that we utilize introductions in a far more responsible manner is that similar policies must become part of the code of practice of every aquatic resource management agency. Furthermore, because more than one political entity is usually affected, decisions concerning introductions must be based on drainage basins, not on political boundaries. This mandates interagency cooperation.

Finally, attitudes toward introductions are changing among most responsible biologists. The rapidly developing aquaculture industry as well as some management agencies, unfortunately, seem to like the old paradigm. Clearly there is a need for management agencies to develop (or redevelop) a conservation ethic. Biodiversity cannot be protected if the same old policies and practices continue in force. Moreover, the best knowledge about how to manage resources does not rest solely in resource agencies. As Hocutt (1984) stated, "...I am constantly amazed at how state fish and wildlife departments ignore the advice of university persons. Every effort is made to separate 'management' philosophy from 'research,' while in fact the two are inseparable."

CONCLUSIONS

Making or allowing introductions of non-native fishes can pose severe threats to the continuity and future of native species biodiversity. The generally accepted paradigm that introductions are beneficial often has been true in agriculture, largely because of the interdependence between cultivated crops or many domesticated animals and humans. Our knowledge of farm and ranch technology greatly exceeds our understanding of aquatic ecosystems, and trying to conclude that they are similar is wrong. Thus, aquatic management strategies that developed with the agricultural introduction paradigm, assuming introductions to be beneficial, have often missed their intended goals and have repeatedly resulted in reducing biodiversity of indigenous species. The challenge to management agencies is to adopt new philosophies and strategies, designed to conserve and enhance native species biodiversity, by minimizing biological alterations of aquatic ecosystems made by agencies and the public with non-native species introductions.

ACKNOWLEDGMENTS

We thank the hundreds of colleagues worldwide who, in addition to our own research over the past 20 years, have helped us understand the complexities of and mistakes (and few successes) with species introductions. For comments on an earlier draft of this manuscript, we are grateful to C. Hubbs, E.P. Pister, and C.R. Robins. We also thank R.C. Szaro for inviting this paper and for the opportunity to bring this message to those who might have been unaware of or have ignored these concerns.

LITERATURE CITED

Acere, T. O. 1988. The controversy over Nile perch, *Lates nilotica*, in Lake Victoria, Africa. *NASA, the ICLARM Quarterly* 11:3-5.

Arratia F. 1978. Comentario sobre la introduccion de peces exoticos en aguas continentales de Chile. *Cien. Forest.* 1:21-30.

Arthington, A. H. 1989. Impacts of introduced and translocated freshwater fishes in Australia. In *Exotic Aquatic Organisms in Asia*, ed. S. S. De Silva, pp. 7-20. Asian Fish. Soc., Spec. Publ. 3.

Barel, C. D. N., R. Dorit, P. H. Greenwood, G. Fryer, N. Hughes, P. B. N. Jackson, H. Kawanabe, R. H. Lowe-McConnel, M. Nagoshi, A. J. Ribbink, E. Trewavas, F. Witte, and K. Yamoaka. 1985. Destruction of fisheries in Africa's lakes. *Nature* (London) 315:19-20.

Behnke, R. J. 1991. America's changing fish fauna. *Trout* Spring 1991:35-38.

Brown,L. R., and P. B. Moyle. 1991. Changes in habitat and microhabitat partitioning with an assemblage of stream fishes in response to predation by Sacramento squawfish (*Ptychocheilus grandis*). *Can. J. Fish. Aquatic Sci.* 48:849-856.

Carlton, J. T. 1987. Patterns of transoceanic marine biological invasions in the Pacific Ocean. *Bull. Mar. Sci.* 41:452-465.

Carlton, J. T. 1992. Dispersal of living organisms into aquatic ecosystems as mediated by aquaculture and fisheries activities. In *Dispersal of Living Organisms Into Aquatic Ecosystems*, eds. A. Rosenfield and R. Mann, pp. 13-46. Maryland Sea Grant Program, College Park, Md.

Coates, D. 1985. Fish yield estimates for the Sepik River, Papua New Guinea, a large floodplain system east of 'Wallace's Line'. *J. Fish Biol.* 27:431-443.

Coates, D. 1987. Consideration of fish introductions into the Sepik River, Papua New Guinea. *Aquacult. Fish. Manage.* 18:231-241.

Coates, D. 1990. Sepik River fish stock enhancement project. Phase one final report and recommendations, Part 2—fish species suitable for stocking and stocking strategies. UN-FAO, Rome. PNG/85/001. Field Doc. 12/b. 37 pp.

Courtenay, W. R., Jr. 1990. Fish conservation and the enigma of introduced species, pp. 11-20. In *Introduced and Translocated Fishes and Their Ecological Effects*, ed. D. A. Pollard, pp. 11-20. Proceedings of the Bureau of Rural Resources 8, Canberra, Australia.

Courtenay, W. R., Jr. 1991. Pathways and consequences of the introduction of non-indigenous fishes in the United States. Unpublished Rept. to Office of Technology Assessment, Congress of the United States. 83 pp.

Courtenay, W. R., Jr. 1993. Biological pollution through fish introductions. In *Biological Pollution: The Control and Impact of Invasive Aquatic Species*, ed. B. N. McKnight, pp. 35-61. Indiana Academy of Sciences, Indianapolis, Ind.

Courtenay, W. R., Jr., D. A. Hensley, J. N. Taylor, and J. A. McCann. 1986. Distribution of exotic fishes in North America. In *Zoogeography of North American Freshwater Fishes*, eds. C. H. Hocutt and E. O. Wiley, pp. 675-698. John Wiley & Sons, New York.

Courtenay, W. R., Jr., and P. B. Moyle. 1992. Crimes against biodiversity: The lasting legacy of fish introductions. *Trans. N. Am. Wildl. Nat. Res. Conf.* 57:365-372.

Courtenay, W. R., Jr., and C. R. Robins. 1975. Exotic organisms: An unsolved, complex problem. *BioScience* 25:306-313.

Courtenay, W. R., Jr., and C. R. Robins. 1989. Fish introductions: Good management, mismanagement, or no management? *Rev. Aquat. Sci.* 1: 159-172.

Courtenay, W. R., Jr., and J. R. Stauffer, Jr., eds.. 1984. *Distribution, Biology, and Management of Exotic Fishes*. Johns Hopkins University Press, Baltimore, Md.

Courtenay, W. R., Jr., and J. N. Taylor. 1984. The exotic ichthyofauna of the contiguous United States, with preliminary observations on intranational transplants. *EIFAC Tech. Pap.* 42:466-487.

Courtenay, W. R., Jr., and J. D. Williams. 1992. Dispersal of exoticspecies from aquaculture sources, with emphasis on freshwater fishes. In *Dispersal of Living Organisms Into Aquatic Ecosystems*, eds. A. Rosenfield and R. Mann, pp. 49-81. Maryland Sea Grant College, College Park, Md.

Deacon, J. E. 1979. Endangered and threatened fishes of the west, pp. 41-64. In *The Endangered Species: A Symposium.* Great Basin Nat. Mem. 3.

Deacon, J. E. 1988. The endangered woundfin and water management in the Virgin River, Utah, Arizona, Nevada. *Fisheries (Bethesda)* 13:18-24.

de Moor, I. J., and M. N. Bruton. 1989. Atlas of alien and translocated indigenous aquatic animals in southern Africa. *S. Afr. Nat. Sci.* Prog. Rpt. 144:1-308.

Echelle, A. A., and P. J. Connor. 1989. Rapid, geographically extensive genetic introgression after secondary contact between two pupfish species (*Cyprinodon*, Cyprinodontidae). *Evolution* 43:717-727.

Elton, C. S. 1958. *The Ecology of Invasions by Animals and Plants.* Chapman and Hall, London.

Gee, J. M. 1965. The spread of Nile perch (*Lates niloticus*) in East Africa with comparative biological notes. *J. Appl. Ecol.* 2:407-408.

Hocutt, C. H. 1984. Toward the development of an environmental ethic for exotic fishes. In *Distribution, Biology, and Management of Exotic Fishes*, eds. W. R. Courtenay, Jr. and J. R. Stauffer, Jr., pp. 374-386. Johns Hopkins University Press, Baltimore, Md.

Hocutt, C. H., R. E. Jenkins, and J. R. Stauffer, Jr. 1986. Zoogeography of the fishes of the central Appalachians and central Atlantic coastal plain. In *The Zoogeography of North American Freshwater Fishes*, eds. C. H. Hocutt and E. O. Wiley, pp. 161-212. John Wiley & Sons, New York.

Hoffman, G. L., and G. Schubert. 1984. Some parasites and diseases of exotic fish. In *Distribution, Biology, and Management of Exotic Fishes*, eds. W. R. Courtenay, Jr. and J. R. Stauffer, Jr., pp. 233-261. Johns Hopkins University Press, Baltimore, Md.

Horwitz, R. J. 1982. The range and co-occurance of the shiners *Notropis analostanus* and *N. spilopterus* in southeastern Pennsylvania. *Proc. Acad. Nat. Sci. Philadelphia* 134:178-193.

Hughes, N. F. 1986. Changes in the feeding biology of Nile perch *Lates nilotica* (L.) (Pisces: Centropomidae) in Lake Victoria, East Africa since its introduction in 1960 and its impact on the native fish community of Nyanza Gulf. *J. Fish. Biol.* 29: 521-548.

Johnson, J. E., and C. Hubbs. 1989. Status and conservation of poeciliid fishes. In *Ecology and Evolution of Livebearing Fishes*, eds. G. K. Meffe and F. F. Snelson, Jr., pp. 301-317. Prentice Hall, Englewood Cliffs, N.J.

Johnson, J. E., and J. N. Rinne. 1982. The Endangered Species Act and southwest fishes. *Fisheries (Bethesda)* 7:2-8.

Laycock, G. A. 1966. *The Alien Animals.* Natural History Press, Garden City, N.Y.

Laycock, G. A. 1992. Aliens. *Wildl. Conserv.* 95(2):60-67.

Lemly, A. D. 1985. Suppression of native fish populations by green sunfish in first-order streams of piedmont North Carolina. *Trans. Am. Fish Soc.* 114:705-712.

Marsh, P. C., and J. E. Brooks. 1989. Predation by ictalurid catfishes as a deterrent to re-establishment of hatchery-reared razorback suckers. *Southwest. Nat.* 34:188-195.

McDowall, R. J. 1984. Exotic fishes: The New Zealand experience. In *Distribution, Biology, and Management of Exotic Fishes*, eds. W. R. Courtenay, Jr. and J. R. Stauffer, Jr., pp. 200-214. Johns Hopkins University Press, Baltimore, Md.

McKay, R. J. 1984. Exotic fishes in Australia. In *Distribution, Biology, and Management of Exotic Fishes*, eds. W. R. Courtenay, Jr. and J. R. Stauffer, Jr., pp. 177-199. Johns Hopkins University Press, Baltimore, Md.

Miller, D. J. 1989. Introductions and extinction of fish in the African Great Lakes. *Trends Ecol. Evol.* 4:56-59.

Miller, R. R. 1961. Man and the changing fish fauna of the American southwest. *Pap. Mich. Acad. Sci. Arts, Lett.* 46:365-404.

Minckley, W. L. 1991. Native fishes of the Grand Canyon region: An obituary? In *Colorado River Ecology and Dam Management*, pp. 124-177. National Academy Press, Washington, D.C.

Minckley, W. L., and J. E. Deacon. 1968. Southwestern fishes and the enigma of endangered species. *Science* 159:1424-1432.

Minckley, W. L., and J. E. Deacon, eds. 1991. *The Battle Against Extinction: Native Fish Management in the American West.* University of Arizona Press, Tucson, Ariz.

Mooney, H. A., and J. A. Drake, eds. 1986. *Ecology of Biological Invasions of North America and Hawaii.* Springer-Verlag, New York.

Moyle, P. B. 1976a. Fish introductions in California: History and impact on native fishes. *Biol. Conserv.* 9:101-118.

Moyle, P. B. 1976b. *Inland Fishes of California.* University of California Press, Berkeley, Calif.

Moyle, P. B. 1986. Fish introductions into North America: Patterns and ecological impact. In *Ecology of Biological Invasions of North America and Hawaii*, eds. H. A. Mooney and J. A. Drake, pp. 27-43. Springer-Verlag, New York.

Moyle, P. B., and R. A. Leidy. 1992. Loss of biodiversity in aquatic ecosystems: Evidence from fish faunas. In *Conservation Biology: The Theory and Practice of Nature Conservation,* Preservation, and Management, eds. P. L. Fiedler and S. Jain, pp. 127-169. Chapman and Hill, New York.

Moyle, P. B., H. W. Li, and B. A. Barton. 1986. The Frankenstein effect: Impact of introduced fishes on native fishes in North America. In *Fish Culture in Fisheries Management*, ed. R. H. Stroud, pp. 415-426. American Fisheries Society, Bethesda, Md.

Moyle, P. B., and J. E. Williams. 1990. Biodiversity loss in the temperate zone: Decline of the native fish fauna of California. *Conserv. Biol.* 4:275-284.

Murphy, B. R., and D. R. Terre. 1984. Genetic conservation of North American walleye stocks: A review of problems and needs. Presented at the 114th annual meeting of the American Fisheries Society, 12-16 August 1984, Ithaca, N.Y. (Abstract).

Paxton, J. R. 1990. Proposed introduction of exotic fishes into the Sepik River System, Papua New Guinea: Potential problems. In *Introduced and Translocated Fishes and Their Ecological Effects*, ed. D. A. Pollard, pp. 158-161. Proceedings of the Bureau of Rural Resources 8, Canberra, Australia.

Philipp, D. P., W. F. Childers, and G. S. Whitt. 1983. A biochemical genetic evaluation of the northern and Florida subspecies of largemouth bass. *Trans. Am. Fish. Soc.* 112:1-20.

Pierce, B. E. 1990. Biological impacts of translocations of North American salmonids, or, How to use the latest U.S. fisheries management techniques to decimate otherwise useful fisheries. In *Introduced and Translocated Fishes and Their Ecological Effects*, ed. D. A. Pollard, pp. 127-135. Proceedings of the Bureau of Rural Resources 8, Canberra, Australia.

Pollard, D. A., and J. J. Burchmore. 1986. A possible scenario for the future of Australia's freshwater fish fauna. In *Limnology in Australia*, eds. P. De Deckker and W. D. Williams, pp. 615-636. CSIRO, Melbourne, Australia.

Pollard, D. A., and P. A. Hutchings. 1990. A review of exotic marine organisms introduced to the Australian region. I. Fishes. *Asian Fish. Sci.* 3:205-221.

Ribbink, A. J. 1987. African lakes and their fishes: conservation scenarios and suggestions. *Environ. Biol. Fish.* 19:3-26.

Sheldon, A. L. 1988. Conservation of stream fishes: Patterns of diversity, rarity, and risk. *Conserv. Biol.* 2:149-156.

Sindermann, C. J. 1986. Strategies for reducing risks from introductions of aquatic organisms: A marine perspective. *Fisheries (Bethesda)* 11:10-15.

Taylor, J. N., W. R. Courtenay, Jr., and J. A. McCann. 1984. Known impacts of exotic fishes in the continental United States. In *Distribution, Biology, and Management of Exotic Fishes*, eds. W. R. Courtenay, Jr. and J. R. Stauffer, Jr., pp. 322-373. Johns Hopkins University Press, Baltimore, Md.

Waples, R. S., G. A. Winans, F. M. Utter, and C. Mahnken. 1990. Genetic approaches to the management of Pacific salmon. *Fisheries (Bethesda)* 15:19-25.

Welcomme, R. L. 1986. International measures for the control of introductions of aquatic organisms. *Fisheries (Bethesda)* 11:4-9.

Welcomme, R. L. 1988. International introductions of inland aquatic species. *FAO Fish. Tech. Pap.* 294:1-318.

Wikramanayake, E. D., and P. B. Moyle. 1989. Ecological structure of tropical fish assemblages in wet-zone streams of Sri Lanka. *J. Zool.* (London) 218:503-526.

17

THE USE OF GENETIC INFORMATION IN ESTABLISHING RESERVES FOR NATURE CONSERVATION

Stephen D. Hopper

Kings Park and Botanic Garden
West Perth, Western Australia
Australia

The establishment of reserves for nature conservation remains one of the enduring strategies for the conservation of global biodiversity, primarily because it is so cost-effective compared with the intensive management required in zoos or botanic gardens for adequate off-site conservation.

Traditionally, the allocation of land for nature conservation has been subservient to other interests. Land considered worthless for agriculture, forestry, industrial, or urban development usually constitutes the bulk of that set aside for conservation.

However, as nature conservation has emerged as a legitimate land use, biologically-based policies focused at the species and community levels of organization have contributed to reserve selection. Thus, centers of species and community richness, representativeness, endemism or rarity have been prominent in the identification of areas deemed suitable for the creation of new reserves for nature conservation (e.g., Burbidge et al. 1990).

The relatively recent emergence of the concept of biodiversity has been a significant stimulus to recognition of the variety of levels of biological organization. Biodiversity does not simply mean numbers of species, but embraces genes and populations within them, as well as the communities, ecosystems and processes of which they are part (Wilson 1988, 1992; Ehrlich and Wilson 1991; Noss 1990).

This perspective focuses on the necessity of an integrated understanding of processes below, above, and at that of species to achieve viable conservation outcomes (Falk and Holsinger 1991). Not only do individual nature reserves occur in a matrix of changing landscapes and ecosystem processes, but also species exist as dynamic metapopulations undergoing constant evolution.

Moreover, species vary significantly in their population genetic architecture and resultant modes of evolution (Grant 1981). From a population genetics viewpoint, there is little value in ensuring that a species is represented in a conservation reserve system without asking what parts of the species population system are essential to ensure ongoing evolutionary and ecological functions.

Here, I wish to explore this theme, drawing on plant conservation experience in Western Australia. We have found that genetics as a discipline can contribute substantially to the identification and management of species and to the allocation of land for reserves for species deemed important (Moran and Hopper 1987; Hopper and Coates 1990).

THE SOUTH-WEST AUSTRALIAN FLORA

The south-west of Australia is a region of significant plant biodiversity with an estimated 8,000 species, 75% of which are endemic (Hopper 1979, 1992). Unlike the equally species-rich California and the Cape Floristic Region of South Africa, south-western Australia is noteworthy in the absence of significant mountains. It is dominated by an extensive low plateau, underlain by Precambrian granites that have formed a continental nucleus of Gondwana above sea level and without glaciers since the Permian, more than 200 million years ago.

As might be expected, soils are highly leached of nutrients through eons of weathering, and form a complex but subtle pattern across the subdued landscapes of the south-west. The flat terrain has resulted in uncoordinated drainage across much of the plateau, so that salts from rain are rarely flushed to the ocean. Instead, they have accumulated in extensive salt-lake systems along ancient broad-valley drainage lines. The salt has surfaced outward from the lakes and drainage lines as replacement of deep-rooted perennial native vegetation by cereal agriculture has elevated saline water tables.

The native flora, dominated by woody perennial and tuberous herbaceous plants, has evolved remarkable adaptations to cope with the soils of the south-west (Pate and McComb 1981; Pate and Beard 1984; Hobbs 1992). It has also endured profound climatic changes, most recently in the late Tertiary and Quaternary (Hopper 1979, 1992). Conditions favoring a predominantly rainforest Tertiary flora were replaced by a more arid mediterranean climate as Australia drifted northwards from Antarctica, and modern weather patterns became established.

Most rainforest taxa became extinct in the south-west (although their descendants can still be seen in eastern and northern Australia), while plants preadapted to cope with drier conditions flourished. Quaternary climatic oscillations associated with the ice ages created substantial pressures and opportunities, especially for the flora occupying transitional country between the remnant high-rainfall forests of the extreme south-west and the deserts inland and to the north and east. With such dynamic opportunities for plant isolation and speciation, it is this transitional country, now the main agricultural wheatbelt, that today carries most of the plant diversity for which the region is renowned.

Consequently, a hallmark of the south-western flora is the large number of species within genera, signalling prolific speciation and recent evolutionary activity (Hopper 1979, 1992; James and Hopper 1981). *Eucalyptus*, for example, has an estimated 300+ species in the region.

Conserving a flora characterized by such pronounced recent speciation is challenging, especially because districts greatest in local endemism coincide with prime agricultural land of the south-west (Hopper et al. 1990). How could evolutionary dynamism and speciation in the flora be maintained in a system of increasingly fragmented nature reserves surrounded by a sea of wheat farms? The exploration of plant population genetic architecture is an essential first step to answer this question.

NATURALLY FRAGMENTED VERSUS CONTINUOUS POPULATION STRUCTURES IN RESERVES

Despite its flat terrain and seeming landscape uniformity, south-western Australia has some habitats that are quite insular and disjunct in distribution. Foremost among these are granite outcrops, which emerge irregularly across the landscape as flat or dome-shaped exposures, some up to many hectares in area and more than 100 m above the surrounding terrain. These outcrops have a rich flora and fauna (e.g., Hopper 1981; Hopper et al. 1982), with many endemic species.

Because most rock outcrops have been spared from clearing for agriculture, they present an opportunity to examine patterns of genetic variation in native species with naturally fragmented population structures. At the same time, other landforms in the south-west are more extensive and continuous, such as valley slopes in the wheatbelt and higher-rainfall southwest. Those in the wheatbelt have been extensively cleared of native vegetation, whereas those in the high-rainfall country remain largely forested for timber production and nature conservation.

Eucalypts dominate most south-western woodland and forest vegetation, and they occur in species-rich kwongan vegetation (equivalent to chaparral) as many-stemmed shrubs or small trees called mallees. Consequently, eucalypts were selected for allozyme studies to compare and contrast population genetic architecture in naturally fragmented versus continuous population systems (Moran and Hopper 1983, 1987; Sampson et al. 1988; Coates and Sokowlowski 1989; Hopper and Coates 1990).

Mallees of disjunct granite outcrops, such as *Eucalyptus caesia* and *E. crucis*, displayed relatively low genetic variation within populations but substantial genetic differences between populations. Continuously distributed populations of forest trees, such as the tall hardwood karri (*E. diversicolor*), showed high genetic variability at each sample site and very little differentiation geographically. The most significant geographical differences in karri were for populations isolated from the main forest belt in relictual sites, such as the Porongurup Range.

These observations translate into clear directions for reserve design, based on the concept of *genetic gap analysis* of Millar and Libby (1991). This simply means examining the geographical spread of detectable genetic variation, overlaying boundaries of existing conservation reserves, and searching for gaps in the reserve system where significant genetic variation of the species in question falls in the gaps between reserves. Gaps can then be assigned priorities for acquisition as reserves according to their contribution to improving the conservation of the species detectable genetic variation.

Hopper and Coates (1990) proposed that for karri, relatively few but large core reserves with high heterozygosity would capture most detectable allozyme variation. This had already been achieved through the creation of large reserves, such as Shannon River National Park. Outliers with unusual alleles, such as at the Porongurups, were also identified as important. Recognizing that allozymes sample only part of karri's genetic variation, it was also stressed that a broad range of habitats occupied by karri in its main range should be included in the large core reserves to cover undetected physiological and ecological variation.

For *Eucalyptus caesia* and *E. crucis* on the insular granite outcrops, replicated geographical coverage, irrespective of population size, was essential to ensure adequate sampling of genetic variation in the species included in the reserve system. For example, a population of only 48 plants at The Humps was the most genetically distinct of all populations of *E. caesia* sampled. An ongoing program of reserve acquisition continues for the rocks on which these species occur, with some significant successes in recent years.

The generality of the patterns of variation found in the naturally fragmented versus continuous population structures of these eucalypts requires additional investigation, but data already in hand on other forest eucalypts are supportive (Moran and Hopper 1987).

Apart from the immediate applications to reserve design, the population genetic data have implications for future management. For example, fragmentation of populations of continuously distributed forest eucalypts would lead to significant loss of genetic variation, with possible inbreeding effects (Charlesworth and Charlesworth 1987), and commence the genetic divergence of previously contiguous populations. Gene flow may have to be enhanced through direct management to maintain levels of heterozygosity (Ledig 1986). Conversely, naturally disjunct species may well be able to continue their evolutionary and ecological processes without direct management intervention if enough of their immediate habitat is retained in the fragmentation process.

RESERVATION OF TRANSILIENT AND HYBRID POPULATIONS

A special category of populations deserving attention in reserve design is those where the evolution of genetic novelty is pre-eminent. Genetically transilient populations (cf., Templeton this volume) and hybrid populations are examples.

Now spanning three decades, the detailed studies of James (1965, 1981, 1982) and James et al. (1990 and references therein) on the Australian granite rock herb *Isotoma petraea* (Lobeliaceae) provide a superb example of genetic transilience. Throughout much of arid central Australia, this species is outbreeding and diploid with normal chromosome behavior at meiosis. In the wheatbelt and adjacent areas of south-western Australia, however, the species becomes inbreeding and forms chromosomal ring structures at meiosis, thus indicative of translocation heterozygosity and a balanced lethal system. Moreover, the number of chromosomes involved in ring formation increases towards the south-western margin of the species range. This is clearly a derived condition. James has demonstrated convincingly that the evolution of this population system commenced at a population on Pigeon Rock, at the junction of the outbreeding population system and the inbreeding south-western derivatives (James et al. 1990). Although *Isotoma petraea* is a common species of no immediate conservation concern, its Pigeon Rock population is small and threatened by cattle grazing. Without the benefit of James' careful population genetic studies, this fundamentally important transilient population may well have become extinct without being noticed.

A population of *Eucalyptus caesia* at Yanneymooning Hill has morphological and allozyme characteristics suggesting that it is also either the ancestral transilient population of the derived subspecies *magna*, or it is a hybrid population of *magna* and subspecies *caesia* (Hopper et al. 1982). In either case, the population is one of special evolutionary activity, and warrants inclusion in the nature reserve system. This outcome was successfully achieved in 1992.

Hybrids may be far more important in plant evolution than most conservation managers appreciate. For example, Grant (1981) estimated that 70% of angiosperms might be of hybrid origin, and proposed that "synagameons composed of hybrid semispecies are often more common than well-defined biological species" and "agamic hybrid complexes are likewise quite common."

A recent policy change in Western Australia, motivated by an appreciation of the evolutionary and ecological role of hybrid plants, has enabled rare and endangered hybrids to be afforded the same special legal protection as endangered species and subspecies.

Managers of the Stirling Range National Park, which contains an unusual concentration of eucalypt and other hybrids (Hopper 1993) and is clearly a site of active evolutionary change, can now focus some of their energies toward ensuring conservation of transilient hybrid populations previously considered unimportant.

The conservation management of hybrids and genetically transilient populations deserves serious consideration, research and policy development in the near future if plant biodiversity and its originating processes are to be conserved.

RESTORATION AND RECOVERY OF SPECIES ON NEW RESERVES

In some cases, choice of the location of a new conservation reserve may be predicated on requirements to restore or recover populations of a species or group of species. This is a relatively new field in Western Australia, but one where it is already evident that knowledge of genetic architecture and breeding system is very important.

Although conventional horticulture and transplantation might work for many species, the establishment of self-maintaining populations may be frustrated with such approaches in the case of species, such as *Isotoma petraea* or *Stylidium crossocephalum* (Stylidiaceae: Coates and James 1979) that form intersterile microspecies isolated by chromosomal sterility barriers. It is of fundamental importance, in the absence of detailed population genetic data, to use local propagating stocks for restoration and recovery programs, especially in floras where unusual genetic systems occur, such as in the south-west or documented for groups such as *Clarkia* and *Gilia* in California.

Even where local propagating stocks are used, and the species has normal chromosomal behavior, a knowledge of population genetic architecture is critical to the long-term success of a recovery program. In the endangered Rose Mallee *Eucalyptus rhodantha*, Sampson (1988) and Sampson et al. (1989, 1990) found that the largest remnant stand of the species produced about 60% outcrossed seed, but that a population of 14 plants in a cleared and grazed area produced only 26% outcrossed seed. Such high levels of inbreeding in a eucalypt were unprecedented, and signalled a species in real distress.

Sampson's elegant study of population genetic architecture in the eucalypt proved invaluable for the development of an appropriate recovery strategy. Management is now in place aimed at planting understory species to attract pollinators, building up numbers of the Rose Mallee to 30 or more in rehabilitated areas and planting seedlings of known genotype in a spatial array to minimize inbreeding.

CONCLUSIONS

These few examples from a remote but biologically interesting corner of the world highlight that the potential for evolution and speciation will be lost in reserve design and conservation management without strategies based on population genetic data.

Informed management requires sound biological knowledge at various levels of the organizational hierarchy, and an ability to integrate such knowledge in a way that leads to maximum conservation outcomes. Although it may be convenient to focus policy and management at the species and community levels, it is biologically naive.

Through understanding the detail and complexity of population genetic architecture in at least some of the world's species deemed of conservation importance, we will be better managers for it.

ACKNOWLEDGMENTS

Tom Ledig facilitated my participation in the symposium and he and Bob Szaro were most patient in dealing with a very late manuscript. Much of the research underpinning the paper was completed while I was on the staff of the Western Australian Wildlife Research Centre, Department of Conservation and Land Management. Ideas expressed herein derive from collaborative research with many colleagues, including Sid James, Gavin Moran, Dave Coates, and Jane Sampson, but the responsibility for their presentation in this form is mine alone.

LITERATURE CITED

Burbidge, A. A., S. D. Hopper, and S. van Leeuwen. 1990. Nature conservation, landscape and recreation values of the Lesueur area. *Environ. Protection Authority Bull.* 424 (Perth).

Charlesworth, D., and B. Charlesworth. 1987. Inbreeding depression and its evolutionary consequences. *Ann. Rev. Ecol. Syst.* 18:237-268.

Coates, D. J., and S. H. James. 1979. Chromosome variation in *Stylidium crossocephalum*. F. Muell. (Angiospermae: Stylidiaceae) and the dynamic coadaptation of its lethal system. *Chromosoma* 72: 357-376.

Coates, D. J., and R. E. S. Sokolowski. 1989. Genetic diversity in karri *Eucalyptus diversicolor*. *Austral. J. Bot.* 37:145-156..

Ehrlich, P. R., and E. O. Wilson. 1991. Biodiversity studies: science and policy. *Science* 253:758-762.

Falk, D. A., and K. E. Holsinger, eds. 1991. *Genetics and Conservation of Rare Plants.* Oxford University Press, New York.

Grant, V. 1981. *Plant Speciation.* 2nd ed. Columbia University Press, New York.

Hobbs, R. J. 1992. *Biodiversity in Mediterranean Ecosystems in Australia.* Surrey Beatty & Sons, Chipping Norton, NSW.

Hopper, S. D. 1979. Biogeographical aspects of speciation in the south west Australian flora. *Ann. Rev. Ecol. Syst.* 10:399-422.

Hopper, S. D. 1981. Honeyeaters and their winter food plants on granite rocks in the central wheatbelt of Western Australia. *Austral. Wildl. Res.* 8:187-197.

Hopper, S. D. 1992. Patterns of plant diversity at the population and species levels in south-west Australian mediterranean ecosystems. In *Biodiversity in Mediterranean Ecosystems in Australia*, ed. R. J. Hobbs, pp. 27-46. Surrey Beatty & Sons, Chipping Norton, NSW.

Hopper, S. D. 1993. Eucalypts. In *Mountains of Mystery A Natural History of the Stirling Range*, eds. C. Thomson, G. Hall, and G. Friend, pp. 71-83. Department of Conservation and Land Management, Perth.

Hopper, S. D., and D. J. Coates. 1990. Conservation of genetic resources in Australia's flora and fauna. In *Australian Ecosystems: 200 Years of Utilization, Degradation and Restoration*, eds. D. A. Saunders, A. J. M. Hopkins, and R. A. How. *Proc. Ecol. Soc. Austral.* 16:567-577.

Hopper, S. D., N. A. Campbell, and G. F. Moran. 1982. *Eucalyptus caesia*, a rare mallee of granite rocks from south-western Australia. In *Species at Risk. Research in Australia*, eds. R. H. Groves and W. D. L. Ride, pp. 46-61. Australian Academy of Science, Canberra.

Hopper, S. D., S. van Leeuwen, A. P. Brown, and S. J. Patrick. 1990. *Western Australia's Endangered Flora.* Department of Conservation and Land Management, Perth.

James, S. H. 1965. Complex hybridity in *Isotoma petraea.* I. The occurrence of interchange heterozygosity, autogamy and a balanced lethal system. *Heredity* 20:341-353.

James, S. H. 1981. Cytoevolutionary patterns, genetic systems and the phytogeography of Australia. In *Ecological Biogeography of Australia*, ed. A. Keast, pp. 763-782. Junk, The Hague.

James, S. H. 1982. The relevance of genetic systems in *Isotoma petraea* to conservation practice. In *Species at Risk: Research in Australia*, eds. R. H. Groves and W. D. L. Ride, pp. 63-71. Australian Academy of Sciences, Canberra.

James, S. H., and S. D. Hopper. 1981. Speciation in the Australian Flora. In *The Biology of Australian Plants*, eds. J. S. Pate and A. J. McComb, pp. 361-381. University of Western Australia Press, Perth.

James, S. H., J. Playford, and J. F. Sampson. 1990. Complex hybridity in *Isotoma petraea.* VIII. Variation for seed aborting lethal genes in the O6 Pigeon Rock population. *Heredity* 66:173-180.

Ledig, F. T. 1986. Heterozygosity, heterosis, and fitness in outbreeding plants. In *Conservation Biology, the Science of Scarcity and Diversity*, ed. M. E. Soulé, pp. 77-104. Sinauer Associates, Sunderland, Mass.

Millar, C. I., and W. J. Libby. 1991. Strategies for conserving clinal, ecotypic, and disjunct population diversity in widespread species. In *Genetics and Conservation of Rare Plants*, eds. D. A. Falk and K. E. Holsinger, pp. 149-170. Oxford University Press, New York.

Moran, G. F., and S. D. Hopper. 1983. Genetic diversity and insular population structure of the rare granite rock species *Eucalyptus caesia.* Benth. *Austral. J. Bot.* 31:161-172.

Moran, G. F., and S. D. Hopper. 1987. Conservation of the genetic resources of rare and widespread eucalypts in remnant vegetation. In *Nature Conservation: The Role of Remnants of Native Vegetation*, eds. D. A. Saunders, G. W. Arnold, A. A. Burbidge, and A. J. M. Hopkins, pp. 151-162. Surrey Beatty & Sons, Sydney.

Noss, R. F. 1990. Indicators for monitoring biodiversity: a hierarchical approach. *Conserv. Biol.* 4:355-364.

Pate, J. S., and J. S. Beard, eds. 1984. *Kwongan—Plant Life of the Sandplain.* University of Western Australia Press, Perth.

Pate, J. S., and A. J. McComb, eds. 1981. *The Biology of Australian Plants.* University of Western Australia Press, Perth.

Sampson, J. F. 1988. The population genetic structure of *Eucalyptus rhodantha* Blakely & Steedman and its allies *Eucalyptus crucis* Maiden and *Eucalyptus lane-poolei* Maiden. Ph.D. thesis, University of Western Australia, Nedlands.

Sampson, J. F., S. D. Hopper, and S. H. James. 1988. Genetic diversity and the conservation of *Eucalyptus crucis.* Maiden. *Austral. J. Bot.* 36:447-460.

Sampson, J., S. D. Hopper, and S. H. James. 1989. The mating system and population genetic structure in a bird-pollinated mallee, *Eucalyptus rhodantha. Heredity* 63:383-393.

Sampson, J., S. D. Hopper, and D. J. Coates. 1990. Eucalyptus rhodantha. Wildlife Management Program No. 4. 34 pp. Department of Conservation and Land Management, Perth.

Wilson, E. O., ed. 1988. *Biodiversity.* National Academy Press, Washington, D.C.

Wilson, E. O. 1992. *The Diversity of Life.* Allen Lane, The Penguin Press, London.

18

STRATEGIES TO REDUCE EROSION OF BIODIVERSITY BY EXOTIC TERRESTRIAL SPECIES

Lloyd L. Loope

Haleakala National Park
Makawao, Maui, Hawaii
U.S.A.

Charles P. Stone

Hawaii Volcanoes National Park
Hawaii National Park, Hawaii
U.S.A.

Human-assisted invasions of biogeographic regions by exotic organisms are a major but under-appreciated cause of global loss of biodiversity. Organisms are moved around at an ever accelerating pace, and become established and thrive with increasing frequency and intensity because of habitat modification and fragmentation (D'Antonio and Vitousek 1992). Taxa that evolved in isolation from each other are forced into permanent and irreversible cohabitation, which can initiate a cascading chain of extinctions (Diamond 1989). As a result, a relatively low, but increasing number of successful invaders are now homogenizing and simplifying biotic communities worldwide (Drake et al. 1989). Single introductions frequently lead to extinctions of more than one native species. Collectively, accidental and intentional introductions are major causes of species loss (Macdonald et al. 1989).

The loss of biodiversity from invasions is most severe on oceanic islands and on natural habitat islands (e.g., subtropical Florida (Austin 1978; Center et al. 1991), South Africa's Cape province (Macdonald et al. 1986)), and is increasing in continental ecosystems from accelerating habitat fragmentation and transport of species (McKnight 1993). Although much irreversible damage has already been done and some further anthropogenic invasion is undoubtedly inevitable, action to slow introductions of exotic species is necessary and possible.

Three examples are presented to illustrate the severity of lost biodiversity from introduction of exotics: (1) the worldwide status of exotic grasses; (2) the increasing decline of forests in the Appalachian Mountains of the southeastern United States caused by introduced insects and diseases; and (3) exotics and the status of the conservation of

Hawaiian biota. We then elaborate on the Hawaiian example, detailing strategies for control of exotics and prevention of their continuing establishment.

EXOTIC GRASS INVASIONS

A recent review (D'Antonio and Vitousek 1992) revealed that introduced grasses are widespread, compete aggressively with native species, and substantially affect ecosystems worldwide. The most significant effects of exotic grasses on ecosystems are from their interactions with fire. Invasion of exotic grasses can set in motion a grass/fire cycle because they provide the fine fuel that initiates and sustains fire. Frequency, area, and intensity of fires often increase as a result of grass invasion. Following grass-fueled fires, exotic grasses recover more rapidly than native species and cause a further increase in susceptibility to fire and progressive impoverishment of biodiversity. Such grass species include the well-studied *Bromus tectorum*, a native of Eurasia, in western North America (Billings 1990); *Hyparrhenia rufa*, *Melinis minutiflora*, *Panicum maximum*, and *Brachiaria* spp. (all African C4 species) in Central and South America; and *Cenchrus ciliaris*, *Melinis minutiflora*, *Pennisetum polystachyon*, and *Ehrharta calycina* in Australia. Through production of 2-3 times as much flammable material as native grasses, *Cenchrus ciliaris* alters Australian watercourses from their historical role as fire barriers to one of fire "wicks," serving as conduits carrying fire to new areas. In Central and South America and Australia, the ecological change often involves conversion of rain forest to grassland, triggering regional climatic change, which in turn accelerates loss of biodiversity.

DECLINE OF APPALACHIAN FORESTS

Species diversity of trees is greater in the temperate deciduous and coniferous forests of the Appalachian Mountains in North Carolina and Tennessee than in any other forest region of the United States (Whittaker 1956). However, exotic diseases and insects are causing an alarming erosion of diversity in these forests and other forests of the eastern United States (Langdon and Johnson 1992), demonstrating that high diversity does not necessarily buffer individual species against catastrophic decline (Table 18.1). The extinctions that follow removal of a tree species from an ecosystem remain mostly undocumented, although the demise of five native microlepidopterans formerly associated with the once-dominant American chestnut has been recognized (Pyle et al. 1981). Dogwood anthracnose is one of the new devastating introduced diseases. Its effects might eventually prove to be as harmful as those of chestnut blight and Dutch elm disease. The host is the eastern flowering dogwood (*Cornus florida*), a beautiful and popular tree and the state tree of many states. The causative agent is *Discula destructiva* (Redlin 1991), a native fungus of Asia which entered the United States with plantings of *Cornus cusa*, an Asiatic dogwood introduced by the Arnold Arboretum in Massachusetts. *Cornus cusa* was widely planted and was only mildly affected by the disease (Daughtery 1993). This disease was first noted in New York in the late 1970s and had spread south to the southern Appalachians by 1987 (Langdon and Johnson 1992).

Table 18.1. Number of native, endemic, and established alien species in major groups in the Hawaiian Islands.[1]

Taxonomic Group	No. of Extant Native Species	No./(%) of Endemic Species	No. of Invasive Alien Species
Flowering plants	970	883 (91%)	800
Ferns and allies	143	105 (73%)	21
Hepaticae (liverworts)	168	*ca.* 112 (67%)	2
Musci (mosses)	233	112 (48%)	3
Lichens	678	268 (38%)	0
Resident land birds	57	44 (77%)[2]	38
Mammals (on land)	1	0	18
Reptiles (on land)	0	0	13
Amphibians	0	0	4
Freshwater fish	6	6 (100%)	19
Arthropods	6000-10000	(98%)	*ca.* 2000
Mollusks	*ca.* 1060	(99%)	9

[1] From Loope and Mueller-Dombois 1989. Native species include both indigenous (found elsewhere) and endemic (found only in Hawaii) species. Alien species were brought to Hawaii by humans.

[2] James and Olson (1991) consider 40-55 historically-known species and 35 fossil species of passerines endemic (i.e. at least 75-90 endemic species). An additional 39 species of non-passerine endemics are known (Olson and James 1991). At least 8 additional fossil forms of passerines are undescribed, for at total of 140 species.

The balsam woolly adelgid, *Adelges piceae* (Arthropoda: Homoptera), was first noted in Nova Scotia and Maine in the early 1900s and had spread to the southern Appalachians by 1950 (Eager 1984). It attacks fir (*Abies*) species and might completely eliminate Fraser fir (*Abies fraseri*), a southern Appalachian endemic. In the Great Smoky Mountains National Park, where Fraser fir is common, all mature trees were killed in the past 30 years. Young trees are attacked and killed at about 20 years of age, when they first produce viable seeds (Langdon and Johnson 1992).

HAWAIIAN BIOTA—EXOTIC AND NATIVE

Flora and fauna of oceanic islands in general and the Hawaiian Islands in particular are well known for their great susceptibility to displacement by invasions of non-native species (Elton 1958; Williamson 1981; Brockie et al. 1988; Diamond 1989). Because of their evolution in relative isolation (over 3,000 km from the nearest continent) and in the absence of many forces shaping continental organisms, ecosystems of the Hawaiian Islands are particularly vulnerable to invasion by exotic species from continents. Numbers of exotic species in the Hawaiian Islands exceed numbers of native species in some taxonomic categories (Table 18.2). More native species have been eliminated in Hawaii than anywhere else in the United States. An estimated 10% of the Hawaiian flora is extinct, and another 40-50% is threatened with extinction (Wagner et al. 1990). Federally listed endangered plants in Hawaii are expected to number 200 within the next two years (HDLNR/FWS/TNCH 1991). At least 70 of 140 species of native birds, known

Table 18.2. Selected exotic forest pests of Eastern North America (from Langdon and Johnson 1992).

Native Host Species	Alien Pest	Original Range of Pest
Fraser fir (*Abies fraseri*); also *A. balsamea*	Balsam wooly adelgid (*Adelges piceae*) • insect	Europe, Asia
sugar maple (*Acer saccharum*); others	Pear thrips (*Taeniothrips inconsequens*) • insect	Europe
American chestnut (*Castanea dentata*); some oaks; chinquapins	Chinese chestnut (*Endothia parasitica*) • fungus	Asia
flowering dogwood (*Cornus florida*); other American *Cornus*	Dogwood antracnose (*Discula destructiva*) • fungus	Probably East Asia
Butternut (*Juglans cinerea*) or white walnut	Butternut canker (*Seriococcus clavigineti-juglandacearum*) • fungus	Unknown; most *Juglans* in Asia, South-Central Europe
Eastern white pine (*Pinus strobus*); also some western five-needled pines	White pine blister rust (*Cronartium ribicola*) • fungus	Europe
Oaks (*Quercus* spp.); many other genera such as *Acer*, *Betula*, *Carya*, *Populus*	Gypsy moth (*Lymantria dispar*) • insect	Europe, North Africa, Asia
American mountain ash (*Sorbus americana*)	European sawfly (*Pristiphora geniculata*) • insect	Northern Europe
Eastern or Canadian hemlock (*Tsuga canadensis*); also *Tsuga caroliniana*	Hemlock wooly adelgid (*Adelges tsugae*) • insect	East Asia
American elm (*Ulmus americana*); other eastern elm species	Dutch elm disease (*Ceratocytis ulmi*) • fungus	Europe, Asia

historically or from the modern fossil record, are believed to be extinct as a result of human activities (James and Olson 1991). Thirty extant Hawaiian bird species are listed as federally endangered, 41% of all listed birds in the United States. Although habitat destruction has been a major cause of extinction (Kirch 1982; Gagne 1988), the introduction of exotic species, especially feral ungulates (goats, pigs, sheep, axis deer), ants (especially *Pheidole megacephala*), parasitoids and plants have contributed substantially to the endangerment and extirpation of species (Cuddihy and Stone 1990).

In spite of much degradation, substantial remnants of ecosystems survive, especially at high elevations and on azonal substrates, and provide the opportunity to retain large samples of Hawaiian biodiversity.

Elimination of Exotics in Hawaii

Because Hawaii is already flooded with exotic plants and animals, the protection of native biodiversity might seem hopeless. This was, in fact, a widespread view among land management agencies in Hawaii in the 1960s and 1970s (Mueller-Dombois 1981). In the 1970s at Hawaii Volcanoes and in the 1980s at Haleakala and Hawaii Volcanoes, the National Park Service demonstrated that native Hawaiian biota can be protected with management (Stone and Loope 1987). Although current management is still not adequate, recent efforts by the Hawaii Division of Forestry and Wildlife, The Nature Conservancy of Hawaii, the U.S. Fish and Wildlife Service, and the National Park Service are aimed at more adequately addressing the problems (Stone and Holt 1990; HDLNR/FWS/TNCH 1991).

Hawaii National Park, which included much of the present day Haleakala and Hawaii Volcanoes National Parks, was established in 1916 to preserve geological and biological resources of the Haleakala volcano on Maui and the more geologically active Kilauea volcano on Hawaii Island. The Haleakala (now 11,400 ha) and Hawaii Volcanoes National Parks (93,000 ha) were given separate status in 1961. Together the two parks account for 50% of the legally protected natural area in the state. Each park includes continuous protected sea-level to mountaintop landscapes that are rare elsewhere in Hawaii, an abundance of endangered, threatened, and rare species, and an excellent sampling of the endemic Hawaiian biota. Until recently, however, both parks were degraded by feral goats (*Capra hircus*) and pigs (*Sus scrofa*) (Stone and Loope 1987).

Goats and Pigs

In both parks, the public hunted feral mammals without significantly reducing their populations from establishment of the parks in 1916 until recently. Although drastic removal of plant cover and acceleration of soil erosion by 15,000-20,000 goats were causing an obvious and profound decline of native fauna and flora, goats were tolerated on state and private lands adjacent to both parks. The parks had to be fenced before goats could be eliminated. In an experimental exclosure in the lowlands of Hawaii Volcanoes in 1968, an endemic leguminous vine (*Canavalia kauensis*) new to science, appeared. It germinated from the seed bank in the area protected from goats. This incident and efforts by the National Park Service gained public support for the funding of fences and complete control of goats in Hawaii Volcanoes by 1980. A long-term plan for monitoring numbers of goats and the response of plants to the management was implemented (Stone et al. 1992) and repeated in the smaller but topographically more diverse Haleakala National Park in the 1980s.

The first step toward control of feral pigs in Hawaii also required a clear demonstration of the damage to native ecosystems through exclosures, analyses of food habits, and other studies. Rooting by pigs had destroyed native vegetation, with certain plant species sought out selectively, and had created open areas subject to invasion by exotic plants. Pigs also dispersed exotic plants by passing seeds in their dung. Techniques for the removal of pigs were tested in fenced units of several hundred hectares. Hunting with dogs and snaring with customized wire cables were the most cost-effective techniques (Stone 1991). The latter was especially effective in remote rain forests of Haleakala's Kipahulu Valley (Anderson and Stone 1993). Studies of pig movements with radiotelemetry were used to determine boundaries of some management units for fencing and to develop strategies for more effective control by hunting and snaring. Effectiveness of control and recovery of the community were monitored (Stone 1991). Although pigs have not been eliminated from the Hawaii Volcanoes and Haleakala National Parks, many fenced management units are free of pigs, and available control techniques are adequate.

Exotic Plants

Control of exotic plants in the Hawaiian national parks requires managers to make choices. In Hawaii Volcanoes National Park, 29 exotic plant species are disruptive; 24 are too widespread to control park-wide by conventional methods. Current strategies (Figure 18.1) include the following (Tunison 1992a): (1) feral ungulates are eliminated before or during control of exotic plants; (2) fire is prevented; (3) distribution of important exotic plant species is mapped; (4) localized exotic plant species are eliminated throughout the park (Tunison and Zimmer 1992); (5) biological control is exercised for some widespread species; (6) all disruptive species are controlled in areas of high ecological value (Special Ecological Areas) (Tunison and Stone 1992); (7) fountain grass (*Pennisetum setaceum*), a potentially devastating (because of its role in the fire cycle) widespread species is confined to infested lowland areas (Tunison 1992b); (8) research into herbicidal control is conducted to develop safe and effective chemical control; (9) the ecology of exotic plant species is studied; and (10) the public is educated in the importance of controlling exotic plants.

Biological control of exotic plants is essential in Hawaii (Gardner 1990). A cooperative state and federal steering committee coordinates biocontrol of the most

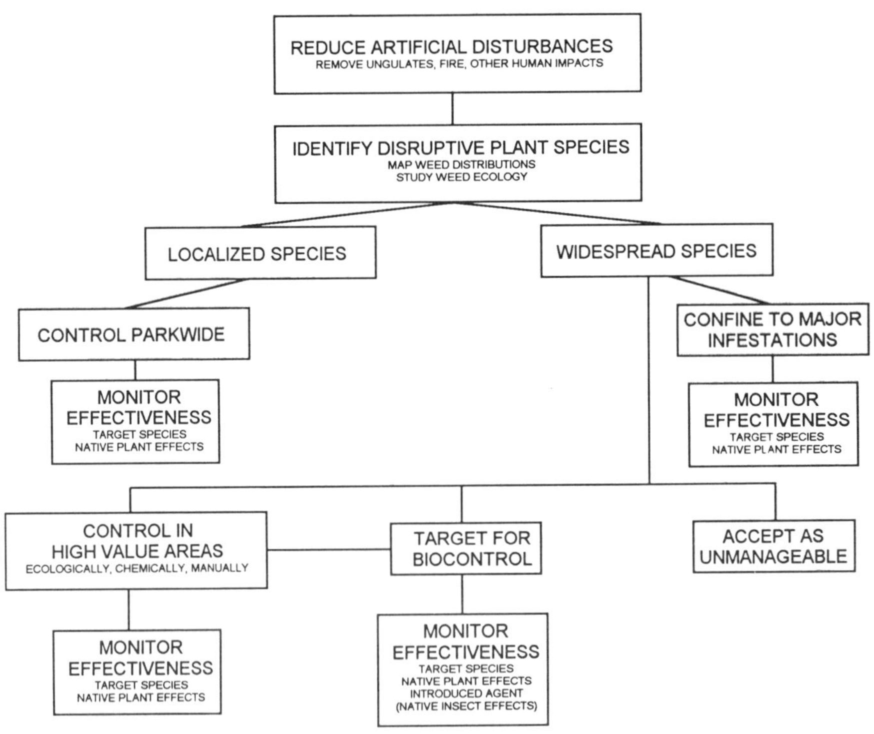

Figure 18.1. Exotic plant control strategies in the alien plant control program at Hawaii Volcanoes National Park (courtesy J.T. Tunison).

widespread and damaging exotic plants that threaten natural areas. Eight plant species are targeted, and insects were released to control four of the species (*Passiflora mollissima*—Passifloraceae, *Clidemia hirta*—Melastomataceae, *Ulex europeus*—Fabaceae, and *Myrica faya*—Myricaceae). An entomologist of the U.S. Forest Service, a plant pathologist of the National Park Service, and scientists of the Hawaii Department of Agriculture are conducting research and exploratory activity in host countries of the target species. Federal entomological research is conducted in an insect quarantine facility at the Hawaii Volcanoes National Park.

The success rate of biological control in Hawaii has been about 50% (including agricultural pests), but costs are high and time lags lengthy (Markin et al. 1992). Other problems include stringent regulations to avoid damage to non-target organisms, difficulties of foreign exploration and study especially in such countries as Colombia and Peru, and conflicts of interest with agriculture. No reasonable alternative to biocontrol exists for some species. In spite of being subjected to much valid criticism (Howarth 1991), primarily for past mistakes involving invertebrate introductions, biological control has had some remarkable successes worldwide and a safe long-term track record for dealing with plant pest problems (Groves 1989).

Preventing Invasions in Hawaii

An average of 20 new immigrant species of invertebrates per year established themselves in Hawaii from 1961 to 1990; about 50% of these are economic pests; about 1 in 20 are considered serious economic pests (TNCH/NRDC 1992). The trend is one of widely fluctuating, but not declining numbers. No one really knows what proportion of the new arrivals damage native biota.

The future of Haleakala and Hawaii Volcanoes National Parks and other protected natural areas as reservoirs of native biodiversity in the Hawaiian Islands might depend more than anything else on the success of keeping new exotic plant and animal species from becoming established. Many aggressive exotic plant and animal species not yet established in Hawaii can, if introduced, exploit and modify habitats not yet threatened by any already established exotic species. Unless combatted with ingenuity and commitment, the insidious threat of exotic species can be expected to proliferate and to inundate all but the most resistant native ecosystems in national parks and other protected areas of Hawaii. In view of the difficulty and expense of controlling invasive species after they are established, preventing establishment of new introductions is not only highly cost effective, but essential.

Recent experience at Haleakala National Park revealed that efforts to stop new exotic species from establishing and becoming widespread are not necessarily doomed to failure. The National Park Service recognizes that the Park (114 km^2 in area), an international biosphere reserve, on an 1,864-km^2 oceanic island, cannot be managed as an island. As with many other conservation areas, the park's survival as a viable conservation unit depends on implementation of regional management. Pigs and goats can be fenced out, but many other exotics cannot. Maui and other Hawaiian islands have to develop and refine interagency cooperation to avoid establishment of aggressive species because no agency or organization has the mandate or responsibility to stop them.

One recent challenge at Haleakala National Park involved an outbreak of rabbits. European rabbits (*Oryctolagus cuniculus*) have been liberated on at least 700 islands throughout the world, and devastation from their establishment has been well documented (Atkinson 1989). Although the Hawaiian Islands are well known to be highly vulnerable to biological invasions, the domestic rabbit had come to be disregarded as a potential invader after more than a century without invasion on a major island. In July 1990 after discovery of a reproducing rabbit population in 25 ha of native shrubland at high elevation (2,075-2,135 m), the administration of the Haleakala National Park initiated monitoring and removal of the rabbits. The population originated from six unwanted cage-reared rabbits that a pet owner released in October 1989. Because of their obvious threat to native biota, the eradication of the feral rabbits was given the highest priority. A total of 93 rabbits was removed from the 25 ha between August 1990 and March 1991 by snaring, shooting, and trapping. Four more rabbits were removed up-slope (elevation 2,135-2,440 m) at distances of 0.6-2.4 km from the main area of infestation. The last one was removed on May 6, 1991. Monitoring of transects (assessment of rabbit pellet presence/absence, age, and abundance), combined with scouting and follow-up of reports of rabbit sightings by visitors and other park employees (concurrently with rabbit control) allowed accurate assessment of numbers and location of remaining rabbits. The park is still faced with the recurring problem of release or escape of pet rabbits (Loope et al. 1992).

Haleakala National Park and its collaborators are attempting to eradicate an invasive tree, *Miconia calvescens*, known as "the plague of French Polynesia." This species grows to 15-18 m and has large, dark green leaves with maroon undersides. It is native from southern Mexico to northern Argentina and Chile, where it grows at elevations of 300-1830 m and seems to be an understory species and an invader of small forest gaps. It was introduced to Tahiti in 1937, but the first published record of its apparent invasiveness there was in 1976 (Birnbaum 1991). Dr. Raymond Fosberg (pers. comm. 1991) of the Smithsonian Institution saw it on the Taravao Plateau in 1971 and concluded that "this is a plant that could destroy Hawaiian forests." Aerial photographs from 1978 showed that the bird-dispersed, shade-tolerant *Miconia* in the forest canopy dominated an area of only 100-200 ha on the Plateau. In early 1983, two hurricanes severely disturbed forests on Tahiti, and by 1989, *Miconia* was present on over 75% of the island and dominated a large part of the canopy (Birnbaum 1991).

Because of its perceived attractiveness as an ornamental, *Miconia calvescens* was introduced in the 1970s by the horticultural industry to at least three Hawaiian islands: Hawaii, Oahu, and Maui. Upon discovery of the plant on Maui in 1990, the National Park Service initiated eradication, with the cooperation of private landowners, although the populations were still 5 km from Haleakala National Park. Founding individuals had grown to 0.3 m in diameter and over 10 m tall and produced abundant seedlings locally. *Miconia calvescens* merits special concern on East Maui because few other plant species are highly invasive in forests above 1,200 m elevation. It is likely to invade such areas, given its native elevational range and its performance in climatically similar Tahiti. It can become established in such dense stands that it can change the landscape and cause massive loss of biodiversity (Gagne et al. 1992; Medeiros and Loope 1992).

Over 13,000 *M. calvescens* plants have been removed from Maui to date by the National Park Service and volunteers. In mid-1992, removal of all known *Miconia* plants seemed possible. However, because of bird dispersal and the fact that the seed bank lasts for at least several years, periodic follow-up for an unknown length of time is necessary.

The invasion of rabbits, *Miconia calvescens*, and other species alerted scientists, managers, and the public in Hawaii to the absence of means that operate to prevent establishment of new aggressive exotic species in the state. The approach at Haleakala National Park has been to publicize these problems, to raise community consciousness and concern, and to remove invading exotics with support from the community. Experience suggests that the public is receptive to a campaign to stop the proliferation of exotic species (Loope 1992). Grassroots efforts, initiated on an emergency basis by necessity, are fully complementary to statewide efforts. No program will succeed without grassroots understanding and support. Experience at Haleakala National Park reinforces the observation by Merton (1978) that successful eradication involves at least four essential ingredients: (1) careful planning, (2) public relations to enlist support and reduce emotional or misinformed criticism, (3) a high level of dedication by project personnel to achieve total eradication, and (4) sustained funding. Merton (1978) stressed that "eradication is often as much a psychological challenge as it is a physical and biological one, and without total commitment by field workers and those financing the project, eradication cannot be achieved." These were, of course, the same essential ingredients for removal of goats and pigs from the Hawaiian parks.

Cooperative Efforts and Current Invasions

Identification of gaps in the legal/quarantine system in Hawaii is in progress as a prelude to increasing effectiveness in preventing entry of invasive exotic species (TNCH/NRDC 1992). Initial interagency cooperation in this exercise was encouraging, but the large number of agencies with responsibilities in this field compound the complexity of perfecting quarantine systems to stop innumerable potential invaders from entering by diverse routes.

Federal agencies in Hawaii are concerned with preventing the introduction of noxious pests into the United States from foreign sources and preventing established pests in Hawaii from reaching the United States mainland. Their work is guided by federal laws and rules that evolved with a focus on protecting large-scale mainland agriculture and enforcing international trade agreements. The U.S. Customs Service, U.S. Fish and Wildlife Service, U.S. Department of Agriculture's Animal and Plant Health Inspection Service, and the U.S. Department of Defense's Military Customs Inspection are the federal agencies most involved in prevention (TNCH/NRDC 1992).

State agencies (primarily the Hawaii Department of Agriculture) have the responsibility of preventing the introductions from the United States mainland that might be damaging in Hawaii and control of newly escaped pests, although the state Department of Land and Natural Resources and federal agencies conduct pest control on the lands they manage. The system is under-funded and understaffed, and its quarantine personnel cannot inspect more than a very small proportion of cargo and other traffic. Penalties for non-compliance with regulations are inadequate. A federal law prohibiting inspection of first class domestic mail leaves that avenue for pests unchallenged. The responses of agencies to perceived mandates have led to a set of programs that is effective in their own jurisdictions, but leaves many gaps and leaks for the entry and establishment of pests. Until now, agricultural pests were targeted and little emphasis was placed on preventing establishment of species that threaten biodiversity. Interisland spread of pests is a major and largely unregulated problem. Multi-agency cooperative planning for major refinement of the quarantine and control system under the auspices of the Hawaii Department of Agriculture is proposed (TNCH/NRDC 1992).

PRINCIPLES FOR MANAGING EXOTICS

Species Can Make a Difference

In the first of a series of important related contributions to the literature (Vitousek and Walker 1989; Vitousek 1990), Vitousek (1986) used the title "Biological invasions and ecosystem properties: can species make a difference?" Vitousek and colleagues confirmed that individual species can modify ecosystem processes by diverse mechanisms. The biological pollution caused by exotic species is not simply a vague, generalized accumulation of exotic taxa that collectively degrade ecosystems and biodiversity. Introduction of individual species can have devastating consequences, including massive extinction. A particularly well-documented example is that of the extinction of 14 endemic taxa of the land snail genus *Partula* on the island of Moorea in French Polynesia by the intentional introduction in 1978 of the snail *Euglandina rosea* as a biological control of the giant African snail *Achatina fulica* (Clarke et al. 1984; Murray et al. 1988). The same *Euglandina rosea* has been intentionally introduced on many Pacific islands, driving many native snail species to extinction (Hadfield 1986; Hadfield and Miller 1992).

Identifying Potential Invaders

In spite of several efforts to identify the life history and the physiological, genetic, and ecological traits of successful invaders, such efforts have little predictive value in forecasting which species can invade a particular community (Crawley 1987; Noble 1989). Predicting potential invaders through observation of which invasive species occur elsewhere in similar habitats does not seem to have received rigorous scrutiny in the literature, but seems to be effective in practice to the extent that it has been used. Although there is "A Geographical Atlas of World Weeds" (Holm et al. 1979) for agricultural weeds, no comparable volume exists for exotic species that invade natural vegetation. Development of such a volume is a high priority, especially for the world's islands.

Common sense prediction is undoubtedly routinely used in many locations vulnerable to invasions. For example, agencies in Hawaii use this approach to prevent the invasion of such exotics as the brown tree snake (*Boiga irregularis*), a highly destructive species on Guam (Savidge 1987). Knowledge of behavior of invasive species in other parts of the world also resulted in prompt eradication of an incipient rabbit population on the island of Maui and control of several invasive plant species (*Cortaderia jubata* and *Verbascum thapsus*, in addition to *Miconia calvescens*). A particularly noteworthy case from mainland United States is that of *Mimosa pigra*, a prickly shrub native to Central and South America, which extensively (450 km^2) invaded wetlands in Australia's Northern Territory, including a portion of the Kakadu National Park (Braithwaite 1989). Soon after this species was discovered in 1984 during the construction of Florida's Turnpike, just outside the Everglades drainage, eradication was initiated, based upon the behavior of the species in similar habitat in Australia.

The Lag Time of Invasions and Opportunities for Eradication

Some invasive species, especially flying insects, spread so rapidly when they reach a new area that timely eradication is never a serious option. With invasive plants and at least some animal groups, however, there is often a lag between the time that species are introduced and the time their invasiveness is perceived to be a problem. Ewel (1986) reviewed the lag phenomenon and gave four possible explanations:

1. Certain exotic plants might persist before extensive ecosystem modification by humans, then explode demographically when triggered by disturbance.
2. Expansion is rapid, but not noticeable or at least not noticed.
3. Development of large populations with adequate reproductive potential for invasion might take several decades. Salisbury (1961) referred to this phenomenon as "infection pressure." A small population may disperse few propagules, and until the right combination of conditions for germination, survival, growth, and reproduction is achieved, the species does not extend its range.
4. Its genetic structure might confine a new colonist to restricted habitats until appropriate genotypes become available through recombination or introgression for expansion into different habitats (Baker 1965). This tentative explanation can have practical consequences; if genetically different individuals of the species are introduced into the same area, opportunities for invasion are enhanced.

A prime example of a damaging exotic species that experienced a lengthy lag time is *Mimosa pigra* in the Northern Territory of Australia. Introduced, probably as a novel ornamental in the 1870s, it underwent a 100-year lag phase as "an occasionally troublesome but minor weed" (Lonsdale et al. 1988). In the 1970s, seeds were water-dispersed by exceptionally heavy seasonal rains to all parts of the Adelaide River floodplain, where overgrazing of wetlands by water buffaloes (*Bubalus*) allowed ready establishment of seedlings. Because seeds were then rapidly dispersed to other drainages, the nearby Kakadu National Park was massively invaded within 10 years (Lonsdale et al. 1988).

Local, Regional, and National Coalitions to Combat Exotics

Bederman (1991) observed that "the U.S. experience in controlling accidental introductions illustrates a common tendency in environmental policy-making: little concrete action until the risk is manifest." The zebra mussel (*Dreissena polymorpha*), which was introduced in ship ballast from Europe into the Great Lakes in 1985 (Garton 1993) and already causes billions of dollars of annual damage (especially by clogging water intake structures), recently raised awareness of the public and of politicians. Problems associated with this mussel caused initiation of a major study of exotic species in the United States by the Congressional Office of Technology Assessment (OTA 1993).

In October 1991, a national conference, "Biological Pollution: the Control and Impact of Invasive Exotic Species," sponsored by the Indiana Academy of Science, brought together about 300 concerned professionals. This conference generated much enthusiasm for action and alarm over exposure to the paucity of serious efforts to control exotic species. National and state budgets for the prevention and control of exotic species are pathetically small. Much of the funding is directed toward dealing with effects rather than dealing with the root of the problem—stopping the flow of new introductions. The public and policy-makers are very poorly informed about the magnitude of the problem and ways of appropriately addressing it. The Indiana Academy of Science plans a second national "Biological Pollution" conference for October 1993 with the hope of aiding development of a national strategy for combatting exotic species problems.

Efforts at dealing with exotic species in Hawaii, some of which are described above, illustrate the need for bringing about essential cooperation and communication among federal, state and local agencies and the public. Florida's Exotic Pest Plant Council (EPPC) seems to be the most experienced and most effective regional organization fostering cooperation among agencies. EPPC focuses on invasive woody exotic species and devotes major efforts toward obtaining funding for biological control and management of *Melaleuca quinquenervia* (Doren and Whiteaker 1991; Doren 1991). Very little was done to coordinate any control of woody exotic weeds in southern Florida prior to formation of EPPC in 1984. Although the current level of cooperation and effort in southern Florida is 15-20 years overdue (Myers 1991), EPPC is now functioning effectively and serves as a prototype for cooperative regional management of exotic species.

Managers and scientists concerned over the loss of biodiversity must learn to communicate and cooperate with a large potential reservoir of allies in combatting exotic invasions, because invasions also threaten agriculture, health (Hertzberg 1992), tourism (e.g., Fritts 1988), and sport fisheries (Courtenay and Stauffer 1984; Wilcove et al. 1992). They must also interact with specialized groups of those same largely allied interests, particularly horticulture (Yee and Gagne 1992), some of which promote policies for the free flow of species. From the standpoint of preserving biodiversity, it would seem that current regulations are balanced in favor of the horticultural and pet industries rather than conservation, because the burden of proof is generally placed on the conservation community to demonstrate that a species is invasive before it can be prohibited. Wilcove et al. (1992) have proposed, for fish introductions, that those proposing an intentional introduction be made to bear the cost of control efforts or unanticipated damages when an introduction produces unexpected harmful results.

Existing Federal Legislation

Westbrooks (1991,1993) described implementation of the Federal Noxious Weed Act of 1974. Congress provided the authority to prevent the introduction and spread of designated foreign weeds (about 100 taxa), but much of the effort is left to the states because adequate federal funding has not been provided. The Animal and Plant Health Inspection Service (APHIS) of the U.S. Department of Agriculture has 18 years of experience in administering this act and has an excellent framework for prevention and control of exotic species through interagency cooperation. The weeds on the list to date are all threats to agriculture, but the same framework can be used to deter foreign weeds that threaten biodiversity in the United States.

APHIS is currently supporting the eradication of several weeds, but the only one that is crucial to natural ecosystems is the effort against *Mimosa pigra* in southeastern Florida. Because of its known potential to invade wetlands similar to the Everglades, this weed received much attention (Kipker 1992). Partial funding is provided by APHIS; the Bureau of Aquatic Plant Management of the Florida Department of Natural Resources in Tallahassee and Royal Palm Beach administers the eradication. In April 1992, a coordinator in the Royal Palm Beach office evaluated the success of the monthly herbicide application by a contractor to known populations at four known sites in Palm Beach, Martin, and Highlands counties; served as a mediator between contractors and private landowners; searched for new *Mimosa pigra* localities on adjacent lands; and coordinated with researchers (Kipker 1992). Simultaneously, researchers at the University of Florida station at Ft. Lauderdale are investigating the biology of the species and assisting in refining control methods. Formidable logistical and biological (seed bank lasts for decades) barriers to immediate success exist (A. Boshell, pers. comm., 1992), but efforts must be sustained as long as a reasonable chance for success exists.

APHIS is the agency mandated to maintain oversight of the flow of potentially damaging species into the United States. For this reason, APHIS recently became involved in an evaluation of pests likely to accompany imported timber from the Soviet Union (USDA 1991). The prospect of Soviet timber imported to idle sawmills in the U.S. Pacific Northwest necessitated a scientific analysis of the tremendous risks of introduction of new forest pests and extensive discussions in the political arena to identify management options (Shannon 1991). The required treatments of imported Soviet lumber were identified, and the sawmill industry must now determine whether economics warrant the drastic treatments required to prevent unacceptable risk of introduction of dozens of new pests.

A NATIONAL STRATEGY FOR EXOTICS

Development of a comprehensive national strategy for exotic species is well beyond the scope of this contribution. The problem is being addressed by OTA and others, as pointed out earlier. It is, however, probably not premature to suggest some components necessary within such a strategy, as follows:

1. Rigorously review the common sense concept of predicting which species will invade a particular region by determining which species invade similar habitats elsewhere in the world.

The concept superficially seems to be sound but seems not to have been rigorously scrutinized in the voluminous recent literature on terrestrial invasions (Mooney and Drake 1986; Groves and Burdon 1986; Drake et al. 1989; Macdonald et al. 1989). With a similar approach, Kohler and Stanley (1984) developed a protocol for evaluating desirability of exotic fish introductions, which requires evaluation of such parameters as present range vs. native range and habitat/niche requirements.

2. Compile existing information and fill information gaps to produce a comprehensive atlas of invasive plant and animal species of the world, with details on habitat (including physical environment) and effects on biodiversity, agriculture, health, and tourism. This should be produced both in hard copy and in a computerized version with more detail.

3. Revise the Federal Noxious Weed Act of 1974 to broaden the authority of USDA-APHIS over quarantine and eradication of invasive plant species that threaten biodiversity. Review other federal legislation about various groups of animal species that threaten biodiversity and develop appropriate legislative changes.

Lists of prohibited organisms would closely correspond with the species with damaging effects on biodiversity somewhere in the world, documented in the comprehensive atlas of invasive plant and animals species. A reasonable principle might be that federal funding normally matches state funding, at least when state agencies conduct the eradication.

4. Develop an appropriate new international convention or conventions to update the 1952 International Plant Protection Convention and other pertinent treaties. Investigate the advantages of developing Bilateral Quarantine Agreements as are proposed in New Zealand (TNCH/NRDC 1992) with countries with which the U.S. trades in agricultural, forest, or other biological products.

5. Bring about, through legislation if necessary, increased interagency cooperation and coordination with the United States quarantine network to prevent entry of invasive organisms that threaten biodiversity. Develop coordinated computerized databases to assist quarantine inspectors in detection and identification, especially those in remote locations.

6. Direct federal, state and local governments to stop indiscriminate planting of invasive exotics. Planting practices by public agencies can make a difference. Nadkarni and Odion (1986) reported that the area of a large wildfire in California chaparral had partially re-burned only two years later. The partial re-burn corresponded to a localized area which had been replanted after the first fire with the exotic annual grass *Lolium multiflorum*. The Illinois Department of Natural Resources recently changed its policy from one of planting exotics throughout the state to one of planting almost nothing but natives (Harty 1993). Other state and federal agencies should assume a similar practice—or at least avoid such past mistakes as the planting of kudzu (*Pueraria lobata*) on roadcuts in the southeastern U.S. or roadside plantings of invasive Pampas grass (*Cortaderia jubata*) in California.

7. Consider incorporating concerns for addressing harm by exotic species in an amended National Environmental Policy Act, thus requiring consideration of effects of a proposed action on biodiversity (Doremus 1991).

8. Encourage regional and local cooperation in addressing exotic species problems, based on patterns in Florida (Exotic Plant Pest Council) and Hawaii. Mandates of federal agencies such as the U.S. Forest Service, U.S. Fish and Wildlife Service, National Park Service, Bureau of Land Management, U.S. Soil Conservation Service, and the military should be strengthened. These agencies should be required to evaluate possible effects of exotic species on biodiversity when planning and implementing projects.

9. Provide appropriate public information and education on the role of exotic species in eroding biodiversity. There is a need to explain the serious damage caused by exotic species to the public. Options for the control of vertebrates were reduced in many areas

because of pressure from animal rights groups. With sensitivity to ethical concerns and full disclosure of the harmful consequences of these exotic animals for biodiversity, land managers can gain support for control much more often than they have in the past (Stone 1992).

LITERATURE CITED

Anderson, S. J., and C. P. Stone. 1993. Snaring to control feral pigs (*Sus scrofa*) in a remote Hawaiian rain forest. *Biol. Conserv.* 63(3):195-201.

Atkinson, I. 1989. Introduced animals and extinctions. In *Conservation for the Twenty-first Century*, eds. D. Western and M. C. Pearl, pp. 54-75. Oxford University Press, New York.

Austin, D. 1978. Exotic plants and their effects in southeastern Florida. *Environ. Conserv.* 5:25-34.

Baker, H. G. 1965. Characteristics and modes of origin of weeds. In *The Genetics of Colonizing Species*, eds. H. G. Baker and G. L. Stebbins, pp. 147-172. Academic Press, New York.

Bederman, D. J. 1991. International control of marine "pollution" by exotic species. *Ecol. Law Quart.* 18:677-717.

Billings, W. D. 1990. *Bromus tectorum*, a biotic cause of ecosystem impoverishment in the Great Basin. In *The Earth in Transition: Patterns and Processes of Biotic Impoverishment*, ed. G. M. Woodwell, pp. 301-322. Cambridge University Press, Cambridge.

Birnbaum, P. 1991. *Exigences et Tolerances de Miconia calvescens a Tahiti*. Centre ORSTOM de Tahiti. Papeete, Polynesie Francaise.

Braithwaite, R. W., W. M. Lonsdale, and J. A. Estbergs. 1989. Alien vegetation and native biota in tropical Australia: the impact of *Mimosa pigra*. *Biol. Conserv.* 48:189-210.

Brockie, R. E., L. L. Loope, M. B. Usher, and O. Hamann. 1988. Biological invasions of island nature reserves. *Biol. Conserv.* 44:9-36.

Center, T. D., R. F. Doren, R. L. Hofstetter, R. L. Myers, and L. D. Whiteaker, eds. 1991. *Proceedings of the Symposium on Exotic Pest Plants*. Technical Report NPS/NREVER/NRTR-91/06. USDI, National Park Service, Washington, D.C.

Clarke, B., J. Murray, and M. S. Johnson. 1984. Extinction of endemic species by a program of biological control. *Pac. Sci.* 38(2):97-104.

Courtenay, W. R., Jr., and J. R. Stauffer, Jr. 1984. *Distribution, Biology, and Management of Exotic Fishes*. The Johns Hopkins University Press, Baltimore, Md.

Crawley, M. J. 1987. What makes a community invasible? In *Colonization, Succession and Stability*, eds. A. J. Gray, M. J. Crawley, and P. J. Edwards, pp. 424-453. Blackwell Scientific Publications, Oxford.

Cuddihy, L. W., and C. P. Stone. 1990. *Alteration of Native Hawaiian Vegetation: Effects of Humans, Their Activities and Introductions*. Univ. Hawaii Coop. Natl. Park Resour. Stud. Unit, University of Hawaii Press, Honolulu, Hawaii.

D'Antonio, C. M., and P. M. Vitousek. 1992. Biological invasions by exotic grasses, the fire cycle, and global change. *Ann. Rev. Ecol. Syst.* 23:

Daughtery, M. 1993. Dogwood anthracnose disease: native fungus or exotic invader? In *Biological Pollution: the Control and Impact of Invasive Exotic Species*, ed. B. N. McKnight. Indiana Academy of Science, Indianapolis, Ind.

Diamond, J. 1989. Overview of recent extinctions. In *Conservation for the Twenty-first Century*, eds. D. Western and M. C. Pearl, pp. 37-41. Oxford University Press, New York.

Doremus, H. 1991. Patching the ark: improving legal protection of biological diversity. *Ecol. Law Quart.* 18:264-333.

Doren, R. F. 1991. Forward. In *Proceedings of the Symposium on Exotic Pest Plants*, eds. T. D. Center, R. F. Doren, R. L. Hofstetter, R. L. Myers, and L. D. Whiteaker, pp. ix-xii. Technical Report NPS/NREVER/NRTR-91/06. USDI, National Park Service, Washington, D.C.

Doren, R. F., and L. D. Whiteaker. 1991. The Exotic Pest Plant Council: an example of interagency cooperation to solve resource related problems. In *Proceedings of the Symposium on Exotic Pest Plants*, eds. T. D. Center, R. F. Doren, R. L. Hofstetter, R. L. Myers, and L. D. Whiteaker, pp. 111-114. Technical Report NPS/NREVER/NRTR-91/06. USDI, National Park Service, Washington, D.C.

Drake, J. A., H. A. Mooney, F. di Castri, R. H. Groves, F. J. Kruger, M. Rejmanek, and M. Williamson, eds. 1989. *Biological Invasions: A Global Perspective*. SCOPE 37. John Wiley and Sons, New York.

Eager, C. C. 1984. Review of the biology and ecology of the balsam woolly aphid in southern Appalachian spruce-fir forests. In *The Southern Appalachian Spruce-Fir Ecosystem: Its Biology and Threats*, ed. P. S. White, p. 37. U.S. Department of the Interior, National Park Service, Research/Resource Management Report SER-71. Atlanta, Ga.

Elton, C. S. 1958. *The Ecology of Invasions by Animals and Plants*. Methuen, London, England.

Ewel, J. J. 1986. Invasibility: lessons from South Florida. In *Biological Invasions of North America and Hawaii*, eds. H. A. Mooney and J. A. Drake, pp. 214-230. Springer-Verlag, New York.

Fritts, T. H. 1988. *The Brown Tree Snake: A Harmful Pest Species*. U.S. Government Printing Office, Washington, D.C.

Gagne, B. H., L. L. Loope, A. C. Medeiros, and S. J. Anderson. 1992. *Miconia calvescens*: a threat to native forests of the Hawaiian Islands. *Pac. Sci.* 46:390-391.

Gagne, W. C. 1988. Conservation priorities in Hawaiian natural systems. *BioScience* 38(4):264-271.

Gardner, D. E. 1990. *Role of Biological Control as a Management Tool in National Parks and Other Natural Areas*. National Park Service, Tech. Report NPS/NRUH/NRTR-90/01. Washington, D.C.

Garton, D. W. 1993. The zebra mussel, *Dreissena polymorpha*: environmental limits and impacts on native organisms in North America. In *Biological Pollution: the Control and Impact of Invasive Exotic Species*, ed. B. N. McKnight. Indiana Academy of Science, Indianapolis, Ind.

Groves, R. H. 1989. Ecological control of invasive terrestrial plants. In *Biological Invasions: A Global Perspective*, eds. J. A. Drake, H. A. Mooney, F. di Castri, R. H. Groves, F. J. Kruger, M. Rejmanek, and M. Williamson, pp. 437-461. SCOPE 37. John Wiley and Sons, New York.

Groves, R. H., and J. J. Burdon. 1986. *Ecology of Biological Invasions*. Cambridge University Press, Sydney, Australia.

Hadfield, M. G. 1986. Extinction in Hawaiian Achatinelline snails. *Malacologia* 27(1):67-81.

Hadfield, M. G., and S. E. Miller. 1992. Alien predators and decimation of endemic Hawaiian tree snails. *Pac. Sci.* 46:395.

Harty, F. M. 1993. How Illinois kicked the exotic habit. In *Biological Pollution: the Control and Impact of Invasive Exotic Species*, ed. B. N. McKnight. Indiana Academy of Science, Indianapolis, Ind.

Hawaii Department of Land and Natural Resources, U.S. Fish and Wildlife Service, and The Nature Conservancy of Hawaii (HDLNR/FWS/TNCH). 1991. *Hawaii's Extinction Crisis: a Call to Action*. Honolulu, Hawaii.

Hertzberg, H. 1992. Summer's blood. *Time* August 10, 1992, pp. 46-48.

Holm, L., J. V. Pancho, J. P. Herberger, and D. L. Plucknett. 1979. *A Geographical Atlas of World Weeds*. John Wiley and Sons, New York.

Howarth, F. R. 1991. Environmental impacts of classical biological control. *Ann. Rev. Entomol.* 36:485-509.

James, H. F., and S. L. Olson. 1991. Descriptions of Thirty-two New Species of Birds from the Hawaiian Islands. Part II. Passeriformes. *Ornithol. Monogr.* 46:1-88.

Kipker, R. 1992. The Status of *Mimosa pigra* in Florida. Florida Department of Natural Resources, Bureau of Aquatic Plant Management, Tallahassee, Fla.

Kirch, P. V. 1982. The impact of prehistoric Polynesians on the Hawaiian ecosystem. *Pac. Sci.* 36(1):1-14.

Kohler, C. C., and J. G. Stanley. 1984. A suggested protocol for evaluating proposed exotic fish introductions in the United States. In *Distribution, Biology, and Management of Exotic Fishes*, eds. W. D. Courtenay, Jr., and J. R. Stauffer, Jr., pp. 387-406. The Johns Hopkins University Press, Baltimore, Md.

Langdon, K. R., and K. D. Johnson. 1992. Alien forest insects and diseases in eastern U.S. National Park Service units: impacts and interventions. *The George Wright Forum* 9:2-14.

Lonsdale, W. M., K. L. S. Harley, and J. D. Gillett. 1988. Seed bank dynamics in *Mimosa pigra*, an invasive tropical shrub. *J. Appl. Ecol.* 25:963-976.

Loope, L. L. 1992. Preventing establishment of new alien species in Haleakala National Park and the island of Maui, Hawaii. *The George Wright Forum* 9:20-31.

Loope, L. L., A. C. Medeiros, W. Minyard, S. Jessel, and W. Evanson. 1992. Strategies to prevent establishment of feral rabbits on Maui, Hawaii. *Pac. Sci.* 46:402-403.

Macdonald, I. A. W., F. J. Kruger, and A. A. Ferrar, eds. 1986. *The Ecology and Management of Biological Invasions in Southern Africa*. Oxford University Press, Cape Town, South Africa.

Macdonald, I. A. W., L. L. Loope, O. Hamman, and M. B. Usher. 1989. Wildlife conservation and the invasion of nature reserves by introduced species: a global perspective. In *Biological Invasions: A Global Perspective*, eds. J. A. Drake, H. A. Mooney, F. di Castri, R. H. Groves, F. J. Kruger, M. Rejmanek, and M. Williamson, pp. 215-255. SCOPE 37. John Wiley and Sons, New York.

McKnight, B. N., ed. 1993. *Biological Pollution: the Control and Impact of Invasive Exotic Species*. Indiana Academy of Science, Indianapolis, Ind.

Markin, G. P., P. Y. Lai, and G. Y. Funasaki. 1992. Status of biological control of weeds in Hawaii and implications for managing native ecosystems. In *Alien Plant Invasions in Native Ecosystems of Hawaii: Management and Research*, eds. C. P. Stone, C. W. Smith, and J. T. Tunison, pp. 466-482. Univ. Hawaii Coop. Natl. Park Resour. Stud. Unit., University of Hawaii Press, Honolulu, Hawaii.

Medeiros, A. C., and L. L. Loope. 1992. Ecology of *Miconia calvescens* and *Cyathea cooperi*, two new invasive plant species in the Hawaiian Islands. *Bull. Ecol. Soc. Am.* 73(2):270.

Merton, D.V. 1978. Controlling introduced predators and competitors on islands. In *Endangered Birds: Management Techniques for Preserving Threatened Species*, ed. S. A. Temple, pp. 121-128. University of Wisconsin Press, Madison, Wisc.

Mooney, H. A., and J. A. Drake, eds. 1986. *Biological Invasions of North America and Hawaii*. Springer-Verlag, New York.

Mueller-Dombois, D. 1981. Understanding Hawaiian forest ecosystems: the key to biological conservation. In *Island Ecosystems: Biological Organization in Selected Hawaiian Communities*, eds. D. Mueller-Dombois, K. W. Bridges, and H. L. Carson, pp. 502-520. U.S. IBP Synthesis Series 15. Hutchison Ross, Stroudsberg, Pa.

Murray, J., E. Murray, M. S. Johnson, and B. Clarke. 1988. The extinction of *Partula* on Moorea. *Pac. Sci.* 42: 150-153.

Myers, R. L. 1991. Are we serious about exotic species control? A symposium conclusion. In *Proceedings of the Symposium on Exotic Pest Plants*, eds. T. D. Center, R. F. Doren, R. L. Hofstetter, R. L. Myers, and L. D. Whiteaker, pp. 255-259. Technical Report NPS/NREVER/ NRTR-91/06. USDI, National Park Service, Washington, D.C.

Nadkarni, N., and D. Odion. 1986. Effects of seeding exotic grass *Lolium multiflorum* on native seedling regeneration following fire in a chaparral community. In *Proceedings of the Chaparral Ecosystems Research Conference*, California Water Resources Center, Report 63, pp. 115-122.

Noble, I. R. 1989. Attributes of invaders and the invading process: terrestrial and vascular plants. In *Biological Invasions: A Global Perspective*, eds. J. A. Drake, H. A. Mooney, F. di Castri, R. H. Groves, F. J. Kruger, M. Rejmanek, and M. Williamson, pp. 301-313. SCOPE 37. John Wiley and Sons, New York.

Office of Technology Assessment (OTA). 1993. *Harmful Non-indigenous Species in the United States*. U.S. Government Printing Office, Washington, D.C.

Olson, S. L., and H. F. James. 1991. Descriptions of Thirty-two New Species of Birds from the Hawaiian Islands. Part I. Non-Passeriformes. *Ornithol. Monogr.* 46:1-88.

Pyle, R., M. Bentzien, and P. Opler. 1981. Insect conservation. *Ann. Rev. Entomol.* 26:233-258.

Redlin, S. C. 1991. *Discula destructiva* sp. nov., cause of dogwood anthracnose. *Mycologia* 83:633-642.

Salisbury, E. J. 1961. *Weeds and Aliens*. MacMillan, New York.

Savidge, J. 1987. Extinction of an island forest avifauna by an introduced snake. *Ecology* 68:660-668.

Shannon, M. J. 1991. Agency role in exclusion/management of invading pests.

Paper presented at Symposium, "Biological Pollution: the Control and Impact of Invasive Exotic Species" IUPUI University Place Conference Center, Indianapolis, October 25-26, 1991.

Stone, C. P. 1991. Feral pig (*Sus scrofa*) research and management in Hawaii. In *Biology of Suidae*, eds. R. H. Barrett and F. Spitz, pp. 141-154. Inst. Nat. Recherche Agron., Toulouse, France.

Stone, C. P. 1992. Humane natural area management in Hawaii. *The George Wright Forum* 9:32-35.

Stone, C. P., L. W. Cuddihy, and J. T. Tunison. 1992. Responses of Hawaiian ecosystems to removal of feral pigs and goats. In *Alien Plant Invasions in Native Ecosystems of Hawaii: Management and Research*, eds. C. P. Stone, C. W. Smith, and J. T. Tunison, pp. 666-704. Univ. Hawaii Coop. Natl. Park Resour. Stud. Unit, University of Hawaii Press, Honolulu, Hawaii.

Stone, C. P., and R. A. Holt. 1990. Managing the invasions of alien ungulates and plants in Hawaii's natural areas. *Monogr. Syst. Bot., Missouri Bot. Gard.* 32:211-221.

Stone, C. P., and L. L. Loope. 1987. Reducing negative effects of introduced animals on native biotas in Hawaii: what is being done, what needs doing, and the role of national parks. *Environ. Conserv.* 14(3):245-258.

The Nature Conservancy of Hawaii and the Natural Resources Defense Council (TNCH/NRDC). 1992. *The Alien Pest Species Invasion in Hawaii: Background Study and Recommendations for Interagency Planning*. Honolulu, Hawaii.

Tunison, J. T. 1992a. Alien plant control strategies in Hawaii Volcanoes National Park. In *Alien Plant Invasions in Native Ecosystems of Hawaii: Management and Research*, eds. C. P. Stone, C. W. Smith, and J. T. Tunison, pp. 485-505. Univ. Hawaii Coop. Natl. Park Resour. Stud. Unit, University of Hawaii Press, Honolulu, Hawaii.

Tunison, J. T. 1992b. Fountain grass control in Hawaii Volcanoes National Park: management considerations and strategies. In *Alien Plant Invasions in Native Ecosystems of Hawaii: Management and Research*, eds. C. P. Stone, C. W. Smith, and J. T. Tunison, pp. 376-393. Univ. Hawaii Coop. Natl. Park Resour. Stud. Unit, University of Hawaii Press, Honolulu, Hawaii.

Tunison, J. T., and C. P. Stone. 1992. Special Ecological Areas: an approach to alien plant control in Hawaii Volcanoes National Park. In *Alien Plant Invasions in Native Ecosystems of Hawaii: Management and Research*, eds. C. P. Stone, C. W. Smith, and J. T. Tunison, pp. 781-798. Univ. Hawaii Coop. Natl. Park Resour. Stud. Unit, University of Hawaii Press, Honolulu, Hawaii.

Tunison, J. T., and N. G. Zimmer. 1992. Success in controlling localized alien plants in Hawaii Volcanoes National Park. In *Alien Plant Invasions in Native Ecosystems of Hawaii:*

Management and Research, eds. C. P. Stone, C. W. Smith, and J. T. Tunison, pp. 506-524. Univ. Hawaii Coop. Natl. Park Resour. Stud. Unit, University of Hawaii Press, Honolulu, Hawaii.

U.S. Department of Agriculture (USDA). 1991. *An Efficacy Review of Control Measures for Potential Pests of Imported Soviet Timber*. Animal and Plant Health Inspection Service. USDA Misc. Publ. No. 1496.

Vitousek, P. M. 1986. Biological invasions and ecosystem properties: Can species make a difference?. In *Biological Invasions of North America and Hawaii*, eds. H. A. Mooney and J. A. Drake, pp. 163-176. Springer-Verlag, New York.

Vitousek, P. M. 1990. Biological invasions and ecosystem properties: towards an integration of population biology and ecosystem studies. *Oikos* 57:7-13.

Vitousek, P. M., and L. R. Walker. 1989. Biological invasion by *Myrica faya* in Hawaii: plant demography, nitrogen fixation, ecosystem effects. *Ecol. Monogr.* 59:247-265.

Wagner, W. L., D. R. Herbst, and S. H. Sohmer. 1990. *Manual of the Flowering Plants of Hawaii*. Bishop Museum and University of Hawaii Press, Honolulu, Hawaii.

Westbrooks, R. G. 1991. Plant protection issues: I. A commentary on new weeds in the United States. *Weed Technol.* 5:232-237.

Westbrooks, R. G. 1993. Regulatory exclusion of federal noxious weeds. In *Biological Pollution: the Control and Impact of Invasive Exotic Species*, ed. B. N. McKnight. Indiana Academy of Science, Indianapolis, Ind.

Whittaker, R. H. 1956. Vegetation of the Great Smoky Mountains. *Ecol. Monogr.* 26:1-80.

Wilcove, D., M. Bean, and P. C. Lee. 1992. Fisheries management and biological diversity: problems and opportunities. *Trans. 57th Ann. Wildlife and Nat. Res. Conf.* 57:373-383.

Williamson, M. 1981. *Island Populations*. Oxford University Press, Oxford.

Wilson, E. O. 1989. Conservation: the next hundred years. In *Conservation for the Twenty-first Century*, eds. D. Western and M. C. Pearl, pp. 3-7. Oxford University Press, New York.

Yee, R. S. N., and W. C. Gagne. 1992. Activities and needs of the horticulture industry in relation to alien plant problems in Hawaii. In *Alien Plant Invasions in Native Ecosystems of Hawaii: Management and Research*, eds. C. P. Stone, C. W. Smith, and J. T. Tunison, pp. 712-725. Univ. Hawaii Coop. Natl. Park Resour. Stud. Unit, University of Hawaii Press, Honolulu, Hawaii.

19

MANAGEMENT OF LAND AND SPECIES RICHNESS IN THE TROPICS

Ariel E. Lugo and Sandra Brown[1]

International Institute of Tropical Forestry
USDA Forest Service
Rio Piedras, Puerto Rico
U.S.A.

Public attention on biodiversity has centered on species extinctions caused by deforestation and other human activities such as hunting and poaching, introduction of exotics (King 1984), and habitat fragmentation (Newmark 1990; Cutler 1991; Saunders et al. 1991). As public concern for the preservation of species increases, pressure builds on governments and natural resource agencies to reverse current trends of natural resource degradation. The polarization between resource management agencies and the public contributes to a situation where well-meaning people and institutions fail to agree on realistic solutions to problems that are fundamental to the survival of our civilization. Such a situation invites propagation of myths, advancement of unrealistic solutions to resource management problems, and adoption of strategies that can harm communities and societies without resolving the fundamental problem of sustaining human activity in the biosphere in an environmentally sound way.

Four notions are prevalent about species richness in the American tropics, these deserving clarification at the outset. First is the idea that the tropics are rich in species because of historically "stable" environmental conditions (Dobzhansky 1950). It is now known that the American tropics have had a long history of both natural and human-induced disturbances (Bush et al. 1992). Human activity has shaped the history and species composition of the lowland vegetation in Central America for about 11,000 years (Gomez-Pompa et al. 1987; Bush et al. 1992). Another notion is the idea that persistent drought created islands of moist forests where species gained refuge through the Pleistocene (Haffer 1969; Simpson and Haffer 1978). This notion has been widely debated (Prance 1982; Gentry 1986) and questioned for Latin America by the works of Bush et al. (1992) and Nelson et al. (1990). Centers of endemism apparently occur as a result of localized climatic or edaphic phenomena or because of artifacts of scientific collecting and rarity of species, and not necessarily as a result of Pleistocene refugia. A third notion is the idea that the new world tropics had a lower species richness than the old world

[1] Present address: U.S. Environmental Protection Agency, Environmental Research Laboratory, Corvallis, Oregon.

tropics (Whitmore 1975). However, the studies of Gentry (1982) and Gentry and Dodson (1987) have shown that species richness in the American tropics can in fact exceed those reported for the old world tropics. Finally, the belief that all organisms reach their maximum diversity in the lowland moist tropics has been revised by Mares (1992) who showed that the richness of mammals peaks in the dry ecosystems which have a greater geographical extension than moist forests.

The dissolution of these myths illustrates that scientific attention to species richness is broader than the current public discussion about extinctions of species. Science is responsible for cataloging biodiversity, explaining how it developed and is maintained, and understanding its ecological functions. Such knowledge puts people in a better position to manage all natural resources sustainably. Unfortunately, the task facing scientists is enormous, and there is not enough information, adequate scientific infrastructure, or the means to satisfy all our needs and curiosity. Moreover, given the growth of molecular biology at the expense of field biology in most of our universities and technical institutions, science is losing its capacity to catalog and study the species richness of the world, public interest in the subject notwithstanding.

Knowledge of species richness issues is inadequate. We still have not cataloged 80% of the earth's species, we have little information on the rate and mechanisms of evolution and extinction of species, and we cannot agree on the ecological role of biodiversity. Most of our focus has been limited to studying a small sub-set of organisms and geographic locations. This limitation in the breadth of study biases theories, hypotheses, and generalizations about the patterns and ecological advantages (if any) of high species richness in an ecosystem.

Because of our ignorance, it is easy for the debate on species extinctions and natural resource management to become emotional and lack a clear focus. A conservation strategy that emerges from this situation is to advocate the preservation of all species (with the possible exception of exotics, pests, vagrants, tramps, super tramps, weeds, parasites, diseases), assume that nature is fragile and intolerant of human activity, consider people as intruders in natural systems, ignore human needs, and maintain the status quo as much as possible.

We believe in preserving **all** species including those in parenthesis listed above, which are normally excluded because of their commonness or their perceived negative ecological impacts. Many times, the value of species is determined by utilitarian or anthropocentric arguments that negate the importance of certain other species. Moreover, we doubt the wisdom of preservation as the **only** focus of the conservation strategy stated above. Survival conditions in the biosphere are being driven by unprecedented rates of global change powered by high rates of human use of energy and natural resources. It is unrealistic to embrace the status quo because it does not exist; environmental change is inevitable. Therefore, research and management activities must be refocused to understanding and managing environmental change (Lugo 1992a, 1994).

To help focus the debate about biodiversity and land use in tropical forests, we first review patterns of species richness that are accepted by most scientists and then address four questions: (1) What is the rate of extinction of species in the tropics? (2) How is land use changing in tropical landscapes? (3) What kinds of ecosystems are being favored by human activity? and, (4) How can people manage biodiversity?

PATTERNS OF SPECIES RICHNESS IN THE TROPICS

Certain patterns of species richness have been demonstrated in the ecological literature (Table 19.1). It is apparent that the distribution and concentration of species in the landscape follows a natural order. Understanding these patterns is critical for intelligent land use and for balancing human needs with the conservation of species. For example, because high species concentrations apparently occur in specialized environments, it is possible to identify and preserve those environments and concentrate human use in locations where the density of species is lower or where the ecosystems are more resilient (Lugo 1991).

Understanding the factors that regulate species richness in the tropics is often addressed from an evolutionary perspective in which scientists try to understand **the origin** of high species diversity. Of equal importance to ecology and of greater importance to ecosystem management is to understand how high species densities are **maintained** in ecosystems over centuries as opposed to millennia for evolutionary phenomena. Odum (1983) developed a hypothesis of energy causality for the maintenance of species diversity in ecosystems and summarized his views in the model shown in Figure 19.1 (see also Lugo 1988a). Currie and Paquin (1987) compiled data for large-scale patterns of tree species richness in the temperate zone and obtained the highest correlation between species richness and solar energy input. They suggested that species richness was limited by available contemporary energy.

The model in Figure 19.1 suggests the following: (1) Species diversity is an ecosystem attribute that develops as a second priority to the basic structural and functional framework

Table 19.1. Patterns of species diversity as shown by scientific observation.

- The diversity of most groups of plants and animals increases toward the tropical latitudes, and within tropical latitudes species diversity increases towards the equator.
- The number of species (plants and animals together) per unit area of land is usually greater in the lowland tropical rain forests and lower in high latitudinal ecosystems.
- Tropical forests have a higher diversity than high-latitudinal forests in all scales of measure: within habitat, between habitats, and landscapes.
- Within a given altitudinal belt in the tropics, the number of tree species increases with increasing rainfall.
- Across altitudinal belts, the number of tree species and bird species decreases with increasing altitude.
- Mammal species richness is higher in the tropical dry biomes than in the moist, wet, or rain forest biomes.
- In islands, the number of animal species increases with the size of the island and with proximity to a continent. If the topography of the island is rugged, the number of species also increases.
- Tropical mountainous areas (particularly high mountains) support a larger number of species (plant and animal) than equivalent temperate mountains because environments range from lowland tropical at the base to alpine conditions at the summit.
- Ecosystems with severe limiting factors such as high salinity flooding, or freezing temperatures have reduced species richness.
- As rain forests recover from disturbance by acute events such as hurricanes or fires, species diversity increases in the disturbed area until a peak is reached after several decades.
- High-yield systems under intensive human management are species-poor, and increase in species richness to values observed in less disturbed and more mature systems when management pressure for high yield is reduced.
- The landscape is not uniform in terms of its species diversity. Often pockets with a large number of species can be interspersed within areas of low species diversity.

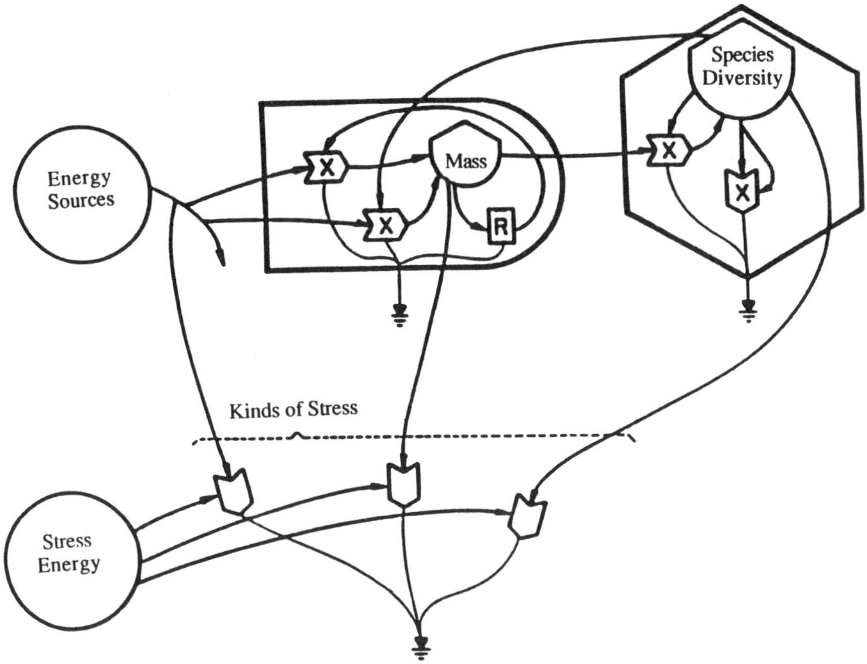

Figure 19.1. Minimodel of the ecological function and maintenance of species diversity in ecosystems (Odum 1983). Stress energy can divert energy from various storages or flows with the same effect, a lowering of diversity. See text for other attributes of this model.

of the system (e.g., production, consumption, and recycling). (2) Developing and maintaining many species per unit area involves an energy cost in the form of organization, behavior, and integration of biotic processes. (3) The energy cost of maintaining species diversity increases as a quadratic function of the number of species, a situation that sets limits on the number of species per unit area. (4) The maximum species diversity in a system will depend on the balance of available energy resources that promote growth and energy drains (stressors) in that particular environment. (5) Species diversity provides a feedback to the productive sector of the system and contributes to increased efficiency of resource use and energy flow through the biota. (6) Species diversity provides redundancy of function and availability of pre-adaptations to cope with environmental uncertainty (Odum 1983; Lugo 1988a). These are controversial and untested hypotheses but they are alternatives to the current paradigms on biodiversity. Odum's model and ideas provide approaches for the management of species richness by considering diversity as an ecosystem attribute similar to biomass or productivity.

Following any kind of disturbance or alteration of the environment (soil, atmosphere, hydrology), succession takes place and a new group of species that function as an ecosystem is established at the site. If the perturbation is below certain thresholds of intensity and spatial scale, usually the same group of species that was present before the perturbation will prevail. But when the perturbation is extreme, new combinations of species form new ecosystems, sometimes with exotic species. Only under the most extreme modifications of the site will species fail to establish a new functioning ecosystem. Evolutionary forces, availability of genetic resources, and site conditions determine the species combinations that will prevail at any site. Odum (1983) termed the process of ecosystem-reorganization following disturbance "self-design" because the biota adjusts to the prevailing conditions through natural processes. Self-design is ecosystem regeneration, but because the term regeneration is used in forestry to mean the establishment of the tree component, we use self-design to emphasize whole ecosystem regeneration.

In a changing global environment, self-design is a critical ecosystem attribute to use as much as possible to our advantage. The combinations of species that develop after disturbance through natural successions will be more persistent due to their better adaptation to prevailing environmental conditions. They will also be cheaper to establish than alternative solutions, based on preconceived notions of species composition. The alternative to self design is artificial regeneration, an expensive human activity. In addition, we lack sufficient knowledge to regenerate biodiversity artificially in all tropical environments.

Because species regeneration work is done by natural processes, self-design is a strategy to pursue in the tropics as we try to rehabilitate the millions of hectares damaged by human activity (Grainger 1988). However, two points should be made clear. (1) After humans modify the landscape, the resulting forests will have native species previously used by people at the site (mostly exotic species) as well as species that were rare in pre-disturbance forests. In some landscapes, the total species inventory is higher after human intervention (Burgess and Sharpe 1981). (2) Human intervention is needed to guide the natural recovery of vegetation to a desired state. The use of natural processes means less expenditure of human effort than in the artificial regeneration strategy. Thus, use of self-design does not imply a hands-off management policy. Sometimes the species pool in an ecosystem is not diverse enough to compensate for certain environmental conditions that humans or catastrophic events impose on the landscape. These situations call for importation of species either by humans or natural dispersal.

RATES OF SPECIES EXTINCTIONS IN THE TROPICS

The loss of species as a result of human activity is a very serious problem because of the irreversibility of the extinction process. The problem is particularly critical for species that are endemic to restricted areas and which are difficult to identify before human activity places them in irreversible danger. Once lost, these species cannot be replaced because populations occur nowhere else.

To estimate rates of species extinctions, scientists usually start with the species-area relationship: $S = CA^z$ where S = number of species, A = area, and Z and C = constants (MacArthur and Wilson 1967). The number of species lost is estimated from average values of Z and rates of deforestation (Lovejoy 1980; Ehrlich and Ehrlich 1981; Simberloff 1986). Figure 19.2 illustrates how this method produces estimates of species extinction for different Z values and rates of deforestation. Lugo (1988b) and Lugo et al. (1993)

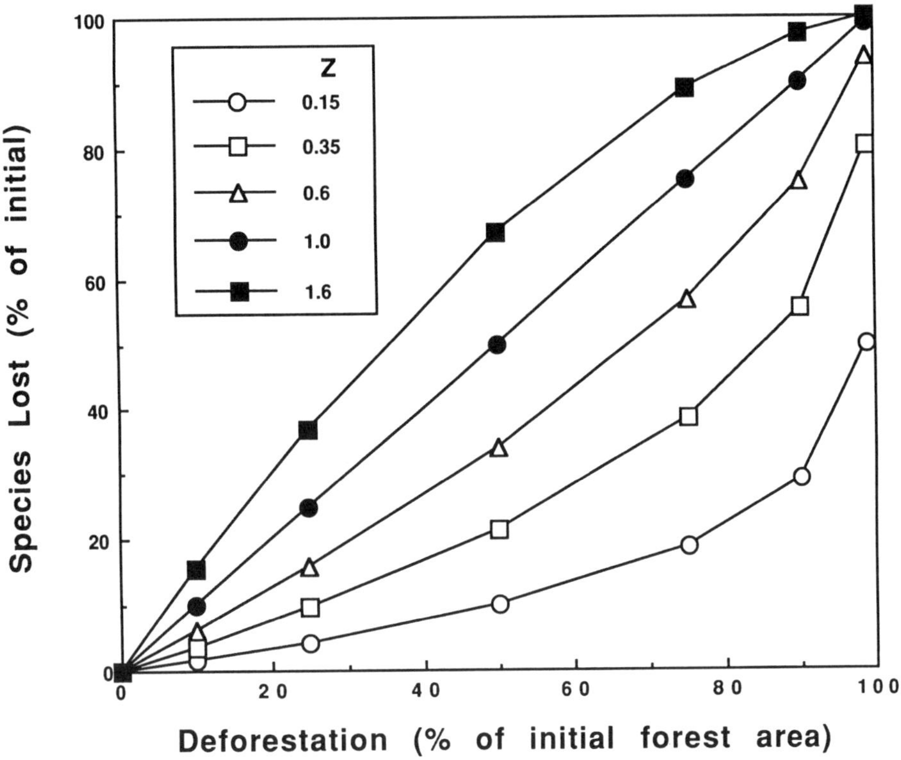

Figure 19.2. Estimates of species lost due to deforestation using different Z values. Species loss rates are predicted by noting the change in species numbers associated with a reduction in forest area and dividing by the time interval of the deforestation event. Higher Z values are associated with steeper species-area relationships.

suggested that this methodology overestimates the rates of extinctions for the following reasons: (1) the method assumes that deforestation results in the complete loss of species at the site; (2) the methodology ignores the resiliency or persistence of species in the landscape (i.e., through dispersal or survival); (3) predictions of extinction rates for specific groups of organisms, such as birds or trees, are higher than empirically observed extinctions in areas where deforestation events have taken place (e.g., in Puerto Rico and Singapore); and (4) human efforts to restore species through reforestation or land rehabilitation are not considered.

Our inability to predict rates of species extinctions accurately is not an excuse to be complacent about the issue. Knowing that careless human activity can cause significant extinctions of species, and that this is inimical to our future survival, should be sufficient to spur us into protective actions. But we also need better understanding of the effects of human activity and more knowledge of the factors that lead to species extinctions so that we can avoid them.

LAND-USE CHANGES IN THE MOIST TROPICS

In spite of a long history of human activity (Bush et al. 1992), the moist tropics still contain a large area of unfragmented mature forests (FAO 1985). However, we have suggested that most of these forests have been or still are subject to human disturbance, often at low intensity (Brown and Lugo 1990, 1992; Brown et al. 1991). Most scientific inquiry regarding tropical biodiversity issues has focused on these so-called mature forests. Less scientific attention has been given to the effects of different land-use practices on species composition. Basic ecological information is lacking on the types of forests in the tropics that are becoming prevalent as a result of human activity (Brown and Lugo 1990).

A simplified diagram of land transformation in the moist tropics is useful for illustrating the types of ecosystems that require more ecological attention (Figure 19.3). Humans transform the landscape with the purpose of obtaining food and resources to sustain economic activity. It is necessary to manage the various states of converted lands

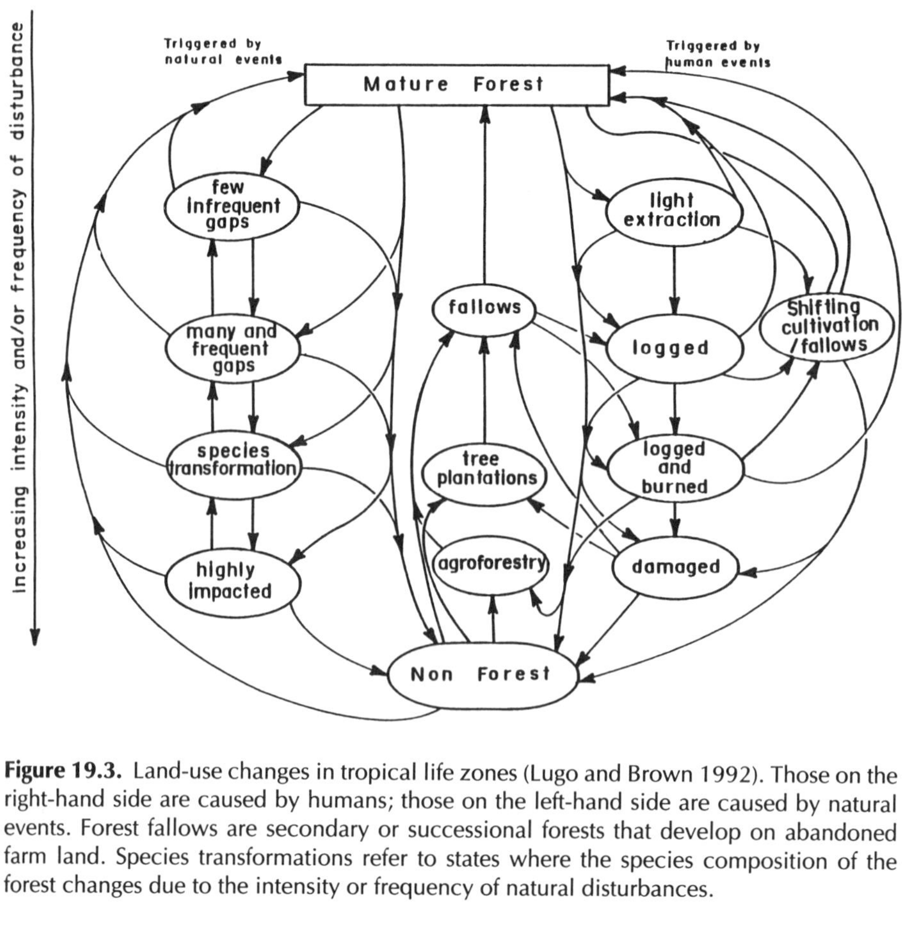

Figure 19.3. Land-use changes in tropical life zones (Lugo and Brown 1992). Those on the right-hand side are caused by humans; those on the left-hand side are caused by natural events. Forest fallows are secondary or successional forests that develop on abandoned farm land. Species transformations refer to states where the species composition of the forest changes due to the intensity or frequency of natural disturbances.

in Figure 19.3 so that people can maximize benefits while protecting the productive capacity of the land as well as its ecological values and resiliency (Brown and Lugo 1990).

Data on the species richness of land uses depicted in Figure 19.3 are lacking. Most of the available data for plant species richness in the tropics are for small sized plots usually in mature forests (Figure 19.4). In contrast, the temperate data set is more evenly spread across all spatial scales (Figure 19.4). In fact, in a global survey of species richness, Rice and Westoby (1983) found that the 0.1-ha plot was the most common in the ecological literature. There is a need to verify the patterns of species richness for large spatial scales and to understand patterns of species richness under a variety of land uses.

The data that we discuss next address these issues for plant species using forest inventories that cover thousands of hectares in contrast to fractions of a hectare. We believe that large-scale inventory data are required to sort out the effects of human activity on the world's species richness. For example, when large areas are considered, Z values are smaller than those obtained from small sized plots (Williamson 1981). This is another reflection of the principle that area considerations affect species richness determinations (Preston 1962; Williamson 1981). Therefore, when addressing environmental problems at global or local scales, it is prudent to base solutions on information collected at the corresponding spatial and temporal scale. We have shown that similar large-scale inventory data are needed to address adequately the issue of forest biomass (Brown and Lugo 1992).

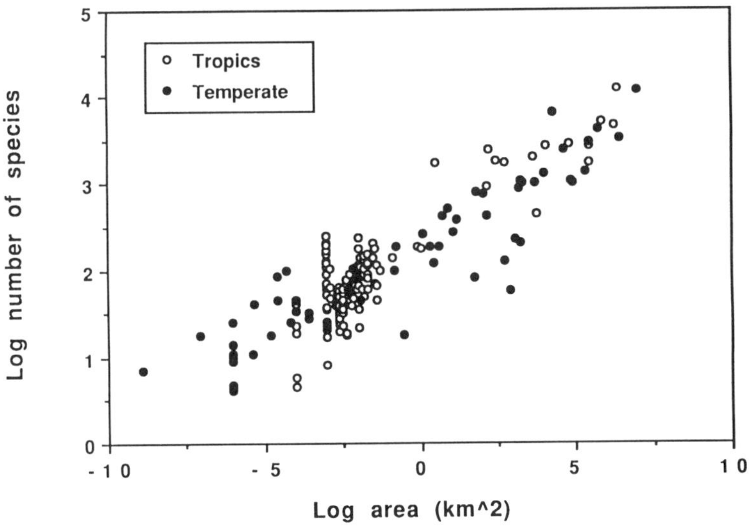

Figure 19.4. Relationship between number of plant species and area (km^2) for tropical and temperate ecosystems. Data are from Centre Technique Forestier Tropical (1969, trees ≥ 10 cm diameter), Clement and Nouvellet (1978, trees ≥ 10 cm diameter), Gentry (1982; contained two tropical data sets: a summary of the literature [trees with diameter ≥ 2-15 cm], and his data for all trees with diameter ≥ 2.5 cm plus vines), Gentry (1984, trees and vines ≥ 2.5 cm diameter), J.P. Veillon (pers. comm., trees ≥ 10 cm diameter), and Williams (1964; global for all flowering plants).

SPECIES-AREA RELATIONSHIPS IN MATURE AND DISTURBED FORESTS

Fitting the species-area model to a global data base of plant species results in a Z value of 0.21 ($r^2 = 0.78$), close to the estimates of Williams (0.20; 1964) and Preston (0.22; 1962) who analyzed a subset of the data in Figure 19.4. As expected, slight differences in the Z value occur if the model is applied separately to the tropics and temperate zone. The corresponding Z values are 0.19 ($r^2 = 0.85$) for the temperate zone and 0.23 ($r^2 = 0.74$) for the tropics. This reflects the greater density of species in the tropics compared to the temperate zone. Using a sub-set of our data, Preston (1962) found Z values of 0.22 and 0.25 for temperate and tropical regions, respectively. He considered 0.27 as the maximum Z value possible for these kinds of species-area curves.

The scatter of points around the regression line of species-area curves most likely reflects differences in environmental factors and degree of human disturbance. The influence of climate on plant species richness is illustrated by the data for Sri Lanka (based on a national forest inventory) where tree species for the same number of plots is considerably lower in the dry zones than in the moist zones (Figure 19.5). A plot of the log of species versus log of number of plots (implies area) shows that the slope (Z value) is higher in moist (0.45) than for the dry (0.37) forests. The maximum number of tree species in Sri Lanka's dry forests is about 50-60 in comparison to about 110 for the moist forests.

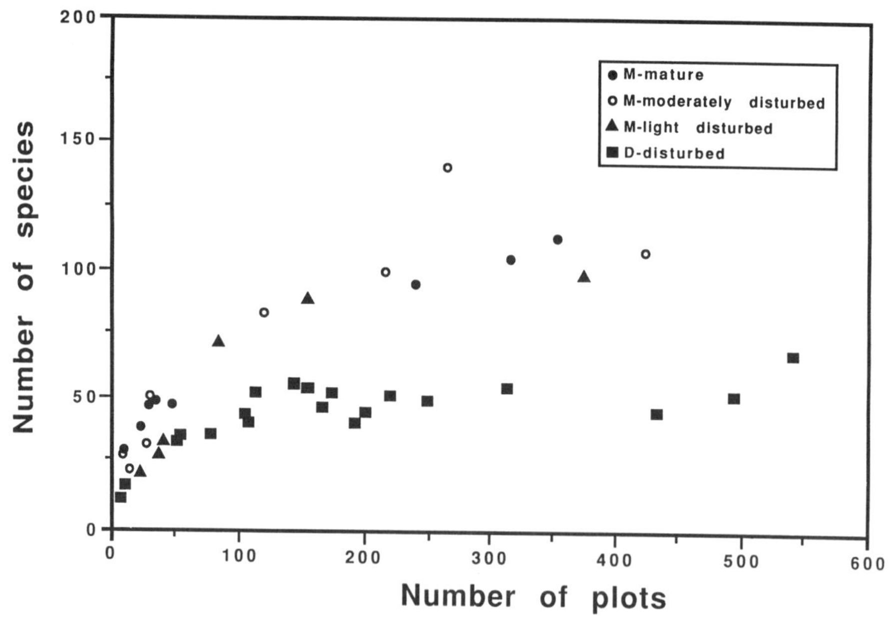

Figure 19.5. Relationship between species (all trees with diameter ≥ 10 cm) and plots in moist (M) and dry (D) climate zones and by degree of disturbance (data from Andrews 1961). Plots were dimensionless and based on the basal area prism method of forest inventory. The inventory was nation-wide and included a total of 37 forest reserves.

The data for the nationwide forest inventory in Sri Lanka suggest that regardless of the degree of disturbance, sampling of the same number of plots gives similar numbers of tree species because all data for moist forests fall on the same curve (Figure 19.5). Even the so called mature forests are most likely disturbed to some degree, often by illicit felling as determined from the description of the forest reserves (Andrews 1961). The other causes of forest disturbance were shifting cultivation and wood harvest.

WHAT TYPES OF POPULATIONS AND ECOSYSTEMS ARE WE SELECTING FOR?

As the era of successional vegetation unfolds (Brown and Lugo 1990), people continue to transform the landscape and in the process select for those ecological attributes of ecosystems that favor short-term human survivorship. These are attributes associated with successional or transitional ecosystems, i.e., high rates of net primary productivity, low biomass, fast turnover of mass and nutrients, small size of organisms, rapid metabolism, and short life spans. Many of these changes are necessary to maintain a higher population density because they maximize use opportunities caused by high yields and adjustment to environmental change caused by fast growth and regeneration. A challenge facing all of us is to divert our activities to the appropriate ecological life zones so that sufficient area is set aside for preservation of mature ecosystems which have contrasting characteristics and sufficient genetic diversity to provide insurance against unpredictable environmental change (Lugo 1991). Figure 19.6 illustrates guidelines for optimal use of life zones that protect species richness and assures human needs can be balanced on a landscape basis. We believe that the time has come to assign a use to every square kilometer on the planet.

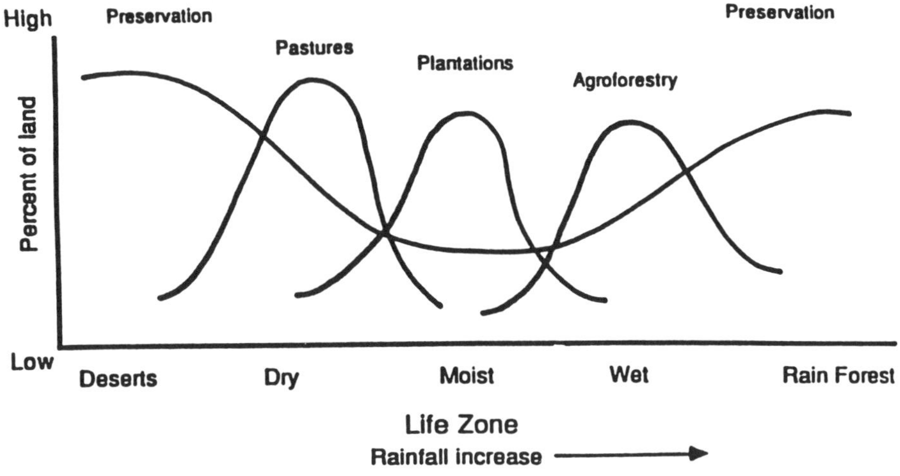

Figure 19.6. Suggested land uses according to type of ecological life zone. Notice that there is preservation and use in all life zones, but the emphasis of the land use changes according to ecological conditions (Lugo 1987).

HOW CAN ENVIRONMENTAL CHANGE BE MANAGED?

Managing the conditions of environmental change is imperative for our well being (Lugo 1992a). However, this activity faces staggering social, technical, and economic obstacles. It is not our objective to discuss these problems nor their root causes. We agree with Schmink (1987) and Hecht and Cockburn (1989) that the roots of the problem of resource use in tropical countries are social and their solution requires greater attention to socioeconomic factors. The lack of adequate institutional infrastructure to deal with resource management and the scarcity of trained human resources compound the situation. This is why only a small fraction of the tropical forest biome (< 5%) is under **any** kind of management (Schmidt 1987). Our focus here is limited to management actions that use ecological information to protect or rehabilitate species richness while also using the land for human benefit. These recommendations can only be implemented if there are institutions with sufficient resources and political support to act on them.

We believe that no matter how degraded a site is, there is always an opportunity to return it to production and some level of ecological health. Obviously the complexity of the job and the time and cost required to achieve a measure of success are directly proportional to the degree of damage inflicted on the land and are dependent on removing of the factors that caused the degradation. The important lesson to be learned is that human activity can be as effective in repairing damaged lands as it is in causing damage. Therefore, we need to be pro-active and optimistic so that the net effect of our actions is beneficial.

Islands in the Pacific, which for millennia have suffered chronic exploitation by humans and in the processes have lost their endemic avifauna and forests (Steadman 1991), are worse-case scenarios from which lessons can be derived. The transformation of formerly forested landscapes to shrubby vegetation, pastures, and damaged sites has been documented for Fiji (Ash 1992), Nauru and the Gilbert Islands (Thaman 1992), and Samoa and Tonga (Whistler 1992). Similar transformations have also occurred in Java (Smiet 1990, 1992). Accounts of the effects of humans on these islands include the initial loss of large endemic animals, the early elimination of natural vegetation from fertile lands, invasion of exotic species (Stoddart 1992), and reduction of biomass and stature of vegetation. In spite of the devastation implied in these accounts, there are glimmers of hope.

First, these islands have been subject to some of the most extreme examples of human exploitation of land. For example, for the Nauru and the Gilbert Islands, Thaman (1992) reports mining, bombing, urbanization, cultivation, and exploitation on islands of areas less than 300 km^2. In addition, the activity took place in small islands generally far from sources of replacement species. Population pressure on the land was 10 to 100 times that for continental areas. In spite of this human assault on the biota, vegetation change has been gradual by species replacement and substitution of original ecosystems by new mixtures of native and exotic species that somehow function, retain some endemic species including vertebrates, and provide the required services and products to the human population (Ash 1992; Thaman 1992). Data from Java show that in spite of the dramatic changes induced by people, the new communities retain a similar tree species composition (Table 19.2), although many of them are probably not the original ones.

Experience in the Caribbean islands suggests that land managers can direct succession to improve the productivity of the landscape and maintain its biodiversity (Wadsworth

Table 19.2. Numbers of tree species (diameter > 10 cm) and areas sampled under a variety of land uses and in different areas of Java (Smiet 1992).

Undisturbed		Slightly disturbed		Moderately disturbed		Heavily disturbed[1]	
Numbers	Area (ha)	Numbers	Area (ha)	Numbers	Area (ha)	Numbers	Area (ha)
57	1.00	78	1.00	31	0.16	72	1.20
10	0.04	27	2.00	44	5.40[1]	74	3.00
9	0.04					74	2.10
						87	2.90

[1] Trees > 12.5 dbh.

and Birdsey 1982; Lugo 1988c). This was accomplished by removing human pressure from some lands, applying silvicultural treatments to abandoned and regenerating forest stands (Table 19.3), and concentrating human populations in cities. These solutions require strong land management institutions and economic opportunity for displaced human populations.

Table 19.3. Forestry measures that might be taken to optimize biodiversity conservation while treating stands for optimal wood production. These suggestions allow production forestry to continue while conserving species. From Frank H. Wadsworth (pers. comm.).

- Secure present public forest estates by strengthening legal protection, identification on the ground, protective patrolling, and establishing support from the surrounding human population by fostering compatible use of the forests to the people.
- Determine through field studies any components of forest diversity subject to special jeopardy (critical species) and their environmental requirements.
- On the basis of such studies foster public or private protection of the full gamut of forest land environments needed to conserve native biodiversity and particularly the critical species, including, if necessary, lands not currently well forested.
- Increase the eficiency of forest product harvesting, processing, and marketing to provide more income for rural participants and to finance protection of additional forests.
- Identify alternative income opportunities for all users of the forest, taking into consideration social and cultural factors.
- Within production forests, exclude removal of products from some ten % of the area, diffusely scattered in small watersheds, to provide a network of natural environments and habitats which can be readily identified on the ground and thus protected from modification.
- Leave unmodified all riparian forests within 20 m of perennial streams less than 10-m wide and within 40 m of wider rivers.
- Prior to timber harvesting, identify environmental components such as keystone species and food-source or nest-site trees which might be crucial to known critical species and, on up to 30 % of the forest area, exclude them from harvesting plans, including assurance of their preservation during harvesting.
- Control intensity of timber harvesting to levels that conserve a forested environment in cutover stands with most of the attributes of the uncut forest. This generally means leaving more than one-half of the basal area.
- Where indicated, limit harvesting seasonally during periods of reproduction of critical species of forest animals.
- In post-harvesting silvicultural refinement, conserve up to 30 % of the forest or of trees that might be needed for the critical species.

Use of tree plantations is another strategy available for rehabilitating lands and species in sites that have lost them. Plantations can serve as foster ecosystems for native species by restoring forest conditions and soil fertility to degraded lands incapable of supporting native forests (Figure 19.7, Lugo 1988c, 1992b; Parrotta 1992, 1993). Under plantation canopies, about one native tree species became established annually, a rate matching that observed in secondary forests (Lugo et al. 1993). The difference is that more species are re-established in secondary forest succession because they start with a higher initial number of species than a plantation which starts with practically zero.

To conclude, managing species is feasible and possible but requires a strong commitment to land management. Land management must focus on a multiplicity of ecosystem states ranging from mature forests under protection to many other stages of land conversion including damaged and repaired lands (Figure 19.8). The strategy is to minimize states associated with low productivity and damaged lands. Reversing these degraded states can be expensive and time consuming particularly when it is not feasible to use natural regeneration processes. Maintaining a productive landscape with a variety of uses and ecosystems is the best strategy to maintain biodiversity (Pimentel et al. 1992).

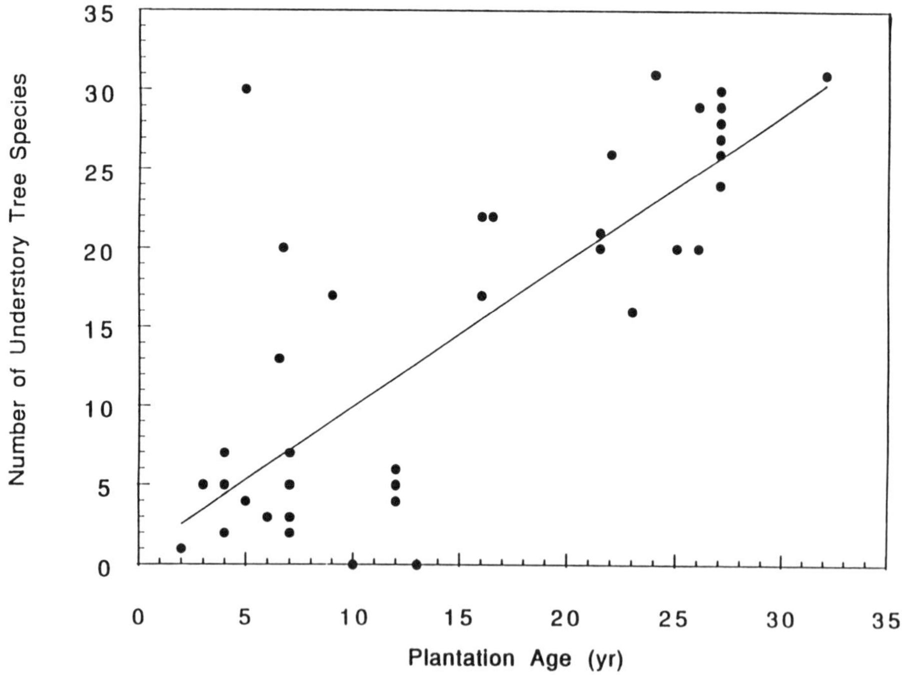

Figure 19.7. Relationship between age of tree plantation on degraded lands in Puerto Rico and India and establishment of native understory tree species on the plantation (Lugo et al. 1993).

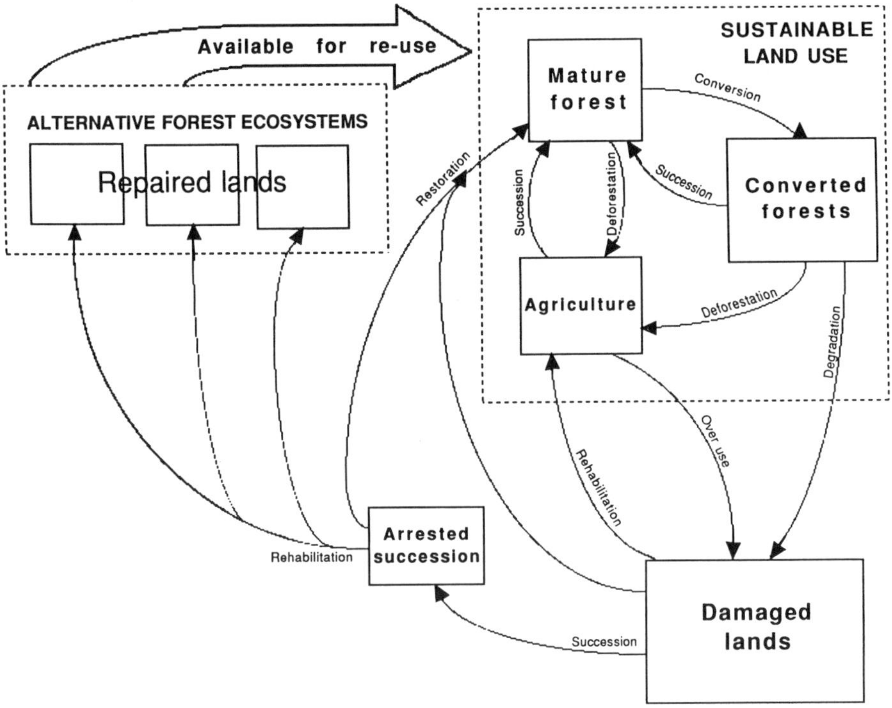

Figure 19.8. Dynamics of land-use change in relation to human activities. A landscape with reduced areas of damaged lands and a management strategy for repairing and restoring ecosystems is the best strategy for maintaining biodiversity.

ACKNOWLEDGMENTS

This work was carried out in cooperation with the University of Puerto Rico. We thank Juan Torres and two reviewers for their comments and suggestions, and Mildred Român and María M. Rivera for their logistic support.

LITERATURE CITED

Andrews, J. R. T. 1961. A forest inventory of Ceylon. Prepared by the Hunting Survey Corporation Ltd., Toronto, Canada in cooperation with the Forest Department, Ceylon. A Canada-Ceylon Colombo Plan Project, Government Press, Ceylon.

Ash, J. 1992. Vegetation ecology of Fiji: past present, and future perspectives. *Pac. Sci.* 46:111-127.

Brown, S. and A. E. Lugo. 1990. Tropical secondary forests. *J. Trop. Ecol.* 6:1-32.

Brown, S. and A. E. Lugo. 1992. Aboveground biomass estimates for tropical moist forests of the Brazilian Amazon. *Interciencia* 17:8-18.

Brown, S., A. J. R. Gillespie, and A. E. Lugo. 1991. Biomass of tropical forests of South and Southeast Asia. *Can. J. For. Res.* 21:111-117.

Burgess, R. L., and D. M. Sharpe, eds. 1981. *Forest Island Dynamics in Man-dominated Landscapes*. Ecological Studies 41. Springer-Verlag, New York.

Bush, M. B., D. R. Piperno, P. A. Colinvaux, P. E. de Oliveira, L. A. Krissek, M. C. Miller, and W.

E. Rowe. 1992. A 14,300-yr paleoecological profile of a lowland tropical lake in Panama. *Ecol. Monogr.* 62:251-275.

Clement, J., and Y. Nouvellet. 1978. Inventaire forestier dans la zone d'Edenzork. Republique du Gabon, Societe Gabonaise de Cellulose. Centre Technique Forestier Tropical, Nogent-sur-Marne, France.

Centre Technique Forestier Tropical(CTFT). 1969. Republique Federale du Cameroun. Inventaire de 100.000 hectares de forest dense dans la region d'Edea. Fascicule I. Nogent-sur-Marne, France.

Currie, D. J., and V. Paquin. 1987. Large-scale biogeographical patterns of species richness of trees. *Nature* 329:326-327.

Cutler, A. 1991. Nested faunas and extinction in fragmented habitats. *Conserv. Biol.* 5:496-505.

Dobzhansky, T. 1950. Evolution in the tropics. *Am. Sci.* 38:209-221.

Ehrlich, P., and A. Ehrlich. 1981. *The Causes of the Disappearance of Species.* Random House, New York.

Food and Agriculture Organization (FAO). 1985. Forest resources 1980. Rome, Italy.

Gentry, A. H. 1982. Patterns of neotropical plant species diversity. *Evol. Biol.* 15:1-84.

Gentry, A. H. 1986. Endemism in tropical vs. temperate plant communities. In *Conservation Biology, the Science of Scarcity and Diversity*, ed. M. E. Soulé, pp. 153-181. Sinauer Associates Inc., Sunderland, Mass.

Gentry, A. H., and C. Dodson. 1987. Contributions of nontree to species richness of a tropical rain forest. *Biotropica* 19:149-156.

Gomez-Pompa, A., J. S. Flores, and V. Sosa. 1987. The 'pet kot': a man-made tropical forest of the Maya. *Interciencia* 12:10-15.

Grainger, A. 1988. Estimating areas of degraded tropical lands requiring replenishment of forest cover. *Internat. Tree Crops J.* 5:31-61.

Haffer, J. 1969. Speciation in Amazonian forest birds. *Science* 165:131-137.

Hecht, S., and A. Cockburn. 1989. *The Fate of the Forest. Developers, Destroyers and Defenders of the Amazon.* Verson, New York.

King, C. 1984. *Immigrant Killers.* Oxford University Press, Melbourne, Australia.

Lovejoy, T. E. 1980. A projection of species extinctions. In *The Global 2000 Report to the President. Entering the Twenty-first Century*, study director G. O. Barney, pp. 328-331. Vol. 2. Council on Environmental Quality, Washington, D.C.

Lugo, A. E. 1987. Tropical forest management with emphasis on wood production. 1987. In *Ecological Development in the Humid Tropics. Guidelines for Planners*, eds. A. E. Lugo, J. R. Clark, and R. D. Childs, pp. 169-189. Winrock International, Morrilton, Ark.

Lugo, A. E. 1988a. Diversity of tropical species: questions that elude answers. *Biol. Internat.* 19:37 pp.

Lugo, A. E. 1988b. Estimating reductions in the diversity of tropical forest species. In *Biodiversity*, ed. E. O. Wilson, pp. 58-70. National Academy Press, Washington, D.C.

Lugo, A. E. 1988c. The future of the forest: ecosystem rehabilitation in the tropics. *Environ.* 30(7):16-20; 41-45.

Lugo, A. E. 1991. Cities in the sustainable development of tropical landscapes. *Nature and Resources* 27:27-35.

Lugo, A. E. 1992a. Managing tropical forests in a time of climate change. In *Forests in a Changing Climate*, ed. A. Qureshi, pp. 336-344. Climate Institute, Washington, D.C.

Lugo, A. E. 1992b. Tree plantations for rehabilitating damaged forest lands in the tropics. In *Ecosystem Rehabilitation*, Vol. 2: Ecosystem Analysis and Synthesis, ed. M. K. Wali, pp. 247-255. SPB Academic Publishing, The Hague, The Netherlands.

Lugo, A. E. 1994. Tropical forests: their future and our future. In *Tropical Forests: Management and Ecology*, eds. A. E. Lugo and C. Lowe, Chapt. 1. Springer Verlag, New York. (in press).

Lugo, A. E., J. Parrotta, and S. Brown. 1993. Reduction of species due to tropical deforestation and their recovery through management. *Ambio* 22(2-3):106-109.

Lugo, A. E., and S. Brown. 1992. Tropical forests as sink of atmospheric carbon. *For. Ecol. Manage.* 54:239-255.

Mares, M. A. 1992. Neotropical mammals and the myth of Amazonian biodiversity. *Science* 255:976-979.

MacArthur, R. H., and E. O. Wilson. 1967. *The Theory of Island Biogeography*. Monographs in Population Biology 1. Princeton University Press, Princeton, N.J.

Nelson, B. W., C. A. C. Ferreira, M. F. da Silva, and M. L. Kawasaki. 1990. Endemism centers, refugia and botanical collections density in Brazilian Amazonia. *Nature* 345:714-716.

Newmark, W. D. 1990. Tropical forest fragmentation and the local extinction of understory birds in the eastern Usambara Mountains, Tanzania. *Conserv. Biol.* 5:67-78.

Odum, H. T. 1983. *Systems Ecology: An Introduction*. John Wiley and Sons, New York.

Parrotta, J. A. 1992. The role of plantation forests in rehabilitating degraded tropical ecosystems. *Agric. Ecosyst. Environ.* 41:115-133.

Parrotta, J. A. 1993. Secondary forest regeneration on degraded tropical lands: the role of plantations as 'foster ecosystems.' In *Restoration of Tropical Forest Ecosystems*, eds. H. Lieth and M. Lohmann, pp. 63-73. Kluwer Academic Publishers, Dortrecht, The Netherlands.

Pimentel, D., U. Stachow, D. A. Takacs, H. W. Brubaker, A. R. Dumas, J. J. Meaney, J. A. S. O'Neil, D. E. Onsi, and D. B. Corzilius. 1992. Conserving biological diversity in agricultural/ forestry systems. *BioScience* 42:354-362.

Prance, G. T., ed. 1982. *Biological Diversification in the Tropics*. Columbia University Press, New York.

Preston, F. W. 1962. The canonical distribution of commonness and rarity: Part 1. *Ecology* 43:185-215.

Rice, B., and M. Westoby. 1983. Plant species richness at the 0.1 ha scale in Australian vegetation compared to other continents. *Vegetatio* 52:120-140.

Saunders, D. A., R. J. Hobbs, and C. R. Margules. 1991. Biological consequences of ecosystem fragmentation: a review. *Conserv. Biol.* 5:18-32.

Schmidt, R. 1987. Tropical rain forest management: a status report. *Unasylva* 39(156):2-17.

Schmink, M. 1987. The rationality of forest destruction. In *Management of the Forests of Tropical America: Prospects and Technologies*, eds. J. C. Figueroa Colòn, F. H. Wadsworth, and S. Branham, pp. 11-30. Institute of Tropical Forestry, Rio Piedras, P.R.

Simberloff, D. 1986. Are we on the verge of a mass extinction in tropical rain forests? In *Dynamics of Extinctions*, ed. D. K. Elliot, pp. 165-180. John Wiley and Sons, New York.

Simpson, B. E., and J. Haffer. 1978. Speciation patterns in the Amazonian forest biota. *Ann. Rev. Ecol. Syst.* 9:497-518.

Smiet, A. C. 1990. Forest ecology on Java: conversion and usage in a historical perspective. *J. Trop. Ecol.* 2:286-302.

Smiet, A. C. 1992. Forest ecology on Java: human impact and vegetation of montane forest. *J. Trop. Ecol.* 8:129-152.

Steadman, D. W. 1991. Extinct and extirpated birds from Aitutaki and Atiu, southern Cook Islands. *Pac. Sci.* 45:325-347.

Stoddart, D. R. 1992. Biogeography of the tropical Pacific. *Pac. Sci.* 46:276-293.

Thaman, R. R. 1992. Vegetation of Nauru and the Gilbert Islands: case studies of poverty, degradation, disturbance and displacement. *Pac. Sci.* 46:128-158.

Wadsworth, F. H., and R. A. Birdsey. 1982. Un nuevo enfoque de los bosques de Puerto Rico. In *Noveno Simposio de Recursos Naturales*, pp. 12-27. San Juan, P.R.

Whistler, A. W. 1992. Vegetation of Samoa and Tonga. *Pac. Sci.* 46:159-178.

Whitmore, T. C. 1975. *Tropical Rain Forests of the Far East.* Clarendon Press, Oxford, England.

Williams, C. B. 1964. *Patterns in the Balance of Nature and Related Problems in Quantitative Ecology*. Academic Press, New York.

Williamson, M. 1981. *Island Populations.* Oxford University Press, Oxford, England.

20

THE DIVERSITY AND MAINTENANCE OF OLD-GROWTH FORESTS

Thomas A. Spies

USDA Forest Service
Pacific Northwest Research Station
Corvallis, Oregon
U.S.A.

Jerry F. Franklin

College of Forest Resources
University of Washington
Seattle, Washington
U.S.A.

The contribution of old-growth forests to biological diversity has only become widely recognized in the last 10 years (Franklin et al. 1981), despite an interest in the ecology of virgin or "wild" forests which has existed among naturalists and ecologists for at least 130 years (Thoreau 1864). On a global scale, the area of old-growth forests has probably steeply declined over the last 2,000 years, although no estimates of past or current amounts of old growth are available. The area of "primary" forest or original forest cover, of which old-growth forests would make up a large part, is estimated to be only 24% of its pre-settled-agriculture amount, and in the United States excluding Alaska, less than 5% of its presettlement amount (Postel and Ryan 1991). In most areas of the world, where old-growth forests are totally gone or exist as fragments in highly altered landscapes, the ecological values of old growth are poorly known. Scientific knowledge of the ecological characteristics of old-growth forests is probably best developed for forests of the Pacific Northwest (Franklin et al. 1981; Ruggiero et al. 1991), and scientific interest in the ecology and conservation and management of old growth is increasing in many parts of the world (Mayer et al. 1987; Kaufmann et al. 1992; Esseen et al. 1992; Botkin and Talbot 1992; Peterkin 1993). In regions lacking old-growth forests, efforts are being made to restore this type of ecosystem (Henderson and Hedrick 1991), typically with very limited information to develop the specific objectives of restoration. Maintenance or restoration of old growth requires a good understanding of forest dynamics and the diversity and ecological roles of all successional stages including old growth. In most biomes, adequate information about the ecological characteristics of old growth does not exist, especially information about how old growth is distinguished from other forest conditions and how it varies across landscapes and vegetation types.

Understanding the ecology of old growth and maintaining its distinctive biological diversity requires a hierarchical perspective. Biological hierarchies have been expressed in different ways (Rowe 1961; MacMahon et al. 1978; Noss 1990) and we use the framework comprised of genes, species and populations, ecosystems, and landscapes (Noss 1990). Placing old growth in a biological hierarchy provides a more unified concept of old growth and demonstrates how human manipulations at one biological level can affect old-growth values at other levels. Our objectives in this paper are to: (1) review the ecological perspectives on old growth at genetic and physiological, population and species, ecosystem, landscape levels; (2) characterize some of the diversity in types of old growth; and (3) discuss options for the maintenance and restoration of old growth.

PERSPECTIVES AND DEFINITIONS

The term "old growth" has many definitions. To some it is synonymous with "virgin" or "primary" forest, i.e., forests that have never been cut or lack evidence of human activities. To others, old growth is the same as "climax" or "late successional" forests where stand composition is relatively stable. To others, old growth represents an "ancient forest" of very old trees which has spiritual, cultural, and/or ecosystem values. In this paper we begin with a very simple and broad definition of old growth as an ecosystem distinguished by old trees not necessarily (1) in a late successional condition and (2) free of evidence of human activity. Within this broad definition we recognize more restrictive definitions that we hope will encompass most of the ecological perspectives on old growth.

Concepts and definitions of old growth have become more refined in recent years with scientific discoveries and changing societal values. In North America old growth was traditionally defined from a timber production viewpoint as a stand of trees beyond a certain age, or rate of growth. Before old forests became a subject of detailed scientific inquiry, there was little interest in precise definitions or in knowing the amount or distribution of old growth. Since the mid-1970s, old growth has taken on more scientific and political significance, and generic and specific definitions or standards have been developed (Old-Growth Task Group 1986; USDA Forest Service 1989) primarily as tools for inventory and mapping. The specific definitions are typically based on minimum levels of structural features such as large or old live and dead trees, features that can be relatively easily measured in extensive ground-based inventories or in some cases with remote sensing (Cohen and Spies 1992). These precise definitions should be distinguished from general characterizations of old growth which provide the ecological basis of the rather arbitrary inventory standards.

Although inventory of old growth requires common and relatively precise definitions, ecological and management characterizations of old growth will be diverse and hierarchical, a reflection of the complexity of ecosystems and diversity of scientific and societal values. Although the concept of old growth has been too narrowly focused in the past on the age and structure of the trees as a timber crop, the nature of old growth at genetic, species, population, ecosystem, and landscape levels is strongly controlled by life histories of cohorts of trees, and tree population phenomena. Landscape and ecosystem characteristics, such as disturbance regimes and the physical environment, however, constrain the population, species, and genetic patterns of old growth.

GENETIC AND PHYSIOLOGICAL PERSPECTIVES

The genetic characteristics of old-growth forests have received relatively little attention because most forest genetics research has been oriented toward genetic traits associated with young trees growing in forest plantations and because of the difficulties of studying genetic characteristics that might take centuries to be expressed. From a genetic perspective, a population of old trees in a forest represents a genetic pool that is the product of a long period of intense competition and selection. However, the oldest or most dominant trees in a forest are not necessarily the most genetically superior for a particular trait, because survivorship can be more related to "accidents" of history and site variation than genetically superior traits which increase reproductive success (Oliver and Larson 1990).

The selective pressures shaping the gene pools of trees vary with the growth stage of the individuals and the population (Namkoong and Conkle 1976). In the establishment stage of forest development, genetic patterns can be strongly affected by microenvironmental heterogeneity (Campbell 1979). Following canopy closure, which signifies the beginning of the thinning stage of stand development, cohort competition for resources, weather, and other macro-site factors appear to affect survivorship and control genetic patterns (Namkoong and Conkle 1976). Genetic patterns created by selective pressures in early stages of stand development might still be evident in old-growth populations (Campbell 1979). Consequently, the genetic structure within a population of old-growth trees is the sum of selection forces acting over centuries. Cohorts of old individuals might be genetically distinct from nearby populations of younger trees (Tiegerstedt et al. 1982), if the regeneration is a product of inbreeding. Selection and competition which occur as the population matures, resulting in the death of inbred individuals, are important processes that reduce inbreeding depression in natural populations. The existence of trees in a landscape that have survived into mature and old-growth stages probably help keep the larger population in the landscape well adapted and genetically healthy. However, genetic structure among tree age cohorts might not differ if levels of individual-tree heterozygosity and outcrossing are high (Neale 1985).

The forces of natural selection are hypothesized to decline following the onset of reproduction for many organisms (Olshansky et al. 1993). However, for many tree species that regenerate infrequently, natural selection pressures probably continue into mature and even old ages. As trees age and stand density decreases, mortality results more from "extrinsic" factors such as fire, wind, insects, and diseases than "intrinsic" factors associated within-stand competition for resources (Oliver and Larson 1990). Consequently, any natural selection during the old-growth stage is probably associated with the ability of the individual to withstand disturbance (e.g., thick bark and decay-resistant wood), regenerate dead or damaged parts (e.g., release of dormant buds below damaged stems), or allocate resources to seed production when highly stressed.

The process of senescence or aging in trees is characterized by increases in physiological stresses resulting from the effects of large size on food and water supply to meristems and the accumulation of defects and injuries from the sources mentioned above. These damages accrue, either rapidly or slowly, to a point at which the tree no longer has enough vital structures to regenerate the dead parts and the tree dies. This is a general pattern of degeneration that is found in all organisms, particularly the large, complex ones with sufficient redundancy of vital structures (Gavrilov and Gavrilova 1993) to allow for a period of decline. In large, long-lived species with decay resistant wood, such as Sequoia and Douglas-fir, this period of senescence, or the old-growth stage, can be centuries long. Given that many disturbances, such as fire, which damage or kill canopy trees, can also create opportunities for regeneration, some natural selection probably occurs for the ability to survive these events by growing large, disturbance-resistant structures and surviving disturbances, at least long enough to produce seeds that could colonize the open habitats created by the disturbance. At extreme ages, however, old individuals have little influence on population gene pools (Pianka 1978) because of their low reproductive value relative to more numerous and vigorous younger age classes of trees.

SPECIES AND POPULATION PERSPECTIVES

It is widely recognized that life history characteristics of the tree species play a major role in determining rates and patterns of change during forest succession and stand development (Noble and Slayter 1977). The rate at which old growth develops, the structure and composition of old growth, and the relative stability of old growth will differ as a function of the age of reproduction, shade tolerance, size at maturity, longevity, and reproductive systems of the tree species of a landscape or region. For example, short-lived tree species such as aspen (*Populus* spp.) can reach maturity and old age in 50 years and be completely gone from a site in 150 years, whereas Douglas-fir can take 150 to 250 years to reach maturity and live for another 700 years. Because of their light requirements, neither species can regenerate in small openings left by the death of individual trees, and typically require relatively large and intense disturbances to regenerate.

The dynamics of aggregates of conspecific tree species is a population-level phenomena, with the population characteristics determined by the life history characteristics of the tree species. A general theoretical model of forest dynamics following major disturbances such a large intense fires or clearcutting has been developed and refined by several investigators (Bormann and Likens 1979; Oliver 1981; Peet and Christensen 1987; Mueller-Dombois 1987). In this model, forest development is driven by population process of trees and can be broken into several stages: establishment, thinning, transition, and steady state (Peet and Christensen 1987). These stages are distinguished by growth rates of dominants, mortality rates and causes, origin of the tree cohorts, and importance of small canopy openings to tree regeneration.

Only Oliver (1981) has used the term "old growth" in a forest classification scheme. In his scheme old growth corresponds to the transition and steady-state stages. More recently, Oliver and Larson (1990) have termed these stages "transition old growth" and "true old growth". In the transition stage, the canopy is dominated by large individuals of the cohort that established following the major disturbance. However, new tree establishment of the same or different species is occurring within the stand and those individuals are growing into intermediate and dominant canopy positions. In the steady-state stage, most or all of the individuals in the initial cohort have died, gap dynamics control growth in the understory and the pattern of canopy tree replacement, and the forest can be viewed as a shifting mosaic of all the stages of stand development.

The general model of forest stand dynamics described above lacks some important details of forest dynamics. Recent research in the Pacific Northwest has indicated that stand dynamics between the end of the thinning stage and the end of the transition stage can be quite protracted and characterized by considerable change (Spies and Franklin 1991) which is obscured by recognizing only thinning and transition stages. In addition, the general model has only focused on the population dynamics of live trees, ignoring the dynamics of dead trees. Dead trees play many important ecological roles (Harmon et al. 1986) in a forest and we propose that the general model of forest dynamics must encompass tree population dynamics over the entire existence of a tree as a biological and ecological structure. Based on these perspectives of forest dynamics, we modify the general model of forest dynamics, with the recognition that it represents an idealized continuum of change.

Establishment Phase

After disturbance new individuals and species establish for some period of time. Dead tree accumulations are typically high and originate in the predisturbance stand and are now declining. The accumulations of dead trees may be relatively low if the predisturbance stand had low numbers or small sizes of live trees.

Thinning Phase

Once canopy closure as occurred, additional establishment might stop or become rare, and the understory might become relatively free of vegetation ("Stem exclusion stage" of Oliver 1981). Mortality resulting from competition among canopy trees eventually leads to increases in understory light levels which allows establishment or growth of surviving individuals of shade tolerant trees, shrubs, and herbs ("Understory reinitiation stage" of Oliver 1981). Accumulations of dead trees are beginning to originate in the post-disturbance stand from thinning mortality although the small stems are typically not persistent. Considerable accumulations of well-decayed dead trees might still occur if carry-over from previous stand was high and decay rates are low.

Mature Phase

During this phase some of the first live structural characteristics of old growth appear, such as relatively large, live trees and local areas of multiple canopy layers. Growth rates of the dominant trees (either single or multiple species) have slowed and trees begin to approach their maximum height and crown width. Canopy gaps that form from the death of dominant trees do not close from lateral growth of surrounding canopy trees. However, gaps are typically smaller and shade tolerant saplings are less restricted to gaps than in the later phases (Spies et al. 1990). Shade tolerant associates might be regenerating throughout the understory and a few may have reached intermediate canopy levels. Mortality occurs less from inter-tree competition and more from other factors, such as disease or windthrow. The mass of dead wood is typically at a low point and most of the dead wood originating in the pre-disturbance stand is gone (Spies et al. 1988).

Transition Phase

This phase is characterized by a transition from a forest dominated by live and dead trees that established following the initial large disturbance to a forest dominated by live and dead trees that established in the understory and gaps of the developing forest. This phase can be subdivided into early and late stages. In the early stage the canopy is still dominated by a relatively dense stand of large old trees from the cohort that established following the major disturbance. The crowns of the old trees have become large, relatively open and frequently flat-topped with large limbs. Trees that regenerated in the understory have typically increased in density and filled-in the intermediate canopy levels. The numbers and mass of dead trees is building in response to mortality of the large dominants and codominants. In the latter stage of this phase, all or nearly all of the live individuals that made up the original cohort have died. However, the population of dead, relatively large tree boles (standing or fallen or both) from this cohort is close to its maximum density and biomass.

Shifting-Gap Phase

During this stage the live and dead structures of the original tree cohort are gone, and the establishment and growth of the population of live and dead trees results from gap-phase dynamics. The biomass of dead trees has declined from its high levels in the previous phase. Tree death might be less synchronous in space and time than in previous phases and the wood of late successional tree species might decay more rapidly so dead wood amounts might be lower. The forest can be viewed as a relatively fine-grained shifting mosaic of all stages of stand development. Slow changes in forest composition and structure can still occur as a consequence of climatic change or changes in mortality rates or establishment rates among tree species.

The general model of forest stand dynamics described above applies primarily to post-disturbance conditions following catastrophic disturbance, in which all or most of the original stand is killed by the disturbance. Analyses of large disturbances, however, frequently reveal that areas of low to moderate disturbance severity occur where portions of the original stand survive (Zackrisson 1977; Morrison and Swanson 1990; Foster and Boose 1992). In cases of patchy disturbances, the forest can behave as combinations of the above stages occurring the same time and space with varying degrees of interaction between cohorts. For example, a new establishment phase can occur beneath the surviving canopy. The new cohort can proceed through various stages while the surviving cohort becomes a remnant population which might disappear slowly or rapidly, leaving a patch of forest in a regeneration, thinning or transition stage. Consequently many stands or landscapes following "major" natural disturbances are composed of an overlapping mosaic of tree population patches of different phases. These mosaics retain some of the characteristics of the previous old-growth stages, such as large live or dead trees, and might return more rapidly to the more complex structure of the transition and shifting gap phases than in the case where the disturbance kills or removes all the trees. In the case of the shifting-gap phase of old growth, in which the tree size structure appears to be a result of fine-grained small gap disturbances, analysis of forest age structure and dynamics frequently reveals that these stands have had a history of larger disturbances that is partially obscured by the patchiness of the larger disturbances, differential growth rates of even-aged groups of species, and/or an overlay of smaller disturbances (Foster 1988).

Not all of the latter stages described above occur in a given forest type and some might be very ephemeral. For example, the shifting-gap phase rarely occurs in forest types such as Douglas-fir because it can take 800 to 1000 years for the original tree cohort to die out and another several hundred years for the tree boles to decay. Stand-replacing fires typically occur before this time (Franklin and Hemstrom 1981; Spies et al. 1988). The transition phases can be very short, e.g., a few decades, in some forest types such as *Populus* and *Alnus* where the tree dominants are short-lived and have high decay rates.

ECOSYSTEM PERSPECTIVES

At the ecosystem level we consider the composition, structure, and function of the biological and physical components of old growth. The development of old-growth plant and animal communities relative to younger stand developmental stages is not well known for most groups of organisms. Tree composition of old-growth or late successional forests have been documented for many temperate forests ecosystems (Nichols 191;3, Cline and Spurr 1942; Jones 1945; Habeck 1968; Muller 1982; Peterkin and Jones 1987; Foster 1988). Far less is know about the compositional uniqueness of old-growth forests for other life forms such as non-tree vegetation, vertebrates, and soil and canopy invertebrates (Anon. 1993; Romme et al. 1992). Recent studies in the Pacific Northwest indicate that most vertebrate and plant species do not show differences in abundance between closed canopy forest developmental stages (Ruggiero et al. 1991). However, in the survey, at least 49 plant (including hypogenous fungi and epiphytic lichens) and animal species did occur more frequently in old growth (transition phase) relative to earlier forest stages (thinning and mature phases). Differences in plant communities between old-growth and younger stages or secondary forests have also been observed in other forest types (Peterkin and Game 1984; Whitney and Foster 1988; Romme et al. 1992; Duffy and Meier 1992). Community differences in invertebrates have also been observed (Heliovaara and Vaisanen 1984; Schowalter 1989).

Although old growth might contain plant and animal specialists, it is typically not the stage of highest species richness during stand development. The establishment stage has the highest diversity for plants and mammals (Brown 1985; Schoonmaker and McKee

1988; Spies 1991). Species richness typically declines in the thinning stage and increases again to intermediate levels in the later stages.

The distinctiveness of the plant and animal communities of old-growth forest can be related to changes in resource levels and the spatial patterns of their availability, habitats and structures, and the poor colonizing ability of many old-growth forest species. For plants, the most dramatic resource change involves light levels and patterns in the understory. Light levels under closed canopies of late successional forests can be as little as 2% of full sunlight (Canham et al. 1990). Consequently, many shade intolerant species do not occur or are rare in old-growth forest stands where the canopy is composed of shade tolerant species. Where canopy openings occur, light levels can be high enough to enable shade intolerant plants to exist within the mosaic of old-growth vegetation. In the darkest areas within these forests, mycotrophic plants that lack chlorophyll might find suitable habitat.

Changes in food resources and habitat conditions drive development of old-growth animal communities. Because understories of forests are low light environments, vegetative and fruit production is typically lower than in more open environments. In old growth, the presence of canopy openings can result in higher forage production and quality relative to younger forests with dense continuous canopies (Alaback 1982; Hanley 1984). The presence of large standing dead and fallen trees and multiple layers of vegetation in old-growth stages provides nesting and roosting sites and hiding and thermal cover for many species. Although, large dead trees are not unique to old growth they are far more common relative to earlier developmental stages (Spies et al. 1988). The presence of relatively tall deep canopies provides for vertical niche differentiation and gradient of microclimate from exposed high, radiation environments at the top of the canopy to buffered environments near the forest floor.

Some of the differences observed between old-growth forests and younger forests might result from the slow rate of dispersal and colonization of suitable sites of many forest organisms, such as herbs which might be dispersed by ants or splashes from raindrops. Consequently, for many organisms old-growth forests are ecosystems which simply have had a longer time to accrue forest species than younger forests (Peterkin and Game 1984; Matlack in press). This suggests that accelerating stand development through silvicultural means will not be able to shorten the long time periods required for the development of some components of these old ecosystems.

The maturation, transition, and shifting gap phases of stand development appear to differ in some ecosystem characteristics from the thinning and establishment stages of natural forests and plantations. Relative to younger stages, these later stages of forest development can have more buffering of microclimates (Chen et al. 1993), greater production of food for some consumer organism groups (Ure and Maser 1982; Huff et al. 1992), higher storage of carbon (Harmon et al. 1990), greater inputs of precipitation (Harr 1982) and higher populations of arthropod predators and organisms beneficial to other ecosystems or successional stages (Schowalter 1989). Soil erosional potential is low in transitional old-growth stands (Swanson et al. 1982). Epiphytic nitrogen-fixing lichens, which can be more abundant in old growth (Lesica et al. 1991), can be sources of nitrogen for the ecosystem (Sollins et al. 1980). Ecosystem biomass typically peaks in the transition phase old-growth development and stays relatively high. In some ecosystem types, such as Douglas-fir-western hemlock, live biomass might decline as the original tree canopy breaks-up and is replaced by a forest of smaller stature (T. Spies, J. Franklin, and Spycher unpubl. data). While net primary productivity might be close to zero in the later stages of forest development, gross primary productivity can be relatively high (Grier and Logan 1977).

Nutrient retention varies through stand development. Highest outputs of nutrients from watersheds occur in the establishment phase and in the shifting-gap phases when net biomass increments are low (Vitousek and Reiners 1975). During the thinning and maturation stages when net biomass increments are high, nutrient outputs are low (Cole et al. 1967). During the transition old-growth phase, nutrient outputs also appear to be low (Sollins et al. 1980), but net biomass accumulation is also low (Grier and Logan 1977). The reason for the retention of nutrients during this phase, despite the low net biomass accumulation, may be net accumulation of detrital components (Sollins et al. 1980; Spies et al. 1988) as the original cohort of trees dies.

LANDSCAPE PERSPECTIVES

The general model of forest dynamics described above has been applied primarily to "stands," i.e., relatively homogeneous patches of vegetation whose areas are typically determined by management considerations (Daniel et al. 1979). In practice, stands usually range from 10s to 100s of ha in size. Although the "shifting-gap" phase of old growth describes a stand mosaic, it is typically applied to a fine-grained, "homogeneous," repeating pattern of different vegetation phases generated by the death of single or small groups of trees.

Old growth cannot be characterized entirely at a single spatial scale, e.g. a stand typical of most forestry cutting units. Vegetation patterns, disturbances and ecological processes occur at both finer and coarser scales (Vitousek and Reiners 1975). In addition, homogeneity is scale-dependent—what appears heterogeneous over small areas might appear homogeneous over large areas. For example, a small watershed can be composed of two very different "stands" based on vegetation structure and development, but over a larger area those two stands can be part of a "homogenous" repeating mosaic of vegetation phases.

In a landscape perspective, old growth is characterized across multiple spatial scales including landscapes (Forman and Godron 1986) and "regional landscapes" (Noss 1983) which are heterogeneous mosaics of interacting ecosystems ranging from 10^1 to 10^7 km^2 in area. Rather than emphasizing homogeneity, as the stand concept does, a landscape perspective emphasizes spatial patterns and heterogeneity. The principles of landscape ecology are not restricted to mosaics of large extent (Forman and Godron 1986), however. Forest mosaics covering small areas (<100 ha) are also composed of interacting ecosystems and can be viewed as "landscapes."

Old growth can be viewed as an element of a larger mosaic of vegetation phases and as the entire dynamic mosaic itself (Mueller-Dombois 1987; Vankat et al. 1991). The first view emphasizes influences of the surrounding forest matrix on the old-growth patches and interactions and movements between the old-growth parts of the mosaic.

For example, microclimate and forest structure of edges of old-growth stands bordering clear cuts might be altered for more than 200 meters into the interior of the stand (Chen et al. 1993).

In the second landscape perspective, the ecological character and function of the entire mosaic is emphasized. Mosaic characteristics, such as kind and proportion of patch types, their sizes, spatial pattern, and stability, affect the behavior of the landscape. In fine-grained mosaics, e.g., shifting-gap old growth, where contacts between patches are frequent because of the high edge to patch area ratio, the taller and more massive older patches can exhibit considerable control over the microclimate, resources, habitat, and function of the younger patches in the mosaic. In coarser-grained mosaics which occur over larger spatial extents, older and younger patches are larger and contacts between patch types are less frequent as a proportion of the mosaic area than in fine-grained mosaics. Consequently, edge effects among stages might have less ecological influence on the dynamics and function of the entire mosaic. However, interactions among patches still occur through processes that operate over greater distances such as dispersal of wide-ranging organisms, disturbance initiation and propagation, and flows of water and materials. In landscapes where old growth exists as patches in a matrix of younger or more homogeneous stand conditions, outbreaks of insects, which originate and spread in a relatively uniform matrix, can threaten the stability of the old-growth islands (Lynch and Swetnam 1984).

The spatial and ecological characteristics of old-growth mosaics will vary as a function of the landscape disturbance regimes including disturbance size, frequency, and intensity. We describe below a classification of old-growth types based on species life history traits and landscape dynamics which might be useful in conserving and managing a broad range of old-growth types.

A LANDSCAPE CLASSIFICATION OF OLD GROWTH

The diversity of old-growth types and ecological perspectives can be organized into a frame-work based on landscape disturbance regimes and species life history characteristics and (Table 20.1). A first division in the classification is based on disturbance patch size which will control the species that can regenerate and the size of area needed to maintain the landscape dynamics. Shade intolerants are restricted to coarse-grained mosaics where light and other resources are typically higher in disturbance patches. Shade tolerants, on the other hand, can regenerate in large or small patches, but tend to be less common in larger patches where they are frequently outcompeted by faster growing intolerants or stressed by environmental extremes.

Examples of coarse-grained old-growth forests include, aspen, lodgepole pine, and Douglas-fir. Examples of fine-grained old-growth forests include those dominated by shade tolerants such as hemlock, beech and maple, and *Picea* and *Abies* spp. Most tree species or groups have the potential to occur as coarse-grained old growth, whereas, the set of species that can occur as fine-grained old growth will be limited to those tolerant of resources levels in small disturbance patches. Waves of old growth can occur as successive populations or communities mature in the same area. Eventually, as small disturbances occur, the coarse-grained old growth will become fine-grained until a major disturbance intervenes.

A further division of old-growth forests can be made along a gradient of tree or patch longevity (Table 20.1), determined by the interplay between environment, physiology, and genetics. The longevity of a patch of trees has both ecological and management implications. The longer a patch persists the greater its influence will be on the soil characteristics, such as build-up of litter layers and chemical and climatic influences on soil development. Patch conditions that persist a long period of time can also provide more stable habitat and be more likely to be found, and used by less mobile organisms.

Examples of short-lived old-growth forests include coarse-grained types, such as aspen and lodgepole pine, and fine grained types, such as balsam fir or grand fir. In these forest types, old-growth conditions can begin in as little as 50 years and the original tree population can die out in less than 100 years. Consequently, old-growth conditions might persist at the patch level for 50 years or less. Intermediate to long-lived trees such as Douglas-fir, eastern white pine and ponderosa pine can dominate a patch for 400 to 1,000 years.

A third division of old-growth forests can be made on the basis of the type of disturbance that regenerates the canopy trees and/or kill the canopy trees. Disturbances that remove the understory vegetation and litter layers will create very different regeneration environments than those that do not disturb the lower strata of the forests. Old-growth forests in which fire is a frequent component typically have less down woody debris than old-growth forests in which fire is infrequent. Old-growth forests and landscapes, such as Ponderosa pine, which are subjected to frequent low intensity fires, experience a shifting mosaic of safe-sites and canopy openings. Fire helps to clear away competing vegetation and reduce coarse and fine litter for pine establishment (Moir and Dieterich 1988). In the absence of fire, fire-sensitive species such as *Abies* spp. become established and eventually replace the pine.

Table 20.1. Classification of old-growth forest types based on landscape structure and dynamics and tree longevity.

I. Coarse-grained forest mosaics (disturbances >> 0.1 ha)
- A. Short-lived trees (maximum age < 250 years)
  - Examples: aspen, red alder, jack pine, black cherry, pin cherry, black spruce
- B. Intermediate to long-lived trees (maximum age >250)
  - Examples: Douglas-fir, ponderosa pine, white pine, tulip poplar, longleaf pine

II. Fine-grained forest mosaics (disturbances << 0.1 ha)
- A. Short-lived trees (maximum age < 250)
  - Examples: balsam fir, white fir, black spruce
- B. Intermediate to long-lived trees (maximum age > 250)
  - 1. Fire-frequent forests
    - Examples: ponderosa pine, longleaf pine, Douglas-fir
  - 2. Fire-infrequent forests
    - Examples: western hemlock, sugar maple

ACTIVE MANAGEMENT FOR OLD GROWTH AT THE PATCH LEVEL

The maintenance and management of old-growth forests requires consideration of both patch (stand) and landscape (mosaic) level perspectives of old growth. Maintenance of old-growth patches requires that disturbances be infrequent enough that old, relatively large live and dead trees can develop. The time it takes for a single patch to develop old-growth characteristics depends upon the life span of the trees, density of the trees, and productivity of the environment.

Several patch-level approaches are available to retaining or accelerating the development of old-growth characteristics, such as multiple tree sizes, relatively large diameter trunks and crowns, and large dead trees (Nyberg et al. 1987; Spies et al. 1991). The first approach is to use silvicultural systems that do not remove all the structural elements of old growth at the time of cutting. Retaining live and dead trees on cutting units, which imitates natural disturbances more than clearcutting, can help to accelerate the redevelopment of transition and shifting gap stages of old-growth ecosystems. A second cohort of trees then becomes established, either artificially or naturally, beneath the surviving canopy. This approach, which is called green tree retention (Franklin 1989), or irregular shelterwood (Smith 1962), leaves the stand in a higher state of structural diversity so that it will take less time to reach the later stages of stand development than using clearcutting. This silvicultural approach is now being implemented in many areas of the Pacific Northwest and some parts of Sweden, where 5 to 40 trees per ha are left from the previous stand.

Selection systems (Smith 1962) that remove single or small groups of trees can also be used to maintain characteristics of shifting-gap old-growth types on some sites. However, selection management can be difficult to apply in practice (Oliver and Larson 1990), and is not a suitable imitation of natural landscape disturbance regimes for all types of old growth.

Where forest stands have already been regenerated through even-aged methods such as clearcutting, and relatively dense young plantations exist, thinning can be applied to make more resources available to the remaining trees and allow them to attain large dimensions sooner than if the thinning process occurs more slowly. Thinnings would be applied in a patchy manner to provide for horizontal and vertical heterogeneity. Several thinnings can be applied during the lifetime of the stand. However, when trees are relatively large, it is important that thinning be applied in a way that retains or allows development of large standing dead and fallen trees.

MANAGEMENT OF OLD GROWTH AT THE LANDSCAPE LEVEL

Old-growth conservation requires a dynamic landscape perspective. The maintenance of old growth within large areas is dependent the rate, size, pattern, and intensity of disturbances. If the frequency of disturbance is too high, some types of old growth, such as Douglas-fir or ponderosa pine, will not develop in a landscape and if the frequency of disturbance is too low, other types, such as aspen, will be lost from landscapes. If disturbances are too small, then shade intolerant types of old growth, such as alder or long-leaf pine, will not occur.

The minimum dynamic area of land (Pickett and Thompson 1978) required to maintain a mosaic containing old growth will vary with the disturbance regime of the landscape and the size of patch needed to regenerate the old-growth species. Size of the area will be a function of the number of patches needed to represent compositional and structural diversity of a landscape and the degree of desired degree of stability in the proportions of the different patch types (Shugart 1984). For example, if 100 patches are needed to maintain a quasi-equilibrium, then an area of 10 ha is needed, without considering edge effects, for a fine-grained old-growth type characterized by 0.1-ha disturbance patches, . Whereas, a quasi-equilibrium of 100 patches of coarse-grained old growth characterized by 1,000-ha fire disturbance patches, would require 100,000 ha. Edge effects (Chen et al. 1993; Matlack in press) and the needs of wide-ranging old-growth associated organisms would also affect the area of landscape needed to maintain old-growth conditions.

Old growth is sensitive to the frequency of disturbance in a landscape. In the case of Douglas-fir, a long-lived species, the old phases can persist in landscapes for 500 to 1,000 years. Consequently, many areas of the mosaic must remain free of severe disturbances for 200 to over 500 years, the typical period of development for the transitional old-growth phase of this species (Spies and Franklin 1991). The frequency of disturbance to maintain an old-growth aspen landscape, by contrast, would have to be 50 to 100 years depending the productivity of the landscape.

IMPLICATIONS TO CONSERVATION OF OLD GROWTH

The diversity of old-growth types means that conservation, management, and restoration strategies must be tailored to the particular type of old growth. Patch level and landscape level plans and management practices that work for one type might not work well for another. For example, growth types maintained by frequent burning, such as ponderosa pine, will require prescribed burning or thinning every decade to maintain their structure and function (Moir and Dieterich 1988).

The feasibility and opportunities for conservation and management of old growth will vary according to type, remaining amounts, and land ownership patterns. In the Pacific Northwest, where old growth has been rapidly declining but some relatively large areas remain on National Forest lands, old-growth conservation strategies have focused on reserving large existing old-growth areas and maintaining connectivity between them for wide-ranging species such as the Northern Spotted Owl (Thomas et al. 1990; Johnson et al. 1991; Anon. 1993). In these conservation plans, the old-growth reserves are large and numerous enough, that regional amounts of old-growth Douglas-fir will probably be stable for many decades despite the large fires characteristic of this region.

Many forest regions do not currently have large areas of remaining old growth or large, uniform land ownership patterns to facilitate implementation of idealized conservation networks. Where landscapes and ownerships patterns currently allow only small old-growth areas (<100 ha), it only will be feasible to grow and maintain fine-grained old-growth types. However, even fine-grained types are subject to infrequent coarse-scale disturbances, so their long-term stability is doubtful. If prescribed fire is not allowed or used, then fire-frequent old-growth types cannot be maintained with any size of landscape. If timber rotations are too short, then long-lived old-growth types will not develop. Conversely, where timber rotations are long or natural disturbances are suppressed, then short-lived old-growth types such as aspen will not be common. Biodiversity planning and management will necessarily involve evaluating tradeoffs between different old-growth types.

Although old-growth types have disappeared from most landscapes or are not feasible to maintain or restore over large areas, it is still possible to sustain some elements of biological diversity associated with old growth. This can be accomplished in several ways.

Maintaining Current Areas of Old Growth

In landscapes where old growth is limited, maintaining existing areas of old growth will typically be the best strategy to providing old-growth habitat in the short term (Harris 1984; Thomas et al. 1990; Johnson et al. 1991). This is particularly true of long-lived old-growth types, which are both relatively stable and take long periods to develop. Silvicultural manipulations to existing stands will only degrade the quality of the old growth, and manipulations to young stands to accelerate development of old-growth conditions will not provide for short-term habitat needs.

Providing for Continued Presence of Old Growth

Long-term strategies for old-growth conservation require identifying areas that will provide future old-growth patches and landscapes. This is required from the stand-point of losses of old growth that are deterministic (forest succession) and stochastic (e.g., wild disturbances). Where short-lived old-growth types are desired, the need to plan disturbances, such as cutting or prescribed burning, will be relatively immediate.

Facilitating Dispersal Among Old-Growth Areas

In most forested landscapes, human needs for wood products or other land-uses, will limit the areas available to develop multi-aged forest mosaics. Old-growth areas will probably be fragments of once-larger expanses and some old-growth species may have difficulty maintaining population levels in isolated fragments. Smaller areas of old growth may still be able to support species specializing in old-growth habitats if the ability of those species to move among old-growth areas is not seriously impaired.

Dispersal can be maintained or facilitated through several means: (1) maintain short distances between old-growth reserves or habitat areas; (2) create a general landscape matrix for the reserves in which habitat conditions favor dispersal and reduce the probability of mortality (Thomas et al. 1990); (3) identify landscape areas of old-growth or similar habitat to serve as corridors and "stepping stones" between old-growth areas (Harris 1984). The relative values of these approaches have received relatively little research attention and should be a high priority for research.

Providing Old-Growth Habitat Elements in Managed Forests

Plantation forestry has traditionally attempted to create relatively uniform, short-rotation forests. This type of managed forest lacks many of the characteristics of old-growth patches and landscapes. Silvicultural practices (Franklin 1989; Spies et al 1991), such as long-rotations, green tree retention, and selection forestry, might be able to maintain old-growth structures and their associated species and functions in managed forest landscapes.

Accelerating the Development of Old-Growth Characteristics

Even where timber objectives are not a concern, silviculture could be used to speed the development of old-growth structures or change the direction of forest succession toward desired old-growth conditions. This might be desirable for landscapes currently lacking old-growth conditions and will require centuries before they develop. For example, many young forest plantations were planted at high densities, which will greatly slow the onset of old-growth structures. Thinning and underplanting in such stands can help to reduce the time for old growth development. In other cases, cutting or prescribed fire could be used to reset the landscape to allow development of coarser-grained or fire-frequent old-growth types.

CONCLUSIONS

Old-growth forests contribute to biological diversity at many levels of the biological hierarchy, from genes to landscapes. Old-growth ecosystems are as diverse as the species and landscapes that mold them. On a given area many stages of forest development will occur over time including several different types and stages of old growth. The types differ in their stability, composition, and stand and landscape structure. All types of old growth share the common element of an aging population or community of trees. The presence of an aging population of trees, the environments created by these trees and the long time-spans following intense disturbances leads to the development of an ecosystem which is distinctive in its composition, structure, and function.

Old growth cannot be maintained without a patch dynamics (landscape) perspective. The rates and sizes of disturbances will determine potential types of old growth in an area. Sustaining elements of biological diversity associated with old growth can be facilitated by: (1) maintaining current areas of old growth; (2) providing areas for replacement of old growth lost to disturbances; (3) facilitating organism dispersal among old-growth areas in landscapes; (4) providing old-growth habitat elements in managed forests; and (5) using manipulations to reduce the time needed to develop old-growth characteristics.

LITERATURE CITED

Alaback, P. B. 1982. Dynamics of understory biomass in Sitka spruce-western hemlock forests of southeastern Alaska. *Ecology* 63:1932-1948.

Anonymous. 1993. Forest Ecosystem Management: An Ecological, Economic, and Social Assessment. Report of the Forest Ecosystem Management Assessment Team. USDA Forest Service. Government Printing Office 1993-793-071.

Bormann, F. H., and G. E. Likens. 1979. *Pattern and Process in a Forested Ecosystem*. Springer-Verlag, New York.

Botkin D. B., and L. M. Talbot. 1992. Biological diversity and forests. In *Managing the World's Forests, Looking for Balance Between Conservation and Development*, ed. N. P. Sharma, pp. 47-74. Kendall/Hunt Publishing Co., Dubuque, Ia.

Brown, E. R., ed. 1985. Management of wildlife and fish habitats in forests of western Oregon and Washington. USDA Forest Service, Pacific Northwest Region, Portland, Ore.

Campbell, R. K. 1979. Genecology of Douglas-fir in a watershed in the Oregon Cascades. *Ecology* 60(5):1036-1050.

Canham, C. D., J. S. Denslow, J. S. Platt, W. J. Runkle, T. A. Spies, and P. S. White. 1990. Light regimes beneath closed canopies and tree-fall gaps in temperate and tropical forests. *Can. J. For. Res.* 20:620-631.

Chen, J., J. F. Franklin, and T. A. Spies. 1993. An empirical model for predicting diurnal air-temperature gradients from edge into old-growth Douglas-fir forest. *Ecol. Modeling* 67:179-198.

Cline, A. C., and S. H. Spurr. 1942. The virgin upland forest of central New England. A study of old growth stands in the Pisgah mountain section of southwestern New Hampshire. *Harvard Forest Bull.* 21.

Cohen, W. B., and T. A. Spies. 1992. Estimating structural attributes of Douglas-fir/western hemlock forest stands from landsat and spot imagery. *Remote Sens. Environ.* 41:1-17.

Cole, D. W., S. P. Gessel, and S. F. Dice. 1967. Distribution and cycling of nitrogen, phosphorus, potassium, and calcium in a second-growth Douglas-fir ecosystem. In *Primary Production and Mineral Cycling in Natural Ecosystems*, ed. H. E. Young, pp. 197-223. University of Maine Press, Orono, Me.

Daniel, T. W., J. A. Helms, and F. S. Baker. 1979. *Principles of Silviculture*. 2nd ed. McGraw-Hill, New York.

Duffy, D. C., and A. J. Meier. 1992. Do Appalachian herbaceous understories ever recover from clearcutting? *Conserv. Biol.* 6:196-201.

Esseen, P. A., B. Ehnstrom, L. Ericson, and K. Sjoberg. 1992. Boreal forests-the focal habitats of Fennoscania. In *Ecological Principles of Nature Conservation. Applications in Temperate and Boreal Environments*, ed. L. Hansson, pp. 252-325. Elsevier, London.

Foster, D. R. 1988. Disturbance history, community organization and vegetation dynamics of the old-growth Pisgah Forest, south-western New Hampshire, U.S.A. *J. Ecol.* 80:79-98.

Foster, D. R., and E. M. Boose. 1992. Patterns of forest damage resulting from catastrophic wind in central New England, USA. *J. Ecol.* 80:79-98.

Forman, R. T. T., and M. Godron. 1986. *Landscape Ecology*. John Wiley and Sons, New York.

Franklin, J. F. 1989. Toward a new forestry. *Am. For.* (November/December):37-44.

Franklin, J. F., and M. A. Hemstrom. 1981. Aspects of succession in coniferous forests of the Pacific Northwest. In *Forest Succession: Concepts and Applications*, eds. D. C. West, H. H. Shugart, and D. B. Botkin, pp. 212-219. Springer-Verlag, New York.

Franklin, J. F., K. Cromack, Jr., W. Denison, A. McKee, C. Maser, J. Sedell, F. Swanson, and G. Juday. 1981. Ecological characteristics of old-growth Douglas-fir forests. Gen. Tech. Rep. PNW-118, USDA Forest Service, Pacific Northwest Forest and Range Experiment Station, Portland, Ore.

Gavrilov, L. A., and N. S. Gavrilova. 1993. Fruit fly aging and mortality. *Science* 260:1565.

Grier, C. C., and R. S. Logan. 1977. Old-growth *Pseudotsuga menziesii* communities of a western Oregon watershed: biomass distribution and production budgets. *Ecol. Monogr.* 47(4):373-400.

Habeck, J. R. 1968. Forest succession in the Glacier Peak cedar-hemlock forests. *Ecology* 41:872-880.

Hanley, T. A. 1984. Relationships between Sitka black-tailed deer and their habitat. USDA Forest Service. Gen. Tech. Rep. PNW 168.

Harmon, M. E., J. F. Franklin, F. J. Swanson, P. Sollins, S. V. Gregory, J. D. Lattin, N. H. Anderson, S. P. Cline, N. G. Aumen, J. R. Sedell, G. W. Lienkaemper, K. Cromack, Jr., and K. W. Cummins. 1986. Ecology of coarse woody debris in temperate ecosystems. In *Advances in Ecological Research*, Vol. 15, eds. A. MacFadyen and E. D. Ford, pp. 133-302. Academic Press, Orlando, Fla.

Harmon, M. E., W. K. Ferrell, and J. F. Franklin. 1990. Effects on carbon storage of conversion of old-growth forests to young forests. *Science* 247:699-702.

Harr, R. D. 1982. Fog drip in the Bull Run municipal watershed, Oregon. *Water Resources Research* 18:785-789.

Harris, L. D. 1984. *The Fragmented Forest*. University of Chicago Press, Chicago, Ill.

Heliovaara, K., and R. Vaisanen. 1984. Effects of modern forestry on northwestern European forest invertebrates: a synthesis. *Acta Forestalia Fennica* 189:5-29.

Henderson, D., and L. D. Hedrick, eds. 1991. Restoration of old growth forests in the Interior Highlands of Arkansas and Oklahoma. Ouachita National Forest and Winrock International Institute for Agriculture Development, Ark.

Huff, M. H., D. A. Manuwal, and J. A. Putera. 1992. Winter bird communities in the southern Washington Cascade Range. In *Wildlife and Vegetation of Unmanaged Douglas-fir Forests*, tech. coords. L. F. Ruggiero, K. B. Aubry, A. B. Carey, and M. H. Huff, pp. 207-220, Gen. Tech. Rep. PNW-GTR-285, USDA Forest Service, Portland, Ore.

Johnson, K. N., J. F. Franklin, J. W. Thomas, and J. Gordon. 1991. Alternatives for management of late-successional forests of the Pacific Northwest. Report to the Agriculture Committee and the Merchant Marine and Fisheries Committee of the U.S. House of Representatives. Washington, D.C.

Jones, E. W. 1945. The structure and reproduction of the virgin forest of the north temperate zone. *New Phytol.* 44:130-148.

Kaufmann, M. R., W. H. Moir, and W. W. Covington. 1992. Old-growth forests: what do we know about their ecology and management in the southwestern and Rocky Mountain regions. In *Old-growth Forests in the Southwest and Rocky Mountain Regions, Proceedings of a Workshop*, tech. coords. M. R. Kaufmann, W. H. Moir, and R. L. Bassett, pp. 1-11. USDA Forest Service GTR RM-213, Fort Collins, Colo.

Lesica, P., B. McCune, S. V. Cooper, and W. S. Hong. 1991. Differences in lichen and bryophyte communities between old-growth and managed second-growth forests in the Swan Valley, Montana. *Can. J. Bot.* 69:1745-1755.

Lynch, A. M., and T. W. Swetnam. 1992. Old-growth mixed-conifer and western spruce budworm in the southern Rocky Mountains. In *Old-growth Forests in the Southwest and Rocky Mountain Regions, Proceedings of a Workshop*, tech. coords. M. R. Kaufmann, W. H. Moir, and R. L. Bassett, pp. 81-99. USDA Forest Service GTR RM-213, Fort Collins, Colo.

MacMahon, J. A., D. L. Phillips, J. V. Robinson, and D. J. Schimpf. 1978. Levels of biological organization: an organism-centered approach. *BioScience* 28:700-704.

Matlack, G. R. In Press. Plant species mobility determines the composition of the understory in a fragmented forest landscape in eastern North America. *Ecology*.

Mayer, H., K. Zurigl, W. Schrempf, and G. Schlager. 1987. Urwaldreste, Naturwaldreservate und schutzenswerte Naturwalder in Osterterreich (Old-growth remnants, natural forest reserves and natural forests of protection-value in Austria). Institut fur Waldbau, Universitat fur Bodenkultur, Wien.

Moir, W. H., and J. H. Dieterich. 1988. Old-growth ponderosa pine from succession in pine-bunchgrass forests in Arizona and New Mexico. *Natural Areas J.* 8:17-24.

Morrison, P. H., and F. J. Swanson. 1990. Fire history in two forest ecosystems of the central western Cascade Range, Oregon. Gen. Tech. Rep. PNW-GTR-254 USDA Forest Service, Pacific Northwest Research Station, Portland, Ore.

Mueller-Dombois, D. 1987. Natural dieback in forests. *BioScience* 37:575-583.

Muller, R. N. 1982. Vegetation patterns in the mixed mesophytic forests of eastern Kentucky. *Ecology* 63:1901-1917.

Namkoong, G., and M. T. Conkle. 1976. Time trends in genetic control of height growth in ponderosa pine. *For. Sci.* 22:2-12.

Neale, D. B. 1985. Genetic implications of shelterwood regeneration of Douglas-fir in southwestern Oregon. *For. Sci.* 31:995-1005.

Nichols, G. E. 1913. The vegetation of Connecticut. II. Virgin forests. *Torreya* 13:199-215.

Noble, I. R., and R. O. Slayter. 1977. Post-fire succession of plants in Mediterranean ecosystems. In *Proc. Symposium on the Environmental Consequences of Fire and Fuel Management in Mediterranean Ecosystems*, eds. H. A. Mooney and C. E. Conrad, pp. 27-63. USDA Forest Service. Gen. Tech. Rept. WO-3, Washington, D.C.

Noss, R. F. 1983. A regional landscape approach to maintain diversity. *BioScience* 33:700-706.

Noss, R. F. 1990. Indicators for monitoring biodiversity: a hierarchical approach. *Conserv. Biol.* 4:355-364.

Nyberg, J. B., A. S. Harestad, and F. L. Bunnel. 1987. "Old growth" by design: managing young forests for old-growth wildlife. *Trans. N. Am. Wildl. Nat. Resour. Conf.* 52:70-81.

Old-Growth Definition Task Group. 1986. Interim definitions for old-growth Douglas-fir and mixed-conifer forests in the Pacific Northwest and California. Research Note PNW-447. USDA Forest Service, Pacific Northwest Research Station, Portland, Ore.

Oliver, C. D. 1981. Forest development in North America following major disturbances. *For. Ecol. Manage.* 3:153-168.

Oliver, C. D., and B. C. Larson. 1990. *Forest Stand Dynamics*. McGraw-Hill, New York.

Olshansky, S. J., B. A. Carnes, and C. K. Cassel. 1993. Fruit fly aging and mortality. *Science* 260:1565-1566.

Peet, R. K., and N. L. Christensen. 1987. Competition and tree death. *BioScience* 37:586-595.

Peterkin, G. F. 1993. *Woodland Conservation and Management*. 2nd. ed. Chapman and Hall, London.

Peterkin, G. F., and M. Game. 1984. Historical factors affecting the number and distribution of vascular plant species in the woodlands of central Lincolnshire. *J. Ecol.* 72: 155-182.

Peterkin, G. F., and E. W. Jones. 1987. Forty years of change in Lady Park Wood: the old growth stands. *J. Ecol.* 75:477-512.

Pianka, E. R. 1978. *Evolutionary Ecology*. 2nd ed. Harper and Row, New York.

Pickett, S. T. A., and J. N. Thompson. 1978. Patch dynamics and the design of nature preserves. *Biol. Conserv.* 13:27-37.

Postel, S., and J. C. Ryan. 1991. Reforming Forestry. In *State of the World, 1991*, ed. L. R. Brown, pp. 74-92. W. W. Norton, New York.

Romme, W. H., D. W. Jamieson, J. S. Redders, G. Bigsby, J. P. Lindsey, D. Kendall, R. Cowen, T. Kreykes, A. W. Spencer, and J. C. Ortega. 1992. Old-growth forests of the San Juan National Forest in southwestern Colorado. In *Old-growth Forests in the Southwest and Rocky Mountain Regions, Proceedings of a Workshop*, tech. coords. M. R. Kaufmann, W. H. Moir, and R. L. Bassett, pp. 154-165. USDA Forest Service GTR RM-213, Fort Collins, Colo.

Rowe, J. S. 1961. The level-of-integration concept and ecology. *Ecology* 42:420-427.

Ruggiero, L. F., L. C. Jones, and K. B. Aubry. 1991. Plant and animal habitat associations in

Douglas-fir forests of the Pacific Northwest: An Overview. In *Wildlife and Vegetation of Unmanaged Douglas-fir Forests*, tech. coords. L. F. Ruggiero, K. B. Aubry, A. B. Carey, and M. H. Huff, pp. 447-462. Gen. Tech. Rep. PNW-GTR-285, USDA Forest Service, Portland, Ore.

Schoonmaker, P., and A. McKee. 1988. Species composition and diversity during secondary succession of coniferous forests in the western Cascade Mountains of Oregon. *For. Sci.* 34(4):960-979.

Schowalter, T. D. 1989. Canopy arthropod community structure and herbivory in old-growth and regenerating forests in western Oregon. *Can. J. For. Res.* 19:318-322.

Shugart, H. H. 1984. *A Theory of Forest Dynamics*. Springer-Verlag, New York.

Smith, D. M. 1962. *The Practice of Silviculture*. 7th ed. John Wiley and Sons, New York.

Sollins, P., C. C. Grier, F. M. McCorrison, K. Cromack, Jr., R. Fogel, and R. L. Fredriksen. 1980. The internal element cycles of an old-growth Douglas-fir ecosystem in western Oregon. *Ecol. Monogr.* 50(3):261-285.

Spies, T. A. 1991. Plant species diversity and occurrence in young, mature, and old-growth Douglas-fir stands in western Washington and Oregon. In *Wildlife and Vegetation of Unmanaged Douglas-fir Forests*, tech. coords. L. F. Ruggiero, K. B. Aubry, A. B. Carey, and M. H. Huff, pp. 111-121. Gen. Tech. Rep. PNW-GTR-285, USDA Forest Service, Portland, Ore.

Spies, T. A., and J. F. Franklin. 1991. The structure of natural young, mature and old-growth forests in Washington and Oregon. In *Wildlife and Vegetation of Unmanaged Douglas-fir Forests*, tech. coords. L. F. Ruggiero, K. B. Aubry, A. B. Carey, and M. H. Huff, pp. 91-121. Gen. Tech. Rep. PNW-GTR-285, USDA Forest Service, Portland, Ore.

Spies, T. A., J. F. Franklin, and M. Klopsch. 1990. Canopy gaps in Douglas-fir forests of the Cascade Mountains. *Can. J. For. Res.* 20:649-658.

Spies, T. A., J. F. Franklin, and T. B. Thomas. 1988. Coarse woody debris in Douglas-fir forests of western Oregon and Washington. *Ecology* 69(6):1689-1702.

Spies, T. A., J. Tappeiner, J. Pojar, and D. Coates. 1991. Trends in ecosystem management at the stand level. *Trans. N. Am. Wildl. Nat. Res. Conf.* 56:628-639.

Swanson, F. J., R. L. Fredricksen, and F. M. McCorrison. 1982. Material transfer in a western Oregon forested watershed. In *Analysis of Coniferous Forest Ecosystems in the Western Unite States*, ed. R. L. Edmonds. US/IBP Synthesis Ser. 14. Hutchinson Ross Publishing Co., Stroudsburg, Pa.

Tiegerstedt, P. M. A., D. Rudin, T. Niemela, and J. Tammisola. 1982. Competition and neighborouring effect in a naturally regenerating population of scots pine. *Silvae Fennica* 16:122-129.

Thomas, J. W., E. D. Forsman, J. B. Lint, E. C. Meslow, B. R. Noon, and J. Verner. 1990. A conservation strategy for the northern spotted owl. USDA Forest Service and USDI National Park Service, Portland, Ore.

Thoreau, H. D. 1864. *The Maine Woods*. Ticknor and Fields, Boston, Mass.

Ure, D. C., and C. Maser. 1982. Mycophagy of red-backed voles in Oregon and Washington. *Can. J. Zool.* 60:3307-3315.

USDA Forest Service. 1989. Generic definition and description of old-growth forests. Report on file at PNW Research Station, Forestry Sciences Laboratory, Corvallis, Ore.

Vankat, J. L., J. Wu, and S. A. Fore. 1991. Old growth forests by design: Applying the concepts of landscape ecology. In *Restoration of old growth forests in the Interior Highlands of Arkansas and Oklahoma*, eds. D. Henderson and L. D. Hedrick, pp. 153-177. Ouachita National Forest and Winrock International Institute for Agriculture Development, Ark.

Vitousek, P. M., and W. A. Reiners. 1975. Ecosystem succession and nutrient retention: A hypothesis. *BioScience* 25(6):376-381.

Whitney, G. G., and D. R. Foster. 1988. Overstorey composition and age as determinants of the understorey flora of woods of central New England. *J. Ecol.* 76:867-876.

Zackrisson, O. 1977. Influence of forest fires on the North Swedish boreal forest. *Oikos* 29:22-32.

21

TRANSLOCATION IN CONSERVATION

Alan R. Templeton

Department of Biology
Washington University
St. Louis, Missouri
U.S.A.

Translocation is the movement by humans of any organism from one location to another. Quite often the movement of organisms is a consequence of unintentional human actions. Such unintentional translocations were numerous in the past five centuries and often harmed the native biodiversity (Atkinson 1989). Other translocations were intentional for a variety of reasons: the desire to bring familiar species into an area being colonized, aesthetics, agriculture, and pest and game management. Most intentional translocations have also harmed the native biodiversity (Atkinson 1989). Despite these numerous examples of harm to native flora and fauna, intentional translocations have and continue to be a tool in many programs with conservation-oriented goals (Greig 1979; Griffith et al. 1989).

This paper is limited to a treatment of intentional translocations for conservation goals. Such goals include the reintroduction of a population into an area from which it was extirpated, the introduction of an endangered population into a new area outside its historical range to protect it from global extinction, the augmentation of natural populations to protect them from extinction through demographic and genetic stochasticity, and the re-establishment of gene flow and dispersal between natural populations to counteract the effects of habitat fragmentation. This paper focuses on the goal of reintroduction, the least controversial use of translocation in conservation (Greig 1979; Griffith et al. 1989).

A four-point program for reintroductions is presented, involving (1) formulating specific goals, (2) preparing for the reintroduction, (3) implementing the reintroduction with an integrated plan, and (4) monitoring and performing follow-up studies of the reintroduced populations. The general principles, questions, and problems of reintroduction at all these stages is illustrated by the reintroduction of collared lizards (*Crotaphytus collaris*) in the Missouri Ozarks.

FORMULATING SPECIFIC GOALS

The primary goal of most reintroductions is the establishment of a self-sustaining population in the area of release. Demographic factors are often the primary determinant of the chances of self-sustainability during the initial generations after release (Lande 1988), but genetic diversity is required for long-term sustainability through evolutionary flexibility in the face of environmental uncertainty. To increase the chances of long-term persistence of the species, an additional goal of many reintroductions should be the preservation of the species genetic diversity. Genetic diversity can exist at many levels: individual heterozygosity, among-individual variation within local populations, and among-population variation within species. In addition, hybridization and interspecific genetic diversity is a complication for many endangered populations (Templeton 1991a; O'Brien and Mayr 1991). Because of these many levels of genetic diversity, the optimal strategy for preserving genetic diversity in a reintroduced population must be determined on a case-by-case basis (Templeton 1991b). This paper is confined to one case only: the collared lizard reintroduction program. The goals of this program are: (1) the establishment of self-sustaining populations of collared lizards in the Missouri Ozarks and (2) the preservation of the genetic diversity of the Ozark collared lizards (Templeton 1990). Until goals are formulated, it is impossible to discuss strategies.

PREPARATION FOR REINTRODUCTION

Before implementing a reintroduction, two necessary prerequisites are (1) to have access to areas appropriate for reintroduction and (2) to have access to individuals for reintroduction. Although these two basic needs seem obvious and even trivial, the bulk of the work of a reintroduction often consists of these two preparatory steps.

Preparing the Habitat for Reintroduction

A reintroduction is impossible without an appropriate area for release of translocated organisms. Because reintroduction involves the translocation of organisms into areas they had previously inhabited, either historical knowledge or inferences from current biogeographical patterns are needed to identify areas of former occupation. Once such areas have been identified, it is necessary to know (or at least have a reasonable hypothesis) why the former local populations went extinct. Remedial action must be taken to prepare the habitat for the translocated individuals to preclude the reintroduced population from sharing the fate of the former population.

Most subspecies of the collared lizard live in the desert or dry prairie regions of the American Southwest. The Ozark populations, therefore, represent both a geographic and an ecological extreme of the species (Sexton et al. 1992). Although no true deserts are in the Ozarks, the collared lizards live in habitats that mimic some, but not all, features of a desert. In the Ozarks, collared lizards live on glades—open habitats characterized by exposed bedrock usually with a southerly or southwesterly exposure (Nelson and Ladd 1981). The thin rocky soils and their extreme pHs (acidic when the bedrock is igneous or sandstone, basic when the bedrock is limestone or dolomite) inhibit the growth of trees. As a consequence, glades are open habitats that are usually exposed to the full force of the sun during the hottest part of the day (Nelson and Ladd 1981).

This creates a hot, xeric, open rocky habitat upon which collared lizards and other desert-adapted organisms can live.

One of the few trees that can invade this open, hot, xeric habitat is the juniper, *Juniperus virginiana* (Ladd 1991). Prior to European settlement, the Ozarks were subjected to frequent fires, often started by the native Amerindian populations (Ladd 1991). These frequent fires prevented the junipers from becoming established on the glades. Since European settlement, the incidence of fires has plummeted, and many former glades have become completely shaded by junipers (Guyette and McGinnes 1987). Once this occurs, other trees can invade and many inhabitants of the glade become extinct (Ladd 1991).

As the role of fire in the Ozarks became increasingly appreciated, fire regimes were re-established on state, federal, and private lands owned or managed by the Missouri Department of Conservation, the Missouri Department of Natural Resources, the Mark Twain National Forest, and The Nature Conservancy. More extensive use of fire is being planned (Nigh et al. 1992). These agencies rejected the notion that, because many of the pre-settlement fires were anthropogenic, the fires were unnatural and forest succession would be better. This notion is based on the false premise that humans are not part of the natural world. Humans are indeed part of the natural world and have had and will continue to have a large effect on most living communities. Conservation programs that ignore current and past human effects are almost certainly doomed to failure. Regardless of views about naturalness, much of the biodiversity of the Ozarks prior to settlement depended on frequent fires, and the suppression of these fires results only in a continuing loss of biodiversity without corresponding gains (Ladd 1991; Nigh et al. 1992).

Under these re-established burning regimes, some glade habitat has been restored. The glade flora returned remarkably well and does not require reintroduction (Templeton et al. 1990). Apparently, relictual plant populations persisted in the shaded environment and/or seed pools provided a natural source of new propagules. Occasionally, collared lizards made a spontaneous comeback (pers. observ.), for example, at Victoria Glade in Jefferson County, just south of St. Louis. This glade is now jointly owned by the Missouri Department of Conservation and The Nature Conservancy. Because of its proximity to St. Louis, this glade has been extensively and frequently studied by a variety of investigators since the late 1940s. Collared lizards were thought to have become extinct on this large glade because none had been sighted since at least 1977 during frequent visits by the author, Owen Sexton of Washington University, Douglas Ladd of The Nature Conservancy, and associated students and workers. However, after several burnings and cuttings, at least 16 collared lizards were observed on Victoria Glade by the author during the spring of 1992 (this number has increased to 27 in 1993). Apparently, the lizards survived in parts of this large, formerly fragmented glade complex at low density, and the population size is expanding now that the habitat has been restored.

The observations at Victoria and in other glades indicate that regular burning restores the glade as suitable habitat for collared lizards. However, glades showing spontaneous comebacks of collared lizards are the exception. Collared lizards are near the top of the glade food chain and, hence, tend to have small population sizes even in open glades (Templeton et al. 1990). Therefore, they are prone to local extinction when the glade habitat degenerates, as indicated by their current patchy distribution across glades in the

Ozarks (Templeton et al. 1990). Moreover, the lizards avoid shaded habitats, and genetic data support the conclusion that even a few hundred yards of forest is sufficient to stop all dispersal (Templeton et al. 1990). Hence, once extirpated from a glade, they are incapable of natural recolonization under current climatic conditions (as discussed in Templeton et al. 1990, the Ozark populations were most likely established during the Ipsothermal Maximum about 6,000 years ago). Consequently, there is a need for the reintroduction of collared lizards as part of an overall restoration of glade communities.

A Source of Individuals for Release

The second basic requirement for a reintroduction is a source of individuals for reintroduction. The choice of such sources is critical to the success of the program and should be done in a thorough and informed fashion.

An examination of past reintroductions indicates that the probability of a self-sustaining reintroduced population increases with increasing geographical proximity of the site of origin of the propagules to the site of release (Greig 1979; Griffith et al. 1989). The usual explanation for this pattern is local adaptation (Templeton et al. 1986); that is, local populations of the species are adapted to the local environmental conditions, and populations from geographical proximal regions are more likely to be adapted to the local conditions at the release site.

That local adaptation is a possibility for collared lizards is strongly indicated by the work of Sexton et al. (1992). Collared lizards are abundant southwest of the Ozarks and are, therefore, an obvious source population. However, Sexton et al. (1992) demonstrated that lizards from the Ozarks display many significant differences from lizards from the American southwest—and some surprising similarities. For example, the total activity season in the northeastern Ozarks is 23.5 weeks in contrast to 24 weeks in central Texas. This similarity is not expected in ectothermic animals living in such different temperature regimes and indicates that the lizards from the Ozarks remain active during cooler temperature conditions than their southwestern conspecifics. Another difference is in how lizards use this activity period. In the northeastern Ozarks, the reproductive season is maximally 7.5 weeks or 32% of the total annual activity season. In central Texas, the maximal reproductive season is 13 weeks or 54% of the total annual activity season. By concentrating their normal reproductive season into a short period in the late spring and early summer, the collared lizards from the Ozarks have the option of a second reproductive season in late summer during years with particularly favorable weather (O. Sexton, pers. comm.). This trait seems to be unique to the Ozark populations.

These and other ecological differences between Ozark and southwestern collared lizards have not yet been shown to have a genetic basis. However, the differences cannot simply be explained as the response of an ectothermic animal in a cooler environment. Moreover, the Ozark lizards are morphologically and genetically distinct from the southwestern lizards and have been evolving as a distinct evolutionary lineage for at least several thousand years (Templeton 1991a; Templeton et al. 1990). In the absence of direct knowledge, it would be safer to assume that these ecological differences represent local adaptation.

To minimize the problems associated with local adaptation, all lizards obtained for reintroduction were captured in other Ozark glades whose habitats had not yet degenerated to the point of causing local extinction. However, even within the Ozarks, the possibility of local adaptation exists. Glades have either acidic or basic soils and differ greatly in their flora (Nelson and Ladd 1981) and insect communities (pers. observ.). Few plants that are endemics to glades live on both glade types, and of those that do, some show ecotypic adaptation to glade type (Ware 1990). There are no obvious ecological differences between collared lizards on acidic versus basic glades (pers. observ.), but detailed studies have not yet been done. Some existing composite glades consist of almost equal parts of sandstone (acidic) and dolomite (basic), such as site number 1 in Sexton et al. (1992). Individual collared lizards move and hunt freely between these two substrate types in such glades (pers. observ.), weakly indicating that these substrate differences are not important to the lizards as long as the habitat is open, rocky, and xeric. However, the importance of local adaptation to glade substrate type is an unanswered question.

Such unanswered questions inevitably arise in almost any reintroduction, and it is critical to articulate clearly such questions before designing the final release strategy. By a clear articulation of such questions, the release program can be designed in such a fashion that the questions will be answered. In this manner, even if the initial reintroductions are failures, much can be learned from them so that future reintroductions will have a greater chance of success. Hence, one need of the collared lizard reintroduction program is to design it to test the null hypothesis of no local adaptation to glade substrate type. Relevant experimental contrasts can be performed only if source populations include lizards captured from both acidic and basic glades and release sites include both acidic and basic restored glades. Hence, both knowledge and questions present constraints upon the strategy of reintroduction.

Another strong predictor of the success of reintroduced populations is the use of wild-caught as opposed to captive-bred individuals (Griffith et al. 1989). Captive-bred individuals are generally inferior because of the importance of learned behaviors (particularly in species with parental care), adaptation to the captive environment, or erosion of genetic variability during captivity (Povilitis 1990; Templeton 1990, 1991b; Varner and Monroe 1990). Strategies exist for avoiding or minimizing these problems (Kleiman 1990; Templeton 1990, 1991a,b; Templeton and Read 1984), but it is better to avoid captive breeding entirely if practical. For collared lizards, captive breeding is also not a good option because Ozark collared lizards are difficult even to maintain in captivity, much less bred. Hence, the source of collared lizards for reintroduction is the remaining natural populations, and the lizards to be translocated are held (but not bred) in captivity only until a sufficient number has been captured for a release.

A final question about the source of individuals for release relates to the goal of preserving genetic diversity. Simply using a large number of founders does not ensure that genetic diversity is preserved. Without a knowledge of how the species' genetic diversity is divided within and between local populations, it is difficult to design a sampling program that ensures preservation of the species' genetic diversity (Templeton 1991b). A survey of genetic variation is the most direct means of obtaining this information, although correlates that indicate population structure can be used if genetic surveys are impractical (Templeton 1991b). Prior to the collared lizard reintroductions, genetic surveys of allozyme variation, mitochondrial DNA restriction site fragment length polymorphisms (RFLPs), nuclear DNA RFLPs, and DNA fingerprinting were carried out on the Ozark lizards. The results revealed little genetic variation within a glade population but extensive genetic differentiation among glade populations—even among glades just a few hundred yards apart (Templeton 1991a; Templeton et al. 1990). In light of this genetic population structure, the strategy of obtaining the lizards for release from the handful of natural glade populations that are still large would miss most of the genetic diversity in the Ozark populations. Instead, the more labor-intensive strategy of sampling animals from many different glades is required for preservation of genetic diversity.

In summary, the above considerations define the source population for the reintroduction of collared lizards as wild-caught individuals from a large number of Ozark glades, representing a mixture of acidic and basic glade substrates.

IMPLEMENTATION OF THE REINTRODUCTION

Once the source populations and release areas have been determined, a strategy for implementing the reintroduction must next be defined, one which takes into account the goals of the program. Such an implementation is rarely simple because the various goals and constraints associated with the source populations and release areas often lead to contradictory ideal actions when regarded in isolation. Hence, at this stage, all the goals and constraints must be considered simultaneously to achieve an integrated release strategy.

For example, how many animals should be released on a restored glade? To minimize demographic stochasticity and genetic drift, the release size should be close to the ultimate carrying capacity. Adult population sizes in eastern Ozark glades tend to be rather small. For example, at Sandy Ridge, a glade with one of the more abundant and reliable collared lizard populations in the eastern Ozarks, the adult population size fluctuated between 21 and 79 individuals in the period 1975-1985 (O. Sexton pers. comm.). Most other glade populations are even smaller. This pattern of small population size on glades is confirmed by the analysis of genetic population structure (Templeton et al. 1990). These observations indicate that release sizes larger than 20 individuals would be wasted effort. Another constraint on founder size stems from the decision to use wild-caught animals. Because collared lizards are difficult to capture and capture rates are very sensitive to weather conditions, it can take much time to capture a large number of lizards. Because the health of wild-caught individuals often deteriorates even if they are held in captivity for a few weeks, small release numbers are better to minimize holding times. A release size of 10 represents a reasonable compromise between these considerations. Such a size is a substantial proportion of the typical glade carrying capacity and yet can be achieved rapidly enough to minimize holding time.

Given a release size of 10, what should the sex ratio be? A 50:50 sex ratio mimics the natural state. However, the initial population growth rate might be enhanced by a female-biased sex ratio. Moreover, males are strongly territorial during the breeding season (Hathaway 1969), which indicates that much of the natural reproduction might be due to a minority of the males. Hence, many released males might make no genetic contribution if males are sufficiently abundant. On the other hand, social dominance is not always a reliable indicator of paternity (Gibbs et al. 1990) and rapid turn-over of dominant males implies that genetic and demographic stochasticity would still be minimized by an even sex ratio even if most reproduction in any one breeding season is accomplished only by a few males. Obviously, information is insufficient on this issue. For the initial releases, a 50:50 sex ratio was adopted because it is the easiest to obtain and mimics the natural situation.

Another consideration is the number of glade populations to sample to obtain 10 individuals for release. To preserve genetic diversity, the decision had been made to sample animals from several glades, as previously discussed. However, this can be done in two fashions. All animals for a given release can be from one glade, but each release would use animals from different glades. Alternatively, each individual release can represent a mixture of animals from several glades. These two strategies are equivalent with regard to overall genetic diversity in the Ozarks, but they differ greatly in their genetic implications for diversity within a single release population. The first strategy has the benefit of mimicking the natural population structure of little genetic diversity in a population from a single glade. It also totally avoids the potential problems associated with outbreeding depression, a phenomenon that can evolve between local populations that are genetically isolated from one another (Templeton et al. 1986), like the Ozark collared lizards. If outbreeding depression exists, hybrids between populations from different glades (or more likely, the backcrosses and F_2) will display depressed fitness, thereby endangering the success of the release (Templeton et al. 1986). On the other hand, a mixed release would create a large within-population genetic variation during the initial generations after release. Hence, a mixed population would have a greater chance of adapting to the local conditions of the restored habitat, which might not be truly restored to a completely natural state.

There are, therefore, legitimate rationales for both pure and mixed glade releases, but a practical constraint also enters into the decision making. As mentioned above, most glades have small population sizes. Hence, taking 10 lizards from a single glade could seriously endanger the viability of that population. It is also impossible to avoid this danger by incubating eggs or raising juveniles in captivity because the lizards cannot be successfully reared in captivity. Because of this concern about the viability of the source population, no more than two lizards were collected from any single glade. This constraint leads to a release strategy consisting of a mixture of animals from different glades, even though it invokes the danger of an outbreeding depression.

The above compromise leads to a general and important point. Natural populations are the products of evolutionary change, and as long as they retain genetic variation, they continue to evolve. Reintroduced populations are no exception. Reintroductions should not attempt to preserve a static entity, but rather an evolving lineage. Evolutionary change should be regarded as an active part of any conservation program that can be used to enhance the chances for success. For example, outbreeding (or inbreeding) depressions, when they are encountered, are not permanent features of populations, but can be eliminated or reduced by natural selection (Templeton 1986; Templeton et al. 1986; Templeton and Read 1984). Viewed from an evolutionary perspective, an outbreeding depression might represent only a temporary and perhaps not very important setback to population growth. Hence, evolutionary change, and the factors that promote or retard it, must be incorporated into the release strategy. In general, when compromises must be made, genetic diversity and the potential for evolutionary change should always have priority over other considerations. Knowledge is insufficient to anticipate all the challenges a reintroduced population will face. If natural or released populations have the potential for evolutionary change, they can evolve solutions to problems such as outbreeding depression or environmental change; without it, their adaptive flexibility is lost. The Earth's biodiversity is the product of past evolution and has never been static. Hence, conservation programs should try to conserve processes (such as evolution) that affect living organisms and ecosystems rather than conserve the current status quo of the living world.

Given that the release will consist of animals from more than one glade, there is still the choice of collecting the release animals only from glades of one pH type or both Ph types (basic or acidic). One option is to capture lizards only from acidic glades for release on restored acidic glades (and likewise for basic glades). This has the benefit of matching yet another ecological variable between source and release sites. Alternatively, one can release lizards captured on both acidic and basic glades onto the same restored glade. The rationale for this is that there is no evidence supporting ecotypic variation of the lizards across glade substrates and, hence, matching lizards according to source and release glade-type is an unnecessary constraint that only increases both labor and holding times for captured individuals. If more conclusive evidence about glade-substrate adaptation in collared lizards were available, future reintroduction programs could be designed to take advantage of this knowledge. From an experimental design perspective, the question of substrate adaptation can be answered by executing one or both of the following release sets (ideally with replication): (1) lizards from acidic glades are released on acidic glades, lizards from acidic glade are released on basic glades, lizards from basic glades are released on acidic glades, and lizards from basic glades are released on basic glades; or (2) lizards from both acidic and basic glades are released on acidic glades, and lizards from both acidic and basic glades are released on basic glades. Ultimately, both sets (1) and (2) should be made, but an initial answer to this question of adaptation to glade substrate can be obtained with the fewest number of releases using the second set (using genetic and ecological data gathered on the release populations to test the hypothesis of substrate adaptation, as described in the next section). Therefore, the initial strategy was to release lizards from both acidic and basic glades according to the second design set.

In summary, the strategy for implementing the reintroduction of collared lizards in glades was determined after consideration of the specific goals of this reintroduction program, the questions needed to be answered, and the practical constraints arising from the decisions concerning source populations. The resulting release strategy is to release five males and five females captured from several different glades (with no more than two individuals coming from the same glade) representing a mixture of acidic and basic rock substrates onto both acidic and basic restored glades. Six such releases were made between 1984 and 1989; three on acidic glades (rhyolite substrate) and three on basic glades (dolomite substrate).

MONITORING RELEASED POPULATIONS

After the release of the populations, a well-designed reintroduction is not completed. The released populations must be monitored to assess the extent to which the goals of the program are met and to answer the questions articulated in formulating the sampling and release strategy. Only with such follow-up studies can we learn from our failures as well as our successes. Conservation biology is a science with a directed agenda: to help conserve biodiversity. Given the rate of loss of biodiversity, management decisions often have to be made with incomplete scientific knowledge. Conservation programs should be implemented to aid in the management of biodiversity and to find answers to questions about biodiversity that can be generalized to other conservation needs. Accordingly, reintroductions should be regarded and designed as both management programs and as scientific experiments.

To achieve these twin objectives in the collared lizard reintroduction program, a blood sample was taken from each lizard prior to release and scored with DNA fingerprinting and other high resolution genetic surveying techniques (see Templeton et al. 1990 for technical details). This provides a baseline for genetic variation in the release population and genetic markers that can be used to address the questions raised previously. After release, each population has usually been resampled once a year to monitor the success of the population, to gather data on individual size and growth rates, to estimate population size and individual survivorship through capture/recapture data, and to obtain blood samples from lizards born into the reintroduced population for genetic monitoring.

In terms of the primary goals (to establish self-sustaining populations and preserve genetic variation), the program has been completely successful so far: all six release populations still exist. These release populations are genetically highly variable within and among themselves.

Besides the primary goals, another important aim of the reintroductions is to answer questions. The specific questions are: what is the optimal sex ratio for release; does outbreeding depression exist and, if so, how long does it persist; and are collared lizards adapted to glade substrate? All of these questions can be addressed by genetic and ecological monitoring of the lizards born into the release populations. By scoring the genotypes of the lizards born in the reintroduced population, the parents of these lizards can be estimated from the baseline genetic survey on the individuals originally released. The proportion of the released animals of each sex that actually reproduced and the variance of male and female reproductive success can be estimated from parentage data, and these estimates in turn determine the optimal release sex ratio. It is important to gather such data for several years after the release because individual lizards can be long-lived (Sexton et al. 1992), and there might be much turnover in male dominance. Genetic markers also allow the ranking of individuals by their degree of hybridity (i.e., the extent to which their ancestry represents mixtures of founders from different glades). By coupling hybrid indices with data on adult size, growth rates, and reproductive success, the hypothesis of an outbreeding depression can be tested (Templeton et al. 1986). Finally, the genetic monitoring allows one to test the null hypothesis that there is no tendency for over or under representation of genes derived from founders on acidic or basic glades. Several markers that are unique to source glade populations are scorable through DNA fingerprinting. These markers are in non-random association with the remainder of the genome in the release populations' gene pools, and such associations are only halved each generation for unlinked markers and decay even more slowly for genes on the same chromosome. Hence, for the first several generations after release, selection for genes conferring adaptation to acidic or basic rock substrates should cause a bias in favor of genetic markers derived from source populations that inhabited the same glade type as the restored glade. The null hypothesis of no bias will be tested for each substrate specific marker within each release population and across replicates.

The data for answering these questions are coming in at a slow but steady pace because of the difficulty of capturing these lizards and because of their long life spans. Accordingly, definite answers to these questions are not yet available but will be forthcoming. Irrespective of outcome, valuable lessons will be learned from the initial release populations. Management and research are, therefore, not incompatible in conservation programs, but rather are inseparable.

CONCLUSIONS

The scope of this paper is limited to translocations of organisms for reintroducing populations into an area from which they had been extirpated. For such cases, it is concluded that an optimal program for reintroductions should involve (1) formulating specific goals, (2) preparing for the reintroduction through habitat restoration, identification of appropriate source populations, and articulation of scientific questions needed to be answered, (3) implementing the reintroduction with an integrated plan that balances the specific goals, practical constraints, and the scientific questions to be answered by the release program, and (4) monitoring and performing follow-up studies on the reintroduced populations so that management and research can be performed simultaneously.

ACKNOWLEDGMENTS

This work was supported by NIH grant R01 GM31571 and the Missouri Department of Conservation. I thank the many graduate students who have helped capture collared lizards over the years, a task that frequently requires going to the hottest places in the Ozarks during the hottest, most humid times of the year. Finally, I thank Dr. Russel Lande and an anonymous reviewer for their comments on an earlier draft.

LITERATURE CITED

Atkinson, I. 1989. Introduced animals and extinctions. In *Conservation for the Twenty-first Century*, eds. D. Western and M. C. Pearl, pp. 54-75. Oxford University Press, Oxford.

Gibbs, H. L., P. J. Weatherhead, P. T. Boag, B. N. White, L. M. Tabak, and D. J. Hoysak. 1990. Realized reproductive success of polygynous red-winged blackbirds revealed by DNA markers. *Science* 250:1394-1397.

Greig, J. C. 1979. Principles of genetic conservation in relation to wildlife management in Southern Africa. *S. Afr. Tydskr. Naturnav.* 9:57-78.

Griffith, B., J. M. Scott, J. W. Carpenter, and C. Reed. 1989. Translocation as a species conservation tool: status and strategy. *Science* 245:477-480.

Guyette, R., and E. A. McGinnes, Jr. 1987. Fire history of an Ozark glade in Missouri. *Trans. Missouri Acad. Sci.* 16:85-93.

Hathaway, L. 1969. *Lizard microdistribution and behavior*. Ph.D. Thesis, Washington University, St. Louis, Mo.

Kleiman, D. G. 1990. Decision-making about a reintroduction: do appropriate conditions exist? *Endangered Species Update* 8:18-19.

Ladd, D. 1991. Reexamination of the role of fire in Missouri oak woodlands. In *Oak Woods Management Workshop*, pp. 67-80. Eastern Illinois University, Charleston, Ill.

Lande, R. 1988. Genetics and demography in biological conservation. *Science* 241:1455-1460.

Nelson, P., and D. Ladd. 1981. Missouri glades — Part I. How many, what kind, and where. *Missouriensis* 3:5-9.

Nigh, T. A., W. L. Pflieger, Jr., P. L. Redfearn, W. A. Schroeder, A. R. Templeton, and F. R. Thompson III. 1992. *The Biodiversity of Missouri*. Missouri Department of Conservation, Jefferson City, Mo.

O'Brien, S. J., and E. Mayr. 1991. Bureaucratic mischief: recognizing endangered species and subspecies. *Science* 251: 1187-1188.

Povilitis, T. 1990. Is captive breeding an appropriate strategy for endangered species conservation? *Endangered Species Update* 8: 20-23.

Sexton, O. J., R. M. Andrews, and J. E. Bramble. 1992. Size and growth rate characteristics of a peripheral population of *Crotaphytus collaris* (Sauria; Iquanidae). *Copeia* 1992:968-980.

Templeton, A. R. 1986. Coadaptation and outbreeding depression. In *Conservation Biology*, ed. M. Soulé, pp. 105-116. Sinauer and Associates, Sunderland, Mass.

Templeton, A. R. 1990. The role of genetics in captive breeding and reintroduction for species conservation. *Endangered Species Update* 8:14-17.

Templeton, A. R. 1991a. Genetics and conservation biology. In *Species Conservation: A Population–Biological Approach*, eds. A. Seitz and V. Loeschcke, pp. 15-29. Birkhauser Verlag, Basel.

Templeton, A. R. 1991b. Off-site breeding of animals and implications for plant conservation strategies. In *Genetics and Conservation of Rare Plants*, eds. D. A. Falk and K. E. Holsinger, pp. 182-194. Oxford University Press, Oxford.

Templeton, A. R., H. Hemmer, G. Mace, U. S. Seal, W. M. Shields, and D. S. Woodruff. 1986. Local adaptation, coadaptation, and population boundaries. *Zoo Biol.* 5:115-125.

Templeton, A. R., and B. Read. 1984. Factors eliminating inbreeding depression in a captive herd of Speke's gazelle (*Gazella spekei*). *Zoo Biol.* 3:177-199.

Templeton, A. R., K. Shaw, E. Routman, and S. K. Davis. 1990. The genetic consequences of habitat fragmentation. *Ann. Missouri Bot. Gard.* 77:13-27.

Varner, G. E., and M. C. Monroe. 1990. Ethical perspectives on captive breeding: is it for the birds? *Endangered Species Update* 8:27-29.

Ware, S. 1990. Adaptation to substrate—and lack of it—in rock outcrop plants: *Sedum* and *Arenaric*. *Am. J. Bot.* 77:1095-1100.

22

STRATEGIES FOR MAINTENANCE AND REPAIR OF BIOTIC COMMUNITY DIVERSITY ON RANGELANDS

Neil E. West

Department of Range Science
Utah State University
Logan, Utah
U.S.A.

Consensus is developing among ecologists that biological diversity will not be effectively conserved in natural reserves alone (Wilcove 1989). Therefore, many are recommending that reserves be complemented by surrounding and connecting semi-natural lands where ecological principles are used to manage for a combination of commodity production and resource conservation purposes (Noss and Harris 1986). The challenge is how to plan and carry out management on such multipurpose lands to minimize loss, maintain, and even enhance biodiversity.

On an area basis, rangelands constitute the largest category of generally non-tilled, but extensively used land across the world (about 47%, Williams et al. 1968). The definition of rangeland has, however, been somewhat muddled. The simplest definition is non-tilled land used primarily for livestock grazing. This is a current use-based definition that includes all kinds of terrestrial ecosystems, be they forest, tall shrubland, grassland, desert or tundra, as long as some domestic animal grazes there now. Even though livestock might not be currently using a terrestrial ecosystem, the land might still bear some imprint of past use related to livestock, e.g., heaths and salt marshes of eastern North America and Europe. There is also some potential to expand domestic animal use. For example, the University of Alaska and Alaskan native peoples are attempting to expand grazing of reindeer, musk oxen, and Icelandic ponies on arctic and subarctic terrain there. Some normally forested lands are also used for livestock grazing for a few years following clear-cutting (e.g., the western white pine belt in northern Idaho). These early seral stages within forests are usually called "transitory" ranges. Thus, a use-based definition puts rangeland into a wide variety of ecosystem types.

Most rangelands occur in grassland, shrub steppe, savanna, open woodland, and semi-desert regions. Thus, most rangeland managers have to become intimately familiar with grassland and desert ecology, a set of knowledge often unfamiliar to foresters, wildlife biologists, and watershed managers. Thus, even where livestock no longer graze such areas, range managers and range scientists comprise the major group of people most adequately trained to understand how such ecosystems are structured and function. Because many readers are probably most familiar with and interested in forests, I see one of my roles as helping to broaden the perspective of this volume by focusing on a sub-set of rangelands, the self-sown vegetation of drier parts of the world and whatever combinations of animals, domestic or wild, occur there.

CHARACTERISTICS OF GRASSLANDS AND DESERTS INFLUENCING COMMUNITY DIVERSITY

Grasslands are where grasses and grass-like plants produce most of the biomass and cover. Other classes of plants, particularly broad-leaved herbs, (which range managers call forbs), provide many more species to the total list of plants, however (Risser 1988). Most of the world's grasslands occur in climates where moderate amounts of precipitation are concentrated during or shortly before the warmest season (Box 1981). The fine, diffuse roots of grass acquire soil moisture shortly after it infiltrates the upper part of the soil. This, plus frequent fires and drought, keep out trees and shrubs. Because of their many basal meristems and much below-ground storage of photosynthates, grasses tolerate close grazing, especially if rest from grazing follows. Large grazers and grasses have thus co-evolved. Without moderate grazing and/or fire, plant litter builds thatch that sequesters nutrients and physically limits vegetative re-growth and seedling establishment (McNaughton et al. 1982). Some grazing and burning are, therefore, necessary to allow optimal light penetration and nutrient cycling. Maximum grassland plant community diversity is usually attained under moderate grazing (West 1993), especially under wetter conditions with a long evolutionary history of herbivory (Figure 22.1). Fossorial mammals such as prairie dogs, pocket gophers, and badgers, provide newly bared spots for plants, especially forbs, to exist in a spatially and temporally fugitive manner. These micro-seral patches are necessary to maintain maximum plant community diversity.

The wetter natural grassland regions are ideally suited for intensive agriculture and have been almost entirely converted into agro-ecosystems. Mesic grasslands remain only on terrain with shallow soils. The bulk of the remaining self-sown grasslands are in much drier rainfall belts, e.g., the short grass prairies of the most western parts of the Great Plains in the United States.

Precipitation in deserts less in amount and more variable in time and space than in grasslands. Succulents survive by taking in water from widespread, near-surface root systems. Shrubs survive drought by tapping water from deep in the profile and shedding leaves and even whole parts of the canopy. Perennial herbs are usually rare, but ephemerals exploit narrow windows of growth and reproductive opportunity after significant storms. Animals either travel to permanent water as needed, get their moisture from food, or spend a lot of time in dormant states. Collectively, this is called the "pulse and reserve" strategy (Noy-Meir 1973, 1974).

True deserts have been relatively little altered by mankind because he does not typically live there permanently (Bahre 1991). Nomadic desert peoples are opportunistic, seldom re-using the same spot until the distant future (Ellis and Swift 1988). Modern settlements are still sparse and subsidized from other zones.

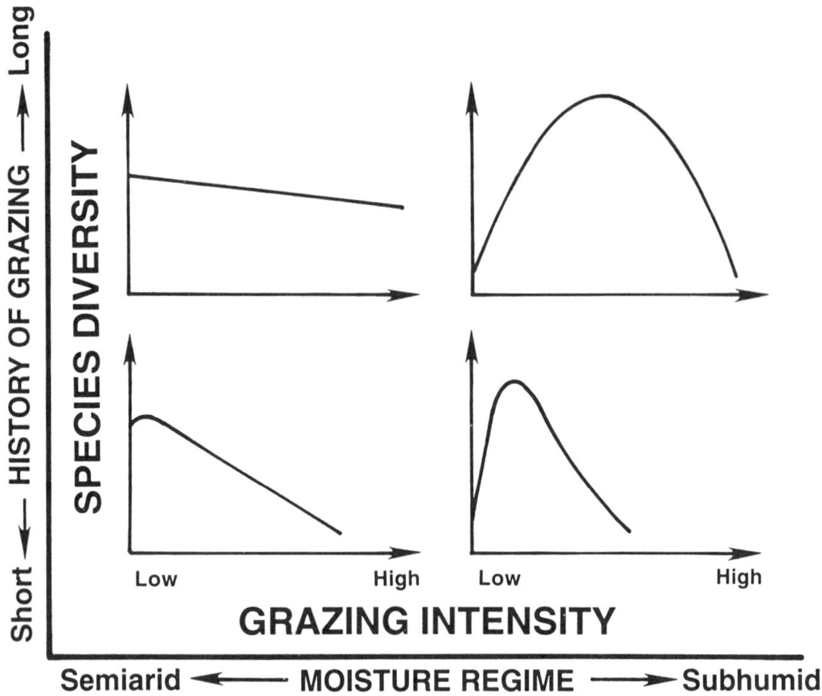

Figure 22.1. Hypothesized relationships between relative plant diversity in grassland communities in relation to grazing intensity along gradients of soil moisture and evolutionary history of grazing (from Milchunas et al. 1988). Republished by permission of the University of Chicago Press.

The major zone of rangeland alteration has been the margin between grasslands and deserts (Bahre 1991). Grasses exist there under marginal conditions (Neilson 1986). These environments are productive enough on the average so that during wetter times they have attracted attempts to graze and farm them. Inevitable droughts have followed and desertification (Schlesinger et al. 1990) has ensued.

Shrub steppes are systems with organismal life and growth forms of both deserts and grasslands. The variation in their climate, with some years of grassland-like climate and some years of desert-like climate, has produced this mix. These fluctuations have also made such ecosystems inherently unstable (Archer and Smeins 1991).

Unrestricted livestock grazing in shrub steppes puts pressure on the perennial herbs, removes fire and tips the competitive balance toward shrubs. Shrubs regenerate quicker under drought conditions and soil disturbance (Neilson 1986; Brown and Heske 1990). Wood harvest in some parts of the world leads to loss of the woody perennials as well. With changed soils and micro-climates, only ephemerals can prevail as a new domain of attraction is formed (Laycock 1991). These changes happened thousands of years ago in the Middle East (Zohary 1973). Similar changes are now happening elsewhere in the world as human populations and material standards of living increase. The resulting simplification, homogenization, and biotic impoverishment will be exacerbated and expanded if global warming ensues.

Biotic Change in Sagebrush Steppe

I do not have the space here to deal with how both local and global environmental changes will influence biodiversity on rangelands at large. This has been attempted elsewhere (West 1993). Many kinds of ecosystems are involved, particularly if the use-based definition of rangeland is employed. I feel it will probably be more productive to illustrate what we are facing and how we can respond via a case study of the sagebrush-steppes of the Intermountain Western United States. I will begin by explaining the past, present and probable future for biotic community diversity in this selected example of an ecosystem type that everyone can agree is rangeland, regardless of preferred definitions.

Definition, Location, Extent, Ownership and Climate

Sagebrush steppe occurs wherever there is or once was a more or less sharing of dominance by sagebrushes (woody *Artemisia*) and bunchgrasses (Tisdale and Hironaka 1981; Blaisdell et al. 1982; West 1983, 1988). This kind of vegetation occurs mostly in the northern part of the Intermountain West (Figure 22.2). It also occurs at higher

Figure 22.2. Map of the sagebrush steppe. Derived from Küchler (1970).

elevations in the south, but the smaller fragments there cannot be mapped at coarse scales. The sagebrush steppe is the most extensive of the North American semi-desert ecosystem types, occupying about 45 x 10^6 ha. About 20% of this type has passed into private ownership, (Yorks and McMullen 1980), about one-half of which has been converted to either dryland or irrigated agriculture over the past 130 years. The approximately 90% remaining in an untilled state lacks irrigation water or is too steep, rocky or shallow-soiled for annual cultivation. The dominant economic use of these rangelands has historically been and continues to be livestock grazing.

The prevailing climate in sagebrush steppe is semi-arid, temperate, and continental (Figure 22.3). Winters are cold enough so that shallow snow packs are common, but snowmelt is usually gentle and soil moisture is recharged so that a spurt of plant growth can take place when temperatures increase in the spring. Summer precipitation is usually not enough to encourage plant regrowth. Early fall precipitation is not dependable and October temperatures usually too cool to allow much regreening of grasses.

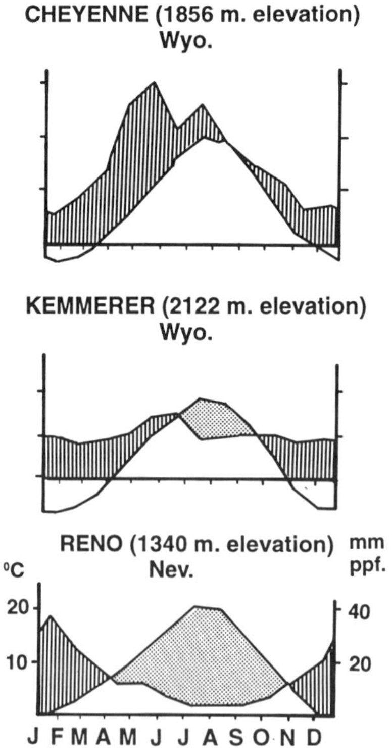

Figure 22.3. Climatic diagrams for two representative stations within the sagebrush steppe (Reno and Kemmerer) and one short grass prairie site (Cheyenne). Average length of summer drought is indicated with stippling, when the curve of mean monthly temperature crosses above the curve of mean monthly precipitation. Note the shift in seasonal precipitation from winter only in the west (Reno) to a late spring—early summer peak in the east (Cheyenne).

Primary Producers

Woody species of *Artemisia* (sagebrushes) are the most characteristic and widespread dominants of lowlands in the Intermountain Region. The progenitors of sagebrush came from Eurasia since late Tertiary and have rapidly undergone radiation into about 12 species (McArthur 1983; Johnson 1987). Furthermore, the major species, *Artemisia tridentata* (big sagebrush), has at least four subspecies or forms that should be recognized if one is to interpret site differences fully (Figure 22.4).

The sizes and degrees of dominance of the several sagebrush species vary greatly with both site and disturbance history. Sagebrush density is generally greater, but stature lower, on more xeric sites. Sagebrush also increases in abundance following excessive livestock grazing combined with lower fire frequency and drought. One can find a continuum of situations, from almost pure stunted sagebrush on shallow, rocky ridgetops to increasing amounts of perennial herbaceous species on more mesic sites.

The floristic diversity of the sagebrush steppe in late seral condition is moderate. On relict sites in central Washington, Daubenmire (1970) found an average of 20 vascular plant species in 1,000 m² plots. Tisdale et al. (1965) found 13 to 24 vascular plant species

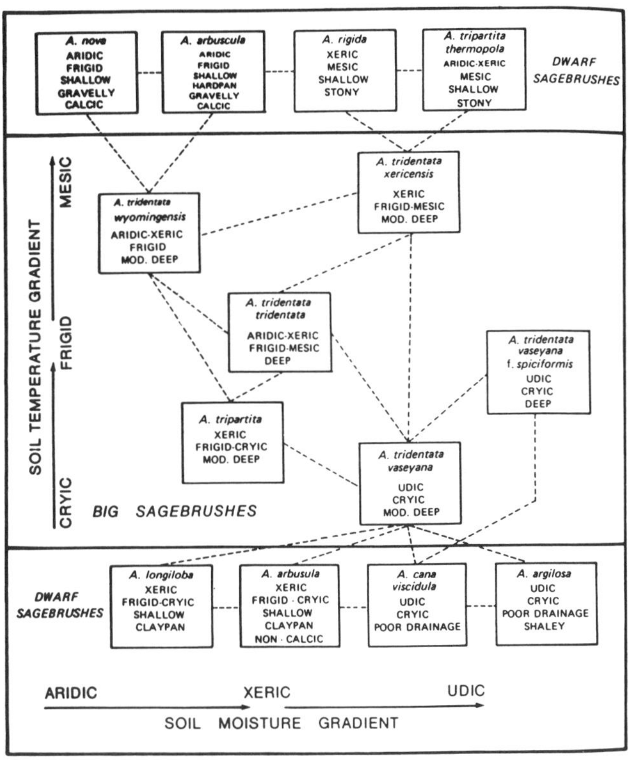

Figure 22.4. Distribution of major sagebrush taxa in relation to gradients of soil temperature and aridity (West 1983, with corrections from more recent literature.)

on three relict stands in southern Idaho. Zamora and Tueller (1973) found 54 vascular plant species in a set of 39 late seral condition stands in the mountains of northern Nevada. Mueggler (1982) found between 24 to 41 vascular plant species in 0.05-ha plots in sagebrush steppe of western Montana.

The vertical and horizontal plant community structures are remarkably similar in all relatively undisturbed examples of this ecosystem type. The shrub layer reaches approximately 0.5 - 1.0 m in height. The shrubs have a cover of about 10-80%, depending on site and successional status. The grass and forb stratum reaches to about 30-40 cm during the growing season. Herbaceous cover also varies widely depending on site and successional status. On relict sites, the sum of cover values usually exceeds 80%, approaching 200% on the most mesic sites (Daubenmire 1970).

The herbaceous life form most prevalent on relict sites are hemicryptophytes (Daubenmire 1975). The proportion of therophytes increases markedly with disturbance. The proportion of geophytes is around 20%. A microphytic crust dominated by mosses, lichens, and algae is commonly found where litter from perennials is not excessive.

Sagebrushes have both fibrous roots that can draw water and nutrients near the surface and a taproot that can feed from deep in the soil profile. Near the end of the growing season for grasses, sagebrushes nocturnally "lift" water from more than 90 cm and excrete it in the upper part of the soil profile (Caldwell and Richards 1990). This "hydraulic lift" can help the grasses stay active longer than possible on their own.

Perennial C_3 grasses associated with *Artemisia* vary greatly throughout the region. The bunch-grasses (*Agropyron spicatum*, *Festuca idahoensis*, *Stipa* spp., *Sitanion hystrix*, *Poa* spp.) dominate the herbaceous layer in the north and western parts of the type. Sod grasses (e.g., *Agropyron smithii*) become more common in the south and east where more growing season precipitation occurs (West 1979). Nearly all herbaceous growth has to occur when the only convergence of warm temperatures and available soil moisture occurs in late spring and early summer.

Total above-ground standing crop phytomass within the sagebrush steppe type varies between about 2,000 to 12,000 kg/ha, depending on site differences, successional status and age of the brush (West 1983). Litter standing crops are about one-half the live non-woody material (West 1985). Below-ground phytomass is similar in magnitude to that above-ground. Annual net above-ground primary production varies between about 100 to 1,500 kg/ha, depending on site, successional status, stand age, and preceding climatic conditions (Passey et al. 1982). Table 22.1 gives an example from a Wyoming Big Sagebrush site in southeastern Oregon. Note that most of the production (annual growth) and standing crops (accumulated over many years) is sagebrush tissues that are mostly woody. The pre-European era livestock grazing-capacity, when shrubs were fewer and grasses more prevalent, was estimated to be 0.83 AUM per ha (McArdle and Costello 1936). Because of loss of perennial herbs and increase in brush from uncontrolled livestock use, this had declined to an average of 0.27 AUM per ha in the 1930s (McArdle et al. 1936), but had improved slightly to 0.31 AUM per ha by 1970 (Forest-Range Task Force 1972).

Consumers

The native vertebrates using this ecosystem type are a mixture of grassland and desert species. Maser et al. (1984) grouped the vertebrates of sagebrush steppe in southeastern

Table 22.1. Crown cover by major species (%) and average annual forage production (kg/ha) in various condition classes for a Wyoming Big Sagebrush site in southeastern Oregon (Hall 1985).

	Condition Classes			
Crown Cover	Potential Natural Community	Late Seral	Mid-Seral	Early Seral
Wyoming Big Sagebrush	12	14	16	18
Bluebunch Wheatgrass	10	7	4	--
Needle-and-thread grass	2	3	2	--
Sandberg's bluegrass forbs	5	7	5	2
Forbs	5	6	7	5
cheatgrass	1	3	7	10
total forage production (kg/ha)	670	450	225	110

Oregon into 16 life forms and related them to vegetation structure and other features of habitat (Figure 22.5). The vertebrate community is more diverse when the vegetation has the greatest structural diversity as sagebrush steppe. Neither shrub-dominated nor grass-dominated situations favors as many different kinds of vertebrates.

Over 1,000 species of insects have been observed on a sagebrush-grass site in southern Idaho (Bohart and Knowlton 1976). Wiens et al. (1991) recently identified 76 taxa of invertebrates on sagebrush alone in central Oregon. Relatively little is known about the habitat preferences, trophic relationships, and other aspects of the roles of invertebrates in this ecosystem type. Only a few (thrips, webworms, grasshoppers, cicadas, aphids and coccids, (Kamm et al. 1978; West 1983) are known to be irruptive and visibly alter vegetation structure.

Decomposers and Nutrient Cycling

Very little is known about microbes and the decomposition process in this ecosystem type. Initial studies of the nitrogen (West and Skujins 1978) and phosphorus (West et al. 1984a) cycles showed that available forms of these elements may limit plant production in wetter than average years. Allelochemics from sagebrush may inhibit some decomposition and nitrogen-cycling processes, perhaps indirectly strengthening their dominance in this ecosystem type (West 1988).

Interactions Between Plants, Animals and Humans

The pristine sagebrush steppe evolved with large browsers, most of which had disappeared by about 12,000 years ago (Mehringer and Wigand 1990). Graminivores were few in the pre-European system (Mack and Thompson 1982; Harper 1986). The small populations of aboriginal hunters and gatherers probably influenced the vegetation only with burning. It took European colonization to change the native vegetation drastically and the wildlife

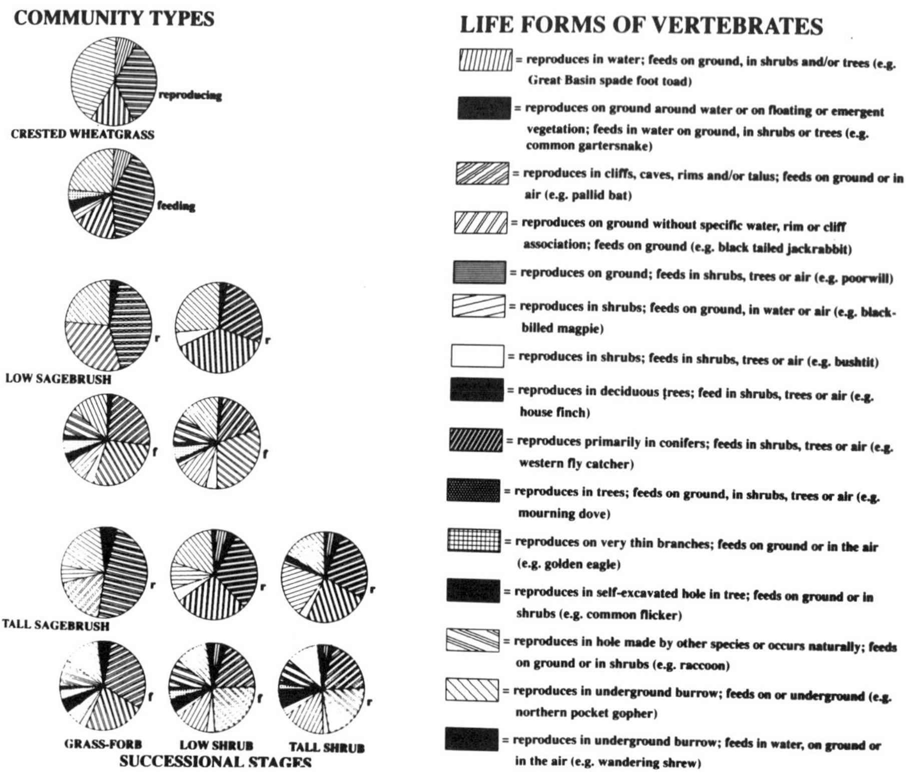

Figure 22.5. Summary of vertebrate wildlife species use of crested wheatgrass, low sagebrush and tall sagebrush/bunchgrass plant communities in various seral conditions. Pie charts above each plant community/structural condition show the relative contributions of 16 possible animal life forms to the vertebrate communities occurring as either feeding or reproducing categories in that plant community type (data from Maser et al. 1985).

habitat it provides (Young 1989). Livestock populations built up rapidly near the end of the 19th century. Griffiths (1902) judged that the grazing capacity of these ranges had been exceeded by 1900. Hull (1976) examined historical documents and concluded that major losses of native perennial grasses and expansion of shrubs took only 10-15 years after a site was first grazed by livestock.

Sagebrushes are long-lived and loaded with chemical defenses against herbivory, making them unpalatable to livestock. The native grasses are extremely palatable, especially when green. They die easily when grazed heavily in the spring (Miller et al., in press). In addition they rarely produce good seed crops or build substantial soil seed reserves (Hassan and West 1986).

The only time the grasses and forbs have an advantage over brush is when sites are burned. Most *Artemisia* species do not resprout after fire, but have to reestablish from seed, thereby greatly influencing successional patterns (Figure 22.6). The fire return interval in the pre-European sagebrush steppe probably varied between 60 and 110 years

(Whisenant 1990). The advent of introduced winter annual grasses, notably *Bromus tectorum* in the 1890s (Mack 1981), led to earlier, more frequent (less than 5 years) and larger fires. After repeated fires, combined with unrestricted grazing (Figure 22.6), native vegetation becomes easily replaced by other, even more noxious introduced annuals, such as medusahead (*Taeniatherum caput-medusae*) and yellow star thistle (*Centaurea solistolisis*). The result is a considerable decrease in plant species and structural diversity and forage production (Billings 1990; Whisenant 1990; Table 22.1).

Wildlife responds dramatically to changes in vegetation structure (Figure 22.5). Habitat for wildlife reproduction is usually more restrictive than habitat for feeding. For example, sage grouse require openings for leks and then riparian habitat with numerous forbs and insects to raise chicks (Call and Maser 1985). The much more prevalent uplands with sagebrush dominance suffice during the rest of the life cycle. Some sagebrushes stand up above the snowpack and are important sources of proteins and lipids for wintering mule deer and sage grouse. A rangeland without these shrubs is inhospitable to these two important native animals.

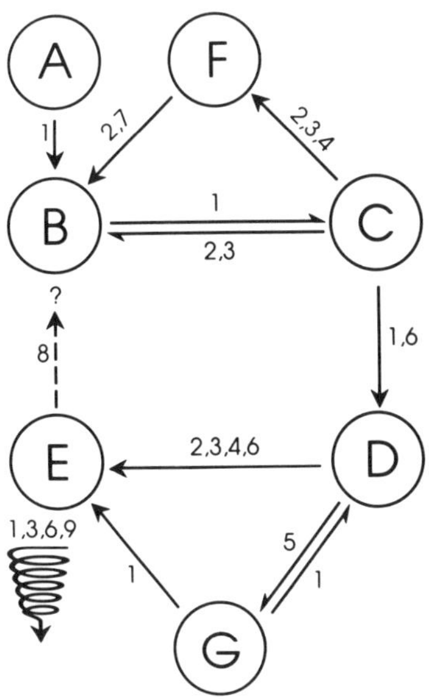

COMMUNITY TYPES

A= Pristine, "complete", original sagebrush steppe, pre-European or relict condition, no livestock influence.

B= Late seral sagebrush steppe, all life-forms present, lightly grazed.

C= Denser, larger sagebrush, remnants of native, perennial herbs remain, moderately grazed.

D= densest, largest sagebrush, understory dominated by introduced annuals, heavily grazed.

E= Cheatgrass/Medusahead, shrubs and native herbs absent, burned after heavy grazing.

F= Native perennial grasses dominate, moderately grazed.

G= Introduced wheatgrasses or ryegrass seedings.

FORCES

1= livestock grazing

2= rest from livestock grazing

3= fire

4= herbicides killed broadleaf plants

5= tillage followed by seeding

6= accelerated soil erosion

7= interseeding

8=restoration

9= further weed invasion

Figure 22.6. Major succession states (circles) and pathways (arrows) of succession for sagebrush steppe (derived from West 1979, with modifications suggested by Laycock 1991).

There has already been a vast replacement of native plant species by European invaders. More is expected, especially if global warming materializes. Controlling fires entirely is an impossibility. Reductions or even complete removal of livestock will not result in a rapid return to the vegetation that occurred before European colonization (Miller et al. in press). Our major means of obtaining greater dependability of forage production and soil protection while at the same time reducing the chance of fire has been to plant Eurasian wheatgrasses and ryegrasses (West 1988). However, this can only be done easily on relatively level sites with deep soils. The remaining rougher terrain awaits more imaginative management.

We can now turn to possible ways to conserve remaining community diversity, alter existing stands, or rehabilitate degraded diversity. Figure 22.6 will be used to guide the following discussion. This figure is a state and transition model (Laycock 1991) thought to accommodate better our current understanding of degradation and successional processes in sagebrush steppe.

Preservation of Relatively Unaltered Ecosystems

Few pristine, relictual areas (circle A in Figure 22.6) remain. Tisdale et al. (1965) describe an example. Less than 1% of the sagebrush steppe that remains has avoided the impact of livestock. All of the sagebrush steppe is further influenced by air pollutants, climatic change and invasion by exotics. Accordingly, most of the late seral sagebrush steppe (circle B in Figure 22.6) has had light livestock use. Even light livestock use puts inordinate pressure on a few highly palatable species of plants ("ice cream"), explaining the lack of a return arrow from B to A.

The lightly grazed, largely intact sagebrush steppe represented by state B in Figure 22.6 largely exists on either gentle terrain far from water or steep terrain closer to water. Cattle are currently the main domestic species using the gentler terrain. In earlier decades sheep utilized the steep areas as well because they were herded, can go farther without water (utilizing snow during the winter), and consume more browse. In some places, feral horses and burros put considerable pressure on such rangeland. About 20% of the remaining sagebrush steppe is estimated to be in state B.

The perceived will of a majority of Americans now is to identify these remaining areas and protect them from being developed. Such areas serve as the "parts catalogue" for restoration efforts. The Gap Analysis Program of the U.S. Fish and Wildlife Service (Scott et al. 1987; Scott 1990) and the various natural heritage programs initiated by the Nature Conservancy are well underway to put these views in action. We can thus expect to see physical modifications (formerly called range "improvements") to be more spatially limited than in the past. The remaining relatively unaltered areas will be consciously protected to provide the later seral condition patches necessary to hold a broader spectrum of all species, and meet the special requirements for some featured species such as sage grouse (Call and Maser 1985). Rangeland managers in the past have strived to reduce the limitations of the land for producing livestock, mainly topography, forage availability and water. For example, trails were constructed into areas where topographic breaks limited livestock access. Natural water was supplemented by development of springs, building stock tanks and small dams, drilling wells, piping and hauling water. Fences were constructed and salt distributed to control livestock movement and institute grazing management systems (e.g., rest-rotation grazing). All these "improvements" were designed to distribute livestock utilization more uniformly across the land. The net result has been progressively more widespread and intensive use of a landscape that has become partially "tamed" from the "wild."

Alteration of Existing Heavily Grazed Stands

Because livestock grazing of native sagebrush steppe favors the unpalatable and especially woody species, dominance becomes concentrated in the few woody plants on areas with a history of heavy livestock grazing, but not fire (State C, Figure 22.6). About 30% of this ecosystem type is estimated to exist currently in this state. Most of these stands can stay stagnated for decades (Rice and Westoby 1978; West et al. 1984b; Sneva et al. 1984; Winward 1991). The dense stands of "excess" sagebrush prevent the herbaceous species from recovering. Such brush-choked stands are usually chosen by both livestock and wildlife managers for manipulation to diversify vegetation structure, enhancing it for livestock production in spots, concentrating livestock use, reducing their pressure elsewhere, while simultaneously advantaging some wildlife species through vegetation modifications via grazing systems, prescribed burning, herbicides, and brush beating or chaining. For example, sheep grazing in the fall, because they consume more sagebrush then (Laycock 1967), can be used to obtain a reversal from State C to State B. Prescribed burning (Harniss and Murray 1973) and herbicides (Evans et al. 1979) were applied to stands with sufficient remnant populations of native herbs to quickly recover following brush kill. Rest from livestock use, such as with a rest-rotation grazing system, will often allow a slower return to State B from State C. Reduction of brush also enhances water yields (Sturges 1977). When phenoxy herbicides are used alone (Evans et al. 1979) or in conjunction with fire, the community becomes native grass-dominated (State F, Figure 22.6). This conversion only slowly returns to State B with judicious grazing and a secondary treatment with prescribed burning. About 5% of the remaining sagebrush steppe is now in State F. Mueggler (1982) found enhanced alpha diversity in sagebrush steppe communities in western Montana following prescribed fire, 2,4-D and brush-beating treatments. Summer fires can damage some of the grasses (Young 1983), but encourage the resprouting rabbitbrushes (*Chrysothamnus* spp.) and horsebrushes (*Tetradymia* spp).

If accelerated soil erosion does not ensue and the fundamental potential of the site does not change, then State C can be maintained or managed toward States B or F. However, as herbaceous plants and litter in the interspaces between shrubs is reduced, soil erosion frequently increases. When the probable threshold of use is exceeded, the site can irreversibly change to one of lesser potential. This explains the single downward arrow between States C and D. This is where the syndrome of desertification is most evident. All the former states can be dealt with via "soft energy" management approaches. Once this threshold is exceeded, however, subsequent management requires expensive, risky, "hard energy" solutions. Unfortunately, it is often easier to get political attention after major damage has been done rather than getting budgets and personnel to plan, monitor and tweak the natural systems at opportune times.

The brush-thickened, introduced annual, native herb depauperate State D comprises about 30% of the current sagebrush steppe. Reduction or removal of livestock only hastens further degradation from State D because livestock remove part of the fuel load and thus reduce the chance of fire destroying the sagebrush.

If insufficient amounts of native herbs remained on sagebrush steppe, the usual land management agency response has been to replace them with introduced wheatgrasses and ryegrasses, especially crested wheatgrass (Asay 1987). Indeed, some large treatment areas have monocultures of Eurasian perennial grasses prevail (State G, Figure 22.6). About 5% of the original sagebrush steppe has already been transferred to State G.

Wildlife biologists have noted declines in the numbers of birds (Olson 1974; Reynolds and Trost 1979, 1981), small mammals (Reynolds and Trost 1980) and large reptiles (Reynolds 1979) on such seedings of introduced grasses. It should be noted, however, that such studies present a worst case scenario because in the crested wheatgrass area were taken in the center of large treatments. No provision for increased diversity near edges (Thomas et al. 1979) was mentioned. Today's more sensitized planners would provide for optimum edge effect and patchiness (McEwen and DeWeese 1987).

Livestock-grazed grass plantings in the sagebrush steppe region will eventually experience shrub re-invasion. Not all brush can now be eliminated by re-treatment. Herbicide use on public lands in the Pacific Northwest has been suspended by judicial decree. Prescribed burning of the coarser, introduced grasses is difficult and leaves patches where the shrubs prevail. There are, therefore, chances to enhance edge effects in areas that were formerly homogenized. As in the untilled native areas, we could enhance wildlife habitat by providing a mix of successional stages or stand conditions, providing both cover and forage for either featured species or total species richness (Maser et al. 1984). For example, some success has been attained in creating alternate leks for sage grouse breeding following disturbances of development (Eng et al. 1979).

Rehabilitation of Burned Sagebrush Steppe

Despite greatly increased attention to fire prevention and control, much of the sagebrush steppe has been burned at least once during the past 3 decades and is now dominated by introduced annuals, mainly grasses such as cheatgrass and medusahead (State E, Figure 22.6). The Bureau of Land Management (M. Pellant pers. comm.) estimates that about 3 million acres of public lands in Idaho, Utah, Oregon, and Nevada are now dominated by cheatgrass and medusahead. Because of their short stature, restricted nutritional characteristics, and susceptibility to recurring fires, such areas are undesirable from all viewpoints. Without nutritional supplementation, livestock can graze there only during the plant growing season. Only the most generalist animals, such as the introduced chukars, horned larks, grasshoppers, and deer mice seem to thrive on the annual grasslands (Figure 22.5). When such areas burn in early summer, soils are bared to wind and water erosion during the convectional storms of summer. The consequent needs for revegetation after fire are increasing.

Land dominated by annuals may provide good watershed protection during years without fire and actually appear to be more productive of total plant tissues than the original sagebrush-native perennial grass and forb combination (Rickard and Vaughan 1988). This might, however, be only a temporary situation based on mineralization of nutrients from the enormous below-ground necromass of the original system, much of which only slowly decomposes in this environment. When these reserves of nutrients and soil organic matter are finally respired away, the annual grasslands could become much less productive. Similar transitions happened in the Middle East several millennia ago (Zohary 1973). Many other, more noxious weeds from that region could find their way here, and we could witness the downward spiral depicted in the lower left hand corner of Figure 22.6.

If society makes the investment in repairing severely damaged sagebrush steppe, creating perennial grass-dominated pastures of much greater productivity of species palatable to livestock, this should compensate for livestock reductions and other management restrictions on States B, C, and F (Figure 22.6).

Rather than allowing the annual grasslands derived from former sagebrush steppe to remain and the land to degrade further, some land managers are attempting to intervene. A joint U.S. Forest Service, Bureau of Land Management, Agricultural Research Service, and University of Idaho program is underway to reduce these threats (Pellant 1990). The most notable component of this effort is the "greenstriping" program most evident in southern Idaho. The basic approach is to begin breaking up the now vast stretches of cheatgrass and other annual dominance that have developed as fires have become earlier, larger and more frequent. Investigators and practitioners are attempting to break the area into smaller, burnable units, especially nearer to cities. The approaches used thus far include planting strips of vegetation that stay green longer than cheatgrass.

Although the introduced wheatgrasses and ryegrasses do stay green longer and burn less readily because of coarser above-ground structure, they are not native. Accordingly, a root-sprouting big sagebrush is seen as a better "keystone" species to put back in this area. A few sagebrushes may actually help sustain introduced perennial grasses by harboring the predators on black grass bugs (*Labops* spp) (Haws 1987). Furthermore, total plant community production can be enhanced (Harniss and Murray 1973) because sagebrushes help trap blowing snow (Sturges 1977) and scattered sagebrushes moderate temperatures (Pierson and Wight 1991) and benefit the re-establishment of native herbs and protect them from excessive utilization (Winward 1991). Sagebrushes also extract nutrients from deep in the soil and recycle them to the surface through litter production (Mack 1977; West 1985).

Whether we can accomplish restoration of sagebrush steppe (the dashed arrow between States E and B in Figure 22.6) is highly questionable. Even where money is less limiting and topsoil is stockpiled on coal strip mines, early results are only partially encouraging (Hatton and West 1987). We will have to know much more about how sagebrush steppe ecosystems are structured and function and obtain vast budgets and more trained personnel before such efforts are routinely successful. It is cheaper to foster land stewardship (manage while in States A, B, C or F of Figure 22.6) than rely on restoration efforts after degradation has taken place (States D, E, G of Figure 22.6).

Guidelines for Rangeland Management Sensitive to Biodiversity

The sagebrush steppe is but one example of rangeland. Whereas it provides illustrations of how biodiversity interacts with land management, other rangeland ecosystems offer other kinds of opportunities to practice management with concerns for biodiversity.

First, tradeoffs must be made between maximum production of livestock and the best possible wildlife habitat, watershed, and soil profile protection in the ecosystem management philosophy. No matter what the manager does or does not do, habitat of some species will be enhanced and that of others simultaneously diminished. Grazing by livestock can be advantageous or disadvantageous to wildlife and other land uses depending on species, uses and values, and their ecosystem context. Retention of species richness, featured species, and stenotypic species cannot usually be simultaneously maximized on the same small piece of land. Such objectives have to be managed for on landscape and regional bases.

Mid-seral and even early seral conditions are not detrimental to all wildlife species. Some species require these conditions to complete their life cycle (West 1993). Maser et al. (1984) found that the greatest species numbers and highest population levels of most featured vertebrate species of sagebrush steppe are in mid-seral condition. Thus, maintenance of disturbances to create early to mid-seral conditions is desirable on some parts of a landscape. This is one reason why total removal of livestock from public rangelands would not necessarily lead to optimal habitat for featured species or species richness. Furthermore, prohibiting livestock grazing would lead to more fine herbaceous fuels and thus hotter, more frequent and larger fires and even more takeover by introduced annual plants in sagebrush steppe as well as many other kinds of adjacent rangelands.

Domestic livestock management directly affects wildlife in two ways: (1) consumption of forage that could be used by wildlife, and (2) alteration of vegetation as it influences escape, thermal and protective cover. Most animal species are more narrowly adapted to vegetation structure for thermal and hiding cover than they are to particular plant species for food (Dealy et al. 1981). Structural diversity of rangeland vegetation also relates positively to wildlife diversity, except if the mosaic is on a scale too small to meet the home range needs of species that require large blocks of uniform vegetation. For example, optimum spacing between stands of big sagebrush and crested wheatgrass for black-tailed jack rabbits requires that the wheatgrass openings be no more 600 m across because rabbit use of wheatgrass occurs within 300 m of the type edge (Westoby and Wagner 1973).

Domestic livestock management indirectly affects wildlife by (1) human and livestock presence, (2) fencing, (3) salting, (4) water developments, (5) roads, (6) trails, (7) predator control and other physical and chemical manipulations, such as prescribed burning, chaining, cabling, root plowing, brush beating, reseeding, herbicides. The latter treatments usually simplify and homogenize habitat structure, but mosaics and edge can be increased with planning and plant species richness enhanced by interplanting in areas with large expanses of currently homogeneous vegetation.

Better planning and management should result in retention of livestock and their use as tools for constructive improvement of wildlife habitat, watersheds, and fire management. Roads can be closed and off-road vehicles prohibited at critical times. Scattered trees could either be retained or planted to provide shade and storm cover for both livestock and wildlife, simultaneously enhancing overall distribution.

Use of electronic sensors on livestock and thoughtful placement of invisible electronic boundaries offer promise to foregoing building more fences and possibly even to remove eventually the existing ones (Fay et al. 1989). We should recognize, however, that fences serve as perch posts for some birds (Graul 1980), and, thus, unintended impacts of fence removal and pole line installation could ensue.

Some people (Roger Banner pers. comm.) feel that there is promise in either selecting domestic animals that naturally spend less time in sensitive areas (e.g., riparian zones) or training them to avoid such areas through adversive conditioning.

Some water sources can be completely or selectively closed off to favor certain species. Development of naturally occurring springs through installation of perforated pipe and water troughs has made more water available for drinking by both ungulates and birds. Unfortunately, the natural wetlands around the original springs have often been highly altered. We can, however, pipe out the overflow and create new fenced out wetlands to replace those altered (Kindschy 1978).

We could even fertilize certain portions of some plant communities to increase and freshen (make more palatable) some areas to draw animals to them to enhance fire control, and strutting grounds. Guzzlers (artificial watersheds) can be built in areas with limited free water for drinking.

CONCLUSIONS

I have demonstrated that application of some knowledge, logic and planning should allow us to continue using rangelands for traditional values as well as provide for preservation and even enhancement of biodiversity. Because of their rarity, desirability in research, and guiding management, most areas that have escaped livestock use thus far should probably be protected to provide maximum landscape diversity and materials for restoration efforts. For the much larger fraction of the rangelands that had and now have extensive use, but are still dominated by native plant species, we should continue to apply our increasing knowledge of individual species responses and community dynamics in devising low input ways to direct succession to desired outcomes. Prescribed burning, behavioral modification of animals, and improved ways of distributing animals without fences should be further developed. If some compromise is deemed possible, livestock grazing systems compatible with wildlife, recreational use, watershed and soil protection can be devised. It is much cheaper and satisfying to prevent such semi-natural areas from slipping over the brink of irreversible trends toward desertification than trying to rehabilitate or restore areas that have already been seriously degraded. We should not let the possibility of artificial recovery prevent us from concentrating on trying to develop sustainable use strategies to begin with.

LITERATURE CITED

Archer, S., and F. E. Smeins. 1991. Ecosystem-level processes. In *Grazing Management: An Ecological Perspective*, eds. R. K. Heitschmidt and J. W. Stuth, pp. 109-139. Timber Press, Portland, Ore.

Asay, K. H. 1987. Revegetation in the sagebrush ecosystem. In *Integrated Pest Management on Rangeland: State of the Art in the Sagebrush Ecosystem*, ed. J. A. Onsager, pp. 19-27. USDA Agricultural Research Service ARS-50.

Bahre, C. J. 1991. *A Legacy of Change: Historic Human Impact on Vegetation in the Arizona Borderlands*. University of Arizona Press, Tucson, Ariz.

Billings, W. D. 1990. *Bromus tectorum*, a biotic cause of ecosystem impoverishment in the Great Basin. In *The Earth in Transition: Patterns and Processes of Biotic Impoverishment*, ed. G. M. Woodwell, pp. 301-322. Cambridge University Press.

Blaisdell, J. P., R. B. Murray, and E. D. McArthur. 1982. Managing intermountain rangelands—sagebrush—grass ranges. USDA Forest Service. Intermountain Forest and Range Expt. Sta. Gen. Tech. Rep. INT-134.

Bohart, G. N., and G. E. Knowlton. 1976. Invertebrates. In *Final Environmental Impact Statement for the Sodium Cooled Class III*, Appendix D-IV, pp. D-47-D-58. Argonne National Laboratory. Design Safety Research Experiment Facility Idaho Nat. Engineer. Lab.

Box, E. O. 1981. *Macro-climate and Plant Forms: An Introduction to Predictive Modelling in Phytogeography*. Dr. W. Junk, The Hague.

Brown, J. H., and E. J. Heske. 1990. Control of a desert-grassland transition by a keystone rodent guild. *Science* 250:1705-1707.

Caldwell, M. M., and J. H. Richards. 1990. Hydraulic left: Water efflux from upper roots improves effectiveness of water uptake by deep roots. *Oecologia* 79:1-5.

Call, M. W., and C. Maser. 1985. Wildlife habitats in managed rangelands—the Great Basin of Southeastern Oregon: Sagegrouse. USDA Forest Service, Pacific Northwest Forest and Range Expt. Sta. Gen. Tech. Rep. PNW-187.

Daubenmire, R. F. 1970. Steppe vegetation of Washington. Wash. Agric. Expt. Sta. Bull. 62.

Daubenmire, R. F. 1975. An analysis of structural and functional characters along a steppe-forest catena. *Northwest Sci.* 49:122-140.

Dealy, J. E., D. A. Leckenby, and D. M. Concannon. 1981. Wildlife habitats in managed rangelands the Great Basin of southeastern Oregon: Plant communities and their importance to wildlife. USDA, Forest Service, Gen. Tech. Rep. PNW-160. Pacific Northwest Forest and Range Expt. Sta., Portland, Ore.

Ellis, J. E., and D. M. Swift. 1988. Stability of African pastoral ecosystems: alternate paradigms and implications for development. *J. Range Manage.* 41:450-459.

Eng. R. L., E. J. Pitcher, S. J. Scott, and R. J. Greene. 1979. Minimizing the effect of surface coal mining on a sagegrouse population by a directed shift of breeding activities. In *The Migitation Symposium*, tech. coord. G. A. Swanson, pp. 464-468. USDA Forest Service, Rocky Mtn. Forest and Range Expt. Sta. Gen. Tech. Rep. RM-65.

Evans, R. A., J. A. Young, and R. E. Eckert, Jr. 1979. Herbicides as a management tool. In *The Sagebrush Ecosystem: A Symposium*, pp. 110-116. Utah State University, Logan, Utah.

Fay, P. K., V. T. McElligott, and K. M. Havstad. 1989. Containment of free-ranging goats using pulsed radio-wave activated shock collars. *Appl. Anim. Behav. Sci.* 23:165-171.

Forest-Range Task Force. 1972. The Nation's Range Resources—A Forest-Range Environmental Study. Forest Resource Report 19, Forest Service, U.S. Department of Agriculture, Washington, D.C.

Graul, W. D. 1980. Grassland management practices and bird communities. In *Management of Western Forests and Grasslands for Nongame Birds*, eds. R. M. DeGraaf and N. G. Tilghman, pp. 38-47. USDA Forest Service Gen. Tech. Rep. INT-86. Intermountain Forest and Range Expt. Sta., Ogden, Utah.

Griffiths, D. 1902. Forage conditions on the northern border of the Great Basin, being a report upon investigations made during July and August, 1902, in the region between Winnemucca, Nevada and Ontario, Oregon. U.S. Bur. Plant Industry Bull. 15.

Hall, F. C. 1985. Wildlife habitats in managed rangelands—The Great Basin of southeastern Oregon: Management practices and options. USDA Forest Service. Gen. Tech. Rep. PNW-189.

Harniss, R. O., and R. B. Murray. 1973. 30 years of vegetal change following burning of sagebrush-grass range. *J. Range Manage.* 26:322-325.

Harper, K. T. 1986. Historical environments. In *Handbook of North American Indians*. Vol. II, Great Basin, ed. W. L. D'Azeredo, pp. 51-63. Smithsonian Institution Press, Washington, D.C.

Hassan, M. A., and N. E. West. 1986. Dynamics of soil seed pools in burned and unburned sagebrush semi-deserts. *Ecology* 76:269-272.

Hatton, T. J., and N. E. West. 1987. Early seral trends in plant communities on a surface coal mine in southwestern Wyoming. *Vegetatio* 73:21-29.

Haws, B. A. 1987. The status of IPM strategies for controlling grass bugs infesting introduced grassland monocultures. In *Integrated Pest Management on Rangeland: State of the Art in the Sagebrush Ecosystem*, ed. J. A. Onsager, pp. 67-72. USDA Agricultural Research Service ARS-50.

Hull, A. C. Jr. 1976. Rangeland use and management in the Mormon West. In *Symposium on Agriculture, Food and Man—A Century of Progress*, pp. 21. Brigham Young University, Provo, Utah.

Johnson, K. L. 1987. Sagebrush types as ecological indicators to integrated pest management (IPM) in the sagebrush ecosystem of western North America. In *Integrated Pest Management on Rangeland: State of the Art in the Sagebrush Ecosystem*, ed. J. A. Onsager, pp. 1-10. USDA Agricultural Research Service ARS-50.

Kamm, J. A., F. A. Sneva, and L. M. Rittenhouse. 1978. Insect grazers on the cold desert biome. In *Proceedings of the First International Rangeland Congress*, ed. D. N. Hyder, pp. 479-483. Society for Range Management, Denver, Colo.

Kindschy, R. R. 1978. Rangeland management practices and bird habitat values. In *Proceedings of a Workshop on Nongame Bird Habitat Management in the Coniferous Forests of the Western U.S.*, tech. coord. R. M. DeGraaf, pp. 66-69. USDA Forest Service, Pacific Northwest Forest and Range Expt. Sta., Gen. Tech. Rep. PNW-64.

Küchler, A. W. 1970. Potential natural vegetation (map at scale of 1:7,500,00). In *The National Atlas of the U.S.A.*, pp. 90-91. U.S. Geological Survey, Government Printing Office, Washington, D.C.

Laycock, W. A. 1967. How heavy grazing and protection affect sagebrush-grass ranges. *J. Range Manage.* 20:206-213.

Laycock, W. A. 1991. Stable states and thresholds of range condition on North American rangelands: a viewpoint. *J. Range Manage.* 44:427-433.

Mack, R. N. 1977. Mineral return via litter of *Artemisia tridentata*. *Am. Midl. Nat.* 97:189-197.

Mack, R. N. 1981. Invasion of *Bromus tectorum* L. into western North America: an ecological chronicle. *Agroecosyst.* 7:145-165.

Mack, R. N., and J. N. Thompson. 1982. Evolution in steppe with few large hooved mammals. *Am. Nat.* 119:757-773.

Maser, C., J. W. Thomas, and R. G. Anderson. 1984. Wildlife habitats in managed rangelands - The Great Basin of southeastern Oregon. The relationship of terrestrial vertebrates to plant communities. Parts 1 and 2. USDA Forest Service, Pacific Northwest Forest and Range Expt. Sta. Gen. Tech. Rep. PNW-172.

McArdle, R. E., and D. F. Costello. 1936. The virgin range. In *Letter from the Secretary of Agriculture to the U.S. Senate—A Report on the Western Range—A Great But Neglected Resource*, H. A. Wallace. Senate Document 199:71-80.

McArdle, R. E., et al. 1936. The Whiteman's toll. In *Letter from the Secretary of Agriculture to the U.S. Senate—A Report on the Western Range—A Great but Neglected Resource*, H. A. Wallace. Senate Document 199:81-116.

McArthur, E. D. 1983. Taxonomy, origin, and distribution of big sagebrush (*Artemisia tridentata*) and allies (sub-genus *Tridentatae*). In *Proceedings of the First Utah Shrub Ecology Workshop*, ed. K. L. Johnson, pp. 3-13. Utah State University, Logan, Utah.

McEwen, L. C., and L. R. DeWeese. 1987. Wildlife and pest control in the sagebrush ecosystem: Basic ecology and management considerations. In *Integrated Pest Management on Rangeland: State of the Art in the Sagebrush Ecosystem*, ed. J. A. Onsager, pp. 76-85. USDA Agricultural Research Service ARS-50.

McNaughton, S. J., M. B. Coughenour, and L. L. Wallace. 1982. Integrative processes in grassland ecosystems. In *Grasses and Grasslands: Systematics and Ecology*, eds. J. R. Estes, R. J. Tyrl, and J. N. Brunken, pp. 167-194. University of Oklahoma Press, Norman, Okla.

Mehringer, P. J., Jr., and P. E. Wigard. 1990. Comparison of late Holocene environments from woodrat middens and pollen: Diamond Craters, Oregon. In *Packrat Middens: The Last 40,000 years of Biotic Change*, eds. J. L. Betancourt, T. R. Van Devender, and P. S. Martin, pp. 294-325. University of Arizona Press, Tucson, Ariz.

Milchunas, D. G., O. E. Sala, and W. K. Lauenroth. 1988. A generalized model of the effects of grazing by large herbivores on grassland community structure. *Am. Nat.* 132:87-106.

Miller, R. J., T. Svejcar, and N. E. West. in press. Implications of livestock grazing the Intermountain Sagebrush Region: Plant Composition. Proc. Symposium AIBS San Antonio, Tex. Society for Range Management, Denver, Colo.

Mueggler, W. F. 1982. Diversity of western rangelands. In *Natural Diversity in Forest Ecosystems*, eds. J. L. Cooley and J. H. Cooley, pp. 211-217. Proceedings of a Workshop, Institute of Ecology, University of Georgia, Athens, Ga.

Neilson, R. P. 1986. High resolution climatic analysis and Southwest biogeography. *Science* 232:27-34.

Noy-Meir, I. 1973. Desert ecosystems: Environment and producers. *Ann. Rev. Ecol. Syst.* 4:25-51.

Noy-Meir, I. 1974. Desert ecosystems: Higher tropic levels. *Ann. Rev. Ecol. Syst.* 5:195-214.

Noss, R. F., and L. D. Harris. 1986. Nodes, networks, and MUMs: Preserving diversity at all scales. *Environ. Manage.* 10:299-309.

Olson, R. A. 1974. Bird populations in relation to changes in land use in Curlew Valley, Idaho and Utah. M.S. thesis, Idaho State University, Pocatello, Idaho.

Passey, H. B., V. K. Hugie, E. W. Williams, and D. E. Ball. 1982. Relationships between soil, plant community, and climate on rangelands of the Intermountain West. USDA Soil Conservation Service, Tech. Bull. No. 1669.

Pellant, M. 1990. The cheatgrass wildfire cycle—are there any solutions? In *Proceedings of a Symposium on Cheatgrass Invasion, Shrub Die-off and Other Aspects of Shrub Biology and Management*, eds. E. D. McArthur et al., pp. 11-18. USDA Forest Service, Gen. Tech. Rep. INT-276, Intermountain Res. Sta., Ogden, Utah.

Pierson, F. B., and J. R. Wight. 1991. Variability of near-surface temperature on sagebrush rangeland. *J. Range Manage.* 44:491-497.

Reynolds, T. D. 1979. Response of the reptile populations to different land management practices on the Idaho National Engineering Laboratory site. *Great Basin Nat.* 39:255-262.

Reynolds, T. D., and C. H. Trost. 1979. The effect of crested wheatgrass plantings on wildlife on the Idaho Nat. Engineer: Laboratory site. In *The Mitigation Symposium: A National Workshop on Mitigating Losses of Fish and Wildlife Habitats*, tech. coord. G. A. Swanson, pp. 665-666. USDA Forest Service, Gen. Tech. Rep. RM-65, Rocky Mtn. Forest and Range Expt. Sta., Ft. Collins, Colo.

Reynolds, T. D., and C. H. Trost. 1980. The response of native vertebrate populations to crested wheatgrass planting and grazing by sheep. *J. Range Manage.* 33:122-125.

Reynolds, T. D., and C. H. Trost. 1981. Grazing, crested wheatgrass, and bird populations in southeastern Idaho. *Northwest Sci.* 55:225-234.

Rice, B., and M. Westoby. 1978. Vegetative responses of some Great Basin shrub communities protected against jackrabbits or domestic stock. *J. Range Manage.* 31:28-34.

Rickard, W. H., and B. E. Vaughan. 1988. Plant communities: characteristics and responses. In *Shrub-steppe: Balance and Change in a Semi-arid Terrestrial Ecosystem*, eds. W. H. Rickard, L. E. Rogers, B. E. Vaughan, and S. F. Liebetrau, pp. 109-179. Elsevier, Amsterdam.

Risser, P. G. 1988. Diversity in and among grasslands. In *Biodiversity*, ed. E. O. Wilson, pp. 176-180. National Academy Press, Washington, D.C.

Schlesinger, W. G., J. F. Reynolds, G. L. Cunningham, L. F. Huenneke, W. W. Jarrell, R. A. Virginia, and W. G. Whitford. 1990. Biological feedbacks in global desertification. *Science* 247:1043-1048.

Scott, J. M. 1990. Preserving life on earth: We need a new approach. *J. For.* March:13-14.

Scott, J. M., B. Csuti, J. D. Jacobi, and J. E. Estes. 1987. Species richness: a geographic approach to protecting future biological diversity. *BioScience* 37:782-788.

Sneva, F. A., L. R. Rittenhouse, P. T. Tueller, and P. Reece. 1984. Changes in protected and grazed sagebrush-grass in eastern Oregon, 1937-1974. Oregon Agric. Res. Sta. Bull. 663.

Sturges, D. L. 1977. Snow accumulation and melt in sprayed and undisturbed big sagebrush vegetation. USDA Forest Service, Rocky Mtn. Forest and Range Expt. Sta. Research Note RM-348.

Thomas, J. W., C. Maser, and J. E. Rodiek. 1979. Wildlife habitats in managed rangelands -The Great Basin of Southeastern Oregon: Edges. USDA Forest Service, Pacific Northwest Forest and Range Exper. Sta., Gen. Tech. Rep. PNW-85.

Tisdale, E. W., and M. Hironaka. 1981. The sagebrush-grass region: A review of the ecological literature. Forest Wildlife and Range Expt. Sta. Contrib. 209. University of Idaho, Moscow, Idaho.

Tisdale, E. W., M. Hironaka, and M. A. Fosberg. 1965. An area of pristine vegetation in Craters of the Moon National Monument, Idaho. *Ecology* 46:349-352.

West, N. E. 1979. Basic synecological relationships of sagebrush-dominated lands in the Great Basin and Colorado Plateau. In *The Sagebrush Ecosystem: A Symposium*, pp. 33-41. Utah State University, Logan, Utah.

West, N. E. 1983. Western Intermountain sagebrush steppe. In *Temperate Deserts and Semi-deserts.* Vol. 5, Ecosystems of the World, ed. N. E. West, pp. 351-374. Elsevier, Amsterdam.

West, N. E. 1985. Aboveground litter production of three temperate semidesert shrubs. *Am. Midl. Nat.* 113:159-169.

West, N. E. 1988. Intermountain deserts, shrub steppes and woodlands. In *North American Terrestrial Vegetation*, eds. M. G. Borbour and W. D. Billings, pp. 210-230. Cambridge University Press, New York.

West, N. E. 1993. Biodiversity on rangelands. *J. Range Manage.* 46:2-13.

West, N. E., R. A. Griffin, and J. J. Jurinak. 1984a. Comparison of phosphorus distribution and cycling between adjacent native and cultivated grass-dominated ecosystems. *Plant and Soil* 81:151-164.

West, N. E., F. D. Provenza, P.S. Johnson, and N. K. Owens. 1984b. Vegetation change after 13 years of livestock grazing exclusion on sagebrush semidesert in central Utah. *J. Range Manage.* 37:262-264.

West, N. E., and J. Skujins, eds. 1978. *Nitrogen in Desert Ecosystems.* Dowden, Hutchinson and Ross, Stroudsburg, Pa.

Westoby, M., and F. H. Wagner. 1973. Use of a crested wheatgrass seeding by black-tailed jackrabbits. *J. Range Manage.* 26:349-351.

Whisenant, S. G. 1990. Changing fire frequencies on Idaho's Snake River Plains; Ecological and management implications. In *Proceedings of a Symposium on Cheatgrass Invasion, Shrub Die-off, and Other Aspects of Shrub Biology and Management*, eds. E. D. McArthur et al., pp. 4-10. USDA Forest Service, Gen. Tech. Rep. INT-270. Intermountain Research Station, Ogden, Utah.

Wiens, J. A., R. G. Cates, J. T. Rotenberry, N. Cobb, B. Van Home and R. A. Redak. 1991. Arthropod dynamics on sagebrush (*Artemisia tridentata*): Effects of plant chemistry and avian predation. *Ecol. Monogr.* 61:299-321.

Wilcove, D. S. 1989. Protecting biodiversity in multiple-use lands: lessons from the U.S. Forest Service. *Trends Ecol. Evol.* 4:385-388.

Williams, R. E., B. W. Allred, R. M. Denio, and H. A. Paulsen, Jr. 1968. Conservation, development and use of the world's rangeland. *J. Range Manage.* 21:355-360.

Winward, A. H. 1991. A renewed commitment to management of sagebrush-grasslands. In *Management in the Sagebrush Steppe*, pp. 2-7. Oregon Agric. Expt. Sta., Special Report 880.

Yorks, T. P., and C. McMullen, eds. 1980. Western shrub and grasslands. Vol. II of the descriptive supplement to FRODAS: An integrated Resource Data Analysis System for the Forest-Range. Natural Resource Ecology Lab, Colorado State University, Ft. Collins, Colo.

Young, R. P. 1983. Fire as a management tool in rangelands of the Intermountain Region. In *Managing Intermountain Rangelands—Improvement of Range and Wildlife Habitats*, compilers S. B. Monsen and N. Shaw, pp. 18-31. USDA Forest Service, Gen. Tech Rep. INT-157, Intermountain Forest and Range Expt. Sta., Ogden, Utah.

Young, J. A. 1989. Intermountain shrub steppe plant communities—pristine and grazed. In *Proceedings of the Western Raptor Management Symposium and Workshop*, ed. B. G. Pendleton, pp. 3-14. National Wildlife Federation, Sci. & Tech. Series No. 12., Washington, D.C.

Zamora, B., and P. T. Tueller. 1974. *Artemisia arbuscula, A. longiloba* and *A. nova* habitat types in northern Nevada. *Great Basin Nat.* 33:225-242.

Zohary, M. 1973. *Geobotanical Foundation of the Middle East.* Gustav Fischer, Verlag, Stuttgart.

23

STRATEGIES FOR ECOSYSTEM-BASED CONSERVATION OF FISH COMMUNITIES

Jack E. Williams[1]

Office of the Director
Bureau of Land Management
Washington, D.C.
U.S.A.

Gary E. Davis

National Park Service
Channel Islands National Park
Ventura, California
U.S.A.

The increasing loss of genetic and species diversity is one of the world's most critical problems and as E. O. Wilson succinctly noted, that loss "is the folly that our descendants are least likely to forgive us." It is not surprising that our concerns for biodiversity traditionally have focused at the level of individual species. We have seldom considered conservation needs or opportunities at the community-level. The escalating loss of species suggests, however, that we need additional conservation strategies. The purposes of this paper are to describe a strategy that maintains aquatic community integrity and to characterize the ecosystem-based management approach necessary to maintain fundamental ecological processes across landscapes and seascapes.

BACKGROUND

Species loss and rarity in freshwater ecosystems of North America is particularly dramatic. Miller et al. (1989) documented the extinction of 40 fish taxa from North America during the past century, 21 of these since 1960. In addition, repeated negative surveys suggest that the Maryland darter (*Etheostoma sellare*) and High Rock Springs tui chub (*Gila*

[1] Present address: Intermountain Research Station, 316 Myrtle Street, Boise, Idaho.

bicolor ssp.) have recently become extinct. The number of North American freshwater fishes classified by the American Fisheries Society as endangered, threatened, or of special concern increased 45% during the 1980s (Williams et al. 1989).

Species losses in aquatic communities of North America exceed those of terrestrial systems (Williams and Neves 1992). In proportion to the entire fauna, fish species loss and rarity in much of western North America is as great or greater than in tropical regions (Moyle and Williams 1990; Williams and Neves 1992).

In some river systems, habitat loss is so pervasive and non-native species so widely established, that a majority of native fishes are threatened or endangered. The fabric that holds communities together has frayed and unravelled. For example, despite many years of intensive "pre-listing" recovery efforts involving the captive propagation and release of many thousands of young fish into the Colorado River drainage, the razorback sucker (*Xyrauchen texanus*) was declared an endangered species in 1991. All of the "big river" fishes of the mainstream Colorado are now officially listed as endangered or threatened. The native fishes of the Colorado River have largely been replaced by dozens of species native to the eastern United States (Minckley 1982; Tyus 1992). Likewise, 46% of the Sacramento River basin's relatively diverse fish fauna is declining or extinct while non-native species continue to spread (Moyle and Williams 1990). Such declines are not restricted to western waters, where fish communities often are depauperate and highly endemic. In the Illinois and Maumee rivers, 67% and 44% respectively, of their diverse and relatively widespread native fishes have exhibited significant declines or total extirpation (Karr et al. 1985).

Existing systems of protected areas have not been adequate to protect North America's native fish diversity. Most parks, refuges, and wilderness areas are designed to protect terrestrial species and landforms. Because of their linear nature, river systems are particularly difficult to protect within most existing preserves (Moyle and Sato 1991). Even river segments protected by large preserves, such as the Colorado River in Grand Canyon or Canyonlands National Parks, are highly vulnerable to invasion by non-native species located outside of, but readily accessible to, the protected areas, and because of influences of upstream dam releases on water quality and flow rates (Williams 1991). Adequate protection of riverine habitats is particularly important because flowing waters, despite their relatively small water volume when compared to other aquatic habitats, contain the vast majority of aquatic biodiversity (Allan and Flecker 1993).

Conservation efforts for marine environments have lagged far behind those for terrestrial and freshwater systems. Less than 1% of the world's marine area is included in an established protected area, and nearly all of the more than 1,000 established marine protected areas allow extractive uses (Salm and Clarke 1984; Kelleher and Bleakley 1992). In the United States, 34 national parks and all of the national marine sanctuaries support thousands of tons of recreational and commercial fishery harvests. Only 10% of the Great Barrier Reef Marine Park in Australia is closed to all fishing. Truly protected areas large enough to assure biological integrity are exceedingly rare in the marine environment, in spite of more than 40 years of growing concern about human impacts on the oceans. Powerful economic interests drive exploitation of marine resources to produce short-term goals at the expense of long-term productivity and biological diversity. An integrated approach to ocean management is imperative as marine resources are exhausted and no new fishing grounds are left. Sustainability of both biodiversity and productivity are inseparable and need to be the highest goal of marine resources management as food production in the sea shifts from a "hunt and gather" mode to mariculture.

Successful conservation of native fish communities, whether in protected areas or in managed landscapes, requires the maintenance of interactions among the organisms as well as between organisms and their environment. A successful strategy for fish conservation requires improvements to our network of protected areas as well as the development of a management framework that maintains aquatic community integrity across managed landscapes.

PROCESSES THAT AFFECT COMMUNITY INTEGRITY

Fish communities are the products of interactions among water quality and quantity, condition of the watershed, and the species themselves. The relative importance of the processes that influence these variables is a function of their intensity and duration. Large flood and other disturbance events might dramatically alter landscapes, create habitat mosaics, and, thereby, shape associated communities. Smaller events might lack the energy to effect such changes. Human activities can alter or negate the effects of natural processes. For example, dams can eliminate large floods downstream and dramatically alter and stabilize water temperature regimes, whereas reduced vegetative cover in watersheds will increase runoff and magnify extremes in flow.

Many of the natural disturbance events affect the interactions among watersheds, riparian communities, and aquatic habitats. The consequences of these interactions are often most visible in the riparian zone, which is one of the most dynamic and productive portions of the landscape (Gregory et al. 1991). An understanding of the processes controlling the connection between aquatic habitats and watersheds, and the results of human disruption of these factors is necessary for proper management.

Processes Controlling Water Quality and Quantity

Floods and drought are often critical and essential forces shaping fish communities. Native fish communities are well-adapted to such natural variation. Floods might be necessary, for example, to remove fine sediments from spawning areas, introduce large woody debris to stream channels, and scour new pool habitats (Swanston 1991). Also, periodic high flows might be critical for transport of nutrients and creation of high turbidity conditions favorable to native fishes (Pflieger and Grace 1987). In the arid southwestern United States, native fishes tend to survive the vagaries of flash flooding that periodically scours their habitats, while non-native fishes introduced into these streams often are eliminated by floods (Minckley and Meffe 1987). Restoration of hydrographs similar to the natural flow regime of the Missouri River appears to be the key for rejuvenation of fish and floodplain plant communities in that system (Hesse and Mestl 1993).

Water chemistry can also be a determining factor influencing fish communities. Annual patterns of water depth fluctuations determine the composition and abundance of fishes in Florida's Everglades (Kushlan 1976) and downstream estuaries (Davis 1978).

Storm events and seasonal variation in freshwater input will alter salinity and, thereby, control the structure and location of fish communities in areas of tidal influence. Thermal spring systems in the western United States are often characterized by high salinity, alkalinity, and/or highly elevated levels of boron, arsenic or other chemicals. These isolated systems support a sparse but highly-specialized fish community dominated by pupfishes (Cyprinodontidae) and certain minnows (Cyprinidae).

Acid rain has altered or eliminated fish communities throughout much of southeastern Canada and the New England states. Approximately 100,000 lakes in Ontario and Quebec are adversely affected by acid precipitation (D. E. McAllister pers. comm.). Similar but more localized effects result from mine drainage. Where acidification does not eliminate all fishes, surviving communities reflect the different buffering qualities of watersheds and varied physiological tolerances of fishes to reduced Ph.

Successful management depends, in part, on restoring the natural hydrologic regime, including low and high flow events, and reconnecting rivers to their floodplains. The differential susceptibility of native and non-native species to environmental stochasticity might be the key to survival of native fishes in many parts of the country.

Processes Controlling Watershed Condition

Processes that substantially alter watershed condition will produce changes in fish communities. Wildfires, although more commonly known for their impacts on terrestrial landscapes, can result in substantial changes to aquatic communities as was demonstrated by the Yellowstone fires of 1988 (Minshall et al. 1989). Fires have their greatest impacts in headwater areas, but can affect sediment suspension, nutrient cycling, availability of woody debris, and types of aquatic species present throughout river systems (Minshall et al. 1989).

Natural erosion processes, caused by floods or other high flows, can be essential to resupply spawning gravel from upstream or to cause succession of riparian communities. However, erosion also can be severe enough to incise channels and render habitat unsuitable in watersheds damaged by poor land-use practices.

Other natural processes that shape fish communities include stochastic events, such as storms, earthquakes, and landslides, or even the activities of beavers. A more through discussion of natural processes and their effects on fish communities is provided by Swanston (1991).

Successful management also depends on determining those natural processes that control communities and how human activities affect them. In many Pacific Northwest streams, large, woody debris is essential to increase habitat complexity and provide spawning, rearing, and overwintering habitat necessary in communities dominated by salmonid fishes (Sedell et al. 1988; Bisson et al. 1992). Fallen trees and root wads create deep scour pools and capture gravel as they move downstream. The presence of large, woody debris also provides substantial benefits in estuaries and shallow ocean areas (Gonor et al. 1988). Processes that influence the presence of large coniferous trees along Pacific Northwest streams, such as timber harvest practices, road building, and forest disease, will in large measure determine the structure and productivity of affected fish communities.

Processes Controlling Population Abundance

The presence of non-native species profoundly affects native fishes, especially in areas of modified habitats (Courtenay and Stauffer 1984). Non-native fishes often are predators that are introduced to improve sport fishing, but often cause declines or extinction of native species (Moyle 1976). Such predators include members of the sunfish (Centrarchidae), North American catfish (Ictaluridae), temperate basses (Percichthyidae), and trout (Salmonidae) families. Baitfishes, such as threadfin shad (*Dorosoma petenense*), golden shiner (*Notemigonus crysoleucas*) and red shiner (*Cyprinella lutrensis*), purposefully introduced to support populations of introduced sportfish or inadvertently introduced by anglers, can become the dominant fish when introduced into novel habitats. Introduced species might also be the source of exotic parasites or diseases that can depress native fishes (Heckmann et al. 1986).

The establishment of introduced fishes tends to be favored in altered habitats. Alteration of habitats, however, might not be obvious. Even the Colorado River in Canyonlands National Park in Utah, which might appear pristine, has been extensively altered by upstream releases from Flaming Gorge Dam that control water flow patterns, temperature, and turbidity. Recent surveys of Cataract Canyon in Canyonlands National Park found that only 18% of the fishes were native to the Colorado River system (Valdez and Williams 1991).

The structure of fish communities can also be altered by the selective removal of certain predators. Many coastal fisheries are sustained by a process of serial depletions. That is, harvest efforts shift from one species to another as each stock is depleted.

Habitat alterations, pollution, and fishing also threaten the integrity and diversity of marine communities. The complexity of marine ecosystems, combined with simplistic, single-species fisheries models and overfishing have doomed major fisheries worldwide to collapse in the 20th century (Clark 1985; Keen 1991). In coastal California, overall fishery landings have been sustained or increased only by fishing in new territories or shifting harvest to new species (Davis and Dugan 1993). Landings for 75% of the nearshore taxa that constituted important fishery elements in the late 1940s had declined more than 60% by the late 1980s. Most harvests peaked in the 1950s and 1960s. Even species such as green and black abalones, which did not enter the fishery until the mid-1950s or late 1960s, had declined dramatically by 1980. Southern California landings of the rapidly growing red sea urchin fishery declined after 1986, and only harvests from new territories in northern California have allowed landings to increase. Species that once provided major value and weight to landings have declined precipitously, such as white seabass (*Atractoscion nobilis*), abalone (*Haliotis* spp.), giant sea bass (*Stereolepis gigas*), and California sheepshead (*Semicossyphus pulcher*) and have been replaced by rock crabs (*Cancer* spp.), red sea urchin (*Strongylocentrotus franciscanus*), and rockfish (*Seabastes* spp.).

Reef fisheries are particularly difficult to manage on a species-specific basis because of the high number of interacting species and the wide variety of gear and fishery objectives. Intense fishing reduces abundance, alters population age structure, and can alter fish community structure, all of which lead to classic signs of overfishing (Pauly 1979; Munro and Williams 1985; Polovina and Ralston 1987). A community approach to marine fisheries management seems to offer the best opportunity to sustain fishery yields and protect biological diversity, but such an approach has not yet been adequately tested (Davis 1989; Bohnsack 1990; Roberts and Polunin 1991). In the few cases where it has been examined in marine systems, community management appears to work well (Buxton and Smale 1989; Ballantine 1991).

The movement of animals within watersheds or between oceans and fresh waters provide essential sources of energy and minerals into headwater streams. Anadromous fishes, such as salmon, provide major releases of biomass and minerals as they move into spawning streams and die. Similar impacts, albeit less dramatic, occur within river systems as a result of upstream spawning runs of trouts, suckers, and other fishes. Even large emergences of aquatic insects, such as mayflies, can be valuable for input of energy to relatively unproductive headwaters. Changes that impede the upstream flow of energy can change fish community structure with a breakdown in community integrity.

Successful management also depends on emphasizing the values of native species and controlling non-native species. In many streams in the western United States, for example, native redband (*Oncorhynchus mykiss* sspp.) and cutthroat (*O. clarki* sspp.) trouts have been eliminated and replaced by poorly-adapted hatchery trout. In addition to numerous detrimental effects associated with introductions of non-native species, there is growing evidence of the negative consequences of large-scale introductions of hatchery stocks of native species (Hilborn 1992).

ADVANTAGES OF COMMUNITY CONSERVATION

Protection of entire communities offers many ecological and economic advantages over species-level conservation. Community management is a proactive approach protecting more than just those species listed as endangered or threatened. Candidate species, other locally rare species, and species not yet recognized as needing special management are conserved, and future listings of such species as endangered or threatened can be avoided.

Many existing refuges tend to freeze present conditions and act more as museums than as functional communities that interact with dynamic landscapes or seascapes (Bohnsack 1990; Moyle and Sato 1991). Conservation of entire, natural aquatic communities allows species interactions and evolution to continue. This occurs because management focuses on ecosystem processes and functions that continue to shape communities, thus promoting dynamic conservation.

Various single-species refuges are plagued by problems associated with maintenance of infrastructure, such as fences, pipes, and valves that supply water (Simons 1987; Baugh and Deacon 1988; Williams and Williams 1989). When natural processes exert their forces, maintenance costs, which are often high in single-species refuges (Williams 1991), are minimized.

Management for entire communities has several economic advantages. First, it minimizes the need to establish single-species refuges because entire assemblages of species are protected before any component species becomes endangered. Second, landscapes managed to conserve viable communities can preempt processes leading to declines in individual species that would require protection under the Endangered Species Act and retain more opportunities to provide a mix of recreation, timber products, and other commodities. On multiple-use lands that contain endangered or threatened species, management options are often more limited, and a greater proportion of available funds must go towards triage efforts to prevent extinctions.

Across much of the managed landscapes, habitats are so degraded that management options are already limited. A nationwide rivers inventory in 1982 found that only 1.8% of all rivers in the contiguous 48 states remain in high quality condition (Benke 1990). Considerable effort will be needed to restore ecological processes that maintain communities, but in the cases of our rivers, riparian areas, and estuaries, this effort will yield dividends beyond conservation of biological diversity. Water quality will improve because of decreasing nonpoint source pollution, and water availability will be more consistent as watersheds increase their storage capacity. Restored riparian functions will decrease erosion and increase the capacity of floodplains to capture silt and absorb energy of high flows.

MANAGEMENT STRUCTURE AND FUNCTION

Most multiple-use land management agencies are structured toward the production of commodities, such as board-feet of timber, livestock forage (AUMs), or recreational user days (see A, Figure 23.1). Within such a structure, programs compete for available

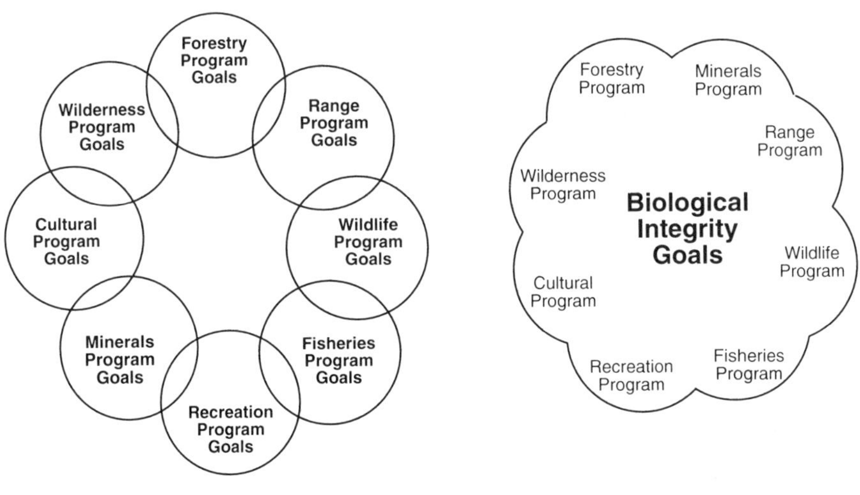

Figure 23.1. Comparison of program organization within hypothetical multiple-use management agencies. Organizational structure **A** shows little integration (overlap) between programs and none among all programs. As program integration increases, as shown in **B**, more emphasis is placed on common management themes. Explicit inclusion of the concept of maintaining or restoring biological integrity as a primary common goal for land management ensures integration among agency programs.

natural resources, for lands for their programs, and for funding. This conflict is reflected in land-use plans, which typically offer separate alternatives that emphasize one commodity over all others. For example, one alternative will provide increased livestock grazing on the landscape, another emphasizes management to increase recreation use, whereas a third emphasizes wilderness designations. As recently noted, during this process of allocating land parcels to predominant uses, we have run out of available land (Williams and Rinne 1992).

Increased integration of agency programs promotes cooperation toward common goals rather than competition for funds and available land to produce commodities (see B, Figure 23.1). An integrated organization facilitates management for biological integrity as a primary common goal. Biological integrity is defined as "the capability of supporting and maintaining a balanced, integrated, adaptive community of organisms having a species composition and functional organization comparable to that of natural habitat of the region" (Karr and Dudley 1981). Such a capability requires that natural processes controlling habitat and community diversity are maintained. Under ecosystem-based, integrated management, the primary goal is to maintain and/or restore biological integrity of rivers, estuaries, rangelands, and forests. Additional productivity above that needed to maintain biological integrity can then be allocated among specific commodity interests.

For ecosystem-based management to be most effective, it should be organized by watershed boundaries. Watersheds are natural links between landscape-scale strategies and local activity planning. Seldom in the past have land management agencies made plans based on land units encompassed by ecosystem boundaries. Land-use planning is organized by administrative units, agency boundaries, county or state borders, or other units that have little correspondence to ecosystem boundaries. Successful ecosystem management, therefore, requires modification of existing planning boundaries. And because a single agency seldom manages entire watersheds, cooperative planning and management among federal and state agencies, and other landowners are necessary.

A new strategy stressing cooperative watershed management is "Bring Back the Natives," a joint effort of the USDA Forest Service, USDI Bureau of Land Management (BLM), and National Fish and Wildlife Foundation. This strategy seeks to restore the biological integrity of entire river systems and their native species on public lands by integrating efforts of the Forest Service and BLM in stream management where both agencies oversee large tracts of land within a common watershed (Williams and Williams 1992). Most "Bring Back the Natives" projects involve streams with headwaters on Forest Service land and downstream reaches on BLM lands. Another key element of these projects is the modification of land management practices, particularly those associated with livestock grazing and timber harvest, which resulted in past declines in aquatic conditions.

Another cooperative Forest Service and BLM strategy, nicknamed PACFish," has been developed in response to the large number of endangered, threatened, or of special concern stocks of anadromous salmonids on public lands. The goal of PACFish is restore freshwater habitat conditions to contribute to the sustained natural production of the natural complex of salmonid species occurring in stream systems. Rather than focusing on the fish themselves, the strategy focuses on the restoration and maintenance of quantified habitat features, such as the number of deep pools per stream-mile, water temperature, and amount of large woody debris in riparian areas. These features are surrogates for watershed health. The PACFish strategy was adopted by the Forest Ecosystem Management Assessment Team (1993) as the riparian and aquatic components of the President's plan for old growth forests in the Pacific Northwest.

In the marine environment, the Great Barrier Reef Marine Park Authority in Australia has utilized a successful strategy that integrates concerns for sustainable development with protection of biological diversity. The Authority has jurisdiction over a large area in which fishing, recreation, and tourism coexist on an environmentally sustainable basis. Conflicts between disruptive or extractive uses and non-extractive uses are resolved by zoning, based on ecological features. The U.S. National Marine Sanctuaries Program is beginning to emulate this model with the recent creation of large marine sanctuaries in the Florida Keys and central California. Internationally, the World Conservation Union (IUCN) has adopted guidelines for establishment of marine protected areas, but an integrated global approach to management of the seas that balances extractive uses with sustainability has not yet been implemented (Kelleher and Kenchington 1991).

In fresh waters, it has become increasingly clear that long-term improvements to fisheries can seldom be achieved through the use of instream structures. Of paramount importance are the contributions of riparian and upland elements to the aquatic condition. Recent surveys in the Pacific Northwest demonstrate that instream structures commonly used to improve fisheries habitats have a high failure rate and are often counterproductive in streams with high sediment loads, high peak flows, or erodible bank materials (Frissell and Nawa 1992). Fisheries conservation is more likely to succeed by focusing on riparian management, watershed integrity, and maintaining interactions between rivers and watersheds, rather than in relying only on modifications to instream habitats (Gregory et al. 1991), where costs invariably exceed benefits, and improvements are temporary at best.

CONCLUSIONS

Recent observers have suggested that existing laws and authorities in the United States are adequate to protect biological diversity if agencies choose to apply them correctly (Williams and Deacon 1991). Karr (1990) noted the sound intent behind endangered species and water resource legislation, but also found a lack of enthusiasm and success in proactive implementation of these laws. An ecosystem-based approach to conservation of biological integrity is not only consistent with legislation, such as the Clean Water Act and Endangered Species Act, but is strongly encouraged by these laws.

The increasing number of rare fishes in the United States and the inability of traditional, species-level conservation efforts to stem this rising tide, suggest that broader efforts will be required to conserve aquatic diversity (Table 23.1). Managed landscapes can play a pivotal role in the conservation of aquatic diversity while still providing for commodity production. Also needed, however, are an increased emphasis on watershed boundaries and river systems in defining protected areas on the landscape, and an increased emphasis on establishing protected areas free of fishery harvest on the seascape.

Our knowledge base is now adequate to maintain successfully natural aquatic communities across multiple-use landscapes. The processes affecting aquatic communities are becoming known (Gregory et al. 1991; Swanston 1991) and are beginning to be incorporated into agency programs (e.g., Gregory and Ashkenas 1990). As Deacon and

Table 23.1. Major features of traditional versus ecosystem-based management. Right-hand column indicates changes necessary to conserve fish communities.

Traditional	Ecosystem-Based
Resource Use Planning	
commodity-based goals	biological integrity goals
administrative boundaries	ecological boundaries
agency-specific planning	multi-agency planning
Management Focus	
species and/or commodities	ecological processes and functions
streams	watersheds
Protected Areas	
allow fishery harvests	restrict fishery harvests
terrestrial-based system	terrestrial, freshwater, and marine systems

Minckley (1991) eloquently stated in their book *Battle Against Extinction: Native Fish Management in the American West*,

> During the past two decades it has become evident that knowledge no longer limits our ability to protect native fishes. Most Endangered Species can be recovered, if we choose.

LITERATURE CITED

Allan, J. D., and A. S. Flecker. 1993. Biodiversity conservation in running waters. *BioScience* 43(1):32-43.

Ballantine, B. 1991. Marine reserves for New Zealand. University of Auckland. Leigh Lab. Bull. No. 25.

Baugh, T. M., and J. E. Deacon. 1988. Evaluation of the role of refugia in conservation efforts for the Devils Hole pupfish, *Cyprinodon diabolis* Wales. *Zoo Biol.* 7:351-358.

Benke, A. C. 1990. A perspective on America's vanishing streams. *J. N. Am. Benth. Soc.* 9:77-88.

Bisson, P. A., T. P. Quinn, G. H. Reeves, and S. V. Gregory. 1992. Best management practices, cumulative effects, and long-term trends in fish abundance in Pacific Northwest river systems. In *Watershed Management: Balancing Sustainability and Environmental Change*, ed. R. J. Naiman, pp. 189-232. Springer-Verlag.

Bohnsack. J. A. 1990. The potential of marine fishery reserves for reef fish management in the U.S. southern Atlantic. So. Atlantic Fishery Management Council, Coastal Resources Div., Contrib. No. CRD/89-90/04.

Buxton, C. D., and M. J. Smale. 1989. Abundance and distribution patterns of three temperate marine reef fish (Teleostei: Sparidae) in exploited and unexploited areas off the southern cape coast. *J. Appl. Ecol.* 26:441-451.

Clark, C. W. 1985. *Bioeconomic Modelling and Fisheries Management*. John Wiley and Sons, New York.

Courtenay, W. R., Jr., and J. R. Stauffer, Jr., eds. 1984. *Distribution, Biology, and Management of Exotic Fishes*. John Hopkins University Press, Baltimore, Md.

Davis, G. E. 1978. Changes in the Everglades National Park red drum and spotted seatrout fisheries 1958-1978: fishing pressure, environmental stress, or natural cycles? In *Proceedings of a Colloquium Gulf States Marine Fisheries Commission*, pp. 81-87.

Davis, G. E. 1989. Designated harvest refugia: the next stage of marine fishery management in California. CalCOFI Rept. 30:53-58.

Davis, G. E., and J. E. Dugan. 1993. Applications of marine refugia to coastal fisheries management. *Can. J. Fish. Aquat. Sci.* 49:in press.

Deacon J. E., and W. L. Minckley. 1991. Western fishes and the real world: the enigma of "Endangered Species" revisited. In *Battle Against Extinction: Native Fish Management in the American West*, eds. W. L. Minckley and J. E. Deacon, pp. 405-413. University of Arizona Press, Tucson,Ariz.

Forest Ecosystem Management Assessment Team. 1993. Forest ecosystem management: an ecological, economic, and social assessment. Report of the Forest Ecosystem Management Assessment Team to President Clinton.

Frissell, C. A., and R. K. Nawa. 1992. Incidence and causes of physical failure of artificial habitat structures in streams of western Oregon and Washington. *N. Am. J. Fish. Manage.* 12:182-197.

Gonor, J. J., J. R. Sedell, and P. A. Benner. 1988. What we know about large trees in Estuaries, in the sea, and on coastal beaches. In *From the Forest to the Sea: a Story of Fallen Trees*, eds. C. Maser et al., pp. 83-112. USDA Forest Service Gen. Tech. Rept. PNW-GTR-229.

Gregory, S. V., and L. Ashkenas. 1990. Riparian management guide: Willamette National Forest. USDA Forest Service, Pacific Northwest Region.

Gregory, S. V., F. J. Swanson, W. A. McKee, and K. W. Cummins. 1991. An ecosystem perspective of riparian zones. *BioScience* 41:540-551.

Heckmann, R. A., J. E. Deacon, and P. D. Greger. 1986. Parasites of the woundfin minnow, *Plagopterus argentissimus*, and other endemic fishes from the Virgin River, Utah. *Great Basin Nat.* 46:662-676.

Hesse, L. W., and G. E. Mestl. 1993. An alternative hydrograph for the Missouri River based on the precontrol condition. *N. Am. J. Fish. Manage.* 13:360-366.

Hilborn, R. 1992. Hatcheries and the future of salmon in the Northwest. *Fisheries* 17(1):5-8.

Karr, J. R. 1990. Biological integrity and the goal of environmental legislation: lessons for conservation biology. *Conserv. Biol.* 4:244-250.

Karr, J. R., and D. R. Dudley. 1981. Ecological perspectives on water quality goals. *Environ. Manage.* 5:55-68.

Karr, J. R., L. A. Toth, and D. R. Dudley. 1985. Fish communities of midwestern rivers: a history of degradation. *BioScience* 35:90-95.

Keen, E. A. 1991. Ownership and productivity of marine fishery resources. *Fisheries* 16(4):18-22.

Kelleher, G. K., and C. Bleakley. 1992. Marine and coastal. In *Regional Reviews*, pp. 1.1-1.16. IVth, World Congress on National Park and Protected Areas, Caracas, Venezuela. IUCN, Gland, Switzerland.

Kelleher, G. K., and R. A. Kenchington. 1991. Guidelines for establishing marine protected areas. Great Barrier Reef Marine Park Authority/UNESCO/IUCN. Gland, Switzerland.

Kushlan, J. A. 1976. Environmental stability and fish community diversity. *Ecology* 57:821-825.

Miller, R. R., J. D. Williams, and J. E. Williams. 1989. Extinctions of North American fishes during the past century. *Fisheries* 14(6):22-38.

Minckley, W. L. 1982. Trophic interrelations among introduced fishes in the lower Colorado River, southwestern United States. *Calif. Fish and Game* 68:78-89.

Minckley, W. L., and G. K. Meffe. 1987. Differential selection by flooding in stream-fish communities of the arid American Southwest. In *Community and Evolutionary Ecology of North American Stream Fishes*, eds. W. J. Matthews and D. C. Heins, pp. 93-104. University of Oklahoma Press, Norman, Okla.

Minshall, G. W., J. T. Brock, and J. D. Varley. 1989. Wildfires and Yellowstone's stream ecosystems. *BioScience* 39:707-715.

Moyle, P. B. 1976. Fish introductions in California: history and impact on native fishes. *Biol. Conserv.* 9:101-117.

Moyle, P. B., and G. M. Sato. 1991. On the design of preserves to protect native fishes. In *Battle Against Extinction: Native Fish Management in the American West*, eds. W. L. Minckley and J. E. Deacon, pp. 155-169. University of Arizona Press, Tucson, Ariz.

Moyle, P. B., and J. E. Williams. 1990. Biodiversity loss in the temperate zone: decline of the native fish fauna of California. *Conserv. Biol.* 4:275-284.

Munro, J. L., and D. M. Williams. 1985. Assessment and management of coral reef fisheries: biological, environmental, and socio-economic aspects. *Proc. Fifth Internat. Coral Reef Congress, Tahiti* 4:544-578.

Pauly, D. 1979. Theory and management of tropical multispecies stocks: a review, with emphasis on Southeast Asian demersal fisheries. *ICLARM Studies and Reviews* 1:1-35.

Pflieger, W. L., and T. B. Grace. 1987. Changes in the fish fauna of the lower Missouri River, 1940-1983. In *Community and Evolutionary Ecology of North American Stream Fishes*, eds. W. J. Matthews and D. C. Heins, pp. 166-177. University of Oklahoma Press, Norman, Okla.

Polovina, J. J., and S. Ralston, eds. 1987. *Tropical Snappers and Groupers: Biology and Fisheries Management*. Westview Press, Boulder, Colo.

Roberts, C. M., and N. V. C. Polunin. 1991. Are marine reserves effective in management of reef fisheries? *Rev. Fish Biol. and Fish.* 1:65-91.

Salm, R. V., and J. R. Clarke. 1984. Marine and coastal protected areas: a guide for planners and managers. IUCN. Gland, Switzerland.

Sedell, J. R., P. A. Bisson, F. J. Swanson, and S. V. Gregory. 1988. What we know about large trees that fall into streams and rivers. In *From the Forest to the Sea: a Story of Fallen Trees*, eds. C. Maser et al., pp. 47-81. USDA Forest Service Gen. Tech. Rept. PNW-GTR-229.

Simons, L. H. 1987. Status of the Gila topminnow (*Poeciliopsis occidentalis occidentalis*) in the United States. Arizona Game and Fish Dept. Spec. Rept. Project E-1.

Swanston, D. N. 1991. Natural processes. In *Influences of Forest and Rangeland Management on Salmonid Fishes and their Habitats*, ed. W. R. Meehan, pp. 139-179. Am. Fish. Soc. Spec. Publ. 19.

Tyus, H. M. 1992. Razorback sucker listed as endangered: a further decline in the Colorado River system. *Endangered Species Update* 9(5&6):1-3.

Valdez, R. A., and R. G. Williams. 1991. Endangered fishes of Cataract Canyon Proc. Desert Fishes Council 20(for 1988):25-26.

Williams, C. D., and J. E. Deacon. 1991. Ethics, federal legislation, and litigation in the battle against extinction. In *Battle Against Extinction: Native Fish Management in the American West*, eds. W. L. Minckley and J. E. Deacon, pp. 109-121. University of Arizona Press, Tucson, Ariz.

Williams, C. D., and J. E. Williams. 1989. Refuge management for the threatened Railroad Valley springfish in Nevada. *N. Am. J. Fish. Manage.* 9:465-470.

Williams, C. D., and J. E. Williams. 1992. Bring Back The Natives: a new strategy for restoring aquatic biodiversity on public lands. *Trans. 57th N. Am. Wildl. Nat. Res. Conf.* 57:416-423.

Williams, J. E. 1991. Preserves and refuges for native western fishes: history and management. In *Battle Against Extinction: Native Fish Management in the American West*, eds. W. L. Minckley and J. E. Deacon, pp. 171-189. University of Arizona Press, Tucson, Ariz.

Williams, J. E., and R. J. Neves. 1992. Introducing the elements of biological diversity in the aquatic environment. *Trans. 57th N. Am. Wildl. Nat. Res. Conf.* 57:345-354.

Williams, J. E., and J. N. Rinne. 1992. Biodiversity management on multiple-use federal lands: an opportunity whose time has come. *Fisheries* 17(3):4-5.

Williams, J. E., J. E. Johnson, D. A. Hendrickson, S. Contreras-Balderas, J. D. Williams, M. Navarro-Mendoza, D. E. McAllister, and J. E. Deacon. 1989. Fishes of North America endangered, threatened, or of special concern: 1989. *Fisheries* 14(6):2-20.

24

CONSERVING PLANT SPECIES DIVERSITY: THE CHALLENGE OF RECOVERY

Bruce M. Pavlik

Department of Biology
Mills College
Oakland, California
U.S.A.

On the eve of the 21st Century, there is crisis and hope. The present crisis of mass extinction, caused by population growth, technology, and resource consumption, is unprecedented in the short history of the human species and in the long history of this planet. Plants, in particular, are bearing the brunt of the crisis. Nearly 20,000 taxa are globally rare or threatened and as many as 60,000 face extinction by the middle of the next century (Table 24.1; IUCN/WWF 1988). Most of these reside in species-rich tropical forests which are being destroyed or degraded at a rate of 11.4 million hectares per year (Wilson 1988). Many, however, are in the anthropogenic landscapes of the north temperate zone. More than 2,000 plant taxa are now threatened in the continental United States (Ayensu and DeFilipps 1978), with 820 in California alone (Skinner and Pavlik 1993). The presummed extinction of 360 plant taxa has been documented in the United States, including Hawaii (Ayensu and DeFilipps 1978), whereas the Center for Plant Conservation estimates that 680 more could become extinct before the year 2000.

The extinction of so many plant species might go unnoticed if it were not for the fact that humans depend upon plant resources for food, medicine, clothing, shelter, and other essential products or elements of culture. Recent studies have demonstrated that humans depend on a wider array of plant species than was previously thought (Myers 1983; Prescott-Allen and Prescott-Allen 1990), thus underscoring the practical importance of conserving plant diversity. Other reasons for minimizing anthropogenic extinction are ecological rather than utilitarian: plants play a fundamental role in the biosphere, because they produce reduced carbon, recycle nutrients, and form the material basis of natural landscapes. There are also philosophical reasons, those that emphasize the "inherent wrongness of destruction" (Ehrenfeld 1988) and seek a new ethic that places humans, non-humans, and whole ecosystems in a long-term, mutually-sustainable context. As conservationists have presented these utilitarian, ecological, and philosophical perspectives, people and their governments have slowly, haltingly, begun to respond.

The crisis of mass extinction and the hope of human response were forcefully brought together by the adoption of the Endangered Species Act of 1973. With that law the conservation of rare plants was formally integrated into the natural resource policies of the United States government. By extending the protection of declining populations and

Table 24.1. Number of plant taxa known to be at risk of extinction or presumed extinct. Taxa include named species, subspecies, and varieties.

	Taxa	Percent of native flora	Source
At Risk (rare or threatened)			
Globally	19,077	8	IUCN/WWF 1988
United States	2,893	13	Ayensu and DeFilipps 1978
California	820	12	Skinner and Pavlik 1993
Presumed Extinct			
United States	360	2	Ayensu and DeFilipps 1978
California	37	1	Skinner & Pavlik 1993

critical habitats to plants, Congress and the Nixon administration called an end to the vertebrate chauvinism (or perhaps "vertism") so pervasive among wildlife laws, regulations, and agencies at the time (Bean 1983; Bartel 1987). This was a major step toward embracing the broader concept of conserving biological diversity regardless of the phylogenetic disposition of organisms. Although there is still much work to be done in developing biodiversity policy, it is heartening to see the Forest Service, the Bureau of Land Management, and other federal agencies attempting to redefine their activities in this regard (Topik 1991; Hastey 1993). Recommendations for strategic biodiversity policy have also been made in California (Jensen et al. 1990, Cochrane 1992) and at the recent Earth Summit in Rio de Janiero.

Despite such policy advances, however, no plant taxa have been fully recovered under the Endangered Species Act (Rohlf 1991; Pavlik 1994a), indicating that conservation is limited by something other than a lack of policy. That something is a management-oriented understanding of rare plants and their habitats so that natural populations can be enhanced or new populations created to reduce the probability of extinction. Enhanced or newly created populations are often required by recovery plans in order to down-list or delist a species under the law. Failure to recover listed species, combined with backlash from frustrated economic interests (landowners, developers, county supervisors) could soon weaken or repeal many well-intentioned but unproductive conservation policies.

Our ability to make policy has far exceeded our ability to implement it. Significant progress has been made in areas where the tools of implementation have been available and effective. For example, sophisticated inventories of rare and common species are now possible using traditional tools from systematics and biogeography combined with computer databases and satellite imagery. Although constrained by time and money, large parcels of habitat have been preserved using conventional financial and legal tools in the name of species conservation. But there has not been sufficient progress toward ensuring that the rare species, even those in national parks and wildlife refuges, will persist into the 21st Century.

The recovery of endangered plants, including the restoration of their critical habitats, might be the greatest challenge facing species-level conservation programs. How can recovery efforts be improved? The purpose of this paper is to examine (1) how small reserves can become more effective, (2) the importance of demographic monitoring, (3) the role of population viability analysis, and (4) recommendations for improving recovery efforts.

HOW CAN SMALL RESERVES BE MORE EFFECTIVE?

Island biogeography provides a theoretical frame-work for designing nature reserves to minimize extinction probabilities and maximize the conservation of biodiversity (Diamond and May 1976; Jensen 1987). Models suggest that the amount of preserved habitat should reflect local species richness, distance from immigration sources, critical resource availability, and the probability of disturbance from beyond reserve boundaries. Some conservationists argue that large reserves with less edge support more species and larger populations for longer periods of time than several small reserves of the same total area (Diamond and May 1976; Terborgh 1976; Wilcox and Murphy 1985). Others argue that plant species richness in particular can be conserved in relatively small areas (Simberloff and Abele 1976a,b; Simberloff and Gotelli 1984; Lahti and Ranta 1985, 1986). The SLOSS (Single Large or Several Small) debate attempted to address fundamental economic and political limits of conservation with sound biological reasoning.

The limitations of this theoretical approach are amply discussed elsewhere (Soulé and Simberloff 1986; Robinson and Quinn 1992), but recently Saunders et al. (1991) underscored the futility of SLOSS arguments from the perspective of actually having to manage existing small or fragmented biological reserves. They point out that regardless of the amount of preserved habitat, most harm originates from the surrounding landscape where managers are unlikely to have much influence. What is needed, they argue, is ecosystem-level management which is specific to the ecological and political constraints of the reserve. Often there is no chance to argue scientifically about the size of a reserve in a checkerboard landscape with rapid development and high real estate values. Size and shape of a reserve often conform to the proposed placement of streets, buildings, and golf courses; only a few grudging concessions can be extracted by government biologists or local activists. In rapidly-developing areas, therefore, the argument should not end with whether or not the reserve will be small (assume the reserve will be small). Long-term, ecosystem management plans must also be adopted to maximize the value of small reserves, with legal and financial responsibility for those plans assigned to developers whose projects create the need and revenue for small reserves.

Is such an approach biologically sound? Reznicek (1987) emphasized that in regions without large tracts of natural land for protection, small reserves would be sound because many plant taxa have life history traits that decrease the probability of extinction in small populations. These include persistent seed banks, great longevity, vegetative propagation, habitat specificity, minimal habitat area per individual, and minimal harm from inbreeding depression. Genetic models have also shown that small plant populations (effective population sizes (N_e) between 25 and 100 individuals) might be better able to maintain their heterozygosity than was previously assumed. After 25 generations of drift and overdominance (greater fitness of heterozygotes at the loci examined), a population with $N_e = 25$ could retain over 92% of its original heterozygosity if selection intensities were moderate (Lesica and Allendorf 1992). Taken together, these demographic and genetic analyses indicate that plants are more tolerant of small population size than animals and better able to maintain long-term viability. This tolerance does not discount the importance of large populations, which are less susceptible than small populations to drift and environmental stochasticity. Instead, tolerance means that small plant populations on small reserves can be viewed as worthwhile nuclei for the enhancement and recovery of target species.

But just how large should a small reserve be for a target species? It should be large enough to (1) allow for significant propagule dispersal within, rather than beyond, the reserve, (2) maintain critical ecosystem processes, (3) contain suitable but presently unoccupied habitat, and (4) accommodate population growth (or at least regeneration) of the target species in the future. These conditions are necessary in order to achieve a basic conservation objective: the area of available habitat should enable target populations to demonstrate self-maintenance by completing critical demographic processes (Pavlik 1994a). If this objective cannot be met in a small reserve under natural conditions, then a greater degree of management intervention is required to compensate for deficits in habitat quantity or quality. Intensive management, paid for by the developer, may be especially necessary in small reserves with a high potential for anthropogenic disturbance or invasion by exotic species. Provisions for the ongoing management of populations and ecosystems must be built into local development projects and into the funding of small reserves as part of a comprehensive mitigation plan (Howald and Wickenheiser 1990).

Is such an intensive, management-oriented approach desirable? Ideally, rare plant populations should be self-maintaining, with little or no requirement for human intervention (Pavlik 1994a). This is more likely to be the case for larger, more heterogeneous reserves that experience natural rather than anthropogenic disturbance, migration, and succession. But given the large number and wide geography of rare plant occurrences in rapidly urbanizing states such as California, small, isolated reserves will be the rule for hundreds of plant taxa. This is especially true at low elevations where natural communities tend to have the greatest number of rare plant occurrences and existing reserves are uncommon (Figure 24.1) and where most land development and conversion take place. Small reserves, therefore, will be of great importance to rare plant recovery, but only if they are managed in the long-term for ecosystem viability (Soulé and Simberloff 1986; Reznicek 1987). With so many populations and entire taxa at stake, sound criteria should be established immediately for managing rare plant ecosystems of various kinds and sizes.

Managing small reserves on the basis of ecosystem viability is easier said than done. Models concerning the viability of single species populations are cumbersome and impractical enough, let alone assemblages of interacting species and microenvironments. Clearly, viability criteria and management techniques need to be established from descriptive studies of intact natural systems and empirical studies of similar but disturbed systems. For example, Taylor (1992) demonstrated the importance of uplands in the maintenance of species richness patterns of vernal pools in California. His results explain the failure of some existing pool reserves, those established on the extent of pool habitat alone, to retain their component species. Without the uplands,

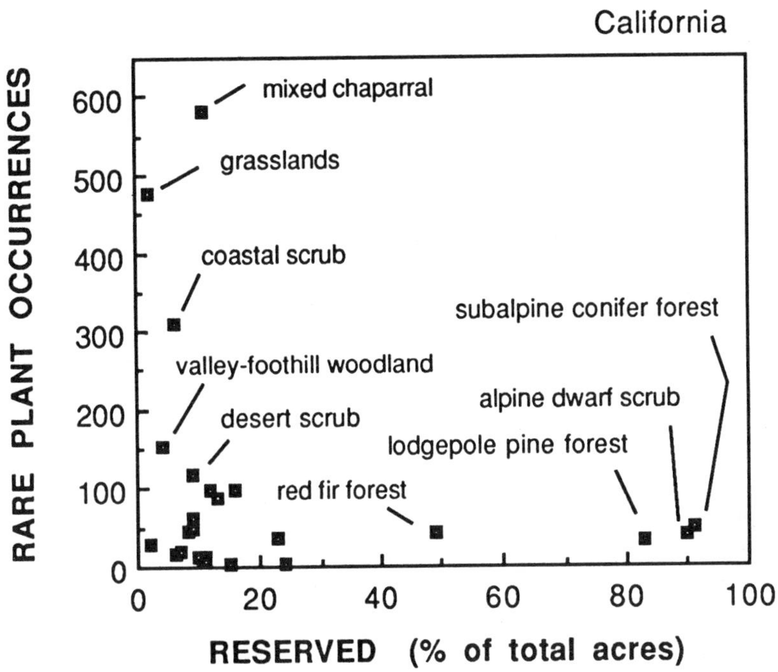

Figure 24.1. Most rare plant occurrences in California are at low elevation in plant communities with less than 10% reserved acreage (parks, research natural areas, wildlife refuges). These communities (e.g., mixed chaparral, coastal scrub) are also intensively developed so many rare plants are at risk of extinction or subject to isolation in small reserves. Data from CNDDB (1992) and Jensen et al. (1990).

intensive management becomes necessary if rare plant populations are to be maintained as part of the pool reserve. In 1987, a vernal pool in Sonoma County, California, habitat for more than 500 plants of the endangered *Blenospermum bakeri*, became a corner lot reserve after the construction of tract housing. In 1992, the *Blenospermum* population consisted of fewer than 10 plants in a competitive sward of introduced grasses (Ann Howald pers. comm.). Except for a 6-foot chain link fence, no management of the pool ecosystem was implemented to offset its small size, isolation, or invasion by exotic species. Neither was there demographic monitoring to explain the decline of *Blenospermum* or to suggest specific actions for halting or reversing the trend. Although another population of this rare plant has been lost, this unfortunate "experiment" verifies the need to build ecosystem management into the process of establishing small reserves if adequate habitat with buffers and corridors cannot be preserved.

In many cases the ability to manage ecosystems or conduct the experiments that reveal how management should be done, is severely constrained by size of the reserve and patterns of land use or resource regulation. In general, smaller reserves have a greater potential for conflict with adjacent developments (e.g., housing, pasture). Without adequate buffer zones, it is difficult to conduct control burns, weed eradication, rodent control, and other activities that might be considered inappropriate for a suburban landscape. Such activities are essential to small reserves because they maintain the patchiness or early seral conditions that rare plants often require, especially in low elevation habitats subject to invasion and domination by weeds (e.g., grasslands, coastal scrub, oak woodlands). This problem can sometimes be overcome by employing small-scale versions of management activities, such as burning in 2 X 2-m sheet metal "burn boxes" which confine flames, increase fire control, and generate small amounts of heat and smoke (Pavlik et al. 1993). But decreasing the scale of management simply increases management intensity; actions become surgical rather than conventional, requiring greater control, accuracy, and effectiveness. To be effective for recovery purposes, small reserves require more intensive management than large reserves of the same taxon.

If intensive, ecosystem-based management is developed for small reserves, who will implement it? With almost 10,000 known rare plant occurrences in California alone (CNDDB 1992), it is easy to imagine reserves outnumbering employed rare plant biologists by several orders of magnitude. The bulk of the work needs to be done by local volunteers, interested members of the California Native Plant Society, and other such organizations under the guidance of a rare plant biologist. The biologist would oversee a system of many small reserves within the regional, developed landscape, and receive salary from mitigation funds. Volunteers would be trained to perform demographic monitoring, restoration, and other routine management tasks, and guided by clear objectives, appropriate techniques, and good relations with the general public.

DEMOGRAPHIC MONITORING IS NEEDED FOR RECOVERY

"Monitoring" is an often-used word that means making measurements or assessments over time. It can apply to any sort of measurement, with any sort of technology, and over any time scale. It can be used to describe research at the level of the cell, population, community, or ecosystem. It can be done within the context of descriptive or experimental work for a great variety of purposes. Despite this tremendous ambiguity, the word "monitoring" is left to speak for itself in environmental impact reports, mitigation agreements, rare plant conservation programs, and recovery plans. Without precisely defining the specific purpose and procedure for monitoring, very little useful information will be generated by the money, time, and effort spent. Actually, poorly-designed monitoring probably does more harm than good because crucial questions go unanswered, actions are delayed, impacts begin to compound, and well-intentioned biologists are frustrated by having to gather irrelevant data.

I recommend that any biologist who is performing such monitoring to stop making estimates of canopy cover and stop shooting another series of photo plots and ask the hard question, "why is this being done?" Or "what impending decision will be guided by the results of this monitoring?" Or in general "what is the purpose of all of this work?" I suspect that in most cases, the answer is "to detect change." But there are at least five reasons why changes in the number of populations, population size, or population area are insufficient and sometimes misleading for managing rare plants (Bradshaw 1981; Davy and Jefferies 1981; Sutter 1986; Pavlik 1994a):

(1) The number of plants counted per year might not provide important insights into the population biology of a target species. Change in the size of a population might not be, in and of itself, a matter of concern to conservation. In fact, great fluctuations in population sizes are the norm for some plant life forms (annuals, herbaceous perennials, Harper 1981). Also, some rare species exist as a complex of metapopulations whose distribution and abundance change rapidly through time (Griggs and Jain 1983; Shaffer 1990; Menges 1990). Metapopulation structure means that short-term trends in population size and number do not adequately portray conservation status or recovery potential. (2) Significant change can be slow and difficult to detect over relatively short periods of time, especially with typical bureaucratic or fiscal constraints. Periods of five to 15 years or more might be required to document genuine trends, during which time employees leave, survey methods change, budgets are cut, and/or interest in the project fades. Even if bureaucratic and fiscal commitment remain high, species with large, natural variations in population size might be beyond hope of recovery by the time a significant reduction is detected. (3) Many rare plant taxa lack "replicate" populations that are needed to determine the significance of detectable change. (4) If change is detected from census data for reproductive individuals, its relationship to the population and to the species as a whole is obscure because seeds, seedlings, and juvenile plants might be better indicators of demographic change. (5) Without knowledge of the underlying biological processes that affect population stability, (e.g., survivorship, fecundity, duration in the seed bank), no meaningful management of the population can be designed, implemented or evaluated.

So why should we monitor rare plants? We should monitor to address a specific question regarding a specific conservation action. In some cases, the question can be related to conducting resource inventories ("which rare species are present on the site?") or to determining changes in risk factors for a particular species ("why is the status of this rare plant changing?"). Answering such questions involves inventory and survey, respectively, which could be regarded as rare plant monitoring in the broadest sense (Palmer 1987; Menges and Gordon 1994). For purposes of inventory and survey, it might be appropriate to count the number of rare plants in a population, to map distributions, to describe habitat relationships with vegetation cover estimates, and to gather many kinds of descriptive data. But if the questions fall into the realm of recovery—the realm of predicting trends and determining limiting factors so that populations can be enhanced or created—then monitoring should employ demographic techniques that follow the fates of individual plants *in situ* (Davy and Jefferies 1981; Travis and Sutter 1986; Pavlik and Barbour 1988; Pavlik 1994a).

The recovery of an endangered species should use demographic monitoring to determine (1) which populations are declining and require management (trend analysis) and (2) what factors are most responsible for the decline (factor resolution) and should be ameliorated (Pavlik 1994a). Although non-demographic forms of monitoring (census, photo surveys) provide a preliminary trend analysis, they might not distinguish trivial change from significant decline. Furthermore, the most effective management prescriptions will come from comparing the performance of individuals in growing as well as declining populations (Menges 1990; Pavlik et al. 1993). Does competition from weeds limit seed production and, therefore, population size of the target species? Do herbivores have a significant effect on seedling survivorship? Obviously, the answers to these questions can only come from field experiments which are well-designed and statistically-empowered. Monitoring, therefore, should have an experimental frame-work for testing a specific, management-oriented hypothesis. Hypothesis testing has taken a back seat to description in most conservation efforts (Freas and Murphy 1990), and I suggest this has been to the detriment of some critically endangered plant populations. Management of those populations, therefore, should be driven by the results of sound experimentation, using demographic monitoring to obtain relevant data in the shortest time possible.

The recovery of rare plant populations is, by its very nature, an experimental as well as a demographic venture (Pavlik 1994a,b). Descriptive or comparative studies can be used to select a reintroduction site, but those studies and selection criteria are not actually tested until propagules are released on site. Alternative management techniques can then be tested by following the fates and demographic performance of those propagules. Habitat manipulation, outplanting potted plants, sowing seeds, and other restoration techniques have population-level effects that must be quantified and evaluated to assess efficacy and success. If certain restoration techniques stabilize a declining population or create a new, self-sustaining population within historic range, then the recovery efforts have been successful. If not, then new techniques or approaches need to be implemented and tested to see if they are effective. In either case, the quality of decisions made during the recovery process depends on the pertinence and reliability of the data. Pertinence and reliability can be enhanced by focusing on empirical studies of population performance with demographic monitoring.

Demographic monitoring does have its drawbacks. Marking seedlings, measuring seed production, estimating mortality in the seed bank are more time-consuming and, therefore, more expensive than a rough census or new photo plot. A greater knowledge of plant ecology and statistics is required so that management personnel may not feel qualified to design, implement, and interpret a demographic program. These drawbacks can be minimized if demographic monitoring is simplified and structured to meet the specific needs of rare plant recovery (Menges and Gordon 1994; Pavlik 1994a). Ultimately, simplicity and structure could be expressed as guidelines for designing demographic monitoring programs based on life history and other important traits of the target taxon. Guidelines in the form of a flow chart or dichotomous key would enable managers to select a few, pertinent population parameters for addressing specific recovery problems. With a strong focus on pertinent data required to test a management-oriented hypothesis, the time spent on a demographic monitoring program is likely to be less than the time spent collecting trivial or possibly misleading data over many years with no positive outcome for the target species.

WHAT IS THE ROLE OF POPULATION VIABILITY ANALYSIS?

Population viability analysis or PVA (Gilpin and Soulé 1986) is a technique used to examine the dynamics of small populations from demographic characteristics measured in the field. With the matrix projection techniques of Leslie (1945) and Lefkovitch (1965), survivorship and fecundity data are integrated into a single, predictive model. PVA has been used on several plant species, including the endangered *Pedicularis furbishae*

(Menges 1990) and two endangered species of *Calochortus* (Fiedler 1987). Size-class (or life cycle stage-class) categories replace age-specific schedules, thus avoiding some of the problems imposed by life table analyses (Menges 1986, 1992). This kind of deterministic modeling is simplistic because demographic characteristics are held constant in the same population over time or between different populations of the same species. Sophistication can be added by sampling several populations during several years or by using stochastic simulation models to predict population trends under non-equilibrium conditions (Menges 1992).

Depending upon implementation, PVA can estimate finite rates of population increase, extinction probabilities, and minimum viable population sizes or model the effects of critically-limiting factors at the population level (sensitivity analysis). Under the right conditions, PVA is among the most powerful tools for meeting conservation objectives. Objectives that are specific to a single taxon or involve PVA on a few, representative taxa tend to be short-term, requiring less than five years to complete (Table 24.2). It is clear that PVA and other forms of demographic analysis (Pavlik 1994a) can now be used to meet short-term, taxon-specific objectives.

If PVA could be done on all of the rare taxa in the flora of a region, a set of long-term conservation goals could also be met. Among them is the determination of endangerment status of the rare plants in a flora, as suggested by Mace and Lande (1991). Taxa with the highest probability of extinction would receive the highest status for legal protection and conservation. This is both logical and desirable, but it is also impractical and unattainable for the following reasons:

(1) The sheer numbers of rare plant taxa, not to mention the numbers of populations of those taxa (minimum of 4 populations/taxon would have to be examined with PVA, Pavlik 1994a), are prohibitive. With so many plants in peril (Table 24.1), there is not enough time to determine the relative status of all rare plants using PVA, even in the smallest regional floras. Menges' (1990) PVA on *Pedicularis furbishae* obtained data on 15 populations during four years, but the preliminary field studies by Menges and his predecessors took several more years. This is fine for some highly endangered taxa, but not for a list of rare plants in a regional flora. As more and more PVAs become available,

Table 24.2. Conservation applications of PVA (population viability analysis). Short-term indicates a time frame of five years or less for PVA to meet objectives for one or a few taxa. Long-term indicates a time frame of 50+ years for PVA to meet objectives in a regional flora.

Taxon-specific, short-term objectives	Flora-specific, long-term objectives
Identifying populations for management (trend analysis)	Determining endangerment status and conservation priorities
Designing management/recovery actions (factor resolution)	Identifying regional preserves to maximize overall conservation of biodiversity
Designing a reserve for a taxon	Predicting regional levels of biodiversity under existing or changing environments
Investigating life history strategies and the most effective means of conserving different life forms	Providing landscape-level management strategies for self-maintenance of rare plant populations

some general patterns might emerge to guide conservation decisions. Hopefully, such patterns would depend on the life history, breeding system, or other important and easily observed traits. But the few existing PVAs for rare plants emphasize the population- and taxon- specific peculiarities associated with rarity and endangerment (e.g., Fiedler 1987; Menges 1990). If that is the case, possibly no model systems identified by PVA could be used to manage or conserve a wider spectrum of unanalyzed taxa.

(2) A large number of rare taxa have life history characteristics that do not lend themselves to PVA. Cryptic clonal growth (e.g., bulbils), long-distance seed dispersal, and extreme longevity in some size/stage classes (so that some transition probabilities during the study period cannot be measured) impose difficulties on the prediction of population viability. Of the 47 rarest plants in the Canterbury region of New Zealand, only about 12 (25%) had life histories that could be readily assessed with PVA (Norton et al. 1991). Species with ong-lived seeds are especially problematic because survivorship within the seed bank needs to be thoroughly assessed and incorporated into the projection matrices. This might exclude most annual species and many short-lived perennials and further limit the application of PVA. In California, these two life history traits are possessed by more than 80% of the rarest taxa (Pavlik and Skinner 1993). Menges (1986), however, suggested that the relevant seed bank data can be obtained from empirical studies in the field. Standard procedures need to be developed for circumventing the technical problems associated with PVA in its present form so that a wider variety of plant populations can be analyzed in this way.

(3) Estimates of minimum viable population size are sometimes based on genetic, rather than demographic analyses. The often cited figure of 500 individuals per population (also used by Mace and Lande 1991) has it origin in studies of bristle characteristics in homogeneous lines of *Drosophila* (Franklin 1980). Plants and many animals might not fit well into the fruit fly model. Menges (1992) reviewed some of the critical plant and animal differences and concluded that plants have (a) small neighborhood sizes, (b) extremely localized genetic differentiation (e.g., ecotypes), (c) greater tolerance of inbreeding and low intrapopulational genetic variation (the often-cited cheetah story might not be relevant), and (d) unusual life history/reproductive features (clonal growth, apomixis, hybridization, polyploidy) that complicate and partially invalidate purely genetic approaches to plant conservation.

Although PVA might apply to a limited number of situations, it can still be of use in rare plant recovery. PVA is one form of demographic monitoring that identifies populations in decline through trend analysis. PVA can also be used for determining the best management prescription through factor resolution. Trend analysis and factor resolution are the two principal activities of demographic monitoring for recovery (Pavlik 1994a). Menges' (1990) study of *Pedicularis furbishae* is the best example of how PVA has been applied to trend analysis and factor resolution on an endangered plant. Only four of the 15 populations examined along the St. John's River were found to be declining; thus, management could precisely focus on a small subset of the species. Declining populations were associated with high canopy cover by woody overstory species, a condition promoted by flood control along the river. Menges concluded that the recovery of *P. furbishae* would be favored by managing the riverine habitat for early seral species with a more natural flooding regime. Such a regime could also be mimicked by work crews thinning or removing the overstory canopy near populations of *P. furbishae*. In this case, PVA was able to identify the populations in need of management and, with some additional data on habitat, a specific management recommendation was formulated. But given the described limitations, managers will find PVA technically cumbersome and impractical under many circumstances. Other approaches to trend analysis and factor resolution presented elsewhere (Pavlik 1994a) are more likely to answer the same basic questions more efficiently and effectively for the management of a wider variety of endangered plant taxa.

RECOMMENDATIONS FOR IMPROVING RECOVERY EFFORTS

Recovery plans for endangered plants require that natural populations be protected within critical habitat so they can maintain themselves over long periods of time. Protected natural populations contain the subtle genetic and ecological characteristics that reflect the evolutionary history of a species. In extreme cases, recovery might also require the creation of new, self-sustaining populations within historic range and the enhancement of small, natural populations *in situ* (Knudsen 1987, Whitten 1990). Created populations reduce the probability of extinction and restore the species to a greater ecological potential. Enhanced natural populations are larger, less likely to lose their genetic variability, and less susceptible to small-scale perturbation.

No plant taxa have, in fact, been recovered and conserved *in situ* by enhancing or creating populations. Despite numerous projects that have attempted to protect (e.g., Elias 1987; CDFG 1991), enhance (Ferreira and Smith 1987; Pavlik and Manning 1993; DeMauro 1994), or create (e.g., Havlik 1987; Simonich and Morgan 1990, Reichenbacher 1990; Olwell et al. 1987, 1990; Parenti and Guerrant 1990; Pavlik et al. 1993) populations, the goal of recovery has remained elusive for a variety of reasons. Natural populations protected in small reserves often lack proactive, ecosystem management. Self-maintenance has rarely been demonstrated for natural, new, or enhanced populations of a target species in the wild. This is particularly true when recovery involves long-lived perennial plants and the use of transplantation to circumvent critical life history phases (Fahselt 1988). The issue of how much genetic variation should be built into a new plant population has often been ignored (Clegg and Brown 1983; Guerrant 1992). Recovery has also not been attained because propagules, appropriate habitat, time, effort, and money are usually in short supply.

Perhaps the most common deficiency, however, comes from a failure to design recovery projects as field experiments within a frame-work of demographic monitoring. Demographic monitoring is needed to detect short-term population trends and to analyze factors that limit population establishment, growth, or self-maintenance. Only experiments can test the predictions of population models such as PVA (Gilpin 1987) and verify an ecological understanding of rare taxa. If demographic monitoring is incorporated into the experiment, evaluation of critical limiting factors and habitat conditions becomes possible. Such an evaluation can determine the success of reintroduction or ecosystem management techniques and indicate other actions to hasten recovery (Harvey 1985; Menges 1986; Pavlik 1994a,b). Without an objective, rigorous measure of success or failure, it is unlikely that any recovery project can be brought to a sound, satisfactory conclusion.

When can one conclude that a plant species has recovered? Recovery plans specify the number and size of viable populations to be achieved, but the scientific justifications can be vague or lacking a theoretical basis. More work is needed to provide useful recovery criteria, especially because rare plant biology, endangerment factors, and legal protection are so varied. It is logical to suggest that the number of viable populations comprising a recovered species should in some way reflect the former area of distribution. Once wide-ranging taxa should have more populations than those that were naturally rare. Although the theory is only partially developed, five populations per taxon seems to be a minimum (see Falk and Holsinger 1991). Also the distribution of populations should correspond to the distribution of genetic variation in a species. Because widespread taxa tend to be more variable than narrowly-distributed taxa (Hamrick et al. 1979; Soltis and Soltis 1991; Karron 1992), the conserved populations of a geographically widespread, but endangered taxon must contain more genetic variation if population viability and evolutionary potential are to be maintained. Furthermore, localized extinction and recolonization are features of many species with complicated metapopulation dynamics, such as *Pedicularis furbishae* (Menges 1990) and *Circium pitcheri* (McEachern and Bowles 1994). If such dynamism is characteristic of a target species, it seems futile to manage every declining population. Species persistence over the entire landscape rather than population persistence at specific sites should be the management goal (Griggs and Jain 1983; Shaffer 1990; Menges 1990).

How can population viability be defined for purposes of recovery? Models that generate extinction probabilities under a variety of demographic and environmental scenarios are providing valuable insights, but relatively few species have been examined with such models. Minimum viable population sizes have been derived for a few animals, ranging between 50 and 2,500 individuals (Mace and Lande 1991). Given (1994) has suggested 500 as a first approximation of minimum viable population sizefor most plant species. However, annual species with outcross mating systems or unusual life history traits or habitat requirements may require larger populations (>1,000) to dampen the effects of environmental and demographic stochasticity (Pavlik 1994b). Long-lived, self-compatible species, on the other hand, might not require large numbers of individuals (<500) to maintain viable populations (Soulé and Simberloff 1986; Reznicek 1987; Lesica and Allendorf 1992).

Until sophisticated models become available for a wider variety of plants, recovery of most species should proceed so that extinction in the near-term is prevented. Perhaps successful recovery should be based on the *in situ* restoration of critical demographic processes in populations of target species, rather than an attainment of a specific, minimum population size (Whitson and Massey 1981; Soulé 1987; Pavlik 1994a). For most plants, these processes include seed or ramet production, dispersal into high quality habitat, germination, and survival to reproductive maturity. All of these can be manipulated to some degree, and all such manipulations can be evaluated for success (whether or not the population has grown or stabilized *and* whether or not all critical demographic processes have been restored) and cost effectiveness (population increase per unit of time, money, and effort). Such an approach was recently proposed for a large number of Britain's most endangered vascular plants (Whitten 1990).

However, approximately 70-80% of endangered plant taxa are composed of five or fewer populations (Brown and Briggs 1991; Guerrant 1992). It will often be necessary, therefore, to create new, self-sustaining populations to decrease the overall risk of extinction. There have been few attempts to create new populations of endangered plants *in situ*, and few of those incorporated demographic monitoring and population genetics into an empirical frame-work allowing for some evaluation of success or cost effectiveness. Standard but flexible protocols for creating populations need to be developed to address demographic and genetic issues. Furthermore, a number of important, practical matters remain unresolved. Do we attempt to create new populations of naturally rare taxa or only of those declining because of human disturbance? How ecologically similar does a reintroduction site have to be to the existing natural site(s)? Should propagules from different sources be mixed to maximize the initial genetic variability of the inoculum? When should transplants from greenhouse propagation be placed *in situ* instead of seeds? Resolving or at least debating these questions can lead to better approaches for creating new populations and successfully meeting the challenge of recovery.

CONCLUSIONS

The challenge of recovery can only be met if rare plant populations and their ecosystems are scientifically managed . Increasingly, those populations and ecosystems will be set aside on smaller-than-optimal reserves bordering on intensively developed lands. The smaller the reserve, the more management will be required, and the financial and legal responsibilities for that management need to be clearly assigned. Small-scale experimentation is an essential component of the management process and demographic monitoring allows managers to address a specific question whose answer will allow a specific recovery action to be taken. Monitoring rare plant populations is, therefore, a frame-work built around a decision-making process—it is never an end unto itself, nor should it be used only to detect change. Population viability analysis (PVA) is one, particularly powerful, form of demographic monitoring that addresses these short-term, taxon-specific questions. Successful recovery projects will be those that rigorously demonstrate an ongoing ability of the target taxon to complete its critical life history processes in natural, enhanced and created populations throughout historic range.

Several management-oriented, technical procedures must be developed so that recovery becomes a standard, rather than exceptional, component of plant conservation programs. The procedures are suggested by the present analysis of small reserves, demographic monitoring, PVA, and the enhancement or creation of new populations for purposes of recovery. These include; (1) practical but effective replication of natural disturbance factors on a small scale if early seral conditions are required by a target taxon, (2) effective and environmentally-benign weed or herbivore control methods, (3) guidelines for designing simple demographic monitoring programs based on life history and other important traits of the target taxon, (4) modification of PVA methods so that the population dynamics of a wider variety of rare plants could be assessed in this way, (5) field tests of PVA and other models that suggest minimum viable population sizes and the minimum number of populations per taxon, (6) methods for genetically formulating a founding population, and (7) a critical distillation of success criteria based on conservation theory and empirical studies from a wide array of rare plants.

ACKNOWLEDGMENTS

I am grateful to Ken Berg, Ann Howald, and Sandy Morey (Endangered Plant Program, California Department of Fish and Game) for their support of my research on endangered plants. The manuscript was much improved by comments from three anonymous reviewers.

LITERATURE CITED

Ayensu, E. S., and R. A. DeFilipps. 1978. *Endangered and Threatened Plants of the United States.* Smithsonian Institution Press, Washington, D.C.

Bartel, J. A. 1987. The federal listing of rare and endangered plants: What is involved and what does it mean? In *Conservation and Management of Rare and Endangered Plants,* ed. T. Elias, pp. 15-22. California Native Plant Society, Sacramento, Calif.

Bean, M. J. 1983. *The Evolution of National Wildlife Law.* Praeger Publishers, New York.

Bradshaw, M. E. 1981. Monitoring grassland plants in Upper Teesdale, England. In *The Biological Aspects of Rare Plant Conservation,* ed. H. Synge, pp. 241-251. John Wiley and Sons, London.

Brown, A. H. D., and J. D. Briggs. 1991. Sampling strategies for genetic variation in *ex situ* collections of endangered plant species. In *Genetics and Conservation of Rare Plants,* eds. D. A. Falk and K. E. Holsinger, pp. 99-119. Oxford University Press, Oxford.

Clegg, M. T., and A. D. H. Brown. 1983. The founding of plant populations. In *Genetics and Conservation: A Reference for Managing Wild Animal and Plant Populations,* eds. C. M. Schonewald-Cox, S. M. Chambers, B. MacBryde and L. Thomas, pp. 216-228. Benjamin Cummings, London.

California Natural Diversity Data Base (CNDDB). 1992. Unpublished data on file. California Department of Fish and Game, Sacramento, Calif.

Cochrane, S. 1992. The outlook for legal recognition and protection of rare plant communities (Abstract). Paper presented at AAAS Pacific Division Meeting, 24 June, 1992, Santa Barbara, Calif.

Center for Plant Conservation (CPC). 1988. CPC survey reveals 690 native U.S. plants may become extinct within 10 years. *Plant Conserv.* 3(4):1.

Davy, A. J., and R. L. Jefferies. 1981. Approaches to the monitoring of rare plant populations. In *The Biological Aspects of Rare Plant Conservation,* ed. H. Synge, pp. 219-232. John Wiley and Sons, London.

California Department of Fish and Game (CDFG). 1991. 1991 annual report on the status of California's State Listed Threatened and Endangered Plants and Animals. State of California, The Resources Agency, Sacramento, Calif.

DeMauro, M. 1994. Development and Implementation of a recovery program for the federally threatened Lakeside daisy (*Hymenoxys acaulis* var. *glabra*). In *Recovery and Restoration of Endangered Species,* eds. M. C. Bowles and C. Whelan. Cambridge University Press, Cambridge (in press).

Diamond, J. M., and R. M. May. 1976. Island biogeography and the design of nature reserves. In *Theoretical Ecology: Principles and Applications,* ed. R. M. May, pp. 163-186. W. B. Saunders, Philadelphia, Pa.

Ehrenfeld, D. 1988. Why put a value on biodiversity? In *Biodiversity,* ed. E. O. Wilson, pp. 212-216. National Academy Press, Washington, D.C.

Elias, T., ed. 1987. *Conservation and Management of Rare and Endangered Plants.* California Native Plant Society, Sacramento, Calif.

Fahselt, D. 1988. The dangers of transplantation as a conservation technique. *Natural Areas J.* 8(4):238-244.

Falk, D. A., and K. E. Holsinger, eds. 1991. *Genetics and Conservation of Rare Plants.* Oxford University Press, New York.

Ferreira, J., and S. Smith. 1987. Methods of increasing native populations of *Erysimum menziesii.* In *Conservation and Management of Rare and Endangered Plants,* ed. T. Elias, pp. 507-511. California Native Plant Society, Sacramento, Calif.

Fiedler, P. L. 1987. Life history and population dynamics of rare and common mariposa lilies (*Calochortus* Pursh: Liliaceae). *J. Ecol.* 75:977-995.

Freas, K. E., and D. M. Murphy. 1990. Hypothesis-based monitoring and management for endangered species recovery. Abstract presented in: Symposium on Recovery and Restoration of Endangered Species (M. Bowles, coordinator), Society for Ecological Restoration, Chicago, Ill.

Franklin, I. R. 1980. Evolutionary change in small populations. In *Conservation Biology: An Evolutionary-Ecological Perspective,* eds. M. E. Soulé and B. A. Wilcox, pp. 134-150. Sinauer Associates, Sunderland, Mass.

Gilpin, M. E. 1987. Minimum viable populations: A restoration ecology perspective. In *Restoration Ecology: A Synthetic Approach to Ecological Research,* eds. W. R. Jordan III, M. E. Gilpin, and J. D. Aber, pp. 301-305. Cambridge University Press, Cambridge.

Gilpin, M. E., and M. E. Soulé. 1986. Minimum viable populations and processes of species extinctions. In *Conservation Biology: The Science of Scarcity and Diversity*, ed. M. E. Soulé, pp. 19-34. Sinauer Associates, Sunderland, Mass.

Given, D. R. 1994. *Principles and Practice of Plant Conservation.* IUCN/WWF, Gland, Switzerland, in press.

Griggs, F. T., and S. K. Jain. 1983. Conservation of vernal pool plants in California, II. Population biology of a rare and unique grass genus, *Orcuttia. Biol. Conserv.* 27:171-193.

Guerrant, E. O. 1992. Genetic and demographic considerations in rare plant conservation. In *Conservation Biology: The Theory and Practice of Nature Conservation, Preservation, and Management,* eds. P. L. Fiedler and S. Jain, pp. 321-346. Chapman and Hall, New York.

Harper, J. L. 1981. The meanings of rarity. In *The Biological Aspects of Rare Plant Conservation,* ed. H. Synge, pp. 189-204. John Wiley and Sons, London.

Harvey, H. J. 1985. Population biology and the conservation of rare species. In *Studies on Plant Demography. A Festschrift for John L. Harper,* ed. J. White, pp. 111-123. Academic Press, London.

Hamrick, J. L., Y. B. Linhart, and J. B. Mitton. 1979. Relationships between life history characteristics and electrophoretically-detectable genetic variation in plants. *Ann. Rev. Ecol. Syst.* 10:173-200.

Hastey, E. 1993. Rare plant conservation on Bureau of Land Management Lands. In *California Native Plant Society's Inventory of Rare and Endangered Vascular Plants of California.* 5th ed., eds. M. Skinner and B. M. Pavlik. California Native Plant Society Special Publication No. 1, Sacramento, Calif.

Havlik, N. A. 1987. The 1986 Santa Cruz Tarweed relocation project. In *Conservation and Management of Rare and Endangered Plants,* ed. T. Elias, pp. 421-423. California Native Plant Society, Sacramento, Calif.

Howald, A. M., and L. P. Wickenheiser. 1990. Mitigation plan annotated outline for endangered plants of California. California Department of Fish and Game, Sacramento, Calif.

International Union for the Conservation of Nature/World Wildlife Fund (IUCN/WWF). 1988. Plant Conservation Programme. IUCN, Gland, Switzerland.

Jensen, D. B. 1987. Concepts of preserve design: What we have learned. In *Conservation and Management of Rare and Endangered Plants,* ed. T. Elias, pp. 595-603. California Native Plant Society, Sacramento, Calif.

Jensen, D. B., M. Torn, and J. Harte. 1990. In Our Own Hands: A Strategy for Conserving Biological Diversity in California. California Policy Seminar Research Report, University of California, Berkeley, Calif.

Karron, J. D. 1992. Patterns of genetic variation and breeding systems in rare plant species. In *Conservation Biology: The Theory and Practice of Nature Conservation, Preservation, and Management,* eds. P. L. Fiedler and S. Jain, pp. 87-98. Chapman and Hall, New York.

Knudsen, M. D. 1987. Recovery of endangered and threatened plants in California: The federal role. In *Conservation and Management of Rare and Endangered Plants,* ed. T. Elias, pp. 461-469. California Native Plant Society, Sacramento, Calif.

Lahti, T., and E. Ranta. 1985. The SLOSS principle and conservation practice: an example. *Oikos* 44:369-387.

Lahti, T., and E. Ranta. 1986. Island biogeography and conservation: a reply to Murphy and Wilcox. *Oikos* 47:388-389.

Lefkovitch, L. P. 1965. The study of population growth in organisms grouped by stages. *Biometrics* 21:1-18.

Lesica, P., and F. W. Allendorf. 1992. Are small populations of plants worth preserving? *Conserv. Biol.* 6:135-139.

Leslie, P. H. 1945. On the use of matrices in certain population mathematics. *Biometricka* 33:183-212.

Mace, G. M., and R. Lande. 1991. Assessing extinction threats: Toward a reevaluation of IUCN threatened species categories. *Conserv. Biol.* 5:148-157.

McEachern, K., and M. Bowles. 1994. Recovery planning for the threatened Great Lakes Thistle *Circium pitcheri* according to a metapopulation model. In *Recovery and Restoration of Endangered Species,* eds. M. C. Bowles and C. Whelan. Cambridge University Press, Cambridge (in press).

Menges, E. S. 1986. Predicting the future of rare plant populations: demographic monitoring and modeling. *Natural Areas J.* 6(3):13-25.

Menges, E. S. 1990. Population viability analysis for an endangered plant. *Conserv. Biol.* 4:52-62.

Menges, E. S. 1992. Stochastic modeling of extinction in plant populations. In *Conservation Biology: The Theory and Practice of Nature Conservation, Preservation, and Management,* eds. P. L. Fiedler and S. Jain, pp. 44-61. Chapman and Hall, New York.

Menges, E. S., and D. R. Gordon. 1994. Three levels of monitoring intensity for rare plant species. *Ecol. Appl.* (in press).

Myers, N. 1983. *A Wealth of Wild Species: Storehouse for Human Welfare.* Westview Press, Boulder, Colo.

Nelson, J. R. 1984. Rare plant surveys: Techniques for impact assessment. *Fremontia* 12(3):19-23.

Norton, D. A., D. R. Given, and B. M. Pavlik. 1991. Revisiting rarity—a botanical perspective on rarity and extinction. Paper presented to New Zealand Botanical Society Conference, 23 November 1991, Auckland, New Zealand.

Olwell, P., A. Cully, and P. Knight. 1990. The establishment of a new population of *Pediocactus knowltonii:* Third Year Assessment. In *Ecosystem Management: Rare Species and Significant Habitats,* eds. R. S. Mitchell, C. J. Sheviak, and D. J. Leopold, pp. 189-193. New York State Museum Bull. 471.

Olwell, P., A. Cully, P. Knight and S. Brack. 1987. *Pediocactus knowltonii* recovery efforts. In *Conservation and Management of Rare and Endangered Plants,* ed. T. Elias, pp. 519-522. California Native Plant Society, Sacramento, Calif.

Palmer, M. E. 1987. A critical look at rare plant monitoring in the United States. *Biol. Conserv.* 39:113-127.

Parenti, R. L., and E. O. Guerrant, Jr. 1990. Down but not out: Reintroduction of the extirpated Malheur wirelettuce, *Stephanomeria malheurensis.* In *Endangered Species Update* 8:62-63.

Pavlik, B. M. 1994a. Demographic monitoring and the recovery of endangered plants. In *Recovery and Restoration of Endangered Species,* eds. M. C. Bowles and C. Whelan. Cambridge University Press, Cambridge (in press).

Pavlik, B. M. 1994b. A framework for defining and measuring success in rare plant reintroductions. In *Restoring Diversity; Ecological Restoration and Endangered Species,* eds. D. A. Falk, P. Olwell, and C. Millar. Island Press, Washington, D.C. (in press).

Pavlik, B. M., and M. G Barbour. 1988. Demographic monitoring of endemic sand dune plants, Eureka Valley, California. *Biol. Conserv.* 46:217-242.

Pavlik, B. M., and E. Manning. 1993. Assessing limitations on the growth of endangered plant populations, I. Experimental demography of *Erysimum capitatum* ssp. *angustatum* and *Oenothera deltoides* ssp. *howellii. Biol. Conserv.* 65:257-265.

Pavlik, B. M., D. Nickrent, and A. M. Howald. 1993. The recovery of an endangered plant. I. Creating a new population of *Amsinckia grandiflora. Conserv. Biol.* 7: 510-526.

Pavlik, B. M., and M. W. Skinner. 1993. Ecological characteristics of California's rare plants. In *California Native Plant Society's Inventory of Rare and Endangered Vascular Plants of California.* 5th ed., eds. M. W. Skinner and B. M. Pavlik. California Native Plant Society Special Publication No. 1, Sacramento, Calif.

Prescott-Allen, R., and C. Prescott-Allen. 1990. How many plants feed the world? *Conserv. Biol.* 4:365-374.

Reichenbacher, F. W. 1990. Reintroduction brings Kearney's Blue Star from extinction's edge. *Plant Conserv.* 5(1):3.

Reznicek, A. A. 1987. Are small reserves worthwhile for plants? *Endangered Species Update* 5(2):1-3.

Robinson, G. R., and J. F. Quinn. 1992. Habitat fragmentation, species diversity, extinction, and design of nature reserves. In *Applied Population Biology,* eds. S. K. Jain and L. W. Botsford, pp. 238-248. Kluwer Academic Publishers, the Netherlands.

Rohlf, D. J. 1991. Six biological reasons why the Endangered Species Act doesn't work—and what to do about it. *Conserv. Biol.* 5:273-282.

Saunders, D. A., R. J. Hobbs, and C. R. Margules. 1991. Biological consequences of ecosystem fragmentation: A review. *Conserv. Biol.* 5:18-24.

Shaffer, M. L. 1990. Population viability analysis. *Conserv. Biol.* 4:39-40.

Simberloff, D., and L. G. Abele. 1976a. Island biogeography and conservation practice. *Science* 193:285-286.

Simberloff, D., and L. G. Abele. 1976b. Island biogeography and conservation: Strategy and limitations. *Science* 193:1032.

Simberloff, D., and N. Gotelli. 1984. Effects of insularization on plant species richness in the prairie-forest ecotone. *Biol. Conserv.* 29:27-46.

Simonich, M. T., and M. D. Morgan. 1990. Researchers successful in transplanting dwarf lake iris ramets (Wisconsin). *Restoration and Manage. Notes* 8(2):131-132.

Skinner, M. W., and B. M. Pavlik, eds. 1993. *California Native Plant Society's Inventory of Rare and Endangered Vascular Plants of California.* 5th ed. California Native Plant Society Special Publication No. 1, Sacramento, Calif.

Soltis, P. S., and D. E. Soltis. 1991. Genetic variation in endemic and widespread plant species: Examples from Saxifragaceae and *Polystichum* (Dryopteridaceae). *Aliso* 13:215-223.

Soulé, M. E. 1987. Introduction. In *Viable Populations for Conservation*, ed. M. E. Soulé, pp. 1-10. Cambridge University Press, Cambridge.

Soulé, M. E., and D. Simberloff. 1986. What do genetics and ecology tell us about the design of nature reserves? *Biol. Conserv.* 35:19-40.

Sutter, R. D. 1986. Monitoring rare plant species and natural areas—ensuring the protection of our investment. *Natural Areas J.* 6(1):3-5.

Taylor, D. W. 1992. Biodiversity patterns and formulation of restoration standards for vernal pools in the Sacramento Valley, California. (Abstract). Paper presented at Southern California Academy of Sciences Meeting, 2 May 1992, Los Angeles, Calif.

Terborgh, J. 1976. Island biogeography and conservation: Strategy and limitations. *Science* 193:1029-1030.

Topik, C. 1991. Forest Service and The Nature Conservancy join forces to conserve biological diversity. *Endangered Species Bull.* 16(4):8.

Travis, J., and R. Sutter. 1986. Experimental designs and statistical methods for demographic studies of rare plants. *Natural Areas J.* 6(3):3-12.

Whitson, P. D., and J. R. Massey. 1981. Information systems for use in studying population status of threatened and endangered plants. In *Rare Plant Conservation: Geographical Data Organization,* eds. L. E. Morse and M. S. Henifin, pp. 217-236. New York Botanical Garden, Bronx, New York.

Whitten, A. J. 1990. Recovery: A proposed programme for Britain's protected species. Nature Conservancy Council, CSD Report No. 1089.

Wilcox, B. A., and D. D. Murphy. 1985. Conservation strategy: The effects of fragmentation on extinction. *Am. Nat.* 125:879-887.

Wilson, E. O. 1988. The current state of biodiversity. In *Biodiversity*, ed. E. O. Wilson, pp. 3-17. National Academy Press, Washington, D.C.

25

CONSERVATION OF DESERT FISHES: SPAIN, AUSTRALIA, AND THE UNITED STATES

John N. Rinne

Rocky Mountain Forest and Range Experiment Station
Southwestern Forestry Sciences Complex
Northern Arizona University
Flagstaff, Arizona
U.S.A.

Walter Ivantsoff and L. E. M. Crowley

School of Biological Sciences, Macquarie University
South Australian Museum
Adelaide
Australia

Javier Lobón-Cervia

Department of Evolutionary Ecology
National Museum of Natural Sciences (CSIC)
Madrid
Spain

Desert areas are characterized by reduced rainfall and few permanent-flowing streams. By definition, these environments appear to be unlikely habitats for fishes. Because streams are shallow and air temperature is high, water temperatures frequently exceed 30°C and can exceed 40°C in the daytime hours (Deacon and Minckley 1974; Naiman and Soltz 1981). Evaporation can elevate salinities to levels several times greater than seawater (Soltz and Naiman 1978). Elevated water temperatures, accompanied by cyclic changes in photosynthesis and respiration, cause oxygen levels may fluctuate widely during any 24-hour period and to approach zero. Streamflow can change from intermittent stagnant pools to raging torrents within seconds (Rinne and Minckley 1991).

Desert fish faunas are commonly characterized by low diversity and high endemicity (Sostoa et al. 1984; Minckley and Deacon 1991; Rinne and Minckley 1991). Species have adopted strategies to survive cycles of flood, drought, and rapidly changing conditions of water quality (Deacon and Minckley 1974; Lob_n-Cervia et al. 1991). Ever-increasing human populations of desert areas in large cities such as Madrid, Spain, Perth, Australia, Phoenix, Arizona, and Las Vegas, Nevada have placed demands on the basic medium which is both at a premium and in which fishes precariously survive—water. Conservation of desert fishes is difficult because of demands for water for agricultural irrigation, power generation, industrial development, and human consumption.

This paper collectively addresses (1) diversity and endemic nature of fishes in three desert countries on three continents in the world, (2) status of endangered faunas, (3) major reasons for loss of diversity, and (4) past, present, and potential future efforts to conserve desert fishes.

DIVERSITY OF FAUNAS

Spain

The fish fauna of Spain is highly endemic and dominated by cyprinids (Table 25.1, Figure 25.1A). Twenty-six of 42 (62%) of those taxa native to Spain are endemic (Sostoa et al. 1984; Icona 1986; Rincon et al. 1990). Twenty-two of 24 native cyprinids are endemic. The majority (22 of 26, 85%) of endemic native taxa of fishes are cyprinids (51% of all native taxa); only two cyprinodontids and cobitids each are likewise native and endemic. Compared to more mesic areas of central and northern Europe, the diversity of fishes in Spain is very low. There is also a general lack of native predatory fishes in the waters of Spain. Only the migratory eel (*Anguilla anguilla*) and the native brown trout (*Salmo trutta*) are piscivores.

More than a dozen species of fishes have been introduced into the waters of Spain (Sostoa et al. 1984; Elvira 1990; Table 25.1). Most of these are large, principally predatory species which have been stocked in the 20th century for sport fishing in the many artificial reservoirs (see below).

Australia

Although the freshwater fish fauna of Australia (Allen 1989) contains 180 species in 30 families, the desert regions (Figure 25.1B) are occupied by only 27 species of native fishes in 12 families (Glover 1982, 1990; Table 25.2). The only cyprinids are the introduced common goldfish and carp. The Lake Eyre drainage contains the greatest number (27) of desert fish species; almost one-half (13) are found only in this drainage basin. All eight native species in the Western Plateau drainages are common to those present in the Lake Eyre system. The Bulloo-Bancannia system of drainages contains seven species; three are common to the above two drainage systems, one to the Lake Eyre system alone, and three endemic to this arid land drainage system. Almost a dozen taxa (40%) of fishes in these three desert areas are considered endangered, vulnerable, or restricted (Harris 1987; Jackson 1992).

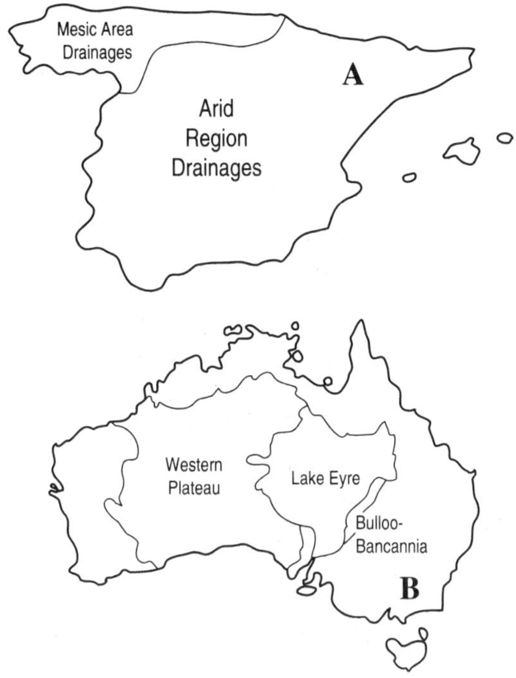

Figure 25.1. Map showing general location of arid, desert areas of Spain (A) and Australia (B). The endorheic Lake Eyre, Western Plateau, and Bulloo-Bacannia arid land drainages.

Five species belonging to three families (Poecilidae, Cyprinidae, and Percidae, Table 25.2) have been introduced into Australian waters either intentionally or by natural dispersion through periodic flooding. Similar to the occurrence of only two native predators in the desert waters of Spain, large, native predators are generally lacking in Australian waters.

United States

Diversity of fishes in deserts of the western United States is low (Pister 1981; Sigler and Sigler 1987; Rinne and Minckley 1991; Table 25.3, 25.4). Most fish taxa in deserts in western North America are of recent origin (Smith 1981). Three major deserts—the Sonoran, Mohave, and Great Basin—comprise the majority of arid landscapes of the western Unites States (Figure 25.2). These deserts are characterized by endemic fish faunas inhabiting aquatic systems ranging from major river systems, through endorehic basins, to minute springs and spring runs obscured within vast desert landscapes.

Sigler and Sigler (1987) list 91 taxa of fishes as established in the arid Great Basin area (Table 25.3). Of these, 44 (48%) have been introduced, mostly for sport fishing, as bait fishes, or as a result of the aquarium trade. Similar to the deserts of Spain and Australia, the Cypriniformes comprise a major component (37 of 47, 79%) of the native

Figure 25.2. Map of the United States indicating general location of the Great Basin, Mohave, and Sonoran desert areas of southwestern United States and northwestern Mexico.

fauna. Livebearers and killifish (Atheriniformes) comprise the next largest group (14 species), however, one-half are introduced species. The absence of large predators, aside from taxa of salmonids in upper elevation, more mesic areas, is evident. The Great Basin area ranks first in the United States in extinction of species of fishes with more than a dozen taxa having been lost in the last century in this arid region (Miller et al. 1989).

The fish fauna of the Sonoran Desert region (Figure 25.2B, Table 25.4) displays a pattern of diversity similar to that of Great Basin deserts (Table 25.3). Faunal diversity, endemicity, and conservation measures of this arid region parallel those of other western North American deserts. Because of our familiarity with this area it will be used to discuss the overall nature and conservation status of desert fishes in the western United States.

Table 25.1. List of native and introduced fishes of Spain, 1992.

Family / Species	Common name	
Petromyzontidae		
Lampetra fluviatilis	European river lamprey	N[1]
Lampetra planeri	European brook lamprey	N
Petromyzon marinus	Sea lamprey	N
Acipenseridae		
Acipenser sturio	Atlantic sturgeon	N
Anguillidae		
Anguilla anguilla	Common eel	N
Clupeidae		
Alosa alosa	Allis shad	N
Alosa fallax	Twaite shad	N
Salmonidae		
Hucho hucho	Danube sturgeon	I
Salmo salar	Atlantic salmon	N
Salmo gairdneri	Rainbow trout	I
Salmo trutta fario	Brown trout	N
Salmo trutta trutta	Brown trout	N
Salvelinus fontinalis	Brook trout	I
Salvelinus namaycush	Lake trout	I
Esocidae		
Esox lucius	Pike	I
Cyprinidae		
Barbus bocagei	Barbel	N-E
Barbus comiza	Barbel	N-E
Barbus graellsi	Barbel	N-E
Barbus haasi	Barbel	N-E
Barbus meridionalis	Barbel	N-E
Barbus microcephalus	Barbel	N-E
Barbus sclateri	Barbel	N-E
Barbus steindachnen	Barbel	N-E
Carassius auratus	Gold fish	I
Cyprinus carpio	European carp	I
Chondrostoma lusitanicum	Nase	N-E
Chondrostoma polylepis polylepis	Nase	N-E
Chondrostoma polylepis willkommii	Nase	N-E
Chondrostoma toxostoma arrigones	Nase	N-E
Chondrostoma toxostoma toxostoma	Nase	N-E
Gobio gobio	Gudgeon	I
Iberocypris palaciosi		N-E
Leuciscus cephalus cabeda	Chub	N-E
Leuciscus cephalus cephalus	Chub	N-E
Leuciscus cephalus pyrenaicus	Chub	N-E
Leuciscus carolitartii	Chub	N-E
Phoxinellus hispanicus	Minnow	N-E
Phoxinus phoxinus	Minnow	N
Rutilus alburnoides	Grey-roach	N-E
Rutilus arcasii	Red-roach	N-E
Rutilus lemmningii lemmingii	Roach	N-E
Scardinius erythrophthalmus	Rudd	I
Tinca tinca	Tench	N

continued

Table 25.1.—*continued*

Family / Species	Common name	
Cobitidae		
Cobitis calderoni	Calderois loach	N-E
Cobitis paludicola	Loach	N-E
Noemacheilus barbatulus	Stone loach	N
Ictaluridae		
Ictalurus melas	Black bullhead	I
Cyprinodontidae		
Aphanius iberus	Fartet	N-E
Valencia hispanica	Sakarug	N-E
Poeciliidae		
Gambusia affinis	Western mosquitofish	I
Atherinidae		
Atherina boyeri	Sand smelt	N
Gasterosteidae		
Gasterosteus aculeatus	Stickleback	N
Cottidae		
Cottus gobio	Bull head	N
Centrarchidae		
Lepomis gibbosus	Pumpkinseed	I
Micropterus salmoides	Largemouth bass	I
Cichlidae		
Cichlasoma facetum	Cichlid	I
Blenniidae		
Blennius fluviatilis	Freshwater blennoid	N

[1]N = Native; I = Introduced; N-E = Native endemic.

Table 25.2. List of fishes from drainages of three arid regions of Australia (Figure 25.1B). All are native, except three introduced species (I).

Family / Species	Common name	
Clupeidae		
Nematolosa erebi	Bony bream	
Retropinnidae		
Retropinna semoni	Australian smelt	
Plotosidae		
Porochilus argentius	Silver tandan	
Neosilurus hyrtlii	Hyrtl's tandan	
Neosilurus sp.1	Dalhousie catfish	
Neosilurus sp.2	Cooper's Creek tandan	
Neosilurus glencoensis	Tellow tandan	
Neosilurus sp.3	Barcoo catfish	
Poeciliidae		
Gambusia affinis	Western mosquitofish	
Gambusia dominicensis	Domingo mosquitofish	I
Gambusia holbrooki	Eastern mosquitofish	

continued

Table 25.2.—*continued*

Family Species	Common name	
Melanotaeniidae		
Melanotaenia splendida tatei	Desert rainbowfish	
Atherinidae		
Craterocephalus eyresii	Lake Eyre hardyhead	
Craterocephalus centralis	Finke River hardyhead	
Craterocephalus dalhousiensis	Dalhousie hardyhead	
Craterocephalus gloveri	Glover's hardyhead	
Pseudomugilidae		
Scaturiginichthys vermielipinnis	Red finned blue-eye	
Ambassidae		
Denariusa bandata	Pennyfish	
Ambassis mulleri	Mueller's glassfish	
Pecrcichthyidae		
Macquaria sp.	Callop	
Macquaria ambigan	Yellowbelly	
Theraponidae		
Leiopotherapon unicolor	Spangled perch	
Amniataba percoides	Banded grunter	
Bidyanus welchi	Welch's perch	
Scortum barcoo	Barcoo grunter	
Eleotridae		
Mogurnda mogurnda	Purple spotted gudgeon	
Mogurnda sp.	Flinders Range gudgeon	
Hypseleotris klunzingeri	Western carp gudgeon	
Hypseleotris sp.1	Barcoo gudgeon	
Gobiidae		
Chlamydogobius eremius	Desert goby	
Chlamydogobius sp.1	Dalhousie goby	
Chlamydogobius sp.2	Elizabeth Springs goby	
Cyprinidae		
Cyprinus carpio	Common carp	I
Carassius auratus	Common goldfish	I
Percidae		
Percia fluviatolis	English perch	I

Table 25.3. Fishes of the Great Basin area (Figure 25.2A) modified from Sigler and Sigler (1987). Listed (i.e., threatened and endangered) fishes are designated by "L."

Family	Species	Common name	
Salmonidae			
	Oncorhynchus nerka	Sockeye salmon (Kokanee)	I
	Prosopium abyssicola	Bear Lake whitefish	N
	Prosopium gemmiferum	Bonneville cisco	N
	Prosopium spilonotus	Bonneville whitefish	N
	Prosopium williamsoni	Mountain whitefish	N
	Salmo aguabonita	Golden trout	I-L
	Salmo clarki	Cutthroat trout	N
	Salmo clarki henshawi	Lahontan cutthroat trout	N-L
	Salmo clarki seleniris	Paiute cutthroat trout	N-L
	Salmo clarki utah	Bonneville cutthroat trout	N-L
	Salmos gairdneri	Rainbow trout	I
	Salmo trutta	Brown trout	I
	Salvelinus fontinalis	Brook trout	I
	Salvelinus namaycush	Lake trout	I
	Thymallus arcticus	Arctic grayling	I
Esocidae			
	Esox lucius	Northern pike	I
Cyprinidae			
	Acrocheilus alutaceus	Chiselmouth	
	Carassius auratus	Goldfish	I
	Ctenopharyngodon idella	Grass carp	I
	Cyprinus carpio	Common carp	I
	Eremichthys acros	Desert dace	N-L
	Gila alvordensis	Alvord chub	N-L
	Gila atraria	Utah chub	N
	Gila bicolor	Tui chub	N-L
	Gila boraxobius	Borax Lake chub	N-L
	Gila copei	Leatherside chub	N
	Gila orcutti	Arroyo chub	I
	Gila robusta jordani	Pahranagat roundtail chub	N-L
	Hesperoleucus symmetricus	California roach	I
	Iotichthys phlegethontis	Least chub	N-L
	Lepidomeda albivallis	White River spinedace	N-L
	Lepidomeda mollispinus pratensis	Big Spring spinedace	N-L
	Notemigonus crysoleucas	Golden shiner	I
	Notropis hudsonius	Spottail shiner	I
	Orthodon microlepidotus	Sacramento blackfish	I
	Pimephales promelas	Fathead minnor	I
	Ptychocheilus oregonensis	Northern squawfish	I
	Relictus solitarius	Relict dace	N
	Rhinichthys cataractae	Longnose dace	N-L
	Rhinichthys osculus	Speckled dace	N-L
	Richardsonius balteatus	Redside shiner	N
	Richardsonius egregius	Lahontan redside	N

continued

Table 25.3. —*continued*

Family Species	Common name	
Catostomidae		
Catostomus ardens	Utah sucker	N
Catostomus clarki	Desert sucker	N
Catostomus clumbianus	Bridgelip sucker	N
Catostomus discobolus	Bluehead sucker	N
Catostomus fumeiventris	Owens sucker	N
Catostomus macrocheilus	Largescale sucker	N
Catostomus platyrhynchus	Mountain sucker	N-L
Catostomus tahoensis	Tahoe sucker	N
Catostomus warnerensis	Warner sucker	N-L
Chasmistes cujus	Cui-ui	N-L
Chamistes liorus	June sucker	N-L
Ictaluridae		
Ictalurus catus	White catfish	I
Ictalurus melas	Black bullhead	I
Ictalurus natalis	Yellow bullhead	I
Ictalurus nebulosus	Brown bullhead	I
Ictalurus punctatus	Channel catfish	I
Loricariidae		
Hypostomus plecostomus	Suckermouth catfish	I
Cyprinodontidae		
Crenichthys baileyi	White River springfish	N-L
Crenichthys nevadae	Railroad Valley springfish	N-L
Cyprinodon diabolis	Devils Hole pupfish	N-L
Cyprinodon nevadensis	Amargosa pupfish	N-L
Cyprinodon radiosus	Owens pupfish	N-L
Cyprinodon salinus	Salt Creek pupfish	N-L
Empetrichthys latos latos	Pahrump killifish	N-L
Fundulus zebrinus	Plains killifish	I
Lucania parva	Rainwater killifish	I
Poeciliidae		
Gambusia affinis	Mosquitofish	I
Poecilia latipinna	Sailfin molly	I
Poecilia mexicana	Shortfin molly	I
Poecilia reticulata	Guppy	I
Xiphophorus helleri	Green swordtail	I
Gasterosteiformes		
Gasterosteus aculeatus	Threespine stickleback	N-L
Percichthyidae		
Morone chrysops	White bass	N
Centrarchidae		
Archoplites interruptus	Sacramento perch	I
Lepomis cyanellus	Green sunfish	I
Lepomis gibbosus	Pumpkinseed	I
Lepomis macrochirus	Bluegill	I
Micropterus dolomieui	Smallmouth bass	I

continued

Table 25.3. *—continued*

Family / Species	Common name	
Centrarchidae *(continued)*		
Micropterus punctulatus	Spotted bass	I
Micropterus salmoides	Largemouth bass	I
Pomoxis annularis	White crappie	I
Pomoxis nigromaculatus	Black crappie	I
Percidae		
Perca flavescens	Yellow perch	I
Stizostedion vitreum vitreum	Walleye	I
Cichlidae		
Cichlasoma nigrofaciatum	Convict cichlid	I
Tilapia zilli	Redbelly tilapia	I
Cottidae		
Cottus bairdi	Mottled sculpin	N-L
Cottus beldingi	Paiute sculpin	N
Cottus extensus	Bear Lake sculpin	N

[1]N = Native; I = Introduced.

Table 25.4. List of native and introduced fishes from the Sonoran Desert region (Figure 25.2B), United States and Mexico (Rinne and Minckley 1991).

Family / Species	Common name
Native species	
Elopidae	
Elops affinis	Pacific tenpounder
Engraulidae	
Clupeidae	
Dorosoma smithi	Pacific shad
Salmonidae	
Oncorhynchus apache	Apache trout
Oncorhynchus chrysogaster	Mexican golden trout
Oncorhynchus gilae	Gila trout
Oncorhynchus mykiss nelsoni	San Pedro Martir trout
Oncorhynchus sp.	Yaqui trout
Cyprinidae	
Agosia chrysogaster	Longfin dace
Campostoma ornatum	Mexican stoneroller
Codoma ornata	Ornate minnow
Cyprinella formosa	Beautiful shiner
Gila cypha	Humpback chub
Gila ditaenia	Sonoran chub
Gila elegans	Bonytail
Gila eremica	Desert chub
Gila intermedia	Gila chub
Gila purpurea	Yaqui chub
Gila robusta	Roundtail chub

continued

Table 25.4. —*continued*

Family / Species	Common name
Cyprinidae *(continued)*	
Gila sp.	Undescribed chub (Rio Yaqui)
Lepidomeda altivelis	Pahranagat spinedace
Lepidomeda albivallis	White River spinedace
Lepidomeda mollispinis	Virgin spinedace
Lepidomeda vittata	Little Colorado spinedace
Meda fulgida	Spikedace
Moapa coriacea	Moapa dace
Plagopterus argentissimus	Woundfin
Ptychocheilus lucius	Colorado squawfish
Rhinichthys deaconi	Las Vegas speckled dace
Rhinichthys osculus subspp.	Speckled dace
Tiaroga cobitis	Loach minnow
Catostomidae	
Catostomus bernardini	Yaqui sucker
Catostomus cahita	Cahita sucker
Catostomus insignis	Sonoran sucker
Catostomus latipinnis	Flannelmouth sucker
Catostomus leopoldi	Leopold sucker
Catostomus wigginsi	Opata sucker
Catostomus sp.	Little Colorado sucker
Pantosteus clarki	Desert mountain-sucker
Pantosteus discobolus	Bluehead mountain-sucker
Pantosteus plebeius	Rio Grande mountain-sucker
Xyrauchen texanus	Razorback sucker
Ariidae	
Ictaluridae	
Ictalurus pricei	Yaqui catfish
Poeciliidae	
Poeciliopsis o. occidentalis	Gila topminnow
Poeciliopsis o. sonoriensis	Yaqui topminnow
Poeciliopsis monacha	Common name unknown
Poeciliopsis monacha-occidentalis	All-female topminnow
Poeciliopsis prolifica	Common name unknown
Cyprinodontidae	
Cyprinodon macularius	Desert pupfish
Cyprinodon sp.	Whitefin pupfish
Cyprinodon sp.	Monkey Spring pupfish
Goodeidae	
Crenichthys baileyi	White River springfish
Mugilidae	
Mugil cephalus	Striped mullet
Cichlidae	
Cichlasoma beani	Sinaloan cichlid
Sciaenidae	
Cynoscion macdonaldi	Totoaba
Centropomidae	

continued

Table 25.4. —*continued*

Family / Species	Common name
	Introduced species
Lepisosteidae	
Amiidae	
Esocidae	
Esox lucius	Northern pike
Clupeidae	
Dorosoma petenense	Threadfin shad
Salmonidae	
Oncorhynchus mykiss	Rainbow trout
Oncorhynchus clarki	Cutthroat trout
Salmo trutta	Brown trout
Salvelinus fontinalis	Brook trout
Cyprinidae	
Cyprinus carpio	Common carp
Ctenopharyngodon idellus	Grass carp
Cyprinella lutrensis	Red shiner
Notemigonus crysoleucus	Golden shiner
Pimephales promelas	Fathead minnow
Richardsonius balteatus	Redshide shiner
Ictaluridae	
Ameiurus melas	Black bullhead
Ameirus natalis	Yellow bullhead
Ictalurus punctatus	Channel catfish
Ictalurus furcatus	Blue catfish
Pylodictis olivaris	Flathead catfish
Centrarchidae	
Lepomis cyanellus	Green sunfish
Lepomis macrochirus	Bluegill
Lepomis microlophus	Redear sunfish
Micropterus dolomieui	Smallmouth bass
Micropterus salmoides	Largemouth bass
Pomoxis annularis	White crappie
Pomoxis nigromaculatus	Black crappie
Percichthyidae	
Morone chrysops	White bass
Morone mississippiensis	Yellow bass
Morone saxatilis	Striped bass
Percidae	
Stizostedion vitreum	Walleye
Cichlidae	
Oreochromis aurea	Blue tilapia
Oreochromis mossambica	Mossambique Tilapia
Tilapia zilli	Redbelly tilapia

Rinne and Minckley (1991) illustrated and discussed native fishes in arid regions of the Sonoran Desert of the southwestern United States and northern Mexico. Of the 41 species of freshwater fishes of this desert region, 10 occur only in Mexico, 9 only in the United States, and 22 are in common. These numbers illustrate the regional nature of the fish fauna between these two countries. Taxa found exclusively in one country or another represent local endemic species.

Thirty four of the 41 (83%) species in the Sonoran Desert region are Cypriniformes—minnows (23) and suckers (11). Similar to the eel in Spain, only one native cyprinid species, the Colorado Squawfish (*Ptychocheilus lucius*), is a large (>1 m) predatory species. The Ahteriniforms (five livebearers and two killifishes) comprise the remaining freshwater taxa. Three taxa have become extinct in this desert region in the past three decades (Miller et al. 1989).

Over the past century almost 140 nonnative species of fishes have been introduced into the waters of Arizona (Rinne in press). Of these, more than 50 have become established, resulting in a "loading" or supersaturation of aquatic habitats in this arid area. Most of these taxa have been introduced for sport, bait, or as biological control agents. In addition, a majority of the species introduced have come from faunal regions characterized by "faunal saturation." In regions such as the Mississippi Basin and eastern coastal regions in North America and tropical Mexico and Africa, species occupying respective lakes and drainages number in the hundreds. By comparison, in the arid regions of the American West and Southwest, fish taxa of drainages number in the tens. Accordingly, many of the introduced species have evolved in "biological adversity" compared to the "biological naivete" evolutions of desert fishes.

Miller (1961) first suggested that the native fish fauna of the Southwest was endangered. A few years later, Minckley and Decon (1968) discussed southwestern fishes and the concept of "endangerment." More recently, Johnson and Rinne (1982) discussed southwestern fishes in context of the 1973 Endangered Species Act. The plight of native fishes in the western United States and the struggle to sustain them in their natural habitats are excellently detailed in Minckley and Deacon (1991). In a recent listing of threatened, endangered, or special concern species by the American Fisheries Society (Williams et al. 1989), 28 of the 41 taxa (68%) discussed and well illustrated by Rinne and Minckley (1991) were presented as listed species.

REASONS FOR LOSS OF DIVERSITY

Spain

Since 1900, Spain has experienced a gradual increase in construction of dams and reservoirs. At present, over 1,000 major reservoirs are functioning on the waterways of Spain to supply electric power and water for industry and domestic uses. Many of these reservoirs were constructed either in the upper reaches of rivers to capture water as it issues from more mesic mountainous regions or near river mouths to capture this resource before it empties into the sea.

The proliferation of artificial structures has affected the native fish fauna in at least three ways. First, they prevented natural movements of migratory native fish species. Second, they significantly reduced spawning and rearing habitats for native taxa through both inundation and qualitative and quantitative alteration of hydrologic regimes. Finally, reservoirs provided suitable habitats for introduction and progressive expansion of populations of large, predatory, nonnative sport species, principally pike (*Esox lucius*) and large mouth black bass (*Micropterus salmoides*).

Although mechanisms of interactions are not completely understood, native species apparently have not been able to withstand predatory impacts not previously experienced (i.e., biological naivete). For example, migration of pike upstream from reservoirs near river mouths has apparently been responsible for extinction of one small endemic minnow, *Rutilius arcasii*, in several streams (Rincon et al. 1990). In addition to the two large predators mentioned above, and similar to the United States deserts, the diminutive Western mosquitofish, *Gambusia affinis*, has negatively impacted the range and number of two endemic Cyprinodonts, *Aphanius iberus* and *Valencia hispanica*. Genetic contamination of native stocks of salmonids has resulted from over-stocking of both brown trout (*Salmo trutta trutta*) and Atlantic salmon (*Salmo salar*).

In addition to impoundments, channelization and increasing industrial and agricultural pollution have negatively altered habitats for native fishes in Spain. Large areas of marshlands, an essential habitat for Cyprinodontiformes, have been lost to fruit orchard development. Tourism-related developments have also negatively impacted this critical aquatic habitat.

In summary, habitat alteration and introductions of nonnative fish species have, in combination, negatively impacted the diversity of the native fish fauna of Spain.

Australia

Fishing by aboriginal people might have been a cause for loss of some faunal diversity in Australian waters. However, since the arrival of Europeans in the arid drainage areas of Australia, there has been increasing pressure on native fishes for recreational and commercial enterprises. Several species in the Lake Eyre drainage are large enough for either type of enterprise. The main target species for recreational fishing is the yellowbelly (*Macquaria ambigua*). Bony bream (*Nematolosa erebi*) and spangled perch (*Leiopotherapon unicolor*) are also being caught. Gill nets are illegal, but they are often used by "recreational fishermen" to catch native species. Use of gill nets to catch yellowbelly commercially in the Lake Eyre drainage threatens the survival of this species (Glover 1990).

Tourism also plays a major role in destruction, alteration, and loss of aquatic habitats and, in turn, the decrease of native fish diversity in Australia. Previous free access to habitats, such as Dalhousie Springs, has resulted in degradation of habitat through swimming and washing with soaps and detergents, fishing, littering, and destruction of vegetation and surrounding banks of the spring and its run.

Grazing of vegetation in shallow aquifers by sheep and cattle has been detrimental to springs in some cases (Ivantsoff et al. 1991; Unmack and Brumley 1991). Although not specifically studied, urination and defecation into shallow pools increase nutrient supply, bringing about increases in green and blue-green algae concentrations.

Utilization of groundwater through untapped bores and open drains for both agricultural irrigation and water for domestic livestock has reduced flows and fish habitat in springs (Glover 1990). At Dalhousie Springs, a 6% decline in flow has occurred in the past decade (Smith 1989).

The introduced Eastern mosquitofish, *Gambusia holbrooki*, and goldfish, *Carassius auratus*, are likely to threaten native fishes and ultimately reduce their diversity through competition and introduction of new pathogens and parasites. As has been reported for the Western mosquitofish in Southwestern United States (Meffe 1986), the Eastern mosquitofish also is an aggressive predator and eats larvae of native species. Where the mosquitofish is abundant in small creeks in the Brisbane area, native species are rare (Milton and Arthington 1983). The Eastern mosquitofish is widespread in some areas of the Lake Eyre drainage. It has been reported recently in the Aramac region where it could pose a threat to the diminutive (25 mm) newly described endemic pseudomugilid fish, the red-finned blue-eye, *Scaturiginichthys vermeilipinnis (Pseudomugilidae)*.

The common carp, *Cyprinis carpio*, has only recently been reported to inhabit a dam on Leigh Creek in northern Australia. Introductions of this species into the waters of Australia have been extremely successful. In Lake Bonney, South Australia, it apparently has displaced the previously abundant hardyhead, *Craterocephalus fluviatilis*, as well as other native fish species.

Western and Southwestern United States

Similar to Spain and Australia, two major impacts separately and in combination have caused a marked, negative impact on the native fish fauna of deserts of western North America: (1) modification of habitat and (2) introduction of nonnative species. Miller (1961) first warned of the effects of humans on the native fish fauna of the American Southwest. Damming, irrigation diversion, and groundwater mining have greatly modified the surface hydrology of these arid environments (Rinne 1991). In Arizona, for example, 70% of the major river courses have been dried or quantitatively and/or qualitatively modified by mainstream dams (Rinne 1991). Introductions of nonnative species for sport, bait, or biological control have tripled the fish fauna of Arizona (Rinne 1991). Much of the success of introductions of nonnative species can be attributed to the controlling of natural, highly variable flow regimes, creating habitats more amenable to taxa introduced from the Mississippi Basin and other eastern drainages.

Only one undescribed species of fishes has become extinct in the waters of Arizona; however, many more have become drastically reduced in range and numbers, despite extensive recovery efforts since passage of the Endangered Species Act of 1973. In California, among 20 species of native fishes, only seven have viable, healthy populations, six are endangered, and seven are extinct (Pister 1981). Here also, the introduction of 52 nonnative fishes has been a major contributor to this marked decline in native freshwater fishes in California waters within the Mohave Desert area.

CONSERVATION EFFORTS AND STRATEGIES

Spain

The political administrative structure and thereby policy in Spain is divided among 17 separate "Comunidades Autonomas." Each of these political entities has its own government and parliament. Accordingly, the responsibility for conservation and management of fauna, flora, and the environment falls to each governing unit. In 1990, because of the increasing concern for biological conservation, a State Law on conservation of fauna and flora was passed. This Law transcended these extant governing bodies and had general applicability for conservation efforts throughout Spain. It was largely based on information in the Red Data Book of the Spanish Vertebrates (Icona 1986). The Law mandated developing specific programs for conservation and recovery of all vertebrates including fishes.

Australia

Each Australian state and territory has its own legislation to control and regulate the taking of fish. Nevertheless, in many instances, little can be done to patrol effectively and to prevent illegal take by nets and recreational fishing in the vast arid zones. Permits are necessary to collect fishes for scientific purposes, but other than near major human population centers, enforcement is virtually impossible.

The Australian federal government is presently considering national and international legislative controls that will protect native aquatic fauna. Courtenay (1989) stressed the importance of adopting legislation to protect aquatic diversity in Australia from introductions and translocations of nonnative species. The Wildlife Protection Act of 1982, although important, addresses only imports and exports of wildlife. Only a few edible and commercially valuable species are listed under this Act, thereby affording little protection to the native fish fauna as a whole.

Several species of aquarium fishes are prohibited by the Wildlife Act, but those species already introduced are not affected. Several states are independently in the process of enacting legislation to prohibit possession of imported fishes. One portion of the Australian Fisheries Act states that "any exotic fish including common carp taken are not to be returned to the water and must be destroyed by the fisherman." The Victorian Fisheries Act of 1968 also declared that "noxious" species of fish are not to be returned to the water. Smaller commercially unimportant species such as *Gambusia* are generally ignored by all except the scientific community.

Recent surveys have been conducted of the mound springs and large warm pools in the arid zone by the Australian and South Australian Museums. Because the habitat at Dalhousie Springs was considered to be "very fragile", Zeidler and Ponder (1989) urged concerned governments to protect this and other existing relatively stable artesian springs systems. In 1985, because of recommendations by these two workers, the region was proclaimed a National Park. Apart from the Dalhousie Springs National Park little is being done to protect natural springs in the arid zone.

Open bores, dams, and reservoirs have been demonstrated to extend the ranges of several native Australian fish species (Glover and Sim 1978; Glover 1979, 1989). However, in South Australia, the Department of Mines has been conducting a program of capping and regulating outflows of bores since 1978. Although some bores remain open, the partial enclosure of drains to facilitate limited watering of livestock has reduced evaporation and depletion of groundwater reserves. This action, however, could have a negative effect on native fishes.

United States

Past conservation efforts for native fishes in desert environments of the United States had its inception in the late 1960s and early 1970s. The National Environmental Policy Act (NEPA) of 1969, the Endangered Species Preservation and Conservation Acts of 1966 and 1969 and, later, the Endangered Species Act of 1973 laid the legislative groundwork for conservation of all threatened and endangered species and their habitats. It was a time when individuals interested in desert fishes as a resource took a stand to preserve species being rapidly and markedly impacted by the demand for water for agricultural and housing developments.

In April 1969, a group of a less than a dozen (*ad hoc*) individuals from several agencies and universities gathered in Death Valley to prevent the potential loss of a group of rare species of desert fishes through pumping of aquifers and subsequent drying desert springs in the Ash Meadows system of southwestern Nevada, adjacent to Death Valley. Out of these efforts, the Desert Fishes Council was founded in November 1976 and now has over 500 members (Pister 1981, 1990, 1991). This group presently addresses the conservation of desert fishes and their habitats throughout the western United States, northern Mexico, and is international in scope. Two decades after founding of the Council, "Battle Against Extinction, Native Fish Management in the American West;" (Minckley and Deacon 1991) was published documenting the struggle of two decades to conserve (and often preserve) native fishes, in arid environments of the American West.

In the past, conservation efforts in the southwestern United States for the recovery of the native fishes have been extensive and have taken one or a combination of approaches. Conservation activities are normally outlined in "Recovery Plans," which are documents for federally listed species drafted by an interagency recovery team. To stem the extensive loss of aquatic habitats through hydrological alteration, most recovery plans for listed fish species include objectives and activities that will secure, maintain, and enhance their habitats (Rinne and Turner 1991).

Removal of nonnative fishes has been a common practice in conservation efforts and is usually accomplished through application with piscicides (Rinne and Turner 1991). Unfortunately, the success of renovation attempts is often reduced because of the size and/or complexity of habitats, flow regimes (in spring systems), lack of security of habitat through private ownership or special management, variable conditions of habitats from year to year, and the ever-present threat of reestablishment of introduced species.

Johnson and Rinne (1982) first expressed the need to move from listing (and protection by law) of native southwestern (US) species as dictated by the 1973 Act to active species recovery programs. Large-scale but genetically sound rearing of fishes in hatchery environments (Rinne et al. 1986), coupled with intensive, long-term reintroduction into what is deemed "favorable habitat" in the wild, also has been used extensively in conservation efforts (Johnson 1985; Simons 1987; Minckley and Deacon 1991). Several such, extensive reintroductions of razorback sucker (*Xyrauchen texanus*) into unrenovated (i.e., treated with ichthyiotoxin) streams and rivers in Arizona (Johnson 1985) failed largely because of predation by nonnative species, such as the piscivorous flathead catfish (*Pilodictus olivaris*) (Marsh and Brooks 1989) and introduced Centrarchidae fishes. Over 10 million razorback suckers were stocked in waters of the Gila River Basin, Arizona in the 1980s; fewer than 120 have been recaptured, and those mostly within two weeks of stocking.

The apparent "successful recovery" of the Gila topminnow (*Poeciliopsis o. occidentalis*) through extensive and widespread introductions similar to efforts for the razorback sucker becomes questionable because of their occurrence in habitats unprotected from nonnative species (Simons et al. 1989). Purchase and management of riparian/stream habitats by private agencies such as The Nature Conservancy have been a boon to desert fishes.

Enhancement of aquatic habitat through in-stream improvements has become a common practice for recovery of salmonids. Indeed, such improvements frequently have been shown to improve trout populations (Rinne 1981). Nevertheless, because such in-stream, site-specific improvement is often conducted without regard to surrounding land use and resultant condition of the supplying watersheds, the probability of failure and loss of in-stream structures increases (Szaro and Rinne 1988; Lafayette and Rinne 1991).

Past management and conservation efforts for native desert fishes in the southwestern United States were often products of necessity, undertaken under crises situations and narrowly focused. In some cases, species of fishes in isolated desert springs left no alternative to a "species-by-species" approach to recovery (Simons et al. 1989). Future conservation efforts must consider the total diversity (i.e., genetic, species, ecosystem, and landscape) of the fauna relative to alteration of aquatic systems (Rinne 1990), Where possible, conservation activities must begin to move away from single-species management toward ecosystems (i.e., river systems) or groups of species (Johnson and Rinne 1982; Rinne 1984; Williams et al. 1985; Rinne 1990). Further, species not currently on official federal lists but on respective state lists, U.S. Fish and Wildlife Service candidate species list, American Fisheries Society special concern list, and U.S. Forest Service sensitive species lists should also be considered for immediate conservation attention.

We must undertake conservation activities for desert fishes where the habitat is secure (e.g., federal, state, and private conservation lands and special management areas) or relatively so and where we can reasonably be most effective. By law (the Endangered Species Act) habitats on federal lands should be afforded ample protection from the negative impacts of agency activities. In practice, such security is not a hard-and-fast rule. Furthermore, conservation objectives must be realistic and achievable rather than idealistic. Finally, cooperative interagency management activities that are synergistic in recovering species should be adopted. Multidisciplinary, holistic approaches incorporating hydrologists, biologists, geneticists, geologists, and botanists who address entire ecosystems, or the biodiversity philosophy of conservation, must be formulated (Pister 1990).

Agency targets and recovery plan objectives, although well-intentioned, often can be detrimental to the recovery of a rare fish over the long term. Although goals in terms of number of populations required for delisting the Apache trout (*Oncorhynchus apache*) and Gila topminnow might be quantitatively achievable, they might be qualitatively unsound (Rinne 1990). The sustainability of these populations indefinitely must be the criterion for delisting species. A recent chemical renovation activity of the Virgin River with piscicide to remove the introduced red shiner (*Cyprinella lutrensis*) and restore the endangered woundfin minnow (*Plagopterus argentissimus*) is a case in point. Placing a higher priority on completion of the operation on schedule rather than insuring that all components of this large-scale, multi-agency activity were in place, resulted in treatment of excessive reaches of the river and negative impacts on other native species. The failure to remove nonnative salmonids in the chemical renovation of Ord Creek to recover the Apache trout strongly suggests that no matter how well-intentioned or planned the project, success is not easily achieved (Rinne et al. 1981). Furthermore, as occurred in the Virgin River and in Ord Creek (Minckley and Mihalik 1981), aquatic invertebrate biodiversity is also likely to be impacted.

Because of occurrences such as the Virgin River, Ord Creek, and numerous past renovations with specific agency goals (Binns 1967; Holden 1991; Rinne and Turner 1991), a philosophy of management should be adopted that considers the ecosystem and its total biodiversity (Williams et al. 1985; Rinne 1990). To be successful, such a philosophy necessitates a multi-agency and multidisciplinary approach (Pister 1990; Rinne 1990). That is, the broader the area a recovery or conservation plan addresses, the greater the probability that more than one agency will be involved in management. Equally important, it is more probable that a larger component of total fish faunal diversity rather than single species will be addressed (Rinne 1984). Recovery teams and plans, such as the Colorado Fishes and the Desert Fishes Recovery teams, nominally address this philosophical approach to conservation of desert fishes. However, because of the extent of the Colorado River Basin (and other major desert rivers), its drastic physical alteration, and multi-agency jurisdictions that divide conservation efforts into upper and lower basin entities, a holistic or ecosystem approach has not been realized.

Future management strategies need to address the multitude of species that are declining but are not yet federally listed. Management and research activities that consider such candidate, vulnerable, restricted, sensitive, or special species lists will become more proactive. For example, the Southwest Region of the Forest Service (Arizona and New Mexico) has more than 230 sensitive vertebrate species, 15 of which are fishes. Unfortunately, much endangered species management and research has adopted a "flagship" philosophy, wherein most budgeting and effort is directed toward a few high-profile species (e.g., bald eagle, grizzly bear, whooping crane, and currently in the Southwest, Mexican spotted owl). Although flagship efforts might indirectly spill over and affect other species needing immediate conservation attention, this is a serendipitous but less desirable way to manage rare aquatic biodiversity and their ecosystems (Rinne and Medina in press). Granted, we cannot direct intensive conservation efforts at 230 species, but through timely conservation we can increase the probability that many of these species will never be listed as threatened or endangered. Furthermore, if we plan and design our conservation efforts on an ecosystem basis, many species, including some that are in need of help, will automatically be addressed.

Much has been accomplished since adoption of legislation facilitating management and conservation of rare species of fishes in the deserts of the western United States. It behooves us to consider and adopt innovative and progressive approaches in future conservation efforts for this regional fish fauna. To do less would discredit and make a mockery of extensive past efforts by State and Federal agencies and private individuals that frequently were necessary crisis management endeavors to secure the very existence of rare fishes. Indeed, it is time to move from "saving" to "recovering" native fishes in the arid West and Southwest (Johnson and Rinne 1982). Many of these fishes occur in often isolated habitats in an arid region characterized by the virtual absence of the basic life medium of fishes and aquatic biota, water. We have come far, yet have far to go.

Only if the overall physical and political aspects of southwestern water use are considered in our conservation efforts will aquatic habitats and native fish biodiversity be maintained.

Great strides have been made in understanding the issues of conservation and perpetuation of biodiversity of fishes in the American West and Southwest since enactment of the Endangered Species Act of 1973. Case histories demonstrate that some success has been achieved, but at great cost in time and money. Although piscine diversity of this xeric region is low, the uniqueness of the taxa dictates continued efforts to conserve biodiversity. Minimal success in conserving native fish biodiversity has been attained chiefly at a species level; genetic and ecosystem (watershed or river drainages) levels only now are emerging as important components of the effort to conserve the native fish fauna of the Southwest. A more encompassing landscape biodiversity (i.e., access multiple ecosystems) approach to conserving desert fishes has the least probability of being realized. The present often disjunct and isolated nature of aquatic ecosystems (drainages), interposed with the rapid development of the arid regions of the western United States render this approach moot at best.

CONCLUSIONS AND RECOMMENDATIONS

Native fish faunas of desert areas of Spain, Australia, and the United States are low in diversity (Tables 25.1-25.4). They are dominated by cypriniform taxa and a lack of large natural predators. All have sustained introductions of nonnative species of fishes, with the greatest impact being in the United States, less in Spain, and least in desert regions in Australia. The United States and Spain have suffered extensive damming of mainstream rivers. As a result, about one-third of all species in North America are listed as threatened, endangered, or of special concern (Williams et al. 1989); many have become extinct (Miller et al. 1989). Similarly, in Australia 34% of 192 species fall in categories of "endangered, vulnerable, potentially threatened, unknown, and restricted" (Pollard et al. 1990). Finally, in Spain the "double dose disadvantage" of loss and modification of habitat and the introduction of nonnative and "exotic" species of fishes is also present (Elvira 1990).

To realize long-term success, where possible the conservation of native fish diversity in Spain, Australia, and the United States must be addressed at all levels (i.e., genetic, species, ecosystem, and landscape) of biodiversity. Understanding and managing genetic diversity of desert fishes is basic to conservation of desert fishes (Meffe 1986; Meffe, this volume). For larger stream and river systems, conservation efforts should address multiple species of fishes and large-scale habitat approaches. However, because of the characteristic aquatic infrastructure of deserts, consisting of isolated, small (yet important) aquatic habitats containing highly-adapted species, this objective is not always feasible. However, because these small habitats are ecosystems, they remain in need of our constant attention. Acquisition of aquatic habitats should proceed on an ecosystem basis to include multiple species, and should be complemented by a program of research and development. Priority should be given to intact and recoverable aquatic systems. These efforts should have the short-term goal of species preservation yet embrace the long-term goal of biodiversity maintenance over the greatest possible range of biota.

The 1973 Endangered Species Act was instrumental in the United States in initiating and effecting conservation of desert fishes in the United States (Williams 1981). However, without many individual efforts and, collectively, those of conservation organizations such as the Desert Fishes Council, the current diversity and status of native, usually rare, desert fishes would be grim at best. The plea for and the means to conservation of this valuable "non-resource" (Ehrenfeld 1976) has been adequately made (Pister 1990; Williams and Miller 1990; Rinne and Minckley 1991; Minckley and Deacon 1991). The Australian Society for Fish Biology was formed in 1971 and is rapidly becoming an important professional society to protect and conserve native fish biodiversity (Pollard 1989). There is no parallel legislation to the United States Endangered Species Act of 1973 in either Spain or Australia; however, such has been encouraged by Courtenay (1989) for Australia. Professional societies in the United States and Australia are encouraged and will ultimately be essential for the conservation of native desert fishes in Spain. Finally, in the words of a noted desert fish conservationist in the United States, we must "recognize that in unity there is strength" and that "only by continuing to work together, researcher and lay conservationist alike, do we retain a possibility of preserving this unique fauna" (Pister 1990).

LITERATURE CITED

Allen, G. R. 1989. *Freshwater Fishes of Australia*. T. F. H. Publications, Inc., Garden City, N.Y.

Binns, N. A. 1967. Effects of rotenone treatment on the fauna of the Green River, Wyoming. *Wyo. Game and Fish Comm. Fish. Tech. Bull.* 1:1-114.

Courtenay, W. R. 1989. Fish introductions and translocations, and their impacts in Australia. In *Introduced and Translocated Fishes and their Ecological Effects*, ed. D. A. Pollard, pp. 171-189. *Aust. Soc. Fish Biol. Publ.* 8.

Deacon, J. E., and W. L. Minckley. 1974. Desert fishes. *Desert Biology*. Vol. 2, ed. G. W. Brown, Jr., pp. 385-488. Academic Press, New York.

Ehrenfeld, D. W. 1976. The conservation of non-resources. *Am. Sci.* 64:648-656.

Elvira, B. Iberian endemic freshwater fishes and their conservation status in Spain. *J. Fish Biol.* 37:231-232.

Glover, C. J. M. 1979. Studies on Central Australian fishes: Further observations and records. Part. I. *Austral. Nat.* 53:58-62.

Glover, C. J. M. 1982. Adaptations of fishes in arid Australia. In *Evolution of Flora and Fauna of Arid Australia*, eds. W. R. Baker and P. J. M. Greenslade, pp. 241-246. Peacock Publications, Fernvale, South Australia.

Glover, C. J. M. 1989. In *Natural History of Dalhousie Springs*, eds. W. Zeidler and W. F. Ponder. South Australian Museum, Adelaide.

Glover, C. J. M. 1990. Fishes. In *Natural History of the North East Deserts*, eds. M. J. Tyler, C. R. Twidale, M. Davies, and C. B. Wells, pp. 189-197. Roy. Soc. S. Austral. Inc., Adelaide.

Glover, C. J. M. and T. C. Sim. 1978. Studies of Central Australian ichthyology. *Austral. Zool.* 19(3):35-44.

Harris, J. H., ed. 1987. Proceedings of the conference on Australian threaten fishes. Austral. Soc. for Fish Biology, Melbourne, 15-16 August 1985. Division Fish Biology, N.S.W. Dept. Agric., Sydney.

Holden, P. R. 1991. Ghosts of the Green River: impacts of Green River poisoning on management of native fishes. In *Battle against extinction—native fish management in the American West*, eds. W. L. Minckley and J. E. Deacon, pp. 55-68. University of Arizona Press, Tucson, Ariz.

Icona. 1986. Lista Roja de los Vertebrados de Espanâ. Pu Minis. Agric. Pesca Y aliment. Madrid, Spain.

Ivantsoff, W., P. Unmack, B. Saeed, and L. E. L. M. Crowley. 1991. A red-finned blue-eye, a new species and genus of the family Pseudomugilidae from central western Queensland. *J. Austral. New Guinea Fishes Assoc.* 6(4):277-282.

Jackson, P. 1992. Australian threatened fishes—1991 supplement. *Austral. Soc. Fish. Biol.* 21(2):20-23.

Johnson, J. E. 1985. Reintroducing the natives: razorback sucker. *Proc. Desert Fishes Council* 13 (1981):73-79.

Johnson, J. E., and J. N. Rinne. 1982. The Endangered Species Act and Southwest fishes. *Fisheries* 7(2):1-8.

Lafayette, R. A., and J. N. Rinne. 1991. Riparian structures and watershed treatments in the Southwest: History, status and management implications—a preliminary report. Water Resour. Research Institute Rep 265: 133-138. New Mexico State University, Las Cruces, N.M.

Lobón-Cervia J., C. Kontañes, and A. de Sostoa. 1991. Influence of the environment upon the life history of gudgem, *Gobio gobio* (L.): a recent and successful colonizer of the Iberian Peninsula. *J. Fish Biol.* 39:235-300.

Marsh, P. C., and J. E. Brooks. 1989. Predation by Ictalurid catfishes as a deterrent to re-establishment of hatchery-reared razorback suckers. *Southwest. Nat.* 34(2):188-195.

Meffe, G. K. 1983. Attempted chemical renovation of an Arizona spring brook for management of the endangered Sonoran topminnow. *N. Am. J. Fish. Manage.* 3:315-321.

Meffe, G. K. 1986. Conservation genetics and the management of endangered fishes. *Fisheries* 11(1):14-23.

Miller, R. R. 1961. Man and the changing fish fauna of the American Southwest. *Pap. Mich. Acad. Sci. Arts. Lett.* 46: 365-404.

Miller, R. R., J. D. Williams, and J. E. Williams. 1989. Extinctions of North American fishes during the past century. *Fisheries* 14(6):22-38.

Milton, D. A., and A. H. Arthington. 1983. Reproductive biology of *Gambusia affinis holbrooki* Baird and Girard, *Xiphorus helleri* (Gunther) and *X. maculatus* (Heckel)(Pisces: Poeciliidae) in Queensland, Australia. *J. Fish. Biol.* 23:23-41.

Minckley, W. L., and P. Mihalik. 1981. Effects of chemical treatment for fish eradication on stream-dwelling invertebrates. *J. Ariz.-Nev. Acad. Sci.* 16:79-82.

Minckley, W. L., and J. E. Deacon. 1968. Southwestern fishes and the enigma of "endangered species." *Science* 159:1424-1432.

Minckley, W. L., and J. E. Deacon. 1991. *Battle against Extinction: Native Fish Management in the American West.* University of Arizona Press, Tucson, Ariz.

Naiman, R. J., and D. L. Soltz, eds. 1981. *Fishes in North American Deserts.* John Wiley and Sons, New York.

Pister, E. P. 1981. The conservation of desert fishes. In *Fishes in North American Deserts*, eds. R. J. Naiman and D. L. Soltz, pp. 411-445. John Wiley and Sons, New York.

Pister, E. P. 1990. Desert fishes: an interdisciplinary approach to endangered species conservation in North America. *J. Fish Biol.* 37:183-188.

Pister, E. P. 1991. Desert Fishes Council: catalyst for change. In *Battle Against Extinction—Native Fish Management in the American West*, eds. W. L. Minckley and J. E. Deacon, pp. 55-68. University of Arizona Press, Tucson, Ariz.

Pollard, D. A., ed. 1989. Introduced and translocated fishes and their ecological effects. *Proc. Austral. Soc. Fish. Biol.* 8:1-181.

Pollard, D. A., B. A. Ingram, J. H. Harris, and L. F. Reynolds. 1990. Threatened fishes in Australia—an overview. *J. Fish. Biol.* 37:67-78.

Rincon, P. A., J. C. Velusio, N. Gonzalez-Sanchez, and E. Pollo. 1990. Fish assemblages in small streams in western Spain. The influence of an introduced predator. *Arch. Hydrobiol.* 118 (1):81-91.

Rinne, J. N. 1981. Stream habitat improvement and native SW trout. *USDA, For. Serv. Res. Note* RM-409:1-4.

Rinne, J. N. 1984. Indicator on diversity of species management: the case of native southwestern fishes. In *Proc. Workshop on Management of Nongame Species and Ecological Communities*, ed. W. McComb, pp. 75-92. University of Kentucky, Lexington, Ky.

Rinne, J. N. 1990. Biodiversity and native fish management: southwestern United States. *Proc. West Assoc. Fish and Wildl. Agencies* 70:152-165.

Rinne, J. N. 1991. An approach to management and conservation of a declining regional native fish fauna: Southwestern United States. In *Wildlife Conservative: Present trends and perspectives for the 21st Century. Prov. Internatl. Symp. Wildl., 5th Internatl. Congr. Zool.*, N. Maruyama et al., pp. 56-60. Tsukuba and Yokohama, Japan.

Rinne, J. N. In press. The effects of introduced fishes on native fishes: Arizona, southwestern United States. *World Fish. Congr.*, May 1992. Athens, Greece.

Rinne, J. N., and W. L. Minckley. 1991. Native fishes in arid lands—A dwindling resource of the desert Southwest. *USDA, For. Serv. Gen. Tech. Rep.* RM-206:1-45.

Rinne, J. N., and P. R. Turner. 1991. Reclamation and alteration as management techniques, and a review of methodology in stream renovation. In *Battle against Extinction: Native Fish Management in the American West*, eds. W. L. Minckley and J. E. Deacon, pp. 219-246. University of Arizona Press, Tucson, Ariz.

Rinne, J. N., and A. L. Medina. In Press. Implications of Multiple Use Management Strategies on Native Southwestern(USA) Fishes. World Fish. Congr., Athens, Greece. May 4-8, 1992.

Rinne, J. N., W. L. Minckley, and J. N. Hanson. 1981. Chemical Treatment of Ord Creek, Apache County, Arizona, to re-establish Arizona trout. *Ariz.-N. Mex. Acad. Sci.* 16:74-78.

Rinne, J. N., J. E. Johnson, B. L. Jensen, A. W. Ruger, and R. Sorenson. 1986. The role of hatcheries in the management and recovery of threatened and endangered species. In *Fish Culture in Fisheries Management*, ed. R. H. Stroud, pp. 271-285. Am. Fish. Soc., Bethesda, Md.

Sigler, W. F., and J. W. Sigler. 1987. *Fishes of the Great Basin.* University of Nevada Press, Reno, Nev.

Simons, L. H. 1987. Status of Gila topminnow (*Poeciliopsis o. occidentalis*) in the United States in 1987. Unpubl. Rep. to U. S. Fish and Wildlife Service, Albuquerque, N.M.

Simons, L. H., D. A. Hendrickson, and D. Papoulias. 1989. Recovery of Gila topminnow: A success story? *Biol. Conserv.* 3(1):11-15.

Smith, M. L. 1981. Late Cenozoic fishes in the warm deserts of North America: A reinterpretation of desert adaptations. In *Fishes in North American Deserts*, eds. R. J. Naiman and D. L. Soltz, pp. 11-38. John Wiley and Sons, New York.

Smith, P. C. 1989. Hydrogeology. In *Natural History of Dalhousie Springs*, eds. W. Zeidler and W. F. Ponder, pp. 27-41. South Australia Museum, Adelaide.

Soltz, D. L., and R. J. Naiman. 1978. The natural history of native fishes in the Death Valley system. *Nat. Hist. Mus. Los Angeles Co., Calif.*, Sci. Series 30: 1-76.

Sostoa, A., F. J. Sostoa, J. Lobón-Cervia, B. Elvira, J. A. Hernando, and M. Avelia. 1984. Atlas Y Distribucion delos Peces de agua dulce de España. El Projecto, methodos Y Resuttudos Prelimanares. *Bol. Ast. Cent.* 13 (25):75-81.

Szaro, R. C., and J. N. Rinne. 1988. Ecosystem approach to management of southwestern riparian communities. *Trans. 53rd N. A. Wildl. Nat. Res. Conf.* 53:502-511.

Unmack, P., and C. Brumley. 1991. Initial observations on spawning and conservation status of the red-finned blue-eye (*Scaturiginichthys vermeilipinnis*). *J. Austral. New Guinea Fishes Assoc.* 6(4):282-284.

Williams, J. D. 1981. Threatened desert fishes and the Endangered Species Act. In *Fishes in North American Deserts*, eds. R. J. Naiman and D. L. Soltz, pp. 447-475. John Wiley and Sons, New York.

Williams, J. E., and R. R. Miller. 1990. Conservation status of North American fish fauna in fresh water. *J. Fish. Biol.* 37:79-86.

Williams, J. E., D. B. Bowman, J. E. Brooks, A. A. Echelle, R. J. Edwards, D. A. Hendrickson, and J. J. Landye. 1985. Endangered aquatic ecosystems in North American deserts with a list of vanishing fishes of the region. *J. Ariz.-Nev. Acad. Sci.* 20:1-62.

Williams, J. E., J. E. Johnson, D. A. Hendrickson, S. Contreras-Balderas, J. D. Williams, M. Navarro-Mendoza, D. E. McAllister, and J. E. Deacon. 1989. Fishes of North America Endangered, threatened, and of special concern: 1989. *Fisheries* 14(6):2-21.

Zeidler, W., and W. F. Ponder. 1989. Conclusions and recommendations. In *Natural History of Dalhousie Springs*, eds. W. Zeidler and W. F. Ponder, p. 125. South Australia Museum, Adelaide.

26

CANOPY TREE RETENTION AND AVIAN DIVERSITY IN THE OREGON CASCADES

Andrew J. Hansen

Biology Department
Montana State University
Bozeman, Montana
U.S.A.

Patrick Hounihan

Forest Science Department
Oregon State University
Corvallis, Oregon
U.S.A.

Forested managers are developing and implementing new strategies to sustain ecological productivity, biodiversity, water quality, and other factors (Kessler et al. 1992). This trend has generated considerable debate on the economic, social, and ecological consequences of these new strategies. A centerpiece of "ecological forestry" in the Pacific Northwest (PNW) United States is the retention of variable densities of live trees in harvest units in an attempt to maintain canopy complexity over the full rotation cycle (Franklin 1989; Gillis 1990). This approach contrasts sharply with the traditional practice of clearcutting all live and dead trees to facilitate the establishment and growth of plantations of shade-intolerant species Douglas-fir (*Pseudotsuga menziesii*).

The rationale for canopy tree retention derives from patterns of disturbance and succession in natural PNW forests. The variable disturbance regimes, large tree sizes, and "legacies" of structures surviving disturbances result in natural forests of all ages having high variability in tree size and canopy layering (Spies and Franklin 1991; Hansen et al. 1991). This structural complexity influences several ecological attributes (Franklin et al. 1981). Retaining canopy complexity in managed forests is hypothesized (Swanson and Franklin 1992; Franklin 1992) to: maintain habitat diversity for forest organisms; promote nutrient cycling; maintain beneficial predator/prey relationships among forest invertebrates; provide refugia and inocula for nonvagile mycorrhizae and invertebrates; provide sources of coarse woody debris for both uplands and streams; and enhance dispersal opportunities for species that avoid forest openings.

Critics of canopy tree retention, on the other hand, argue that it increases harvest and regeneration costs, threatens human safety, promotes the spread of forest diseases, and inhibits the growth of regenerating Douglas-firs (Atkinson 1992). Clearly, trade-off analyses are needed on the relative costs and benefits of this silvicultural strategy.

The effects of canopy retention in managed forests on bird communities have yet to be examined in the region. Ecological theory, however, suggests that the relationship between canopy density and avian diversity should be strong. The habitat niche hypothesis maintains that structurally complex forests contain more habitat niches and should thus support more bird species than do more homogeneous forests (MacArthur and MacArthur 1961; Urban and Smith 1989). This hypothesis, which has been found to prevail in natural forests in the region (Ruggiero et al. 1991), suggests that bird diversity is positively correlated with canopy tree retention.

An alternative hypothesis asserts bird diversity is inversely related to canopy density because of food availability (Hansen et al. in press). According to this hypothesis, net primary production (NPP) in closed-canopy conifer stands is mostly fixed as wood and unpalatable conifer leaves. In open-canopy stands, relatively more of the NPP is available to consumers because it is fixed in the form of palatable hardwood and herb leaves, buds, fruits, seeds, and flowers. The higher level of available energy (in plants and invertebrates) and greater number of energy pathways in open canopy forests should support higher bird diversity (Begon et al. 1986). Possibly both habitat niche diversity and energetics influence bird communities, resulting in complex relationships between canopy tree density and bird diversity.

Regardless of the underlying ecological processes, forest managers want to know how canopy tree retention influences biodiversity. Which species and communities respond to canopy tree density and are those responses positive or negative? Do thresholds exist where small changes in tree density result in large changes in biodiversity? What densities and size-class distributions of tree retention best accomplish specific biodiversity and other objectives? Such knowledge can help managers design forests to achieve their management goals better.

In this paper we report on a correlative study of bird response to a gradient of canopy tree retention levels in managed forests at higher elevations in the west Cascades of Oregon. Specific questions were: (1) Do significant relationships exist among canopy tree density and attributes of bird communities?, (2) How strong are these relationships relative to the effects of other habitat measures?, (3) Are the associations between canopy tree density and birds positive or negative, linear or nonlinear?, (4) What size classes of retained trees are most closely associated with variation in the abundance of individual bird species?

METHODS

Study Sites

Habitat structure and breeding birds abundance were measured across 16 managed stands representing a gradient of canopy tree retention levels. The stands were within the McKenzie and Sweet Home Ranger Districts of the Willamette National Forest in western Oregon. Fourteen stands were between Highway 126 to the west and the Mt. Washington

Wilderness to the east. The other two stands were approximately 15 km to the north near the junction of Highway 126 and Highway 20.

This area is on the western edge of the High Cascades Province (Franklin and Dyrness 1973), a high-elevation plateau of volcanic origin (late Pliocene and Pleistocene epochs) that straddles the Cascade crest. Pleistocene glaciers covered the area. Soils are immature, derived from glacial deposits and volcanic ejecta. The climate here is cool and moist, with a mean annual precipitation of 2190 mm, mean January temperature of -1.7° C and a mean July temperature of 14° C (Franklin and Dyrness 1973). A snowpack persists over much of the winter and spring. The study area lies between the *Tsuga heterophylla* and *Abies amabalis* Zones (Franklin and Dyrness 1973). Dominant trees are Douglas-fir, western hemlock (*Tsuga heterophylla*), grand fir (*Abies grandis*), noble fir (*Abies procera*), and silver fir (*Abies amabilis*). Common understory plants are chinquapin (*Castanopsis chrysophylla*), Pacific dogwood (*Cornus nuttallii*), and cascara buckthorn (*Rhanus purshiana*). These stands were generally established 100 to 150 years ago after intense wildfires. Shrub biomass within the study sites appeared to be substantially less than at lower elevations to the west. This might reflect the effects of elevation, snowpack, soils, or wildfire history.

The stands were 9 ha or greater in area, at elevations of 731-1,189 m, and varied in aspect (Table 26.1). Each stand had been subjected to timber harvest. Four sites were

Table 26.1. Attributes of stands sampled in this study.

Location[1]	TRI Number[2]	Area (HA)	Elevation (M)	Aspect[3]/ Slope[4]	Harvest Year	Harvest System[5]	Number of Plots
15-7-8	7109-123	13	914	NE/25	1987	1	6
15-7-31	7110-138	19	914	SE/15	1989	1	6
15-6-25	7110-133	9	762	0	1987	1	6
15-6-25	7110-244	12	884	0	1987	1	6
15-7-27	7110-50	17	1189	W/10	1988	2	6
15-7-28	7110-275	17	1066	W/10	1988	2	8
16-7-8	7112-147	13	1098	SE/12	1988	3	6
15-7-15	7109-137	15	1146	W/15	1985	3	7
14-6-1	3503-95	16	1098	E/15	1987	4	6
14-6-14	3502-64	20	1112	SE/15	1986	4	7
15-7-19	7109-212	13	914	SE/5	1987	4	6
16-7-4	7110-263	13	1113	0	1986	4	6
15-7-15	7109-52	13	1128	W/10	1979	5	6
15-7-19	7109-84	35	914	N/25	1982	5	6
15-6-35	7110-128	19	731	SW/15	1985	5	6
15-6-25	7110-18	82	792	0	1982	5	8

[1] Township (south) - Range (east) - Section

[2] USDA Forest Service Total Resource Inventory Number

[3] Cardinal direction

[4] Percent slope

[5] 1- Clearcut (no retention); 2 - Clearcut (light retention); 3 - Clearcut (moderate retention); 4 - Shelterwood; 5 - Commercial thin.

See text for explanation.

clearcut, four were clearcut with retention of 2-14 trees per ha, four underwent shelterwood cuts where a mean of 19 trees per ha were retained, and four stands were commercially thinned to a mean of 87 trees per ha. The retention and shelterwood units were harvested between 1985 and 1989, and most were burned and planted with Douglas-fir; the trees retained were 30 cm diameter at breast height (dbh) or larger. The commercial thins were done between 1979 and 1985, and the majority of trees retained were 10-30 cm dbh. In all stands, the trees retained were dispersed, rather than clumped.

Bird and Habitat Sampling

The abundance of breeding birds was sampled using the Variable Circular-Plot method (Reynolds et al. 1980). Plot centers were placed 100 m apart and at least 75 m from stand edges. Six to eight plots were located in each stand, depending on the size of the stand. Censuses began each day at dawn and continued for no more than four hours. Observers walked to a plot center, waited two minutes, and then recorded all birds seen and heard during an eight-minute period. Variables recorded for each bird were: species, distance class to bird at first detection, distance class to bird at nearest detection, distance class of bird from the nearest edge of the stand, and type of stand neighboring that stand edge. Flagging was placed 40 m from each plot center along the four cardinal directions to aid in estimating distance class to a bird. Each plot was censused five times during the period May 15 - June 31, 1991. Observers rotated among plots and stands to minimize bias. The common name, scientific name, and code for each bird species sampled are listed in Table 26.2.

Habitat measurements were centered on the bird census plots. The variables measured involved topography, tree density, understory cover, and canopy cover. Descriptions of the variables and the sampling scheme are presented in Table 26.3 and Figure 26.1. The measurements were made during July and August 1991.

Data Analyses

Only birds registered within 50 m of plot centers were included in the analyses. This prevented overlap in the areas covered from adjacent plots. An analysis by Spencer (1993) in habitats similar to the study area revealed that the songs of all bird species in our area can be detected within 50 m. We calculated relative abundance for each species as the number of individuals registered within the 50-m radius plot at either the first or nearest detection. The results for each species were averaged over plots within a stand and across censuses and are reported as mean number of registrations/ha/census. Total bird abundance was the sum for all species sampled.

Analyses of habitat associations for individual bird species were done only for species with 9 or more registrations. All bird species detected were considered for community-scale bird-habitat analyses. These community variables were: relative abundance of all species combined (ALL), bird species richness (RICHNESS), Shannon's diversity index (SHANNON) and Hill's N2 (HILL'S), a diversity index for the relatively abundant species (Ludwig and Reynolds 1988).

Table 26.2. Bird species sampled in this study.

Species Code	Common Name	Latin Name	Registrations[1]
AMRO	American robin	*Turdus migratorius*	12
BRCR	brown creeper	*Certhia americana*	10
CBCH	chestnut-backed chickadee	*Parus rufescens*	6
CHSP	chipping sparrow	*Spizella passerina*	18
DEJU	dark-eyed junco	*Junco hyemalis*	176
DOWO	downy woodpecker	*Picoides pubescens*	6
DUFL	dusky flycatcher	*Empidomax oberholseri*	5
EVGR	evening grosbeak	*Coccothraustes vespertinus*	8
FOSP	fox sparrow	*Passerella iliaca*	2
GCKI	golden-crowned kinglet	*Regulus satrapa*	22
GRJA	gray jay	*Perisoreus canadensis*	3
HAFL	Hammond's flycatcher	*Empidonax hammondii*	14
HAWO	hairy woodpecker	*Picoides villosus*	16
HTWA	hermit/Townsend's warbler[2]	*Dendroica occidentalis* and *D. townsendi*	50
HOWR	house wren	*Troglodytes aedon*	5
MCWA	McGillivray's warbler	*Oporornis tolmiei*	9
OSFL	olive-sided flycatcher	*Contopus borealis*	2
PISI	pine siskin	*Carduelis pinus*	6
PUFI	purple finch	*Carpodacus purpureus*	1
RBNU	red-breasted nuthatch	*Sitta canadensis*	9
RBSA	red-breasted sapsucker	*Sphyrapicus ruber*	2
RUHU	rufous hummingbird	*Selasphorus rufus*	5
SOSP	song sparrow	*Melospiza melodia*	1
STJA	Steller's jay	*Cyanocitta stelleri*	9
TOSO	Townsend's solitaire	*Myadestes townsendi*	1
TRSW	tree swallow	*Tachycineta bicolor*	2
WCSP	white-crowned sparrow	*Zonotrichia leucophrys*	10
WEBL	western bluebird	*Sialia mexicana*	11
WEFL	western flycatcher	*Empidonax difficilis*	6
WETA	western tanager	*Piranga ludoviciana*	20
WIWA	Wilson's warbler	*Wilsonia pusilla*	1
WIWR	winter wren	*Troglodytes troglodytes*	1
YRWA	yellow-rumped warbler	*Dendroica coronata*	5

[1] Number of registrations within 50 m of a plot center during the 5 censuses.

[2] These two species overlap and hybridize in the study area and are very difficult to distinguish by song Consequently they were lumped in this study.

Tree density data were averaged across subplots within each plot. Means and standard deviations were then calculated among plots within each stand and used in the bird habitat analyses. Similarly, understory and canopy cover were averaged among subplots within each plot and the mean calculated among plots within each stand. The four tree

Table 26.3. Habitat attributes used in the analyses of bird habitat relationships. See Figure 26.1 for a depiction of the sampling design.

Attribute	Code[1]	Description
Elevation	ELEV	Elevation above sea level (m) at plot center
Slope	SLOPE	Average slope (%) within 20 m of plot center as measured with a clinometer
Aspect	ASPECT	Average aspect (cardinal direction) within 20 m of plot center as measured with a compass
Tree density	TOT2	Number of trees (per ha) 10-30 cm diameter at breast height (dbh) within 4 5-m radius subplots placed 20 m from plot center
	TOT3	Number of trees (per ha) 30-50 cm dbh as sampled in TOT2
	TOT4	Number of trees (per ha) 50-90 cm dbh as sampled in TOT2
	TOT5	Number of trees (per ha) >90 cm dbh as sampled in TOT2
	TOT2-5	Number of trees (per ha) >10 cm dbh as sampled in TOT2
	TOT45	Number of trees (per ha) >50 cm dbh as sampled in TOT2
	SHADE2	Number of shade-tolerant (grand fir) trees (per ha) 10-30 cm dbh as sampled in TOT2
	SUN2	Number of shade-intolerant (Douglas-fir and noble fir) trees (per ha) 10-30 cm dbh as sampled in TOT2
Canopy cover	CANCOV	Percent cover of all vegetation above 2 m in height measured with the moosehorn technique at 4 points 20 m from plot centers
Understory	UNDCOV	Percent cover of vegetation <2 m in height estimated visually within 4 1-m subplots distributed around each of 4 points 20 m from plot centers

[1] Habitat codes followed by "L" in the text denote a log transformation. Habitat codes followed by "SD" in the text denote standard deviation.

size classes that were >10 cm dbh were aggregated into the variable TOT2-5. We refer to this variable as canopy tree density. Habitat variables were generally not included in the bird habitat analyses if they: had a sample size of 9 or fewer non zero values; had distributions that differed significantly from normal (data transformations were used where helpful); were strongly correlated with other habitat variables; or were judged not have ecological relevance to the bird species. The variables used were: ELEV, ASPECT, UNDCOV, and a log transformation of TOT2-5 (denoted as TOT2-5L). For all analyses, relationships were considered statistically significant at the 0.05 level.

Stepwise linear regression was used to determine the relative amount of variation in each of the bird variables associated with each habitat variable (ELEV, ASPECT, UNDCOV, AND TOT2-5L). The shapes of the relationship between TOT2-5 and each bird variable were determined by fitting a linear model and two nonlinear models, the natural growth function (y=a+b*(1-exp(-b*x))) and the logistic function (y=a/(1+b*exp(-c*x))). The model with the tightest fit was selected as the best descriptor of the relationship. A log transformation of TOT2-5 was not used in this analysis because we wanted the plots of bird abundance on tree density to be readily interpretable by forest managers. Plots of these curves enable the reader to identify thresholds where small changes in tree density are associated with large changes in bird abundance.

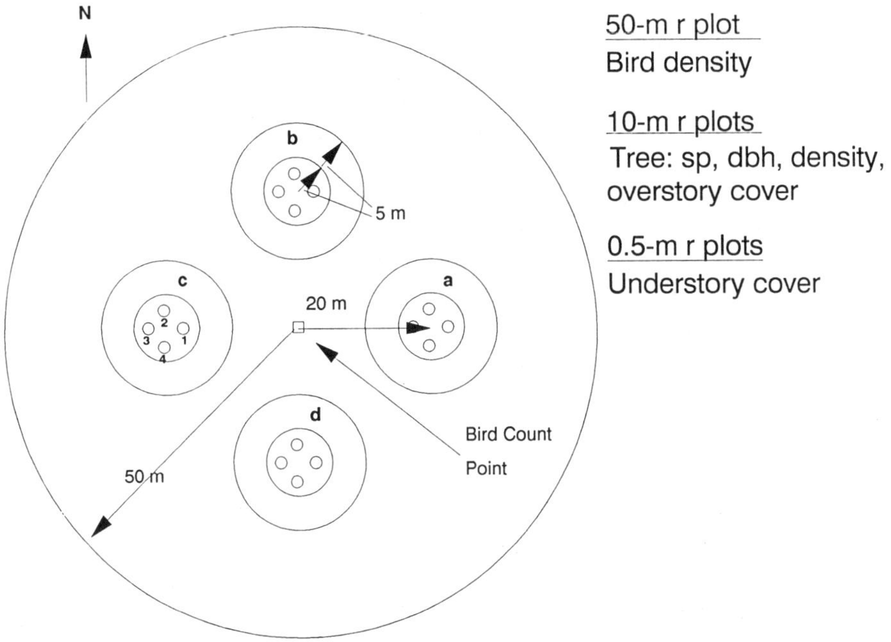

Figure 26.1. Depiction of the vegetation sampling protocol used at each plot within a study stand. See Table 26.3 for a list of the variables sampled.

The final analysis asked which combination of tree density variables (mean and variance of the tree size classes) were most associated with variation in each of the bird variables. Stepwise linear regression (p-value to enter=0.05) was used to select habitat variables that were significantly related to variation in each of the bird variables. The habitat variables included mean and standard deviation of TOT2-5, TOT3, TOT45, and TOT2 differentiated into shade tolerant species (SHADE2) and shade intolerant species (SUN2). This was done to differentiate between habitats with shaded and open understories. The analysis was done separately for untransformed variables and log-transformed variables, and the best models (highest R^2) are reported here.

An assumption of linear regression is that the error term is normally distributed. We tested this assumption for the variable sets selected in the stepwise procedure and found that the residuals for all but one of the species (brown creeper) did not differ significantly from a normal distribution.

RESULTS

A total of 33 species of birds were tallied in the study plots (Table 26.2). Fourteen of these had 9 or more registrations and were subjected to species-level statistical analyses. Species associated with forest canopies and boles were relatively abundant in the sample. Understory associates other than dark-eyed junco were sparse.

Tree species with more than 9 non-zero values in the > 10 cm dbh classes included Douglas-fir, western hemlock, grand fir, and noble fir. Correlation analyses showed that most vegetation variables were not biased in distribution relative to elevation and aspect. None of the variables was correlated with ASPECT and only TOT45 was significantly associated with ELEV (Table 26.4). CANCOV, UNDCOV, and several tree density variables were positively related. The correlation was especially strong between CANCOV and TOT2-5. For this reason, CANCOV was not included in the bird habitat analyses.

The abundances of 7 bird species were significantly related to the major habitat variables (Table 26.5), particularly tree density. TOT2-5L was associated with most of the variation in abundance for 6 of the 7 bird species. This was also true for the community-level variables, RICHNESS, SHANNON, and HILL'S. Total bird abundance (ALL) was most strongly related to UNDCOV.

Nonlinear regression models provided better fits than linear models for the relationships between each of the bird variables and TOT2-5. Both the directions of the slopes and shapes of the curves differed among species. American robin and dark-eyed junco abundances were negatively related to tree density, and their abundances dropped precipitously between clearcuts and stands with only a few canopy trees per ha (Figures 26.2 and 26.3). The American robin feeds on the ground in open stands, and the dark-eyed junco forages and nests in well developed understories below open canopies.

The abundance of the canopy gleaner hermit/Townsend's warbler increased proportionally with tree density (Figure 26.4). This trend began at the origin and continued to the highest tree densities. Golden-crowned kinglet, also a canopy gleaner, exhibited a threshold relationship. It was absent from stands with fewer than 25 trees per ha and increased proportionally with tree density in stands with higher stem densities (Figure

Table 26.4. Pearson correlation coefficients and p-values among habitat variables.

	Elev	Aspect	Cancov	Undcov	Tot2-5	Tot2	Tot3	Tot45	Shade2
Sun2	-0.43	0.27	0.81	0.58	0.93	0.98	0.55	-0.20	0.84
	0.09	0.31	0.00	0.02	0.00	0.00	0.03	0.46	0.00
Shade2	-0.48	0.44	0.95	0.72	0.89	0.91	0.47	-0.09	
	0.06	0.09	0.00	0.00	0.00	0.00	0.07	0.74	
Tot45	0.57	0.00	0.18	0.00	0.12	-0.21	0.15		
	0.02	1.00	0.50	1.00	0.66	0.44	0.59		
Tot3	-0.19	0.32	0.58	0.29	0.66	0.56			
	0.49	0.23	0.02	0.28	0.01	0.02			
Tot2	-0.49	0.31	0.87	0.63	0.95				
	0.05	0.24	0.00	0.01	0.00				
Tot2-5	-0.31	0.33	0.94	0.63					
	0.24	0.22	0.00	0.01					
Undcov	-0.28	0.26	0.74						
	0.30	0.34	0.00						
Cancov	-0.30	-0.39							
	0.26	0.13							
Aspect	0.09								
	0.73								

Table 26.5. Results of stepwise regressions of bird species community attributes against the habitat variables ELEV, ASPECT, UNDCOV, AND TOT2-5L.

Species[1] Code	Variables Included	Partial R^2	Overall F-Value	Overall Prob > F
AMRO	TOT2-5L	.47	12.4	.003
DEJU	TOT2-5L	.34	7.1	.018
GCKI	TOT2-5L	.50	14.3	.002
HAFL	TOT2-5L	.55	23.7	.0001
	ELEV	.23		
HTWA	TOT2-5L	.67	27.4	.0001
	ELEV	.12		
MCWA	UNDCOV	.26	5.0	.043
RBNU	TOT2-5L	.39	8.8	.010
ALL	UNDCOV	.45	11.7	.004
RICHNESS	TOT2-5L	.59	19.1	.0001
	UNDCOV	.12		
	ELEV	.12		
SHANNON	TOT2-5L	.76	29.8	.0001
	ELEV	.06		
HILL'S	TOT2-5L	.68	30.4	.0001

[1] Bird and habitat codes are defined in Tables 26.2 and 26.3 and in the text. The significance level for allowing a habitat variable to stay in the model was 0.05. Only statistically significant models are shown. Insignificant models resulted for chipping sparrow, hairy woodpecker, house wren, Steller's jay, white-crowned sparrow, western bluebird, and western tanager.

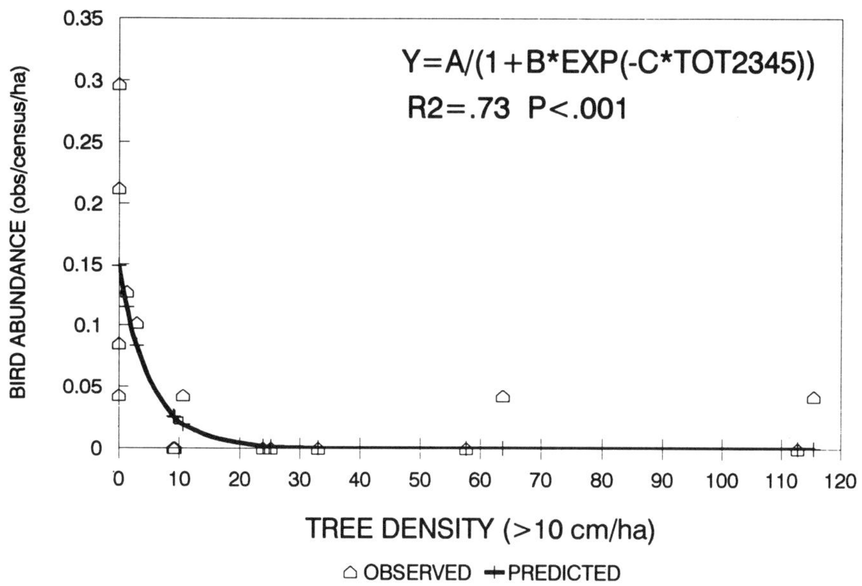

Figure 26.2. Data points and nonlinear regression line describing the relationship between canopy tree density and the American robin.

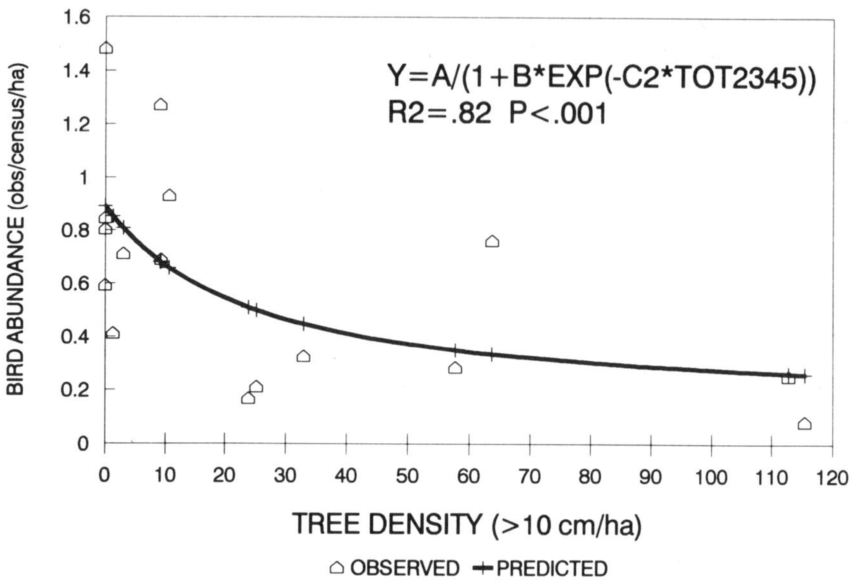

Figure 26.3. Data points and nonlinear regression line describing the relationship between canopy tree density and the dark-eyed junco.

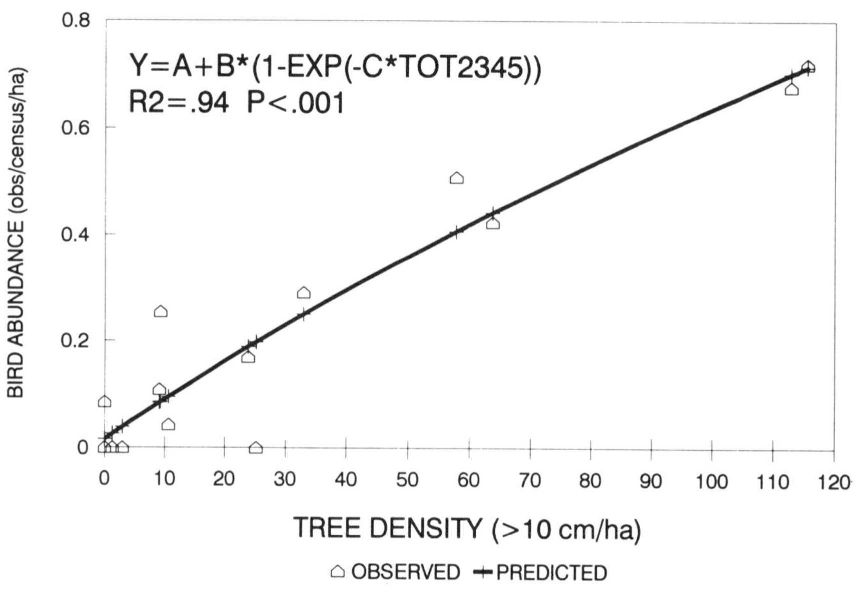

Figure 26.4. Data points and nonlinear regression line describing the relationship between canopy tree density and the hermit/Townsend's warbler.

26.5). Hammond's flycatcher, a sallying insectivore, showed a similar relationship (Figure 26.6). The curve for brown creeper, which forages on the bark on tree boles and branches, was sigmoidal, with the lower asymptote at about 10 trees per ha and the upper asymptote at about 30 trees per ha (Figure 26.7). Red-breasted nuthatch, also a bark forager, had a logistic relationship with tree density (Figure 26.8). Overall, these habitat associations were quite strong: tree density accounted for more than 70% of the variation in abundance for 6 of the 7 bird species with significant models (Figures 26.2-26.8).

The bird community measures were also tightly related to tree density. Total bird abundance increased proportionally with TOT2-5 (Figure 26.9). The three measures of diversity, RICHNESS, SHANNON, and HILL'S, each had logistic curves, reaching asymptotes at about 15-30 trees per ha (Figures 26.10-26.12). Tree density was associated with 96% to 99% of bird diversity. Forest structural complexity is a possible factor underlying the relationships between bird diversity and tree density. We found that an index of structural complexity (mean number of tree size classes per stand) was also related to TOT2-5 (Figure 26.13). This index was significantly correlated with bird diversity (SHANNON) (n=16, R=0.70, p<0.003).

Bird species differed in their associations with the various tree size classes. TOT2-5 was the first variable selected in the stepwise regressions for American robin, dark-eyed junco, golden-crowned kinglet, and hermit/Townsend's warbler (Table 26.6). Red-breasted nuthatch was associated with variation in tree density (TOT2-5SD). Brown creeper, hairy woodpecker, and Steller's jay were significantly related to the mean and variance of large trees. Shade-tolerant trees 10-20 cm dbh and large trees was associated with most of the variation in Hammond's flycatcher density.

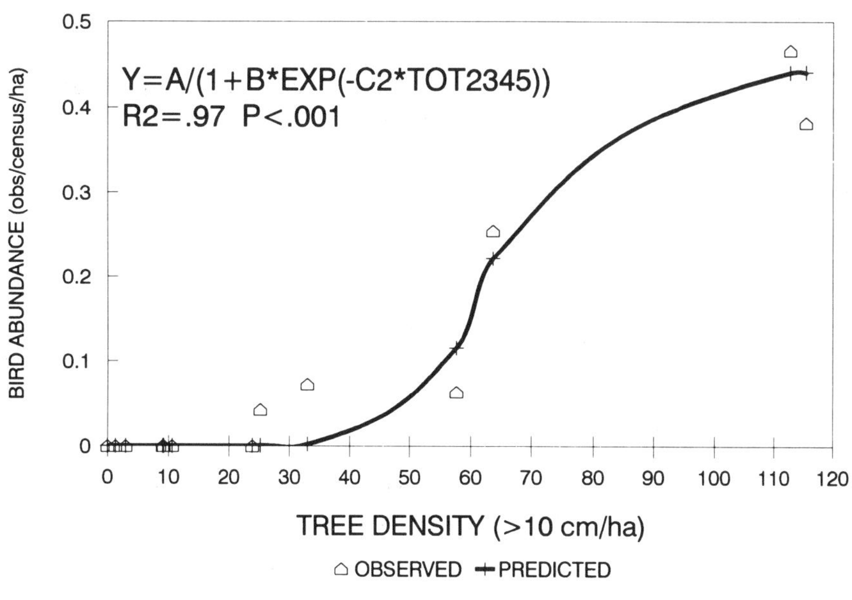

Figure 26.5. Data points and nonlinear regression line describing the relationship between canopy tree density and the golden-crowned kinglet.

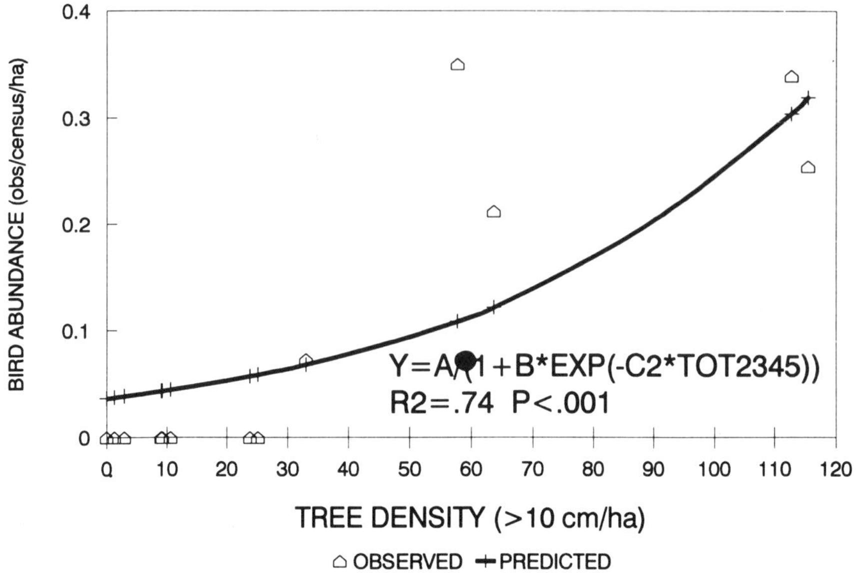

Figure 26.6. Data points and nonlinear regression line describing the relationship between canopy tree density and the Hammond's flycatcher.

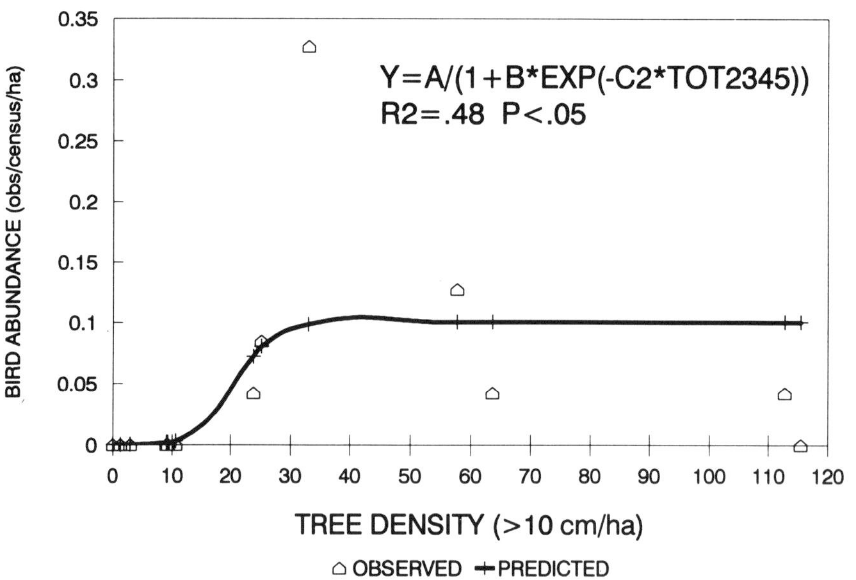

Figure 26.7. Data points and nonlinear regression line describing the relationship between canopy tree density and the brown creeper.

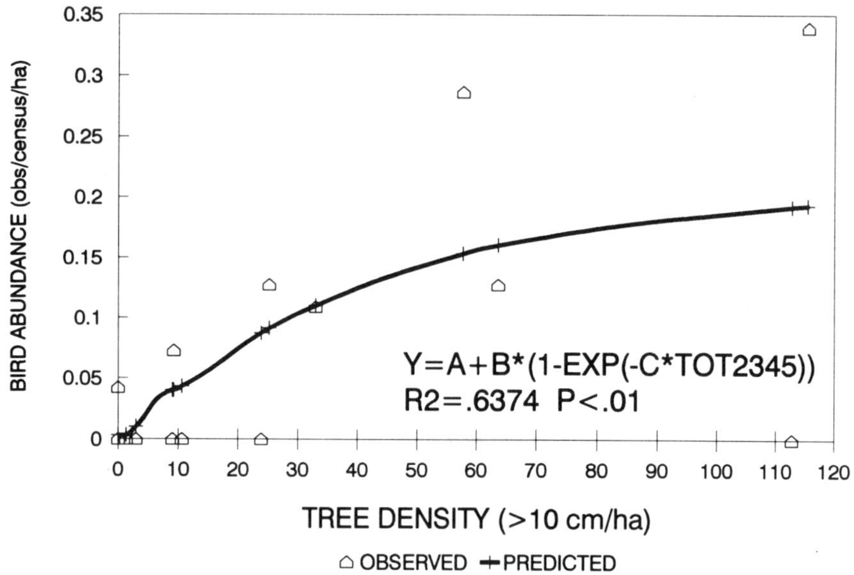

Figure 26.8. Data points and nonlinear regression line describing the relationship between canopy tree density and the red-breasted nuthatch.

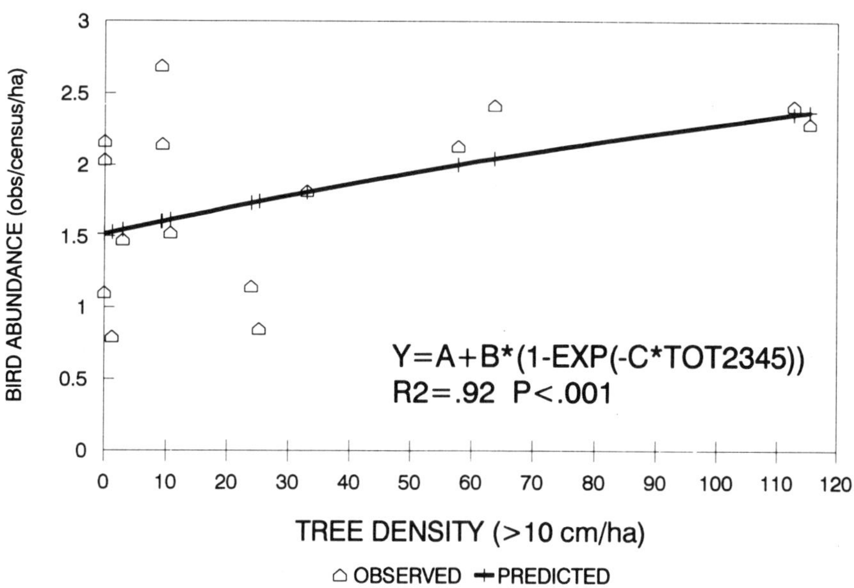

Figure 26.9. Data points and nonlinear regression line describing the relationship between canopy tree density and total bird abundance.

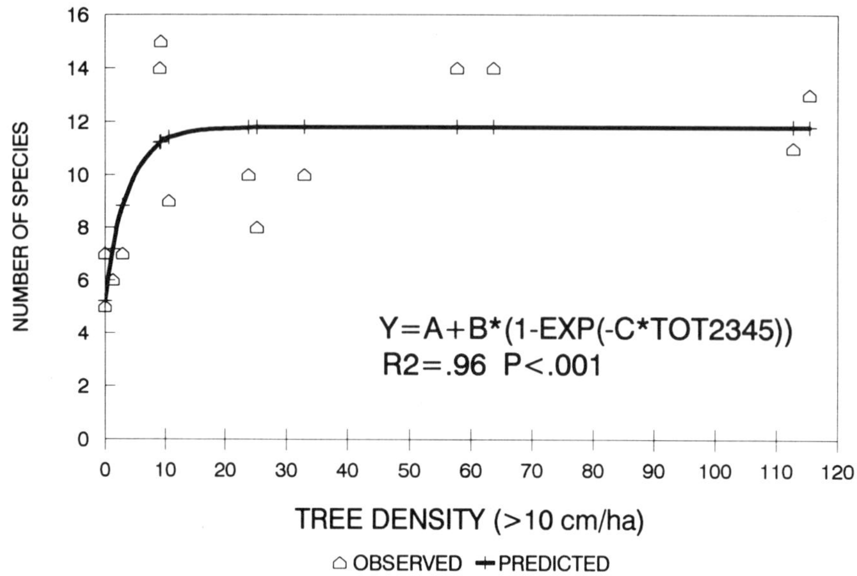

Figure 26.10. Data points and nonlinear regression line describing the relationship between canopy tree density and bird species richness.

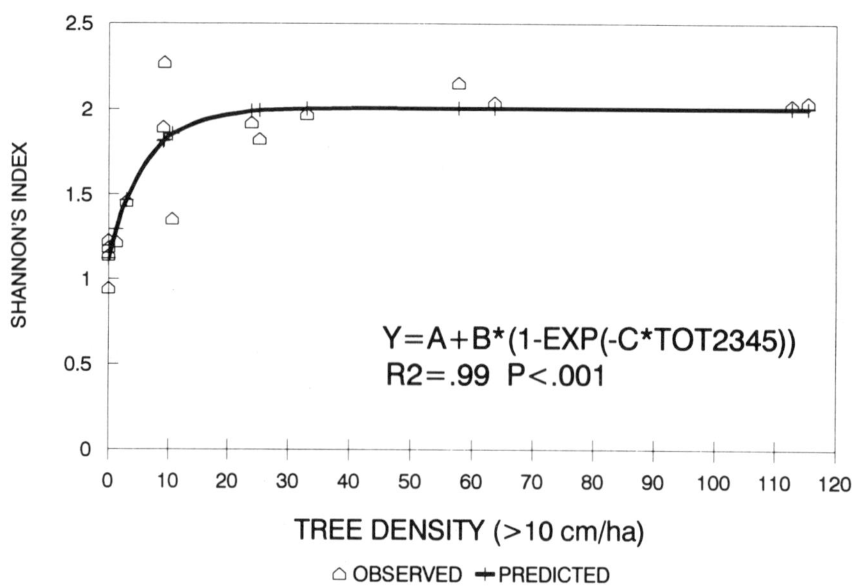

Figure 26.11. Data points and nonlinear regression line describing the relationship between canopy tree density and bird diversity (Shannon's index).

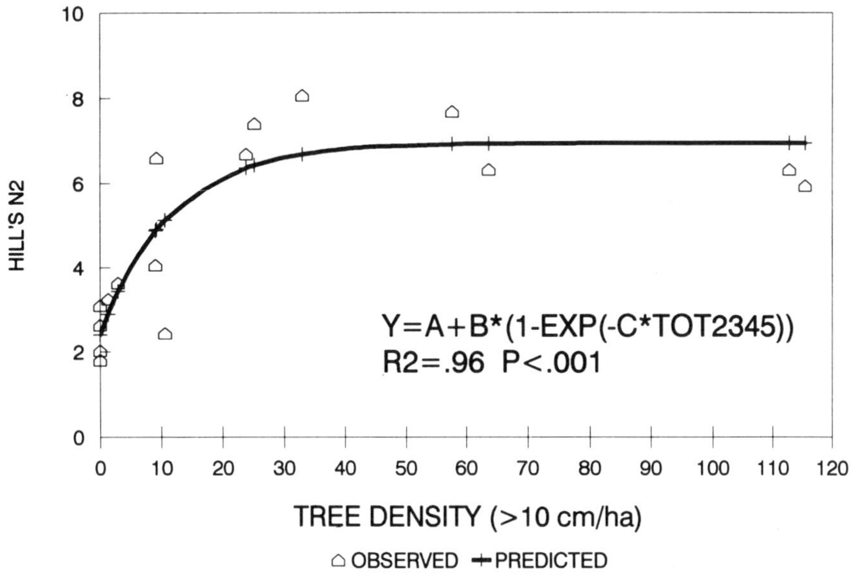

Figure 26.12. Data points and nonlinear regression line describing the relationship between canopy tree density and bird diversity (Hill's N2).

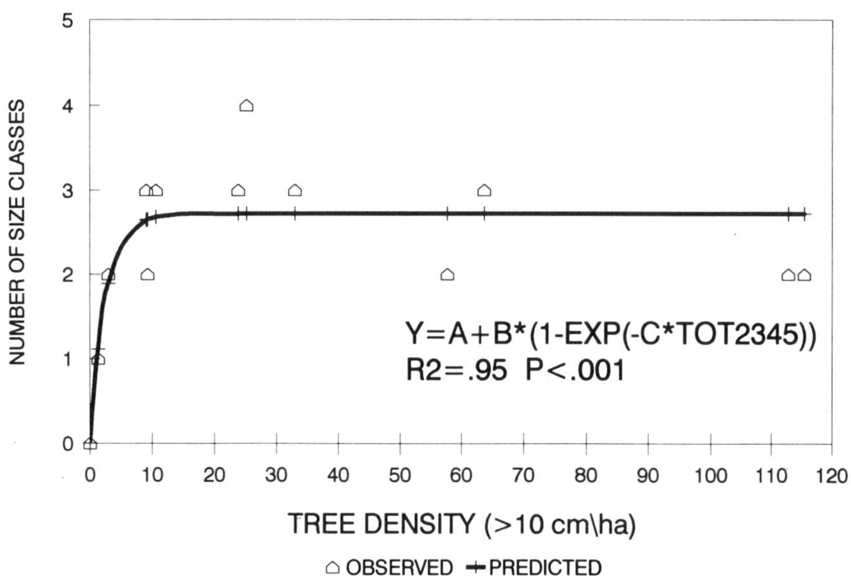

Figure 26.13. Data points and nonlinear regression line describing the relationship between canopy tree density and the number of canopy tree size classes present.

Table 26.6. Results of stepwise regressions of bird species abundance against the mean and standard deviations of each of the tree density variables.

Species[1] Code	Variables Included	Overall R^2	Overall F-Value	Overall Prob > F
AMRO	TOT2-5L	.56	8.3	.005
BRCR	TOT45, TOT45SD	.79	24.6	.0001
DEJU	TOT2-5L	.34	7.1	.018
GCKI	TOT2-5, TOT2-5SD	.94	97.3	.0001
HAFL	SHAD2, TOT45	.97	237.8	.0001
HAWO	TOT45L	.33	6.7	.020
HTWA	TOT2-5	.90	123.6	.0001
RBNU	TOT2-5SD	.59	20.1	.0002
STJA	TOT45SD	.37	8.3	.01

[1] Bird and habitat codes are defined in Tables 26.2 and 26.3. The significance level for allowing a habitat variable to stay in the model was 0.05. Only statistically significant models are shown.

DISCUSSION

The results suggest that the density of canopy trees in managed forest stands strongly influences breeding bird abundance and diversity. Significant relationships were found between tree density variables and 13 of the 18 bird species and community variables with sufficient sample sizes for analysis. Moreover, the density of trees greater than 10 cm dbh (TOT2-5) was associated with more of the variation in the bird variables than were the other major habitat variables (elevation, aspect, and understory cover).

The latter finding should not be interpreted as evidence against the potential importance of these other habitat variables. The study sites were selected to represent a gradient of tree densities and to control for or randomize elevation and aspect. Another study surveying a broader range of elevations found elevation to be a strong correlate with bird richness (Huff and Raley 1991). It is likely that some of our results reflect the fact that most of our sites were at relatively high elevations (see below).

Canopy tree density (TOT2-5) was associated with a surprisingly large proportion (48% to 97%) of the variation in bird abundance for the bird species with significant models (Figures 26.2-26.8). The relationships were even stronger (89% to 99%) for the community variables ALL, RICHNESS, SHANNON, and HILL'S (Figures 26.9-26.12). For comparison, previous studies in the region have generally found that habitat variables were associated with less than 50% of the variation in bird abundance (e.g., Gilbert and Allwine 1991; Manuwal 1991). Thus, our results suggest that in this study area tree density is an important determinant of habitat quality for the bird community and several individual species.

Species were individualist in their associations with tree density. As expected, species requiring open-canopy habitats were negatively associated with tree density, while closed-canopy species showed positive associations. American robin, for example, was only abundant in clearcuts, likely because overstory shading inhibits food availability for this ground-foraging omnivore. Similarly, canopy- and bole-foraging species were most abundant in dense stands probably because the high canopy volumes and dense boles conferred foraging and nesting opportunities.

All of these relationships between birds and tree density were best described by nonlinear models. These curves revealed interesting thresholds where small changes in tree density were associated with large differences in bird abundance. The abundances of golden-crowned kinglet and Hammond's flycatcher, for example, were positively associated with tree density only above a threshold of about 25 trees per ha. The relationship between bird abundance and tree density for brown creeper became positive at about 10 trees per ha and plateaued above about 30 trees per ha. It is difficult to speculate on the ecological factors that may explain these thresholds. For example, golden-crowned kinglet and hermit/Townsend's warbler have similar foraging habits, yet hermit/Townsend's warbler showed very different responses to stands with few canopy trees (cf. Figures 26.4 and 26.5).

The bird community variables were all positively associated with tree density. Total bird abundance varied at lower tree densities but was consistently high above about 30 trees per ha. The high abundance of hermit/Townsend's warbler, golden-crowned kinglet, and Hammond's flycatcher in the densest stands accounts for this trend. Bird richness and diversity increased with tree density up to a threshold and then leveled off. The low bird abundance and diversity values in stands with few trees were due to the relative paucity in the study area of species specializing in open-canopy habitats.

Bird species differed in their response to the size classes of the retained trees. The abundances of the open-canopy specialists American robin and dark-eyed junco were inversely related to tree density in general, regardless of size-class distribution. Similarly, canopy-gleaning species (golden-crowned kinglet and hermit/Townsend's warbler) probably benefit from increased canopy volume and thus were positively associated with canopy tree density regardless of the size-class distribution. On the other hand, birds that forage and nest in larger trees with furrowed bark (brown creeper, hairy woodpecker) were associated with the density of trees > 50 cm dbh. Interestingly, Hammond's flycatcher was associated both with shade-tolerant trees 10-30 cm dbh and with large trees, suggesting it requires habitats with complex, multi-layered canopy configurations. Knowledge of these habitat associations can allow managers to design stands to favor particular groups of species (see below).

Ecological Mechanisms

What ecological processes underlie these positive associations between tree density and bird diversity and abundance? The results provide partial support for the hypothesis that habitat niche diversity is a fundamental determinant of bird diversity (MacArthur and MacArthur 1961). An index of habitat diversity (tree size-class richness) increased logistically with tree density and was positively correlated with bird diversity (SHANNON). This correlation was not significant, however, when the four sites with no canopy trees were omitted from the analysis. Thus, the analysis simply revealed that stands with 1 or more of the tree size classes had higher bird diversity than stands with 0 tree size classes. A larger sample over a greater range of structural configurations is needed to better understand the relationships among canopy tree retention, habitat niche diversity, and bird diversity.

The data did not support the hypothesis (Hansen et al. in press) that bird diversity is highest in open-canopy stands because of increased available energy. In fact, understory vegetation and understory birds were poorly represented across all the study stands. These findings conflict with other studies from lower elevations, where Turner and Long (1975) found that productivity of conifer leaves and understory plants was higher in young open-canopy stands than in closed-canopy forests. Also, Hansen et al. (unpubl. data) found in the Oregon Coast Range that bird richness and abundance did not differ significantly between 5- to 10-year-old clearcuts and 90- to 130-year-old natural forest. The patterns of bird diversity and abundance reported here may be unique to the High Cascades Province. The high elevation, heavy snow packs, soils, or fire history appear to have suppressed the shrub communities and caused the understory bird community to be depauperate.

This conclusion suggests that vegetation and bird response to silvicultural treatments vary among topographic and geomorphic settings. The relative influences on bird communities of niche habitat diversity and stand energetics likely vary across the region. Studies across a range of environmental settings are needed to better understand these ecological relationships and to allow managers to tailor silvicultural strategies to local conditions.

Limitations of the Study

The study has various limitations that necessitate additional research on the topic. Our sample of stands was relatively small in number and included a very limited range of structural configurations within each tree-density class. For example, all units with light to moderate retention contained only trees >30 cm dbh, but those with the heaviest retention contained trees 10-50 cm dbh. The study involved only stands 2-11 years past harvest and it reveals little about older stands. Our study area was high in elevation and on soils that are somewhat unique in the region. Only birds were studied; they might or might not reflect biodiversity patterns for other taxonomic groups. Finally, birds were sampled for only one breeding season, and we do not know how much variability can be expected among years. We also do not know how birds may use these stands during other times of the year. Future studies should endeavor to sample a greater range of structural configurations, forest age classes, topographic and geomorphic settings, and taxonomic groups.

The study only hinted at the sorts of ecological processes linking forest structure and biodiversity. Studies are needed to test the habitat niche hypothesis, the stand energetics hypothesis, and other possible hypotheses relevant to the question.

A final limitation is that the study dealt only with the stand level. Landscape-scale factors such as stand diversity, grain size, juxtapositioning, and connectivity are known to influence vertebrate habitat quality and population viability. Difficult as these studies are to design and execute, landscape-scale research is needed to determine how canopy tree retention should be distributed across stands and landscapes to maintain viable populations of vertebrate species.

CONCLUSIONS AND IMPLICATIONS FOR MANAGEMENT

This study is one of the first in the PNW to quantify the responses of birds to the increasingly common forestry practice of canopy tree retention. The study has several implications for management. Most importantly, the results provide evidence that canopy tree retention strongly influences bird species and community attributes. The statistical associations we found between the bird attributes and canopy tree density may be the strongest animal habitat relationships yet documented in the region. This suggests that canopy structure is a fundamental driver of various ecological properties, including species diversity. Clearly, canopy tree retention is an important silvicultural tool for achieving a more ecological forestry.

Furthermore, the study revealed specific canopy tree densities that appear to be thresholds for influencing bird populations and communities. Such data should enable managers to design stands for specific biodiversity goals. For example, bird diversity reached a plateau in stands with about 20-25 trees per ha (Figs. 11, 12). Retaining fewer trees than this would probably reduce bird diversity, but retaining more trees would not necessarily increase bird diversity. Similar data for wood production, harvest costs, sedimentation rates, and other responses could be used to select tree retention levels that best satisfy diverse management goals.

The individualistic responses of bird species to canopy retention levels emphasize the importance of setting specific management objectives. Managing for "biodiversity" is a hollow goal. Any management action will benefit some species and hinder others. Managers should carefully evaluate which species or community attributes they are most concerned with and design silvicultural strategies accordingly (Hansen et al. 1993).

Where the goal is to maintain habitats for all native bird species, our results indicate that a range of canopy tree densities and size-class distributions should be maintained across the landscape. This would accommodate open-canopy specialists as well as species requiring structurally-complex stands. However, determining the distribution of tree densities and size classes that best promote various biodiversity goals will require further work.

ACKNOWLEDGMENTS

We thank the Blue River, McKenzie, and Sweet Home Ranger Districts, Willamette National Forest, for providing study sites. John Cissel and the staff at the Blue River Ranger District also aided with study site selection. The habitat survey protocol was adapted from work by Kevin McGarrigal. Joseph McLaughlin, Robyn Spencer, and Eric Horvath helped with the field work. Helpful comments on the manuscript were provided by Kevin Gutzwiller, Tom Sabin, and Robyn Spencer, and Jane Thomas. The study was funded by the USDA Forest Service New Perspectives Program, Pacific Northwest Forest and Range Experiment Station, the Cascades Center for Ecosystem Management, and the COPE program (College of Forestry, Oregon State University). This is Paper 2914 of the Forest Research Laboratory, Oregon State University.

LITERATURE CITED

Atkinson, W. 1992. Silvicultural correctness: The politicalization of forest science. *Western Wildlands* 17(4):8-12.

Begon, M., J. L. Harper, and C. R. Townsend. 1986. *Ecology: Individuals, Populations, and Communities*. Sinauer Associates, Inc., Sunderland, Mass.

Franklin, J. F. 1989. Toward a new forestry. *Am. For.* (Nov/Dec):37-44.

Franklin, J. F. 1992. Scientific basis for new perspectives in forests and streams. In *Watershed Management: Balancing Sustainability and Environmental Change*, ed. R. Naiman, pp. 25-72. Springer Verlag, New York.

Franklin, J. F., and C. T. Dyrness. 1973. *Natural Vegetation of Oregon and Washington*. Oregon State University Press, Corvallis, Ore.

Franklin, J. F., K. Cromack, W. Denison, A. McKee, C. Maser, J. Sedell, F. Swanson, and G. Juday. 1981. *Ecological characteristics of old-growth Douglas-fir forests*. USDA Forest Service Gen. Tech. Rep. PNW-GTR-118, Portland, Ore.

Gilbert, F. F., and R. Allwine. 1991. Spring bird communities in the Oregon Cascade Range. In *Wildlife and Vegetation of Unmanaged Douglas-fir Forests*, eds. L. F. Ruggiero, K. B. Aubry, A. B. Carey and M. H. Huff, pp. 145-160. USDA Forest Service Gen. Tech. Rep. PNW-GTR-285, Portland, Ore.

Gillis, A. M. 1990. The new forestry: An ecosystem approach to land management. *BioScience* 40:558-562.

Hansen, A. J., T. A. Spies, F. J. Swanson, and J. L. Ohmann. 1991. Conserving biodiversity in managed forests. *BioScience* 41:382-392.

Hansen, A. J., S. L. Garman, D. L. Urban, and B. Marks. 1993. An approach for managing vertebrate diversity in multiple-use landscapes. *Ecol. Appl.* 3:481-496.

Hansen, A. J., R. Spencer, A. Moldenke, and A. McKee. In Press. Ecological processes linking forest structure and avian diversity in western Oregon: A test of hypotheses. In *Biodiversity, Temperate Ecosystems, and Global Change*, ed. T. Boyle. Springer Verlag, New York.

Huff, M., and C. Raley. 1991. Regional patterns of diurnal breeding bird communities in Oregon and Washington. In *Wildlife and Vegetation of Unmanaged Douglas-fir Forests*, eds. L. F. Ruggiero, K. B. Aubry, A. B. Carey, and M. H. Huff, pp. 177-206. USDA Forest Service Gen. Tech. Rep. PNW-GTR-285, Portland, Ore.

Kessler, W. B., H. Salwasser, C. W. Cartwright, Jr., and J. A. Caplan. 1992. New perspectives for sustainable natural resources management. *Ecol. Appl.* 2(3):221-225.

Ludwig, J. A., and J. F. Reynolds. 1988. *Statistical Ecology: A Primer on Methods and Computing*. John Wiley and Sons, New York.

MacArthur, R. M., and J. W. MacArthur. 1961. On bird species diversity. *Ecology* 42:494-498.

Manuwal, D. A. 1991. Spring bird communities in the southern Washington Cascade Range. In *Wildlife and Vegetation of Unmanaged Douglas-fir Forests*, eds. L. F. Ruggiero, K. B. Aubry, A. B. Carey and M. H. Huff, pp. 161-176. USDA Forest Service Gen. Tech. Rep. PNW-GTR-285, Portland, Ore.

Reynolds, T. T., J. M. Scott, and R. A. Nussbaum. 1980. A variable circular-plot method for estimating bird densities. *Condor* 82:309-313.

Ruggiero, L. F., K. B. Aubry, A. B. Carey, and M. H. Huff, eds. 1991. *Wildlife and Vegetation of Unmanaged Douglas-fir Forests*. USDA Forest Service Gen. Tech. Rep. PNW-GTR-285, Portland, Ore.

Spencer, R. 1993. Bird communities in managed conifer stands in the Oregon Cascades: Habitat associations and nest predation. M.S. Thesis, Oregon State University, Corvallis, Ore.

Spies, T. A., and J. F. Franklin. 1991. The structure of young, mature, and old-growth forests in Oregon and Washington. In *Wildlife and Vegetation of Unmanaged Douglas-fir Forests*, eds. L. F. Ruggiero, K. B. Aubry, A. B. Carey, and M. H. Huff, pp. 91-110. USDA Forest Service Gen. Tech. Rep. PNW-GTR-285, Portland, Ore.

Swanson, F. J., and J. F. Franklin. 1992. New forestry principles from ecosystem analysis of Pacific Northwest forests. *Ecol. Appl.* 2(3):262-274.

Turner, T., and J. N. Long. 1975. Accumulation of organic matter in a series of Douglas-fir stands. *Can. J. For. Res.* 5:681-690.

Urban, D. L., and T. M. Smith. 1989. Microhabitat pattern and the structure of forest bird communities. *Am. Nat.* 133(6):811-829.

IV

MAINTAINING BIODIVERSITY ON A LANDSCAPE SCALE

27

CONSERVING BIODIVERSITY IN MANAGED LANDSCAPES

Kenton R. Miller

Biological Resources and Institutions
World Resources Institute
Washington, D.C.
U.S.A.

This paper will explore some key technical and policy implications of a shift in the scope and scale of resource management. Emerging scientific knowledge and local experience suggest that biodiversity will be retained to the extent that whole regions are managed cooperatively among protected areas, farmers, foresters and other neighboring land users.

BACKGROUND

Traditionally, wild species and ecosystems have been managed through the establishment of national parks, national forests, wildlife refuges, and other types of protected areas. These have been complemented by programs to protect and rescue endangered species by holding selected plants and animals in off-site facilities, such as zoos, game ranches, botanic gardens and aquaria, and by storing germplasm in seed banks and tissue culture collections.

The continuing and accelerating loss of biotic diversity in all parts of the world (Myers 1979; Reid and Miller 1989; McNeely et al. 1990) can be attributed to an array of causes related to human activities and lifestyles. Science and policy analysis are pointing to two factors, however, that suggest a need to reconsider current approaches to conservation and the management of biological resources.

First, two internationally negotiated agreements stemming from the Earth Summit (in Rio de Janeiro, 1992) call for a transformation in the management of natural resources—that biodiversity be maintained, and that biological resources be used in ways that are both sustainable and equitable. The debate has shifted from a restricted consideration of how to protect wild flora, fauna, and microbial life, to the tripartite goal to protect, systematically document and understand, and properly employ nature. Thus, expansion of the *scope* of biodiversity management is called for, one that will embrace the various concepts and activities associated with saving, studying, and using nature and biological resources.

Second, conservation biology and landscape ecology, and the broad experience of societies and communities around the world, are showing that the geographic *scale* of biodiversity management has been too restricted. Beyond the frontiers of core protected areas, and in addition to the projects to rescue endangered species, management programs need to encompass whole regions that provide the space and habitat elements required by the species and wild communities of the area, including opportunities for movement among protected wildland sites. Within this larger context, biodiversity management can include not only protection of the biota but also steps to know and understand nature and to ensure that whatever uses are made of it are sustainable and are equitably shared with local, national, and international human communities.

For the most part, existing parks and reserves are simply too small to address this expanded scope if taken in isolation from the surrounding landscape. Many populations of species stand little chance of survival in sufficient numbers to ensure their genetic strength and adaptability to changes in climate and land-use patterns. Long-term maintenance of species and their genetic variation will require cooperative efforts across entire landscapes to ensure that adequate populations of plants, animals, and microorganisms can be maintained, migrate, and adapt. A new land-use paradigm is emerging, variously referred to as "ecosystem management" (Agee and Johnson 1988; Slocombe 1992), "landscape management" (Noss 1983; Turner 1989), and "bioregional management" (Sale 1985; Berg 1986). Each of these and other associated concepts carry distinct meaning and connotation for different academic and social groups. This paper will employ the term "bioregional management" to denote a land/water territory the extent of which is determined by a combination of ecological, social, and economic considerations.

This reorientation opens the possibilities to safeguard our biotic wealth more scientifically, and also to address inequities that have been experienced by local residents. For example, people and neighbors are often displaced by the establishment of protected areas, greatly restricting their traditional access to grazing, firewood, hunting and fishing, water, and other natural resources. Furthermore, few opportunities for employment or other benefits come to these same individuals and communities (WRI/IUCN/UNEP 1992).

By expanding the scale of management to include protected areas, corridors (Csuti 1991), and the intervening land uses, regional approaches can include as partners the neighboring farmers, foresters, rural villages, local universities and research stations, indigenous communities, industry, and NGOs as members of the larger team of collaborators. Each can contribute to achieving a patch-work quilt of landscape diversity that maintains diversity, fosters inventory and research, expands appreciation and understanding as to how local ecosystems work, and the production of goods and services of economic value. Experience around the world demonstrates this possibility (K. Miller unpubl. data).

Landscape ecology provides the initial basis for determining appropriate scales for bioregional management. However, scale must also include what residents consider to be their "homeland," and what economic capacity is required to make a regional economy viable. Communities can negotiate conservation plans that consider their options for maintaining the biotic content, ecological structure, and function of their landscape. They can weigh the factors of environmental change that have the potential to alter significantly and affect their region. They can integrate protected areas and off-site facilities within the mosaic of farms, harvested forests and wildlife, and fished freshwater and marine environments.

Thus, my argument is that two fundamental shifts can greatly enhance the maintenance of biodiversity: shift the *scope* of conservation practice to embrace more comprehensive goals and objectives, and shift the *scale* of conservation practice to embrace larger land and water areas that are determined by ecological, social and economic considerations. Such larger and more complex practices are referred to here as "bioregional management regimes." Important technical and policy issues need to be addressed, however, before bioregional management regimes can be expected to succeed.

COMPREHENSIVE SCOPE FOR CONSERVATION

The *Global Biodiversity Strategy* (WRI/IUCN/UNEP 1992) proposes that the goals and operational objectives of biodiversity management should be sufficiently broad to include the set of components that link the biota with human intellectual development and human use. That is:

> **Save**—maintain ecological composition (species and genetic variation), structures (habitats, seral stages) and functions (processes such as nutrient and detritus cycles, pollination, migration), prepare for change (in climate and that caused by alterations in land uses and catastrophic events), and maintain human practices that have selected species and genetic variation, and shaped landscapes (selected resource practices and traditional cultures).

To save biodiversity implies establishing the full range of protected areas and off-site facilities; restoring areas to extend effective boundaries of key sites or to reclaim degraded sites for wildlife conservation; reintroducing extirpated species, and expanding plant and animal populations (Cade 1988; Jordan 1988). It also implies establishing management schemes at bioregional scales that provide for a mosaic of diverse habitats including farms, harvested forest, game areas, fishery zones, and corridors.

> **Study**—establish research, inventory, experimental, and outreach activities to gain understanding, appreciation, and awareness of biodiversity values and options.

To study biodiversity implies (1) comprehending its content, extent, role, and values, to provide options for experimentation and demonstration of sustainable approaches to nature, forest, wildlife, watershed, grazing, agricultural, and fisheries management, (2) inventorying to learn the biological content, its distribution and attributes, (3) monitoring and analyzing trends, and (4) developing information services to support public awareness and appreciation, extension, and informed decision-making.

> **Use Sustainably and Equitably**—redirect and improve on-going land and water management practices that utilize biodiversity and biological resources to approaches that are sustainable and equitable.

The third dimension of biodiversity conservation involves the reform of current approaches to forest, agriculture, grazing, wildlife and fisheries management to meet two conditions. First, biological resources should be managed to yield material, aesthetic, cultural, spiritual and ecosystem goods and services for present use while maintaining the diversity of species, genetic traits, habitats, and ecosystems to safeguard long-term future values and options. Second, policies should be formulated that provide incentives to explore unknown biota and expand intellectual and economic opportunities, with special consideration given to providing options to local communities for improved livelihoods. And, third, the costs and benefits derived from these practices should be distributed equitably. Notwithstanding the particular needs of local residents and indigenous communities, the legitimate interests of members of society that live elsewhere, along with future generations, must also be considered.

LANDSCAPE SCALES FOR BIODIVERSITY MANAGEMENT

Scientists make the case that biodiversity will be maintained successfully to the extent that land and water territories of the appropriate scales are managed as regional units (Noss 1983; Soulé 1991; Harris and Atkins 1991). Why? Five fundamental reasons can be deduced from ecological, social, and economic considerations:

1. **Ecological Composition.** Intensive programs to manage populations of particular species and wildland communities contribute to the welfare of those components of ecosystems, but such focused approaches can only pretend to deal with a relatively small portion of the ever-growing list of endangered species (Reid 1992). Not all species and their genetic variation will be found within established or potential networks of protected areas. Although several nations have committed 10% and more of their entire

Table 27.1. Percent of National Land Area Protected, Selected Countries.[1]

Country	Percent of Land Protected
Botswana	17.2
Cameroon	4.4
Chad	0.1
Cote d'Ivoire	6.3
Rwanda	12.4
Belize	3.2
Canada	5.0
Costa Rica	11.9
Haiti	0.3
Panama	17.2
Argentina	4.6
Brazil	2.4
Ecuador	37.7
India	4.1
Indonesia	9.3
Japan	6.4
Bulgaria	1.2
Greece	0.8
United Kingdom	18.9

[1] *From WRI,* World Resources Report, 1992-93

territories to protected area regimes (WRI 1992; WCMC 1992) (Table 27.1.), many wild species and varieties occupy niches and have home ranges that extend beyond reserve boundaries. Most reserves are too small to contain all habitat requirements. Peninsulas and other odd boundary patterns encourage qualitative degradation through edge effects (Noss 1983).

2. **Ecological Structures.** Not all ecosystem types, including old growth, second growth, pioneer stages, and flood plains, are found in existing or potential protected areas. Some landscapes are exceptionally diverse and any attempt to cover the range of habitats within one or more protected areas would imply excessive withdrawal of land and water resources from other human purposes, along with major impacts upon local societies and economies. Furthermore, most protected areas have been selected and managed to meet objectives other than, or in addition to, the conservation of biodiversity (See Table 27.2.).

These areas may well have been located to conserve watersheds, to protect scenic wonders, and to maintain prime recreation sites. Sites key for biodiversity goals may have been unrecognized or undervalued at the time of establishment.

Many key habitat and ecosystem types, including patches of old-growth forest and wetland, are found on farms and forest holdings, often in private, community or corporate hands. These patches need to be managed as components of the overall landscape under arrangements that are equitable to land owners.

3. **Ecological Functions.** Hydrological flows, nutrient cycles, migratory patterns, and other ecological processes typically span scales greater than the area contained within most protected areas. Entire watersheds from mountain sources to shorelines require management strategies that address quantity and quality of stream flow. This involves control of pollution and contaminants, preservation of wetlands, erosion control, and mitigation of impacts from land uses including grazing regimes that often contribute excessive nutrient fluxes to soils, streams and inshore habitats, as exemplified by the impact of sheep excreta upon the Great Barrier Reef in Australia's Queensland (W. Craik pers. comm.).

The routes utilized for the daily and seasonal movement of insects, birds, and mammals are open to risk as wildlands and waters between protected areas are converted to other human-oriented purposes. These movements are vital for pollination and dispersal. Climate change can be expected to cause range shifts for all organisms (Peters and Lovejoy 1992). Mitigation strategies typically include the establishment of corridors (connectivity) to provide forest or other suitable cover (Soulé 1991). The Florida Panther project illustrates a state-wide approach to provide habitat for a wide-ranging endangered species (Harris and Atkins 1991). In Central America, local government agencies, NGOs, and Wildlife Conservation International are proposing a multi-country network of corridors among the region's national parks and reserves that will extend from Colombia to Mexico (Marynowski 1992).

4. **Adapting to Change.** Forests, prairies, wetlands, coral reefs, and other environments are constantly undergoing change. Within a forest stand, individual trees fall, open gaps to sun light, and initiate competition among plants for this space. Within each area, communities of plants, animals, and micro-organisms are passing through (seral) stages.

Gradual processes of change are interrupted by periodic, often catastrophic impacts,

Table 27.2. Protected Areas Management Categories[1].

	IUCN Cat. I-III		IUCN Cat. IV-V		Total	
	Number	Area (ha)	Number	Area (ha)	Number	Area (ha)
Africa	260	88,722,877	381	35,918,296	641	124,641,173
North & Central America	610	170,344,290	1,073	91,415,737	1,683	261,760,027
South America	289	58,190,622	291	56,182,497	580	114,373,119
Asia	410	35,397,425	1,762	66,025,886	2,172	101,423,311
Europe	289	8,056,879	1,635	32,031,759	1,924	40,088,638
Former Soviet Union	175	23,908,331	38	465,995	213	24,374,326
Australia & S. Pacific	443	67,872,385	494	16,481,489	937	84,353,874
Antartica	12	220,649	1	36,700	13	257,349
Total	2,488	452,713,458	5,675	298,558,359	8,163	751,271,817

[1] Protected areas fall into two main groups. In *strictly protected areas* (such as scientific reserves, national parks, national monuments, and wildlife sanctuaries) natural landscapes dominate. These are characterized by relative freedom from exotic species, cultivation, and human settlement. In *extractive protected areas* (such as national forests, hunting and fishing zones, and protected rural landscapes) limited harvesting of natural resources is allowed, generally under government control. Protected areas are given a great variety of names by the nations establishing them, but IUCN has classified these sites into five categories according to their management objectives.

Strictly Protected Areas

1. *Strict Nature Reserves.* Generally smaller areas where the preservation of important natural values with minimum human disturbance are emphasized.

2. *National Parks.* Generally larger areas with a range of outstanding features and ecosystems that people may visit for education, recreation, and inspiration as long as they do not threaten the area's values.

3. *Natural Monuments.* Similar to National Parks but usually smaller areas protecting a single spectacular feature or historic site.

Extractive Protected Areas

4. *Habitat and Wildlife Management Areas.* Areas managed to protect and utilize wildlife species.

5. *Protected Landscapes.* Areas consisting of publicly or privately owned lands that may be subject to resource extraction--including farms, forests, freshwater areas, and coasts—and their associated human settlements, where the objective is to maintain the quality of the overall landscape, harmonious human interaction with it, and the biodiversity it contains.

Source: IUCN/CNPPA, 1990; World Conservation Monitoring Center, U.K.

such as by storms and geological processes that significantly alter community composition and structure (Hunter et al. 1988). Within a larger landscape, such processes of change promote a mosaic-like pattern of patches, each featuring successive stages of biotic development. The organisms of such a landscape interact with adjacent patches and often, through daily movements and migration, with patches at some considerable distance. (See Figure 27.1.)

A central strategy for addressing the process of change is to establish landscape configurations that provide the set of habitat options required for survival of all organisms. In addition to protected areas that will predictably cover between 5 and 10% of a particular region, the major part of the larger landscape can be managed through forestry, agriculture, grazing, fisheries, and wildlife regimes that mimic to some extent the dynamic patchwork quilt of nature. Flexibility is key. The nature and extent of perturbations, such as wind, flood, or earthquake, are often unpredictable.

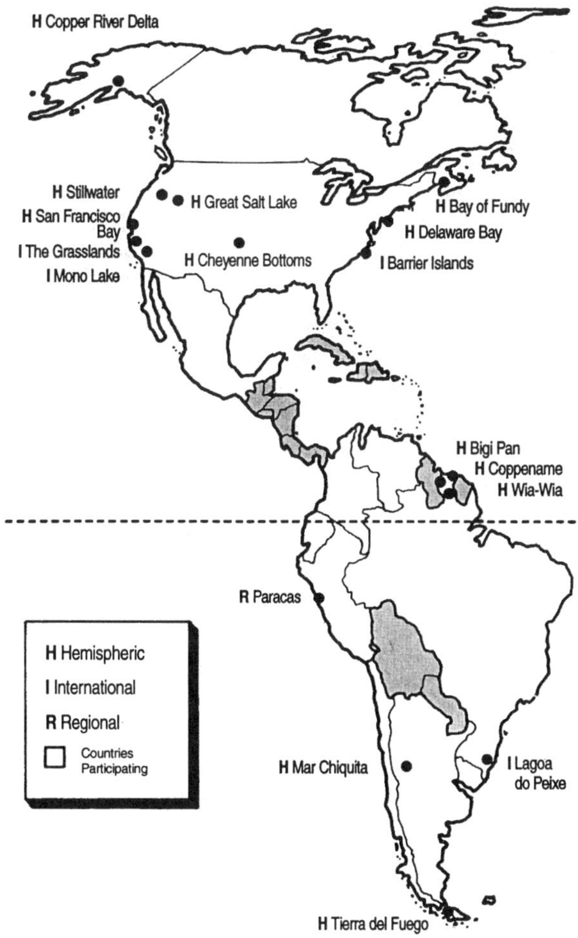

Figure 27.1. International Cooperation on Shorebird Conservation. The Western Hemisphere Shorebird Reserve Network (WHSRN) was established in 1985 in response to declining shorebird populations and wetlands. The Network brings together wildlife agencies, land owners, private conservation groups, and others to solve conservation challenges. Membership in WHSRN is voluntary: management decisions and priorities remain the prerogative of the land administrator. To be included in the network, a site must meet certain biological criteria. Hemispheric Reserves are used by at least 500,000 shorebirds annually (or 30 percent of the flyway population). International Reserves host at least 100,000 shorebirds annually or 15 percent of a flyway population. Regional reserves host 20,000 shorebirds annually (or 5 percent of a flyway population). Endangered Species Reserves are of critical importance to the survival of one or more endangered shorebird species. As of September 1991, a total of 17 reserves have been included in the network, protecting 30 million shorebirds and 4 million acres of land.

5. **Human Settlements and Land Use.** Human cultures have selected and domesticated plant and animal varieties that are found in agricultural, pasture, forest, urban environments, and others. Some traditional cultures continue land-use practices that maintain progenitor races of plants and animals of considerable value to humans world-wide. Some varieties and environments were established through human enterprise back several millennia. It is feared that some of these varieties will remain available only to the extent that traditional cultivation practices are continued (CLADES 1991). The territorial units of land and water needed to embrace the range of species and genetic traits, ecosystems, ecological functions, and the space within which to adapt to change extend well beyond scales that can reasonably be contained in protected areas. This factor of scale, coupled with the recognition of the role of human settlements and land uses in the development and maintenance of varieties and landscapes of importance to human interests, argue for their inclusion in bioregional management programs.

Moreover, in most parts of the world people are already settled in the matrix of the landscape, in between the existing reserves, and have established ownership or traditional rights of access to natural resources. From a biodiversity conservation perspective, human settlements and land uses represent a significant opportunity for achieving the objectives of biodiversity management. For example, to establish and maintain a dynamic patchwork of land uses and successional stages would be economically and technically impossible to administer and implement without residents who, under some cooperative arrangement, seek these same (or somewhat modified) uses of their resources in self interest.

For all these reasons it can be postulated that biodiversity is at once a biological resource and a cultural resource: humans have evolved with it, have modified it while utilizing its components, and their presence is now clearly critical either as a force for degradation and impoverishment, or as force for restoration and long-term maintenance. What is implied here is the development of a "biodiversity economy" where human values and behavior seek to respectfully maintain nature, get to know and appreciate it, and take advantage of its many goods and services, in the process of which gainful employment is generated.

TECHNICAL ISSUES FOR REGIONAL MANAGEMENT

Successful bioregional management regimes will depend to large extent upon the development of clear guidelines for addressing the following technical issues:

1. **Scale.** How can the "right scale" be determined? What criteria can be used to determine the appropriate size of regions to be managed? At the risk of oversimplification, the issue of how to determine scale can be summarized briefly in terms of ecological, social and economic criteria:

Ecologically, scale is a function of the geographic land/water territory employed by the target community or communities of plant, animal and microbial life, the pathways required to allow for migration and adaptation to change (Waller 1991), and the specific goals of the conservation strategy (Noss 1991).

Socially, the extent of the region of a human community is determined by the cultural history of the group, its perception of homeland, and the patterns of use to which local natural resources are employed (Robinson 1983; Alexander 1990).

And economically, scale with respect to nature conservation is viewed as a function of effectiveness in achieving management objectives and economic cost (S. Repetto and A. Solow unpubl. data). Current research focuses on optimal reserve size, that for which the marginal cost of preserving target species is just equal to the marginal benefit of preservation derived by society. Costs include site acquisition, construction, maintenance, and management. Benefits, which are more difficult to quantify, are derived from principles of conservation biology. (See Tisdell and Broadus 1989 for economic work on marine parks.) Economic scale is also critical to ensure that local industry, land use, and tax base are quantitatively sufficient, and appropriately flexible and dynamic to sustain the livelihoods and infrastructure of the region's residents.

2. **Protected Areas.** Three issues are of particular concern: How can sites for biodiversity management be more rigorously selected? What category of management (Table 27.2.) is most useful to achieve the selected objectives? And, how can management plans be prepared and implemented to ensure the adequate incorporation of biodiversity considerations into the overall operation of the protected area?

Needed are more rigorous selection procedures to identify candidate sites that possess biotic content, structures, and functions relevant for the maintenance of particular communities, habitats and ecosystems (Margules et al. 1988; Vane-Wright et al. 1991; Scott et al. 1991).

The choice of a management category can play a fundamental role in orienting the formulation of objectives and activities within each area. The IUCN system of protected area categories (IUCN/CNPPA 1990)—scientific reserves, national parks, natural monuments, wildlife reserves and landscape or multiple-use reserves—provides a simplified model to assist wildland planners in determining management categories according to the objectives to be pursued. (See Table 27.2.) At issue is the need to formulate management approaches for each site on a case by case basis in terms of objectives and interventions required.

Although technically the choice of category should be determined by the analysis of management objectives (Miller 1980; IUCN/CNPPA 1990), the selection of national park, national forest, wildlife sanctuary or other approach for public or private reserves is most often predetermined by the legislative mandate of the agency responsible or mission of the private organization. Thus, even though all institutions involved in nature conservation might be mandated to protect species and sites, some will also carry mandates to extract resources, foster recreation and tourism, protect water sources, manage game, and maintain natural and cultural features of importance for indigenous communities. Conflicts arise from attempts by managers to find the "right mix" of objectives, and the trade-offs inherent in multiple-output approaches to area administration (O'Toole 1988; Aplet et al. 1993).

Management of new sites can be planned and implemented according to clearly focused objectives among which the protection of species and genetic traits, inventory and research, and scientifically-guided human use are taken as central options. Plans for existing areas can evolve to incorporate biodiversity considerations that might portend changes in zoning, use patterns and management practices.

3. **Connectivity.** Corridors can be established to connect sites across the landscape (Noss 1987). In some cases, corridors can be established using public domain wildlands. In most cases, however, land and water areas will already be under agricultural, grazing,

forest, or fishing uses by residents, usually under private ownership. Several important technical issues arise: the cost for purchase of lands for corridors can be disproportionately high relative to wild land where the areas are in agricultural or other high market value uses. Establishing institutional arrangements with other public and private land owners can be complex and require incentives and trade-offs that require considerable political capital and organizational creativity.

Perhaps most evasive at this time, however, is to determine the relative value of various land uses to connect biologically the disbursed protected core zones in the landscape. Will patches of agriculture, tree plantations, pasture, coffee or cacao serve the purpose, or is it necessary to change land use and perhaps ownership, and to initiate restorative processes? Also, what is the appropriate corridor width, how should the edges of the corridors be managed, and what is the best pattern of patches within the matrix of the corridor and surrounding lands? Obviously, the implications of these questions will have fundamental impact not only upon biodiversity conservation but also upon human settlement and access to and use of natural resources.

4. **Off-site Facilities.** Botanic gardens, zoos, aquaria, germplasm banks and tissue libraries are taking important steps to collaborate with in-situ programs for biodiversity conservation (IUCN/WWF 1989; IUCN 1993; IUDZG/CBSG (IUCN/SSC) 1993; IBPGR 1991). At issue is how the potential of these facilities can be utilized strategically, especially given their limited size and capacity (Conway 1988). Beyond their traditional roles in public awareness, education, and research, these facilities have demonstrated their value to biodiversity conservation. Examples include the experience of captive breeding and reintroduction of the Arabian Oryx, the Golden Lion Tamarind, and the White Rhino (Conway 1988). Examples from the plant side include the Florida torreya, the Arizona Agave, and the *Penstemon barrettiae* in Oregon (Falk 1992).

Strategically, this potential could be planned and coordinated to maintain minimum viable populations by keeping some threatened individuals in the wild and translocating others into captivity, complemented by the necessary interventions to maintain genetic viability. Reproductive research and other experimental work can take further advantage of these facilities. Seed banks and tissue libraries can provide a near permanent reference system for the living materials found in a region.

5. **Restoration.** Among the tools and technologies central to biodiversity management will be restorative procedures for application both within protected areas and across bioregions. Science and technology for restoration are desperately needed. Protected areas that are too small could be expanded by reclaiming adjacent lands, social considerations permitting. Habitats within reserves could be modified to favor threatened populations. Corridors could be added to ensure the possibility of migration, daily movements, and adaptation to global change.

Across the landscape, degraded areas can be restored with native organisms to favor diversity and meet other objectives, such as erosion control and water quality (Galli and Janio 1991). The matrix of patches in distinct cover types and at varying successional stages, and viable population sizes of selected species can be managed through such interventions as the scientific use of fire and animal culling (de Vos et al. 1983).

6. **Knowledge.** Considerable effort has gone into the formulation of research priorities for the conservation of biodiversity (Soulé 1991; NRC 1992). The above cited technical issues can be resolved to the extent that science provides clear guidance. Resource

managers, community and business leaders will need to know which sites in their region, because of their respective composition, structure and function, will require special handling. What management interventions will serve? Which land uses are sustainable. In many cases, answers to these and other questions will depend upon research and experimentation within the particular bioregion.

7. **Status and Trends.** What indicators can be employed to track the conditions and trends of biodiversity, and of the achievement of biodiversity management goals? Are existing policies having the desired effects on biodiversity? What problems are beginning to arise that call for various management activities or interventions?

Twenty-two indicators have been proposed for use in tracking various characteristics of biodiversity (Table 27.3.). Even though there is no consensus on how "biodiversity" should be measured, the WRI/World Conservation Union (IUCN)/Bariloche Foundation team has offered a working set of indicators, 13 of which deal with wild species and genetic diversity, four with community diversity, and five with domesticated species. These can be used to understand better how well the goals of biodiversity conservation are being met at local, national, regional or global levels (Reid at al. 1993).

8. **Information.** What information is needed by the various constituency groups (political and community leaders, business, scientists, resource managers, citizenry) to be able to make pondered decisions about land use? Although information at the macro scale is being developed and widely distributed (see, for example, WCMC 1992), local levels of detail are seldom available in forms that serve local decision making needs. Furthermore, the characteristics of local demand for information is little understood. Options for addressing this subject will be further developed below.

9. **Extend, Demonstrate, and Build Awareness.** Even with viable options for the sustainable use of biological resources ready for application to particular regions, there remains at issue how these options can be demonstrated and promoted for field application. The implication is that each bioregion should include facilities and programs for experimentation and demonstration, along with services for extension and public awareness, a subject also treated more fully below.

POLICY ISSUES FOR REGIONAL MANAGEMENT

Shifts in scope and scale also raise a series of policy issues:

1. **National Policies, Plans, and Strategies.** Moving bioregional management concepts from paper to the field requires a clear policy context. At the national level, government policies are needed that establish national goals for biodiversity conservation and give unambiguous direction to government agencies. National strategies and plans can provide vision for bioregional efforts. These instruments need to offer specific objectives and incentives, identify biodiversity assets, propose mechanisms to coordinate interagency programs and public and private initiatives, and set priorities. Efforts to prepare case-history based guidelines are underway by a UNEP-sponsored team headed by WRI in cooperation with the World Conservation Union (report expected in 1994). The preparation of National Conservation Strategies (Halle and Furtado 1987) and UNEP-sponsored Country Studies are attempts in this direction (UNEP 1992).

2. **Integration of Tools and Technologies.** Parks, private reserves, botanic gardens

Table 27.3. Indicators of Biodiversity Conservation.

Indicator	Biodiversity Conservation Concerns		
	Genetic Diversity	Species Diversity	Community Diversity
Wild Species' and Genetic Diversity			
1. Species richness (number, number per unit area, number per habitat type)	X	X	
2. Species threatened with extinction (number or percent)	X	X	
3. Species threatened with extirpation (number or percent)	X	X	
4. Endemic species (number or percent)	X	X	
5. Endemic species threatened with extinction (number or percent)	X	X	
6. Species risk index	X	X	
7. Species with stable or increasing populations (number or percent)	X	X	
8. Species with decreasing populations (number or percent)	X	X	
9. Threatened species in protected areas (number or percent)	X	X	
10. Endemic species in protected areas (number or percent)	X	X	
11. Threatened species in ex situ collections (number or percent)	X	X	
12. Threatened species with viable ex situ populations (number or percent)	X	X	
13. Species used by local residents (number or percent)	X	X	
Community Diversity			
14. Percentage of area dominated by non-domesticated species		X	X
15. Rate of change from dominance of non-domesticated species to domesticated species		X	X
16. Percentage of area dominated by non-domesticated species occurring in patches greater than 1,000 sq km		X	X
17. Percentage of area in strictly protected status		X	X
Domesticated Species Diversity			
18. Accessions of crops and livestock in ex-situ storage (number or percent)	X		
19. Accessions of crops regenerated in the past decade (percent)	X		
20. Crops (livestock) grown as a percentage of number 30 y ears before	X		
21. Varieties of crops/livestock grown as a percentage of number 30 years before	X		
22. Coefficient of kinship or parentage of crops	X		

and experiment stations exist in many regions. At issue is how to reverse typical "institutional fragmentation" and develop common strategies for the formulation of objectives, the selection of sites, species and genetic traits to receive priority attention, and other factors, that are fully integrated with the overall bioregional program? What incentives are needed to promote this integration? How can the established barriers among these communities be relaxed and new partnerships formed? Specifically, how can policies, plans and strategies be developed that bring farmers, foresters, religious and indigenous leaders, local offices of governmental agencies, NGOs and citizens, business and industry together to formulate common goals and a shared plan of action?

3. **Congruency of Jurisdictions.** Governmental agencies that administer natural resources, health, transportation, and other sectors implement their activities through field districts and offices. Seldom, however, do these territories conform to the landscape regions that would be defined by ecological, economic, and social criteria. In fact, the districts of individual agencies seldom align one with another. Thus, how can public programs and investments be expected to collaborate with and support regional initiatives?

4. **Partnership of Constituencies and Interests.** Stewardship of the biotic wealth of a region will depend in great part upon the commitment of local citizens and business and industry to their "homeland." How can this stewardship be fostered?

(a) *Empowerment of Participants.* At issue is the distribution of power. What is the best balance between central control through law, regulation and budget versus local control through decentralized approaches to decision-making and distribution of resources? History suggests that no single solution is applicable to all cases. Experience of the Great Barrier Reef Marine Park Authority demonstrates how central authority can be exercised creatively to stimulate community and industry cooperation, self regulation, and restraint (W. Craik and D. Clasen unpubl. data). The example of La Amistad National Park and Biosphere Reserve illustrates how government policy and a clear vision can provide the context within which local groups can formulate a program to their mutual interest (CI/OAS 1990).

(b) *Apportionment of Ownership and Access.* In many regions, the involvement and long-term commitment of people and industry to a common plan of action will depend upon the degree to which their own roots are firmly tied to the ground. Assured access to natural resources (water, lumber, firewood, hunting and fishing, grazing and tourism) is key to some constituencies, whereas outright ownership can be vital to others (native communities). Lack of resolution of conflicts relative to access and ownership can thwart any plans to achieve regional management.

(c) *Open Negotiation Process.* Ideally, mechanisms should be established whereby all constituencies can come to the table and negotiate goals and means, cooperative arrangements, compensation for opportunity costs, and a fair share in the benefits. In the process of preparing a management plan for the Amistad Biosphere Reserve, the program fostered negotiation among local indigenous groups, the national electrical and water institutions, commercial and subsistence farmers, the national park agency, and other groups with interest in the region.

5. **Leadership.** Achieving regional management is ultimately a political process. At issue is who will provide the leadership to formulate locally the vision, the willingness to come to the table, the free flow of information, and the horse trading that must result in developing a common plan? The Great Barrier Reef suggests the value of an initial and continuing role for a central authority working with local community, governmental and industrial groups. La Amistad benefits from a full-time resident project leader who works with the many varied interests. Both cases emanate from national legislation, yet employ a range of local committees and working groups to examine thematic issues, propose options, and track results.

6. **Capacity, Training, and Education**. Finally, how can the capacity to do all this

be developed? Regional management initiatives will require a shared awareness of the problems and the opportunities, inventory of assets, data management, information management, research in a variety of fields, monitoring, and the capacity and attitudinal disposition to negotiate solutions. Schools will need to add teaching materials and course work to their curricula that feature regional issues and include field trips to local sites and activities. Ideally, all residents of the bioregion, including all hierarchies of administration, will need to be involved. Few new physical facilities will be required, but investments, incentives and compensatory arrangements will all need budgets. This will no doubt call for reapportionment of tax revenues, new "green" taxes, targeted governmental allocations, bond drives, private grants, industry-supported training opportunities, and other creative financial mechanisms.

ADDITIONAL ISSUES

Two additional ingredients can contribute to achieving bioregional management, both of which have been tried and tested in other contexts:

Regional Information Centers

The experience of the National Biodiversity Institute (INBio) of Costa Rica (Sandlund 1991), a public interest private organization that is promoting the biological survey of Costa Rica, the systematic screening of its biota for potential products, and the equitable attribution of benefits from royalties back to the country's conservation programs and employment base, demonstrates how such an institution can "broker" information to interests groups in a particular bioregion (e.g., Guanacaste) and foster dialogue in their common interest. In Colombia, the Pro-Sierra Nevada Foundation project in the Sierra Nevada region (J. Mayr pers. comm.), established a series of modest village centers where local indigenous and rural residents meet and discuss community interests. Biodiversity has been introduced into the debate in the context of overall community need for sustainable resource use.

Arguably, a "regional information center" is a key ingredient of any regional project. Depending upon the case, such centers could be installed in pre-existing facilities (the post office, town hall, tribal commons, or NGO headquarters) and cost little. Data base and information management capacity could be donated by local business and industry. The point is, only by having local residents fully informed and aware of their own challenges and options can group effort be achieved and a vision shared.

Regional Experimental and Demonstration Stations

Each region could benefit from a facility that develops, demonstrates and extends models and examples of how people can direct their efforts towards sustainable livelihoods in that particular region. Such mechanisms as biosphere reserves and existing range, forest, and agricultural stations could serve this purpose with some refocusing of goals and methods. In pragmatic terms this can be approached through cooperative arrangements with local universities, farmer, fishery and forestry cooperatives and associations, and industry. Linked to the regional information centers, these demonstration facilities could serve to develop and demonstrate scientifically and technologically based land-use practices that incorporate goals and activities related to biodiversity management. Ultimately, a network of such facilities, spread across biomes, could provide the ecological and economic basis for the formulation of policies, incentives, and guidelines that are consistent with local conditions and potentials. The experience of the US forest, range and agricultural research system can offer considerable insight here. Similarly, the experience of the Tributary Area Development Program of the Tennessee Valley Authority is particularly relevant as an approach to halt erosion, reforest watersheds, restore farmland, pasture, and woodlands, and promote the livelihoods of rural communities in that region.

LITERATURE CITED

Agee, J. K., and D. R. Johnson. 1988. *Ecosystem Management for Parks and Wilderness.* University of Washington Press, Seattle, Wash.

Alexander, D. 1990. Bioregionalism: science or sensibility. *Environ. Ethics.* 12:161-173.

Aplet, G. H., N. Johnson, J. T. Olson, and V. A. Sample. 1993. *Defining Sustainable Forestry.* Island Press, Washington, D.C.

Berg, P. 1986. *Growing a Like-Place Politics.* Sierra Club, San Francisco, Calif.

Cade, T. J. 1988. Using science and technology to reestablish species lost in nature. In *Biodiversity,* ed. E. O. Wilson, pp. 279-288. National Academy Press, Washington, D.C.

CLADES. 1991. *Status and Trends in Grass-Roots Crop Genetic Conservation Efforts in Latin America.* University of California, Berkeley, Calif.

Conservation International and Organization of American States (CI/OAS). 1990. *Estrategia para el Desarollo Institutional de la Reserva da la Biosferia " La Amistad. "* A Report to the Government of Costa Rica, San Jose, Costa Rica.

Conway, W. 1988. Can technology aid species conservation? In *Biodiversity,* ed. E. O. Wilson, pp. 263-268. National Academy Press, Washington, D.C.

Csuti, B. 1991. Introduction: conservation corridors: countering habitat fragmentation. In *Landscape Linkages and Biodiversity,* ed. W. E. Hudson, pp. 81-90. Island Press, Washington, D.C.

de Vos, V., R. C. Bengis, and H. J. Coetzee. 1983. Population control of large animals in the Kruger National Park. In *Management of Large Animals in African Conservation Areas,* ed. N. Owen-Smith. Haum, Pretoria, Republic of South Africa.

Falk, D. A. 1992. From conservation biology to conservation practice for protecting plant diversity. In *Conservation Biology; The Theory and Practice of Nature Conservation Preservation and Management,* eds. P. L. Fiedler and S. K. Jain, pp. 397-431. Chapman and Hall, New York and London.

Galli, L. F., and C. G Janio. 1991. *Recuperacao de Areas Degradadas atraves do Reflorestamento com Especies Nativas—A Experiencia da Companhia Energetica de Sao Paola (CESP).* Mimeo.

Halle, M., and J. Furtado. 1987. The role of national conservation strategies in attaining the objectives of the World Conservation Strategy. In *Conservation with Equity: Strategies for Sustainable Development,* eds. P. Jacobs and D. A. Munro. IUCN, Gland, Switzerland.

Harris, L. D., and K. Atkins. 1991. Faunal movement corridors in Florida. In *Landscape Linkages and Biodiversity,* ed. W. E. Hudson, pp. 117-134. Island Press, Washington, D.C.

Hunter, M. L., Jr., G. L. Jacobson, Jr., and T. Webb. 1988. Paleoecology and the course-filter approach to maintaining biological diversity. *Conserv. Biol.* 2:375-385.

International Board for Plant Genetic Resources (IBPGR). 1991. *Geneflow.* IBPGR, Rome, Italy.

IUCN. 1993. Parks for life. Report to the IV World Congress on National Parks and Protected Areas. IUCN, Gland, Switzerland.

IUCN and WWF. 1989. *The Botanic Gardens Conservation Strategy.* IUCN, Gland, Switzerland.

IUDZG/CBSG (IUCN/SSC). 1993. *The World Zoo Conservation Strategy; The Role of the Zoos and Aquaria of the World in Global Conservation.* Chicago Zoological Society, Brookfield, Ill.

Jordan, W. J. 1988. Ecological restoration; reflections on a half-century of experience at the University of Wisconsin Madison Arboretum. In *Biodiversity*, ed. E. O. Wilson, pp. 311-316. National Academy Press, Washington, D.C.

Magules, C. R., A. O. Nicholls, and R. L. Pressey. 1988. Selecting networks of reserves to maximize biological diversity. *Biol. Conserv.* 44:11-37.

Marynowski, S. 1992. Paseo pantera: the great american biotic interchange. In *Wild Earth: Wildlands Project* (Special Issue), pp. 71-74. Cenozoic Society, Inc., Canton, N.Y.

McNeely, J. A., K. R. Miller, W. V. Reid, R. A. Mittermeir, and T. B. Werner. 1990. *Conserving the World's Biological Diversity.* World Resources Institute, World Conservation Union, Conservation International, Washington, D.C and Gland, Switzerland.

Miller, K. R. 1980. Planificación de Parques Nationales: Ejemplos y casos de America Latina. FEPMA, Madrid.

Myers, N. 1979. *The Sinking Ark: A New Look at the Problem of Disappearing Species.* Pergamon, Oxford, England.

National Research Council (NRC). 1992. *Conserving Biodiversity: A Research Agenda for Development Agencies.* National Academy Press, Washington, D.C.

Noss, R. 1983. A regional landscape approach to maintain diversity. *BioScience.* 33:700-706.

Noss, R. 1987. Protecting natural areas in fragmented landscapes. In *Natural Areas J.* 7(1):2-13.

Noss, R. 1991. Landscape connectivity: different functions at different scales. In *Landscape Linkages and Biodiversity*, ed. W. E. Hudson, pp. 27-39. Island Press, Washington, D.C.

Office of Technology Assessment (OTA). 1987. *Technologies to Maintain Biodiversity.* OTA-F-330. Government Printing Office, Washington, D.C.

O'Toole, R. 1988. *Reforming the Forest Service.* Island Press, Washington, D.C.

Peters, R. L., and T. E. Lovejoy. 1992. *Global Warming and Biological Diversity.* Yale University Press, New Haven, Conn.

Reid, W. V. 1992. The United States needs a national biodiversity policy. In *Issues and Ideas.* World Resources Institute, Washington, D.C.

Reid, W. V., J. A. McNeely, D. B. Tunstall, D. A. Bryant, and M. Winograd. 1993. *Biodiversity Indicators for Policy-Makers.* World Resources Institute and World Conservation Union, Washington D.C and Gland Switzerland.

Reid, W. V., and K. R. Miller. 1989. *Keeping Options Alive: The Scientific Basis for Conserving Biodiversity.* World Resources Institute, Washington, D.C.

Robinson, J. L. 1983. *Concepts and Themes in the Regional Geography of Canada.* Talonbooks, Vancouver, Canada.

Sale, K. 1985. *Dwellers in the Land: The Bioregional Vision.* Sierra Club, San Francisco, Calif.

Sandlund, O. T. 1991. *Costa Rica's INBio: Towards Sustainable Use of Natural Biodiversity.* NINA, Trondheim, Norway.

Scott. J. M., B. Csuti, and S. Caicco. 1991. Gap analysis: assessing protection needs. In *Landscape Linkages and Biodiversity*, ed. W. E. Hudson, pp. 15-26. Island Press, Washington, D.C.

Slocombe, D. S. 1992. *Integrating Park and Regional Planning through an Ecosystem Approach.* Paper presented at the IV World Congress on National Parks and Protected Areas, Caracas, Venezuela, 10-21 February, 1992.

Soulé, M. E. 1991. Theory and strategy. In *Landscape Linkages and Biodiversity*, ed. W. E. Hudson, pp. 91-104. Island Press, Washington, D.C.

Tisdell, C., and J. M. Broadus. 1989. Policy issues related to the establishment and management of marine reserves. *Coastal Manage.* 17:37-53.

Turner, M. G. 1989. Landscape ecology: the effects of pattern on process. *Ann. Rev. Ecol. Syst.* 20:171-197.

United Nations Environmental Programme (UNEP). 1992. *Convention on Biological Diversity.* UNEP document No. 92-5509. Nairobi, Kenya.

Vane-Wright, R., C. J. Humphries, and P. H. Williams. 1991. What to protect? Systematics and the agony of choice. *Biol. Conserv.* 55:235-254.

Waller, D. M. 1991. Introduction to conserving biodiversity: a unified approach. In *Landscape Linkages and Biodiversity*, ed. W. E. Hudson, pp. 3-13. Island Press, Washington, D.C.

World Conservation Monitoring Center (WCMC). 1992. *Global Biodiversity: Status of the Earth's Living Resources*. WCMC, Cambridge, U.K.

World Resources Institute (WRI). 1992. *World Resources, 1992-93.* World Resources Institute, Washington, D.C.

WRI/IUCN/UNEP. 1992. *Global Biodiversity Strategy: Guidelines to Save, Study and Use the Earth's Biotic Wealth Sustainably and Equitably.* WRI, IUCN, UNEP, Washington, D.C., Gland, Switzerland, and Nairobi, Kenya.

28

BEYOND PROTECTED AREAS: CHANGING PERCEPTIONS OF ECOLOGICAL MANAGEMENT OBJECTIVES

Walter V. Reid

World Resources Institute
Washington, DC
U.S.A.

In 1955, a small patch of mature forest at Rutgers University, Hutcheson Memorial Forest, was set aside as a nature preserve to maintain a sample of New Jersey's natural ecosystems (Botkin 1990). Three decades later Japanese honeysuckle, Norway maple, and Chinese tree of heaven had invaded, the dominant oaks had ceased regenerating, and the sub-dominant sugar maples were outcompeting the oaks. Clearly, the management strategy, minimizing human intervention, was failing to achieve the desired objective. But was the objective even attainable? Or more precisely, did the objective of maintaining a "natural ecosystem" make any sense at all?

For decades, the continuum from "natural" to "disturbed" ecosystems has underlain ecological management goals. At one end of the spectrum, human disturbance is minimized or allowed only when necessary to recreate natural conditions (e.g., controlled burning). At the other end, ecosystems are managed for extractive uses. Management objectives for natural or semi-natural ecosystems are to maintain biodiversity, maintain ecological services such as flood control and pollution filtration, and to meet demands for recreation or tourism. For extractive protected areas, primary objectives include the production of timber, crops, or livestock.

But advances in the sciences of ecology, paleobiology, and conservation biology are calling into question the very meaning of a "natural" ecosystem, traditionally considered to be an ecosystem undisturbed by humanity. In eastern North America, for example, Native Americans' use of fire had modified ecosystems for millennia before Europeans arrived. Excluding human intervention today might produce a different unnatural state, but it certainly could not produce a "natural" state. And even though it might be possible to recreate historic disturbance regimes, how could it be determined whether the resulting system was "natural?" Ecological systems are dynamic, changing continually through time. Thus, we can be sure that whatever community existed 500 years ago is not what would exist today under "natural" conditions, but we have no objective means of determining what today's "natural" state would be.

Many of today's objectives for ecological management were developed at a time when ecologists believed that communities were comprised of relatively stable assemblages of species, that population sizes tended toward equilibrium levels, and that humans played a relatively minor role in shaping the structure of communities. Today, scientists have modified or abandoned these assumptions; yet, their policy implications remain enshrined in management practices.

Clearly, it is time to rethink the fundamental concerns of ecological management. This paper explores the heritage of our current ecological management concerns as they are typically applied to public land management in the United States and examines how well those concerns hold up in view of changes in both ecological science and conservation philosophy. It briefly reviews several other possible management objectives, including "ecological health" and "biological integrity," and proposes a different objective more in keeping with current understanding of ecology and evolutionary biology.

A LINGUISTIC AND MANAGEMENT LEGACY

The roots of current land management policies in the United States trace back to two schools of conservation thought formed in the late 1800s and early 1900s. On the one hand, such founders of the wilderness movement as George Perkins Marsh, Henry David Thoreau, and John Muir saw conservation as an alternative to development (Worster 1977). To them, the value of conservation was spiritual, not economic. When Muir lost the battle for preserving Hetch Hetchy valley, for example, he wrote: "These temple destroyers, devotees of ravaging commercialism, seem to have a perfect contempt for Nature, and instead of lifting their eyes to the God of the mountains, lift them to the Almighty Dollar" (Fox 1981, p. 142).

In contrast, Gifford Pinchot and Theodore Roosevelt saw conservation in utilitarian terms. For Pinchot, conservation ensured the continued production of valuable biological resources. Foreshadowing current debates over sustainable forestry and the value of non-timber forest products, Pinchot wrote: "The purpose of Forestry, then, is to make the forest produce the largest possible amount of whatever crop or service will be most useful, and keep on producing it for generation after generation of men and trees" (Pinchot 1947, p. 31).

Two objectives of public land management grew out of these contrasting views of conservation. In the tradition of Thoreau and Muir, the National Park system was established to "conserve the scenery and the natural and historic objects and the wildlife therein, so that they can be enjoyed by this and future generations" (Congress of the United States 1916, p. 73). Pinchot's philosophy was expressed in the establishment of the Forest Reserve system, designed to maximize the productivity of those major resources in which people had a clear, direct, and immediate interest (Worster 1977). In this same tradition, wildlife refuges were created to maximize the production of game species for hunters, often through the use of intensive predator-control programs.

By the middle of the 20th century, however, the sharp distinction between utilitarian and non-utilitarian conservation goals began to fade with the growth of the science of ecology. Ecological science was beginning to take hold early in the century. Leopold's

(1933) book *Game Management*, for example, provided a scientific underpinning for early wildlife management efforts. But the 1940s and 1950s saw the field drop its natural history orientation and become a more empirical and experimental science. Led by the detailed empirical field studies of Elton (1924), Lack (1954), and others, and MacArthur's (1955, 1972) application of basic mathematical models to these field observations, ecology became a deductive science.

One of the first achievements of the revitalized science of ecology was to provide strong empirical support to the notion of a "balance of nature." Both population models and long-term datasets supported the idea that population sizes tended toward equilibrium values, and suggested that communities exhibited characteristics like "stability" and "resilience" (Odum 1953; MacArthur 1955). Guided by a mechanical metaphor of ecological dynamics, human disturbance could be seen to upset the stability of an ecological system. Thus, a "natural" state could be achieved simply be removing human disturbance. Without today's paleobiological evidence, which demonstrates the dynamic nature of community composition, the notion that biological communities represented stable associations of co-evolved species was a strong one.

If nature existed in a balance, then the consequences of disturbing that balance were obviously of concern. Conservationists and scientists argued that the disturbance of an ecosystem must be kept within bounds or the system could collapse (Holling 1973). Moreover, the preservation of species diversity emerged as one of the key elements thought to maintain ecosystem stability. Backing up intuition with a simple mathematical model, MacArthur showed that the appealing concept of a "web of life" made scientific sense (MacArthur 1955). The greater the diversity of species in an ecosystem, the more stable the system was likely to be. (Although ecologists had discarded this idea of a direct relationship between diversity and stability by the 1970s (May 1973; Goodman 1975), it is probably too useful for environmentalists ever to reject.)

The combination of scientific support for the "balance of nature" and the apparent risks of destabilizing ecosystems thus cast the preservation of natural ecosystems and wilderness in a very different light. Now, protecting habitats and ecosystems from human disturbance was not only in keeping with the spiritual values of Muir, but made sound ecological sense as well. Where predators had been the target of extensive control programs in the early part of the century, their protection could now be justified by their critical role in ecosystem function (Worster 1977). Wilderness areas became more than recreation areas; they were now seen as some of the "healthiest" of ecosystems and thus best able to provide such ecological services as clean water, flood control, and habitat for commercially valuable species.

This emergence of what Worster (1977, p. 284) refers to as the "Age of Ecology" helped coalesce the utilitarian and non-utilitarian conservation rationales. This evolution of conservation philosophy is well-reflected in both the life and writings of Aldo Leopold. Raised in Pinchot's school of utilitarian conservation, by 1947 Leopold had changed his own beliefs to a biocentric view of conservation that saw value in nature for nature's sake. But Leopold stressed the ecological underpinnings of both utilitarian and biocentric views of conservation: "a system of conservation based solely on economic self-interest is hopelessly lopsided. It tends to ignore, and thus eventually to eliminate, many elements in the land community that lack commercial value, but that are (as far as we know) essential to its healthy functioning. It assumes, falsely, I think, that the economic parts of the biotic clock will function without the uneconomic parts" (Leopold 1966, p. 251). Conserving natural systems and maintaining natural ecological processes was good not only for the spirit; but it was also an essential means of avoiding ecological degradation that would undermine human development.

The arrival of the Age of Ecology set the stage for the development of the concept of "protected areas." Instead of viewing National Parks as an alternative to development, they could be seen as one of many different types of public land management serving development needs. At one end of the continuum, "strictly protected areas" like National Parks were kept free of all human disturbance and maintained in a natural state to meet one set of conservation objectives (recreation, species protection), while at the other end protected areas were managed for such extractive uses as timber harvest.

At the Second World Conference on National Parks held in 1972, this new view of National Parks serving development, rather than simply protecting the "heritage" of a nation, was still the minority (Hales 1989). But by the time of the third conference in 1982 (now called the World National Parks Congress) the Protected Area concept was moving into the mainstream. The pioneering work of Dasmann (1973) and Miller (1975) had provided the conceptual basis for establishing protected area "categories," and the publication of the World Conservation Strategy (IUCN/UNEP/WWF 1980) placed protected area conservation squarely in the mainstream of "sustainable development." The Third World National Parks Congress, entitled "Parks for Sustainable Development," led to the adoption of an eight-category system of protected areas by IUCN (IUCN Commission on National Parks and Protected Areas 1984). And by 1992, the Congress was renamed the World Congress on National Parks and Protected Areas.

Implicit in the concept of protected areas is the idea that the continuum of management objectives represents differing levels of permissible human intervention and, thus, differing approximations to a "natural" ecosystem. Yet, where the definition of a "natural ecosystem" seemed straightforward given assumptions of equilibrium co-evolved communities, that assumption was being overturned by ecologists just as managers were succeeding in translating it into policy.

Thirty years ago, science supported the notion that a natural ecosystem could be defined on the basis of its appearance. This is clearly seen in the "Leopold Report" published in 1963 (Advisory Board on Wildlife Management 1963). This report, requested by Secretary of the Interior Stewart Udall and developed by a committee chaired by A. Starker Leopold (Aldo Leopold's son), provided recommendations for game management in National Parks. The report recommended that the primary goal of wildlife management should be to maintain or recreate the biotic associations in each park "as nearly as possible in the condition that prevailed when the area was first visited by the white man" (Advisory Board on Wildlife Management 1963, p. 4). The report stated that "A national park should represent a vignette of primitive America," and that "Above all other policies, the maintenance of naturalness should prevail" (Advisory Board on Wildlife Management 1963, p. 4, 9.).

Prior to this report's release, Park management had focused largely on protecting species from human disturbance, including hunting and fire. (Predator control was common in the early part of the century but had largely ceased by the 1960s in National

Parks.) The Leopold report, in contrast, argued that this management practice created unnatural conditions since fire was common under "natural" conditions and since ecosystems were already disturbed through the extinction of large predators or the fragmentation of habitat. The report drew on the assumption that a balance of nature did exist and proposed that management interventions should be designed to restore that balance. The report suggested that the state of the system when Europeans first arrived could be considered "natural" and adopted as a target.

Although maintaining an ecosystem in the condition that it existed when Europeans first arrived might still represent a rational management objective, we now know that it is anything but natural. In the case of National Parks, the choice of the reference point for recreating a "natural" system could not have been worse. In advance of the white man's first visit to the regions that became our national parks, diseases that swept across the continent after the first contact with early explorers had devastated Native American populations (Crosby 1986). Thus, neither the ecological communities nor the apparent interaction between humans and these communities could have been considered natural when Europeans first settled.

Second, the ecological justification for recreating a previous set of ecological conditions has been eroded by the findings of the dynamic changes in community composition (Hunter et al. 1988) and the "non-equilibrium" state of communities. Thus, even if it were possible to define the "natural" community that existed in 1600, it is a safe bet that it would be quite different today, with or without humans on the continent. This consideration has led to revised National Park Service policies under which "change [will] be recognized as an integral part of the functioning of natural systems. The National Park Service will not seek to preserve natural systems in natural zones as though frozen at a given point in time" (NPS 1988).

And while modern science called into question these traditional goals of management, modern conservation philosophy now calls into question the terminology. The very name "protected area" is a throw-back to early conservation philosophy that viewed conservation as an alternative to development, not a component of development. The term "protected area" makes sense when wild ecosystems are seen as being protected from development, or as being protected from uses that are inappropriate or incompatible with their management objectives. But *any* land, public or private, is "protected" in this sense; certain uses are allowed while others are not. The term conveys the message that barriers exist between the resource and society, at a time when it is urgent that people become aware that these resources are as important to their economic and social well-being as is a farm or a factory.

BEYOND PROTECTED AREAS

Advances in ecological science not only held implications for the objectives of protected area management, but also for the relevant scale of management. For more than two decades, environmental policy-makers have placed considerable emphasis on designing protected area "systems," based on the assumption that a patchwork of land uses and management objectives could, through careful planning, provide the various services needed from ecosystems. However, many important ecological processes take place on a scale exceeding the "patch size" that results from typical protected area system designs. This has always been obvious in the case of migratory species, and both the design and management of protected area systems have incorporated ways to protect these species.

More recently, however, advances in techniques for determining minimum viable populations (MVPs) for species have demonstrated the inadequacy of many protected areas for maintaining viable populations, particularly of larger vertebrates, and studies of the potential impacts of climate change on ecological systems have focused attention on the inadequacy of traditional protected areas to "protect" species migrating in response to changing climates.

One implication of the large scale required for ecological management is that the management of "unprotected" land is often as important for the conservation of biodiversity and the maintenance of important ecological processes as the management of protected areas. This is particularly clear for ecological services, because, for example, pollutant filtration and clean water is as valuable to a community downstream from a timber plantation as it is to a community downstream from a National Park. Large areas are needed also for biodiversity conservation to maintain populations of some wild species. Moreover, a significant fraction of the diversity that we seek to maintain is a product of domestication and, thus, found outside today's protected areas.

Finally, growing recognition of "cumulative effects" calls into question the effectiveness of relatively independent management objectives for different lands within a patchwork of land uses. Given the interconnection of ecological systems, some integration of management objectives is essential. Maintaining native fish populations in a protected river, for example, requires compatible upstream agricultural and forestry practices.

To address these concerns, managers have begun to look beyond protected areas to seek ways of coordinating management inside and outside of protected areas to meet ecological objectives. For example, increasing emphasis has been placed on the establishment of biological "corridors." Such corridors can help ensure that several protected areas, each too small to maintain viable populations of certain species, can function as a larger unit and maintain both species populations and important ecological processes.

A second, more process-oriented solution has been to establish bioregional management mechanisms. In theory, such mechanisms provide a means of coordinating management on public and private lands to meet broader ecological management objectives at minimum cost (or maximum benefit) to private landowners. Numerous obstacles to effective bioregional management exist, but in the absence of any good alternatives, such approaches are increasingly being tried around the world (WRI/IUCN/UNEP 1992).

A third mechanism for moving conservation beyond protected areas involves policy incentives that can modify private land use decisions. Whereas several decades ago the purchase of private land by conservation groups or by the National Park Service or Fish and Wildlife Service was the mechanism of choice, today planners and conservation groups have recourse to a wide array of conservation incentives, including conservation easements, tax incentives, land conversion fees, and development fees.

As ecosystem managers increasingly look beyond the boundaries of protected areas, a new vocabulary will be needed to describe these regions. Terms like "Ecologically Sensitive Areas" (McNeely et al. 1990) or "Ecological Reserves" (Reid 1992), for example, might better describe regions managed primarily to conserve biodiversity, "Timber Production Areas" for regions managed for timber production, and so forth. If a collective term is needed to encompass all types of wildland management, that is, the management of areas comprised primarily of non-domesticated species, then the term "Wildland Management Area" would suffice, or perhaps "Wild Ecosystem Management Area" to cover terrestrial, freshwater, and marine systems. Both in practice and in name, we must seek to emphasize the linkage of protected areas with society rather than their isolation from humanity.

OBJECTIVES FOR ECOLOGICAL MANAGEMENT

Thus, fundamental changes have taken place in our understanding of ecological systems over the four decades since Leopold wrote *A Sand County Almanac*. The scale at which we now approach ecological management is far greater than it was in Leopold's day. Increasingly, policy-makers are viewing human uses of resources within the context of regional and national conservation needs rather than relegating conservation only to the domain of national parks and protected areas.

Perhaps most significantly, however, the very objective of conservation has become more elusive in recent years. The Age of Ecology had put the strength of science behind Leopold's (1966, p. 262) credo that "a thing is right when it tends to preserve the integrity, stability, and beauty of the biotic community. It is wrong when it tends to do otherwise." Given the assumption of the balance of nature and the reliance on mechanical metaphors of stability and integrity, it was relatively easy to see the human modification of ecosystems as necessary evils that had to be kept to a minimum lest the stability of these systems be disrupted.

But if communities are not in balance and no "natural" state can be defined, then what should the overarching concern of ecological management be? Intuition suggests that there must be some fundamental concern that can guide the relationship between humanity and ecological systems, some basic objective that addresses our underlying sense that the biosphere's capacity to support life, including humanity, must be maintained and that it must contribute to sustainable development.

Increasingly, that underlying concern is being seen to be the concept of "ecosystem health" or "biological integrity" (Noss 1990). Biological integrity is defined as the ability to support and maintain "a balanced, integrated, adaptive community of organisms having a species composition, diversity, and functional organization comparable to that of natural habitat of the region" (Karr 1991, p. 69). An ecosystem can be considered healthy when its capacity for self-repair when perturbed is preserved and minimal external support for management is needed (Karr 1991).

Ecosystem health and biological integrity are macro-descriptors that define a state of nature against which to measure degradation. They appear to open the way to developing biological indicators that can reveal the basic condition of our ecological systems. And they appear to provide a basic objective of ecological management that will subsume other conservation needs. For example, Walker (1992, p. 18) suggests that "The best way to minimize species loss is to maintain the integrity of ecosystem function."

Unfortunately, the conceptualization of biological integrity and ecosystem health as ecological management objectives fits poorly with ecological reality. Although it is entirely appropriate for political or educational reasons to seek to maintain a "healthy ecosystem," just as we seek a "strong economy," the former provides no more guidance to the biological manager than the latter provides to the economist.

Ecologists seem to be bent on exhausting every possible mechanical and medical metaphor before recognizing the unique nature of ecological systems. Concepts, such as inertia, elasticity, amplitude, hysteresis, and stability (Westman 1978) have no doubt helped ecologists frame questions about ecological processes, but none has ever stood the test of time as a biological indicator. Yet, these concepts still live on and guide policy, even though their meaning is unclear. Walker (1992, p. 22), for example, notes that "The worrisome cost of decline in biodiversity, particularly to politicians who may be held accountable, is the threat of a collapse in the 'stability' of ecosystems (whatever that means)."

Just as ecologists sought measures of stability and failed, it seems that the metaphors of health and integrity are doomed to failure as well. Establishing when an ecosystem is "sick" or disintegrating is entirely subjective. Is the eastern deciduous forest more healthy today than it was before climate change led to the migration of oak and hemlock into the community 10,000 years ago? Did the loss of the chestnut decrease the biological integrity of the system?

A sense of objectivity can be achieved if we assume that any change from the "natural condition" is evidence of declining health or loss of biological integrity, but this founders on the same shoals as traditional management objectives. How would we objectively define a "natural" state? Recognizing the difficulty of defining a natural ecosystem by its composition, several attempts have been made to redefine it based on "process." Prescott-Allen (1991) defines a natural ecosystem as one where since the industrial revolution (1750) human impact (a) has been no greater than that of any other native species, and (b) has not affected the ecosystem's structure. Botkin (1990, p. 157), stresses that we must appreciate that change is "natural, desirable, and acceptable," implying that natural ecosystems are those undergoing natural rates of change. But in either case, by recognizing man's role in "nature" and in particular by giving man a role in maintaining "natural" ecosystems, the very definition of "nature", that part of the material world unaffected by man, is contradicted.

The continuing pursuit of ecological management objectives related to ecosystem health and biological integrity reflects an undying faith among ecologists that we can ultimately determine what "makes the system tick." Guided throughout our careers by Leopold's admonition "to keep every cog and wheel is the first precaution of intelligent tinkering" (Leopold 1966, p. 190), we continue to search for the mainspring even though it is now clear that the system does not tick at all— it undergoes both sudden and gradual change through both deterministic and chaotic processes.

Clearly, efforts to prescribe objectives for ecological management in terms of the "state" of ecological systems will not succeed simply because ecological systems are neither static nor balanced. But what are we left with? Although we might not be able to define the state of the system, we certainly still must pursue some management objective that will ensure that the regional and global biota maintains its capacity to support life. And we can be relatively certain of the targets of ecological management even if we have difficulty defining the objective: We clearly want to maintain ecological services important to humanity, and we want to maintain biodiversity to keep options open for future generations.

One overarching objective could be to preserve the *status quo*. Given that we have little ability to predict how a change in an ecological system will affect ecological services and biodiversity conservation, we could simply acknowledge that any change carries with it a risk, and a good risk-averse strategy is to do our best not to change the existing system. In a sense, this approach would be functionally no different than traditional objectives of maintaining natural ecosystems. We would accept whatever now exists as what we should maintain because it is providing valuable services.

The problem with the "if it ain't broke, don't fix it" approach is, of course, that the system is already broken. Declining biodiversity, declining soil fertility, water pollution, and air pollution all suggest that the *status quo* is insufficient.

Alternatively, we could state the management objective in terms of maintaining ecological processes rather than maintaining particular ecological states. For example, we could seek to establish "evolutionary preserves," areas in which natural evolutionary processes continue unabated. A process approach would also frame objectives in terms of ensuring that valuable ecological services are not degraded. But again, how would we judge what is an impairment of the process and what is not? Certainly, evolution would be more rapid in a fragmented habitat: so should an evolutionary preserve be large to slow species extinction, or small to promote speciation? And how should the goal of maintaining ecological services address the impacts of climate change, which will presumably fundamentally alter the location of various services from ecological systems.

The pursuit of an overarching ecological management objective is doomed to failure because of the unpredictability of the behavior of ecological systems and because of our lack of knowledge of those aspects of the behavior of ecological systems that are predictable. These two characteristics ensure that objectives related to ecological states, ecological processes, or macrodescriptors combining states and processes, will fall short of the more basic concern. We know enough to be sure that there is nothing ecologically profound about what we have called "natural systems" (and they cannot be defined anyway), but we will never know enough to be sure what the "best" state is or what the impacts of change will be.

This suggests that the objective of ecological management should not be ecological, but social: the overarching objective of ecological management should be to maximize human capacity to adapt to changing ecological conditions. This might seem overly-anthropocentric, but it is entirely consistent with a biocentric outlook. To maximize human capacity to adapt to changing conditions, we must also maximize life's (or, the biosphere's) capacity to adapt to change.

In practice, the adoption of such a management objective would have relatively little effect on current management targets. To adapt to change, humanity needs the diversity from which new innovations can be created and needs productive ecosystems that provide the biological and economic capital to invest in making those innovations. Thus, maintaining biodiversity is a prerequisite for maximizing human ability to respond to changing conditions, as is maintaining the productivity of agricultural systems, the yield from forests and fisheries, clean water, and clean air. Similarly, the three requirements of biodiversity conservation stated in the *Global Biodiversity Strategy*—save it, study it, and use it sustainably and equitably—are all key elements of a strategy to maximize human adaptability (WRI/IUCN/ UNEP 1992).

But, the adoption of such a management objective would have two effects. First, it would change the justification used for management goals. Rather than having recourse to concepts of biological integrity, ecosystem health, or ecological stability, justifications would be cast in terms of economic, social, and ethical needs and responsibilities. We would justify the maintenance of species within a protected river system because we have an ethical responsibility to future generations, not to maintain the system's "health." Or, we might seek to maintain those species because we know that their loss would impair important ecological services (or, because we do not know enough to judge whether it would), not because we are worried about the biological "integrity" of the system.

Second, such an objective would place well-deserved emphasis on the fundamental importance of change as a characteristic of ecological systems and as a focus of management concern. The objective implies that a premium should be placed on policies geared toward flexibility and adaptation to change, rather than direct regulation geared toward static targets. For example, why not make greater use of development fees to reduce the rate of loss of habitats that are valuable for maintaining biodiversity and ecological services? Perhaps all development within five meters of sea level, or within "ecologically sensitive areas" as identified through Gap Analysis, should be assessed a fee (which could be used to subsidize development in areas with less value for biodiversity or ecological services). Similarly, greater use of public funds for purchasing conservation easements, and defraying lost property taxes for local communities, can help management move beyond protected areas.

CONCLUSIONS

For over three decades, a coalescence of utilitarian and non-utilitarian conservation philosophy, guided by now-dated ecological theory, has helped push conservation goals into the mainstream of development planning. But this coalescence was based in part on ecological misunderstandings. Using ecological theories based on ecosystem stability, biological integrity, and the assumption that natural ecosystems provide higher quality services, wilderness preservation could be justified on the very same economic grounds as resource exploitation. But both economic reality and ecological knowledge changed dramatically over those decades. Today, an economic case for wilderness conservation hardly needs to rest on ecological theories and, at the same time, the ecological support for preservationist goals has weakened considerably.

Abandoned by ecologists who can no longer provide firm scientific support for strict preservation, "deep ecologists" have reclaimed the moral high ground and reasserted the spiritual value of wilderness and the value of wilderness for its own sake. There is indeed value in leaving portions of the earth surface untouched by deliberate human management. The value can be defended on both scientific grounds, by providing a baseline against which the effects of human intervention can be measured, and philosophical: "in wildness is the preservation of the world" (Thoreau, cited in Chase 1987). There is no need to justify such areas as "natural." (Indeed, they are likely to become among the least "natural" as exotics invade and habitat fragmentation and changes in disturbance regimes take their toll on native species.)

Recognizing these changes, it is time to adopt a new objective for ecosystem management: maximizing the human capacity to adapt to changing ecological conditions. There is no point in continuing to pursue a "correct" state of ecological systems as an objective of management when it is clear that ecosystems are both dynamic and subject to random changes. There is every reason, however, to seek to manage ecological systems, including areas left untouched by humanity, to ensure that humanity will be capable of responding to change.

ACKNOWLEDGMENTS

I would like to thank Jeffrey McNeely and two anonymous reviewers for providing helpful comments on an earlier draft of this manuscript, and Kathleen Courrier for her help in editing the manuscript.

LITERATURE CITED

Advisory Board on Wildlife Management. 1963. *Wildlife Management in the National Parks.* A. S. Leopold (chairman). Appointed by Secretary of the Interior Udall. March 4. U.S. Department of Interior, Washington, D.C.

Bergstrom, J. C., and H. K. Cordell. 1991. An analysis of the demand for and value of outdoor recreation in the United States. *J. Leisure Research* 23:67-86.

Botkin, D. B. 1990. *Discordant Harmonies.* Oxford University Press, New York.

Chase, A. 1987. *Playing God in Yellowstone: The Destruction of America's First National Park.* Harcourt Brace Jovanovich, New York.

Congress of the United States. 1916. U.S. Statutes at Large, vol. 28:73. (The National Park Service Act.)

Crosby, A. W. 1986. *Ecological Imperialism.* Cambridge University Press, Cambridge.

Dasmann, R. F. 1973. A system for defining and classifying natural regions for purposes of conservation. *IUCN Occasional Paper* 7:1-18.

Elton, C. 1924. Periodic fluctuations in the numbers of animals: their causes and effects. *Brit. J. Exper. Biol.* 2:119-163.

Fox, S. 1981. *John Muir and His Legacy: The American Conservation Movement.* Little, Brown, and Co., Boston, Mass.

Goodman, D. 1975. The theory of diversity-stability relationships in ecology. *Quart. Rev. Biol.* 50:237-266.

Hales, D. 1989. Changing concepts of National Parks. In *Conservation for the Twenty-first Century,* eds. D. Western and M. Pearl, pp. 139-144. Oxford University Press, Oxford.

Holling, C.S. 1973. Resilience and stability of ecological systems. *Ann. Rev. Ecol. Syst.* 3:33-50.

Hunter, M. L., Jr., G. L. Jacobson Jr., and T. Webb III. 1988. Paleoecology and the coarse-filter approach to maintaining biological diversity. *Conserv. Biol.* 2:375-385.

International Union for the Conservation of Nature and Natural Resources, United Nations Environment Programme, World Wildlife Fund (IUCN/UNEP/WWF). 1980. *World Conservation Strategy*, Morges, Switzerland.

IUCN Commission on National Parks and Protected Areas (IUCN). 1984. In *National Parks, Conservation, and Development: The Role of Protected Areas in Sustaining Society*, eds. J. A. McNeely and K. R. Miller, pp. 47-55. Smithsonian Institution Press, Washington, D.C.

Karr, J. R. 1991. Biological integrity: A long-neglected aspect of water resource management. *Ecol. Appl.* 1:66-84.

Lack, D. 1954. *The Natural Regulation of Animal Numbers*. Oxford University Press, London.

Leopold, A. 1933. *Game Management*. Scribner, New York.

Leopold, A. 1966 [first published in 1949]. *A Sand County Almanac: With Essays on Conservation from Round River*. Ballantine Books, New York.

McNeely, J. A., K. R. Miller, W. V. Reid, R. A. Mittermeier, and T. B. Werner. 1990. *Conserving the World's Biodiversity*. World Resources Institute, Washington, D.C.

MacArthur, R. H. 1955. Fluctuations of animal populations, and a measure of community stability. *Ecology* 36:533-536.

MacArthur, R. H. 1972. *Geographical Ecology*. Harper and Row, New York.

May, R. M. 1973. *Stability and Complexity in Model Ecosystems*. Princeton University Press, Princeton, N.J.

Miller, K. R. 1975. Guidelines for the management and development of national parks and reserves in the American humid tropics. In *Proceedings of a Meeting on the Use of Ecological Guidelines for Development in the American Humid Tropics*, pp. 94-105. Caracas 20-22 February 1974. IUCN, Switzerland.

Noss, R. F. 1990. Can we maintain biological and ecological integrity? *Conserv. Biol.* 4:241-243.

Odum, E. P. 1953. *Fundamentals of Ecology*. Saunders, Philadelphia, Pa.

Pinchot, G. 1947. *Breaking New Ground*. Harcourt Brace Jovanovich, New York.

Prescott-Allen, R. 1991. Classification of ecosystem conditions. In *Caring for the Earth: A Strategy for Sustainable Living*, p. 34. IUCN, Gland Switzerland.

Reid, W. V. 1991. Toward a national biodiversity policy. *Iss. Sci. Technol.* 8(3):59-65.

USDI National Park Service (NPS). 1988. *Management Policies*. National Park Service, Washington, D.C.

Walker, B. H. 1992. Biodiversity and ecological redundancy. *Conserv. Biol.* 6:18-23.

Westman, W. E. 1978. Measuring the inertia and resilience of ecosystems. *BioScience* 28:705-710.

World Resources Institute, International Union for the Conservation of Nature and Natural Resources, United Nations Environment Programme (WRI/IUCN/UNEP). 1992. *Global Biodiversity Strategy: Guidelines for Action to Save, Study and Use Earth's Biotic Wealth Sustainably and Equitably*. World Resources Institute, Washington, D.C.

Worster, D. 1977. *Nature's Economy*. Cambridge University Press, Cambridge.

29

INTEGRATION OF *IN SITU* CONSERVATION OF PLANT GENETIC RESOURCES INTO LANDSCAPE AND REGIONAL PLANNING

Gordon Brent Ingram

Department of Forest Resources Management
University of British Columbia
Vancouver
CANADA

Conservation is a social process, steeped in values and socially determined priorities and acceptable margins of risk. Although the past decade has seen a progressive politicization of the use and transfer of germplasm and the natural ecosystems that support respective populations, the policy dimenions of the relationships between information, genetic conservation priorities, and subsequent requirements for protected area design and management have barely been considered. This paper lays the basis for a theory of environmental planning for the *in situ* conservation of genetic resources.

I begin with the still largely theoretical question of the integration of conservation of wild plants with genetic resources into networks of protected areas involving a range of other concerns for habitat as well as for values and resources related to local subsistence, recreation, and heritage. My argument is that the *in situ* conservation of genetic resources is, inherently, a reflection of social concerns of germplasm and landscape-oriented interventions for the persistence of genotypes and alleles. There are, however, some crucial technical components in such programmes to minimize genetic erosion. *In situ* conservation of plant genetic resources must be based on ecogeographical surveys that correlate environmental factors to intra-specific variability. Current networks of protected areas, which support populations of genetic resources, are largely unorganized and *ad hoc* for those specific purposes and will require careful review and coordination over the coming decade if there is to be adequate conservation in terms of needs for germplasm. Such networks will necessarily overlap with and complement other operations for the conservation of biological diversity and natural areas. Special kinds of monitoring and management prescriptions are necessary.

Five functional categories of spatial zones for genetic resource conservation are considered: cores, buffers, transitions, corridors, and barriers. Three examples are explored: the Sahel of west Africa, crabapple on the north coast of British Columbia, and the rich human forests of islands off of New Guinea.

It is with the *in situ* conservation of genetic resources that the complex linkages required between social values, economic policy, land management, ecology, and biology become prohibitive for determination of land management prescriptions through reductionist science. The "biodiversity" paradigm (Wilson 1988) has radically transformed the formerly *ad hoc* vision of conservation of genetic variability along with the growing interest in utilization of germplasm from wild populations in the improvement of crops and livestock. The result is a set of rapidly intensifying social and economic pressures for *in situ* conservation and a lag in the necessary theory and techniques.

The status of most populations with genetic resources is still poorly known, and the development of new theory and techniques for determining conservation requirements and documentation of wild populations is necessary. A number of technical issues must be addressed before biosphere reserves and other internationally and nationally recognized protected areas can become effective for the conservation and procurement of genetic resources. Desired levels of conservation of less common genes should be determined. Acceptable levels of conservation and margins of security for loss of diversity should be set. Access to and ease of procurement of the genetic resources of wild species must be effectively regulated and in some cases expanded. Regulation and expansion of the systems of distribution (Kloppenburg and Kleinman 1988) of the germplasm taken from protected populations are inevitable.

An additional purpose of this paper is to take stock of what we know and, based on cursory evaluations of the various types of programmes, to explore ways to integrate better the various concerns for a range of plant and animal populations into regional and national decision-making frameworks (WRI, IUCN and UNEP 1992). It is worthwhile to explore the possibilities, for one or two decades into the future, when new vehicles for more comprehensive conservation as the *Convention on Biological Diversity* (UNEP 1992) have been instituted. It is necessary to begin exploring typologies of *in situ* conservation as based on different factors related to particular gene pools, genetic erosion factors, sites and regions, and demands for germplasm and then to identify needs for new theory, technique, and approaches to decision-making.

We can envision a world with a growing number of programmes of surveys, evaluation, protected areas, monitoring (Dallmeier et al. 1992), and germplasm procurement for particular plant and animal species and sites of interest. *In situ* conservation of genetic resources represents a fertile cusp between conservation biology and environmental planning which has been barely explored (Ingram 1990b). It is necessary to develop a more dedicated set of conservation planning "tools" (Hoose 1981) as part of better identification of options for intervention in the midst of continued losses of populations and for increased access to the germplasm of the populations that remain. Perhaps more importantly, we must better situate the concerns for genetic resources within the pantheon of the conservation of biological diversity and the inevitable trade-offs and compromises of decisions over land.

By linking utilization of biological resources with funding for conservation, the *Convention on Biological Diversity*, and whatever follows, the landscape of conservation has been fundamentally transformed. However, we cannot let the various funding crumbs for reserves, which are generated from genetic resources, deter us from a more holistic vision of conservation of biological diversity and respective transitions to bioregionalism (Cheney 1989; Alexander 1990), reinhabitation, and sustainability (Redclift 1987).

IN SITU CONSERVATION OF GENETIC RESOURCES AS DRIVEN BY SOCIAL PRESSURES AND POLICY CHOICES

Why be concerned with the persistence of genotypes and alleles beyond that necessary for the actual survival of species? Maintenance of certain levels of intra-specific variation is key to a population's fitness. When we maintain as many genotypes as possible, we are, at the very least, preserving additional possibilities for utilization of adaptations, either in wild populations or as domesticates. When we maintain as many alleles as possible, we are preserving possibilities for expressions of certain traits, such as disease resistence (Browning 1991), which might otherwise be lost.

The difference between maintenance of fitness and certain levels of security for genotypes in certain environments or alleles at certain levels of occurrence can be great in some cases, or relatively minor in others, depending on the reproductive biology and genetic architecture of the species, as well as the needs and pressures for conservation from a range of interest groups, regions and over various "operational scales" (Delcourt and Delcourt 1992) as related to space and time. Under their framework, the two operational scales, at least as related to the management of edges and successional mosaics which are central to *in situ* conservation of genetic resources, are the "microscale" "1 yr to 500 yr; 1 m^2 to $10^6 m^2$ [100 ha]" and the "mesoscale" "500 yr to 10,000 yr; $10^6 m^2$[1 km^2] to $10^{10} m^2$ [10^4 km^2]."

An additional set of conservation criteria for intra-specific variation, which may not be crucial to a population's continued fitness, is termed "desired diversity" (Namkoong 1983).

Before proceeding, we should consider the notion of the genetic resource and how this label is evolving. A population, whether wild, feral, weedy, or domesticated, is, in itself, a genetic resource when it is recognized to have the potential to provide germplasm through human intervention to enhance a different domesticate or managed wild population. The germplasm in transfer between donor and recipient populations can be seed, sperm, eggs, or cultured material as well as DNA which is injected into living cells. A substantial portion of the species in a region might have some significance as genetic resources as human history unfolds and new technologies invariably make transfers easier. In addition, countless plant species and associated species might have potentials as new domesticates and for environmental restoration.

A species or a population is a genetic resource when it is identified but, similarly, aspects of the genes and associated gene pool can become part of the resource complex when they too are recognized as being valuable. What maintains genetic diversity and our highly fluid subset of genetic resources are complex ecosystems and habitat and successional mosaics across regions. When we transfer seed or even a gene, we are, in effect, transferring oddly incongruous points from one ecosystem into another. When indulging in isolated DNA, the most reduced form of the genetic transfers so far devised, it is easy to forget that there is never a clean transfer. In the case of seeds, there are associated microbes, and sometimes even insects. There are numerous examples, particularly from the humid tropics, of transfers that require such mobile links as vertebrate pollinators so that the new population can be viable. In defining and conserving a genetic resource, particularly involving wild species, the seam between the genetic agent, the actual germplasm, and the context is becoming more fluid and, in terms of ownership and funding, increasingly contentious.

The groups with various interests in and uses for genetic resources are expanding rapidly. Historically, there have been plant breeders and then the actual purveyors of reproductive material such as seed houses. Today, groups of consumers have identified a stake in continued access to products such as drugs from wild populations like Taxol from the Pacific Northwest tree, *Taxus brevifolia*. Groups within production and service sectors advocated increasing access to a growing number of products. If a tiny fraction of the income generated by biotechnology from material from protected areas is actually funnelled back into conservation, we might be able to add protected area administrations and the public which enjoys these areas as a powerful group of indirect beneficiaries.

The proliferation of scientific, industry, and citizen groups associated with the actual genetic transfers and the subsequent experimentation and development now involves both the public and the private sectors. There are the surveyors and collectors, the technicians of the transfer and storage operations, and the evaluators and the breeders. Except in the maintenance of some material from recognized populations, there is no assurance that respective economic and scientific interests and priorities will be in concert.

Groups that are often neglected are subsistence and traditional users with their own interests in particular species and ecosystems. Across a region and within a community is a great range of user types, interests, and priorities; these can even be at odds with the global marketplace and even the liberalism of cosmopolitan "societies" (Dasmann 1975; Ingram 1989). Between and within groups of consumers, purveyors, and subsistence communities, there are bound to be differences in valuation of particular populations and genes, conservation priorities, and acceptable levels of conservation security, risk, access, and costs. This is a microcosm of the complexities of the global marketplace in the 21st century. As genetic resources are increasingly valued, interests will proliferate as will potential points of cooperation and conflict. The currents within this milieu have a direct bearing on what is effectively conserved *in situ*.

IDENTIFICATION AND VALUATION OF GENETIC RESOURCES

Identification procedures for genetic resources have barely been developed. In a sense, any living thing is a potential donor, although only a minute fraction of the species and populations on Earth will ever prompt such levels of monitoring and transfer. Only a minute fraction of the material transfered will be worked into general distribution as new plant varieties and breeds of animals. The following is an example of the ongoing and open-ended nature of determination of genetic resources for food crops.

> IBPGR has formulated a list of crop-collecting priorities to guide conservation activity. In general, the factors considered are the level of risk to the crop and its wild relatives; the crop's economic and social importance; the materials that need collecting; the needs of plant breeders; and the quality of present collections. The list is occasionally updated, and is made both by crop and by region, but there is nothing sacred abouth IBPGR's list. It is based on educated guesses. And they are probably the best guesses that can be made, even if sometimes reflecting a bias toward industrialized countries needs." (Fowler and Mooney 1990: 79).

These species and genera in these lists include the better documented crops. If an expanded list of important potential food plants were compiled (Prescott-Allen and Prescott-Allen 1990), which included species currently only of subsistence importance as well as more distantly related populations that can only contribute genetic material to a food gene pool through laboratory means, the number of "priority crops" would be much greater.

Traditional crop material is defined here as pre-20th Century varieties and "land races" (Williams 1988) which are diverse collections of primitive types of domesticated material and related weeds. A bag of seed from one field of land races is highly diverse and this allows for some production, through the various adaptations, even in the worst years. On the other hand, the seeds of modern varieties are relatively uniform. This "primitive" material is sometimes called "folk seed" (Fowler and Mooney 1990), and the term is perferred by some as it acknowledges the important role farmers have played in selecting, breeding, and distributing crop varieties.

Wild relatives of crops are species that can be used to contribute to the development of new crop varieties. Three general categories of these biological resources correspond to genetic proximity to respective crops. Wild species that can be easily bred with a crop are, therefore, in the *primary gene pool.* Wild species that can be bred with a crop under special conditions are in the *secondary gene pool.* Wild species that can only contribute to the improvement of a crop through genetic engineering techniques are in the *tertiary gene pool.* The trend in biotechnology is to by-pass the genotypes of primary and secondary gene pool populations and to isolate a wide range of material in a technologically driven expansion of the tertiary gene pool. This shift does not make the more closely related wild, weedy, and feral populations less valuable because it is in studying phenotypic expression, *in situ*, and relating it to laboratory evaluation of genotypes that desired expressions can be isolated. The primary and secondary gene pool populations become valuable as much as "libraries" and "*in situ*" laboratories as sources of material for transfers.

The necessary conservation measures differ greatly between traditional varieties and wild relatives of crops because populations evolve within natural ecosystems, whereas land races and primitive crops are products of human practices, modified habitats, transformative social relationships, and the resulting agro-ecosystems (Bedigian 1991). In terms of reproduction, crop and related wild species can be divided into two very different groups: annuals and perennials. Seeds of annuals can be conserved through cold storage. For many perennials, non-ecosystem-based, *ex situ* conservation is not workable because of short-lived seed, the diminishing limitations of meristem culture, and reproduction requirements that are difficult to meet in field genebanks and laboratories (Ingram 1984).

Eight centres of crop origin were identified by Vavilov (1951) who assumed these to be centres of diversity. These areas were further described by Harlan (1976). Some scientists now believe there are twelve centres of diversity (Fowler and Mooney 1990). Crops have dispersed from these areas to create secondary centres of diversity.

There is not necessarily a perfect fit between centres of origin and centres of species and intra-specific diversity for wild relatives of crops. For both mango (*Mangifera* spp.) and citrus (involving *Citrus* spp. and related genera), the centres of domestication were further west toward India while the centre of species diversity extends from Burma through peninsular Malaysia to Sumatra and Borneo. The implications of the location of the centres of diversity are that the great majority of crop genetic resources are in the tropics and in countries with relatively ineffective programs of protected area planning and management.

Recent work on the structure of crop populations (Zimmerer and Douches 1991) and on wild relatives of crops describes a fine mosaic of genotypic variation. This genetic diversity is difficult to capture and maintain even in networks of protected areas and comprehensive genebank collections (Brush 1991). For long-term breeding needs, a wider range of plant characteristics is not of current interest to breeders.

How will each interest group deal with the open-ended nature of setting priorities for surveying, conservation, and evaluation? We can be assured of haphazard responses from the global marketplace of products and information. There will inevitably be the accelerating "generushes" related to particular traits, species, and entire commodity groups: searching for drought-resistence genotypes in times of climatic shifts or rapid screening of a large number of wild relatives for a gene that is resistant to a blight or epidemic. The implications for land management are that a growing number of populations will be considered worthwhile for new protected areas, expansions of current reserves, and intensified monitoring and management.

Significant differences exist between the genetic resources of food crops and other groups, such as plant species with value as forage, medicinals, or in environmental restoration and the wild relatives of livestock. The implications for identification and valuation of wild, weedy, and feral populations are that delineation of a target group for research and conservation is the result of the knowledge base related to taxonomy and the perceived short- and long-term demand for improved or introduced types. In these non-crop groups, the knowledge, political, and capital bases are far more modest.

ECOGEOGRAPHICAL SURVEYS OF GENETIC RESOURCES: LINKAGES BETWEEN TAXONOMIC AND INVENTORY DATA AND PRIORITIES IN CONSERVATION PLANNING

The core of any *in situ* conservation effort is actually the long-term programme of inventory and monitoring. Initial inventories of populations involving collection of field data related to phenotypic expression and biophysical and spatial factors can be termed "ecogeographic surveys" (Ingram 1990a). The sub-text of an ecogeographic survey is the "second-guessing" of crucial relationships between two things. First, there is the ascertaining of the pattern of genetic variability which only occurs when site and laboratory data are correlated. Secondly, a survey explores possible correlations between life histories and landscape trajectories (Hansen et al. 1992), which are particularly vulnerable to disruption from particular segments of cause-effect linkages within the disturbances matrices (Forman and Godron 1986) to ascertain the cumulative impacts of genetic erosion. Vulnerability in this context relates to either habitat loss or habitat alterations that could eventually have a significant impact on the frequencies of certain alleles across landscape units and regions.

These surveys involve subsequent evaluation of sampled material in laboratories and field genebanks through techniques, such as electrophoresis. Material is often "grown out" in controlled conditions and some sense emerges of the overall genetic variation across distributions of each species.

It is usually necessary to record ecogeographic and cultural data at scales involving the site and the habitat isolate to the landscape and the regional meta-population. Site information for each sample is necessary as is more contextual information, such as biogeography which relates to patterns of gene flow. The types of ecological information of relevance will vary with ecosystem structure and respective successional and phenotypic factors. Some information will be spatial and some will be descriptive and biophysical. Usually a host of social-use factors and related selection pressures exists. There might already be some differentiation between phenotypes, and possibly genotypes, along the lines of the units within the cultural landscape (Rapoport 1992). Information on the presence and extent of introduced species is crucial.

Two divergent formats and approaches to ecogeographic surveying have emerged: for an entire species, genus, or gene pool across entire distributions or regions and multi-gene pool surveys for particular biotic districts and regions (Ingram 1990a). The single gene pool surveys have tended to be reductionist and autecological in emphasis, whereas the multi-gene pool surveys have often been more biogeographical in outlook.

The design of a survey programme involves a number of structural decisions as related to:

1. the taxonomic scope of the gene pool(s) as related to the target species;
2. the spatial scope in terms of realms, regions, bioregions (WRI, IUCN and UNEP 1992), political boundaries and districts;
3. the minimum number of samples to be taken and the criteria for sampling;
4. the minimum size of the samples to be taken; and
5. the extent of the data collection and subsequent documentation.

Corollaries exist between the number of samples and sampling sizes and the "where and how much" discussions for planning protected areas.

Of the surveys that have been undertaken over the last decade, there still is a tremendous problem with documentation in terms of the availability of the site information taken for each sample. In many of the *ex situ* settings to which sampled germplasm is deposited, there is very little motivation (or resources) to maintain the kinds of data bases that would allow for intensive use of ecogeographic data in genetic evaluation. This is one of the sleeping problems that will emerge as the *Convention on Biological Diversity* and other accords on access, distribution, and profit sharing of genetic resources are instituted.

In considering the needs for ecogeographic surveying and the subsequent linkages to programmes of *in situ* conservation, it is necessary first to ascertain the political and administrative interests, and the priorization of related questions for particular gene pools. The following are examples of some of the variables of these surveys.

THE SAHEL SURVEY AND THE *PENNISETUM* GENE POOL

The 1984 African famine intensified the concern in crop breeding for drought resistance at a time when there was also increased interest in expansion and sustainability of agriculture in semi-arid regions. Pearl millet (*Pennisetum* spp.), a west African crop, was the subject of research and sampling of germplasm. The surveying was organized by the International Board for Plant Genetic Resources of the Consultative Group on International Agricultural Research and the Food and Agricultural Organization of the United Nations and the cooperating institutes were Sahelian, United States (USDA),

Indian, and French. There were also similar efforts for crops, such as sorghum and okra, in the Sahel. It became evident that a large portion of the variability within the primary gene pool was still in the area of origin: the northern and central band of the Sahel extending from Senegal to Chad. There was particular concern for locating rural landscapes with relatively low human densitities and with high probabilities of introgression across land use-related ecotones involving wild and weedy material and land races (Benz et al. 1990).

The Sahel is a vast region and, to use time and resources effectively for surveying populations in wild and weedy crop gene pools, a range of other species in crop and forage gene pools of global significance was identified. Various sites across a range of ecological zones, countries, regions, and districts were visited (Ingram 1985, 1986b; Ingram et al. 1987) with a vision of a long-term effort to inventory and monitor a range of traditional crops and wild species of importance for food and fodder (Ingram 1986a).

Much of the Sahel is flat and homogeneous with environmental variation largely microscaled. Areas with steeper gradients, such as mountains and marshlands, are of interest particularly in light of the archaeological evidence of patterns of settlement and land use in the region (Munson 1976). An extensive multi-gene pool survey was undertaken in the Aïr Mountains of north Niger (Ingram 1990a), a mountain range which, because of the patterns of rain capture, has maintained Sahelian ecosystems as the Sahara moved southward. But even with only modest grazing pressures, there were extensive areas with dessication and destruction of woodland and savanna vegetation (Ingram 1985; Ingram et al. 1987). In such settings of rapid genetic erosion, human responses, such as recently innovated irrigated gardens, become the cultural refugia for formerly riparian and seasonal wetland vegetation.

The resources required to carry out such a research programme across a region are formidable. The required political cooperation is daunting especially in wartime. Sampling of viable seed is problematic under the best of conditions and almost impossible in times of drought. The collection of site-data (Ingram et al. 1987) is difficult except for the highly trained personnel especially in extreme heat and in 14-hour field days. Local institutes were highly skeptical of such ambitious approaches. But without such accelerated efforts, sampling and monitoring will not be sufficiently ahead of losses to inform conservation planning.

THE CRABAPPLE GENE POOL ON THE NORTHERN COAST OF BRITISH COLUMBIA

Burnaby Island, *Skwa-ikungwa-i*, is part of the Gwaii Haanas National Park Reserve on the Queen Charlotte Islands, *Haida Gwaii*, and, before 1989, was under considerable pressure for conventional clearcut logging. A survey of the island's biological diversity was initiated in the mid-1980s and included concerns for species of traditional significance and genetic resources (Ingram 1989). One species with relatively well-recognized genetic resources, and with some significance to European breeding programmes for apple (*Malus* spp.), is crabapple (*Pyrus fusca*), which was traditionally called *k'anlha* by the Skidegate Haida (Turner and Levine 1971).

The north Pacific coast of North America is relatively depauperate, biologically, because the glacial withdrawl has occurred within the last 10,000 years. However, the islands are remarkable in terms of their rugged landforms (Swanson et al. 1988), the complex biogeoclimatic zonation, ancient coniferous forests (Pojar and Broadhead 1984), and some oddly disjunctive occurrences of plant species (Roemer and Ogilvie 1983).

Four terrestrial, biogeoclimatic zones have been identified in the Burnaby Island area: coastal western hemlock zone, *CWHg*, the superhumid coastal cedars-pine-hemlock zone, *CCPH*, (Pojar and Annas 1980), mountain hemlock zone, *MH*, and alpine tundra. Within these zones are a number of associations (Lewis 1982). Over 90% of Burnaby Island is dominated by coniferous forest. A number of physiognomic types occur in the Burnaby Island area: saltmarsh, beach strands, Sitka spruce forest near shores, alluvial meadow forest, deciduous, red alder woodland, alpine, various associations of western hemlock, Sitka spruce and red cedar, topogenous bogs, fens and marshes, and forests of yellow cedar and mountain hemlock at higher elevations.

The crabapple extends widely across a number of biophysical gradients, although it is usually associated with more geomorphic as well as successional ecotones in the following zones. The crabapple extends from the upper edges of saltmarshes to various lowland forest types (British Columbia Ministry of Forests 1988) including: *Q1*—mixed old growth-moss association (*Tsuga heterophylla- Thuja plicata-Picea sitchensis-Vaccinium* spp.-*Rhytidiadelphus loreum*) comprised of climax lowland forest, dominated by western hemlock, red cedar and Sitka spruce; *Q4*—alluvium Sitka spruce-moss association (*Picea stichensis-Tsuga heterophylla-Rhytidiadelphus loreus-Plagiomnium insigne)*; *Q7*—pine-yellow cedar scrub (*Pinus contorta-Chamaecyparis nootkatensis-Carex* spp.-*Sphagnum* spp.), a complex of shrub and bog associations which extend into the mountain hemlock biogeoclimatic zone; and *Q9*—lowland, colluvial association (*Picea sitchensis- Alnus rubra-Huperzia selago-Pogonatum alpinum),* an "erosional disclimax" (Lewis 1982) and the colluvial phase of *CWHg-Q1*.

The Haida population in the Burnaby Island area was concentrated along a shallow channel with extensive saltmarshes called *K'iit*. The most historically complete and site-specific descriptions of the Haida villages are provided by Swanton (1905a,b) who noted one town occupied before 1870: *£a'nadágAña* translating to "Bad-talk-town." Virtually all of the crabapple trees in the area were "owned" and managed by well-ascribed, matrilineal family groups (Turner and Levine 1971). Therefore, given the extensive modification of lowland areas, as related to Haida settlements, and traditional patterns of utilization, these "wild" populations can be considered forms of cultural artifacts. With the removal of the pressures for clearcut logging several years ago, the threats to these populations are from tourism, introduced black-tailed deer (*Odocoileus hemionus*), and further disintegration of the cultural landscapes which were largely abandoned after the recurring small pox epidemics and resettlement efforts of the late 19th century.

The Pacific crabapple is slow-growing and occurs in relatively low population densities especially in terms of the trees of the Pacific northwest. This species, however, is remarkable in its range of occurrences and the respective biogeoclimatic sub-zones and ecotones. Global warming could shift altitudinal zonation on this highly complex island landscape at such a rapid rate that regeneration declines below viable levels of resilience. In the shorter term, the introduced deer have no predators and their populations continue to increase to levels where all low lying deciduous vegetation is continuously browsed. A more extensive survey of the impacts of introduced deer should include monitoring for crabapple regeneration and mortality from browsing.

A MULTI-GENE POOL RECONNAISSANCE OF FERGUSSON ISLAND, PAPUA NEW GUINEA

Linking the concerns for genetic resources and their conservation *in situ* as part of larger programmes for preservation of primary rain forest has been poorly explored. This survey of the forests of an island in southeastern New Guinea, with high levels of biological richness, was an effort to develop effective inventory and planning practices (Ingram 1989, 1991). Several wild species in the following gene pools were identified including banana, *Musa* spp.; sugar cane, *Saccharum* spp.; mango, *Mangifera* spp.; taro, *Colocasia* spp.; and citrus including *Microcitrus* spp., *Citrus* sp., *Monanthocritrus* sp., and *Wenzelia* spp. These were considered as part of a list of "focal species" (Ingram 1989) to guide the choice and design of protected areas and the relationships with subsequent management.

In such a setting, with so little inventory data and only tentative taxonomic identifications possible, the focus shifts to strategic areas of species richness and types of mosaics that might be sufficiently diverse to support a wide range of genotypes. But even for forest species, it is difficult to generalize about such mosaics. If a specimen of a target species is found, it is still very difficult to generalize about respective distributions, life history, and ecology. The few samples that are found can be subjected to extensive laboratory scrutiny if facilities exist for genetic analysis. But such evaluations usually require a large number of samples and individuals and few of these are possible in the low densities usually occurring in primary rain forest.

A survey such as this is almost more of an object lesson in the conservation of the biodiversity of humid forests. Initial inventory data are non-existent, weak, or suspect. Occurrences are at such low densities as to make any extrapolation on ecological relationships questionable. Few herbarium samples are available for the training of collectors and surveyors and few taxonomists who can find the time to identify specimens at the species level. Even if species are successfully identified, and this usually requires flowering in the case of perennials, the technical possibilities for sampling and *ex situ* conservation are limited without extensive evaluation. If material is successfully "grown out," most likely through meristem culture, the species might well require other aspects of its ecosystem for reproduction. At the same time logging companies are clamouring to remove as much marketable timber as possible and the numbers of swidden farmers continue to grow.

PROTECTED AREAS AND LAND MANAGEMENT IN RESPONSE TO GENETIC EROSION

In referring to the erosion of crop gene pools, the Keystone International Dialogue (1991) noted, "Loss continues today at a rapid rate. As we do not know how much total diversity once existed, it is impossible to quantify the losses."

Unfortunately, there is currently very little monitoring or a funding base for these activities, even where baseline studies exist. Several causal agents fuel genetic erosion or "genetic impoverishment" (Myers 1988), and these have intensified over the last decade. The rates of genetic erosion tend to be the greatest in areas with the most fertile and most easily mechanized agricultural lands and near urban centres and markets. It is in impoverished and marginal areas, such as mountainous uplands, where traditional varieties are still grown and are sufficiently relied upon to allow for conservation of traditional varieties. The crop gene pools that have had the most losses have been primarily those that are subjects of active breeding programs (Fowler and Mooney 1990).

With strategies for *in situ* conservation of genetic resources (Ingram 1987; FAO 1991) across a site or region, measures do not function to maintain an aspect of human experience of wildland but rather to preclude and counter largely cumulative impacts (Dickert and Tuttle 1985). These measures can usually be labeled as either land acquisition or management but can involve a complex set of human activities which, while idealizing the maintenance or reconstruction of relatively "intact" ecosystems, often appear as uncoordinated and *ad hoc*. Genetic erosion is often so rapid and pervasive that the key to conservation programmes is simply in responding to particular cause-effect linkages. Such a reactive approach is fundamentally incomplete from a bioregional or "reinhabitation" perspective but might suffice to maintain acceptable levels of intra-specific variation while freeing up energy for a broader vision of regional environmental management.

An initial framework for understanding genetic erosion in wild, weedy, and feral populations can emphasize models of fragmentation and extinction at finer scales (Merriam and Wegner 1992). However, in considering intra-specific variation, some contradictory management objectives can emerge in contrast to maintenance of fitness. Maintenance of higher levels of genetic variability, such as frequency of rarer alleles or even persistence of a phenotype which might reflect a genotype, requires protection of populations across gradients of active selection. Additional reservoirs, with population sizes sufficiently large to support alleles for which there is currently little selection, are also necessary. This gradient/reservoir dynamic can underly nearly any genetic conservation intervention and requires the integration of concepts of landscape ecology including ecotones, matrices and successional mosaics (Forman and Moore 1992) with laboratory analyses subsequent to surveying and linkages with the (limited) options for land management.

EFFECTIVENESS OF CONSERVATION OF GENETIC RESOURCES

The conservation of genetic resources and the on-site maintenance of particularly valuable populations have been an integral part of most cultures since the shift from gathering to agriculture. However, the pressures for loss of intra-specific variability and the stakes in utilization of germplasm for economic purposes have greatly increased. This decade will see a great proliferation of objectives and approaches for *in situ* conservation as a result of the increasing numbers of species recognized as genetic resources, increasing possibilities for successful transfers of alleles, and intensifying genetic erosion compounded by human population growth, forest clearing, agricultural expansion, and climate change (di Castri and Hansen 1992).

In this context, the following review of the types of *in situ* conservation programmes is more historical and in the past tense. Virtually all of the current programmes are only partically effective and are bound to be integrated into more comprehensive regional programmes for biodiversity.

As noted by Schonewald-Cox and Stohlgren (1988), "Merely designating areas or establishing laws does not guarantee the protection of biological and genetic diversity."

Virtually all protected areas and landscapes have some populations of some species of economic importance which are more or less adequately protected at least in the short-term. However, in virtually all protected areas, substantially increased monitoring and management programmes will be necessary to avert losses of rarer genes and other potential genetic resources. Three general levels of *in situ* conservation of genetic resources are described below.

Random *in situ* conservation of genetic resources is largely unplanned coverage through ecosystem conservation. When protected areas that have been designed or planned without recognition of the occurrence of genetic resource species include such populations, adequate coverage within reserve boundaries for maintenance of viable populations cannot be assumed. The scale for ecosystem conservation is often at the regional level and is often too broad to consider an array of species and finely scaled distribution data. Consequently, the extent of actual conservation is modest especially because narrow amplitudes of ecological requirements for particular species exist.

Ecosystem-level *in situ* conservation of genetic resources requires the planning and design of protected areas with use of lists of species with genetic resources and respective distributional data. Implications of principles of biogeography and conservation biology must be considered. The scale is often district-wide (Ingram 1989) where more specific decisions of about species and sites can be made. Many of the "genetic reserves" (Jain 1975) involve this level of conservation. Management for particular species and associated successional phases is usually necessary. Species that are monitored and managed under programmes of ecosystem coverage can also have ecosystem-level conservation.

Allelic-level *in situ* conservation of genetic resources involves on-going monitoring, management, and procurement for particular levels of conservation for specific "functional population units" (Solbrig 1991). The scale of such planning and management is site-specific. To attain this level of conservation, population viability thresholds must be set with prescriptions for maintenance of intra-specific variation and rarer alleles. Natural and well-protected populations have constant flux of gene frequencies with some of these alleles becoming rare or disappearing. In protected areas, the natural and human-induced dwindling of populations can cause the narrowing of the base of variation and subsequent loss of potentially valuable genetic resources. To maintain rarer alleles or possible adaptive complexes associated with certain environments and selection factors, additional requirements for larger and sometimes additional populations must be determined. It is only at this level of genetic conservation that management can be set to maintain levels of desired diversity.

With species that do not involve a concern for maintenance of potential genetic resources, random *in situ* conservation is often adequate and ecosystem-level *in situ* conservation is needed for the long-term. For *in situ* conservation for crop genetic resources, both traditional varieties and wild species, ecosystem-level *in situ* conservation is always necessary to assure capture of desirable minimum percentages of alleles, and allelic-level *in situ* conservation is necessary for long-term security and germplasm procurement.

Concerns for genetic resources have begun to be worked into broader national park management. However, rarely are there complete lists of plant species and distribution maps even in developed countries. Management is too often oriented to other ecological concerns or for the management of recreation. Biosphere reserves will play an increasingly key role in the conservation and utilization of wild relatives of crops because of institutional potentials for monitoring and international exchange of information (Ingram 1990b). The concept has not been effectively utilized for traditional crops. The network is very new (Batisse 1982; Verhnes 1989), and most biosphere reserves are still poorly inventoried, with only a small portion of the total number of species with potentials for providing genetic material having been identified.

The biosphere reserve concept is based around the vision of a set of land- use zones with one or more "core" protected areas surrounded by well-managed, primarily wildland areas and then surrounded by a heavily modified "transition" zones. By monitoring these transitions from natural baselines to human-modified ecosystems, important data for environmental management can be generated. Over the last decade, the role of biosphere reserves for the conservation of biodiversity has increased (UNESCO 1984). Unfortunately, the management requirements for particular populations with genetic resources have rarely been related to the central concepts of cores, buffers, and transitional areas, although great potential exists (Gregg 1991). Until recently, traditional agriculture was largely suppressed within core areas even where indigenous species were involved.

Extensive debates on the viability of *in situ* conservation of traditional crops have taken place (Altierri 1988; Brush 1988, 1991; Wilkes 1991). Almost all of the current "in farm" crop conservation programs have been initiated by small institutes and non-governmental organizations. In Ethiopia, the Seeds of Survival program is working with the Plant Genetic Resources Centre in Addis Ababa to re-introduce traditional varieties and land races to farmers. During droughts, farmers were forced to eat their stocks of seeds. The new response has been to make extensive collections *ex situ*, re-introduce land races where there is interest, establish research plots, monitor productivity, and establish training and support programs for farmers.

The traditional agroforestry system practiced on Fergusson Island, Papua New Guinea serves as an example of *in situ* conservation of traditional varieties of crops within a particularly biodiversity-rich setting (Flavelle 1991). Because of low human populations, mountainous terrain, and distance to market, cash crop ventures have so far been unsuccessful. The system revolves around yams cultivation, principally of *Dioscorea esculenta* and *D. alata*. These crops are the basis of the subsistence economy, but it is their cultural importance that might ultimately prevent them from being replaced by introduced food crop species.

Yams seeds are inherited through the matrilineage, are exchanged as gifts at funeral feasts and other occasions, and are the focus of magic ritual and myths. Other tubers that farmers have grown for generations but which hold less status than yams include: sweet potato, *Ipomea* sp., *Colocasia esculenta*, and manihot, *Manihot escuelenta*. Prominent food trees in the system include mango, *Mangifera* sp., *Citrus* sp., *Szigium* sp., *Carica* sp., banana, *Musa* sp., *Arctocarpus* sp., *Ficus* sp., *Cocos nucifera*, and sago, *Metroxylon* sp. Every time that a community opposes intrusion of the cash economy and especially such environmentally degrading activities as some forms of logging, it is a form of *in situ* conservation. Such approaches might not be permanent but the record of such traditional conservation measures suggests that they can often be more effective over the longer-term than the measures of an unstable government.

One of the most impressive examples of *in situ* conservation of crops in an internationally monitored protected area is that of Niger's Aïr and Ténéré National Nature Reserve. However, in recent years order has declined (Newby and Wilson 1993). A range of traditional and contemporary garden types are persisting as well as primitive, traditional and more recent crop material for the gene pools of sorghum, *Sorghum* spp., pearl millet, *Pennisetum* spp., barley, *Hordeum* sp., and wheat, *Triticum* sp. A number of key agroforestry gene pools are also being protected including *Acacia* spp.; a close wild relative to cultivated olive, *Olea* sp.; and the important African fruit tree, *Ziziphus* spp. The area is at an ecological crossroad in terms of material and farming practices from North Africa and the Sahel. But, regardless of the support from maintenance of these gardens within the framework of protected area management, progressive desertification, due largely to climatic change, could destroy prospects for long-term *in situ* conservation.

People with a history of living in and off the forest have knowledge about the location and natural history of wild populations of crops, their near relatives, and their sustainable management. Local people have a central role in surveying and monitoring wild populations (Ingram 1990c). The persistence of local knowledge and sustainable traditional forest-use practices over generations require intact and yet evolving cultural systems. Today, most such cultural systems are under assault or are being more subtly undermined by market forces.

Provincial park and nature reserve designations capture only a tiny portion of the wild gene pool of crops. In Amazonia, extractive reserves, designed to be compatible with needs of local people, conserve fragments of forest and the wild gene pools of Brazil-nut, *Bertholletia excelsa*, rubber, *Hevea* spp., cacao and other important crops (Smith and Schultes 1990). Local control and management of extractive reserves are necessary, but are inadequate for sufficient conservation of adequate levels of rarer but, nonetheless, potentially valuable, alleles. Local people need incentives to maintain wild populations and to minimize growing reliance on production of cash crops.

GENERAL PRINCIPLES FOR *IN SITU* CONSERVATION

It would be highly disadvantageous at this phase in the development of conservation biology and landscape ecology to generalize about the nature of the requirements for *in situ* conservation of genetic resources as part of broader regional programmes for the conservation of biological diversity. Many of us are familiar with the almost farsical misreading and misuse of the numbers 50/500 in early discussions on population viability (Frankel and Soulé 1981). However, it is definitely possible to say that requirements for genetic resources will be "different" than other kinds conservation measures if only because the uses and socially related standards for more altruistic and recreationally oriented programmes of protected areas are rooted in separate human needs. However, if we really can construct a bioregional ethic and aesthetic, it should have a place for appreciation of maintenance of a comprehensive set of habitats, edges, gradients, and respective selection factors across regions.

It is also possible to describe the following trends in requirements for *in situ* conservation, in terms of land management, as they emerge from ecogeographical surveys and subsequent and ongoing evaluation.

1. If we are dealing with a gene pool with high significance to the economies of particular regions, it is worthwhile to conserve even the rarer alleles. To maintain these alleles, larger and more numerous populations are usually necessary than those required simply to maintain fitness.

2. In terms of the small number of populations that are wild and feral relatives of livestock, virtually all of the current alleles, even the ones expressed in ways that are currently deleterious to respective populations, are worth maintaining. To assure the persistence of these alleles, culling practices must be seriously reviewed, and many population sizes must be effectively increased through combined *in situ* / *ex situ* strategies.

3. Landscape diversity is "good" but only to a point. Every set of species in particular regions has amplitudes that allow a wide range of populations and habitat and successional isolates and sufficiently large populations to maintain rarer alleles as well as sufficient resilience. We cannot assume that, across a region or landscape unit and for a suite of genetic resources, a median position will be possible. This problem manifests itself in decisions over management of mosaics. Some populations, genotypes and rarer alleles are invariably better protected than others, and this reflects priorities, the state of the (spatial) biogeographic and ecological knowledge as well as the nature of larger gaps in environmental data bases.

4. As concerns for germplasm increase and requirements for conservation are refined, concerns will grow for an expanding set of elements of respective ecosystems, such as mobile links (Nabhan and Norman 1992) and keystone mutualists.

5. Virtually all *in situ* conservation requires interventions and management prescriptions at multiple scales in terms of time and space, and this, in turn, requires linked strategies corresponding to sites, habitat isolates, landscape units, districts, and regions.

6. In planning and management of landscapes, five functional categories for strategies of *in situ* conservation are recognized:

   a. natural cores,
   b. natural and culturally modified buffers,
   c. transitional areas of cultural landscapes involving gradients of modern and traditional, social and technological impacts,
   d. corridors for gene flow, and
   e. barriers for the control of gene flow and the invasion of species.

7. *Ex situ* conservation and evaluation are necessary for any sustained protection of the genetic variability population *in situ*, but might not necessarily be geared to the central questions for land management. Consequently, virtually all *ex situ* programmes need to be reviewed and expanded because there is elaboration on land management questions for particular populations, species, sites and regions.

8. Virtually all *in situ* conservation involves conservation of cultural landscapes either as artifacts and anomolous isolates or as part of traditional human communities under threat. Therefore, every intervention for *in situ* conservation has economic, social,

and cultural impacts that can either support or undermine such communities. These communities often have the only expertise for the necessary fine-scaled management. The pattern of their adaptations to changing land-use patterns and resource availability, which are partially influence by *in situ* interventions, has a direct bearing on options for regional strategies (Nabhan et al. 1991). Essential to effective *in situ* conservation of genetic resources must be viable forms of co-management, which require directing of the global economic benefits of genetic resources back to local operations (Dasmann 1991; WRI, IUCN and UNEP 1992).

9. Because of the dual nature of traditional and global interests in maintenance of genetic resources and the dynamics of co-management that are emerging in most networks of protected areas, it is necessary to observe the principle of multiple and, sometimes conflicting, sets of management objectives for particular sites and populations. Effective *in situ* conservation of these resources is as much about building social, political, and administrative frameworks to resolve these contradictions and respective conflicts as is the design of reserve boundaries.

GUIDELINES FOR MONITORING, HYPOTHESIS TESTING, AND ADAPTIVE MANAGEMENT

Genetic resources are, by their nature, dynamic, fleeting, and unpredictable. Their requirements, as we are coming to perceive them, are profoundly humbling to the environmental planner and land manager. We will require vastly increased "protection" in the coming decades, but this must be based as much on various "inquiring systems" as networks of protected areas. It is with genetic resources that perhaps we can say that our networks of reserves are only as effective as our knowledge bases or at least our abilities to translate the second-guessing of our ignorance into prudent decision-making. Reductionist science does not give us much help, and we are left to relate our correlations of ecogeographic data to laboratory evaluations and then to a myriad of possible cause-effect linkages. How do we begin this open-ended and daunting enterprise for a particular region, park, reserve network or gene pool? The following are some key principles.

1. Concepts of genetic diversity, as they relate to definitions of genetic resources and risk of loss, need to be expanded and clarified within broadly oriented research programmes on biological diversity (Solbrig 1991).

2. It is necessary to clarify relationships between selection factors, ecologically and geographically related variation and both prioritites for conservation and crop breeding, this should be for particular plant groups as related to life histories, reproduction, and environmental heterogeneity. Literature review and field research phases must follow.

3. The pool of methodologies for surveying wild species with genetic diversity is still remarkably undeveloped (Ingram 1990a), and various approaches must be tested and evaluated. Major revisions between surveying phases are inevitable.

4. Documentation practices for collection of germplasm and respective site and regional data, involving both biophysical and cultural factors, are currently poor and highly variable. Even though every species, population, and context is unique, there needs to be some standardization for particular plant and population types, ecosystem and landscape types, and types of genetic erosion.

5. Given the extreme importance of a wide range of information, a distribution system for survey and collection data on genetic resources must be developed by and for larger regional land management and genetic resource institutes. An *in situ* conservation strategy is in no small part structured by the nature of these information systems.

6. Linkages and methodologies for *ex situ* evaluation and the "growing out" of material from wild populations must be reconsidered in the light of expanding pools of information on respective ecosystems and selection factors. This is particularly important to expanded use of germplasm from humid tropical ecosystems which might thrive only with complex sets of conditions. The nature of evaluations and the extent of their implications for *in situ* conservation interventions will change with the extent of sampling and evaluation techniques and the subsequent information base.

7. In the design of new protected areas as well as in management plans for cores and buffers, models are necessary to account for population viability, prescribed levels of conservation of rarer genes, and selection factors across successional mosaics. Such models are best computerized and have spatial aspects that could be integrated into conservation planning through geographic information systems.

8. The theories of domestication for many crop groups and gene pools have gone through paradigm shifts that have direct bearings on priorities for conservation of traditional varieties and primitive and weedy populations. *In situ* or in-farm conservation components of larger gene pool conservation and screening strategies should reflect careful review of hypotheses of crop evolution and genetic erosion.

FRAMEWORKS FOR INTEGRATION OF CONCERNS FOR GENETIC RESOURCES INTO PLANNING FOR CONSERVATION OF BIOLOGICAL DIVERSITY

The previous principles were applied for the species and scales in the three examples. A framework can be based on the principle of the relatively *ad hoc* and scattered nature of necessary for cores, buffers, transitional areas, corridors, and barriers. These functions can be integrated through relatively subtle manipulation of landscape mosaics providing that human activity can be sufficiently constrained and that regional and global shifts are not too rapid and extreme.

With the Sahel Survey and surveying for the *Pennisetum* gene pool, the precise nature of the cumulative impacts of genetic erosion, as they derive from desertification and human pressures, become the key set of research questions for successful networks of genetic reserves and other managed areas. If, for example, the recent shifts in loss of vegetation are less permanent and involve shorter cycles (Malo and Nicholson 1990), reserves might be smaller as connectivity and the management of populations across ecotones are more viable. But if we are looking at much longer-term changes in the region's biogeography, a few large preserves, primarily the Aïr Mountains reserves, are possible. Management would have to focus on these few large areas and be highly site-specific in steep, complex areas with mosaics of small landscape units. The scale of the strategies and the organization of the interventions would differ radically depending on the model for "desertification" that is accepted.

Some corollaries with the cultural diversity of the Sahel also exist. If we accept that plant domestication involved small, scattered points with selection factors from pastoralism, such cultural landscapes, indeed cultural interfaces, are of crucial importance for *in situ* conservation. If we assume a more pastoral, northern Sahel and a largely dry-land farming in central and southern Sahel, maintenance of gradients of land use within these zones might be more important. Again the scales of strategies and interventions, and the nature of their organization across national boundaries, vary radically with species and regional landscapes.

With the crabapple gene pool in the example from the northern coast of British Columbia, populations are "protected" within the context of the South Moresby National Park Reserve but will be viable only as genetic resources over a range of landscapes and with viable populations, if specific management plans are prescribed. Threats from climate change can only be countered at international levels.

The more immediate threats from deer, which invaded only in the 1950s and whose numbers continue to climb, require fencing. A fenced area becomes a kind of fragment or restoration-zone isolate. A range of trade-offs exists between a larger number of small-fenced sites and a smaller number of large-fenced areas. However, the reality of fencing in such steep terrain is that the technical constraints of construction and maintenance are as crucial to the survival of any regenerating populations as the fine points of capture of gradients of biophysical factors.

Trampling from tourism, in synergy with browsing, could prove to have a similarly lethal effect on shore populations of crabapple. *In situ* conservation measures might best be organized around design programmes involving trails and signs to keep people away from key or vulnerable populations and points within the mosaic.

In the multi-gene pool reconnaissance of Fergusson Island, Papua New Guinea, it is highly debatable if information on the location of certain individuals of genetic resource species should provide the basis for choice of protected areas even after years of field work. Inventorying and conservation can best focus on the stability-related refugia for species associated with primary rain forest and areas of more tightly situated and high-contrast edges for other species. Individual trees must be identified and marked and certain buffers should be related to the spatial patterns of reproduction of particular species.

The best way to minimize risk is to slow the cutting of primary forest and to diminish maximum block size while increasing the minimum dimensions of forest fragments. An alternative might be to alter logging practices, as related to timber removal or road engineering, to lessen overall fragmentation and other impacts. To integrate timber harvesting into cultural mosaics of swidden and gathering, the control of invasions of alien grasses is necessary.

CONCLUSIONS

The integration of concerns for the *in situ* conservation of genetic resources will require some of the most complex conservation frameworks for choice and compromise in regional planning, landscape design, and land management. Inevitably heightening conflicts will occur between different interest and user groups as related to germplasm and sites. Such integration must allow for a wide range of objectives and acceptable levels of conservation security and access to germplasm and must function to resolve respective conflicts to acquire and manage natural and culturally modified territory. Management will necessarily be co-management with parallel concerns related to the interests of both global biotechnology and traditional users.

The core of a regional programme for the *in situ* conservation of genetic resources is a long-term inventory, research, and monitoring effort that in turn informs adaptive management. As well as ascertaining initial conservation requirements, such research must allow for hypothesis testing on an indefinite basis as related to particular species, populations, sites, and landscape units.

The face of *in situ* conservation at regional levels will be dominated by a range of overlapping concerns for a growing multitude of species and sites with dynamic aggregations of functional spatial units such as cores, buffers, transitions, corridors, and barriers. These functions will be implemented through designs and prescriptions that are mainly conceived and expressed between the scales of 1:1,000 to 1:50,000. In addition, these functions will need to be supported by a range of economic, administrative, and political policies that elaborate on broader frameworks of the conservation of biological diversity and channelling the benefits of biotechnology back to the sources of germplasm.

ACKNOWLEDGMENTS

I am grateful to these friends who have generously shared their perspectives and information over the last decade: Trevor Williams, Vernon Heywood, Hugh Synge, Joan McPherson, and Louis Olivier.

LITERATURE CITED

Alexander, D. 1990. Bioregionalism: science or sensibility. *Environ. Ethics* 12:161-173.

Altieri, M. 1988. Rethinking crop genetic resource conservation: a view from the south. *Conserv. Biol.* 3(1):77-79.

Batisse, M. 1982. The biosphere reserve: a tool for environmental conservation and management. *Environ. Conserv.* 9(2):101-111.

Bedigian, D. 1991. Genetic diversity of traditional sesame cultivars and cultural diversity in Sudan. In *Biodiversity: Culture, Conservation and Ecodevelopment*, eds. M. L. Oldfield and J. B. Alcorn, pp. 25-35. Westview, Boulder, Colo.

Benz, B. F., L. R. Sánchez-Velásquez, and F. J. Santana Michel. 1990. Ecology and ethnobotany of *Zea diploperennis*: preliminary investigations. *Maydica* 35:85-98.

British Columbia Ministry of Forests. 1988. *Biogeoclimatic Zones of British Columbia 1988* (map). British Columbia Ministry of Forests Research Branch, Victoria, B.C.

Browning, J. A. 1991. Conserving crop plant-pathogen coevolutionary processes *in situ*. In *Biodiversity: Culture, Conservation and Ecodevelopment*, eds. M. L. Oldfield and J. B. Alcorn. pp. 59-85. Westview, Boulder, Colo.

Brush, S. 1988. Rethinking crop genetic resource conservation. *Conserv. Biol.* 3(1):19-29.

Brush, S. 1991. A farmer-based approach to conserving crop germplasm. *Econ. Botany* 45(2):153-165.

Cheney, J. 1989. Postmodern environmental ethics: ethics as bioregional narrative. *Environ. Ethics* 11:117-134.

Dallmeier, F., M. Kabel, and R. Rice. 1992. Methods of long-term biodiversity inventory plots in protected tropical forest. In *Long-Term Monitoring of Biological Diversity in Tropical Forest Areas: Methods for establishment and inventory of permanent plots*, ed. F. Dallmeier, pp. 11-46. UNESCO, Paris.

Dasmann, R. 1975. Difficult marginal environments and the traditional societies which exploit them. *Survival Intern. Rev.* II(1):11-15.

Dasmann, R. F. 1991. The importance of cultural and biological diversity. In *Biodiversity: Culture, Conservation and Ecodevelopment, eds.* M. L. Oldfield and J. B. Alcorn, pp. 7-15. Westview, Boulder, Colo.

di Castri, F., and A. J. Hansen. 1992. The environment and development crises as determinants of landscape dynamics. In *Landscape Boundaries: Consequences for Biotic Diversity and Ecological Flows*. eds. A. J. Hansen and F. di Castri, pp. 3-18. Springer-Verlag, New York.

Delcourt, P. A., and H. R. Delcourt. 1992. Ecotone dynamics in space and time. In *Landscape Boundaries: Consequences for Biotic Diversity and Ecological Flows*, eds. A. J. Hansen and F. di Castri, pp. 19-54. Springer-Verlag, New York.

Dickert, T. G., and A. E. Tuttle. 1985. Cumulative impact assessment in environmental planning: a coastal wetland watershed example. *Environ. Impact Assess. Rev.* 5:37-64.

Flavelle, A. 1991. A traditional agro-forestry landscape on Fergusson Island, Papua New Guinea. Thesis submitted for an M.Sc. in Forestry. On file University of British Columbia, Vancouver, Canada.

Food and Agriculture Organization of the United Nations (FAO). 1991. *Strategies for the establishment of a network of in situ conservation areas*. Commission on Plant Genetic Resources. Fourth Session, April 1991.

Forman, R. T. T., and M. Godron. 1986. *Landscape Ecology*. John Wiley and Sons, New York.

Forman, R. T. T., and P. N. Moore. 1992. Theoretical foundations for understanding boundaries in landscape mosaics. In *Landscape Boundaries: Consequences for Biotic Diversity and Ecological Flows*, eds. A. J. Hansen and F. di Castri, pp. 236-258. Springer-Verlag, New York.

Fowler, G., and P. Mooney. 1990. *Shattering: Food, Politics and the Loss of Genetic Diversity*. University of Arizona Press, Tucson, Ariz.

Frankel, O. H., and M. E. Soulé. 1981. *Conservation and Evolution*. Cambridge University Press, New York.

Gregg, W. P. 1991. MAB Biosphere Reserves and conservation of traditional land use systems. In *Biodiversity: Culture, Conservation and Ecodevelopment*, eds. M. L. Oldfield and J. B. Alcorn, pp. 274-294. Westview, Boulder, Colo.

Hansen, A., D. L. Urban, and B. Marks. 1992. Avian community dynamics: The interplay of landscape trajectories and species life histories. In *Landscape Boundaries: Consequences for Biotic Diversity and Ecological Flows*, eds. A. J. Hansen and F. di Castri, pp. 170-195. Springer-Verlag, New York.

Harlan, J. R. 1976. Genetic resources in wild relatives of crops. *Crop Sci.* 16: 329-333.

Hoose, P. M. 1981. *Building an Ark: Tools for the Preservation of Natural Diversity Through Land Protection*. Island Press, Covelo, Calif.

Ingram, G. B. 1984. *In situ Conservation of Plant Genetic Resources: The Scientific and Technical Basis*. FAO, Rome.

Ingram, G. B. 1985. Report on a survey of some wild relatives of crops and forage species in Niger, 1984, AGPG: IBPGR 85/147. Report to the Republic of Niger, the Food and Agricultural Organization of the United Nations, and the International Board for Plant Genetic Resources. On file, FAO, Rome.

Ingram, G. B. 1986a. Sahel survey documents plant genetic resources. *Nature and Resources* 22(1&2): 45-49.

Ingram, G. B. 1986b. Report on a survey and sampling mission to the southern Air Mountains, in central Niger, November and December 1986. AGPG: IBPGR 86/183. Report to the Republic of Niger, the Food and Agricultural Organization of the United Nations, and the International Board for Plant Genetic Resources. On file, FAO, Rome.

Ingram, G. B. 1987. Conservation of wild plants in crop gene pools and their intraspecific variation: Current needs and opportunities in the moist forest of SE Asia. In *The Conservation and Management of Endangered Plants and Animals*. Proceedings of the Symposium on the Conservation and Management of Endangered Plants and Animals, Bogor, Indonesia, June 1986, eds. C. Santiapillai and K.R. Ashby, pp. 63-85. South-East Asia Center for Tropical Biology, Bogor, Indonesia.

Ingram, G. B. 1989. Planning district networks of protected habitat for conservation of biological diversity: A manual with applications for marine islands with primary rainforest. Ph.D. dissertation, University of California, Berkeley. University Microfilms International, Ann Arbor, Mich.

Ingram, G. B. 1990a. Multi-gene pool surveys in areas with rapid genetic erosion: an example from the Air Mountains, northern Niger. *Conserv. Biol.* 4(1):78-90.

Ingram, G. B. 1990b. The management of biosphere reserves for the conservation and utilization of genetic resources: the social choices. *Impact Sci. Society* 158:133-141.

Ingram, G. B. 1990c. The need for knowledge from indigenous communities in planning networks of protected habitat for the conservation of biological diversity: Three island settings. In *Ethnobiology: Implications and applications*. Proceedings of the First International Congress on Ethnobiology (Belem, Brazil 1988). Part 2, ed. M.J. Plotkin, pp. 87-105. Goeldi Museum, Belem, Para, Brazil.

Ingram, G. B. 1991. Habitat, visual and recreation values and the planning of extractive development and protected areas: a tale of three islands. *Landscape and Urban Planning* 21:109-129.

Ingram, G. B., A. Jackson, S. Tostain, and H. Moss. 1987. Mission to southern Air Mountains, north central Niger-site and sample date. AGPG: IBPGR 86/183. Report to International Board for Plant Genetic Resources, the Food and Agriculture Organization of the United Nations and the Republic of Niger.

Jain, S. K. 1975. Genetic reserves. In *Crop Genetic Resources for Today and Tomorrow*, eds. O. H. Frankel and J. G. Hawkes, pp. 379-396. Cambridge University Press, London.

Keystone International Dialogue Series on Plant Genetic Resources. 1991. *Oslo Plenary Session: Final Consensus Report: Global Initiative for the Security and Sustainable Use of Plant Genetic Resources. Third Plenary Session*. Keystone Center, Keystone, Colo.

Kloppenburg, J. R., and D. L. Kleinman. 1988. Seeds of controversy: National property versus common heritage. In *Seeds and Sovereignty: The Use and Control of Plant Genetic Resources*, ed. J. R. Kloppenburg, pp. 173-203. Duke University Press, Durham, N.C.

Lewis, T. 1982. The ecosystems of Tree-Farm License 24-Queen Charlotte, Islands, B. C. with maps 103B/6—6, 103B/6-7, 103B/6-11, 103B/6-14. On file, Western Forest Products, Vancouver, B. C.

Malo, A. R., and S. E. Nicholson. 1990. A study of rainfall and vegetation dynamics in the African Sahel using normalized difference vegetation index. *J. Arid Environ.* 19(1):1-24.

Merriam, G., and J. Wegner. 1992. Local extinctions, habitat fragmentation, and ecotones. In *Landscape Boundaries: Consequences for Biotic Diversity and Ecological Flows*, eds. A. J. Hansen and F. di Castri, pp. 150-169. Springer-Verlag, New York.

Munson, P. J. 1976. Archaeological data on the origins of cultivation in southwestern Sahara and their implications for west Africa. In *Origins of African Plant Domestication*, eds. J. R. Harlan, J. M. J. de Wet, and A. B. L. Stemler. Mouton, The Hague.

Myers, N. 1988. Draining the gene pool: The causes, course, and consequences of genetic erosion. In *Seeds and Sovereignty: The Use and Control of Plant Genetic Resources*, ed. J. R. Kloppenburg, pp. 90-113. Duke University Press, Durham, N.C.

Nabhan, G. P., D. House, H. Suzan, W. Hodgson, L. Hernández, and G. Malda. 1991. Conservation and use of rare plants by traditional cultures of the U. S. / México borderlands. In *Biodiversity: Culture, Conservation and Ecodevelopment*, eds. M. L. Oldfield and J. B. Alcorn, pp. 127-146. Westview, Boulder, Colo.

Nabhan, G. P., and D. Norman. 1992. A proposal for natural area designation in the Tumacacori Mountains of the Coronado National Forest as a genetic reserve to protect plants of economic importance. On file, Native Seeds / Search, Tucson, Ariz.

Namkoong, G. 1983. Preserving natural diversity. In *Genetics and Conservation*, eds. C. M. Schonewald-Cox, S. M. Chambers, B. MacBryde, and W. L. Thomas, pp. 317-334. Benjamin / Cummings, Menlo Park, Calif.

Newby, J., and A. Wilson. 1993. People in blue: The Tuareg of Niger. In *The Law of the Mother: Protecting Indigenous Peoples in Protected Areas*, ed. E. Kempf. Sierra Club Books, San Francisco, Calif.

Pojar, J., and R. M. Annas. 1980. Coastal cedars-pine-hemlock (CCPH) -a new climatic zone for British Columbia? *Bot. Soc. Am. Misc. Ser. Pub.* 158:90.

Pojar, J., and J. Broadhead. 1984. The green mantle. In *Islands At The Edge: Preserving the Queen Charlotte Islands Wilderness*, eds. Islands Protection Society, pp. 49-72. Douglas & McIntyre, Vancouver, B. C.

Prescott-Allen, R., and C. Prescott-Allen. 1990. How many plants feed the world? *Conserv. Biol.* 4(4):365-374.

Rapoport, A. 1992. On cultural landscapes. *Traditional Dwellings and Settlements Rev.* III(11):33-47.

Redclift, R. 1987. *Sustainable Development: Exploring the Contradictions.* Routledge, London.

Roemer, H. L., and R. T. Ogilvie. 1983. Additions to the flora of the Queen Charlotte Islands on limestone. *Can. J. Bot.* 61:2577-2580.

Schonewald-Cox, C., and T. J. Stohlgren. 1988. Wilderness and the protection of genetic diversity. In *Wilderness Benchmark: Proceedings of the National Wilderness Colloquim*, Tampa, Florida. USDA Forest Service, Southeastern Forest Experiment Station General Technical Report SE-51. Asheville, N.C.

Smith, N., and R. E. Schultes. 1990. Deforestation and shrinking crop gene-pools in Amazonia. *Environ. Conserv.* 17(3):227-234.

Solbrig, O. T. 1991. *From Genes to Ecosystems: A Research Agenda for Biodiversity.* Report of a IUBS-SCOPE-UNESCO workshop, Harvard Forest, Petersham, Mass. International Union of Biological Scientists, Cambridge, Mass.

Swanson, F. J., T. K. Kratz, N. Caine, and R. G. Woodmansee. 1988. Landform effects on ecosystem patterns and processes. *BioScience* 38(2):92-98.

Swanton, J. R. 1905a. Contributions to the Ethnology of the Haida. in *American Museum of Natural History Memoirs* (1893-1930)5 (1):1-300. G. E. Stechert, New York.

Swanton, J. R. 1905b. *Haida Myths and Texts.* Skidegate Dialect. Bureau of American Ethnology Bulletin 29, Smithsonian Institution, Washington, D.C.

Turner, N. J., and R. D. Levine. 1971. The ethnobotany of the Haida Indians of the Queen Charlotte Islands, British Columbia. Report on file University of British Columbia Botanical Garden, Vancouver, B. C.

UNESCO. 1984. The Action Plan for Biosphere reserves. *Nature and Resources* XX(4):1-12.

United Nation Environment Programme (UNEP). 1992. Convention on Biological Diversity 5 June, 1992. version: Na. 92-7807. On file, UNEP, Nairobi.

Vavilov, N. I. 1951. *The Origin, Variation, Immunity and Breeding of Cultivated Plants: Selected writings.* Translated from Russian by K. S. Chester. Ronald Press, New York.

Vernhes, J. R. 1989. Biosphere reserves: The beginnings, the present, and the future challenges. In *Proceedings of the Symposium on Biosphere Reserves, Fourth World Wilderness Congress*, September 14—17, 1987, Estes Park, Colo. U.S. Department of the Interior, National Park Service, Atlanta, Ga.

Wilkes, G. 1991. *In situ* conservation of agricultural systems. In *Biodiversity: Culture, Conservation and Ecodevelopment*, eds. M. L. Oldfield and J. B. Alcorn, pp. 86-101. Westview, Boulder, Colo.

Williams, J. T. 1988. Identifying and protecting the origins of our food plants. In *Biodiversity*, ed. E. O. Wilson, pp. 240-247. National Academy Press, Washington, D.C.

Wilson, E. O. 1988. The current state of biological diversity. In *Biodiversity*, ed. E. O. Wilson, pp. 3-18. National Academy Press, Washington, D.C.

World Resources Institute (WRI), The World Conservation Union (IUCN), and United Nations Environment Programme (UNEP). 1992. *Global Biodiversity Strategy: Guidelines for Action to Save, Study, and Use of Earth's Biotic Wealth Sustainably and Equitably*. World Resources Institute, Washington, D.C.

Zimmerer, K. S., and D. S. Douches. 1991. Geographical approaches to crop conservation: the partitioning of genetic diversity in Andean potatoes. *Econ. Bot.* 45(2):176-189.

30

A TRIAD APPROACH TO LAND-USE ALLOCATION

Malcolm L. Hunter, Jr.

Department of Wildlife
College of Forest Resources
University of Maine
Orono, Maine
U.S.A.

Aram Calhoun

Department of Plant Biology and Pathology
University of Maine
Orono, Maine
U.S.A.

Decisions about how to use land are a fundamental component of natural resource management. Should this valley be dammed to create a reservoir? Should this forest be set aside as a park? Often these decisions are made one at time—should *this* valley be dammed?—, but sometimes a broader temporal and spatial perspective will lead to the evaluation of issues, such as cumulative impact and regional context. One such broad-scale issue is the overall allocation of land to different uses. How much land do we need for agriculture, lumber, recreation, or wildlife?

As with many decisions made in a political arena, the answers often become extremely polarized. Environmentalists advocate setting aside as much land as possible by arguing that too much land has already been converted to human use, and that we should curb our resource consumption and, thus, the demand for more land. Resource developers advocate minimal restrictions on land use by arguing that a growing human population and demands for a higher standard of living, especially among the impoverished people of the developing world, necessitate that more land be allocated to meet human needs. Government employees and policy makers often end up refereeing these arguments and negotiating compromises. The compromises usually involve dividing the "pie" among competing demands; in some cases, they involve promoting intermediate forms of land use in which several compatible uses are accommodated on the same tract.

It is our contention that all forms of land use have validity: (1) intensive commodity production areas, (2) areas with little or no resource use by people except low-intensity recreation, and (3) areas in which modest resource use is allowed while ecological values are carefully protected. We call this three-part system a *triad*, a term first proposed by Seymour and Hunter (1992) (Figure 30.1). The term should not be confused with a triage in which casualties (e.g., war victims or endangered species) are classified as beyond help, requiring help, or not requiring help. We recognize that in reality the intensity of land use is a continuum from no use at all to very intensive human use. We prefer the model of a triad because the concept of a continuum is too easily reduced to its two poles, and we think that polarized constructs lead to divisive discussions.

It is noteworthy that the three different types of land use correspond well with the three major environmental ethics described by Callicott (1990). Romantic-Transcendental Preservationists like John Muir would see the reserves as nature's untrammeled temples. Gifford Pinchot and other Resource Conservationists would be content with plantations as long as they are efficient and sustainable. People who share Aldo Leopold's Evolutionary-Ecological Land Ethic would find scope for being "member-citizens of the land community" in the multiple-use areas.

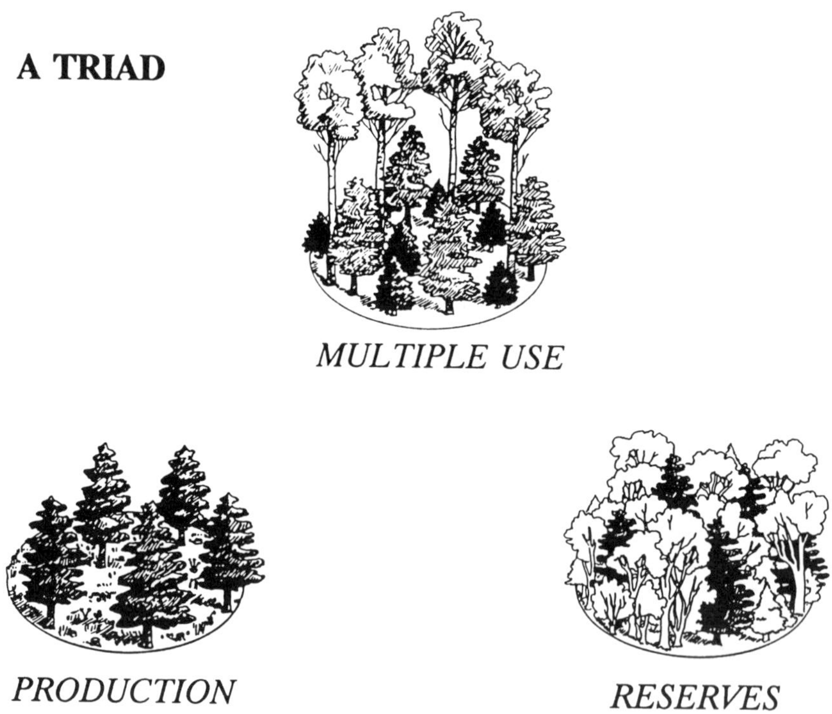

Figure 30.1. A Triad Approach to Land-use Allocation.

Examining current land-use allocation patterns from the triad perspective can reveal imbalances that, if rectified, could improve land use from both an ecological and economic perspective. In this paper, we will first illustrate the triad model by examining some land-use allocation patterns in forest, grassland, aquatic, and wetland ecosystems. Secondly, we will discuss some economic, environmental, and scale issues that arise when land-use allocation patterns are changed.

PART I
FORESTS

The starkly visible contrasts between forests in a pristine state and those that are being intensively manipulated for wood production make forests an excellent example of the polarization that often marks arguments about land use. For many industrialists, an ideal forest is a plantation in which production is maximized by making silviculture a woody analogue of agriculture. To take an extreme example, physicists from Princeton University have calculated that the entire demand for energy in the United States could be supplied from woody biomass plantations and, better yet, these plantations would sequester atmospheric carbon as rapidly as it is generated (R. Williams pers. comm.; also see Cook et al. 1991). The drawback is that it would require converting virtually all of our forests to willow (*Salix*) and aspen (*Populus*) crops grown on 5 to 7-year rotations. At the other extreme, a few environmentalists believe that virtually all forests should be wilderness parks. One member of Preserve Appalachian Wilderness has suggested that 95% of the forests in the eastern United States should be protected from timber cutting (J. Sayen pers. comm.).

These two positions can be viewed as the "intensive use" and "no use" components of the triad model. The "extensive" or "intermediate use" part of the triad is also well represented in forestry; it is the multiple-use doctrine that was the mainstay of foresters for many years and which, with modern refinements, has been resurrected as New Forestry (Franklin 1989). By multiple-use forestry, we mean forestry in which maintenance of biodiversity, protection of water resources, and similar goals stand coequal with timber production. We do not include forestry in which timber production is dominant and these other values are incidental by-products. In the language of New Forestry, it is forestry that views forests as ecosystems which can be carefully managed for limited commercial production, rather than as commodity farms.

To restate our original proposition, we believe that all three types of forest management have validity. Intensively managed forests make sense particularly if they are on productive sites and near mills and roads. If additional intensively managed forests are needed, then obvious candidate sites would be forests that are now managed at a low level of intensity and abandoned agricultural land. Many people view intensive forestry, with its monocultures grown on short rotations and clearcuts, as an environmental calamity, but there is a positive side: by obtaining maximum production on minimal areas, more land can be available for other purposes that are incompatible with timber production (Seymour and Hunter 1992).

Large tracts of unmanipulated forests are needed to protect fully all the ecological values associated with forests. Forests that harbor rare features, such as habitat for endangered species or old growth, are obvious candidates. Because we know so little about the distribution of species, notably insects, fungi and other poorly known taxa, concerns for biodiversity dictate that forest tracts be set aside simply because they are good examples of a particular type of ecosystem and are likely to encompass inconspicuous life forms (Hunter 1990). Some would argue that all of the values maintained in unmanipulated forests can also be obtained in forests that are carefully harvested. We believe that this viewpoint is excessively optimistic, some would say arrogant, about human understanding of ecosystems.

Finally, there are many forests where local needs make multiple-use forestry most appropriate. In some cases, two or more values will be considered of equal importance; in some cases management for a specific goal (e.g., creating optimal habitat for a game species) can be accomplished without compromising some other values. For example, ruffed grouse (*Bonasa umbellus*) and aspen management can be easily integrated on the same tract (Gullion n.d.). Multiple-use forestry will be particularly appropriate for many government agencies that must meet a multi-faceted mandate and for many owners of small tracts who want multiple values from a limited area.

It is easy to see that forestry fits into a triad classification quite well, but to see if a triad perspective offers any useful insights for improved natural resource management we will consider two contrasting case histories from New Zealand and Maine.

New Zealand

In New Zealand, the allocation of government forests (79% of all forests) has become extremely polarized in recent years (Ministry of Forestry 1988). In 1988, approximately 90% of the government forests were composed of native species (primarily *Nothofagus* beech and podocarps); by 1990, virtually all of these had been set aside in parks and other reserves where there will be no timber harvesting. The exceptions are some small state-owned native forests that are still being managed for podocarp (45,000 ha) and beech (80,000 ha) production. In 1988, just under 10% of the government forests were plantations of exotic species (largely *radiata* pine and some eucalypts), and this figure will decrease because most government plantations are being sold to the private sector (New Zealand Forestry Corporation 1989). Privately owned forests (21% of the total) were 35% plantations and 65% indigenous forests; much of the privately owned indigenous forest is not managed for timber production.

This extreme polarization might be understandable, given all the ecological values associated with native forests and the fact that harvest rotations for native trees are about 250 years versus a few decades for the exotic species. On the other hand, the native podocarps produce valuable timber and could be managed on a sustainable basis. Thus, it seems probable that many people would favor a greater allocation for managing native trees on long rotations if most of the productive lowland podocarp forests had not been previously converted to agriculture and plantations. Given the rarity of these forests, however, New Zealanders have decided that they want the large majority of them protected. If they had been thinking in terms of a triad model before most of the lowland podocarp forests had been converted, perhaps the allocation pattern would have been different.

Maine

In Maine the picture is entirely different. Early in the history of settlement, nearly all of the land was sold to private interests and now only 4% (280,000 ha) remains in state and federal ownership (Powell and Dickson 1984). Consequently, less than 2% of Maine's forest (<130,000 ha) have been placed in reserves that are protected from timber harvesting. Only one of these reserves is of substantial size: Baxter State Park (<75,000 ha) was repurchased from timber companies by a former wealthy governor as his gift to the people of Maine. On the other hand, only a small portion of the state's forests (probably 2-10% depending on the exact definition) are intensively managed because abundant natural regeneration usually obviates the need for planting and concomitant forms of intensive silviculture (Seymour and Hunter 1992). Roughly 90% of the state's forests are subject to extensive management, some of which could be characterized as "take the best and leave the rest." In terms of a descriptive interpretation of the triad model, these forests would fall in the intermediate category, but for many of them, it would be a euphemism to call their management multiple-use and they would not fit a prescriptive application of the triad model.

This dominance of low-intensity forestry has led some silviculturists to propose that we meet current demand for timber on a smaller land base by intensifying production. Specifically, Seymour and McCormack (1989) estimated that by intensifying management, current levels of production could be maintained on 20-30% of the forest land base. In a completely laissez-faire economy, such an increase in management could increase overall productivity, but there is an alternative that many foresters favor. For every hectare moved from low-intensity management into high-intensity production, we could move three hectares out of production and into reserve status. It would be neither feasible nor desirable to subject all of Maine's low-intensity forests to such trading, but if done wisely and in moderation, it could ameliorate many conflicts between environmentalists and industrialists. Thus for Maine, a triad perspective highlights some possible changes in allocation that most parties would find desirable.

GRASSLANDS

Grasslands are subtle ecosystems in which the patterns that change through time and space are not obvious to inexperienced observers. Perhaps this is the reason why the advent of cattle and plows has been far less controversial than the influx of axes to forests. Perhaps this is why there are so few grassland reserves in the world. This oversight has been recognized in some situations. In the United States, the need for some large prairie reserves and the problems of overgrazing and subsidized grazing fees on federal land have attracted some attention. In several areas, notably the African Sahel, desertification, the degradation of grasslands into deserts, is a widely recognized problem. Clearly, there are important issues concerning our use of grasslands; does the triad model offer any useful insights?

The use patterns for grasslands readily fit a triad classification. First, at the "no human use" end of the continuum there are grassland reserves, albeit few and small, where biodiversity maintenance and similar goals are dominant. The scarcity of such reserves can probably be attributed to their lack of dramatic scenery and other amenities that make most parks attractive recreational areas. (The savanna reserves of Africa with their spectacular complement of large mammals are a notable exception to this generalization.)

Second, intermediate levels of use characterize large areas of rangeland that are used for livestock grazing but are not intensively managed. Provision of water, control of unpalatable plants such as sagebrush (*Artemisia tridentata*), fencing and rotational grazing schemes, and other management activities can have profound effects on grassland ecosystems, but they still remain grassland ecosystems.

Third, intensive management of grassland ecosystems comes with cultivation, particularly growing domestic grasses like corn (*Zea mays*) and sorghum (*Sorghum bicolor*) for human or livestock consumption. Intensive pasture management (e.g., seeding and fertilizing forage crops) can also effectively convert a grassland ecosystem into a cultivated ecosystem and, thus, be considered intensive management.

The tradeoffs that might be made among different parts of a grassland triad are apparent when one compares the area required to support livestock under extensive versus intensive management. (These are expressed as the hectares required to support an animal unit, one 450-kg cow, for a year, AUY, or a month, AUM). A survey of recommended stocking rates for rangeland in the United States suggested 2 to 45 ha/AUY, depending primarily on rainfall and fertility (Holechek et al. 1989). Actual stocking rates on most rangeland probably fall in the upper half of this range, or beyond. For example, grazing permits and leases on all U.S. Bureau of Land Management holdings in 1990 allowed 13,553,835 AUMs on 64,327,350 ha or 57 ha/AUY (USDI 1990). It is difficult to estimate equivalent figures for maintaining livestock on intensively managed pastures because these operations are often more seasonal and more integrated with feeding of harvested grains than rangeland management. A survey of 123 Iowa beef farmers found that they provided an average of 1 ha of pasture per cow for an average grazing season of 188 days per year (Iowa State University Extension 1991). A Connecticut study found that 1 ha of pasture managed for continuous grazing could provide 470-641 days of grazing per Animal Unit; 1292-1302 days could be provided under short-duration, rotational grazing (R. Hough pers. comm.). If these figures are reasonably representative then intensive pasture management can lower area needs by at least an order of magnitude. Finally, the feed required to maintain a 450 kg cow for a year could be grown on 0.7 ha hay, 0.3 ha corn, 1.0 ha barley, or 1.4 ha oats. (See the appendix 30.1 for methods used to make these estimates.)

Thus, it is apparent that theoretically one could shift livestock production away from rangelands and towards pastures and feedlots and trade ten or more ha of rangeland set aside as reserves for every hectare used for grain production or pastures. There are some obvious problems with this approach, notably economic and ecological feasibility issues, such as how to obtain sufficient water. We will return to these in the second part of the paper.

The desirability of making some shifts is apparent when one compares: (1) the 66 million ha allocated to growing feed grains and hay in the United States in 1991 (corn 30 million ha; oats 4 million ha; barley 3 million ha; sorghum 4 million ha; and hay 25 million ha; USDA 1991); (2) the estimated 92 million ha of grassland used for rangeland in the conterminous United States; and (3) the estimated 1 million ha of grasslands occurring in National Park Service and U.S. Fish and Wildlife Service lands in the conterminous United States. (See the appendix 30.1 for methods used to extract estimates 2 and 3 from data of Crumpacker et al. 1988.) In terms of the triad model, these figures translate into 41.5% intensive, 57.9% intermediate, and 0.6% reserves. If one uses a broader definition that includes shrublands and woodland-grassland mosaics (see appendix 30.1), the distribution becomes: 66 million ha (22.7%) intensive, 220 million ha (75.7%) intermediate and 4.6 million ha (1.6%) reserves. To take another example of the dearth of grassland reserves, Crumpacker et al. (1988) estimated that five different types of mesquite savanna have a total area of 17,636,834 ha in the United States, none of which is found on any type of federal agency or Indian land.

AQUATIC ECOSYSTEMS

Aquatic ecosystems provide many resources for society—water for irrigation, human consumption, and power generation, transportation corridors, waste disposal systems, and others—, but in this paper we will focus on just one issue, fisheries, that is readily analyzed from the triad perspective. First, aquaculture clearly represents the intensive component of the triad model by creating cultivated aquatic ecosystems for producing fisheries products. Second, most aquatic ecosystems are managed for their fisheries resources to some degree, thus constituting extensive ecosystem management. These activities can involve direct manipulation of the fisheries (e.g., regulation of harvest and stocking) or management of the ecosystem in general (e.g., water pollution control and regulating water levels). Finally, there are aquatic ecosystem reserves in which fisheries harvesting is strictly prohibited, but they are rare even among aquatic ecosystems that lie within the boundaries of terrestrial reserves. People expect to be able to fish within a park, probably because they perceive catching fish as less intrusive than hunting birds and mammals. This distinction may exist because people seldom see fish in their natural environment and do not value them except for sport and food. Coral reefs, where snorklers can see spectacular fish and which are occasionally off-limits to fishing, are "the exception that proves the rule."

The choice of generating fisheries products by managing aquatic ecosystems intensively versus extensively is similar to the livestock-grassland issue described above, but is more extreme for two reasons. First, aquatic ecosystem reserves are even rarer than grassland ecosystem reserves. (Unfortunately, we do not have data on the extent of aquatic ecosystems reserved from fishing and can only offer this general statement.) Second, the land-use efficiencies of aquaculture production are even more dramatic than the efficiencies of intensive livestock production. Annual fisheries production in the

United States (primarily fish, molluscs, and crustaceans) averages roughly 27 kg/ha in coastal marine ecosystems and 1.1 kg/ha in freshwater ecosystems (see appendix 30.1 for calculations). In contrast, aquaculture production figures are usually in the hundreds of kilograms per hectare for systems in which the only major addition is fertilizer, and when fish are fed food produced externally (notably fish meal, soybean meal, and grain), production figures typically reach thousands of kg/ha (Lovell 1989). Indeed, given that channel catfish (*Ictalurus punctatus*) gain 0.84 g of body weight for every g of food consumed, compared to 0.48 for chickens and 0.13 for cattle (Lovell 1989), the idea of using aquaculture to relieve pressure on rangeland merits discussion.

With about 92% of fisheries production in the United States derived from species harvested in the wild and 8% from aquaculture (World Resources Institute 1990), there is scope for expanding aquaculture to allow establishment of some aquatic reserves with no net loss of fisheries production. Obviously, such tradeoffs are complex because metric tons of food do not tell the whole story: a catfish reared by a fish farmer is not easily compared to a wild trout caught by an angler for a whole suite of reasons. Nevertheless, such complexities do not belie the desirability of establishing more aquatic reserves. There are at least three reasons why more aquatic reserves are needed. First, as already stated, aquatic reserves are very rare. Second, fisheries exploitation can profoundly influence the entire structure of aquatic ecosystems because many utilized species have keystone ecological roles (Liss and Warren 1980, Northcote 1987). Third, many freshwater species are threatened with extinction (e.g., Moyle and Williams 1990; Minckley and Deacon 1992). Direct overexploitation of endangered aquatic species might not be a common problem, but overexploitation of aquatic organisms in general contributes to a whole syndrome of problems that degrade aquatic ecosystems.

WETLANDS

Wetlands include a diversity of ecosystems ranging from permanently flooded marshes to temporarily flooded forests, and they represent an even wider spectrum of human values. For some people, they are quintessential wastelands; for many conservationists, they have become a rallying point, the temperate analogue of tropical rain forests. In the United States, government programs have reflected these changing values. Through the early 1980s, the emphasis was "reclaiming" wetlands (i.e., making them non-wetlands), largely by subsidizing farming activities. The U.S. Fish and Wildlife Service publication in 1979 of "Classification of wetlands and deep-water habitats of the United States" (Cowardin et al. 1979) signalled the beginning of a change as national attention turned to classifying, inventorying (U.S. Fish and Wildlife National Wetlands Inventory), and evaluating wetlands (Wetland Evaluation Technique) as natural resources rather than as liabilities. However, concomitant with the rising interest in maintaining wetlands for their natural functions and values (sediment and erosion control, flood desynchronization, pollution abatement, and wildlife habitat) has been pressure from development interests to deregulate wetlands. Can the triad perspective lend insight into wetland management strategies, thus striking a balance between these two interests?

Like other ecosystems, wetlands can be classified in terms of the triad system. First, areas managed for rice (*Oryza* spp.) and cranberry (*Vaccinium macrocarpon*) production constitute cultivated wetlands intensively managed for commodity production. Biomass production of cattails (*Typha* spp.), peat mining, and management of wetlands for sludge spreading and tertiary wastewater treatment can also constitute intensive management of wetlands. Second, managing swamps for timber production and marshes for waterfowl and shellfish production usually involve less dramatic changes to the ecosystem, notably water level regulation, and can be considered extensive management. Finally, many wetlands are relatively free of human manipulation because regulations inhibit their use, because they are in reserves, or because no one has decided to use them yet.

Although wetlands fit a triad classification, the kind of land-use tradeoffs suggested by a triad-based analysis of forest, grassland, and aquatic ecosystems might not make sense for wetlands for two reasons. First, the major intensive and extensive uses of wetlands are not readily shifted to a different level of intensity. For example, rice and cranberries cannot be commercially produced extensively (with the minor exception of harvesting wild rice, *Zizania aquatica*); maximizing timber production in a wetland would involve draining it; and growing mallards (*Anas platyrhynchos*) in pens is a poor substitute for wild duck production. Second, the relative scarcity of wetlands in most regions and the history of wetland loss argues against allowing any additional wetlands, whether they are currently unmanipulated conterminous or extensively managed, to be converted to intensive uses, especially because most of the values associated with wetlands are not provided by intensively used wetlands. It is estimated that 87 million ha of wetlands in the pre-European settlement conterminous United States dropped to 41 million ha in the mid-1980s, with over 22 states experiencing losses greater than 50% (Dahl and Johnson 1991). Given that less than 5% of the land area of the conterminous United States is wetland and over half of threatened and endangered species are found in wetlands, further degradation of wetlands would be a significant loss of biodiversity.

A no-net-loss wetland policy does not mean that there can be no additional intensively used wetlands; they could still be artificially created on upland sites. For the remaining wetlands, the key policy decision may be deciding which wetlands can be used by landowners for some economic return, such as timber production, without significantly compromising their ecological values, and which ones cannot. For those that cannot be used by their owners, we need to explore mechanisms, such as property tax abatement, that will offset some of the economic disadvantages accruing to landowners maintaining wetlands for the public good.

PART II
SOME ECONOMIC ISSUES

The tradeoffs suggested by a triad approach to land allocation would not work very well in a laissez-faire economy. Without government intervention, each tract of land would be exploited as intensively as possible, given economic constraints like the cost of doing business and the current market for the products being generated. Of course, governments always regulate the economy in general and the cost of doing business in particular. If society decided to shift the balance of land-use allocations, it could do so through zoning and other forms of regulations, tax incentives and disincentives, and other forms of direct and indirect subsidies. For example, if a government determined that lack of investment capital was the major hurdle to intensifying management on a smaller land base (this will often be the case), then it could offer low-interest rate loans. On the other side of the coin, areas under low-intensity use could be purchased by the government from willing sellers and made into reserves. Funds to support these expenditures could be generated by higher stumpage and grazing fees on public land and higher property taxes on corporate land. Simply tightening environmental regulations on multiple-use lands while relaxing regulations on intensively managed lands would shift the costs of doing business to encourage more production on a smaller area.

As major landowners (the United States government owns 32.1% of the land), governments can also profoundly influence the balance of land-use allocations by management decisions on their own land. For example, if a government were to allocate a large portion of its land to reserves, this could increase the price for some commodities and encourage more intensive use of private land.

An implicit assumption of these potential tradeoffs between extensive management and intensive management is that the level of commodity production would remain about the same; it would just be obtained from a different land area. Economists might argue that production should be increased as much as feasible; for example, we could capture a larger share of the world's lumber market with increased production. Environmentalists might argue that we could reduce production levels by recycling and other efficiencies; for example, we could reduce meat production by eating less meat and more grains. Again, a balance must be struck. We suggest that keeping production levels stable is a simple and reasonable compromise for most situations. Obviously, it will be a challenge meeting the food needs of the world's increasing human population by efficiencies such as better storage and distribution systems, but there is enormous scope for improvement.

SOME ENVIRONMENTAL ISSUES

The tradeoffs between obtaining resources over a limited area through intensive management versus over a larger area through dispersed management have environmental ramifications that have to be weighed. For example, obtaining resources over a large area might require more transportation and consequently more fossil fuel use and road construction. Conversely, intensive management often uses large quantities of fossil fuels in the process of fertilizing, irrigating, and controlling competitive species. Many management practices involve adding toxins to the environment, notably pesticides. Pesticides are more frequently used in intensive management, but when they are used in extensive management, the consequences might be greater because they are distributed over a wider area and more biota are exposed to them. This might be why foresters are more severely criticized for using herbicides and insecticides than farmers are, even though they use far less.

Environmental issues highlight a limitation intrinsic to the simplicity of the triad model: by lumping a wide range of land- use practices into one category, multiple-use, practices that range from environmentally benign to environmentally destructive fall into one category.

As with economic issues, people hold a plurality of environmental values that need to be considered. Most people might agree that environmental degradation should not severely compromise human health, but on other issues such as aesthetics and the well-being of various non-human species (e.g., vertebrates versus invertebrates versus plants), significant differences of opinion exist.

An Important Caveat

In stating that intensive management has a role in wise land use, we must explicitly exclude management regimes that are so intensive that they are not sustainable. Practices that significantly compromise soil fertility are the best examples; they are well known to agriculturists and are becoming of concern in forestry circles too (Maser 1988). Some environmentalists might argue that intensive management can never be sustainable, but this is a matter of definitions. We would reply that backyard vegetable gardens are intensive management and most are sustainable, while there are many examples of unsustainable multiple-use.

COST-BENEFIT ANALYSES

Although timber, livestock, or fish production could be obtained on fewer hectares by substituting intensive management for extensive management on some portion of the land base, such substitutions are not necessarily desirable. They need to be carefully analyzed from an economic cost-benefit perspective which internalizes environmental and social issues. Such analyses might indicate that the desirable substitution rates are quite different from those outlined here. For example, forest plantations designed to be environmentally benign and sustainable indefinitely might produce wood at a rate only twice as great as extensively managed forests, unlike the 3:1 ratio suggested here for Maine forests. Undertaking such analyses is a significant task, and lies well beyond the scope of this paper.

SOME SCALE CONSIDERATIONS

Given that the triad model suggests that land uses should be balanced, then we must consider the scale at which this balancing act should occur. Conservation biologists usually advocate looking beyond property lines to make management decisions in the largest context possible. On the other hand, to take a specific example, it would not be economically or ecologically sound to have all of our forest reserves in the old-growth stands of the Pacific Northwest, all our plantations in a southeastern wood basket, and all our multiple-use forests in the northeast and midwest where high population densities encourage us to wring every product out of every hectare. (This scenario is exaggerated, but there is some basis for it; many USDA-Forest Service foresters outside of the Pacific Northwest have complained that in effect the spotted owl (*Strix occidentalis*) is raiding the timber on their district.) Issues of economic distribution and balance can usually be evaluated at the scale of an individual state or county. Ideally, ecological distribution and balance would be evaluated using ecological units such as physiographic regions or watersheds; in practice, political boundaries are likely to be used for ecological evaluations too.

THE EXTENT OF CHANGE

Some people might believe that, given socioeconomic and political inertia, it would be difficult to effect any significant land use changes even if most parties agreed that they would be wise. Certainly such changes cannot be accomplished quickly, in months or even a few years, but dramatic changes can occur over reasonably short periods. The change in New England from a predominantly agricultural landscape to one that is mostly forested is a recent, well-known example (Irland 1982). An even more rapid replacement of agricultural ecosystems by natural ones might take place in much of Europe during the next few decades because of the new agricultural policies of the European Community. Beyond the obvious economic issues, another major barrier to changing land-use patterns is psychological: resource developers who consider wilderness and other reserves a waste of land, and environmentalists who consider heavy-handed manipulations of ecosystems a travesty.

CONCLUSIONS

The intensity of land use varies in a continuum from no human manipulations to management that is so intensive that natural ecosystems are replaced by artificial or cultivated ecosystems. This continuum can be conceptualized as a triad model in which land uses are characterized as: (1) intensive commodity production areas, (2) areas with little or no resource use by people, or (3) areas in which modest resource use occurs while ecological values are carefully protected. We argue that all three types of land use have validity and we present examples from forest, grassland, wetland, and aquatic ecosystems that suggest that a different balance among these uses could improve resource conservation. For example, in a region where multiple use forest management is dominant, wood production could be intensified on some areas, allowing other areas to be set aside as reserves with no net effect on production.

ACKNOWLEDGMENTS

We wish to thank Bob Hough and Bob Seymour for conversations that contributed significantly to the preparation of this paper. Tom Barron, Bob Hough, Bill Krohn, Lynn Maguire, David Norton, Bob Seymour, and David Wilcove reviewed the manuscript. This is Maine Agricultural Experiment Station Publication No. 1724.

LITERATURE CITED

Callicott, J. B. 1990. Whither conservation ethics? *Conserv. Biol.* 4:15-20.

Cook, J. H., J. Beyea, and K. H. Keeler. 1991. Potential impacts of biomass production in the United States on biological diversity. *Ann. Rev. Energy Environ.* 16:401-431.

Cowardin, L. M., V. Carter, F. C. Golet, and E. T. LaRoe. 1979. Classification of wetlands and deepwater habitats of the United States. U.S. Fish and Wildlife Service, FWS/OBS-79/31, Washington D.C.

Crumpacker, D. W., S. W. Hodge, D. Friedley, and W. P. Gregg, Jr. 1988. A preliminary assessment of the status of major terrestrial and wetland ecosystems on federal and Indian lands in the United States. *Conserv. Biol.* 2:103-115.

Dahl, T. E., and C. E. Johnson. 1991. Wetlands, status and trends in the conterminous United States, mid-1970's to mid-1980's. U.S. Fish and Wildlife Service, Washington, D.C.

Franklin, J. F. 1989. Toward a New Forestry. *Am. For.* Nov.-Dec.:37-44.

Gullion, G. W. n.d. Managing woodlots for fuel and wildlife. Ruffed Grouse Society, Coraopolis, Penn.

Holechek, J. L., R. D. Pieper, and C. H. Herbel. 1989. *Range Management.* Prentice Hall, Englewood Cliffs, N. J.

Hunter, M. L. 1990. Coping with ignorance: the coarse-filter strategy for maintaining biological diversity. In *Balancing on the Brink of Extinction*, ed. K. Kohm, pp. 266-281. Island Press, Washington, D.C.

Iowa State University Extension. 1991. 1991 State livestock enterprise summaries for Iowa: Beef. Iowa State University Extension, Ames, Ia.

Irland, I. C. 1982. *Wildlands and Woodlots: The Story of New England's Forests.* University Press of New England, Hanover, N. H.

Klopatek, J. M., R. J. Olson, C. J. Emerson, and J. L. Joness. 1979. Land-use conflicts with natural vegetation in the United States. *Environ. Conserv.* 6:191-199.

Kuchler, A. W. 1964. Potential natural vegetation of the conterminous United States. American Geographical Society, New York.

Liss, W. J., and C. E. Warren. 1980. Ecology of aquatic systems. In *Fisheries Management*, eds. L. A. Nielsen and R. T. Lackey, pp.41-89. John Wiley & Sons, New York.

Lovell, T. 1989. *Nutrition and Feeding of Fish.* Van Nostrand Reinhold, New York.

Maser, C. 1988. *The Redesigned Forest.* R. & E. Miles, San Pedro, Calif.

Minckley, W. L., and J. E. Deacon, eds. 1992. *Battle against Extinction: Native Fish Management in the American West.* University of Arizona Press, Tucson, Ariz.

Ministry of Forestry. 1988. The forestry sector in New Zealand. Ministry of Forestry, Wellington, New Zealand.

Moyle, P. B., and J. E. Williams. 1990. Biodiversity loss in the temperate zone: decline of the native fish fauna of California. *Conserv. Biol.* 4:275-284.

National Research Council. 1984. Nutrient requirements of beef cattle. 6th ed. National Academy Press, Washington D.C.

New Zealand Forestry Corporation. 1989. Sale of state owned forests in New Zealand. New Zealand Forestry Corporation, Wellington, New Zealand.

Northcote, T. G. 1987. Fish in the structure and function of freshwater ecosystems: a "top-down" view. *Can. J. Fish. Aquatic Sci.* 45:361-379.

Powell, D. S., and D. R. Dickson. 1984. Forest statistics for Maine, 1971 and 1982. USDA Forest Service Resource Bulletin NE-81. Broomall, Penn.

Seymour, R. S., and M. L. Hunter, Jr. 1992. New Forestry in eastern spruce-fir forests: principles and applications to Maine. *Maine Agric. Exper. Stat. Misc. Publ.* 716. Orono, Me.

Seymour, R. S., and M. L. McCormack, Jr. 1989. Having our forest and harvesting it too: the role

of intensive silviculture in resolving forest land-use conflicts. In *Forest and Wildlife Management in New England*, eds. R. D. Briggs, W. B. Krohn, J. G. Trial, W. D. Ostrofsky, and D. B. Field, pp.207-213. *Maine Agric. Exper. Stat. Misc. Rep.* 336. Orono, Me.
U.S. Department of Agriculture (USDA). 1991. Agricultural statistics, 1991. U.S. Government Printing Office, Washington, D.C.
U.S. Department of the Interior (USDI). 1990. Public land statistics. Bureau of Land Management, U.S. Department of the Interior, Washington D.C.
World Resources Institute. 1990. World Resources, 1990-91. Oxford University Press, New York.

Appendix 30.1

Estimating Areas Needed for Cattle Feed Production

To estimate the area of land required to grow enough feed to support a cow, we used 2% of body weight as an average daily requirement for dry matter intake (National Research Council 1984). (The exact figure depends on whether a cow is lactating or pregnant, its breed, the season, and other factors.) This translates into 9 kg of hay for a 450 kg cow daily or 3,285 kg per animal unit year (AUY). The equivalent figures for grains are lower because of differences in nutrient content: corn meal 1,985 kg/yr, corn silage 2,581 kg/yr, barley 2,150 kg/yr, oats 2,442 kg/yr (National Research Council 1984). Average yield figures for these crops for the United States, 1980-89, were: hay 5,274 kg/ha, corn grain 6,639 kg/ha, corn silage 29,608 kg/ha, oats 2,010 kg/ha, barley 2,607 kg/ha (USDA 1991). We multiplied these figures by their % dry matter (30% for silage; 85% for the others) and divided the results by the feeding requirement figures to calculate how many hectares it takes to support an AUY growing different feeds: 0.73 ha hay, 0.29 ha silage corn, 0.35 ha grain corn, 0.97 ha barley, or 1.43 ha oats. These figures are presented as alternatives, but in practice most cattle are fed on a combination of feeds.

Estimating grassland areas. Grassland area estimates were based on potential natural vegetation types (PNVs) defined as grasslands by Kuchler (1964). Crumpacker et al. (1988) provide data on the total area of each PNV type occurring on different types of federal land, and we assumed that all grasslands managed by the National Park Service and U.S. Fish and Wildlife Service were reserves. Grasslands owned by the Fish and Wildlife Service open to livestock grazing would decrease this estimate; grassland reserves owned by state governments, other federal agencies, and private conservation groups would increase this estimate. We have used data from only the conterminous United States to avoid the issue of whether or not tundra is rangeland. To estimate total area of grassland available for livestock, we multiplied estimates from Crumpacker et al. (1988) of the total area of each PNV by the estimate of Klopatek et al. (1979) of the percentage of each PNV that remained in a natural or semi-natural condition, then subtracted the area estimated to occur in reserves. (The condition estimate was unavailable for one PNV type, and we substituted the average value from the other 25 types.) Because shrublands and certain types of woodlands are also used for livestock grazing, we repeated the analysis using all PNV types defined by Kuchler as grassland, shrubland, or grassland-forest mixtures. This broader definition also included alpine areas, deserts, and some wetlands that occur at a large scale (e.g., Everglades and Thule marshes).

Estimating fisheries catch. Data collated by the World Resources Institute (1990) give the annual catch of marine fisheries products in the United States as 5,075,000 metric tons (1985-87 average) and the area of the continental shelf to a depth of 200 m as 1,870,700 km^2, yielding a production estimate of 2.7128 tons/km^2. The corresponding figures for freshwater systems are 73,300 tons (1984-86 average) and 650,306 km^2, giving an average of 0.11 tons/km^2. These estimates should be considered very approximate, perhaps accurate only to the nearest order of magnitude, because of issues such as recreational landings, offshore fishing, and coarseness of the area estimates.

31

MAINTAINING BIODIVERSITY IN A MANAGED LAND AND SEASCAPE : THE GREAT BARRIER REEF, AUSTRALIA

Graeme Kelleher and Wendy Craik

Great Barrier Reef Marine Park Authority
Townsville, Queensland
Australia

"The environment does not exist as a sphere separate from human actions, ambitions and needs". Gro Harlem Brundtland (1987)

"Biodiversity—is the variety of the world's organisms, their genetic diversity and the collection that they form. It can be considered to be the total of the world's species, genes and ecosystems" (IUCN 1992). Biodiversity conservation not only should be considered as a matter of principle, but because of its utility, its importance to all life systems, its provision of economic benefits, and its contribution to the quality of human life.

The "Global Biodiversity Strategy" (WRI/IUCN/UNEP 1991) built on "Caring for the Earth" (IUCN/UNEP/WWF 1991) links biodiversity conservation with human needs. Some of its recommendations include:

- declaration of national sovereignty over genetic resources to ensure wise management of biological resources.
- promotion of policies that are consistent with, rather than counter to ecologically sustainable use of resources, e.g., agricultural policies that do not promote pesticide use and unsustainable farming practices.
- promotion of incentives for local involvement in biodiversity conservation—ecologically sustainable use at a local, individual level.
- strengthened and expanded protected area systems within bioregional planning frameworks.
- establishment of a global early warning system

The Great Barrier Reef Marine Park, comprising one of the world's most biologically diverse areas, is in essence a strategy for biodiversity conservation, which to date, appears to have been relatively successful. Over-exploitation of the Great Barrier Reef has largely been prevented. Through the application of a broadly based management strategy to a large marine ecosystem which caters for use, appreciation and enjoyment, while protecting a range of specific areas and organisms a balance between human need and environmental protection has been struck.

This paper describes the world's largest Marine Park, how it is used and how it is managed as a large marine ecosystem and how that contributes to the maintenance of biodiversity. The pressures that it must sustain and the methods used so far to ensure that it is not progressively degraded are outlined. These pressures come not only from within the Marine Park, but from governments, from users of land adjacent to the Marine Park, and from private interest groups. These pressures are increasing rapidly but so too is our knowledge of the states and processes that sustain the Great Barrier Reef. The changes that can be expected in the next decade are described, with suggested approaches to ensuring that development of the Marine Park is sustainable.

BIODIVERSITY OF THE GREAT BARRIER REEF

The Great Barrier Reef is unique, and the commitment of the Australian people to its conservation is great. Biologically, the Great Barrier Reef supports one of the most diverse ecosystems known. It has developed over 500,000 years on the northeast Continental Shelf of Australia.

The area extends approximately 2,000 km along the eastern coast of Queensland, Australia, from just north of Fraser Island in the south (24°30'S) to the latitude of Cape York in the north (10°41'S), and covers an area of 348,700 km^2 on the continental shelf of Australia. It is acknowledged as an area of great natural beauty and wonder. The unique environment of the Great Barrier Reef, its size and diversity, have been recognized world-wide and led to its inscription in October 1981 on the UNESCO World Heritage List.

The Great Barrier Reef is not a continuous barrier but a broken maze of coral reefs, some with coral cays. Some 2,900 individual reefs, including 760 fringing reefs, lie within the formally defined area known as the Great Barrier Reef Region. These reefs range in size from less than 1 hectare to more than 100 km^2, and in shape from flat platform reefs to elongate ribbon reefs. In some places the reefs are separated by channels of no more than 200 meters width, while elsewhere they are as much as 20 kilometers apart. Most reefs are submerged, with some being exposed at very low tide.

The individual reefs of the Great Barrier Reef are composed of the accumulated remains of plant and animal calcium carbonate skeletal material, supporting a veneer of living plants and animals. There are some 300 reef islands or cays; 87 of them are permanently vegetated. There are about 600 continental or high islands, often with fringing reefs around their margins.

The Great Barrier Reef is an outstanding example of biotically structured ecosystems of high diversity dependent on the structuring organisms. In addition, there is evidence that many species of structuring organisms (in this case coral) are themselves dependent on one another, with some reefs acting as critical sources of larvae and others as sinks. The Marine Park, and areas immediately adjacent to it, contain a broad range of mangroves and 14 species of seagrasses. Disruption of conditions for structuring organisms would have negative consequences for a vast array of dependent species.

Corals

A total of about 450 species of coral in 60 genera are believed to be on the Great Barrier Reef. The majority belong to the order Scleractinia which have calcareous skeletons. The remainder are members of the sub-class Octocorallia (soft corals, sea whips and sea fans) which do not produce calcareous skeleton, and the members of the Class Hydrozoa, two families of which, while producing a calcareous skeleton, are structurally very different from the Scleractinian corals. Many corals have a variety of growth forms which relate to the hydrological regime and exposure of the location in which they develop. Some examples of growth forms and characteristic genera are:

- Branching corals, typified by the genera *Acropora* and *Pocillopora*
- massive brain corals, which are generally rounded structures, exemplified by *Porites*, *Platygyra* and Leptoria
- plate-like corals, which include a large number of genera and a wide range of shapes. Characterized by *Acropora*, *Montipora*, and *Tubinaria*
- encrusting corals, which grow over other organisms as well as reef detritus, characterized by *Montipora*; and
- mushroom corals, which are unattached single polyps with large central mouths of which *Fungia* is the most common.

Other Invertebrate Fauna

The other important members of the Reef fauna include foraminifera, molluscs (amphineurans, gastropods, lamelibranchs, scaphopods, cephalopods), echinoderms, crustaceans, polychaete worms, and ascidians. Many of these species are essential to the kinetic processes of the Reef because they penetrate and break up coral and algal structures, contributing large quantities of detritus to the Reef mass. The molluscan reef fauna is extremely varied. Over 4,000 species of molluscs have so far been collected from the Great Barrier Reef. Some of the better known are trochus shells which are found in coral rubble and on coral and rocky reefs, helmet shells which are readily visible on the sea floor and coral sands of the Reef, and the variously colored species of cowrie shell.

Perhaps the most conspicuous lamelibranchs on the Reef are the giant clams of the family Tridacnidae.

The Reef supports a variety of echinoderms, for example, starfish, sea urchins, and holothurians (bêche-de-mer).

Fishes

Approximately 1,500 species of fishes are known from the Great Barrier Reef area, these exhibiting a variety of size, shape, color, and behavior. There are pelagic (open-water living) species, such as black marlin *(Makaira indica)* and mackerel *(Scomberomorus* sp.*)* which move through the Reef area. Large and colorful demersal (bottom-living) species, such as coral trout *(Plectropomus* sp.), sweetlip *(Lethrinus* sp.), red emperor *(Lutjanus sebae)*, snapper Lutjanidae, and cod Serranidae, spend much of their lives in and around the reefs. The small bright-colored home-ranging fishes, such as butterfly fish (Chaetodontidae) and damsel fish (Pomacentridae), spend most of their adult lives in a restricted territory of a few square meters of a reef.

Sea Turtles

Six species of sea turtles occur in the Great Barrier Reef area: the green *(Chelonia mydas)*, leatherback *(Dermochelys coriacea)*, flatback *(Chelonia depressa)*, loggerhead *(Caretta caretta)*, hawksbill *(Eretmochelys imbricata)* and Pacific Ridley *(Lepidochelys olivacae)*. The Great Barrier Reef provides nesting sites of world importance for the green and loggerhead turtles. All turtles are protected species under Commonwealth and Queensland legislation, and are listed in Appendix 2 of the Convention on International Trade in Endangered Species 1973 (CITES).

Cetaceans

Twenty-four species of cetaceans (whales, dolphins, and porpoises) are known to occur in the waters of the Great Barrier Reef. Of the five species of great whales present in the Reef Region seasonally, the humpback whale *(Megaptera novaeangliae)* is the most commonly seen. The population of humpback whales migrating up the Australian east coast to calving and breeding grounds in the warm waters of the Great Barrier Reef and Coral Sea is estimated at 1,000—1,200 animals. Other cetacean species present are the rare beaked and straptoothed whales *(Mesoplodon* sp.) seen occasionally as stragglers from the open sea, transient pods of melonheaded *(Peponocephala* sp.), pilot *(Globiocephala* sp.) and false killer *(Pseudorca* sp.) whales and species such as irrawaddy (Orcaella sp.), bottlenose *(Tursiops* sp.), spinner *(Stenella* sp.) and humpback *(Sousa* sp.) dolphins which might be resident in shallow inshore areas and off the outer reefs. Shallow inshore areas are extremely important to another marine mammal, the dugong *(Dugong dugon)*, for which the Great Barrier Reef populations form a significant component of the world population. All these marine mammals are protected species.

Birds

The cays and continental islands of the area support 242 species of birds. These include 40 species of sea birds, of which 21 have breeding colonies within the area. Of the 202 species of land birds, breeding sites for 109 have been recorded (Kikkawa 1976).

Some of the earliest research into the Great Barrier Reef took the form of ornithological expeditions, which recorded, amongst the larger sea birds, the reef heron *(Egretta sacra)*, osprey *(Pandion haliaetus)*, pelican *(Pelecanus conspicillatus)*, frigate birds *(Fregata* sp.) and gulls *(Larus novaehollandiae)*. Shearwaters *(Puffinus* sp.) are also common.

Algae

The Great Barrier Reef supports a wide range of fleshy algae, many of which are small and inconspicuous but which are highly productive and heavily grazed by turtles, fish, molluscs and sea urchins. Calcareous algae play a key role in reef building. There are two major groups, Chlorophyta (green algae) and Rhodophyta (red algae), and a third less important group is Phaeophyta (brown algae). The main calcareous genera of Rhodophyta are *Lithothamnion, Lithophyllum, Porolithon, Amphiroa* and *Amansia.* These genera have closely textured skeletal structure and grow as encrustations and are, thus, largely responsible for the survival of the reef mass during severe and continuous wave attack. *Lithothamnion* and related forms are most common on the algal rim of the reef, but are less common on the middle and outer reef flat.

The genera of Chlorophyta, such as *Halimeda, Penidilus* and *Acetabularia,* play a minor role in shielding the reef against wave action, but are important in contributing to sediment. Although *Halimeda* flourishes in almost every part of the reef, *Acetabularia* is restricted to the reef flat, and *Penicilus* occurs only on the inner reef flat.

Seagrasses

Seagrasses, found patchily throughout the Great Barrier Reef area, are an important food source for grazing animals, such as fish and turtles, and are important as prawn nursery areas. In some of the inshore coastal areas, extensive seagrass beds are important feeding grounds for the dugong. There are also offshore seagrass beds of unknown extent.

PRODUCTIVITY

Many areas of high biological productivity are found in the Great Barrier Reef Region. They may be linked to localized upwelling related to high tidal velocities in reef passages. The average nutrient level in the waters of the Reef Region tends to be low. External influences upon nutrient levels could have significant effects in terms of biological productivity.

Unlike other coral reef systems in the world, the Great Barrier Reef has not been intensively used for human subsistence. Apart from some very small areas, the Great Barrier Reef is still in pristine condition relative to the majority of reef areas and has not been unduly affected by human activity.

HUMAN USE OF THE GREAT BARRIER REEF

The estimated value of Reef-related activities (on the Reef and on the adjacent mainland) approximates AUS$1,500 million per annum. Commercial fishing and tourism, recreational pursuits including fishing, diving and camping, traditional fishing, scientific research and shipping all occur within the Great Barrier Reef Marine Park.

Resort tourism is the largest commercial activity in economic terms. In 1991/92, an estimated 243,800 visitor trips resulted in 1,080,000 visitor-nights spent on the 18 island resorts and the visitors spent approximately AUS$188.6 million at the resorts (QTTC 1992). Resort guests make extensive use of reefs and waters for recreational activities, including fishing, diving and snorkeling, water sports, sightseeing and reef-walking, and some collecting.

The popularity of the Reef and adjacent coast region as a tourist destination increased forty-fold over the period from the 1940s to 1980 and is continuing to increase (ATIA 1984). Recently, tourist facilities have been built directly on reefs. The last ten years has seen the introduction of many large, stable, high-speed catamarans providing day trips to islands and outer reefs. The latest development is wave-piercer catamarans which provide fast, more comfortable transport.

There is conflict between some groups of people who use the Reef (for example, between commercial and recreational fishers). There is also conflict between people who wish to exploit the Reef and those who wish to see it maintained in its pristine state forever. Some uses of parts of the Reef have already reached levels which may fully exploit the productive capacity of the system. Bottom trawling for prawns is an example. Run-off from islands and the mainland contains suspended solids, herbicides, pesticides, nutrients, and other materials. They have effects on the Reef, the magnitude of which is not yet known, but we are working on it.

This description also applies to other reef systems throughout the world's tropical seas. The need for and the difficulties of managing uses so that they are sustainable forever are also common. Perhaps the system of management which has been developed on the Great Barrier Reef could be applied elsewhere, although the acceptability of any management system is likely to be diminished where there are very high levels of usage and economic dependence on reef areas, for example in many parts of Asia. It is ironic that these are the areas that are most in need of management.

THE GOAL AND AIMS OF THE AUTHORITY

The Great Barrier Reef Marine Park Authority has derived a primary goal and a set of aims from the provisions of the Act and recognition of the political, legal, economic, sociological, and ecological environment in which it operates. The Authority believes that any use of the Reef or associated areas should not threaten the Reef's essential ecological characteristics and processes, such as biodiversity. Activities depending on the Reef's renewable resources should generally be held at or below maximum sustainable intensities indefinitely. This belief has led the Authority to adopt the following primary goal:

> To provide for the protection, wise use, understanding and enjoyment of the Great Barrier Reef in perpetuity through the development and care of the Great Barrier Reef Marine Park.

However, not only the physical aspects of the Reef need to survive. If the Reef is to be protected, administrative arrangements also must be durable.

Failure of the Authority would not necessarily or even probably be followed by the creation of new, more effective arrangements.

In Australia the major determinant of administrative survivability of organizations like the Authority is public support. In the long run, government support flows from it.

Recognizing that the Authority and the Marine Park concept already have a degree of public support, the Authority must act in ways which sustain or increase that support. What are those ways? It seems clear that the ground work has been well established in the Act through the formal requirements for public participation, the provisions for a Consultative Committee, the composition of the Authority itself and its functions, as well as the ability to perform those functions in association with Queensland or its agencies.

Generally, the public is likely to continue to support the Marine Park and the Authority if the primary goal is being achieved efficiently. For this to occur, the public will have to be aware of what the Authority and its day-to-day management agencies are doing and the way they are doing it, the effectiveness and costs of their programs and the reasons for them, and to the extent practicable, to be involved in the establishment and management of the Marine Park. A set of aims has been derived from this and related observations. They are subordinate to the primary goal and must be read in conjunction with it and with each other.

THE AUTHORITY'S AIMS

Aim 1: "To protect the natural qualities of the Great Barrier Reef, while providing for reasonable use of the Reef Region."

Aim 2: "To involve the community meaningfully in the care and development of the Marine Park."

Aim 3: "To achieve competence and fairness in the care and development of the Marine Park through the conduct of research, and the deliberate acquisition, use and dissemination of relevant information form research and other sources."

Aim 4: "To provide for economic development consistent with meeting the goal and other aims of the Authority."

Aim 5: "To achieve management of the Marine Park primarily through the community's commitment to the protection of the Great Barrier Reef and its understanding and acceptance of the provisions of zoning, regulations and management practices."

Aim 6: "To minimize the costs of caring for and developing the Marine Park consistent with meeting the goal and other aims of the Authority."

Aim 7: "To minimize regulation of, and interference in, human activities consistent with meeting the goal and other aims of the Authority."

Aim 8: "To achieve its goal and other aims by employing people of high calibre, assisting them to reach their full potential, providing a rewarding, useful and caring work environment and encouraging them to pursue relevant training and opportunities."

Aim 9: "To make the Authority expertise available nationally and internationally."

Aim 10: "To adapt actively the Marine Park and the operations of the Authority to changing circumstances".

Taken together, this goal and these aims provide a policy framework for evaluation of proposed programs and actions. We believe that much of the success of the Marine Park is the result of applying these policies. We believe further that they can (and should) be applied in natural resource management anywhere, not only in Australia. Much of the thrust of policy is communication with and involvement of the public in the development and care of the Marine Park. Education and information programs are therefore critically important elements of the Authority's work.

THE MARINE PARK AND THE ZONING SYSTEM

The Great Barrier Reef Marine Park is not a National Park. It is a multiple-use protected area, fitting the definition of Category VIII of the classification system used by IUCN, the World Conservation Union (IUCN 1982). It also meets the criteria for selection and management as a Biosphere Reserve (Category IX), although it has not been formally proposed or established as one. The Reef was inscribed on the World Heritage List in 1981 as a natural site (Category X).

Through the use of zoning, conflicting activities are separated, areas are provided which are suitable for particular activities and some areas are protected from use. Levels of protection within the Park vary from almost complete absence of restriction on activity in some zones to zones within which almost no human activities are permitted. The only activities that are prohibited throughout the Park are oil exploration, mining (other than for approved research purposes), littering, spearfishing with SCUBA, and the taking of large specimens of certain species of fish.

In the zoning plans which have been developed so far, there are three major categories of zones. They are:

1. Preservation zones or Scientific Research zones

Equivalent to IUCN Category I, Scientific Research/Strict Nature Reserve. The onlyhumanactivity permitted is strictly controlled scientific research. These can provide refuge for species and communities.

2. Marine National Park zones

Equivalent to IUCN Category II, National Park. The major uses permitted are scientific, educational and recreational. Again, these contribute significantly to biodiversity conservation.

3. General Use zones

Equivalent to IUCN Categories IV, Managed Nature Reserve and VI, Resources Reserve. Uses are held at levels that do not jeopardize the ecosystem or its major elements. Commercial and recreational fishing are generally permitted, although bottom trawling is prohibited in one of these two zones.

The zones are fixed during the life of a zoning plan (generally five to seven years). They are complemented by generally smaller areas that give special protection from time to time to animal breeding or nesting sites, to sites in general use and other zones which are required to be protected to allow appreciation of nature free from fishing or collecting and to sites suitable for scientific research.

Because of a dramatic increase in use of the Marine Park by tourists, the existing zoning system, which focuses on fishing, is proving inadequate. There is increasing competition for tourism use of particular sites. Usually these sites are near major areas of coastal development (e.g., Cairns or Townsville) or have particular attributes which make them suitable for tourism—the Whitsunday Islands, for example.

A major constraint in zoning for tourism has been that many tourists like to observe the natural qualities of the Great Barrier Reef undisturbed by fishing. Modern technology allows them to do this from semi-submersible vessels and from underwater observatories. If carried out with care these activities can have very little if any effect on the Reef's ecosystem. They are, in other words, compatible with the Authority's goal.

A major innovation that was introduced in the rezoning of the Cairns Section of the Marine Park was the creation of No-structures sub-zones. These sub-zones overlie approximately 22% of the reefs in the Section. Their principal purpose is to ensure that a proportion of those parts of the Reef that are near centers of human population and are therefore subject to heavy human use do not end up dotted with permanent or semi-permanent structures. The public feels very strongly about this issue. Vanclay (1988) found that 77% of tourists interviewed in a survey were opposed to more development on the Great Barrier Reef. It parallels the public concern about excessive coastal development on land.

Tourism is allowed to occur under permit within all except preservation and scientific research zones, that is, in 99.8% of the Marine Park.

A spatial regulation management measure that relates particularly to tourism is the ability to declare Reef Appreciation Areas or Special Management Areas in zones in which fishing and collecting are normally permitted. These provisions allow up to 20% of the area of a reef to be declared a "look, don't take" area to complement the provision of tourism facilities.

All tourist programs and facilities within the Marine Park require a permit. This requirement allows the Authority to assess each proposed tourist operation individually in terms of its suitability.

Tourism programs and other activities requiring a permit are assessed in relation to the following criteria:

(a) the objective of the zone; and
(b) the need to protect the cultural and heritage values held in relation to the Marine Park by traditional inhabitants and other people; and
(c) the likely effect of granting permission on future options for the Marine Park; and
(d) the conservation of the natural resources of the Marine Park; and
(e) the nature and scale of the proposed use in relation to the existing use and amenity, and the future or desirable use and amenity, of the relevant area and of nearby areas; and
(f) the likely effects of the proposed use on adjoining and adjacent areas and any possible effects of the proposed use on the environment and the adequacy of safeguards for the environment; and
(g) the means of transport for entry into, use within or departure from the zone or designated area and the adequacy of provisions for aircraft or vessel mooring, landing, taking off, parking, loading and unloading; and
(h) in relation to any structure, landing area, farming facility, vessel or work to which the proposed use relates:
    (i) the health and safety aspects involved, including the adequacy of construction; and

(ii) the arrangements for removal upon the expiration of the permission of the structure, landing area, farming facility or vessel or any other thing that is to be built, assembled, constructed or fixed in position as a result of that use; and

(i) the arrangements for making good any damage caused to the Marine Park by the proposed activity; and

(j) any other requirements for ensuring the orderly and proper management of the Marine Park.

Additional specific requirements must be considered for traditional fishing and research permits.

Concern to prevent unacceptable ecological impact is paramount in the Authority's management of tourism development. The types of activities which may be associated with reef based tourism operations and which may have biophysical impacts include; discharge of waste, litter and fuel; physical damage to reefs from anchors, people snorkeling, diving and reef walking; disturbance of fauna (especially seabirds); over-fishing or collecting. All of these can be managed to some extent by education, incentive, design, prohibition or limitation.

Proponents of large scale developments or those which have the potential to produce impacts may be required to prepare Environmental Impact Statements or Public Environment Reports. Through environmental impact assessment, the potential impacts can be identified and steps taken to prevent or mitigate the impacts. Often the prevention of impacts is a matter of adopting the appropriate design and operational guidelines.

Another dimension of the tourism impacts issue focuses on amenity—the impacts of tourism on other tourists and users of the Great Barrier Reef. The opposition to more commercial development on the Great Barrier Reef in the future has been recorded (Vanclay, 1988). Although what was meant by development was not defined by many respondents, this response revealed a disquiet about levels of crowding and tourist facility hardware which may occur in the future.

The amenity issue is important at two scales. Overcrowding, reef degradation, or an inappropriate mix of facilities at a particular reef site may detract from the experience of tourists. The Authority has had to address this issue a number of times where new operators have applied for permits to operate at sites already being used by an existing operator. The solutions have included time sharing, delineating operating areas, encouraging use of alternative sites and limiting the number of operators and people visiting a particular site. Research has been undertaken on one occasion with the aim of resolving conflict between two different types of tourist operation wishing to use the same site (e.g., seaplane landing sites in relation to swimmers and snorkelers at Green Island).

On the larger scale, there is a need to ensure that the appropriate mix of tourism developments, which provide a range of recreation opportunities, is provided in an area within reasonable access from major towns and that some areas remain free from the "hardware" associated with tourist facilities—as discussed earlier in relation to No-structures subzones.

The size of the Great Barrier Reef is such that the concept of overdevelopment from tourism can hardly be currently considered a serious threat to conservation of the Great Barrier Reef as a whole. There are around 3,000 reefs and islands within the Great Barrier Reef Region. There are 18 islands with resorts, and 19 reef sites are regular day trip destinations, with permanently moored tourist platforms, while a larger number of sites are visited by intermittent operations. However, the fact that concentrations of people in particular areas and the presence of tourist facilities may detract from the experience of some users is of concern to the Authority. The offshore Cairns area and the Whitsundays are two sites where tourism pressure needs careful consideration.

The Authority is addressing this issue by undertaking strategic planning exercises in areas of increasing tourism demand and undertaking research into identifying the important factors which affect amenity and social carrying capacity. Jointly with the Queensland Department of Environment and Heritage, responsible for Island National Parks and Queensland Marine Parks, the Authority is developing "area statements" and "management plans" for groups of islands and reefs and individual islands and reefs respectively. These plans, which will involve periods of public comment, will identify existing and recommended activities and acceptable levels of impact to attempt to ensure a range of recreational experiences is available and that the cumulative effect of activities is addressed before acceptable levels of impact are exceeded through a series of individual permit decisions. Within each "area" a range of experience "settings" is proposed to ensure that tourism does not preclude alternative opportunities.

To overcome some of the difficulties outlined above and to ensure all interest and user groups and management agencies are working to a common agenda, the Authority is coordinating the development of a 25 Year Strategic Plan for the Great Barrier Reef World Heritage Area with the objective of all groups creating a shared vision of how they would like the Area to be in 25 years time, and developing strategies to achieve that. Details of the process can be found in Raymond and Craik (1992).

LINKAGE WITH THE LAND

It has always been the aim of the Authority to ensure that the Great Barrier Reef is managed as a single ecosystem including all the waters of the Great Barrier Reef Region and the 900 islands within its outer boundaries.

This has largely been achieved through the following mechanisms:

- co-ordination of policy and action between the Federal Government and the State (Queensland) Government in a 4-person Ministerial Council;
- the State Government being represented by one member on the 3-person Authority (the Chairman represents the Federal Government and a third member is independent);
- day-to-day management of the waters and islands is carried out by a single management group, with costs shared by the two governments;
- the policies of the Authority (described previously); and
- very strong commitment of the Authority and Queensland agencies to work together.

The problem of the effects on the Great Barrier Reef of human activities on the mainland are more intractable. Probably the major issue is run-off from farm land of nutrient enriched water. However, even in this case high levels of cooperation in research have been achieved between the Authority, farmers' organizations and state Government agencies responsible for primary industry. Our experience has been that if cooperation is achieved in carrying out research into a problem then that cooperation is likely to extend into defining and applying ways of curing the problem.

FUTURE CHALLENGES

Apart from tourism, there are three major areas of challenge in our task of ensuring that the Great Barrier Reef is sustained into the 21st Century. They are:

1. Achieving sustainable fishing and collecting;
2. Understanding the crown-of-thorns phenomenon; and
3. Protection of the Great Barrier Reef from pollution.

Sustainable Fishing

In common with nearly every major world fishery, the Great Barrier Reef prawn fishery is over capitalised. It is going through the now familiar process of declining catch-effort ratios. This problem is being addressed in two ways.

Through the zoning system, some areas of the Marine Park are closed to trawling. These amount to about 20% of the total area. They act as reference areas, so that the long term effects of trawling can be determined. They also ensure that the species that occupy them are fully protected.

The second approach is being taken by the fishing industry itself. It consists of traditional fisheries practices including seasonal closures, the protection of nursery areas and attempts at limiting the total effort expended in the fishery.

We are fairly confident that these two complementary approaches will together ensure that any damage to the Great Barrier Reef system from trawling can be repaired. Additionally, the Authority is working with the commercial fishing industry in two major studies of the effects of line fishing and trawling on the reef ecosystem. A major multi-agency research program has commenced into these issues.

Crown-of-thorns Starfish

The Federal Government and its agencies have already spent more than $5 M on research into the crown-of-thorns starfish. Not only are we convinced that effort must continue to be applied to this issue at least until we have unequivocally answered the question of whether human activity is affecting the intensity and frequency of population outbreaks.

If it were scientifically demonstrated that human activity is a major factor, then the Authority would react in two ways. First, it would reconsider its policy of limiting control of population numbers to areas of particular scientific or tourist value. Second, it would have to move strongly to ensure that the human activities which were shown to be causative factors were modified.

If, on the other hand, scientific evidence were to accumulate that outbreaks have not changed in intensity or frequency since Europeans arrived in Australia, the Authority would maintain its existing policies. Those policies are to limit control of outbreaks to areas of particular value to tourism or to science and to continue to study scientifically the whole phenomenon and its role in the Reef system.

The sooner we can resolve this issue, the better. Although it is true that crown-of-thorns research answers many scientific questions of a general nature, scarce research funds could be more effectively spent on answering these general questions if crown-of-thorns starfish were not an issue. Further, it is conceivable that postponement of resolution of this issue might lead to progressive degradation of the Great Barrier Reef, requiring hundreds of years to recover.

Pollution Control

Monitoring has shown that the levels of toxic chemicals in virtually the whole Great Barrier Reef Region are very low—close to the limit of detectability. Such chemicals are not likely to be a problem in the medium term but monitoring will be continued.

There is an ever present risk of an oil spill in the Great Barrier Reef. The precautions that have already been taken to minimize this risk include the requirement that all vessels carrying oil or toxic cargoes and all vessels over 70 meters in length using the northern part of the inside passage of the Great Barrier Reef Region and Hydrographers Passage should be piloted and the development of an oil spill contingency plan entitled: "REEF PLAN". This plan is a component of the National Plan to Combat Pollution of the Sea by Oil. In the event of a spill, an On-scene Coordinator is appointed to take direct charge of operations to combat the pollution incident. He is supported by an Operations Controller who is responsible for all administrative actions in support of the On-scene Coordinator. The On-scene Coordinator is advised by a Committee of Experts on action to be taken and on materials and methods best suited for combatting a specific oil pollution situation. The Great Barrier Reef Marine Park Authority is a member of this Committee. We have developed a unique program on Apple Macintosh computers which incorporates bathymetric and environmental information which will eventually cover the whole Region and a predictive capacity for the movement of spilled oil in various weather conditions.

In the light of experiences in other parts of the world it would be naive to expect even the best oil spill contingency plan to prevent damage to the Marine Park by an oil spill. The Authority will continue to work with other agencies in reducing the risk of oil spills and in improving the capacity for a response to such events. A recent expert workshop concluded that further development and application of Differential Global Positioning Systems and electronic mapping offered the best prospects of significantly reducing the risks of oil spills.

A more insidious, but potentially no less serious, concern for the future is increases in nitrogen and phosphorus in waters of the Great Barrier Reef. Work on the Great Barrier Reef, in the Great Barrier Reef Aquarium and in other parts of the world has shown that reef building corals are very vulnerable to increases in the levels of these nutrients. Monitoring has shown that the porosity of some near-shore reefs of the Great Barrier Reef has been increasing over the past few decades due to increase in phosphorus levels. The levels of nitrogen in some parts of the Great Barrier Reef at times exceed significantly levels which have been shown to cause death in branching corals permanently subjected to them.

The Authority received a major grant from the Federal Government to investigate this issue two years ago. This has allowed the establishment of a reef-wide monitoring program and a research program that aims to develop a complete understanding of the origins of these nutrients so that corrective action can be taken if necessary.

We believe that protection of the Great Barrier Reef from increasing nutrient levels may be the greatest challenge facing the Authority in the next two decades.

CONCLUSIONS

The Great Barrier Reef Marine Park is an example of the practical application of the principles defined in the World Conservation Strategy and its successor "Caring for the Earth." It can be seen as a model for development of the kind described in the report of the World Commission on Environment and Development—"Our Common Future." The Marine Park embodies many of the recommendations in the Global Biodiversity Strategy.

Our experience has shown how difficult will be the maintenance of biodiversity and the attainment of sustainable development throughout the world. Unlike many other countries, we are blessed in Australia with a comparatively small population, great natural resources, and a strong emotional and intellectual commitment by the people to the protection of the Great Barrier Reef. Even with these advantages, the Authority is continually struggling to prevent insidious degradation of the Great Barrier Reef ecosystem.

At national and international levels, the adverse effects on marine ecosystems of human activities on land must be recognized and actions taken to reduce them. This will require degrees of cooperation higher than have been attained historically between governments, between levels of government and between agencies within government. Integrated coastal zone management will be an essential precursor to ecologically sustainable use of the world's marine and terrestrial coastal environments.

LITERATURE CITED

ATIA. 1984. *Data review of reef related tourism. 1946-1980.* GBRMPA Research Publication.

Brundtland, G. H. 1987. World Commission on Environment and Development. In *Our Common Future*, p. xi. Oxford University Press, Oxford.

International Union for the Conservation of Nature and Natural Resources (IUCN). 1982. *Categories, Objectives, and Criteria for Protected Areas.* Commission on National Parks and Protected Areas.

IUCN/UNEP/WWF. 1991. *Caring For The Earth: A Strategy for Sustainability.* IUCN, UNEP and WWF.

Kikkawa, J. 1976. The Birds of the Great Barrier Reef. In *Biology and Geology of Coral Reefs*, vol. 3, Biology 2, eds. O. A. Jones and R. Endean, p. 282. Academic Press, New York.

Queensland Tourist and Travel Corporation. 1992. *Queensland Visitor Survey 1991/92.* QTTC, Brisbane.

Raymond K., and W. Craik. 1992. *The Great Barrier Reef World Heritage Area Strategic Plan. Participative Regional Planning.* Florida Keys National Marine Sanctuary First Annual Conference, Coral Reef Coalition, March 19-22, 1992, Key West, Fla.

Vanclay, F. 1988. Tourist perceptions of the Great Barrier Reef. Unpublished report to the Great Barrier Reef Marine Park Authority.

WRI/IUCN/UNEP. 1991. *Biodiversity Strategy and Action Plan.* Draft prepared by WRI, IUCN and UNEP, Washington, D.C.

32

DIVERSIFIED AGROFORESTRY SYSTEMS: BUFFERS FOR BIODIVERSITY RESERVES, AND LANDBRIDGES FOR FRAGMENTED HABITATS IN THE TROPICS

Jiragorn Gajaseni

Department of Biology, Faculty of Science
Chulalongkorn University
Bangkok
Thailand

Rodrigo Matta-Machado and Carl F. Jordan

School of Ecology
University of Georgia
Athens, Georgia
U.S.A.

Agroforestry is a type of land management in which woody perennials are planted on the same land management units as agricultural crops or animals in either a spatial arrangement or a temporal sequence and with ecological and economic interactions between the different components (Fernandes and Nair 1986). The specific sites where agroforestry sometimes is practiced have different names, depending on the region: examples are *conuco* in Venezuela, *ch_cara* in Brazil, *home gardens* in parts of Asia, and *farm* in the United States. For convenience in this paper, we will use the generic name "agroforestry system" for any site in which agroforestry is practiced. Here we provide a review of the evidence of how agroforestry systems can be beneficial for conserving biodiversity.

HISTORICAL PERSPECTIVE

The practice of interplanting crops and trees is as old as agriculture itself. Evidence from excavations at Ban Chieng, Northeastern Thailand indicate that humans cultivated plants in the area for centuries (Gorman 1969). Systems involved combinations of trees, root crops, and seed crops. They are thought to be the origin of regional agriculture (Gorman 1971). Primitive hunting, fishing, and gathering also was important.

The earliest important crops of Southeast Asia were root crops such as taro (*Colocasia esculenta*) and yam (*Dioscorea alata*), both of which preceded rice (*Oryza sativa*). Other crops of great importance to the cultivators were perennials, such as bananas (*Musa* spp.), papayas (*Carica papaya*), coconut palms (*Cocos nucifera*), jackfruits (*Artocarpus* spp.), mangos (*Mangifera indica*), and durians (*Durio zibethinus*) (Pelzer 1978). Archeological evidence from Thailand indicates that it was one of the first rice-growing regions of the world (Saengwichien 1974).

With the advent of the moldboard plow and other agricultural advances, agriculture tended to move away from the traditional systems, at least in the temperate zones, and among large scale farmers in the tropics. But while mixed cropping of trees and crops was forgotten by the industrial farmer, the traditional systems have continued to be practiced by small-scale farmers and by native forest dwellers. Even today many cultivators imitate natural systems by planting a mixture of root crops, herbs, climbers, shrubs, trees, and grain crops, such as maize and upland rice (Harris 1972). Surrounding rural villages might be fields cultivated in a variety of systems, thus reflecting the natural and social environments.

BENEFITS OF AGROFORESTRY

Many recent books on agroforestry emphasize nutrient cycling, prevention of erosion, and control of weeds and insect pests as advantages of agroforestry systems (Beets 1982; Huxley 1983; Gomez and Gomez 1983; FAO 1984; Goodland et al. 1984; Francis 1986; Altieri 1987; Am. Soc. Agronomy 1988; Nair 1989; Kartasubrata et al. 1990; Macdicken and Vergara 1990; Young 1989). For example, MacDicken and Vergara (1990) listed the following beneficial characteristics of agroforestry as opposed to even-aged monocultures of annuals: improved micro-climate, enhanced nutrient cycling, increased soil fertility and soil organic matter, reduction in wind, disease barrier, weed and Pest control, increased economic sustainability, increase in soil porosity and aeration, protection of soil from rainfall impact, erosion control and terrace effect, larger volume of soil exploited, decreased fertilizer requirement, and control of leaching.

BIODIVERSITY AND AGROFORESTRY SYSTEMS

Little attention has been given to the role that agroforestry might play in protecting biodiversity. In light of current concern for loss of biodiversity, especially in the Tropics, the possible role that agroforestry could have in conserving tropical biodiversity is important. Theoretically, agroforestry systems should be better than simple monocultures for conservation of biological diversity because agroforestry systems are structurally more diverse. Structural diversity of a plant community correlates with the diversity of animals that inhabit the community (Urban and Smith 1989). Forest structure determines habitat factors such as microclimate, food abundance and cover that affect the fitness of organisms (Cody 1985).

Altieri et al. (1987) demonstrated that the arthropod diversity in an agroecosystem is a function of its complexity. However, there have been few studies on vertebrate wildlife in simple versus complex agricultural systems. Research into wildlife in forest plantations or agroforestry is extremely rare.

Pimentel et al. (1992) discussed the idea that biological diversity in agricultural and forestry systems can be best conserved by maintaining plant and habitat diversity. They stated that "increasing the diversity of physical habitats increases the diversity of associated plants and other organisms."

Hanski and Tiainen (1988) asked "why population ecologists, who are clearly interested in the role of habitat structure in population ecology, have not used more widely the research opportunities which are incidentally created by agriculture?" "One explanation," they stated, " may be the very intensive management of most agro-ecosystems that appears to leave very little scope for natural population dynamic processes."

BIODIVERSITY AND AGRICULTURAL SIMPLIFICATION

Although few studies have documented increase in biodiversity as a result of substituting agroforestry systems for crop monocultures, several papers address the decrease in biodiversity resulting from the modern trend toward large-scale agriculture and away from smaller mixed systems.

Hanski and Tiainen (1988) found evidence for simplification of biological communities in agricultural systems resulting from changes from mixed farming early in this century to recent large-scale monocultures in Finland. Best (1990) showed that the disappearance of a patchwork of small diversified fields in Iowa and their replacement by unbroken monocultures resulted in a sharp decline in the populations of ring-necked pheasants. He felt that if increased crop diversification could be coupled with a reduction in the size of crop fields, the advantages for birds and other wildlife would be twofold: benefits from a more diverse assemblage of crops, and increases in the relative amounts of edge, with its associated benefits. Remnants of heterogeneous landscapes with native vegetation on farms are often reservoirs for the disappearing biodiversity (Hobbs and Wallace 1991).

Windbreaks and shelterbelts between agricultural fields in the United States shelter a greater biodiversity than surrounding fields (Johnson and Beck 1988; Capel 1988). In Germany, Mader (1988) found 4.1 times more species of arthropods in hedgerows and woodlots than in surrounding fields. Many species were beneficial, for example, spiders that preyed on crop pests. Farmland hedges in England are important for nesting success of birds (Lack 1988) and for breeding of mammals. Of the 28 species of British lowland mammals, 14 are common in hedgerows and breed there (Dowdeswell 1987).

BIODIVERSITY AND BUFFER ZONES

The higher biodiversity in agroforestry systems as compared to annual monocultures is of interest to certain groups, such as hunters who see the more complex vegetation as cover for their game, and to farmers who use integrated pest management instead of chemical control. However, agroforestry systems are not suitable habitat for many species that are threatened by tropical deforestation. Often, rare and endangered species live exclusively in old-growth habitats, and for these, agroforestry systems would not be suitable. Of the 116 species of non-flying mammals in the Malay peninsula, 31% seem to be totally dependent on primary forests for their existence, and cannot adapt to secondary habitat; 77% of the flying mammals seem to be dependent on forest (Hoi-Sen 1978). Thus, patches of structurally more complex vegetation, such as agroforestry systems, cannot protect all tropical species. Strict preserves are needed.

Reserves of virgin forest in the Tropics are often difficult to protect, because the wildlife in the reserves and the humans in the societies surrounding the reserve usually are antagonistic. However, for several reasons, a buffer zone consisting of agroforestry plots could help ease conflicts. Edges between forest and field are typically dominated by generalist species, able to breed and forage in a variety of habitats (Wilson and Diver 1991). Game species used by peasant farmers are often generalists that could find cover in agroforestry systems and the surrounding area. With a buffer zone that supported generalist species, there would be less incentive for the hunters to enter the primary forest set aside for the protection of rare and endangered species. There would be more incentive to hunt in and around the agroforestry systems where game species might be pests on the crops.

Another benefit would be the reduction of the "edge effect," which often occurs around areas of closed canopy forest abutting open land. When primary forest is cut down and converted to pasture or to field crops, the influence of the open area does not stop at the boundary of the forest. For example, students working at the Biological Dynamics of Forest Fragments (formerly the Minimum Critical Size of Ecosystem Project sponsored by Instituto Nacional de Pesquisas de Amazonia, and the Smithsonian Institution) near Manaus, Brazil found changes in micro-climate extending at least 100 meters into the forest. This, in turn, put severe stress on several rain forest tree species, and reduced the biodiversity of closed-canopy rain forest fauna. If the edge of the primary forest were to be surrounded by agroforestry systems instead of completely open pasture or cropland, the edge effect inside the forest would be much reduced.

Agroforestry systems also could act as "landbridges" between fragments of primary forest. In fact, a problem with the Minimum Critical Size of Ecosystem Project was the fact that secondary forest grew up between some of the experimental reserves. This secondary forest served as a bridge through which species could repopulate an area in which the species previously had been depressed (Bierregaard et al. 1992).

The effectiveness of such buffers for biodiversity could be monitored by studies of key species. Some birds might particularly useful as indicators of the effectiveness of management strategies that promote habitat diversity (Szaro 1986). Crops for such agroforestry systems have been found by surveying home gardens and cultivation sites in remote tropical areas (Iltis 1988) and by studying use of plants of indigent peoples (Plotkin 1988).

ADVERSE ASPECTS OF BIODIVERSITY IN AGROFORESTRY BUFFER ZONES

The idea that agroforestry in buffer zones should be promoted to conserve biodiversity in interior forests is appealing, but from the farmer's viewpoint, increased biodiversity can be bad as well as good. Although birds living in shelterbelts or hedges can be predators on insects in the fields, they can also be crop pests (Wilson and Diver 1991). Corridors maintained as wildlife refuges can have important costs (Simberloff and Cox 1987).

They might transmit contagious diseases, fires, and other catastrophes, and they might increase exposure of animals to predators, domestic animals, and poachers. Corridors also bear economic costs, because they take land out of production.

In the design and establishment of agroforestry systems, care must be taken to preclude the impact of species that might harm instead of benefit the native biota. For example, exotic trees have damaged native ecosystems in South Africa (Dennill and Donnelly 1991 for *Acacia longifolia*, and Zimmermann 1991 for *Prosopis* spp.).

IMPROVING BIODIVERSITY THROUGH AGROFORESTRY: CASE STUDY IN THAILAND

In Thailand, taungya reforestation is practiced in a system of forest villages (Gajaseni 1992). These are villages sponsored by the Government and the Forest Village Industry of Thailand. Peasants who live in the villages are allowed to plant their crops amongst the trees for the first few years until the canopy of the trees closes. In most of these forest village plantations, the agroforestry system is quite simple. There is usually just one species of tree, and only a few agricultural crop species.

To examine the biodiversity in these taungya systems, questionnaires were distributed to the officer-in-charge of each of the 140 villages in Thailand. With the help of village officers and workers who were intimately acquainted with each site, a list of species was compiled. Then master lists were made for each of the five major regions of Thailand (Table 32.1.). The table illustrates the relatively high diversity of fauna harbored by these agroforestry systems. Despite the simplicity of the taungya systems, they appear to act as effective land bridges between native forests of the region.

However, the government is having problems retaining peasants at the forest villages. In the northern region of Thailand, upland rice and sorghum is cultivated between the teak seedlings for only two years, but the teak is not ready for harvesting for 40-60 years. Consequently, the peasant farmer must move his subsistence plots every two years. Because of the problems this causes, many peasants leave the villages and resume shifting cultivation. Others stay in the village, but hunt and gather forest products in forest reserves.

Table 32.1. Number of species of faunal types found in agroforestry systems of the five major regions of Thailand.

Region	Forest type	Birds	Amphibians	Reptiles	Mammals
Northern	Mixed deciduous, teak	50	11	27	33
Western	Dry evergreen, mixed deciduous	30	8	5	13
North-eastern	Dry evergreen, deciduous, dry dipterocarp	28	8	16	11
Eastern	Tropical evergreen	34	8	18	16
Southern	Rain evergreen	23	9	18	13

To remedy this situation, Gajaseni and Jordan began an experimental taungya plantation in the Mae Moh Basin, near Lumpang, northern Thailand, in which a wide variety of fruit trees are interspersed among the teak trees. These trees begin to produce at about the time the cultivation of grains must terminate. The fruit will increase the economic sustainability of the system. With a greater economic income from the taungya system, peasant farmers should be more inclined to remain with the system, and to refrain from hunting and gathering in forest reserves.

CONCLUSIONS

When tropical forests are harmed by commercial logging, shifting cultivation, and conversion to pasture, ecosystem structure is simplified. Simplification of ecosystem structure reduces biological diversity. Forest simplification eliminates many plant species, and the animal species dependent on those species disappear.

To prevent loss of biodiversity in the Tropics, the human activity that results in simplification should stop. However, this is not likely to happen. Human demands for forest products, or the conversion of forest land to agricultural land usually takes priority over forest preservation. Replacing shifting cultivation, which destroys forest reserves with sustainable agroforestry systems that form a buffer around the reserve, can reduce the destruction in two ways.

1. Agroforestry systems are better able to *sustain* a supply of vegetables, grains, and protein for the local farmer, thus relieving the pressure to advance farther into the forest.

2. Agroforestry systems in a buffer zone around a reserve for biodiversity relieve the edge effect that occurs when open fields or range extends right to the border of the forest, thereby effectively increasing the size of the reserve.

LITERATURE CITED

Altieri, M. A. 1987. *Agroecology*. Westview Press, Boulder, Colo.

Altieri, M. A., F. J. Trujillo, and J. Farrell. 1987. Plant-insect interactions and soil fertility relations in agroforestry systems: implications for the design of sustainable agroecosystems. In *Agroforestry: Realities, Possibilities and Potentials*, ed. H. L. Gholz, pp. 89-110. Nijhoff Pubs., Dordrecht, The Netherlands.

American Society of Agronomy. 1988. *Agroforestry Land-Use Systems*. Nitrogen Fixing Tree Association, Waimanalo, Hawaii.

Beets, W. C. 1982. *Multiple Cropping and Tropical Farming Systems*. Westview Press, Boulder, Colo.

Best, L. B. 1990. Sustaining wildlife in agroecosystems. In *Farming Systems for Iowa: Seeking Alternatives*, pp. 43-49. Leopold Center for Sustainable Agriculture, Iowa State University, Ames, Iowa.

Bierregaard, R. O., T. E. Lovejoy, V. Kapos, A. Augusto dos Santos, and R. W. Hutchings. 1992. The biological dynamics of tropical rainforest fragments. *BioScience* 42: 859-866.

Capel, S. W. 1988. Design of windbreaks for wildlife in the Great Plains of North America. *Agric. Ecosyst. Environ.* 22/23:337-347.

Cody, M. L. 1985. An introduction to habitat selection in birds. In *Habitat Selection in Birds*, ed. M. L. Cody, pp. 4-46. Academic Press, Orlando, Fla.

Dennill, G. B., and D. Donnelly. 1991. Biological control of *Acacia longifolia* and related weed species (Fabaceae) in South Africa. *Agric. Ecosyst. Environ.* 37: 115-135.

Dowdeswell, W. H. 1987. *Hedgerows and Verges*. Allen and Unwin, London.

FAO. 1984. Sistemas Agroforestales en America Latina y el Caribe. Oficina Regional de la FAO para America Latina y el Caribe, Santiago, Chile.

Fernandes, E. C. M., and P. K. R. Nair. 1986. An evaluation of the structure and function of tropical homegardens. *Agric. Syst.* 21:279-310.

Francis, C. A. 1986. *Multiple Cropping Systems*. MacMillan, New York.

Gajaseni, J. 1992. Socio-economic aspects of taungya. In *Taungya: Forest Plantations with Agriculture in Southeast Asia*, eds. C. F. Jordan, J. Gajaseni, and H. Watanabe, pp. 36-62. C. A. B. International, Wallingford, U.K.

Gomez, A. A., and K. A. Gomez. 1983. *Multiple Cropping in the Humid Tropics of Asia*. International Development Research Center, Ottawa, Ontario, Canada.

Goodland, R. J. A., C. Watson, and G. Ledec. 1984. *Environmental Management in Tropical Agriculture*. Westview Press, Boulder, Colo.

Gorman, C. F. 1969. Haobinhian: Pebble-tool complex with early plant association in Southeast Asia. *Science* 163:671-673.

Gorman, C. F. 1971. Subsistence patterns in Southeast Asia during the Late Pleistocene and Early Recent period. *World Archeol.* 2:217-240.

Hanski, I., and J. Tiainen. 1988. Populations and communities in changing agro-ecosystems in Finland. *Ecol. Bull.* 39:159-168.

Harris, D. P. 1972. The origin of agriculture in the tropics. *Am. Sci.* 60:180-193.

Hobbs, R., and K. Wallace. 1991. Remnant vegetation on farms is a valuable resource. *West. Australia J. Agric.* 32:43-45.

Hoi-Sen, Y. 1978. Natural habitats and distribution of Malayan mammals. In *BIOTROP*, Special Publ. No. 8, pp. 3-9. SEAMEO-BIOTROP, Bogor, Indonesia.

Huxley, R. A. 1983. *Plant Research and Agroforestry*. International Council for Research in Agroforestry, Nairobi, Kenya.

Iltis, H. H. 1988. Serendipity in the exploration of biodiversity. In *Biodiversity*, ed. E.O. Wilson, pp. 98-105. National Academy Press. Washington, D.C.

Johnson, R. J., and M. M. Beck. 1988. Influences of shelterbelts on wildlife management and biology. *Agric. Ecosyst. Environ.* 22/23:301-335.

Kartasubrata, J., S. S. Tjitrosomo, and R. C. Umaly, eds. 1990. *Symposium on Agroforestry Systems and Technologies*. Bogor, Indonesia, 19-21 Sept. 1989. *BIOTROP*, Special Publ. No. 39. SEAMEO BIOTROP, Box 17, Bogor, Indonesia.

Lack, P. C. 1988. Nesting success of birds in trees and shrubs in farmland hedges. *Ecol. Bull.* 39:191-193.

Mader, H. J. 1988. Effects of increased spatial heterogeneity on the biocenosis in rural landscapes. *Ecol. Bull.* 39:169-179.

MacDicken, K. G., and N. T. Vergara. 1990. *Agroforestry Classification and Management*. John Wiley and Sons, New York.

Nair, P. K. R. 1989. *Agroforestry Systems in the Tropics*. Kluwer, Dordrecht, The Netherlands.

Pelzer, K. J. 1978. Swidden cultivation in Southeast Asia: historical, ecological and economic perspectives. In *Farmers in the Forest*, eds. P. Kundstadter, E. C. Chapman, and S. Sabhasri, pp. 271-285. University of Hawaii Press, Honolulu, Hawaii.

Pimentel, D., U. Stachow, D. A. Takacs, H. W. Brubaker, A. R. Dumas, J. J. Meany, J. A. S. O'Neal, D. E. Onsi, and D. B. Corzilius. 1992. Conserving biological diversity in agricultural/ forestry systems. *BioScience* 42:354-362.

Plotkin, M. J. 1988. The outlook for new agricultural and industrial products from the tropics. In *Biodiversity*, ed. E. O. Wilson, pp. 106-116. National Academy Press, Washington, D.C.

Saengwichien, S. 1974. *The Pre-history of Thailand.* Mahidol University Press, Thailand.

Simberloff, D., and J. Cox. 1987. Consequences and costs of conservation corridors. *Conserv. Biol.* 1:63-71.

Szaro, R. C. 1986. Guild management: an evaluation of avian guilds as a predictive tool. *Environ. Manage.* 10:681-688.

Urban, D. I., and T. M. Smith. 1989. Micro-habitat pattern and the structure of forest bird communities. *Am. Nat.* 133:811-829.

Wilson, R. J., and S. G. Diver. 1991. The role of birds in agroforestry systems. In *Proceedings of the Second Conference on Agroforestry in North America*, ed. E. Garrett. pp. 256-273. School of Natural Resources, Springfield, Mo.

Young, A. 1989. *Agroforestry for Soil Conservation.* CAB International, Wallingford, U.K.

Zimmermann, H. G. 1991. Biological control of mesquite, *Prosopis* spp. (Fabaceae) in South Africa. *Agric. Ecosyst. Environ.* 37: 175-186.

33

INTERIOR AND EXTERIOR LAND USE AS FACTORS IN LANDSCAPE BIODIVERSITY FROM CASE STUDIES

Kerry J. Dawson

School of Environmental Design
University of Georgia
Athens, Georgia
U.S.A.

The problems of depreciative human behavior on conservation lands have been repeatedly addressed in recreation literature (Miner 1988). Unfortunately, attempts to derive fully a theory of causation for social predation have not been as successful as theoretical attempts in biogeography (Samdahl 1982). This situation has left voids for both fields in that there are factors of biogeography that significantly affect causation and there are factors of causation that can be very significant for biogeography.

Social predation can be the most important factor in nature reserve biogeography if use is intense and management is weak. If we heed the words of Brian Wilkes (1977) that everyone is a consumptive user of public lands, then it is only the type and degree of use that needs to be assessed. The difficulty is that the current state of knowledge about the type and degree of impinging use is inadequate for the theoretical underpinnings that social impacts on biogeography must have, if we are to address fully the management needs of nature (Stankey 1976).

The purpose of this paper is to present five case studies as projects that address the problems of interior and exterior land-use pressures as negative factors in conservation of biodiversity. Then, by drawing the conclusions to which each case illustrates, a generalization can be made as the basis of a cogent theory supporting the necessity for providing solutions to the problems of human land-use impacts on biodiversity. The five case study sites are the Ocmulgee Wildlife Management Area (Georgia USA), Elkhorn Regional Park (California USA), Fuchs Hammock Natural Area (Florida USA), National Parks and Conservation Areas (Peru), and National Parks and Conservation Areas (Indonesia).

LAND USE AS A BARRIER TO BIODIVERSITY

In the late 1970s, planning for the Ocmulgee WMA in central Georgia was underway. The extensive river swamps of the big bend of the Ocmulgee River offered isolation to much of the river bottomland. An important issue in land acquisition and management was the adjacent upland areas that provide biodiversity and the ability to buffer boundary incursions by human activity. As part of the planning process, two transects were run parallel to the river to measure land-use barriers (Figure 33.1). The transects averaged one-half mile and two miles from the river channel. Two types of barriers were noted: structural (fences, roads) and physical (open fields, channelized side streams). Generally, as the barriers increased, studies by the State Department of Natural Resources indicated that biodiversity was reduced and human access increased (Dawson et al. 1978). For the author, this work highlighted the impact of interior land use on reserve biodiversity.

INTERIOR LAND USE AT ELKHORN REGIONAL PARK

In the early 1980s, a habitat restoration effort on California's Sacramento River furthered the research on interior land use as a factor in site biodiversity (Dawson et al. 1983). Elkhorn Regional Park is located seven miles north of the California State Capitol in the floodplain of the Sacramento River. The Sacramento River lies in a flood plain, 50 miles wide and 200 miles long, known as the Sacramento Valley. Draining the northern portion of California's great Central Valley, the Sacramento River begins near Mount Shasta, and flows for 300 miles to the Sacramento-San Joaquin Delta and from there into San Francisco Bay. It drains the watersheds of the Pit, McCloud, Yuba, Feather, and American rivers. Elkhorn Regional Park is located between the Feather and American forks.

Because Elkhorn is located inside the flood levees of the Sacramento, flooding occurs in the park for six months of the year. The native forest in the park is important to California because water development projects have eliminated 99% of the native riparian vegetation along California's rivers and streams. Elkhorn is important to avian species because 95% of the fauna present in the riparian habitat is avian. The riparian forest at Elkhorn represents a critical home for many riparian-dependent species including the yellow-billed cuckoo, western flycatcher, willow flycatcher, common yellowthroat, blue grosbeak, and lazuli bunting.

A native landscape restoration plan for Elkhorn Regional Park began as part of the overall master plan for the Park. Park development was funded by a grant from the 1976 California State Parks Bond Act. An additional grant was provided by the California Department of Boating and Waterways. Riparian restoration was planned as an integral part of all phases of development, including a picnic area, restrooms, caretakers' residences, riparian interpretive overlooks, and boardwalks. Approximately 90% of the site is set aside as a nature reserve.

The primary factor influencing the present make-up of vegetative patterns at Elkhorn Regional Park is past land-use practice. The original raising of the levees along the Sacramento River prevented the river from flowing over former floodplains and concentrating water at high levels. Additionally, snow melt from the Coast Range and Sierra Mountains was contained in a system of reservoirs and bypass floodways which kept water at high levels within the floodplain for months longer than normal. The effect has been to eliminate traditional riparian vegetation, such as Oregon ash (*Fraxinus latifolia*) from past ecological niches because natural floodplain hummocks are inundated

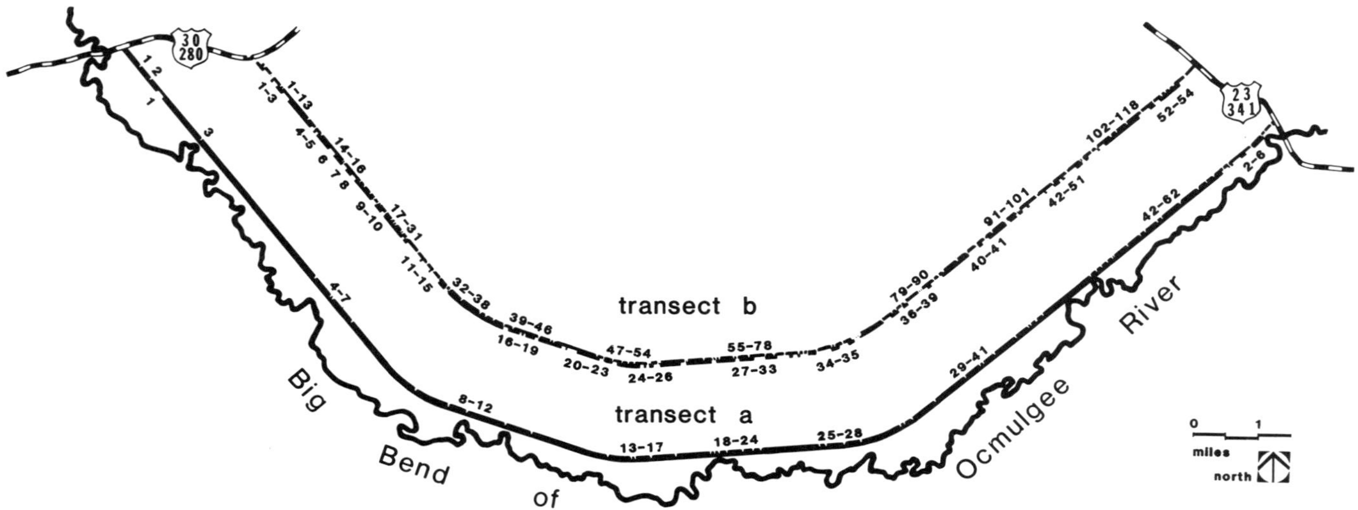

Figure 33.1. Transects register land use barriers (roads, fences, utility easements) on the Ocmulgee Wildlife Management Area (Georgia).

continuously. Although old-growth Oregon ash exists on some natural levees at Elkhorn, no reproduction is occurring. There are also areas where clear cutting has occurred in the past, for uses such as a gunnery range during the Spanish-American War; these now tend to be closed-canopy young cottonwoods. Diverse old-growth riparian forests at Elkhorn tend to be disturbed by past cutting for steamboat fuel and recent selective firewood cutting by local people. Other land-use disturbances include native American foraging, a past ferry, the construction of River Road, agricultural water transport over the levees, levee construction, a past electric trolley line, and introduced exotics.

Elkhorn Research Model

Four major questions formed the basis of the land-use research: (1) what is the avian composition of the riparian forest at Elkhorn Regional Park; (2) what avian species nest at Elkhorn and what are the requirements for nesting; (3) how are avian species using the canopy complex at Elkhorn; and (4) what is the canopy complex and how has past land use affected its present state. The hypothesis to be tested was that, although the forest appeared similar in all areas, if random samples were taken, specific portions of the forest would be more conducive to the habitat for riparian-dependent avian species, that these areas would represent what is truly "natural" at Elkhorn, and that a "naturalized" vegetative model could be formed to guide planting design for proposed restoration areas.

The specific methodology of the avian-use survey was to select statistically random transects through the site. Ten transects varying from 75 to 120 m in length were cut, all 50 m wide. They were walked at a constant speed (estimated 0.5-1 km/hr) at randomly selected hours and all species were recorded along with their activity. Worksheets were used to record data. The worksheet divided the forest into six basic habitats: upper, middle, and lower canopy; overhead; river bank; and over water. Abundance was categorized into high, moderate, and low utilization.

The Park harbors several avian feeding guilds, but most fall into six distinctive categories: piscivores, carnivores, omnivores, insectivores, herbivores-frugivores, and nectarivores. In terms of species composition, 123 species of birds have been recorded on a regular basis from 1965 to 1985. Of these, 102 were found during the census period and 25 additional species were listed as incidental. The most common type of bird at Elkhorn is the insectivore. Of the 123 regularly occurring species, 42 are insectivorous. Their numbers fluctuate as insect numbers fluctuate. During migration insectivores are the dominant bird group, but by winter most have moved out of the area when seed-eating herbivores predominate.

Because species diversity varies from season to season, during winter when seed-eating species dominate, the most abundant genera include *Pipilo*, *Carpodacus*, *Carduelis*, *Zonotrichia*, and *Melospiza*. Insectivores and nectarivores are poorly represented. Woodpeckers remain common in the parkway because of the increase of insects sheltered in the canopy. The parkway is on the eastern boundary of the Pacific flyway, and as a result, several species of waterbirds were found flying overhead. These species are not natural components of the forest but their passage through the area shows the importance of the Sacramento River (they are also by far the most numerous bird group in the

Central Valley during this period). Large flocks of blackbirds, water pipits (*Anthus spinoletta*) and cedar waxwings (*Bombycilla cedrorum*) were also seen moving in the area.

In March the waterfowl give way to the summer visitors. During this period the forest experiences a vegetational bloom when insect numbers dramatically increase. Migrating warblers, vireos, flycatchers, tanagers, and swallows take advantage of the bloom. Several summer visitors arrive and start to set up territories by April 1. These include Swainson's hawk (*Buteo swainsoni*), western kingbird (*Tyrannus verticalis*), ash-throated flycatcher (*Myiarchus cinerascens*), and black-headed grosbeak (*Pheucticus melanocephalus*).

From late May through July, bird activity and species diversity decrease, and nesting species are rearing young. Of the 123 regularly occurring species at Elkhorn, only 38 have been confirmed as nesting there. Another 6 species may nest but have not been confirmed. Of the 38 nesting species, 15 are cavity nesters. Woodpeckers provide nesting holes not only for themselves but for other species as well. Old-growth dead cottonwoods are the favored locations for cavities (Balda 1975) but are limited in numbers.

Histograms of vegetative structure utilization were produced; these compared bird preferences along transects. For selected cavity-nesting species, comparisons in abundance are made between east-side and west-side transects (Figure 33.2). Horizontal

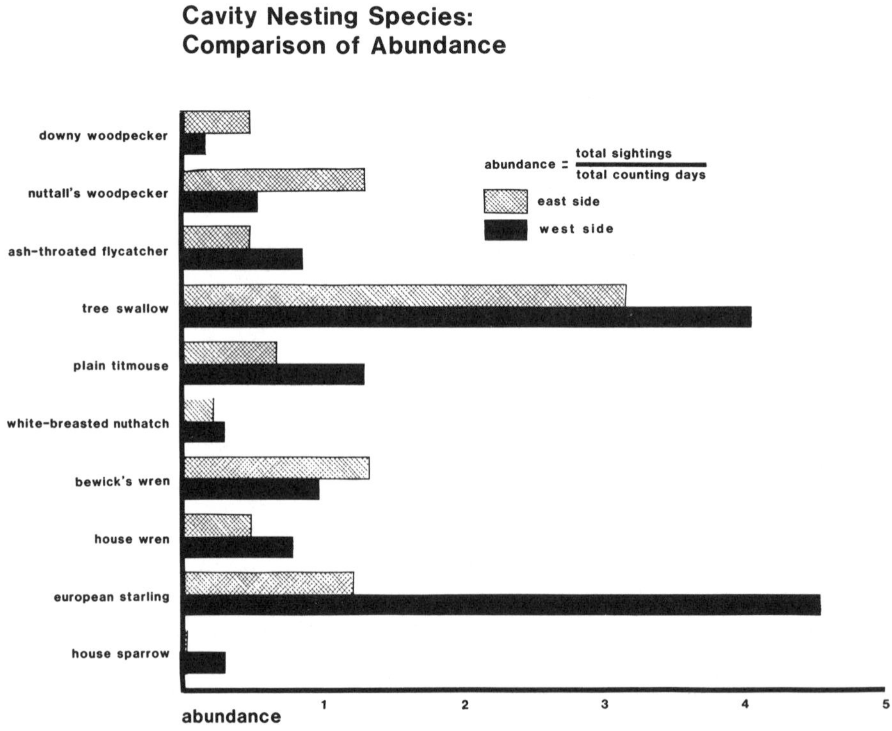

Figure 33.2. Abundances of birds in sample transects at Elkhorn Regional Park.

distribution of vegetative structure was inventoried within random 25-m^2 plots aligned within the 50-m avian transects (Figure 33.3). Vertical distribution of vegetative structure was inventoried along the center line of each transect. Both horizontal and vertical distributions were recorded for composition, diversity, and distribution.

The riparian woodlands of Elkhorn consist of two distinct areas: on the west side of the parking lot, a mature but disturbed area; on the east side of the parking lot, a young stand of dense woodlands. The west side is a more established woodland with a distinctive canopy cover. The upper canopy ranges in height from 22 to 25 m averaging 22.8 m. Major species composition include *Populus fremontii* (Fremont cottonwood), *Quercus lobata* (valley oak), *Platanus racemosa* (California sycamore), and *Salix gooddingii* (black willow). The species composition is stratified in layers from the riverbank to the main levee. Generally, *Populus fremontii* is found scattered throughout the forest with the largest individuals confined to the sloughs and river edge. Bordering the riverbank is a distinct band of *Platanus racemosa*, corresponding to the higher elevation of the natural levee adjacent to the river's edge.

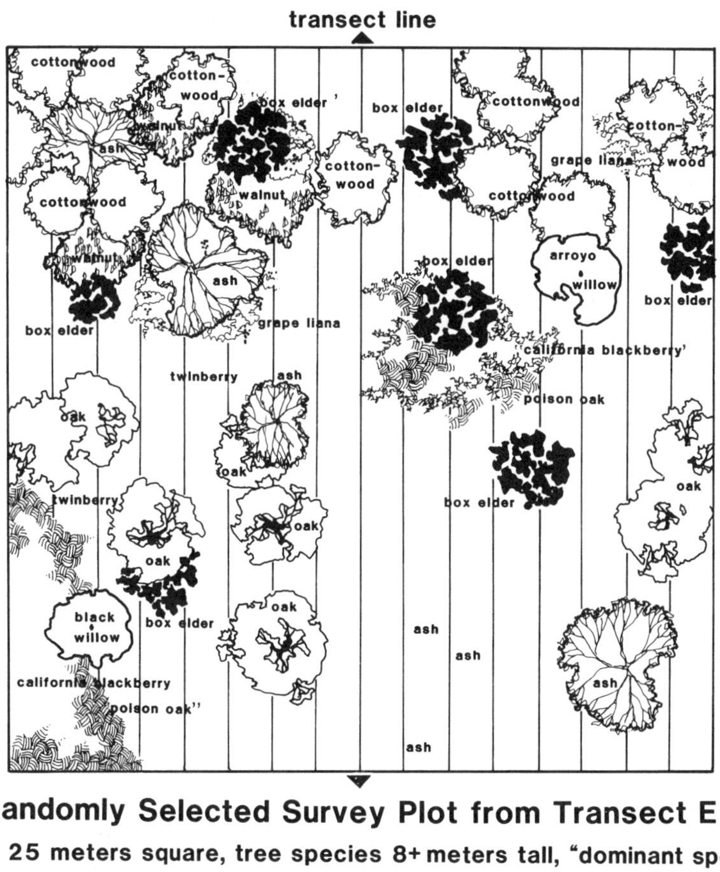

Figure 33.3. Vegetative plots at Elkhorn Regional Park.

With an elevation of over 30 feet, the natural levee of the west side is flooded less than the surrounding floodplain; *P. racemosa* has adapted accordingly. *Q. lobata* also grows only on the higher elevations, and the outer fringes of the forest. *S. gooddingii* is restricted to sloughs and to the river's edge where it often grows to 25 m. The upper canopy cover averages 42%, but it can be very dense, to 100% cover. The number of dead trees on the west side is high (with densities over 5% per acre on the west). These tend to be older trees that have succumbed to root or crown rot and, as will be noted later, are of great importance to the native wildlife.

The middle canopy of the west side is quite diverse and at times 100% cover is common. It is generally 4 to 8.5 m in height, averaging 6.3 m. Here, saplings of the dominant upper canopy species are found, but the middle canopy has its own distinct dominant species. The most common include *Fraxinus latifolius* (Oregon ash), *Juglans hindsii* (California black walnut), *Sambucus mexicana* (elderberry), *Acer negundo* (box elder), and several *Salix* spp. The introduced *Robinia pseudoacacia* (black locust) is restricted to the west side, and is only common in one dense stand outside the levee.

The lower canopy of the west side is generally an impenetrable mass of vegetation 2 to 3 m tall. Characteristically, it includes lianas that grow from the ground into the upper and middle canopy where they form a dense cover over much of the western forest. Many of the species dominant as lower canopy species are also the major lianas of the forest, i.e., *Rhus diversiloba* (poison oak), *Vitis californica* (California grape), and *Rubus vitifolius* (Himalayan blackberry). Other dominant species include *Rubus ursinus* (California blackberry) and *Rosa californica* (California wild rose). The intertwining and trellising of these vines form a curtain, giving the forest a subtropical appearance and the common name of "Sacramento gallery forest." A distinctive lower and middle canopy shrub is *Cephalanthus occidentalis* (Button willow). It occurs along the slough and river's edge, and its fragrant flowers whiten these sloughs during the summer.

In contrast to the west side, a large section of the east side is not a fully mature forest. Sections that were clearcut during the 1950s have only recently recovered. Although not yet fully mature, the trees reach a height of 22 m. The main difference in the upper canopy in most areas is that it is dense, averaging 60% cover and allowing very little light to reach the forest floor. Edges of the east side are more similar to the west side with dense liana growth and similar species composition. The east side is virtually inundated during winter and as a result, no *P. racemosa* are present. *Q. lobata* grows in small stands on levees scattered throughout the east side. The incidence of dead mature trees is less on the east side, where there is a more even canopy continuum. Large dead trees exist near the slough on the east side, providing important nesting holes for many birds.

The middle canopy of the east side contains basically the same species as the west but is somewhat less diverse. The canopy height is higher, 5.5 to 9.0 m, but about the same average density (41 %). Again, *Salix* spp. are restricted to levees along the riverbank and sloughs. *C. occidentalis* is not as widespread on the east side.

The most apparent difference between east and west sides is in the amount of ground cover. The east side is not as impenetrable as the west and open patches of ground are common. Only where light penetrates to the forest floor does dense growth of *Rhus* and *Rubus* occur. Along the sloughs and 25 m out on either side, the ground cover becomes quite thick and 100% cover is common. Surprisingly, the height of the ground cover is about the same, averaging 2.5 m. Fewer patches of *Rubus vitilius* and fewer introduced species (and their individuals) are present on the east side.

Elkhorn Research Results

The goal at Elkhorn was to inventory the forests to determine the abundance of the critically dependent, native avian species, as well as undesirable species (e.g., brown-headed cowbirds and starlings). Major findings of the project generally indicate that vertical distribution of vegetation on the site plays a role equal to horizontal distribution. Areas of the site with diverse, sparse, old-growth timber and dense understories are least desirable for dependent species because the parasitic brown-headed cowbird tends to have more cover. Areas with dense canopies and light understory vegetation seem most desirable. Because almost one-half of the dependent species are insectivores and another one-quarter belongs to other guilds that feed insects to their young, insect-harboring habitats must play an important part in restoration. Figure 33.3 is the random plot most indicative of the desired habitat both for vegetative cover and insect cover; this plot became the model for restoration plans.

Of particular interest at Elkhorn was the obvious influence of past interior land use on present vegetative structure. Elkhorn communities are not truly natural because of intensive land-use influence of present flood control and past clearing. Any attempt to restore the forest to a pre-settlement type would fail, because continued land-use pressure would quickly work to return a naturalized state. These interior land-use pressures are further complicated by the biodiversity influences of exterior land use practices, such as boat traffic on the river and agricultural and urban development on the uplands.

EXTERIOR LAND-USE INFLUENCES AT FUCHS HAMMOCK

To examine further the major importance of exterior land-use pressures on biodiversity identified in the Elkhorn study, research was conducted at Fuchs Hammock Natural Area in Florida.

It is a common misconception that most nature reserves are truly natural. In fact, most have been invaded by some exotic species and most have been altered by land-use practices. In 1970, The Nature Conservancy arranged for the purchase of a natural area in South Florida to conserve a representative sample of the West Indian hammock communities still sprinkled across western Dade County near Everglades National Park. Known as Fuchs Hammock, the purchase area covered only 10.1 ha, making it a relatively small reserve (Fix 1969). As is the desirable practice in most Nature Conservancy purchases, a cooperative public agency responsible for future management existed and gradual transfer of the property to the Metropolitan Miami-Dade County Parks and Recreation Department occurred. Landscape architects on the planning and design staff developed conservation and management plans to establish one of Miami-Dade's first nature reserves, and administrators were satisfied that the cause of native landscape conservation had been served. Unfortunately, the fight for the conservation of Fuchs Hammock had only begun.

When the value of the property was first recognized in the late 1940s, hammocks were still common in private ownership in South Florida. Agriculture was well established on drained marl soils previously occupied by wetlands, but much of the original pinelands and hammocks had been bypassed. That was ended by the post-World War II housing boom that focused on the bypassed woodlands.

As with other hammocks associated with Florida's everglades, Fuchs Hammock is a mixed hardwood sub-tropical forest. One-third of the reserve is disturbed scrubland from previous limestone mining operations. The remainder has an oak overstory to the south and west and a mixed pine overstory to the northeast. The canopy is a closed dome at roughly 20 m with shading at 75-90%. Hammocks are considered to be the climax community of South Florida for uplands without fire.

In the early 1900s, the pinelands and glades surrounding Fuchs Hammock covered several thousand hectares but orchard development on the north and south reduced the contiguous area to 40.5 ha by 1950. A housing project on the east was well established by 1960, and to the west, the most recent addition is an estate, further separating the main hammock from the western natural area.

From 1968 through 1988, Fuchs Hammock was studied for impacts on its most commercial resources: epiphytic orchids, ferns, and bromeliads (Dawson 1991). Surrounding land use was monitored and sample plots established to determine the relationship between predation on these commercial resources and reserve size, boundaries, adjacent land use, and facility design. Complete inventories of all vegetation were monitored early in the period, with sample plots randomly selected on host branches that were 3.7 m long and a minimum 15.2 cm diameter. Host trees were randomly selected from clusters related to eight land-use categories (A through H in Figure 33.4). Recordings of land-use changes, police records, facility changes, and impacts on sample plot populations occurred periodically throughout the monitoring period.

The intent of the survey was to determine the relationship between adjacent land use and the intensity of human predation of the epiphytic plants. Generally, after an initial increase in plants because of suspected stocking, a serious decline in indicator plants occurred as the reserve was publicized in the press during the acquisition years from 1968 to 1978. The trend ended in the late 1970s when the reserve was fenced and police patrols became more frequent.

Tree branches overhanging the estate boundary faced heavy predation during the acquisition year but by 1988 had fully recovered from predation (Figure 33.5). Stocking was suspected. Trees bordering an adjacent natural area fared very well during the survey period as did the trees adjacent to the estate. This was attributed to buffering and remoteness. Trees adjacent to neighboring agriculture had lessened impact as the edge thickened. The central trail showed the most improvement after the reserve was fenced, but was still a popular destination for poachers. Trees adjacent to a subdivision also improved greatly after fencing, but also remained popular to poachers. Branches overhanging neighboring agricultural fields showed a steady decline in numbers of epiphytes because of easy access for theft. Over 60% of the surveyed plants were stolen from branches overhanging the subdivision.

The purchase of Fuchs Hammock brought the area to the attention of the general public for the first time. Human predation began to increase with establishment of the Fuchs Hammock County Nature Reserve, subsequent opening of the area to guided

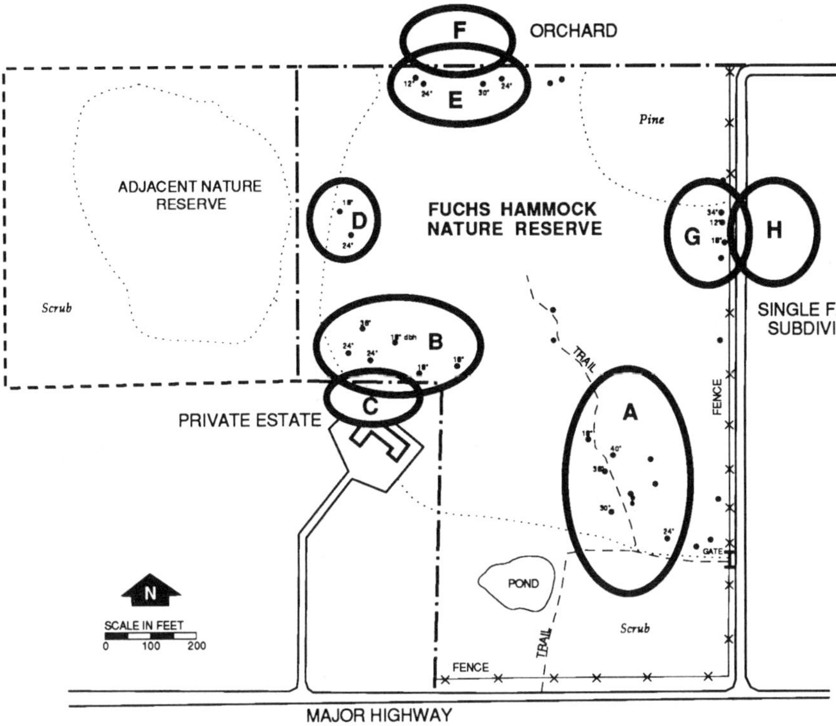

Figure 33.4. Land-use map and transect plots (A-H) at Fuchs Hammock.

tours, and feature articles in the print media. Theft of flora and fauna was both swift and relentless. Letters of complaint from The Nature Conservancy to the county were well received, and ordinances, sign placement, and reduction of access ensued. Short of stationing a ranger on the site, the county gave increasing attention to the property.

Although education reaches many people and ordinances and signs reach additional individuals, the struggle to manage an area such as Fuchs Hammock is difficult because of accessibility. For vulnerable species of a small reserve to survive in the human-occupied environment, buffers or proximity to other landscape reserves are vitally important. Large conservation areas are subject to the impacts of human activities both within and peripheral to their borders.

NATIONAL PARKS AND CONSERVATION AREAS IN PERU

Experience at Fuchs Hammock indicates that some degree of isolation might be necessary for the successful preservation of a small reserve's biodiversity in the face of human impacts. However, large-scale land-planning in Peru and Indonesian Irian Jaya (see below) points to the need for some measure of both accommodation and control of expected human access and activity, exterior as well as interior. In 1984, a large-scale national

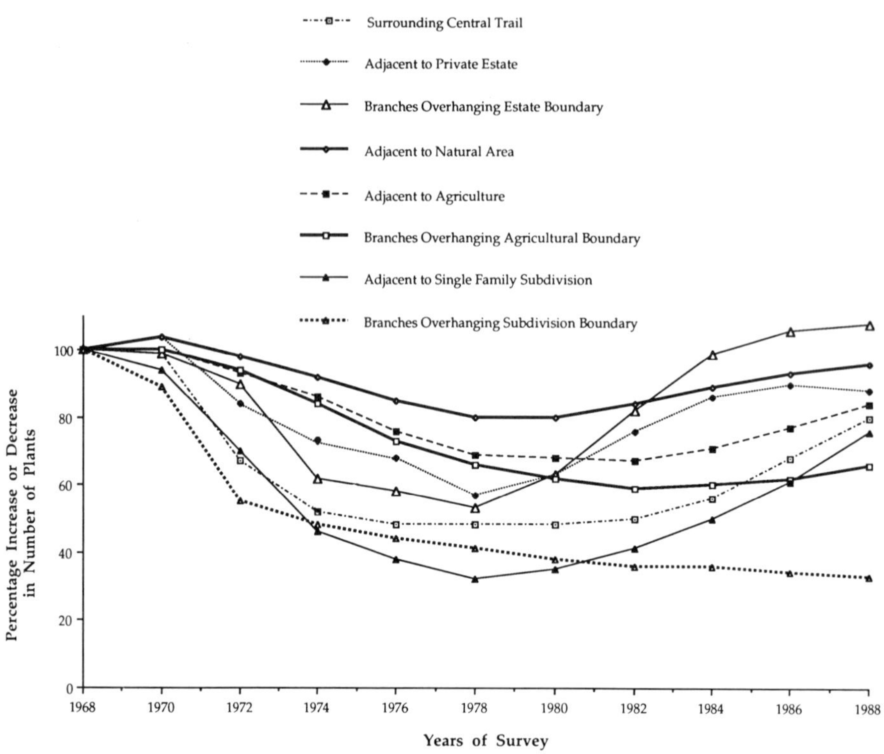

Figure 33.5. Percentage increases or decreases in indicator plants relative to each of eight land use categories, Fuchs Hammock.

park planning effort in Peru dealt adequately only with interior land-use impacts (Figure 33.6). Although it was readily recognized that exterior pressures might be of even greater importance, existing claims of land tenure made consideration difficult to pursue (Dawson 1987).

SUSTAINED BIODIVERSITY IN IRIAN JAYA

In 1991, in a project by the government of Indonesia on sustainable development surrounding conservation lands, both external and internal impacts of land use were assessed and integrated in final conservation plans. The Sustainable Agricultural Development Project in Irian Jaya has the general objective of strengthening the role of small land-holder agriculture in Irian Jaya through technical assistance, conservation, and environmental monitoring. It also has the specific objective to demonstrate the capability of sustainability through appropriate tree crop, field crop, social forestry, livestock development, and expansion of the environmental programs of the Institute of Ecology at Cendrawasi University in Manokwari (Dawson 1992).

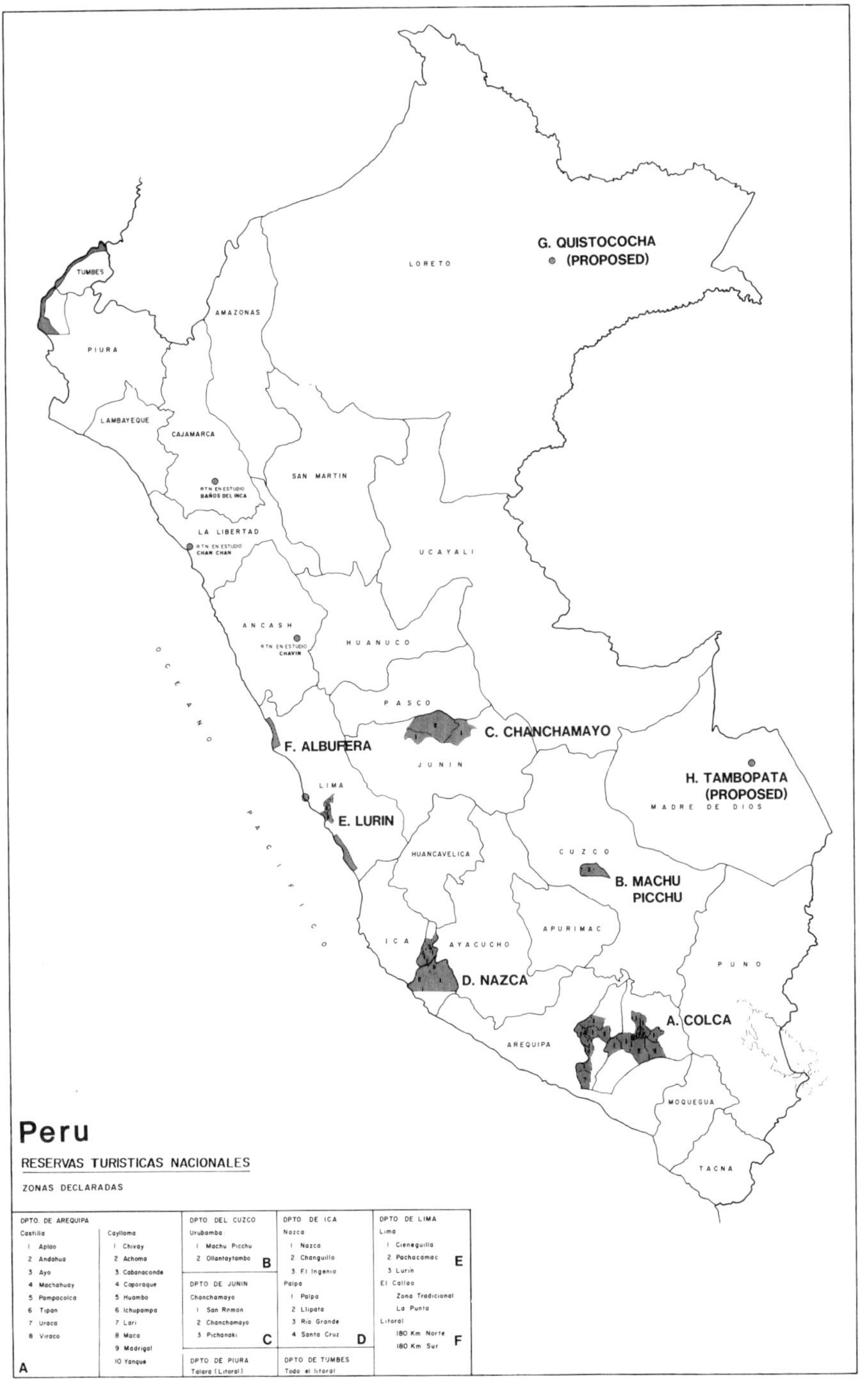

Figure 33.6. Six Peruvian national parks with master plans for eco-tourism during 1984.

Project Components

The project has three principal components: small land-holder development, environmental management through the strengthening of ecological study centers, and environmental management through conservation. Twelve subprojects are in the small land-holder development component, and these are scattered throughout the province (Figure 33.7). Three of the small land-holder subprojects are simple animal health clinics at existing government facilities.

Three veterinary labs are proposed that would be new facilities with staff housing. The remaining six subprojects are agricultural extension projects covering three kecamatans (townships). Extension projects in certain kecamatans will include small land-holder development in field crops. Tree crops, livestock, and social forestry will also be the focus, along with small extension facility improvements, such as the addition of a seed drying platform to the existing seed introduction center in Wamena.

Generally, the small land-holder component of the project is seen as a contribution to environmental quality rather than a negative impact. The purpose of the veterinary labs and animal clinics are to control disease and improve health. The labs will probably be built at existing government complexes, so no removal of native landscape or

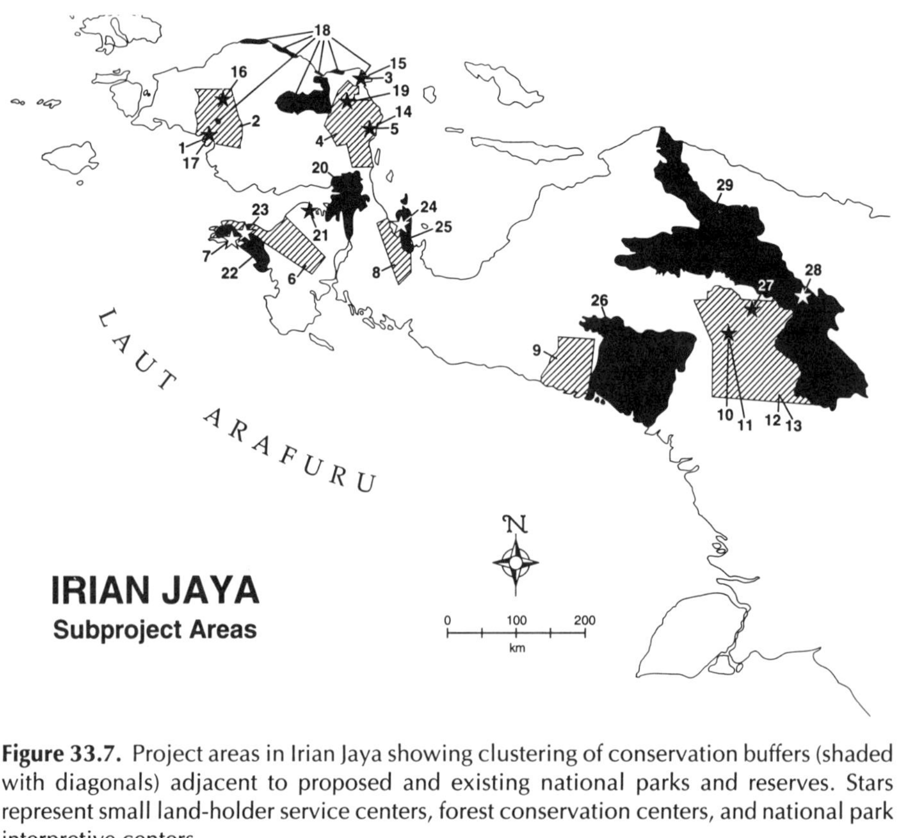

Figure 33.7. Project areas in Irian Jaya showing clustering of conservation buffers (shaded with diagonals) adjacent to proposed and existing national parks and reserves. Stars represent small land-holder service centers, forest conservation centers, and national park interpretive centers.

Class A agricultural soils will occur. Impacts are considered minimal because the technology to be employed will reduce erosion, improve soil fertility, and raise health standards. Because these activities will be oriented toward the small land-holder, they will be small in scale.

Environmental Management

The Environmental Management component of the project focuses on the Manokwari campus of Cendrawasi University and three proposed and one existing farming system field station. It is intended to strengthen environmental management for all the provinces. A new Institute of Ecology lab will bring state-of-the-art environmental monitoring to Irian Jaya. The existing farming station at Ransiki now covers 110 ha and will be expanded to focus on environmental monitoring, research, and field trials to improve the sustainability of agriculture in the coastal strand areas. The new Highlands Farming Station will most likely be established on 7 ha of land now owned by the Indonesian Government a few km northwest of Wamena. The foothill and valley farming station is planned for Ayamaru because of critical environmental conditions and easy access in the bird's head region to Manokwari. The swamps and Riverine field station at Teminabuan are also located for critical environmental conditions. Approximately 10 ha are available near Ayamaru in several locations, but the challenge at Teminabuan will be to find 10 ha of Sago Forest that a local clan would be willing to share. All three proposed farming stations and the existing station at Ransiki will require new administrative headquarters and housing units. The environmental impacts from the expansion and establishment of these stations are considered low because their primary purpose is to monitor environmental management in the respective farming systems and to investigate improved methods in sustainability for all of Irian Jaya.

Conservation and Land-use Buffering

The Environmental Management component dealing with conservation has evolved through several factors. First, Irian Jaya is heavily forested and much of the native population depends on the forest for their swidden agricultural practices. Second, the physiography of Irian Jaya does not lend itself to large-scale agricultural development, so a forest-dependent economy should be the future strategy for sustainability. Third, Irian Jaya has one of the most impressive tropical rain forests in the world, and with the steady growth of eco-tourism and lumber prices, forest conservation will most likely be the wisest economic use. Fourth, there is a need to integrate agricultural planning with conservation areas, and the involvement of local, largely agrarian communities in conservation (Figure 33.8).

The conservation component involves the development of nature reserve management plans for the Bird's Head Conservation Cluster, Greater Bintuni Bay, Fakfak Mountain, Wasior Mountain, Lorentz National Park, and the Greater Mamberamo River Region. These areas are adjacent to the proposed agricultural development areas and will be directly affected by their land-use influences (Figure 33.7). The component also includes the establishment of interpretive centers with rest houses and staff housing at

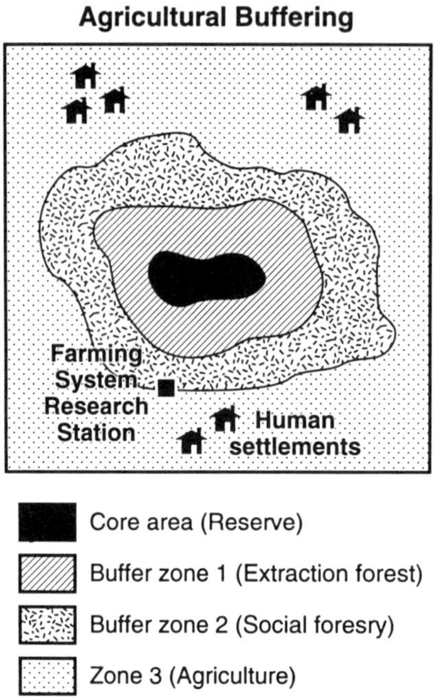

Figure 33.8. Reserve buffering through a four-zone concept: central reserve, extraction forest, social forest, and agricultural development.

the Arfak Reserve, Fakfak Reserve, and the Pass Valley Protection Forest on the summit of the Wamena-Jayapura Road. Lastly, the conservation component includes forest conservation centers for Mangroves at Babo, for lowland forest on the Mamberamo River, and for marine conservation along the Cendrawasi-protected area coast. These forest conservation centers will focus on environmental monitoring and research into the ecology of the native forests of Irian Jaya. These projects will not remove primary rain forest and should be located near existing roads although a village access road is possible. Again, impacts of the conservation component are thought to be minimal because the primary purpose of this activity is to preserve native culture and biodiversity. These conservation centers and previously discussed extension centers will have a primary responsibility to study and manage the land-use system to monitor encroachment on biodiversity.

CONCLUSIONS

Size, configuration, and habitat conditions are recognized generally as the primary design determinants of landscape reserves. Considering the pressure that human society places on isolated nature reserves, it seems imperative that reserves be large enough to contain representative habitats and internal sources of colonization. They must also be buffered by land-use zoning, connected by landscape corridors or clustered in an area where a high percentage of the surrounding land is zoned for conservation and can act as a medium of exchange between reserves.

Small, isolated reserves with separate populations might be necessary because of economics or to avoid inbreeding, wildfires, and disease. Even so, such areas must still be buffered and protected in some way. The most appropriate mean is the police power of local government to zone land surrounding the reserve for conservation and sustainability. Zoning for levels of activity, from management for renewable natural resources to light agriculture, has to be considered on a case-by-case basis. However, tough restrictions on permanent intensive human occupation of land surrounding reserves have to be enforced.

Some people have argued that humans should be viewed as an integral part of nature; therefore, the recreational use of and proximity to nature reserves are to be expected. This is an interesting philosophy and the study of humans as integral with nature is surely a needed area of research. But, we have to realize that vandalism (especially against species not well represented in the justice system), predation, and developments are increasing steadily and, for the time being, the more land buffering human activity from nature reserves, the better in the long run (Samdahl et al. 1992). Because most reserve boundaries are determined administratively rather then ecologically, modeling of perimeter and interior social influences must become as important as modeling perimeter ecological influences (Schonewald-Cox 1988). This was evident at both the Ocmulgee Wildlife Management Area and at Elkhorn Regional Park.

The study of Fuchs Hammock showed that predation on valued resources is much more intense than on other resources (Pittman 1989). Therefore, the intensity of management after the acquisition of a nature reserve with valued resources must match the intensity of aberrant behavior that is evident during its establishment (Sutter 1986). In field work at Fuchs Hammock Nature Reserve in Florida, there were 17 species of orchids identified throughout a twenty year survey period. Of those, six have been extirpated from the reserve due to human activity. The unavoidable conclusion is that the primary extinction factor within Fuchs Hammock has been human predation, not ecological factors.

In Peru, the primary goal of conservation was achieved but at the expense of biodiversity in impacted boundary regions. In Irian Jaya, conservation planning was more successful because an additional goal has been to accommodate land-use buffering. This strategy has been accepted by the Province of Irian Jaya, the Indonesian Government, and the Asian Development Bank. This is important in Irian Jaya because there is still time to influence land-use decisions because development has not yet arrived at the boundary of most conservation areas. If land-use buffering occurs in all areas that still have undisturbed potential, the goals of conserving biodiversity will be enhanced.

LITERATURE CITED

Balda, R. P. 1975. Vegetation Structure and Breeding Bird Diversity. In *Proceedings of the Symposium on Management of Forest and Range Habitats for Nongame Birds.* USDA Forest Service, Gen. Tech. Rep. WO-1:59-80.

Dawson, K. 1978. Ocmulgee Wildlife Management Area Expansion Studies. Heritage Trust Program Report, State of Georgia Archives, Atlanta, Ga.

Dawson, K. 1983. Habitat reclamation in a riparian landscape. In *The Landscape: Critical Issues and Resources*, pp. 1-10. Conference of Educators in Landscape Architecture, Utah State University, Logan, Utah.

Dawson, K. J. 1987. Wilderness Study Comparisons in Two Countries. In *Designation and Management of Park and Wilderness Reserves*, eds. E. E. Krumpe and P. D. Weingart, pp. 67-79. University of Idaho, Moscow, Idaho.

Dawson, K. J. 1991. Subtropical Natural Areas and Social Pressure. In *Healthy Environments*, pp. 109-116. Environmental Design Research Association, Oaxtepec, Morelos, Mexico.

Dawson, K. 1992. Irian Jaya Sustainable Development Project. Vols. 1-4. Government of Indonesia, Ministry of Agriculture, Jakarta, Indonesia.

Fix, J. 1969. Unique hammock preserved. *Fla. Wildl.* 11:20-21.

Miner, C. 1988. Understanding vandalism helps prevent it. USDA, *Forest Research West* 11:71Q

Pittman, C. 1989. Foiling raiders or lost orchids. Santa Rosa, Calif. *Press Democrat*, p. D2, July 22.

Samdahl, D., H. Christensen, and R. Clark. 1982. Prevention and control of depreciative behavior in recreation areas: managerial concerns and research needs. *Forest and River Recreation.* Misc. Publ. 18-1982:52-54. University of Minnesota Agricultural Experiment Station, St. Paul, Minn.

Schonewald-Cox, C. 1988. Boundaries in the protection of nature reserves. *BioScience* 38(7):470,486.

Stankey, G. 1976. Crowding in Parks and Wilderness. *Design & Environ.* :4.

Sutter, R. D. 1986. Monitoring rare plant species and natural areas: ensuring the protection of our investment. *Natural Areas J.* 6:3-5.

Wilkes, E. 1977. The myth of the non-consumptive user. *Can. Field Nat.* 91:343-349.

34

CHANGES IN BIODIVERSITY OF THE WADDEN SEA AND ADJACENT HABITATS: THE NETHERLANDS, GERMANY, AND DENMARK

W. J. Wolff

DLO—Institute for Forestry and Nature Research
and Department of Nature Conservation
Agricultural University
Wageningen
The Netherlands

Here I describe the changes in the biodiversity of the Wadden Sea when humans were able to interfere with natural processes during the last few thousands years. Because few data are available for periods longer ago than about 100 years, my discussion necessarily concentrates on fairly recent developments. It also emphasizes anthropogenic changes over natural changes and ends with measures for the restoration of biodiversity.

The Wadden Sea is a shallow coastal sea (8,000 km^2) or rather an extended series of estuaries along the coasts of The Netherlands, Germany, and Denmark (Figure 34.1). Its tidal basins are sheltered from the North Sea by a series of sandy barrier islands. In the Wadden Sea about 50% of the area consists of bare tidal flats and about 5% of salt marshes. The Wadden Sea in its present position is only about 5,000 years old (Wolff 1983). Hence the time-scale of my discussion relates to this age, and does not take into account the slow processes of plate tectonics and biological evolution.

For a discussion of the ecological development of the Wadden Sea, one must distinguish between natural and anthropogenic developments. Sometimes, however, the boundary is obscure because some changes were initiated by humans, but subsequently have been subject to natural processes. An example is the introduction of the Japanese alga, *Sargassum muticum*. After its introduction along the French coasts of the English Channel from anthropogenic transport of imported oysters, the alga dispersed naturally along the shores of the Channel and the North Sea. For example, some years ago, I observed large amounts of *Sargassum* drifting in the German Osterems estuary between the islands of Borkum and Juist. The quantities were so large that no known growing place of the species along the German, Dutch, or Belgian coasts could have provided them. This observation and our knowledge of the residual currents in the North Sea and the ecology of *Sargassum* imply that this alga in the Osterems drifted from the shores of the Channel all the way to the German Wadden Sea, demonstrating how effective natural

Figure 34.1. The Wadden Sea. The belt of marshes, tidal flats and shallow water in between the barrier island and the coast of the mainland is called Wadden Sea.

processes can be. In the meantime *Sargassum muticum* has established itself permanently along the shores of the Wadden Sea.

The major natural processes to be considered for this review are (1) coastal development under the influence of geomorphological and hydrological processes and in relation to sea level rise, and (2) biogeographical processes leading to the establishment and extinction of species.

The coastal development determines whether a Wadden Sea can exist and, if so, how it looks. The biogeographical processes determine which species inhabit this Wadden Sea. Superimposed on these two major processes, sedimentation, erosion, soil development, plant succession, predator-prey interactions, and many other processes play their part on a local scale. All these secondary processes cause only modifications of the outcome of the two former major processes. The major anthropogenic processes are listed in Table 34.1.

In the following sections I address the natural and man-induced processes in the geomorphological, physical, chemical, and biotic changes in the Wadden Sea.

GEOMORPHOLOGICAL CHANGES

The natural geomorphological development of the Wadden Sea is the result of the dynamic balance between two processes. Sediment transport by rivers and tidal currents tend to fill in the Wadden Sea, whereas rise of sea level because of tectonic subsidence and eustatic sea level rise tends to enlarge the Wadden Sea (Wolff 1983).

In a completely natural situation, a Wadden Sea landscape shows the following sequence of landscape types from the North Sea to the higher inland areas (Figure 34.2):

- beaches and dunes of a barrier island
- salt marshes of a barrier island
- tidal flats and channels
- salt marshes of the mainland coast
- brackish marshes
- freshwater marshes and swamps
- peat moors
- high-lying soils with forests and other dry vegetation types.

Table 34.1. Anthropogenic processes causing changes in the Wadden Sea.

1. Geomorphological changes
    - **(a)** Dike-building across estuaries and tidal channels
    - (b) Reclamation, embankment of tidal flats and marshes
    - (c) Construction of training walls along tidal channels
    - (d) Sand extraction for construction purposes
    - (e) Soil subsidence due to gas extraction
2. Physical changes
    - **(a)** Change of tidal currents and amplitude due to engineering projects
    - (b) Change of temperature due to discharge of cooling water
    - (c) Sea level rise due to climate change
    - (d) Changes of temperature, wind, and precipitation due to climate change
3. Chemical changes
    - **(a)** Changes in salinity due to engineering projects
    - (b) Changes in pH and calcium availability due to reclamation of inland peat moors
    - (c) Pollution
    - (d) Eutrophication
4. Biotic changes
    - **(a)** Decrease of population size and extinction due to hunting
    - (b) Decrease of population size and extinction due to fisheries
    - (c) Introduction of species
    - (d) Increase of population size due to cultivation

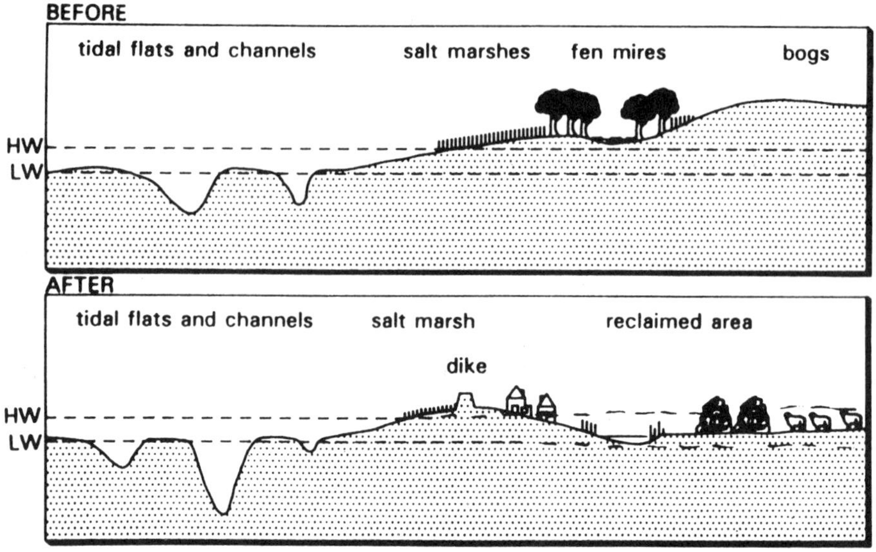

Figure 34.2. Cross-section through the Wadden Sea landscape before human influences started and at present.

In the past, with rising sea level, these belts of different types of landscapes moved inland (Wolff 1983) and would continue to do so if humans had not interfered. At the same time, sediments derived mainly from the barrier islands and the shallow sea floor in front of them were imported into this moving Wadden Sea. The outcome of these two processes is dependent on the amount of available sediment, the tidal range, and the slope of the coast. In some cases, sea level rise was not quick enough to keep pace with sedimentation that completely filled the tidal basin and closed the coast line. This has been the case, for example, south of Den Helder in The Netherlands, where the geological record still shows a fossil Wadden Sea a few meters below soil level (Pons and Wiggers 1959-1960). Elsewhere, for example in Nordfriesland in Germany, the tidal basins enlarged. The coastal processes have a dynamic balance; sometimes sea level rise predominates, whereas in another period sedimentation does.

Until about 1,000 years ago, humans hardly interfered with the geomorphological development of the Wadden Sea. Although the population of the coastal area was fairly dense, people lived on naturally elevated areas and on artificial mounds and did not seriously interfere with the tidal movements. At that time, however, the invention of dike-building introduced a major change (Wolff 1992). In the beginning, only rather small areas in the highest parts of the coastal region were surrounded by low dikes. Soon, however, these areas were connected by other dikes, and at the end of the 12th century most of the vegetated landscape was separated from the tidal Wadden Sea by a contiguous system of dikes. Outside these dikes, large salt marshes still occurred. Gradually, however, these and newly accreting salt marshes were reclaimed and turned into farmland. Thus, the connection between the Wadden Sea and the adjacent belt of brackish and fresh wetlands was broken. At the same time, transports of matter with the tidal and river currents and migrations of animals became more difficult.

This separation of the Wadden Sea from its coastal marshes and peat moors might well have been the most important intervention in its history. However, because it was long ago, one is inclined to neglect this major change of the Wadden Sea ecosystem and to consider the present-day Wadden Sea as being in a normal condition.

Human use of the coastal marshes for agriculture with or without embankment leads to lowering of the level of the land. Drainage compacts the soil because of water loss and, in peat areas, because of wastage of soil from oxidation. Thus, within a few centuries, the vast, high-lying peat moors along the Wadden Sea disappeared and turned into low-lying agricultural landscapes (Edelman 1974; Hoffmann 1980). Such landscapes became very vulnerable to storm surges. With sufficient subsidence the land could be flooded after one of the rather frequent dike bursts (Wolff 1992). Medieval technology was not sufficiently developed to reclaim the lost territory if the newly created flood basin was too large. The strong tidal currents prevented closure of the gap and the land had to be given up. Such events occurred all along the Wadden Sea; the Halligen area in Germany (Figure 34.1) is probably the clearest example. I emphasize that these major losses of land were basically the result of medieval peat reclamation and agriculture, aided only in some cases by digging of peat for fuel or salt production (Borger 1975).

When a tidal basin was created in the described way, deposition of sediments began immediately, and gradually such basins were filled again (Wolff 1983, 1992). This sedimentation process was accompanied by embankment of the newly accreted saltmarshes. The Dollard estuary at the border of Germany and The Netherlands (Figure 34.3) is a good example.

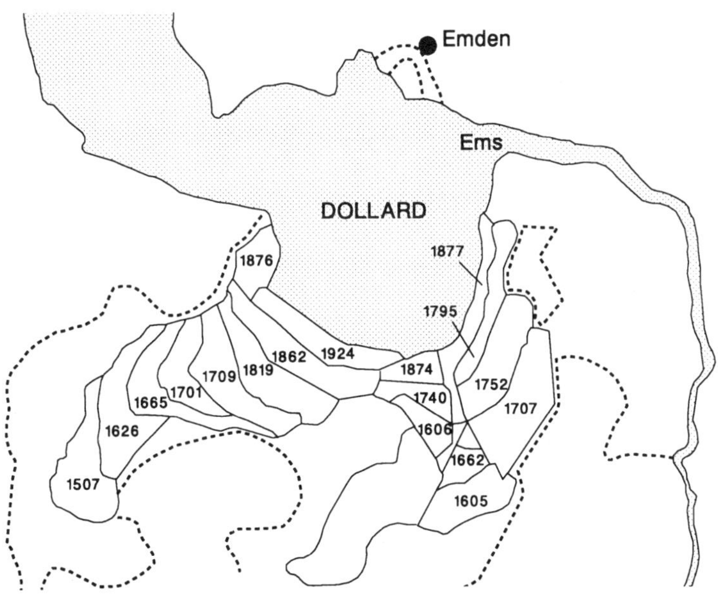

Figure 34.3. Extent of inundation of the Dollard area after medieval dikebursts and subsequent embankments.

For centuries humans embanked and reclaimed salt marshes above mean high tide level. In the 19th century, more difficult reclamations could be undertaken (Wolff 1992). Thus, in 1848, low-lying tidal flats were reclaimed the first time. Still later, entire tidal inlets were separated from the Wadden Sea. The largest one was the 3,700 km^2 Zuiderzee in 1932, but other recent examples are the Lauwerszee (90 km^2) in 1969 and the Nordstrander Bucht (33 km^2) in 1987 (Figure 34.1). The result is a very marked and strict separation between the tidal Wadden Sea on the one hand, and the non-tidal land and freshwater areas on the other hand. Tidal movements nowadays can penetrate only the hinterland along the estuaries of the Ems, Weser, Elbe, Eider, and Varde Aa rivers (Figure 34.1). Along the Dutch coast, no place is left where the tides move inland.

PHYSICAL CHANGES

The engineering works of the past have had an influence on the tides and the tidal currents. Usually the changes were only slight, but in the case of the closure of the Zuiderzee in The Netherlands (Figure 34.1), the tidal amplitude in the remaining part of the Wadden Sea increased by about 50% (Wolff 1983). This change in the tidal dynamics may have been an additional factor in the disappearance of seagrasses (*Zostera marina*) from the Dutch Wadden Sea (Van den Hoek et al. 1979).

The effects on water temperature of the discharge of cooling water into the Wadden Sea and its estuaries are only slight and hardly effect ecological changes (Essink and Wolff 1978).

CHEMICAL CHANGES

Because of the construction of dikes and dams, important chemical changes occurred in the Wadden Sea. One change was that the gradual transition between the sea and the land became much more abrupt. Gradual transitions of salinity with a variety of brackish water types disappeared, and concomitant ecological conditions of these different types of waters vanished. This is especially true for the types of brackish water with only slow or slight changes of salinity, such as the former Zuiderzee in The Netherlands. Thus, several characteristic brackish-water species, such as the Zuiderzee crab *Rhithropanopeus harrisii tridentatus* and the nudibranch *Corambe batava*, disappeared from the Wadden Sea area (Dankers et al. 1981). Today the Wadden Sea is largely a high salinity area, and brackish water occurs only around the major sluices and in the estuaries of the rivers Ems, Weser, Elbe, Eider, and Varde Aa (Figure 34.1). In these areas, salinity fluctuates rapidly and widely under the influence of tidal currents, river discharge, and wind effects (Wolff 1983).

The reclamation of the inland peat moors ended the discharge of very acid water. This might have had an effect on sedimentation, for example, the change from sedimentation of clays poor in lime to clays rich in lime in the Dollard estuary (Wolff 1983).

The most important chemical changes are related to pollution and eutrophication. The volumes edited by Essink and Wolff (1978) and Laane and Wolff (1984) and the paper by Wolff (1988) give extensive reviews of these chemical changes. The general conclusions are that the plant nutrients phosphorus and nitrogen strongly increased; that heavy metals are decreasing after an earlier peak; that several organic micropollutants, mainly chlorinated hydrocarbons, show high concentrations; and that pollution by organic wastewater is decreasing. Very recently, discharges of phosphorus by the Rhine River decreased profoundly, leading to strongly decreasing phosphorus concentrations in the western part of the Wadden Sea (Boddeke and Hagel 1991).

BIOTIC CHANGES

Natural Colonization and Extinction of Species

Although natural changes in distribution of plant and animal species are the rule rather than the exception (Hengeveld 1989), it is hard to demonstrate that natural introductions or extinctions of wild species have occurred in the Wadden Sea area. Many species became extinct and many others colonized the area, but human interference was nearly always involved. In fact, I do not know of any natural extinction or colonization in the Wadden Sea area.

Introduction of New Species by Human Intervention

Among the many species known to have been introduced from abroad into the Wadden Sea area by human intervention (Table 34.2; incomplete list), most were introduced inadvertently. Some species, such as the cord-grass *Spartina townsendii* and the Japanese oyster *Crassostrea gigas*, were introduced deliberately for, respectively, stimulation of accretion and oyster farming. In most cases, nothing indicates that the original Wadden Sea ecosystem was upset because of the new species. Only *Spartina* created far-reaching changes in the vegetation of salt marshes (Beeftink 1977).

Most species from other continents arrived in unknown ways. For some, for example the Chinese mitten crab *Eriocheir sinensis* and the New-Zealand barnacle *Elminius modestus*, transport in ballast water or on the hull of ships is reasonably well established. The same mode of transport may be suspected for many other species (Wolff 1983). Another mode of introduction is transport on imported shellfishes. This is, of course, true for the shellfish species themselves, such as the Japanese oyster *C. gigas*, but also for the species attached to the shells, such as the Japanese alga *Sargassum muticum*.

Other new species were able to colonize the Wadden Sea because their habitat was introduced there. The best example is the construction of artificial rocky shores like breakwaters, moles, and stone-covered dike slopes in places where no rocky shores existed. Many rocky-shore species, such as the kelp *Laminaria saccharina*, colonized these newly created habitats (Den Hartog 1959).

One can conclude that the Wadden Sea today lodges several species of algae and invertebrates which did not occur there before. They did not greatly change the ecosystems, but their addition is also not considered to have enhanced the value of the Wadden Sea habitats because of their often worldwide distribution.

Table 34.2. Some species introduced into the Wadden Sea area because of human intervention. After Wolff (1973), Den Hartog and Van der Velde (1987), Essink (1985) and Essink and Kleef (1988).

Taxa / Species	Origin
Algae	
Asterionella japonica	Pacific
Codium fragile	Pacific
Sargassum muticum	Pacific
Vascular plants	
Spartina townsendii (S. anglica)	America - England
Coelenterata	
Cordylophora caspia	Caspian Sea
Garveia franciscana	America
Mollusca	
Corambe batava	America
Crepidula fornicata	America
Crassostrea gigas (C. angulata)	Pacific
Teredo navalis	?
Ensis directus	America
Petricola pholadiformis	America
Mya arenaria	America
Polychaeta	
Marenzelleria viridis	America
Crustacea	
Balanus improvisus?	?
Elminius modestus	Pacific
Eriocheir sinensis	Pacific
Rhithropanopeus harrisii	America
Tunicata	
Molgula manhattensis	America
Styela clava	America

Extinction and Reduction of Species by Human Intervention

Human activities have had such a strong effect on the Wadden Sea ecosystem that several species became extinct. Because the cause was identified or at least suspected in many cases, the original conditions can be restored and the species reintroduced in several cases.

Exploitation

Wild plants and animals can be exploited on a sustainable basis. This occurs in the Wadden Sea in the forms of salt marsh grazing, fisheries for cockles, shrimps, and various species of fishes, and hunting of some duck species. However, especially among the larger species, overexploitation probably occurred several times in the past.

The fossil and archeological records reveal that grey whales (*Eschrichtius gibbosus*) once occurred in the North Sea. The grey whale is now extinct in the Atlantic Ocean and its adjacent seas and occurs only in the Northern Pacific Ocean (Mitchell 1974; Mitchell and Mead 1977). De Smet (1976) presents evidence for medieval whaling in the North Sea and the English Channel and believes that both the grey whale and the Atlantic right whale (*Eubalaena glacialis*) might have been hunted. He assumes that these slow species had been exterminated in the North Sea and in the English Channel by the Middle Ages. De Smet (1976), therefore, suggests that the grey whale once used the Wadden Sea as a breeding ground, just as it now uses the coastal waters of Baja California.

Subfossil and archeological records of grey seals (*Halichoerus grypus*) are rather common, and this species was probably hunted in the past. Because they stay on dry land for several weeks, the pups were an easy prey, and, hence, were probably overexploited in the Wadden Sea (Reijnders 1978). This provides a likely explanation for the absence of grey seals from the Wadden Sea until quite recently.

The decline of the harbour seal (*Phoca vitulina*) in The Netherlands provides an example of how quick overexploitation of marine mammals can reduce their populations. In the 1950s, virtually all juveniles were killed because of their fur, resulting in a population decrease of about 60-70% in about 10 years (Van Haaften 1974; Reijnders 1976).

Slow reproducing large birds are also vulnerable to exploitation or deliberate killing. This might explain why several species which have a large breeding distribution elsewhere in the Northern Hemisphere do not occur in the Wadden Sea area, although the area seems suitable. Examples are little egrets (*Egretta garzetta*), white-tailed eagles (*Haliaetus albicilla*), ospreys (*Pandion haliaetus*), black kites (*Milvus migrans*), and Caspian terns (*Hydroprogne caspia*). According to Pliny (Voous 1960), Dalmatian pelicans (*Pelecanus crispus*) also occurred in the Wadden Sea area. We might also wonder about the possibility that flamingos (*Phoenicopterus ruber*) existed there too.

Smaller birds became the victim of exploitation mainly because their eggs were collected by the coastal human population. Although a few colonies (Eijerland-Texel, Rottumeroog, Sylt?) were exploited on a sustainable basis (Swennen 1982), many other colonies and populations of species breeding as single pairs disappeared. This exploitation of eggs explains why at the beginning of this century eider ducks (*Somateria mollissima*), lesser black-backed gulls (*Larus fuscus*), herring gulls (*L. argentatus*), common gulls (*L. canus*), and avocets (*Recurvirostra avosetta*) were quite rare along the shores of the Wadden Sea (Leege 1905).

Fish species, especially the larger, slowly reproducing species, might also have become rare because of exploitation. This probably applies to the thornback ray (*Raja clavata*), the sting ray (*Dasyatis pastinacea*), and the dogfish (*Scyliorhinus caniculus*). These declines are not necessarily the result of fisheries in the Wadden Sea, but of intensive bottom-trawl fisheries in the entire North Sea, where these species are mainly incidental catches.

A clear example of overexploitation in the Wadden Sea is the disappearance of the Wadden Sea oyster (*Ostrea edulis*) beds. Oyster fisheries once flourished in the Dutch Wadden Sea and in Nordfriesland in Germany, but completely disappeared because of overfishing. Thus, the classic study of an ecosystem by Moebius (1877) can never be repeated because its subjects are exterminated.

Side-effects of Fisheries

Reise (1982) and Reise et al. (1989) documented the change in the composition of the benthic fauna in the Nordfriesland Wadden Sea in Germany. Long-lived species, such as several species of molluscs, gave way to opportunistic species, mainly polychaetes. This pattern is not restricted to Nordfriesland, however, but is also recognizable in Dutch coastal waters (Horst 1883-84; Wolff 1973; Dekker 1989). Reise (1982) attributes this change mainly to fisheries in the deeper parts of the Wadden Sea and especially those employing heavy bottom gear.

These changes represent not only a change in species, but also a change of ecosystems (Figure 34.4). Formerly, oyster beds, *Sabellaria* reefs, and "meadows" of hydroids were easily recognizable systems with a characteristic flora and fauna. These features hardly or no longer exist in the Wadden Sea.

The effects of fisheries in the North Sea on the population of some larger fish species were already mentioned. Not yet mentioned are the effects of fisheries in the rivers on several anadromous species. These fisheries probably had an effect on populations of species such as the sturgeon (*Accipenser sturio*), the salmon (*Salmo salar*), the allis shad (*Alosa alosa*), and the houting (*Coregonus oxyrhynchus*). However, today such effects of fisheries are masked by the effects of pollution and of regulation of rivers.

Effects of Habitat Loss

Some species disappeared because their habitats were eliminated. One example is the Zuiderzee race of the herring (*Clupea harengus*) which lost its spawning grounds

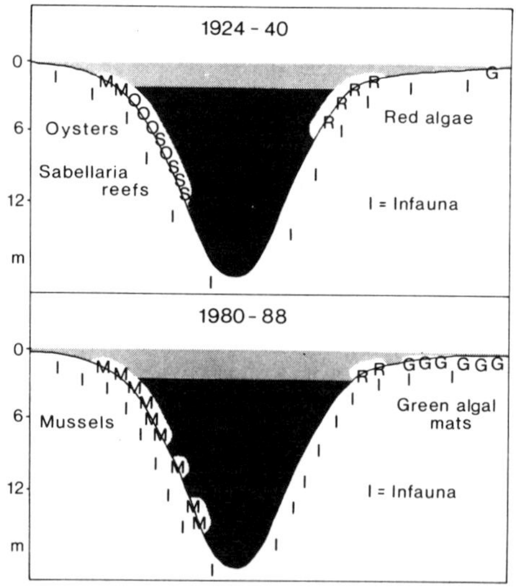

Figure 34.4. Ecosystems in the German Wadden Sea in 1924-40 and in 1980-88. After Reise et al. (1989).

because of the closure of the Zuiderzee. Within a few years it became virtually extinct (Havinga 1954). At the time of the closure of the Zuiderzee, the seagrass beds (*Zostera marina*) in the remaining part of the Wadden Sea also disappeared and never reappeared. With the disappearance of their seagrass habitat, several animal species also disappeared from the entire Wadden Sea: the sea stickleback (*Spinachia spinachia*), the deep-snouted pipefish (*Syngnathus typhle*), and the gastropods *Rissoa membranacea* and *Lacuna vincta*.

Effects of Pollution

Pollution has attracted much attention, but to date its effects have been less far-reaching than those of some other human interventions in the Wadden Sea. No species is known with certainty to have disappeared from the Wadden Sea because of pollution.

Oil pollution is a regular phenomenon in the Wadden Sea and sometimes results in the mortality of tens of thousands of birds. Thus far, however, there are no indications that it influences the population size of any characteristic Wadden Sea bird.

Chlorinated hydrocarbons, however, have had strong effects on the populations of many birds in the Wadden Sea, especially in The Netherlands but also in Germany (Koeman 1971; Becker and Erdelen 1987). Species such as the eider duck, the spoonbill (*Platalea leucorodia*), and the Sandwich tern (*Sterna sandvicensis*) decreased very markedly in the 1960s, and started to increase again only when the cause of the pollution was removed. However, after 25 years, the tern populations have not yet returned to their original levels.

Well known also is the harm of polychlorinated biphenyls on seals (Reijnders 1986). Since the 1960s, the Dutch population of harbor seals showed a continuous decline because of PCB effects until the decrease was balanced in the 1980s by immigration of seals from Germany (Figure 34.5). At its lowest level, only 15% of the original Dutch seal population was left. Whether the harbor porpoise (*Phocoena phocoena*), another marine mammal, is extinct in the Wadden Sea because of PCB pollution is presently being investigated by several research groups.

Pollutants are also strongly suspected to harm the condition of fishes, but effects on population size are unknown. Effects of micropollutants on invertebrates have not been shown to occur in the Wadden Sea area, although negative effects of antifouling paints containing tributyltin are suspected.

In the estuaries and at the outfall of sewage pipelines oxygen shortages from untreated sewage removed benthic fauna from considerable areas of tidal flats (Essink 1984).

Increase of Species by Cultivation

At present, blue mussels (*Mytilus edulis*) and Japanese oysters are cultivated in the Wadden Sea. Mussel cultivation especially changes the environment, and, thus, promotes a number of other species. Eel (*Anguilla anguilla*) and starfish (*Asterias rubens*) can be considered as examples. It has also been suggested that eider ducks increased because of mussel cultivation.

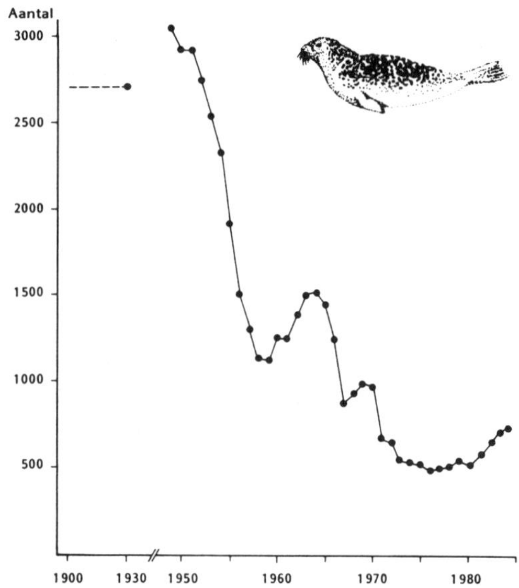

Figure 34.5. Decline of the harbor seal population in the Dutch Wadden Sea.

Increase of Species by Eutrophication

In the Dutch Wadden Sea, eutrophication is obvious and effected increases of several species of microalgae (Cadée 1984) and of several benthic animal species (Beukema and Cadée 1986). Although eutrophication reduces species diversity, extinctions of species from eutrophication are not known in the Wadden Sea.

EFFECTS OF CONSERVATION

Conservation in the Wadden Sea dates back to 1909 when the bird colonies at the tiny island of Jordsand, at that time part of Germany, were protected. This ceased during the 1st World War. Under Danish rule, the protection was restored in 1922. Protection of the equally small island of Griend in the Netherlands followed in 1917. The protection of the island of Mellum in Germany started in 1924. Thus, the original conservation in the Wadden Sea was aimed at the protection of the most important breeding colonies of coastal birds. After the Second World War the coastal salt marshes with lower densities of birds gradually became protected too. Protection of the Wadden Sea itself started in Germany with the reserve "Wattenmeer östlich Sylt" in 1937. However, this and other large reserves established in the German Wadden Sea in later years lacked organized management and, hence, were hardly effective. The Netherlands protected the Dutch Wadden Sea by application of the Nature Conservation Act in 1981; at the same time an organized management was implemented. Denmark protected its part of the Wadden Sea in 1979 as a game reserve and in 1982 as a nature reserve. Three national parks in Germany followed in 1985, 1986, and 1990. Thus, virtually the complete Wadden Sea acquired protected status.

Protection of breeding colonies of birds has been very successful. Breeding populations of almost all species of coastal birds have shown an exponential increase since the beginning of this century (Swennen 1982) (Figure 34.6).

The conservation of salt marsh has preserved some beautiful landscapes at the barrier islands, and is now producing interesting and biologically rich vegetation along the mainland coast (Dijkema and Wolff 1983).

Protection of the Wadden Sea itself has not yet produced obvious favorable ecological changes. Perhaps the return of the grey seal as a breeding species can be viewed as such. At least two major reasons account for the lack of favorable changes: the short period of conservation and the shift from local harm such as overexploitation, which can be counteracted by the creation of reserves, to widespread harm such as pollution by chlorinated hydrocarbons. Control of harm is not exercised by reserves any longer, but requires a strict environmental policy in entire watersheds (e.g., of the rivers Rhine and Elbe).

Indeed, controls effect beneficial ecological changes. After the widespread mortality of seabirds in the western part of the Wadden Sea from the discharge of pesticides at the Rhine estuary (Koeman 1971), measures that decreased these discharges effected a strong increase of the bird populations (Smit and Wolff 1981; Swennen 1982).

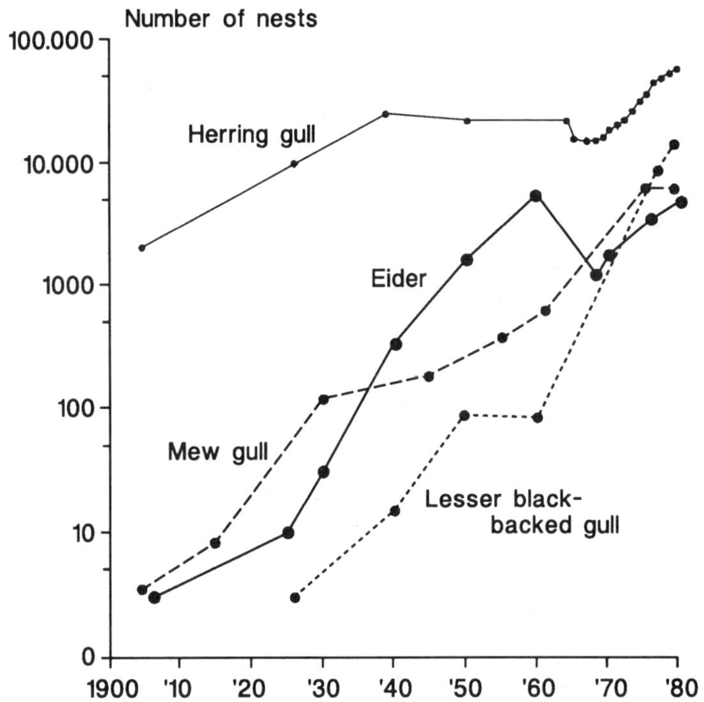

Figure 34.6. Increase of some species of birds breeding in the Dutch Wadden Sea area since about 1900. After Swennen (1982).

Measures to reduce the discharge of organically enriched wastewater resulted in lasting recolonization by benthic invertebrates of tidal flat areas earlier depopulated because of frequent absence of oxygen (Essink 1984).

PAST AND FUTURE MANAGEMENT OF BIODIVERSITY IN THE WADDEN SEA

The Wadden Sea is a hydrographical and ecological entity, but is in three different sovereign states. This makes management of biodiversity in this area a challenging endeavor, and its results can serve as an example for other areas.

Protection of the Wadden Sea in all three countries started as private initiatives (Wolff 1983). Private conservation organizations bought or rented small areas for nature reserves. However, because the larger part of the Wadden Sea is public property, this approach was unlikely to protect the entire Wadden Sea. Therefore, a private National Society for the Preservation of the Wadden Sea and a working group of concerned scientists were created in The Netherlands in 1965. These two groups intended to influence Dutch government policy to gain better protection of the Wadden Sea. In this they and other private conservation organizations, were successful. They prevented the construction of several embankments and other interventions, and, in 1976, the Dutch government published a policy for the protection of the Wadden Sea. Protection was realized in 1981.

In the meantime contacts with Germany and Denmark had been established, both by the private organizations and by the Dutch government. This led to the protection of the Wadden Sea in Denmark in 1979. In Germany, the situation was more complicated because other countries can only address the German federal government whereas the issue of nature conservation is considered by the individual federal states. However, under pressure of private organizations the German Wadden Sea also gained protected status in 1985 and following years.

Looking back, the private conservation organizations clearly played a key role in the conservation of the Wadden Sea. They developed the first initiatives, generated ideas, and kept the government policies moving. An important factor in this was international cooperation between private organizations, strongly stimulated by the World Wildlife Fund since 1978. This resulted in an International Coordination Team, an international Advisory Committee, and in concerted activities.

A major breakthrough occurred when the joint private organizations urged the governments of the three countries to proclaim a joint declaration on the protection of the Wadden Sea in 1982. Since then, protection of the Wadden Sea gained much more impetus. Since 1986, nearly the entire Wadden Sea has obtained protected status. International cooperation between government organizations has developed strongly, for example, in the form of meetings of ministers, conferences of managers, and biannual scientific symposia. A major role is now played by a Common Wadden Sea Secretariate, financed by the three countries and located at Wilhelmshaven, Germany. Among other things, it publishes an international Wadden Sea Newsletter.

It has to be stressed that all of the described developments were the result of constant interaction of private and public organizations.

Although much has been achieved to date, the situation certainly is not yet ideal. The three countries differ in culture, in traditional use of their part of the Wadden Sea, in economic interests in the Wadden Sea area, and in their legal systems. This leads to different accents in protective regulations which can be used, for example, by fishermen or oil companies to reach their different objectives. One way is to shift commercial activities from one country to another, but it is also possible that economic interests in one country point to the situation in one of the other countries to convince their government that regulations should be changed. Hence, much work work is needed before the Wadden Sea can be managed as one coherent biosphere reserve.

At a meeting of the conservation ministers of the three countries in 1991, a policy (Anonymous 1991) was proposed to the ministers by 59 non-governmental conservation organizations from all three countries, stating among other things that a common Wadden Sea policy is needed, that common goals should be set for protection, that an agreement should be reached about the boundaries of the Wadden Sea, that environmental management should be based on watershed management, and that a common Wadden Sea Council should be set up to develop a truly international approach to conservation.

As a final goal the private organizations feel that the Wadden Sea should be managed as one international park extending over three different countries and, in one of these, over at least three different states.

CONCLUSIONS

Naturally the Wadden Sea is a dynamic landscape showing continuous and large changes of its landscapes and ecosystems. However, no examples are known of species extinctions because of this dynamic nature.

During about 1,000 years, the Wadden Sea has seen very great changes because of human interventions. The coastal landscape were mutilated, and tidal inlets and coastal wetlands were severed from the coastal sea. What remains is only a part of an originally much larger coastal system.

From this amputated landscape, several types of ecosystems and many species of mammals, birds, and fishes and some invertebrates disappeared or at least decreased in abundance and distribution because of human intervention, especially exploitation and pollution. New species established themselves in the Wadden Sea area, but these are mainly algae and invertebrates. In general, losses have been incurred by large, long-lived species, whereas additions are often associated with small species with opportunistic life-histories.

Conservation has been successful only in restoring lost biodiversity when changes in the ecosystem were due to exploitation. However, not all exploitation effects have been counteracted by conservation measures. Moreover, many other losses of biodiversity cannot be restored by local conservation but need corrective measures on a watershed, European, or even worldwide scale.

Protection of the Wadden Sea is now increasingly a joint responsibility of the three countries. To reach an international park status, however, much work remains to be done.

LITERATURE CITED

Anonymous. 1991. The common future of the Wadden Sea. A report by the World Wide Fund for Nature. WWF, Husum.

Baerselman, F., and F. W. M. Vera. 1988. Natuurontwikkeling, een verkennende studie. SDU, 's-Gravenhage.

Becker, P. H., and M. Erdelen. 1987. Die Bestandsentwicklung von Brutvögeln der deutschen Nordseeküste 1950-1979. *J. Ornithol.* 126:29-51.

Beeftink, W. G. 1977. The coastal salt marshes of western and northern Europe: an ecological and phytosociological approach. In *Wet Coastal Ecosystems*, ed. V. J. Chapman, pp. 109-155. Elsevier, Amsterdam.

Beukema, J. J., and G. C. Cadée. 1986. Zoobenthos responses to eutrophication of the Dutch Wadden Sea. *Ophelia* 26:55-64.

Boddeke, R., and P. Hagel. 1991. Eutrophication of the North Sea coastal zone, a blessing in disguise. Internat. Council Explor. Sea. C.M.1991/E:7.

Borger, G. J. 1975. De Veenhoop. Een historisch-geografisch onderzoek naar het verdwijnen van het veendek in een deel van West-Friesland. Thesis, Amsterdam.

Cadée, G. C. 1984. Has input of organic matter in the western part of the Dutch Wadden Sea increased during the last decades? *Neth. Inst. Sea Res. Publ. Ser.* 10:71-82.

Dankers, N., H. Kühl, and W. J. Wolff. 1981. *Invertebrates of the Wadden Sea*. Balkema, Rotterdam.

Dekker, R. 1989. The macrozoobenthos of the subtidal western Dutch Wadden Sea. I. Biomass and species richness. *Neth. J. Sea Res.* 23:57-68.

Den Hartog, C. 1959. The epilithic algal communities occurring along the coast of The Netherlands. *Wentia* 1:1-241.

Den Hartog, C., and G. van der Velde. 1987. Invasions of plants and animals into coastal, brackish and fresh water of The Netherlands. *Proc. Royal Neth. Acad. Sci.*, B 90:31-37.

De Smet, W. M. A. 1976. Evidence of whaling in the North Sea and the English Channel during the Middle Ages. In *Scientific Consultation on Marine Mammals*. Bergen, Norway.

Dijkema, K. S., and W. J. Wolff, eds. 1983. *Flora and Vegetation of the Wadden Sea Islands and Coastal Areas*. Balkema, Rotterdam.

Edelman, T. 1974. Bijdrage tot de historische geografie van de Nederlandse kuststrook. *Rijkswaterstaat Publ.* 14:1-82.

Essink, K. 1984. Discharge of organic waste into the Wadden Sea: local effects. *Neth. Inst. Sea Res. Publ. Ser.* 10-1984:165-178.

Essink, K. 1985. On the occurrence of the American jack-knife clam Ensis directus (Conrad, 1843) (Bivalvia, Cultellidae) in the Wadden Sea. *Basteria* 49:73-80.

Essink, K., and H. L. Kleef. 1988. Marenzelleria viridis (Verrill, 1873) (Polychaeta; Spionidae): a new record from the Ems estuary (The Netherlands/Federal Republic of Germany). *Zoöl. Bijdr.* (Leiden) 38:1-13.

Essink, K., and W. J. Wolff. 1978. *Pollution of the Wadden Sea*. Balkema, Rotterdam.

Havinga, B. 1954. Vissen. In *Veranderingen in de flora en fauna van de Zuiderzee (thans IJsselmeer) na de afsluiting in 1932*, ed. L. F. de Beaufort, pp. 253-267. Ned. Dierk. Ver., Den Helder.

Hengeveld, R. 1989. *Dynamic Biogeography*. Cambridge University Press, Cambridge.

Hoffmann, D. 1980. The geological development of the North-Frisian Islands. In *Geomorphology of the Wadden Sea Area*, eds. K. S. Dijkema, H.-E. Reineck, and W. J. Wolff, pp. 85-95. Balkema, Rotterdam.

Horst, R. 1883-84. Verslag omtrent onderzoekingen op de oester en de oestercultuur betrekking hebbende, uitgebracht door de Commissie van het Zoölogisch Station. *T. Ned. Dierk. Ver.*, Suppl. 1:1-673.

Koeman, J. H. 1971. Het voorkomen en de toxicologische betekenis van enkele chloorkoolwaterstoffen aan de Nederlandse kust in de periode van 1965 tot 1970. Thesis, Utrecht.

Laane, R. W. P. M., and W. J. Wolff, eds. 1984. The role of organic matter in the Wadden Sea. Proceedings 4th Int. Wadden Sea Symp., Texel, The Netherlands, 1-3 November 1983. *Neth. Inst. Sea Res. Publ. Ser.* 10-1984.

Leege, O. 1905. *Die Vögel der ostfriesischen Inseln nebst vergleichender Uebersicht der im südlichen Nordseegebiet vorkommenden Arten*. Haynel, Emden.

Mitchell, E. D. 1974. Present status of northwest Atlantic fin and other whale stocks. In *The Whale Problem*, ed. W. E. Schevill, pp. 108-109. Harvard University Press, Cambridge, Mass.

Mitchell, E. D., and J. G. Mead. 1977. History of the grey whale in the Atlantic Ocean. *Proc. Second Conf. Biol. Marine Mammals*, San Diego.

Moebius, K. 1877. *Die Auster und die Austernwirthschaft*. Wiegandt, Hempel and Parey, Berlin.

Pons, L. J., and A. J. Wiggers. 1959-60. De holocene wordingsgeschiedenis van Noord-Holland en het Zuiderzeegebied. *T. Kon. Ned. Aardrijksk. Gen.* 76:104-152; 77:1-57.

Reijnders, P. J. H. 1976. The harbour seal population in the Dutch Wadden Sea: size and composition. *Neth. J. Sea Res.* 10:223-235.

Reijnders, P. J. H. 1978. De Grijze zeehond in het Waddengebied. *Waddenbulletin* 13:500-502.

Reijnders, P. J. H. 1986. Reproductive failure in common seals feeding on fish from polluted coastal waters. *Nature* 324:456-457.

Reise, K. 1982. Long-term changes in the macrobenthic invertebrate fauna of the Wadden Sea: Are polychaetes about to take over? *Neth. J. Sea Res.* 16:29-36.

Reise, K., E. Herre, and M. Sturm. 1989. Historical changes in the benthos of the Wadden Sea around the island of Sylt in the North Sea. *Helgoländer Meeresunters* 43:417-433.

Smit, C. J., and W. J. Wolff, eds. 1981. *Birds of the Wadden Sea.* Balkema, Rotterdam.

Swennen, C. 1982. De vogels langs onze kust. In *Wadden, Duinen, Delta*, ed. Anonymous, pp. 78-100. Pudoc, Wageningen.

Van den Hoek, C., W. Admiraal, F. Colijn, and V. N. de Jonge. 1979. The role of algae and seagrasses in the ecosystem of the Wadden Sea: a review. IN *Flora and Vegetation of the Wadden Sea*, ed. W. J. Wolff, pp. 9-118. Balkema, Rotterdam.

Van Haaften, J. L. 1974. Zeehonden langs de Nederlandse kust. *Wet. Meded. Kon. Ned. Natuurhist. Ver.* 109:1-36.

Voous, K. H. 1960. *Atlas van de Europese Vogels*. Elsevier, Amsterdam.

Wolff, W. J. 1973. The estuary as a habitat. An analysis of data on the soft-bottom macrofauna of the estuarine area of the rivers Rhine, Meuse, and Scheldt. *Zoöl. Verh.* (Leiden) 126:1-242.

Wolff, W. J., ed. 1983. *The Ecology of the Wadden Sea.* Balkema, Rotterdam.

Wolff, W. J. 1988. Impact of pollution on the Wadden Sea. In *Pollution of the North Sea. An assessment*, eds. W. Salomons, B. L. Bayne, E. K. Duursma, and U. Förstner, pp. 441-454. Springer-Verlag, Berlin.

Wolff, W. J. 1992. The end of a tradition. 1000 years of embankment and reclamation of wetlands in The Netherlands. *Ambio* 21:287-292.

35

CONSERVING BIOLOGICAL DIVERSITY THROUGH ECOSYSTEM MANAGEMENT

Hal Salwasser

School of Forestry
The University of Montana
Missoula, Montana
U.S.A.

James A. Caplan and Charles W. Cartwright[1]

USDA Forest Service
Washington, D.C.
U.S.A.

Arlene T. Doyle

Western Washington University
Huxley College of Environmental Studies
Bellingham, Washington
U.S.A.

Winifred B. Kessler

University of Northern British Columbia
Prince George, British Columbia
Canada

Bruce G. Marcot

Pacific Northwest Forest Experiment Station
Portland, Oregon
U.S.A.

Lawrence Stritch

Shawnee National Forest
Harrisburg, Illinois
U.S.A.

This paper describes an ecosystem approach to multiple-use management that features conservation of biological diversity integrated with other uses and values of wildlands.

[1] Present address: USDA Forest Service, South Western Region, Albuquerque, New Mexico.

The approach evolved from a USDA Forest Service program known as New Perspectives for Managing the National Forest System. We discuss concepts and aims of multiple-use ecosystem management and New Perspectives, describe a goal-oriented strategy for conserving biological diversity and give three examples of how goals for conserving specific aspects of biological diversity were integrated with ecosystem management on national forests in the United States. We conclude with a discussion of changes needed to further the development and implementation of ecosystem management as a process to improve relationships between humans and land.

CONSERVING BIODIVERSITY THROUGH MULTIPLE-USE ECOSYSTEM MANAGEMENT

Multiple-use management of wildlands and natural resources is a broad concept. It includes any and all land-uses or management practices that aim to meet more than one specific purpose or goal. For example, a wetland ecosystem managed for waterfowl production, water purification and recreation is a multiple-use site even though waterfowl conservation might be the dominant theme for management. Likewise, a wilderness or natural area that provides limited human recreational uses, regulated grazing by domestic livestock, fishing and hunting is also a multiple-use area. Conversely, a tree farm or grazing pasture that does not include considerations for other resource uses or values, such as aesthetics, species diversity or wildlife habitat, is outside the concept of multiple-use.

The key criterion for multiple use is not whether extractive use of resources occurs but, rather, that a diversity of relatively compatible uses and values is promoted. Thus, while a tree farm that promotes only the production of wood fiber is a single-use forest, a forest managed for diverse tree species, wood products, understory shrubs and herbs for wildlife and fall colors is a multiple-use forest. Individual patches of habitat, farms and ranches can be managed for single resource use. But at geographic scales larger than sites or patches, such as watersheds, landscapes and regions, virtually all places provide a mix of resource values and uses; that is, they are managed for multiple values and uses.

Conservation of biological diversity—the perpetuation and wise use of the variety of life and its many processes—can be partially achieved through multiple-use management of aquatic and terrestrial ecosystems. The key word in this proposition is "partially." Multiple-use encompasses a diverse array of goals, some of which might not directly result in the conservation of biological diversity. Furthermore, other actions in addition to multiple-use management are needed to conserve the full variety of life and its processes, such as intensive production of natural resources to meet burgeoning human needs, strict protection of the most sensitive natural areas, resource conservation, gene banks and captive breeding. And to be more effective, traditional practices of multiple-use management, such as clearcutting forests on erodible soils and grazing domestic livestock in wetlands and riparian areas, need modification to make conservation of biological diversity an integral part of sustaining desired conditions of ecological systems for their multiple uses and values.

To address needed changes, an ecosystem perspective on multiple-use management evolved during the late 1980s and early 1990s (Salwasser et al. 1992). Ecosystem management is broader than traditional multiple-use management because it emphasizes a more comprehensive integration of ecological, social and economic considerations in resource management and applies the result over a variety of temporal and spatial scales. It also has as a fundamental goal the maintenance of diversity in life forms, ecological processes and human communities, economies and cultures. Traditional multiple-use management tends to focus on producing sustainable yields of specific resources and uses with consideration of diversity secondary (Gale and Cordray 1991).

Traditional multiple-use management has been successful in restoring and sustaining certain aspects of biological diversity. For example, the restoration of diverse and productive forests and native vertebrates in the eastern United States during the past century was accomplished through application of multiple-use concepts (MacCleery 1992). However, sustained-yield resource management, when taken to an extreme focus on selected resources such as timber, game wildlife or commercial and sport fisheries, can simplify ecosystems, reducing biological diversity. Loss of old-growth primary forests and their associated diversity through forestry that aimed only for sustained yields of wood is but one example.

An ecosystem perspective on multiple-use management expands the sustained yield concept beyond single resources to whole ecosystems and the multiple-use concept to a wider array of values and uses. Ecosystem management encompasses the production of resources for human use, as any reasonable conservation strategy must. But resource production under ecosystem management derives from desired ecosystem conditions of diversity, long-term productivity and resilience rather than drives resulting ecosystem conditions. Expected yields of selected resources and uses, thus, must be commensurate with the goal of sustaining desired ecosystem conditions.

Ecosystem Management as A Process for Conserving Biological Diversity in Multiple-use Lands

Ecosystem management is how the Forest Service will integrate conservation of biodiversity into its long-standing multiple-use mandate. A full spectrum of land-use classes and resource management practices, ranging from preservation to sustainable production, will be used to restore and sustain diverse, healthy, productive ecosystems. Four principles guide the practical development of ecosystem management (adapted from Robertson 1992):

Protect the land by restoring and sustaining the integrity of its soils, air, waters, biological diversity and ecological processes, thereby sustaining what Aldo Leopold (1949) called healthy land, what some now call ecosystem integrity.

Meet the needs of people who depend on resources of the land for food, fuel, shelter, livelihood and inspirational experiences.

Improve the well-being of communities, regions and nations through efficient and environmentally sensitive production and conservation of natural resources such as wildlife, wood, water, fisheries, minerals, energy, forage for domestic animals and recreation opportunities.

Seek balance and harmony between people and land with equity between interests, across regions and through generations, meeting this generation's resource needs while maintaining options for future generations to also meet their needs.

Ecosystems as a Land and Resource Management Concept

An ecosystem is a community of organisms working together with its environment as an integrated unit. It is both an abstract scientific concept and a real thing. For any plant or animal, including humans, an ecosystem is its home (Berry 1987; Sahtouris 1989; Rowe 1990). A rotting log is the ecosystem for a slime mold, the planet the ecosystem for all life on earth. Ecosystems can be named, delimited and mapped. But they are not independent entities. And they do not stay the same over time. All ecosystems have flows of energy, species and materials moving between them: they are interconnected. All ecosystems have limits set by their biological and physical conditions or states. And every ecosystem is a subset of a larger system.

All ecosystems contain dynamic processes. They change over space and time in response to inputs of energy, new species, natural events, internal growth and development processes and to how people treat the land (Burgess and Sharpe 1981; Shugart 1984; Waring and Schlesinger 1985; Botkin 1990). Ecosystems are always changing, whether people cause the change or not. But most ecosystems are now influenced in some way by human activities or the artifacts of those activities.

The capability of an ecosystem to sustain desired conditions of diversity, ecological services and resource uses and values is a result of its climate, soils, topography, biota, natural processes and how large the ecosystem is. In general, the larger the ecosystem, the more diversity, resilience and productivity it can sustain. Ecosystem capabilities determine what is possible in a human time frame, such as a decade to a generation (any longer than that and the basic capabilities of ecosystems are also likely to change).

Ecosystem management must operate within bounds set by these ecosystem traits. An inventory of current ecosystem conditions and trends can help us determine what conditions are likely in the near future. The needs and aspirations of the people who depend on ecosystems in a particular area for their well-being determine the desired current and future conditions of those ecosystems. Obviously, desired conditions cannot lie outside the bounds of what is there now or what is possible given ecological capabilities, financial resources or technologies. The differences between existing and desired future conditions of ecosystems identify possible management objectives, that is what it will take to sustain or restore ecosystem conditions to their possible and desired states and flows of resource uses and values.

Ecosystems in a Landscape Perspective

Landscapes are the working unit for ecosystem management. A landscape is a large area composed of many different kinds of ecosystems. It has repeatable patterns of habitats, physical features and human influences (Forman and Godron 1986). Landscapes are large enough that it is possible to integrate the protection and management of ecosystems at site, stand and watershed scales. Because of their large size, landscapes often involve multiple land tenures or multiple zones of different land-use classes.

Landscape patterns result from both enduring, slow-changing features of nature (e.g., soils, climate and topography) and more dynamic patterns of biotic communities, ecological processes and disturbances that shape short-term temporal and spatial change. When we look at the earth from an airplane, we see a snapshot of a landscape at a point in time. If we had numerous snapshots, representing repeated fly-overs, spaced several decades apart, we would see that vegetation patterns and human influences on the landscape change from one photo to another, like a kaleidoscope.

Disturbances superimposed on long-term patterns and processes in ecosystems set the context for the temporal diversity of life and its changes over time. In some places the pattern is a patchwork of different kinds of ecosystems: in one example, stand-replacement fires create a mosaic of forests and openings at the landscape scale; in another example, the pattern might be a fairly continuous forest cover. But a view from inside a continuous forest might still reveal a diverse ecosystem containing trees varying in age from very young to very old, multiple canopy layers, and a profuse understory wherever gaps exist in the canopy. Ecosystem dynamics in such a forest might result not from fires but from winds or the actions of age and diseases on individual trees. When global or regional climate change is sufficient, landscape patterns change.

Biodiversity in an Ecosystem Perspective

The variety of life and its many processes in ecosystems determines ecological capabilities. This variety is known by the term biodiversity (Keystone Center 1991). It includes variation and variability in genes, species, plant and animal communities and the many processes through which they are all interconnected through space and time.

The biological diversity of ecosystems has ecological, economic, educational and aesthetic values. It is key to the productivity and sustainability of basic life support systems. It provides numerous current and future resources for human well-being. It provides opportunities for better understanding the myriad of relationships between people and their sources of existence. And it contributes greatly to the beauty and wonder of the world we live in. Biological diversity also has an ethical element: how well we conserve biological diversity demonstrates our respect for other forms of life and our commitment to the well-being of future generations.

Scientists do not fully know the ecological roles or potential values of biological diversity. Nor do they know what all the parts are or understand all the processes that keep ecosystems functioning (Wilson 1993). It is not likely that they ever will. But complete knowledge is not necessary to understand that retaining the natural parts and processes of biodiversity is important for the future health and productivity of all ecosystems (Leopold 1949). How to do this in the face of a growing human population is one of humanity's biggest challenges in the coming decades.

A Goal-oriented Conservation Strategy for Biological Diversity

Incorporating the conservation of biological diversity into multiple-use ecosystem management requires actions aimed to achieve time- and place-specific objectives for conditions of species, biological communities and ecosystems. A strategic framework for such actions and objectives was developed for federal lands through a national policy dialogue (Keystone Center 1991). Although recommendations of the Keystone dialogue are not federal policy as of 1993, they have been adapted here to offer resource managers and scientists a framework of specific and measurable goals. To guide actions on-the-ground, these goals must be reflected in land-use allocations, performance standards in land and resource management plans and working guidelines for project activities in specific places. To succeed in actually conserving major parts of biological diversity, the goals must be accomplished.

1. *Recover and conserve formally listed threatened or endangered species.* Listed species are the most vulnerable officially recognized elements of biodiversity. A net decline in the number of listed species in the area covered by a plan or program is an ideal goal. Achieving it requires protection of existing populations and habitats of listed species and restoration of them if necessary.

2. *Provide for viable populations of native plant and animals species.* Species whose demographic or habitat trends are negative but not yet to the point of endangerment might be the next most vulnerable elements of biological diversity. Some such species might even be more vulnerable than officially listed species. An ideal goal is to secure the places and functions of all native species in regional ecosystems before they reach the point where formal listing as a threatened or endangered species calls into play the extreme measures of protecting species under crisis conditions (Salwasser 1991). This can be achieved by managing habitats, human activities and wild populations of plants and animals to assure that populations of native species are viable and well-distributed throughout their geographic ranges. This requires a combination of actions to protect, restore and enhance sufficient kinds, amounts, qualities and distributions of sub-populations and their habitats. Especially important in achieving population viability is the perpetuation of multiple, interconnected, demographically resilient local populations; the characteristic genetic variation of the entire species; and the full range of the species' roles in ecological processes. Principles and methods of conservation biology (Soulé and Wilcox 1980; Soulé 1986, 1987) are useful in this task.

3. *Maintain a viable network of native biological communities and ecosystems.* Rare, unique or sensitive biological communities or successional stages (often highly productive sites, riparian areas and mature or old-growth successional stages) are likely to be vulnerable elements of biological diversity in certain landscapes. To conserve these communities, lands, human activities and habitats must be managed to assure that a network of representative native biological communities is maintained across the landscape. This can involve ecological restoration in some cases. Especially important are communities or assemblages of species that are rare or imperiled in the region or nation (Jenkins 1988). The matrix conditions of a landscape that contains a network of protected sites should also provide resources for native species to the degree that this is possible, especially conditions needed for normal movement of plants and animals throughout the landscape and a full range of ecological processes characteristic to the area. Where this is not possible, the network of sites may need specific connections between them. The sites and connections must be sufficiently large and diverse to accomplish their intended purposes.

4. *Maintain structural diversity.* Natural elements of structural diversity, such as snags, caves, fallen trees and seeps, provide habitats for many species that would not occur in an area without them. These elements can be jeopardized by intensive human uses, such as fuelwood gathering, heavy livestock grazing, clearcutting and water diversions.

Maintenance of these elements of structural diversity both within patches and across landscapes will assure their roles in sustaining desired conditions of ecosystem diversity, productivity and resilience from site to regional geographic scales (Franklin 1988).

5. *Sustain genetic diversity*. The genetic variation of intensively managed wild plant and animal populations can decline if sufficient attention is not paid to the effects of human selection for various traits. To maintain genetic diversity, populations and habitats must be managed to sustain natural levels of genetic variation within and among populations and the genetic integrity of representative and extreme populations (Ledig 1986; Millar 1987).

6. *Produce and conserve resources needed by people*. Human well-being ultimately depends on natural resources. People will obtain those resources from somewhere. The challenge is to produce them in ways that do not lead to undesired environmental effects at local, regional or global scales. If resources can be produced in ways that reduce human pressures on biological diversity in other places, then resource production zones can have a positive overall effect on biodiversity conservation (Lugo and Brown this volume). To accomplish this, high productivity sites such as flat ground with deep loamy soils, and featured species such as pines, firs, oaks, elk, and trout should be managed with state-of-the-art efficiency in certain places to sustain the production of resources needed by people, thus meeting human needs with minimal impacts on more fragile sites and sensitive species.

7. *Protect ecosystem integrity—soils, waters, biota and ecological processes*. Any human activity has some effect on lands, waters or biota. Ideally, these effects can be minimized through sensitivity to ecosystem integrity. Actions that are known to degrade site conditions or long-term ecosystem diversity, productivity or resilience should be avoided if possible or mitigated promptly when not. To do this, managers should employ the natural restorative powers of ecosystems, considering the kinds, amounts and distribution of living and dead organic matter retained in ecosystems for long-term diversity, productivity and resilience following resource harvest. For ecological sustainability, this is as important as considering how much productivity of the system is removed for human uses.

8. *Restore and renew degraded ecosystems*. Biological communities, waters and soils that have been damaged by natural events or past human actions should be under restoration and renewal programs, embracing the concepts and methods of restoration ecology and management (Cairns 1986, 1988a,b,c, 1989; Jordan et al. 1987; Bonnicksen 1988; Jordan 1988).

There is, of course, more to the conservation of biological diversity in multiple-use lands than identified in this framework, but we believe these actions are a reasonable start on a comprehensive conservation program.

EXPLORING NEW PERSPECTIVES

The Human Dimension

The reason a multi-faceted or ecosystem perspective is needed for conserving biological diversity through multiple-use management is simple. Continued growth in human populations and increases in their production, use and disposal of resources are not matched by corresponding growth in the land base available to meet those demands under traditional resource management approaches while sustaining desired levels of environmental quality (Silver and DeFries 1990). Managers of wildlands and natural resources throughout the world, thus face a dual challenge that grows in difficulty with each passing year: provide people with the resources needed to sustain their lives and well-being while minimizing the impacts of resource production and use on the diversity, productivity and resilience of the ecosystems from which those resources are taken or used (LeMaster 1992; Reid et al. 1992; United Nations 1992).

Forests and woodlands provide an example of this challenge. Forests and woodlands now cover an estimated 31% of the planet's terrestrial surface (4.1 billion ha according to the World Resources Institute 1990). This is about 66% of the forested area that existed prior to the industrial and public health revolutions of several centuries ago. Meanwhile, the number of humans has grown by 11 times, from an estimated 500 million people to about 5.5 billion.

In per capita terms, this means that each person living on the earth had an average of about 12 ha of forest resource in 1750, whereas in 1990, each had only about 0.75 ha. For the United States, the corresponding numbers are: 45 ha of forest per person in 1600 down to 1.2 ha in 1990 (Salwasser et al. 1992). Meanwhile, global use of industrial wood from these forests has increased at an average 2% per year for the past 40 years (Haynes and Brooks 1991). This does not account for fuelwood use, which is approximately equal to industrial roundwood use on a global scale.

The United States, not surprisingly, is a major force in global wood use and its impacts on forests (U.S. citizens use about 33% of the world's annual production of industrial roundwood). The impact of U.S. wood use has been reflected in several events and issues during the 1980s that caused the USDA Forest Service to explore new perspectives for managing the complex lands and resources that comprise the National Forest System.

Global issues included uncertainties associated with climate change, loss of biological diversity and the growing human population. National issues included controversy over logging and forest regeneration methods (mostly clearcutting and even-aged forestry) and declines in forest health because of pollution, drought, insects, fire suppression and past management practices. Controversy over subsidies to the development of public resources (including grazing fees and timber sales whose financial returns do not cover administration costs), loss of old-growth forests and a growing number of endangered species augmented poor conditions of public rangelands, rising demands by people for all natural resources and declining domestic supplies of resources for which demand was creating increased foreign dependencies such as oil and timber as other major national issues. At local scales, concerns for soil productivity, aesthetics of land management practices, water quality, the vitality of communities that depend on public land resources for livelihoods and jobs, large and largely uncontrollable wildfires and new and conflicting uses of public lands fueled the flames of change.

The Forest Service translated these issues into four reasons for exploring new perspectives (Overbay 1992):

- People need and want a wider array of uses, values, products, and services from public lands than in the past, especially, but not limited to, the amenity values and environmental services of healthy, diverse lands and waters;

- New information and a better understanding of ecological processes highlight the role of biological diversity as a factor in sustaining the health and productivity of ecosystems and the need for integrated ecological information at various spatial and temporal scales to improve management;
- People outside the Forest Service want more direct involvement in the process of making decisions about public resources; and
- The complexity and uncertainty of natural resources management call for stronger teamwork between scientists and resource managers than has heretofore been practiced.

Aims of New Perspectives

The Forest Service chartered New Perspectives in 1991 to accomplish five goals: (1) learn how to sustain better the diverse and productive ecological systems; (2) better integrate the different aspects of land and resources management; (3) improve the effectiveness of public participation in resource decision-making; (4) continue building partnerships between forest users and forest managers; and (5) strengthen teamwork between researchers and managers.

All people interested in or affected by management of the National Forest System were invited to participate in shaping the results of New Perspectives and the agenda of work was purposefully left open to allow for new ideas. Work was carried out through more than 200 field research and demonstration projects and dozens of workshops, conferences and symposia held in partnership with universities, professional societies and various conservation and forest-user groups.

Changes in the stewardship of national forests and grasslands would have significant effects on economic, social and environmental goals of the United States. By the late 1980s, national forests and grasslands comprised about 8.5% of the land area of the nation, 77 million ha distributed in 42 states and Puerto Rico. Nearly 70% of this area was forested, about 54 million ha. Timber management was planned for 43% of the forested area: 23 million ha or 30% of the entire land area of the system. The other 57% of forested area, 31 million ha, was devoted to non-timber uses and values. Grazing by domestic livestock occurred on 26% of the total land area, about 20 million ha. Nearly 18% of the total area had been designated by Congress for wilderness preservation, 13.8 million ha. An additional 26% of the land was un-roaded and not planned or scheduled for resource developments, nearly 20 million ha. Thus, 44% of the National Forest System, about 24 million ha, was managed primarily for natural values, ecological processes and low-impact human recreational uses. The entire 77 million ha of the National Forest System is open for various recreational uses. And all of it is wildlife habitat and watershed.

Because of the extent, area and distribution of the National Forest System, these lands are among the nation's richest and most productive public assets for conserving biological diversity and providing citizens with various resource uses and values. For example, more than 70% of the Wild and Scenic River System in the lower 48 states and 84% of the nation's Wilderness Preservation System in those states occur in national forests or grasslands. About 70% of the vertebrate species native to the nation occur in the National Forest System, including all of the large mammals. About 30% of federally listed threatened or endangered species find some of their last remaining habitats in these lands. The forests hold nearly one-half of the nation's standing volume of softwood sawtimber and can provide a sustainable yield of 60 to 80 million m^3 of roundwood per year, about 20% of the nation's annual consumption of softwood timber. More than 40% of all outdoor recreation that occurs on federal lands each year occurs in the national forests or grasslands, nearly twice the recreation use of the national parks. And minerals extracted from these lands generated $4.9 billion in private sector revenues in 1990.

MOVING TOWARDS ECOSYSTEM MANAGEMENT

In the summer of 1992, the USDA Forest Service announced that, as a result of New Perspectives work, ecosystem management would be used to sustain desired conditions of ecosystem diversity, productivity and resilience for the multiple uses and values of national forests and grasslands. Ecosystem management is a process. It is not a goal. The goals for ecosystem management derive from ecological capabilities of the land together with legal mandates and public needs and aspirations.

The Forest Service has a Congressional mandate to manage lands and resources entrusted to its care under the concepts of multiple use and sustained yield, without impairing the long-term productivity of the land. The Forest Service also has a legal mandate to conserve threatened or endangered species and to "provide diversity of plant and animal communities...to meet overall multiple-use objectives (NFMA 1976).

Recent research has shown that biological diversity is important to the long-term productivity and resilience of ecosystems, i.e., the land. Combining this knowledge with the agency's legal mandate, the Chief of the Forest Service, defined ecosystem management as follows: "an ecological approach will be used to achieve the multiple-use management of the national forests and grasslands by blending the needs of people and environmental values in such a way that the national forests and grasslands represent diverse, healthy, productive, and sustainable ecosystems" (Robertson 1992).

Field managers and scientists must carry out the new policy on ecosystem management through changes in agency regulations, national program direction, amendments to integrated forest plans, field projects that carry out the direction in those plans, research programs and cooperative endeavors with conservation partners in universities, other government agencies and the private sector. Most of this work remains to be done throughout the decade of the 1990s.

Working Guidelines for Ecosystem Management

At this point, however, some working guidelines for ecosystem management have evolved from New Perspectives projects. These guidelines remain open for refinement and are presented here to show the state-of-the-art as of 1992.

1. *Work Within the Scope of Natural Processes that Shape Landscape and Ecosystem Conditions.* Plans and field projects should employ the ecological capabilities and natural

processes of different ecosystems, maintaining as much diversity as possible and minimizing the energy costs of management to sustain or restore desired ecosystem conditions and functions.

Natural disturbances such as fires, floods, droughts and storms are major forces which shape ecosystems and landscape patterns. These processes are the context within which long-term management strategies must be developed to sustain desired conditions of ecosystem diversity and productivity. Special care is needed to protect wildland soils and waters, particularly in sensitive areas such as steep slopes, wetlands, riparian zones, fragile sites and rare species' habitats.

Planners and managers should always think about scale effects, both spatial and temporal, at least one scale higher and one scale lower than what you're working on and at least several generations into the future, more and longer if possible. A good motto is to think complex, model simple and maintain options.

2. *Focus on End Results—Desired Future Ecological and Social Conditions* and the land-use classes and management actions that will best attain them. Use landscapes as a basic unit for planning and managing ecosystems to meet specific objectives both desired future ecological conditions and desired economic and social goals while reconciling conflicts between competing uses and values.

3. *Coordinate Strategies for Conservation of Shared Resources.* Many natural resource issues and concerns cross-jurisdictional lines. Examples include migratory fish and wildlife, wide-ranging endangered species, long-term regional timber supplies, air quality and water flows. Regional-scale ecosystems are logical units in which to coordinate land uses and management actions to achieve desired conditions regarding these resources.

Complementary roles for different land tenures, including the legitimate rights of private land holders, can be blended by using existing authorities (Salwasser et al. 1987) or concepts, such as Biosphere Reserves (Greg and McGean 1985) and landscape linkages (Harris and Gallagher 1989).

4. *Get People Involved* in all aspects of public resource decision-making so that managers will know their needs and views; so that people will understand their personal responsibilities, what is possible, and what the relative tradeoffs are; and to obtain informed consent on the course of action selected. Use consensus-building and negotiated problem-solving (Wondolleck 1988) as primary approaches to dispute resolution. People who are affected by resource management and conservation strategies must feel a strong commitment to being part of the solution.

5. *Integrate Information and Technology*, such as ecological classifications, inventories, data management systems, and predictive models, and use them routinely in landscape-scale analyses and conservation strategies. Agencies and affected interest groups and enterprises should contribute to common inventories of the basic conditions of soils, waters, and biota and share data and other information as appropriate to their missions and property rights. Inventories of biological diversity in the United States should build from the foundation of state Heritage Programs (Jenkins 1988) and multi-resource inventories conducted by various state and federal agencies. They should allow prudent choices to be made based on realistic assessments of needs and priorities for investment and protection actions (e.g., Scott et al. 1987, 1991).

6. *Integrate Management and Research* to improve the scientific basis of ecosystem management. Agencies, universities and affected interest groups and enterprises should

cooperate in long-term, interdisciplinary ecosystem research and development. Managers need practical tools and methods for planning and evaluating the expected effects of management options. They also need expanded choices for sustainable harvest and management of resources.

7. *Revitalize Conservation Education and Interpretation.* Agencies, universities and affected interest groups and enterprises should cooperate in comprehensive programs of interpretation, education, and demonstration of ecosystem management. The result should be a better understanding among the citizenry about the effects of personal actions in sustaining desired ecosystem conditions and better support for the complementary roles played by different agencies and ownerships in overall conservation strategies.

8. *Develop, Monitor, and Evaluate Vital Signs of Ecosystem Health.* Agencies, universities and affected interest groups and enterprises should cooperate in identifying and monitoring carefully selected indicators of ecosystem health and diversity, including conditions and trends of valued resources. Monitoring should be guided by the use of decision analysis tools (Maguire 1988, 1991) to ensure that the most vital information is being collected in useable quality and in a timely fashion for the specific purpose of adapting management based on new information (Holling 1978; Walters 1986).

A Larger Role for Research

Research has a significant role to play in ecosystem management, including the use of scientific methods in understanding the basic capabilities of different ecosystems; discerning the needs and wants of people; describing ecologically, economically and socially sound management goals; and designing monitoring systems that will facilitate periodic adaptation to new knowledge (National Research Council 1990; Lubchenko et al. 1991). However, the role for science must reflect the fact that scientists are not the only source of information for solutions to difficult political and social choices. For example, scientifically unique or perfect answers do not exist for how a balance of goals and practices for ecosystem management should be struck. Human values, preferences and aspirations are crucial adjuncts to scientific information in public resource policy-making.

The role of science in ecosystem management is, thus, to help define what is possible and what is desired: to shed light on how to best attain a desired set of conditions or benefits and to help people understand the estimated costs, benefits and consequences of alternative courses. To fulfill this role effectively, social, biological and physical sciences must be integrated to reflect the complexity of how ecosystems actually function.

CASE EXAMPLES OF CONSERVING DIVERSITY THROUGH MULTIPLE-USE ECOSYSTEM MANAGEMENT

The following three examples show how some of the principles and guidelines for an ecosystem approach to multiple-use management were applied to meet specific objectives for conservation of biological diversity at regional, landscape and site scales on national forests in the early 1990s.

A Regional Strategy for Wildlife Population Viability: The Spotted Owl

The first case is a regional strategy for maintaining population viability of a threatened wildlife species: the northern spotted owl (*Strix occidentalis caurinus*) in Oregon, Washington and California. The strategy addresses biodiversity goals for recovery of listed species, viable populations of native species, genetic diversity, structural diversity and a network of biological communities. It also shows the use of ecosystem management guidelines for coordination on shared resources, integration of management and research, and monitoring of ecosystem indicators.

Beginning in the early 1980s, a series of strategies for conserving habitat for viability of northern spotted owl populations were adopted by various state and federal resource management agencies. The strategies had in common the protection of certain kinds of habitats in an interlinked network throughout the range of the subspecies (Salwasser 1988). With time, the strategies included more habitat, in larger blocks, distributed at closer spacing, with specific standards for ecosystem conditions outside of designated habitats for the owls.

The strategy proposed for use in the early 1990s (Thomas et al. 1990) was designed to provide sufficient kinds, amounts and locations of suitable habitat to conserve numbers and a distribution of the subspecies deemed adequate that would ensure its persistence for a century or longer. Management direction consisted of maps that showed locations of habitat reserves and a set of management standards that explicitly defined relationships among reserve locations, sizes and spacing. They also set direction for managing dispersal habitats for juvenile owls. Overall, habitat management was designed to support a population of adequate size and dispersion to minimize risks from environmental variations and catastrophic events, to maintain connectivity for demographic and genetic flows between reserves, and to maintain the current range of environmental conditions throughout the species' range. The policy on spotted owl habitat conservation was in flux at the time this book went to press. Nevertheless, fundamental aspects of the owl conservation strategy warrant description, keeping in mind that the specifics may well have changed with time.

Size of Reserves.—In the 1990 strategy, 69 habitat reserves, each large enough to maintain a local interacting population of at least 20 spotted owl breeding pairs, were designated for protection. The reserves circumscribed locally high densities of owl pairs and some of the better existing nesting, roosting and foraging habitats. Thirty-five additional reserves, each capable of sustaining 2-19 breeding pairs, were identified where it was not possible to have larger reserves. All existing habitat within the reserves was to be maintained and young forests were to be regrown into old forest habitat conditions.

The goal for each habitat reserve was to provide the capability to support at least 20 spotted owl pairs. This goal was derived from empirical analysis of spotted owl populations, evaluations of studies of other raptor species, particularly in island situations, and simulation modeling of "persistence likelihoods" of variously sized clusters of spotted owl pairs. Where field conditions might not provide for 20-pair reserves, the next largest reserve was designated.

Distribution.—Habitat reserves were spaced edge-to-edge, no farther apart than the dispersal capability of two-thirds of juvenile spotted owls observed with radio-tracking. This spacing was designed to help assure that spotted owls could interact among the reserves throughout their range. Redundancy in the number of reserves and their interconnectedness should help increase the likelihood that local populations would remain interlinked despite unplanned loss of habitats from natural catastrophes such as extensive crown fires and wind storms. Reserves were built around suitable habitats that were already protected in national parks, wilderness areas and areas that will not be scheduled for habitat disturbance activities. This was intended to reduce short-term economic effects of the reserve system.

Habitat management standards called for reserves to be linked by a more or less even distribution of dispersal habitat in the forest matrix between the reserves. Spotted owls can disperse through younger forests than those within reserves designed to support reproduction. Such dispersal habitat can be provided over at least 50% of the land area between reserves. This would provide for widely distributed, random dispersal as juvenile northern spotted owls do not seem to use travel corridors *per se*.

Management standards also called for additional owl-pair sites to be conserved in smaller reserves within areas of concern. Areas of concern were geographic locations where highly fragmented habitat and land ownership patterns cause undesired breaks in the distribution of habitat and spotted owl pairs. The purpose of providing for habitat in areas of concern was to maintain some demographic linkage between potentially isolated populations.

Results.—About 2.4 million ha of spotted owl habitat was to be protected in the National Forest System in the range of the northern subspecies. It was estimated that habitats on all land ownership combined could support more than 1,400 breeding pairs. As of 1992, about 630 breeding pairs had been confirmed to occupy designated reserve areas. All administrative units were to follow the same management direction and employ the same monitoring standards.

Economic Considerations.—The spotted owl conservation strategy in 1990 did not call for preserving all existing nesting, roosting and foraging habitat, even though the northern spotted owl was a federally-listed threatened species. This was because social and economic considerations—potentially severe economic impacts of taking old forest habitats out of timber production on national forests—were one consideration in building and implementing the strategy. Social costs of preserving all individuals of the subspecies were deemed too high by policy makers at the time. This has subsequently changed and may yet change again. But one thing remains constant: controversy that still swirls around the spotted owl is centered on the issue of risk to owls versus risk to human livelihood.

Implementation.—Implementation of the spotted owl conservation strategy occurs through regional planning guidelines for consistent designation of habitat reserves and management of dispersal habitat. An integrated program of inventory, monitoring and research on spotted owl populations and habitat relationships is helping to identify habitats for conservation, inventory habitats and populations, track population trends, test assumptions of the strategy and test silvicultural methods for providing habitat. The monitoring program is designed to test specifically and refine management standards, habitat definitions and our understanding of population dynamics. An important element of the monitoring program is communication of the costs, benefits and necessity of the conservation strategy, both within and outside the Forest Service.

The spotted owl/old-growth forest conservation strategy will provide habitat for many other old-growth and older-forest associated species and ecological communities. The need to integrate spotted owl conservation into the overall framework of ecosystem management in the range of the species was identified by Johnson et al. (1991). It is now underway through amendments to forest plans and project decisions.

Ecosystem Management at Landscape Scales: The Bridger-Teton National Forest

The second case illustrates conservation of species and community diversity through integration of land and resource management at the landscape scale. The Bridger-Teton National Forest contains 1.4 million ha of public land in the state of Wyoming. It is located south and east of Yellowstone National Park and is part of the larger Yellowstone region. The Bridger-Teton National Forest contains the headwaters of four major river systems and is home to most of North America's large ungulate species. The recovery strategy for grizzly bear (*Ursos arctos*) serves as an ecological integrator for shared management of different land-use jurisdictions of region, similar to the role played by the spotted owl in the Pacific Northwest.

The Bridger-Teton has been one of the most controversial national forests in the nation. Controversies have come from many sources: threatened and endangered species such, as bald eagle (*Haliaeetus leucocephalus*) and grizzly bear, close proximity to two internationally recognized national parks, public desires for wilderness preservation (0.6 million ha formally designated as of 1986), and considerable potential for economic developments in the form of minerals, oil and gas and outdoor recreation.

The Forest Service is required to prepare a comprehensive land and resource management plan for each National Forest. The Bridger-Teton National Forest began planning in 1983, publishing a Draft Plan and Environmental Impact Statement (EIS) in 1986. It was met with a storm of public controversy, drawing about 20,000 comments from over 9,000 respondents in 49 states. People voiced strong concerns about almost every aspect of the Draft Plan and EIS.

The Final Plan and EIS were released in 1989 and a Record of Decision was published early in 1990 after a 45-day public review period. Intense lobbying went on during the review period, but, once the Record of Decision appeared, the lobbying ended. Seven appeals were filed. One was dismissed. One was settled through negotiation. The Chief of the Forest Service answered the other five appeals by upholding the planning process and decision in all respects. The difference between Draft and Final plans is the difference between a single-resource oriented, technocratic planning process and an integrated, participatory approach to shaping multiple-use management direction.

The Draft Plan.—The draft forest plan had limited public participation and displayed a single desired future condition for the entire national forest. About one page in length, the statement provided a summary of broad goals for resources but no specific information about the agency's intent for the future conditions of specific lands or resources. This discussion was left for a set of "management prescriptions"—a list of permitted management practices and resource uses that constituted the "zoning" direction for future forest management. The management prescriptions were developed by a Forest Service interdisciplinary team with modest public review until the draft plan was published.

In the draft plan, management prescriptions were applied broadly to resources based on an econometric model called FORPLAN which used mainly to present net value determinations, other-resource "constraints," and timber-production-limit coefficients.

The FORPLAN model described timber-harvest areas non-spatially. Management prescriptions were based on FORPLAN outputs and on local knowledge mainly articulated by interdisciplinary team members, including a Wyoming Game and Fish department representative. Sometimes resources in contiguous areas were considered during prescription development, and sometimes, particularly for "suitable" and "scheduled" timber-harvest areas, they were not. With this approach, landscape-level environmental designs, including for example, a migration-corridor design for the Jackson Elk Herd, were not particularly feasible.

The Jackson elk herd (*Cervus canadensis*) roams from its summer habitat in Yellowstone National Park south through Grand Teton National Park (western routes) and the Bridger-Teton National Forest (eastern routes) to its wintering grounds in the National Elk Refuge north of Jackson. There, elk attract the attention of many thousands of visitors each year. Over the years, logging and its resulting increased visibility of elk to people together with increased hunting access in the area south of the Teton Wilderness had reduced the portion of the elk population using eastern migration routes. In the absence of wolves (*Canis lupus*) and relatively unhunted by people, the numbers of elk using the western migration route though Grand Teton National Park increased. Consequently, winter habitat along the western route declined in quality, supplemental feeding programs for wintering elk increased and the potential for disease and winter starvation grew.

During both the draft and final forest planning processes, individual citizens, biologists from the Wyoming Game and Fish Department (the state agency responsible for the elk herd), and National Elk Refuge managers wanted to create a more secure passage for elk through the National Forest while continuing to use hunting to reduce elk numbers. That way, the herd would have the greatest chance of remaining viable and healthy.

In the planning process, the area scheduled for further road building and timber harvest was used as an indicator of future elk habitat conditions. The location of areas scheduled and modification of management practices were designed to secure elk habitat and help sustain or restore forage productivity along elk migration corridors.

The Final Plan.—In the final forest plan, desired future conditions were stated more explicitly. They were tied to actual lands and resources and provided specific management direction. Management objectives were also more explicit with each Desired Future Condition statement representing a commitment to attain a compatible "bundle" of management objectives. In this way, each Desired Future Condition "vision" expressed an array of desired values and uses for each area and that multi-dimensional vision was clearly and tightly connected to management objectives and direction for specific landscapes.

Environmental design of management alternatives proceeded differently in the final plan as well. The design process brought together interested people from the community with the Forest Service interdisciplinary team and Ranger District staff (front-line field managers). All parties labored together over light tables to design alternatives, using Geographic Information System maps, high-altitude photo maps and the "land wisdom" of all participants. This effort, as all other deliberations of the interdisciplinary team, was fully open and accessible to public participants at all times.

As each environmental design was finished, resource-output and condition data resulting from Desired Future Condition assignments were put into the FORPLAN model.

The computer was then used to simulate the expected future performance of each alternative, identifying present net value and other economic indicators as one of many characteristics. Virtually no "constraints" were used in this model because they had already been incorporated into assumptions about Desired Future Condition performance. In that way, team members and public participants designed management scenarios first, outside the computer model, then used the model to assess the likely results of the scenarios.

Because public interest groups were intensely involved in designing the final plan, their interests in such issues as management of elk migration routes were given careful and explicit attention. Differing interests could contribute equally, commercial interests gaining as much consideration as interests concerned about environmental protection. For example, by identifying desired future conditions for elk habitat and restricting surface effects of oil and gas development to achieve those conditions, participants shared access to resources they deemed important in the forest.

The preferred alternatives in the draft and final plan addressed the needs of the Jackson elk herd in a different way. The final plan secured a larger area for elk migration corridors. Better habitat for elk means potential prosperity for commercial and recreational activities dependent on elk hunting and viewing, ski and snowmobile touring and sales of photos and other mementos. The final plan also supported the goals of other federal, State, and local neighbors for elk and the many values and uses of them.

In the final plan, land-use assignments were made to 17 different Desired Future Condition packages. About 5% of the forest is zoned for resource production emphasis, principally energy and timber. About 2% is zoned for wild rivers and domestic water supplies. About 40% is zoned for wilderness management, which has grizzly bear recovery as a high priority. About 2% is designated for grizzly bear recovery outside of wilderness areas. A special designation of 1% was made for environmental education use by the Teton Science School. The remaining 50% of the forest area is zoned for wildlife emphasis with only those resource uses compatible with wildlife diversity allowed. Habitat for elk and grizzly bears along with native fisheries dominated overall land-use allocations in the final plan. Yet substantial resource development and use was still accommodated.

Because mutually compatible objectives were "bundled" together in Desired Future Condition packages, integrated direction governing land occupancy and use allows, among other opportunities, a flow of wood fiber to local mills, access for motorized recreationists and adequate opportunities for oil and gas exploration, leasing and development.

The Bridger-Teton National Forest planning process directly and indirectly addressed all eight biodiversity goals from the Keystone dialogue. Furthermore, it developed many of the ecosystem management guidelines subsequently adopted in the New Perspectives program of the Forest Service.

Restoring Site-scale Diversity: The Shawnee Prairie Barrens

The third example illustrates restoration of biological diversity at a local or site scale within the context of forest plan direction for multiple-use management at the landscape scale. Shawnee National Forest managers in southern Illinois are restoring barrens, a southeastern United States savanna-like natural community, by reintroducing fire, removing invading trees and shrubs and controlling exotic species. They are doing this through extensive partnerships and collaboration with a wide array of people and groups.

The Setting: Simpson Township Barrens.—The Simpson Township Barrens lie within the Shawnee Hills Natural Division of the Shawnee National Forest's Vienna Ranger District. The area contains limestone prairie barrens within a landscape of sandstone-oak woodlands. Both the barrens and woodlands support rare plant communities. Sensitive species include the heart-leaved plantain (*Plantago cordata*), superb lily (*Lilium superbum*), and orange coneflower (*Rudbeckia fulgida*). A rare lizard, the six-lined racerunner (*Cnemidophorus sexlineatus sexlineatus*), is also present.

The Shawnee National Forest Plan recognized that limestone barrens have unique ecological, scientific, and social values. In the Simpson Township Barrens, these values were being lost to: (1) encroachment by woody plants, (2) invasion by exotic plants and (3) siltation from road drainage runoff. Aerial photographs from 1938, Public Land Survey notes from 1806 and field research confirmed that the oak woodland was historically more open and the prairie barrens more extensive than in 1988.

To restore native plant communities to the Simpson Township Barrens, scientists and managers developed a strategy to eliminate competing plant species. Prescribed fire was used in 1988 and 1990 to restore 180 acres in a mosaic of barrens and woodlands. The sandstone-oak woodland was thinned with chainsaws and hand tools to restore the original vegetation mosaic. Exotic white and yellow sweet clovers (*Melilotus alba* and *M. officinalis*) were pulled by hand and exotic black locust (*Robinia pseudoacacia*) was removed. Drainage problems were corrected with the Johnson County Highway Department installing a new culvert. Interpretive signs were installed and a parking area and pull-out were built to enhance the visitor's ability to enjoy the restored area. The project became the model for a program of other similar projects throughout the Forest.

Partnerships with People.—Partnerships were a key component of the Simpson Township Barrens Project. The environmental analysis for the Project was completed by the Forest Service in partnership with the Illinois Department of Conservation's Division of Natural Heritage. Prescribed fire and tree and shrub removal were completed with financial assistance from the Illinois Chapter of The Nature Conservancy, labor and equipment from the Division of Natural Heritage and labor from the Illinois Department of Corrections' Dixon Springs Work Camp. The Johnson County Highway Department corrected drainage problems and constructed the pull-off parking area. Boy Scout Troop 41, from Vienna, Illinois, built a cedar-rail fence and removed trash and blowdown debris. Interpretive signs were provided by the Illinois Nature Preserves Commission, the Illinois Chapter of The Nature Conservancy, the Division of Natural Heritage—Illinois Department of Conservation, Kathy Phelps (a local artist) and the Shawnee National Forest.

Flexibility for Creative Approaches.—The Project integrated goals for vegetation, fire, wildlife and recreation into a landscape-scale "ecosystem management" package to carry out forest plan direction. Prescribed fire was used to restore the area's ecological integrity. Reintroduction of fire favors species such as Cooper's and Sharp-shinned hawks (*Accipiter cooperii* and *A. striatus*) and shade-tolerant plants. Interpretive signs describe the area's vegetation, fire-dependency, wildlife and present management, translating ecological restoration into recreational opportunities.

Partnerships Between Managers, Researchers and Educators.—The Shawnee National Forest engaged numerous outside scientists and resource specialists in developing the Simpson Township Barrens Project. Botanists from the Illinois Native Plant Society and the Department of Conservation provided technical assistance during the planning stage. Researchers from Southern Illinois University's plant biology department reviewed and assisted forest botanists/ecologists in implementing a fire history study of the older oaks present on the site.

Results: Meeting the Challenges of Sustainable Resource Management.—There has been a dramatic restoration of biological diversity in the Shawnee National Forest due to the Simpson Township Barrens Project. Wildflower populations have increased and limestone barrens have expanded. Several rare plants that had been extirpated have now returned. The exotic white and yellow sweet clovers and black locust have been decreased to where annual management actions now exclude them from the restored areas. Interpretive signs and the collaborative nature of the project have given the residents of nearby Simpson, Illinois, a profound sense of pride and ownership in the area. The Forest is now carrying out a full program of similar projects to restore the native diversity of Illinois natural heritage.

SOME NEW TOOLS FOR ECOSYSTEM MANAGEMENT

These cases, along with other New Perspectives projects, integrated specific objectives for conservation of biological diversity into desired future conditions for ecosystems which must provide for multiple resource uses and values. The cases show the importance of taking actions at several geographic scales, sites to regions at a minimum. They also illustrate some key tools for ecosystem management. We highlight three of them: the concept of desired future condition, participatory decision-making and geographic information systems. These tools reflect the importance of new thinking, new alliances and new technologies in ecosystem management.

New Thinking: Desired Future Condition

Desired future condition is a key organizing concept for integrating multiple objectives at different geographic and time scales (36 CFR 219; Caplan 1992). The concept evolved out of traditional resource goal-setting approaches that focused mainly on single-resource production and impact mitigation. Traditional approaches often viewed considerations for environmental protection as constraints on resource production rather than as key elements in sustaining desired ecosystem conditions. Viewing land health concerns as "constraints" is untenable in ecosystem management. Instead, conditions of land health and diversity are what enable sustainable resource production.

The word "desired" in this concept refers to the attainment of values shared by participants: in the case of public resources, the owners and managers of the land. Desires can be for past conditions as well as those yet unattained. "Future" addresses temporal aspects of shared values, that some (old-growth forests, for example) can only be nurtured or attained over long periods of time. "Condition" refers to the capacity of land to provide different environmental services, resource uses or values. "Condition" also refers to the resilience of land to stress-events. "Condition" assumes knowledge of existing conditions based on sound inventories, knowledge of the potential capabilities of the ecosystem to respond to various management strategies and practices, based on sound ecological classifications and interpretations, and some degree of knowledge about how to get from where conditions are to where people want them to be.

A description of "desired future condition" helps define and answer several concerns, such as the need to sustain desired resources and the potential for desired ecosystem responses to global climate change. Desired future condition also allows landscape-scale considerations to be woven into larger-scale, bioregional decisions and policies, such as those meant to sustain birds and butterflies that migrate from one continent to another during the course of their annual cycles.

People define and describe a desired set of conditions, then plan for resource-management practices that will most likely attain those conditions. In a programmatic sense, "practices" include standards, guidelines and other management direction needed to attain goals and objectives for an area. Practices can include vegetation-harvest patterns, prescribed fire, stream-bank restoration and other actions that change resource conditions. By focusing on desired future conditions, participants are able to put management practices in context, ideally replacing internecine battles over methods with higher levels of agreement over outcomes.

New Alliances: Participatory Decision-making

Participatory decision-making is another key to ecosystem management. Ecosystem managers must use planing processes that are relatively "free," "fair," "open" and "tangible" for all participants. Involvement of non-agency people in decision-making must become a regular, day-to-day part of public natural resource management.

Planning processes must minimize participant time and financial costs, yet allow access for anyone interested in or affected by the decisions. They must offer the greatest possible information and interaction to people and give participants a sense of achievement, something tangible to take home and work with.

Participatory decision-making is sometimes referred to as "fishbowl" decision-making: all actions to be decided are clearly visible to participants.

New Technologies: Geographic Information Systems

One of the best technical tools for involving people in ecosystem management is Geographic Information Systems (GIS). GIS gives people new insights into resource relationships and conditions, present and future. In many of today's applications, GIS integrates site-sampled resource data with the sorting and analysis capabilities of computers to produce visual displays and maps of ecosystem attributes.

Two limitations of GIS, however, are incomplete data and data that might not have been gathered with spatial attributes in mind. The historical tendency was to collect data as indices or scaler interpretations of conditions rather than simple, basic descriptions of ecological conditions. Another limitation is that protocols for field sampling call for random, often stratified plot placement across landscapes to assure statistical rigor. With such approaches, boundaries that must be highly accurate for GIS to work effectively remain "fuzzy."

When GIS is enhanced with ties to remote-sensing systems, it gains spatial accuracy. Boundaries of landscape components become clearer and are readily incorporated into GIS data and displays. In addition, field-plot data can be used to glean more information from high-altitude photo reconnaissance or satellite information. This is done by using plot information as the basis for defining resource values represented by the spectral information in each pixel of remotely sensed images. Data bases and maps are generated based on locations of like spectral "signatures" with greater than 90% boundary and resource accuracy in most cases.

In some cases, field sampling must be done to investigate the attributes of pixels that earlier sampling "missed." Global Positioning Systems equipment is used to make sure each pixel's field location is found. If remote-sensing-supported GIS are further tied to cumulative, iterative and adaptive models, they are capable of helping people answer and display many "what if" questions about possible future ecological conditions. As the models input and output data with GIS, the landscape patterns and resource attributes of various resource-management alternatives can be generated and used to support public decisions.

SOME CLOSING THOUGHTS ON NEW PERSPECTIVES AND ECOSYSTEM MANAGEMENT

New Perspectives was a set of projects designed to give people a chance to try some new and some old thinking about how to carry out the management of lands and resources in the National Forest System: preferably to find better ways to blend ecological, economic and social considerations in sustaining desired ecosystem conditions in addition to the flows of uses and products people need from the land. Five themes were emphasized in these projects:

- Sustain diverse and productive ecological systems;
- Integrate the different aspects of land and resource management, research, and conservation;
- Improve the effectiveness of public participation in resource decision-making;
- Build partnerships between forest users and forest managers; and
- Strengthen teamwork between researchers and resource managers.

In the two years that New Perspectives was used by the Forest Service to stimulate new thinking, new technologies and new alliances, it was not possible to build all five themes into every plan or project. This is partly because large organizations, guided by historic plans and priorities, are inherently slow to change. Furthermore, some scientists and managers were reticent to adopt "untested" ideas, especially the social dimensions of ecosystems. And the social aspects of ecosystem management are inherently more difficult to work with than are the technical aspects. Nevertheless, progress was made in exploring the shifting from a resource production focus, with biodiversity conservation as a "constraint," to an ecosystem focus for multiple-use management that includes biodiversity conservation an integral goal.

Sustaining desired ecological, economic and social conditions in ecosystems that are managed for multiple purposes, such as the National Forest System and other public lands in the United States, is a major challenge. But it is not an impossible task if people realize that no single objective can dominate ecosystem management at all geographic scales or even at the same site for all times. Success in sustaining desired ecosystem conditions will depend on having scientifically sound, economically feasible and socially acceptable strategies for achieving combinations of ecological and social goals. For example, it will depend on meeting specific objectives for viability of native species and biological communities such as spotted owls, grizzly bears, elk and tall-grass prairies. It will depend on meeting specific objectives for the characteristics of landscapes such as habitat conditions that are aesthetically pleasing and allow for free movement of plants and animals over time. It will depend on coordination among people responsible for species or resources that transcend administrative boundaries such as the spotted owl, large predators, eagles, migratory birds and mammals and timber and mineral resources. It will depend on practical management standards for the desired characteristics of distinct patches in a landscape, such as diversity of species, structures, and functions provided by snags, fallen trees, riparian areas, prairie barrens, and commensal interactions among species. It will also depend on better integration of research with management, especially in monitoring conditions pertinent to objectives.

These and other actions to sustain desired ecosystem conditions in multiple-use lands are changing traditional approaches to multiple-use management. Some of these changes might result in higher management costs or in less access to resources for people who have high dependencies on those resources for subsistence or economic well-being. On the other hand, they might also result in lower management costs or in resource management being carried out in areas previously considered to be "off-limits" to human use of resources. Only site-specific assessments can show what the nature of the changes might be. In any case, changes in public land uses are often difficult to establish and implement both politically and logistically. But they are tractable if people decide they are willing to make them.

The examples cited above illustrate just three cases of an ecosystem approach to multiple-use management: a regional strategy for habitat conservation to secure population viability for spotted owls by tying different land-use classes together across multiple jurisdictions; an integrated planning process to achieve multiple resource benefits in a large geographic area through social consensus for both economic and biological issues; and ecological restoration that rebuilds both natural biodiversity and community responsibility for conservation of diversity at the landscape and site scale. There is, of course, much more to ecosystem management and to conserving biological diversity than these three cases could illustrate.

We cannot over-emphasize that the learning process on ecosystem management and biodiversity conservation is far from complete. In fact, ecosystem management as a process for sustaining diverse, healthy and productive land has just begun. To date, we have some principles, guidelines, tools, a dozen major research and development programs and several hundred practical demonstrations to draw from. These will expand in our continuing pursuit of new knowledge and technologies.

Finally, an ecosystem perspective on sustaining desired conditions of diversity and productivity in multiple-use lands has potential merit. But it is not sufficient by itself to sustain a harmonious relationship between people, land and resources. Regardless of how well ecosystem management works, other actions will be needed to bring people and land into a better harmony.

NEEDED: A GLOBALLY RESPONSIBLE CONSERVATION ETHIC

Foremost among these actions, North Americans must become more conservative in how they (we) produce, use and dispose of natural resource products. American behaviors regarding consumption and waste of resources are major forces of change in the global ecosystem (Silver and DeFries 1990). Americans need to renew their conservation ethic to bring balance to the complementary roles of managing ecosystems, producing resources and conserving resources (Postel and Ryan 1991). A recent public opinion survey found that, despite an economic recession, the American public was prepared to make the commitments such an ethic entails (The Roper Organization 1992).

Sustaining diverse, productive and resilient ecosystems in the United States is important. But it must be balanced by a commensurate change in how and where Americans get and dispose of their resources. The potential off-site effects of protecting ecosystems in one place, such as old-growth forests or wildlife refuges in certain regions of the United States, while continuing profligate use of resources that are produced in other places, such as oil from the middle east or timber from Canada, are not well understood by many people.

Ecosystems are ultimately interconnected. The potential for interregional and international transfers of the economic, ecological or social effects of where resources are produced and where they are used highlight the veracity of Garrett Hardin's (1985) comment that "it is not possible to do only one thing in an ecosystem." Resource policy in the United States, if it is to truly reflect an ecosystem concept, needs to pay as much attention to the off-site effects of actions or inactions taken to protect nature or produce resources as to on-site effects. Most, if not all, environmental impact assessments ignore these off-site effects of decisions. And we need to think more about long-term dynamics in ecosystems. Things do not stay fixed or in the same place over time. Again, we turn to Garrett Hardin (1985). His key ecological question, "and then what?" is a clue to how citizens and ecosystem managers must think when they think they are closing in on a simple solution to a complex problem.

Needed: A New Model for Conservation Science

Finally, to shape ecosystem perspectives on land and resource management, especially on the linkages between ecological, economic and social factors that an ecosystem view implies, social, biological and physical sciences must become better integrated (NRC 1990; Lubchenko et al. 1991). We cannot learn how to sustain diverse, healthy and productive ecosystems by pursuing only traditional disciplinary sciences whether they be oriented to biological, physical or social disciplines.

Ecosystem perspectives on sustainable resources management have the capability to bring forth a new model for developing the scientific basis of conservation: interdisciplinary teams of researchers working hand-in-hand with managers, educators and citizens to address both short and long-term dynamics in the many dimensions of relationships between people and the land. The working principles for such a model, known as adaptive management, have been evolving for nearly 20 years (Holling 1978; Walters 1986). It is time to make the adaptive management model standard operating procedure for sustaining diverse, healthy, and productive ecosystems.

LITERATURE CITED

36 C.F.R. 219. 1982. National Forest Management Act Land and Resource Planning. *Code of Federal Regulations.*

Berry, W. 1987. *Home Economics.* North Point Press, San Francisco, Calif.

Bonnicksen, T. M. 1988. Restoration ecology: philosophy, goals, and ethics. *Environ. Professional* 10:25-35.

Botkin, D. B. 1990. *Discordant Harmonies: a New Ecology for the Twenty-first Century.* Oxford University Press, New York.

Burgess, R. L., and D. M. Sharpe, eds. 1981. *Forest Island Dynamics in Man-dominated Landscapes.* Springer-Verlag, New York.

Cairns, J., Jr. 1986. Restoration, reclamation, and regeneration of degraded or destroyed habitats. In *Conservation Biology: the Science of Scarcity and Diversity*, ed. M. E. Soulé, pp. 465-484. Sinauer Associates, Inc., Sunderland, Mass.

Cairns, J., Jr., ed. 1988a. *Rehabilitating Damaged Ecosystems.* 2 vols. CRC Press, Boca Raton, Fla.

Cairns, J., Jr. 1988b. Restoration ecology: the new frontier. In *Rehabilitating Damaged Ecosystems*, ed. J. Cairns, Jr., pp. 1-11. Vol. 1. CRC Press, Boca Raton, Fla.

Cairns, J., Jr. 1988c. Restoration of damaged ecosystems and opportunities for increasing diversity. In *Biodiversity*, ed. E. O. Wilson, pp. 333-343. National Academy Press, Washington, D.C.

Cairns, J., Jr. 1989. Restoring damaged ecosystems: is predisturbance condition a viable option? *Environ. Professional* 11:152-159.

Caplan, J. 1992. Striding into elephant country: exploring new ground for planning and management in protected areas.*George Wright Society Forum* 8(4):44-54.

Cartwright, C. W. 1992. New perspectives for national forest managers. *George Wright Society Forum* 8(4):55-60.

Forman, R. T. T., and M. Godron. 1986. *Landscape Ecology.* John Wiley and Sons, New York.

Franklin, J. F. 1988. Structural and functional diversity in temperate forests. In *Biodiversity*, ed. E. O. Wilson, pp. 166-175. National Academy Press, Washington, D.C.

Gale, R. P., and S. M. Cordray. 1991. What should forests sustain? Eight answers. *J. For.* 89(5):31-36.

Gregg, W. P., Jr., and B. A. McGean. 1985. Biosphere reserves: their history and their promise. *Orion* 4(3):41-51.

Hardin, G. 1985. *Filters against Folly: How to Survive despite Economists, Ecologists, and the merely Eloquent.* Penguin Books, New York.

Harris, L. D. 1984. *The Fragmented Forest: Island Biogeography Theory and the Preservation of Biotic Diversity.* University of Chicago Press, Chicago, Ill.

Harris, L. D., and P. B. Gallagher. 1989. New initiatives for wildlife conservation: the need for movement corridors. In *Preserving Communities and Corridors*, ed. G. MacKintosh, pp. 11-34. Defenders of Wildlife, Washington, D.C.

Holling, C. S., ed. 1978. *Adaptive Environmental Assessment and Management.* John Wiley and Sons, New York.

Hunter, M. L., Jr. 1990. *Wildlife, Forests, and Forestry.* Prentice Hall, Englewood Cliffs, N.J.

Jenkins, R. E., Jr. 1988. Information management for the conservation of biodiversity. In *Biodiversity*, ed. E. O. Wilson, pp. 231-239. National Academy Press, Washington, D.C.

Johnson, K. N., J. F. Franklin, J. W. Thomas, and J. Gordon. 1991. Alternatives for management of late successional forests of the Pacific Northwest. A report to the Agriculture and Merchant and Marine Fisheries Committees of the U.S. House of Representatives. Scientific Panel on Late Successional Forest Ecosystems, Washington, D.C.

Jordan, W. R., III. 1988. Ecological restoration: reflections on a half-century of experience at the University of Wisconsin-Madison Arboretum. In *Biodiversity*, ed. E. O. Wilson, pp. 311-316. National Academy Press, Washington, D.C.

Jordan, W. R., III, M. E. Gilpin, and J. D. Aber, eds. 1987. *Restoration Ecology*. Cambridge University Press, Cambridge.

Kessler, W. B. 1992. A tale of two paradigms: multiple use and ecosystem management. *George Wright Society Forum* 8(4):13-20.

Keystone Center. 1991. Final consensus report of the Keystone Policy Dialogue on Biological Diversity on Federal Lands. Keystone, Colo.

Ledig, F. T. 1986. Heterozygosity, heterosis, and fitness in outbreeding plants. In *Conservation Biology: the Science of Scarcity and Diversity*, ed. M. E. Soulé, pp. 77-104. Sinauer Associates, Inc., Sunderland, Mass.

LeMaster, D. C. 1992. Living with the land. *George Wright Society Forum* 8(4):2-12.

Leopold, A. 1949. *A Sand County Almanac and Sketches Here and There*. Oxford University Press, New York.

Lubchenko, J., A. M. Olson, L. B. Brubaker, S. R. Carpenter, M. M. Holland, S. P. Hubbell, S. A. Levin, J. A. MacMahon, P. A. Matson, J. M. Melillo, H. A. Mooney, C. H. Peterson, H. R. Pulliam, L. A. Real, P. J. Regal, and P. G. Risser. 1991. Sustainable biosphere initiative. *Ecology* 72(2):371-412.

MacCleery, D. W. 1992. Condition and trends of U.S. forests: a brief overview. USDA Forest Service, Timber Management Staff, Washington, D.C.

Maguire, L. A. 1988. Decision analysis an integrated approach to ecosystem exploitation and rehabilitation decisions. In *Rehabilitating Damaged Ecosystems*, ed. J. Cairns, Jr., pp. 105-122. CRC Press Boca, Raton, Fla.

Maguire, L. A. 1991. Using decision analysis to manage endangered species. In *Challenges in the Conservation of Biological Resources: a Practitioner's Guide*, eds. D. J. Decker, M. E. Krasny, G. R. Goff, C. R. Smith, and D. W. Gross, pp. 139-152. Westview Press, San Francisco, Calif.

Millar, C. 1987. The California forest germplasm conservation project: a case for genetic conservation of temperate tree species. *Conserv. Biol.* 1:191-193.

National Research Council (NRC). 1990. *Forestry Research: a Mandate for Change*. National Academy Press. Washington, D.C.

Overbay, J. C. 1992. Ecosystem management. USDA Forest Service photocopy. Washington, D.C.

Postel, S., and J. C. Ryan. 1991. Reforming forestry. In *State of the World 1991*, proj. dir. L. R. Brown, pp. 74-92. W. W. Norton and Co., New York.

Reid, W., C. Barber, and K. Miller, eds. 1992. *Global Biodiversity Strategy: Guidelines for Action to Save, Study, and Use Earth's Biotic Wealth Sustainably and Equitably*. World Resources Institute, Washington, D.C.

Robertson, F. D. 1992. Ecosystem management of the national forests and grasslands. June 4, 1992 Memo to Regional Foresters and Station Directors, USDA Forest Service, Washington, D.C.

Roper Organization Inc. 1992. Natural Resource Conservation: Where Environmentalism is Headed in the 1990s. National Environmental Forum Survey conducted for the Times Mirror Magazines, New York.

Rowe, J. S. 1990. *Home Place: Essays on Ecology*. NeWest Publishers Ltd., Edmonton, Alberta, Canada.

Sahtouris, E. 1989. *Gaia: the Human Journey from Chaos to Cosmos*. Pockett Books, New York.

Salwasser, H. 1988. Managing ecosystems for viable populations of vertebrates: a focus for biodiversity. In *Ecosystem Management for Parks and Wilderness*, eds. J. K. Agee and D. R. Johnson, pp. 87-104. University of Washington Press, Seattle, Wash.

Salwasser, H. 1991. In search of an ecosystem approach to endangered species conservation. In *Balancing on the Brink of Extinction: the Endangered Species Act and Lessons for the Future*, ed. K. A. Kohm, pp. 247-265. Island Press, Washington, D.C.

Salwasser, H., D. W. MacCleery, and T. A. Snellgrove. 1992. New perspectives for managing the U.S. National Forest System. Tech. Rep. USA, North American Forestry Commission, USDA Forest Service, Washington, D.C.

Salwasser, H., C. M. Schonewald-Cox, and R. Baker. 1987. The role of interagency cooperation in managing for viable populations. In *Viable Populations for Conservation*, ed. M. E. Soulé, pp. 159-174. Cambridge University Press, New York.

Sample, V. A. 1991. *Land Stewardship in the Next Era of Conservation*. Gray Towers Press, Milford, Pa.

Scott, J. M., B. Csuti, J. D. Jacobi, and J. E. Estes. 1987. Species richness: a geographic approach to protecting future biological diversity. *BioScience* 37(11):782-788.

Scott, J. M., B. Csuti, and F. Davis. 1991. Gap analysis: an application of geographic information systems for wildlife species. In *Challenges in the Conservation of Biological Resources: a Practitioner's Guide*, eds. D. J. Decker, M. E. Krasny, G. R. Goff, C. R. Smith, and D. W. Gross, pp. 167-180. Westview Press, San Francisco, Calif.

Silver, C. S., and R. S. DeFries, eds. 1990. *One Earth One Future: Our Changing Global Environment*. National Academy Press, Washington, D.C.

Society of American Foresters. 1991. *Biological Diversity in Forested Ecosystems: a Position Statement of the Society of American Foresters*. Bethesda, Md.

Soulé, M. E. 1986. Conservation biology and the real world. In *Conservation Biology: the Science of Scarcity and Diversity*, ed. M. E. Soulé, pp. 1-12. Sinauer Associates, Inc., Sunderland, Mass.

Soulé, M. E., ed. 1987. *Viable Populations for Conservation*. Cambridge University Press, New York.

Soulé, M. E., and B. A. Wilcox, eds. 1980. *Conservation Biology: an Evolutionary-ecological Perspective*. Sinauer Associates, Inc., Sunderland, Mass.

Thomas, J. W., E. D. Forsman, J. B. Lint, E. C. Meslow, B. R. Noon, and J. Verner. 1990. A conservation strategy for the northern spotted owl. Interagency Scientific Committee to address the conservation of the northern spotted owl. Government Printing Office, Portland, Ore.

United Nations. 1992. Adoption of agreements on environment and development. United Nations Conference on Environment and Development. Rio de Janeiro, Brazil.

Walters, C. 1986. *Adaptive Management of Renewable Resources*. MacMillan Publishing Co., New York.

Waring, R. H., and W. H. Schlesinger. 1985. *Forest Ecosystems: Concepts and Management*. Academic Press, Inc., New York.

Wilson, E. O. 1988. *Biodiversity*. National Academy Press. Washington, D.C.

Wilson, E. O. 1993. *The Diversity of Life*. W. W. Norton and Co., New York.

Wondolleck, J. M. 1988. *Public Lands Conflict and Resolution: Managing National Forest Disputes*. Plenum Press, New York.

World Commission on Environment and Development. 1987. *Our Common Future*. Oxford University Press, New York.

36

CONSERVATION OF BIODIVERSITY AT THE LANDSCAPE SCALE

Reed F. Noss

Department of Fisheries and Wildlife
Oregon State University
Corvallis, Oregon
U.S.A.

Biodiversity is a multi-scale phenomenon. Unlike traditional wildlife, fisheries, and endangered species programs, the conservation of biodiversity is not solely a species-level problem. Rather, biodiversity exists and must be conserved at genetic, species, and ecosystem levels of organization (OTA 1987) and across many spatial and temporal scales. Some authors have suggested the *landscape* as a meaningful spatial scale at which biodiversity can be recognized, monitored, and managed (Noss 1983, 1990a; Harris 1984; Forman and Godron 1986; Turner 1989; Keystone Center 1991). Landscape diversity can also be considered directly as a higher-order expression of biodiversity. Whereas ecosystem diversity involves merely the number and relative abundance of habitats or communities within a defined area, landscape diversity considers the spatial arrangement of habitats and communities.

Forman and Godron (1986: 11) defined a landscape as "a heterogeneous land area composed of a cluster of interacting ecosystems that is repeated in similar form throughout." Similarly, Urban et al. (1987: 119) characterized a landscape as "a mosaic of heterogeneous land forms, vegetation types, and land uses." These definitions suggest that landscapes have a *pattern* and that this pattern consists of repeated habitat components—for example, patches, corridors, and matrix (Forman and Godron 1986)—that occur in various shapes, sizes, and spatial interrelations. Landscapes are usually considered to be areas at least several kilometers across (Forman and Godron 1986), involving thousands to millions of hectares.

Landscape diversity, then, is pattern diversity (Scheiner 1992). The pattern of habitat components across a landscape has a dominant effect on the distribution of species that are associated with those habitats. Because human activities often change landscape patterns, they have impacts on biodiversity that ripple through other levels of biological organization, affecting species composition and abundances, gene flow, and ecosystem processes. If a forest landscape is fragmented into small patches, for example, those patches may experience a drier microclimate than the original forest, an increased susceptibility to wind-throw, a loss of forest interior species, a reduction of genetic diversity in the remaining populations, and an invasion by weedy and exotic species

(Burgess and Sharpe 1981; Harris 1984; Franklin and Forman 1987). These problems cannot be solved patch by patch, but only across all patches and their matrix. Hence, the landscape is a convenient and useful spatial scale at which to manage and restore land for conservation (Noss 1983; Turner 1989).

Landscape ecology is a broad and interdisciplinary field with contributions from ecology, other biological sciences, geography, landscape architecture, regional planning, sociology, engineering, hydrology, and many other fields (Forman and Godron 1986). My field is applied conservation biology, and I will concentrate here on topics in landscape ecology that are of special relevance to conservation practice. In particular, this paper will explore four landscape-scale topics that have contributed substantial insights to conservation biology over the last few decades: (1) the concept of landscape diversity, (2) disturbance ecology and patch dynamics, (3) fragmentation ecology, including metapopulation dynamics, and (4) reserve network design.

LANDSCAPE DIVERSITY

Most of the science of ecology has been concerned with what goes on inside ecosystems, for instance, habitat use by songbirds in an oak-hickory forest or competition for light among plants in a seral meadow. Species richness or diversity, when measured in ecological studies, has been mostly *alpha* diversity (within-habitat diversity) rather than the diversity of species along environmental gradients (*beta* or between-habitat diversity) or across a geographic region (*gamma* or regional diversity). Note that all three kinds of diversity (Whittaker 1972) are recognized at only the species level of organization; diversity of genes or ecosystems generally was ignored until very recently (Wilson 1988).

Ecologists studying wide-ranging species, such as migratory ungulates or large carnivores, have had to address questions of habitat use at much broader spatial scales than most investigators of population or community ecology, who focused on relatively small organisms and homogeneous sites. Research on wide-ranging animals has been directed at single species rather than entire biotas; diversity is not an issue in such studies. Nevertheless, many topics in wildlife research, including use of habitat corridors by migrating animals and the importance of habitat interspersion and edge effects, date back many decades (Leopold 1933; Stoddard 1936; Lay 1938; Baumgartner 1943; Thomas 1979) and are directly relevant to landscape ecology and conservation.

An important step toward recognizing conservation goals at levels of biological organization above the species was the emergence of *representation* (or representativeness) as a conservation criterion (Usher 1986). For example, two committees of the Ecological Society of America, one of which later split from the Society and evolved into The Nature Conservancy, sought to identify and preserve representative examples of all the natural communities of North America (Shelford 1926, 1933 Kendeigh et al. 1950-51). The Nature Conservancy recognizes the inventory and protection of natural communities as a *coarse filter* that complements the more familiar fine filter of rare species preservation (Noss 1987a; Hunter 1991). A guess is that protection of representative examples of all natural communities in a region might protect 85-90% of the species without having to worry about each species individually. Species that fall through the pores of the coarse filter, including narrow endemics and wide-ranging animals that use many habitat types, would presumably be captured by the fine filter (TNC 1981-84).

But does the fine filter-coarse filter combination, as practiced by The Nature Conservancy, capture all of biological diversity? Or do we need a landscape-scale coarse filter? Consider the concept of environmental gradients. As Gleason (1926) and his followers recognized decades ago, species do not always gather neatly into communities, but instead are distributed along environmental gradients according to their individual physiological requirements and tolerances. These autecological distribution patterns may be modified by interactions with other species, but the community is hardly an organism in the sense of a tightly integrated, coevolved unit that functions to perpetuate itself. It is variable across both time and space.

Traditional classification systems for plant communities employ a standard of homogeneity in sampling; ecotones, habitat mosaics, and disturbed areas are avoided in search of ideal types that can be classified conveniently (Noss 1987a). Finding homogeneous stands of vegetation can be difficult or arbitrary in some landscapes. Yet Daubenmire (1968), among others, instructed ecologists to search for units of vegetation showing little gradient and to stay away from ecotones. Obviously, communities defined in this way do not provide a completely accurate portrayal of species distribution patterns and habitat variation in a landscape. Furthermore, reserves set up around homogeneous community types may fail to protect species dependent on disturbance patches, habitat mosaics, ecotones, or other heterogeneous environments. Because many species of vertebrates and invertebrates—among them holometabolous insects, amphibians, ungulates, and bears—require multiple habitat types to fulfill their life histories (Hansson 1979; Harris 1984; Werner 1986; Noss 1987a), the community-level coarse filter may neglect many species. Different species perceive and use the environment at different scales, so it is worthwhile to take a hierarchical approach and consider ecosystems within ecosystems (Allen and Starr 1982; Noss and Harris 1986).

Climate change is another factor that makes community-level coarse filters less than adequate for biodiversity conservation in many cases. Just as species generally respond individualistically to environmental gradients in space (Gleason 1926), in most cases they also seem to respond independently to changes in time. As climate in North America warmed after the Pleistocene, for example, tree species migrated at different rates and along different routes (Davis 1981), such that community composition has been continuously changing over time. Most modern plant communities are ephemeral and do not reflect long periods of coevolution among species. Therefore, as Hunter et al. (1988) reasoned, community types defined by species composition may not be useful as coarse filters when climate is changing rapidly; instead, physical environments defined by topography, substrate, and other factors may be better coarse filters. A landscape-level coarse filter would seek to represent various environmental gradients and mosaics that are functionally interconnected (Noss 1987a). A topographically and physically diverse landscape, unfragmented by clearcuts, roads, or other dispersal barriers, may allow species to shift their distributions in response to changing temperature and moisture regimes.

Perhaps the best way to represent all ecosystems would be to encompass complete, unfragmented environmental gradients in landscape-scale reserves. That is, a reserve (or a network of connected reserves) includes the highest to lowest and wettest to driest sites, across the full spectrum of substrates, soils, and aspects (Noss 1987a, 1992a). Large, landscape-scale reserves are also better able to encompass natural disturbance regimes and provide a relatively stable mosaic of seral stages and associated populations (Pickett and Thompson 1978; Shugart and West 1981).

A true landscape-level coarse filter, then, is one that captures environmental gradients and mosaics and the full suite of species and processes associated with this heterogeneity. Such a coarse filter targets landscapes ranging between tens of thousands and millions of hectares. A suite of small, separate reserves established on the basis of the conventional species-level fine filter or community-level coarse filter cannot be expected to represent all of biodiversity or to maintain viable populations of wide-ranging species (Noss and Harris 1986; Hunter 1991). The Gap Analysis project of the U.S. Fish and Wildlife Service (Scott et al. 1993) will be useful for landscape-level conservation because it recognizes major vegetation types, as opposed to highly specific plant associations, and identifies large areas of concentrated species richness across gradients.

Whether functional landscapes must be protected as reserves or can be maintained by appropriate management, while allowing for many human uses, is one of the greatest controversies in modern conservation. For instance, both natural disturbances and forestry practices produce mosaics of seral stages on the landscape. But are these different kinds of mosaics equivalent with respect to their influences on species composition and ecological processes? Modern humans have not yet re-learned how to be part of nature or discovered how to emulate natural processes with active management. We will probably never learn enough to justify confidence in total manipulation. In any case, large protected areas always will be needed as benchmarks or control areas for experiments in land management (Leopold 1941). Because we are experimenting with whole landscapes, landscape-sized control areas are required (Noss 1991).

DISTURBANCE ECOLOGY AND PATCH DYNAMICS

One notable generalization of modern ecology is that moderate levels of disturbance maximize the diversity of habitats and species in a landscape (Connell 1978; Huston 1979; White 1979; Pickett and White 1985; Petraitis et al. 1989). The effects of disturbance on diversity are most striking at the landscape scale, where patches ranging from individual treefall gaps to large burned areas and blowdowns can be recognized. Because disturbances are typically patchy in time and space, a landscape can be viewed as a shifting space-time mosaic of patches in various stages of recovery from disturbance (Watt 1947; Bormann and Likens 1979; Baker 1989). Without disturbance, many landscapes would be relatively homogeneous and depauperate in species.

The large spatial scale of many natural disturbances is a strong argument for landscape-scale conservation. A community-level coarse filter would often target areas smaller than the size of a single disturbance. Research Natural Areas (RNAs), for example, are established on federal lands to represent examples of various plant community types. However, 93% of the RNAs on national forests in 1989 were smaller than 1,000 ha and the remaining 7% were less than 5,000 ha (Noss 1990b). In comparison, the Yellowstone fires in 1988 burned 570,000 ha of the Greater Yellowstone Ecosystem, including 400,000 ha of the National Park (Christensen et al. 1989). What if the Yellowstone landscape was represented only by RNAs?

Disturbance is, therefore, an important consideration for determining optimal reserve size, a long-debated issue in conservation biology. Whenever possible, individual reserves should be large, not only to encompass entire environmental gradients and viable populations, but also to allow disturbance regimes to operate unimpeded and maintain a mix of seral stages (Pickett and Thompson 1978; Noss 1983, 1992a; Baker 1992). Reserves that are small relative to the spatial scale of disturbance may experience radical fluctuations in the proportions of different seral stages over time, which in turn threaten populations that depend on certain stages. Many nature reserves are smaller than the area that may be disturbed by a single wildfire or windstorm and, therefore, are vulnerable to catastrophic losses.

In an early paper linking landscape-scale patch dynamics to conservation issues, Pickett and Thompson (1978:34) defined a "minimum dynamic area" as "the smallest area with a natural disturbance regime, which maintains internal recolonization sources, and hence minimizes extinction." The basic idea here is that a nature reserve should be large enough so that only a small part of it is disturbed at any one time. Another requirement is that a source of colonists (i.e., a reproducing population of the same species) exists within dispersal distance so that populations can be reestablished on disturbed sites. In theory, a minimum dynamic area should be able to manage itself and maintain habitat diversity and associated species with little or no human intervention. Shugart and West (1981) estimated that landscapes must be some 50-100 times larger than average disturbance patches in order to maintain a relative steady-state (quasi-equilibrium) of habitats. In a steady-state landscape, the proportions of different seral stages in the overall landscape are almost constant over time, even though the sites occupied by various seral stages change. A steady state may never be reached in some types of ecosystems, such as those regularly experiencing large catastrophic fires (Baker 1989; Sprugel 1991).

The lesson of the minimum dynamic area concept for conservation is that, to maintain biodiversity, all natural seral stages in a landscape should be perpetuated over time and dispersal distances between similar habitats should be surmountable by native species. A steady state of seral stages need not be maintained, for it may rarely occur in nature (Pickett and White 1985; Sprugel 1991). The steady-state concept is useful in the sense that reserves large enough to be close to steady state may experience lower extinction rates than reserves with habitat conditions that fluctuate wildly over time. Larger landscapes buffer the effects of disturbance on diversity of habitats and species (Shugart and Seagle 1985).

Aside from considerations of patch size and seral stage distribution, other aspects of disturbance regimes (Sousa 1984; Pickett and White 1985; Baker 1992) have to be appraised for land management. A striking example of how changes in disturbance frequency and seasonality affect biodiversity is provided by the longleaf pine-wiregrass ecosystem of the southeastern coastal plain. Longleaf pine (*Pinus palustris*) and wiregrass (*Aristida stricta*) are the dominant and ecologically pivotal species in the overstory and herb layer, respectively, in an ecosystem that once covered over 60% of the uplands in this region (Croker 1979). This community type, according to a recent calculation (Ware et al. 1993), has been reduced by over 98% since presettlement time. Logging of longleaf pines was a major cause of the decline, but the failure of the community to recover can be blamed largely on fire suppression, including active suppression and habitat fragmentation, the latter producing barriers to the spread of fires that once covered tens of thousands of hectares (Means and Grow 1985; Noss 1988, 1989).

Fires in the longleaf pine-wiregrass ecosystem are not normally stand-replacing fires; rather, they are low-intensity ground fires that recur at intervals of 2-5 years (Means and Grow 1985; Myers 1985). Germination of longleaf pines is dependent on bare mineral soil, exposed by fire. Both longleaf pine needles and wiregrass are highly flammable and promote frequent fire, probably as a way of excluding fire-sensitive competitors (Platt et al. 1988; Clewell 1989). Fire suppression allowed the invasion of natural, open-structured longleaf pine forests by dense second-growth hardwoods, which have now replaced longleaf pines over much of the Southeast. At least 191 rare and endemic plant taxa are associated with wiregrass in the Southeast (Hardin and White 1989), but sites protected from fire are less rich floristically than their well-burned counterparts (Walker and Peet 1983). Longleaf pine stands also become unsuitable for red-cockaded woodpeckers when a dense understory develops after fire suppression (Jackson 1971). Fire also interacts with environmental gradients to enrich biodiversity at a landscape scale. In the Gulf coastal plain of Florida, Alabama, and Mississippi, herb-bog communities (also known as pitcher plant bogs, seepage slopes, and savannas) are sites of high species richness and endemism (Folkerts 1982; Noss 1988). Fires sweeping down from longleaf pine flatwoods prune back wetland shrubs that would otherwise encroach up the slope-moisture gradient; through this process, fire maintains the open bog community and its unique collection of species (Wolfe et al. 1988). If any of the key elements in this ecotonal ecosystem —fire, hydrology, or the wiregrass matrix that promotes the spread of fire—are disrupted, the bog community is gradually destroyed. Suppression of fire, ditching, and mechanical site preparation after clearcutting (which ruins the wiregrass matrix; Clewell 1989) are responsible for much of the loss of this marvelous expression of landscape diversity. Restoration of sites where wiregrass was destroyed is improbable with current techniques, in part because wiregrass is extremely slow to recolonize lost ground (Means and Grow 1985; Clewell 1989; Duever 1989; Noss 1989).

FRAGMENTATION PROBLEMS

Habitat fragmentation is also usually a landscape-level phenomenon. It was perceived as a conservation problem only when ecologists started taking a broader view of the environment than usually afforded by community-level studies. When attention is focused narrowly on what goes on inside biological communities or other habitat patches (*content*), much of what goes on around them (*context*) may be missed. If context is ignored in conservation decisions and the pattern and structure of the surrounding landscape changes radically, patch content too will be altered by edge effects and other processes (Noss and Harris 1986). Because fragmentation has been reviewed extensively in the recent literature (e.g., Burgess and Sharpe 1981; Noss 1983; Harris 1984; Wilcove et al. 1986; Saunders et al. 1991), I only touch on some of the critical issues here.

Of the many perceivable trends in modern landscapes, fragmentation is one of the most common and obvious. In fact, fragmentation is a seemingly unavoidable accompaniment of human population growth, development, agriculture, and resource extraction. Fragmentation is often said to have two components, a reduction in total area of some habitat type (or perhaps of all natural habitat in a landscape) and an apportionment of the remaining area into small, isolated pieces (Wilcove et al. 1986). The second component is fragmentation per se, but is almost always accompanied by total habitat loss. In Figure 1, each of the first six trends (from the top) can be a form of habitat loss; the last trend, toward an increased number of species requiring listing as endangered or threatened, is a consequence of the others.

Smaller patches and isolated patches (Figure 36.1) lead to a reduction in populations of species sensitive to fragmentation. The sensitivity of species often can be predicted on the basis of autecological characteristics. A species might survive in a fragmented landscape if (1) it can live and reproduce in the matrix such as agricultural land that surrounds remnant habitat patches, (2) it can maintain viable populations in remnant patches, which may be possible for many plants and animals with small home ranges, or (3) it can integrate many patches either into individual home ranges or into an interbreeding population, both of which depend on movement or dispersal between patches. A species incapable of pursuing any of these three alternative strategies will probably go extinct in a fragmented landscape.

In a study in Florida (Noss and Labisky 1988), terrestrial vertebrates predicted to be sensitive to fragmentation and known or suspected to be declining in developed landscapes, shared one or more of the following autecological or situational

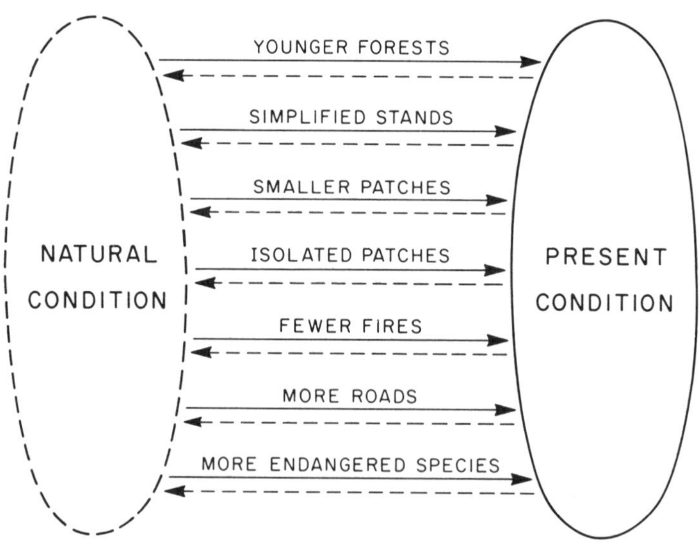

Figure 36.1. The shift from natural condition to present condition for a hypothetical managed forest landscape, as exemplified by seven trends. Dashed arrows in reverse direction to trends indicate the potential for restoration (from Noss in press).

characteristics: (1) initial rarity because of low population density or limited geographic distribution, (2) large home range or territory size, (3) ground-nesting or ground-dwelling habits, making them vulnerable to predation by raccoons, opossums, armadillos, blue jays, or other opportunistic predators that thrive in fragmented landscapes, (4) dependent on fire-maintained habitats such as longleaf pine forests that change radically when fire frequency is reduced by habitat isolation or active suppression, (5) existence in habitats prone to human disturbance or development such as favored homesites, or (6) subject to exploitation or persecution, including some game species (black bear) and unpopular species killed on sight (large snakes).

These findings probably have considerable generality. If we consider the interaction of any of these characteristics with the conversion of a landscape into small, isolated patches, population decline is a predictable consequence. Conservation of such species must consist of mitigating both primary trends (human population growth and development fragmenting natural habitat) and secondary trends (proliferation of opportunistic species, fire suppression, increased human access to natural areas). Retaining habitat corridors or other means of connectivity may be the single most promising mitigation measure in such landscapes (Noss and Harris 1986).

One major lesson from fragmentation studies is that the process is not nearly as simple as early island biogeographic models (MacArthur and Wilson 1967; Diamond 1975) suggested. First of all, fragmentation is a process. It does not appear right away as fragmentation, but as gap formation or perforation of the landscape. At this stage the landscape matrix is still natural habitat. Gradually the gaps get bigger or more numerous, they coalesce, until eventually the matrix is no longer natural. This sequence of fragmentation has been documented in many landscapes. For example, in Cadiz Township, Wisconsin, average forest patch size declined from 8724 ha in 1831, to 36.9 ha in 1882, 13.8 ha in 1902, 7.4 ha in 1935, and 5.8 ha in 1950 (Curtis 1956). The familiar figure (Figure 36.2) from Curtis fails to show the first stage in the fragmentation process (perforation, which must have occurred quickly between 1831 and 1882), but clearly shows the increasing isolation and smaller size of patches over time, accompanied by an overall reduction in forest area.

The trend of decreasing forest patch size and increasing edge is occurring today in many regions with heavy logging. Between 1972 and 1987, the average size of forest patches in two ranger districts of the Willamette National Forest, Oregon, decreased by 17%, the amount of forest-clearcut edge doubled, and the amount of forest interior at least 100 m from edge declined by 18% (Ripple et al. 1991). In the Olympic National Forest, Washington, more than 87% of the old growth in 1940 was in patches larger than 4000 ha; in 1988, only one old-growth patch larger than 4000 ha remained and 60% of the forest was in patches smaller than 40 ha in size. Of the old growth remaining in 1988, 41% was within 170 m of edge and, therefore, vulnerable to blowdown and other edge effects (Morrison 1990).

One important way in which fragmentation of real landscapes differs from the classic island biogeographic model has to do with matrix quality. Real islands are surrounded by a matrix (water) unsuitable to most terrestrial species, whereas habitat islands are surrounded by a potentially wide variety of habitats. Each kind of matrix differs in its suitability for species that live in the habitat islands and for species of the matrix itself. Habitat suitability must be considered not only for living and reproducing, but also for

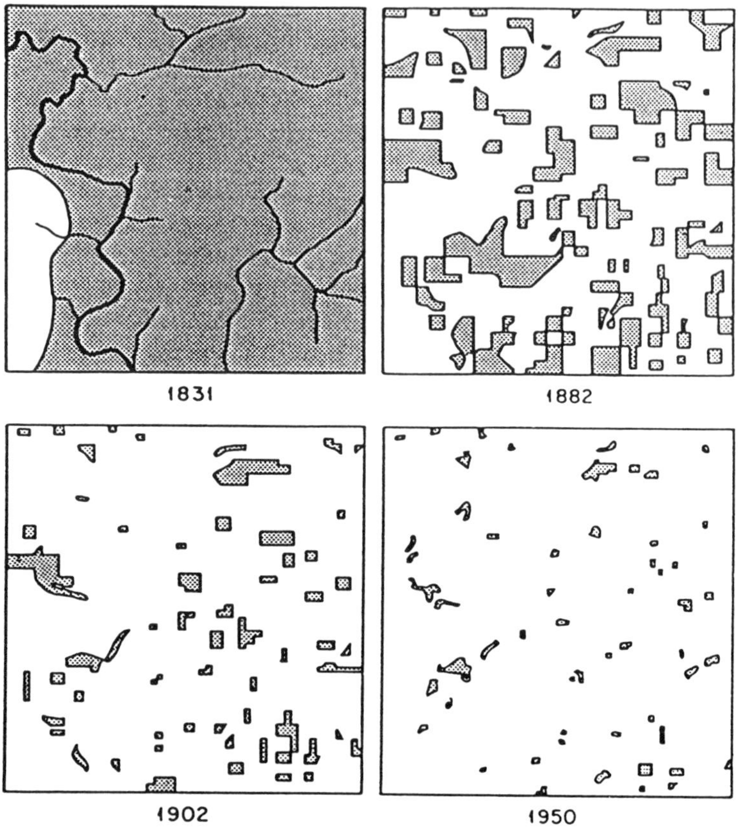

Figure 36.2. Change in forest cover of Cadiz Township, Green County, Wisconsin, from 1831 to 1950. Note that the landscape matrix changed rapidly from forest to farmland between 1831 and 1882. (From Curtis 1956, University of Chicago Press).

moving between patches. For a species that requires remnant habitat patches to persist in a fragmented landscape, movement of individuals between patches must be great enough to balance extirpation from local patches (Wiens 1989). Thus, the study of metapopulation dynamics (Gilpin and Hanski 1991) is of special relevance to fragmented landscapes. Some species, especially those of early successional habitats, are adept at dispersing between isolated patches of suitable habitat. Species of late successional habitats, such as flightless invertebrates in old-growth forests, are usually poorer dispersers and are more vulnerable to the isolating effects of fragmentation (den Boer 1990; Moldenke and Lattin 1990). Aquatic habitats can also become fragmented, with dire consequences for population persistence. Dams not only block access of migratory fishes to upstream areas, but prevent recolonization of stream segments where local populations have been extinguished due to natural or human causes (Moyle and Leidy 1992).

These observations suggest that matrix quality may be the most important consideration for determining the effects of fragmentation in many landscapes. In general, dispersal is easier through habitats similar in structure to the habitat in which a species lives (Wiens 1989). For probably most late-successional species, patches of natural forest in a matrix of tree plantations of various ages are functionally less isolated than forest patches surrounded by agricultural fields, fresh clearcuts, shopping malls, or housing subdivisions. Harris (1984) recommended that old-growth patches in the Pacific Northwest be surrounded by a series of long-rotation stands harvested on different schedules. Recently, management under the banner of "New Forestry" (Franklin 1989; Gillis 1990) is designed to retain coarse woody debris and mature green trees (perhaps 30% of those before cutting) in harvest units, in contrast to conventional clear-cutting and site preparation, which produce structurally barren sites. Unlike a shelterwood cut, New Forestry intends to leave green trees through subsequent rotations. A site harvested under New Forestry might resemble a site after a natural fire or windstorm more than it resembles a clearcut. Retention cuts are expected to mitigate impacts of harvest to soil and soil organisms and provide habitat for many species. Sloppy harvest areas also may be less of a barrier than traditional clearcuts to dispersing animals that require cover, such as the pine marten (Koehler and Hornocker 1977). However, the ability of various fragmentation-sensitive species to travel through New Forestry harvest units, compared with traveling along intact corridors, is unknown. Research to determine what constitutes adequate dispersal habitat for various species (and, conversely, what constitutes movement barriers) is urgently needed.

A question of great importance to land managers is how a fragmented landscape differs from a naturally disturbed landscape. If differences can be minimized—that is, if natural disturbance regimes can be mimicked by management—native biodiversity might be maintained along with the production of commodities. A paradox is evident in this situation: natural disturbance is considered "good" by many ecologists, yet human disturbance in the form of fragmentation is considered "bad." Are there important differences between fragmentation and natural disturbance mosaics? Natural and anthropogenic disturbance regimes obviously differ because they have different effects on species and ecological processes (Harris 1984; Hansen et al. 1991). Some of the important underlying differences seem to be the barren internal structure of anthropogenic openings (such as clearcuts) that are unlike naturally disturbed patches, the greater contrast between adjacent patches (which produces edge effects), and the presence of roads and other avenues of invasion by weedy species, sedimentation, and disruptive human activities in fragmented landscapes. These differences, then, are a consequence of trends in landscape change (Figure 36.1).

RESERVE NETWORK DESIGN

Just as threats to biodiversity can be fruitfully considered at a landscape scale, so can solutions to these problems. Conservation of biodiversity requires management of entire landscapes, often called "ecosystem management" (Agee and Johnson 1988), not just protection of individual reserves (Harris 1984; Brussard 1991; Saunders et al. 1991). In the near term, political resistance to setting aside land solely for biodiversity will make it necessary for multiple-use lands, including those managed largely for timber or livestock forage, to serve a major share of the conservation role. Undoubtedly, most species will persist in the matrix of multiple-use land in most regions. However, not all species will persist, and conservation must consider the needs of species sensitive to human disturbance (Diamond 1976). New reserves still need to be established and the location of these areas is extremely important if all ecosystems are to be represented adequately and viable populations of all native species maintained. This is especially true because the record of multiple-use management for protecting native biodiversity so far is quite poor (Grumbine 1990; Noss 1992a). Therefore, core reserves managed with the protection of biodiversity as the fundamental goal form the backbone for any regional land conservation system; multiple-use lands can then serve as surrounding buffers.

Elsewhere (Noss and Harris 1986; Noss 1987b, 1992a), I have described the basic elements of a regional land conservation system: core reserves, multiple-use buffer zones, and corridors or other means of connectivity (Figure 36.3). This model borrows from the familiar biosphere reserve model of the Man and the Biosphere program (UNESCO 1974; Hough 1988; Batisse 1990; Dyer and Holland 1991) and the multiple-use module idea that applies these concepts at various spatial scales (Harris 1984; Noss and Harris 1986). The goal is an interconnected system of strictly protected areas, surrounded by lands for human activities that are compatible with conservation, and linked in some way that provides for functional connectivity of populations and processes across the landscape.

The idea of planning reserve networks and other land uses at a regional landscape scale has caught on with environmental groups and to some extent with land-managing agencies in the U.S. (for a single-species plan based on landscape and biogeographic principles, see the conservation strategy for the northern spotted owl prepared by the Interagency Scientific Committee; Thomas et al. 1990). However, although many plans have been proposed, no plan focused on maintaining overall biodiversity in a region has

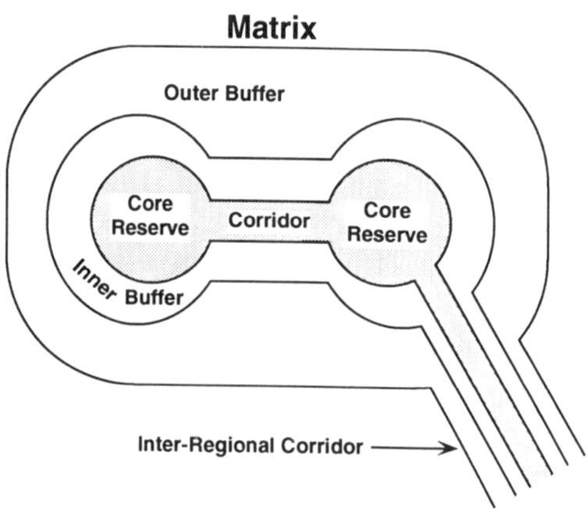

Figure 36.3. A regional reserve network, consisting of core reserves, connecting corridors or linkages, and buffer zones. Only two core reserves are shown, but a real system may contain many reserves. Inner buffer zones would be strictly protected, whereas outer zones would allow a wider range of compatible human uses. In this example, an interregional corridor connects the system to a similar network in another region (from Noss 1992a).

been implemented. Hence, we have no test of the model (Figure 36.3 or similar models) to determine how well it really works. Key unknowns include: (1) the total area needed in core reserves and various buffer zones to represent all ecosystems adequately and maintain viable populations of all native species, (2) the types and intensities of human uses permissible in buffer zones, (3) necessary corridor width and the factors that influence width considerations, (4) the quality of habitat needed to provide adequate connectivity for entire faunas and floras in changing environments. Because so much uncertainty exists and answers to important questions may be a long time coming, it is important to be conservative and err on the side of preservation (Noss 1993).

CONCLUSIONS

None of the significant developments in conservation biology reviewed in this chapter would have been possible if biologists had stuck to their conventional focus on individual species, narrowly defined communities, and separate sites. A continually expanding perspective in space, time, complexity, and ethics is needed to deal with many of the most pressing problems in conservation in the coming decades (Noss 1992b).

Because rigorous experiments at the landscape scale are difficult to conduct and interpret, many questions posed here may never have satisfactory answers. Yet delaying action until data are sufficient is unconscionable (Soulé 1986). Conservation biologists would do well to proceed with a synthetic case-study approach for conservation planning, using the best available information, logical inference, and scientific opinion on how to protect biodiversity in a region of interest. The process of adaptive management (Holling 1978; Walters 1986), which depends on monitoring the consequences of alternative treatments, should be used to adjust and refine management. Finally, I suggest that our conservation goals be ambitious and ethical, with full knowledge that we may be the last generation of scientists with the opportunity to help avert a mass extinction.

ACKNOWLEDGMENTS

I thank Tom Crow, Dennis Glick, Mike Scott, Bob Szaro, and an anonymous technical editor for comments on an earlier draft of this manuscript.

LITERATURE CITED

Agee, J. K., and D. R. Johnson. 1988. *Ecosystem Management for Parks and Wilderness.* University of Washington Press, Seattle, Wash.

Allen, T. F. H., and T. B. Starr. 1982. *Hierarchy: Perspectives for Ecological Complexity.* University of Chicago Press, Chicago, Ill.

Baker, W. L. 1989. Landscape ecology and nature reserve design in the Boundary Waters Canoe Area, Minnesota. *Ecology* 70:23-35.

Baker, W. L. 1992. The landscape ecology of large disturbances in the design and management of nature reserves. *Landscape Ecol.* 7:181-194.

Batisse, M. 1990. Development and implementation of the biosphere reserve concept and its applicability to coastal regions. *Environ. Conserv.* 17:111-116.

Baumgartner, L. 1943. Fox squirrels in Ohio. *J. Wildl. Manage.* 7:193-202.

Bormann, F. H., and G. E. Likens. 1979. *Pattern and Process in a Forested Ecosystem*. Springer-Verlag, New York.

Brussard, P. F. 1991. The role of ecology in biological conservation. *Ecol. Appl.* 1:6-12.

Burgess, R. L., and D. M. Sharpe, eds. 1981. *Forest Island Dynamics in Man-dominated Landscapes*. Springer-Verlag, New York.

Christensen, N. L., J. K. Agee, P. F. Brussard, J. Hughes, D. H. Knight, G. W. Minshall, J. M. Peek, S. J. Pyne, F. J. Swanson, J. W. Thomas, S. Wells, S. E. Williams, and H. A. Wright. 1989. Interpreting the Yellowstone fires of 1988. *BioScience* 39:678-685.

Clewell, A. F. 1989. Natural history of wiregrass (*Aristida stricta* Michx., Gramineae). *Natural Areas J.* 9:223-233.

Connell, J. H. 1978. Diversity in tropical rain forests and coral reefs. *Science* 199:1302-1310.

Croker, T. C. 1979. The longleaf pine story. *J. For. Hist.*, January 1979:32-43.

Curtis, J. T. 1956. The modification of mid-latitude grasslands and forests by man. In *Man's Role in Changing the Face of the Earth*, ed. W. L. Thomas, Jr., pp. 721-736. University of Chicago Press, Chicago, Ill.

Daubenmire, R. 1968. *Plant Communities: A Textbook of Plant Synecology*. Harper and Row, New York.

Davis, M. B. 1981. Quaternary history and the stability of forest communities. In *Forest Succession*, eds. D. C. West, H. H. Shugart, and D. B. Botkin, pp. 132-153. Springer-Verlag, New York.

den Boer, P. J. 1990. The survival value of dispersal in terrestrial arthropods. *Biol. Conserv.* 54:175-192.

Diamond, J. M. 1975. The island dilemma: Lessons of modern biogeographic studies for the design of natural preserves. *Biol. Conserv.* 7:129-146.

Diamond, J. M. 1976. Island biogeography and conservation: Strategy and limitations. *Science* 193:1027-1029.

Duever, L. C. 1989. Research priorities for the preservation, management, and restoration of wiregrass ecosystems. *Natural Areas J.* 9:214-218.

Dyer, M. I., and M. M. Holland. 1991. The biosphere-reserve concept: Needs for a network design. *BioScience* 41:319-325.

Folkerts, G. W. 1982. The Gulf Coast pitcher plant bogs. *Am. Sci.* 70:260-267.

Forman, R. T. T., and M. Godron. 1986. *Landscape Ecology*. John Wiley and Sons, New York.

Franklin, J. F. 1989. Toward a new forestry. *Am. For.* Nov./Dec.:37-44.

Franklin, J. F., and R. T. T. Forman. 1987. Creating landscape patterns by cutting: Ecological consequences and principles. *Landscape Ecol.* 1:5-18.

Gillis, A. M. 1990. The new forestry. *BioScience* 40:558-562.

Gilpin, M., and I. Hanski. 1991. *Metapopulation Dynamics: Empirical and Theoretical Investigations*. Linnean Society of London and Academic Press, London, England.

Gleason, H. A. 1926. The individualistic concept of the plant association. *Bull. Torrey Bot. Club* 43:463-481.

Grumbine, R. E. 1990. Viable populations, reserve size, and federal lands management: A critique. *Conserv. Biol.* 4:127-134.

Hansen, A. J., T. A. Spies, F. J. Swanson, and J. L. Ohmann. 1991. Conserving biodiversity in managed forests. *BioScience* 41:382-392.

Hansson, L. 1979. On the importance of landscape heterogeneity in northern regions for the breeding population densities of homeotherms: A general hypothesis. *Oikos* 33:182-189.

Hardin, E. D., and D. L. White. 1989. Rare vascular plant taxa associated with wiregrass (*Aristida stricta*) in the Southeastern United States. *Natural Areas J.* 9:234-245.

Harris, L. D. 1984. *The Fragmented Forest: Island Biogeography Theory and the Preservation of Biotic Diversity*. University of Chicago Press, Chicago, Ill.

Holling, C. S., ed. 1978. *Adaptive Environmental Assessment and Management*. John Wiley and Sons, New York.

Hough, J. 1988. Biosphere reserves: myth and reality. *Endangered Species Update* 6:1-4.

Hunter, M. L. 1991. Coping with ignorance: The coarse-filter strategy for maintaining biodiversity. In *Balancing on the Brink of Extinction: The Endangered Species Act and Lessons for the Future*, ed. K. A. Kohm, pp. 266-281. Island Press, Washington, D.C.

Hunter, M. L., G. L. Jacobson, and T. Webb. 1988. Paleoecology and the coarse-filter approach to maintaining biological diversity. *Conserv. Biol.* 2:375-385.

Huston, M. 1979. A general hypothesis of species diversity. *Am. Nat.* 113:81-101.

Jackson, J. A. 1971. The evolution, taxonomy, distribution, past populations and current status of the red-cockaded woodpecker. In *The Ecology and Management of the Red-cockaded Woodpecker*, ed. R. L. Thompson, pp. 4-29. Bureau of Fish and Wildlife and Tall Timbers Research Station, Tallahassee, Fla.

Keystone Center. 1991. Final Consensus Report of the Keystone Policy Dialogue on Biological Diversity on Federal Lands. The Keystone Center, Keystone, Colo.

Kendeigh, S. C., H. I. Baldwin, V. H. Cahalane, C. H. D. Clarke, C. Cottam, W. P. Cottam, I. M. Cowan, P. Dansereau, J. H. Davis, F. W. Emerson, L. T. Haig, A. Hayden, C. L. Hayward, J. M. Linsdale, J. A. MacNab, and J. E. Potzger. 1950-51. Nature sanctuaries in the United States and Canada: A preliminary inventory. *Living Wilderness* 15(35):1-45.

Koehler, G. M., and M. G. Hornocker. 1977. Fire effects on marten habitat in the Selway-Bitterroot Wilderness. *J. Wildl. Manage.* 41:500-505.

Lay, D. 1938. How valuable are woodland clearings to birdlife? *Wilson Bull.* 50:254-256.

Leopold, A. 1933. *Game Management*. Charles Scribners Sons, New York.

Leopold, A. 1941. Wilderness as a land laboratory. *Living Wilderness* 6(July):3.

MacArthur, R. H., and E. O. Wilson. 1967. *The Theory of Island Biogeography*. Princeton University Press, Princeton, N.J.

Means, D. B., and G. Grow. 1985. The endangered longleaf pine community. *ENFO Report* 85(4):1-12.

Moldenke, A. R., and J. D. Lattin. 1990. Dispersal characteristics of old-growth soil arthropods: The potential for loss of diversity and biological function. *Northwest Environ. J.* 6:408-409.

Morrison, P. H. 1990. *Ancient Forests on the Olympic National Forest: Analysis from a Historical and Landscape Perspective*. The Wilderness Society, Washington, D.C.

Moyle, P. B., and R. A. Leidy. 1992. Loss of biodiversity in aquatic ecosystems: Evidence from fish faunas. In *Conservation Biology: The Theory and Practice of Nature Conservation, Preservation, and Management*, eds. P. L. Fiedler and S. K. Jain, pp. 127-169. Chapman and Hall, New York.

Myers, R. L. 1985. Fire and the dynamic relationship between Florida sandhill and sand pine scrub vegetation. *Bull. Torrey Bot. Club* 112:241-252.

Noss, R. F. 1983. A regional landscape approach to maintain diversity. *BioScience* 33:700-706.

Noss, R. F. 1987a. From plant communities to landscapes in conservation inventories: A look at The Nature Conservancy (USA). *Biol. Conserv.* 41:11-37.

Noss, R. F. 1987b. Protecting natural areas in fragmented landscapes. *Natural Areas J.* 7:2-13.

Noss, R. F. 1988. The longleaf pine landscape of the Southeast: Almost gone and almost forgotten. *Endangered Species Update* 5(5): 1-8.

Noss, R. F. 1989. Longleaf pine and wiregrass: Keystone components of an endangered ecosystem. *Natural Areas J.* 9:211-213.

Noss, R. F. 1990a. Indicators for monitoring biodiversity: A hierarchical approach. *Conserv. Biol.* 4:355-364.

Noss, R. F. 1990b. What can wilderness do for biodiversity? In *Preparing to Manage Wilderness in the 21st Century*, ed. P. Reed, pp. 49-61. General Technical Report SE-66. USDA Forest Service, Southeastern Forest Experiment Station, Asheville, N.C.

Noss, R. F. 1991. Sustainability and wilderness. *Conserv. Biol.* 5:120-122.

Noss, R. F. 1992a. The Wildlands Project: Land conservation strategy. *Wild Earth* (Special Issue):10-25.

Noss, R. F. 1992b. Issues of scale in conservation biology. In *Conservation Biology: The Theory and Practice of Nature Conservation, Preservation, and Management*, eds. P. L. Fiedler and S. K. Jain, pp. 239-250. Chapman and Hall, New York.

Noss, R. F. 1993. Sustainable forestry or sustainable forests? In *Defining Sustainable Forestry*, eds. G. H. Aplet, N. Johnson, J. T. Olson, and V. A. Sample, pp. 17-43. Island Press, Washington, D.C.

Noss, R. F., and L. D. Harris. 1986. Nodes, networks, and MUMs: Preserving diversity at all scales. *Environ. Manage.* 10:299-309.

Noss, R. F., and R. F. Labisky. 1988. Sensitivity of Vertebrates to Development in Four Upland Community Types in Northern Peninsular Florida. Report to Florida Game and Fresh Water Fish Commission, Tallahassee, Fla.

Office of Technology Assessment (OTA). 1987. *Technologies to Maintain Biological Diversity.* OTA-F-330, Government Printing Office, Washington, D.C.

Petraitis, P. S., R. E. Latham, and R. A. Niesenbaum. 1989. The maintenance of species diversity by disturbance. *Quart. Rev. Biol.* 64:393-418.

Pickett, S. T. A., and J. N. Thompson. 1978. Patch dynamics and the design of nature reserves. *Biol. Conserv.* 13:27-37.

Pickett, S. T. A., and P. S. White. 1985. *The Ecology of Natural Disturbance and Patch Dynamics.* Academic Press, Orlando, Fla.

Platt, W. J., G. W. Evans, and S. L. Rathbun. 1988. The population dynamics of a long-lived conifer (*Pinus palustris*). *Am. Nat.* 131:491-525.

Ripple, W. J., G. A. Bradshaw, and T. A. Spies. 1991. Measuring forest landscape patterns in the Cascade Range of Oregon, USA. *Biol. Conserv.* 57:73-88.

Saunders, D. A., R. J. Hobbs, and C. R. Margules. 1991. Biological consequences of ecosystem fragmentation: A review. *Conserv. Biol.* 5:18-32.

Scheiner, S. M. 1992. Measuring pattern diversity. *Ecology* 73: 1860-1867.

Scott, J. M., F. Davis, B. Csuti, R. Noss, B. Butterfield, C. Groves, J. Anderson, S. Caicco, F. D'Erchia, T. C. Edwards, J. Ulliman, and R. G. Wright. 1993. Gap analysis: A geographical approach to protection of biological diversity. *Wildl. Monogr.* 123:1-41.

Shelford, V. E., ed. 1926. *Naturalist's Guide to the Americas.* Williams and Wilkins, Baltimore, Md.

Shelford, V. E. 1933. Ecological Society of America. A nature sanctuary plan unanimously adopted by the Society, December 28, 1932: The preservation of natural biotic communities. *Ecology* 14:240-245.

Shugart, H. H., and S. W. Seagle. 1985. Modeling forest landscapes and the role of disturbance in ecosystems and communities. In *The Ecology of Natural Disturbance and Patch Dynamics*, eds. S. T. A. Pickett and P. S. White, pp. 353-368. Academic Press, Orlando, Fla.

Shugart, H. H., and D. C. West. 1981. Long-term dynamics of forest ecosystems. *Am. Sci.* 69:647-652.

Soulé, M. E. 1986. Conservation biology and the "real world." In *Conservation Biology: The Science of Scarcity and Diversity*, ed. M. E. Soulé, pp. 1-12. Sinauer Associates, Sunderland, Mass.

Sousa, W. P. 1984. The role of disturbance in natural communities. *Ann. Rev. Ecol. Syst.* 15:353-391.

Sprugel, D. G. 1991. Disturbance, equilibrium, and environmental variability: What is "natural" vegetation in a changing environment? *Biol. Conserv.* 58:1-18.

Stoddard, H. L. 1936. *Management of Wild Turkey.* Trans. North Am. Wildl. Conf. 1:352-356.

The Nature Conservancy (TNC). 1981-1984. Heritage Operations Manual. Unpublished report, Arlington, Va.

Thomas, J. W., ed. 1979. Wildlife Habitats in Managed Forests: The Blue Mountains of Oregon and Washington. USDA Forest Service Agricultural Handbook No. 553, Washington, D.C.

Thomas, J. W., E. D. Forsman, J. B. Lint, E. C. Meslow, B. R. Noon, and J. Verner. 1990. A conservation strategy for the northern spotted owl. USDA Forest Service, USDI Bureau of Land Management, USDI Fish and Wildlife Service, and USDI National Park Service, Portland, Ore.

Turner, M. G. 1989. Landscape ecology: The effect of pattern on process. *Ann. Rev. Ecol. Syst.* 20:171-197.

UNESCO. 1974. Task Force on Criteria and Guidelines for the Choice and Establishment of Biosphere Reserves. Man and the Biosphere Report No. 22, Paris, France.

Urban, D. L., R. V. O'Neill, and H. H. Shugart. 1987. Landscape ecology. *BioScience* 37:119-127.

Usher, M. B., ed. 1986. *Wildlife Conservation Evaluation.* Chapman and Hall, London, England.

Walker, J., and R. K. Peet. 1983. Composition and species diversity of pine-wiregrass savannas of the Green Swamp, North Carolina. *Vegetatio* 55:163-179.

Walters, C. J. 1986. *Adaptive Management of Renewable Resources.* McGraw-Hill, New York.

Ware, S., C. C. Frost, and P. Doerr. 1993. Southern mixed hardwood forest (the former longleaf pine forest). In *Biodiversity of the Southeastern United States: Lowland Terrestrial Communities*, eds. W. H. Martin, S. G. Boyce, and A. C. Echternacht, pp. 447-493. John Wiley and Sons, New York.

Watt, A. S. 1947. Pattern and process in the plant community. *J. Ecol.* 35:1-22.

Werner, E. E. 1986. Amphibian metamorphosis: Growth rate, predation risk, and the optimal size at transformation. *Am. Nat.* 128:319-341.

White, P. S. 1979. Pattern, process, and natural disturbance in vegetation. *Bot. Rev.* 45:229-299.

Whittaker, R. H. 1972. Evolution and measurement of species diversity. *Taxon* 21:213-251.

Wiens, J. A. 1989. *The Ecology of Bird Communities.* Vol. 2. Processes and Variations. Cambridge University Press, New York.

Wilcove, D. S., C. H. McLellan, and A. P. Dobson. 1986. Habitat fragmentation in the temperate zone. In *Conservation Biology: The Science of Scarcity and Diversity*, ed. M. E. Soulé, pp. 237-256. Sinauer Associates, Sunderland, Mass.

Wilson, E. O., ed. 1988. *Biodiversity.* National Academy Press, Washington, D.C.

Wolfe, S. H., J. A. Reidenauer, and D. B. Means. 1988. An Ecological Characterization of the Florida Panhandle. Biological Report 88(12). U.S. Fish and Wildlife Service, National Wetlands Research Center, Slidell, La.

V

POLICY AND SOCIAL CONSIDERATIONS

37

DEVELOPING PUBLIC EDUCATION PACKAGES: A U.S. NATIONAL PARK SERVICE PERSPECTIVE

Gary W. Mullins and Michael D. Watson

School of Natural Resources
The Ohio State University
Columbus, Ohio
U.S.A.

Developing public education packages is a useful technology for maintaining biodiversity. Such packages of education materials fulfill well-documented needs. For example, a report on biodiversity from the Office of Technology Assessment (1987) noted:

> Conveying the importance of biological diversity requires formulating the issue in terms that are technically correct yet understandable and convincing to the general public. To undertake the initiative will require not only biologists but also social scientists and educators working together.

In another report, the U.S. National Park Service Director's Task Force on Biological Diversity asserted that the public is unaware of the difficulty and complexity of protecting biodiversity (Schoenwald-Cox 1987). That report continued by stressing education as one strategy for maintenance of biodiversity in national parks. A NPS workshop on biodiversity in 1990 (Dottavio et al. 1990) resulted in a call to "improve communications with the public about the importance of preserving biodiversity, the challenges of doing so, and the elements needed for success."

This paper focuses on a case study of efforts by the U.S. National Park Service (NPS) to communicate biodiversity and other critical resource issues, both to its staff and to the various clientele they serve.

BACKGROUND

Interpretation, environmental education, and a plethora of other terms are used to delineate specific types of ecological communications. Ecological communications is defined as the array of public education strategies used by resource management agencies and conservation organizations to communicate with their various clientele. The delineation in types of communications is based for the most part on the audience receiving the information. Scientific information available to be delivered is similar for most audiences, and most ecological communicators have access to a relatively full array of communication techniques. The difference in the various types of ecological communications is that the messages are packaged with different objectives using appropriate language and technologies to target those messages most effectively to specific audiences. For example, school-based environmental education is considered to be part of the formal education system. Here the level of learning and messages are prescribed by teachers and school administrators for students. On the other hand, non-formal environmental education or interpretation is elective learning that occurs in parks, forests, biological reserves, zoos, and like settings. Similar environmental messages can be presented in recreational and formal educational settings, but recreating audiences tend to function in leisure frames of mind and time-scales, and are not seeking formal learning experiences. Additionally, informal environmental information is available through channels such as television broadcasts, books, conservation newsletters, or a family outing to explore a nearby creek.

The array of formal, non-formal, and informal environmental education opportunities for conveying messages related to critical resource issues such as biodiversity is vast. Ultimately the aim is to match media and message to a targeted audience for a desired set of responses. These desired responses range from learning environmentally friendly behaviors to people simply reflecting on their personal relationship with nature.

Understanding of message content by communicators is critical. The NPS Director's Task Force on Biological Diversity (Schoenwald-Cox 1987) asserted that the National Park Service is not fully aware of the difficulty and complexity of protecting biodiversity. In the focus of this paper, the concern is for the capabilities within NPS to understand biodiversity problems, issues, and solutions to the extent that these can be effectively communicated to various NPS clientele. Before interpreters can interpret and educators can educate, they must understand the breadth and complexity of their subject matter as well as they understand their communication media and clientele.

One focus within the National Park Service is the education of its staff, as well as education of its clientele. Since 1988, NPS has engaged in a series of activities related to interpreting effectively critical resource issues, namely acid deposition and air quality, biodiversity, and global climate change. These programs are based on the idea that an internal needs assessment of the NPS staff is prerequisite to success. Based on findings from needs assessments, efforts were made to meet the needs of NPS communicators to help them more effectively prepare to communicate with their audiences.

EARLY EXPERIENCES—"CLEARING THE AIR"

The National Park Service, along with other resource management organizations, is faced with increasing pressures on resources. In addition to sponsoring on-going science programs entailing inventory, research, and monitoring, NPS addresses these issues through resource management, protection, and public education.

In 1988, various NPS divisions and departments, the U.S. Man and Biosphere (MAB) program, and other cooperators instituted an acid deposition and air quality interpretive effort entitled "Clearing the Air." The program was based on a need perceived by the NPS director, administrators, science program managers, and interpreters. Questions about acid rain were being asked by visitors, and some related interpretive programming was being conducted. Given the impact and potential impact of deteriorating air resources on NPS landscapes, and the agency's existing air quality research program, a public need to understand risks to their ecological and cultural heritage was assumed.

Researchers from The Ohio State University and NPS scientists from the Great Smoky Mountains National Park conducted an assessment of current related NPS interpretive activities and of NPS personnel's needs for support in interpreting acid deposition and air quality issues in national parks. Interpreters and researchers in the Great Smoky Mountains, Sequoia, and Kings Canyon National Parks studied systems for incorporating acid deposition and air quality messages into existing park interpretive themes. Given the survey feedback from the field, input from units involved in the effort, and visitor input from study parks, interpretive materials were developed. This first public education package included sets of reference slides, a slide program, fact sheets, program idea sheets, and a video; all were distributed to national parks. NPS service-wide, regional, and park-level training programs, in both content and process, were conducted as part of that effort beginning in 1988 and continues today.

A follow-up assessment showing relatively high marks for utility and quality of the acid deposition and air quality materials and training was reported by Mullins and Watson (1992). Portions of the package have been widely distributed both nationally and internationally. "Clearing the Air" was referenced in several national media and has been included in educational materials, such as the Southern Appalachian Regional Man and Biosphere Program's "Our Backyard Biosphere Reserve" (Tassier et al. 1988). Acid deposition and air quality interpretive messages have been most readily integrated into the themes of parks which are impacted most from deteriorating air resources, and/or have on-going air quality research programs. Parks with Man and Biosphere status, in general, are also strong participants in related interpretive efforts.

"Clearing the Air" provided valuable experience for the NPS Biological Diversity Interpretive Program which followed in 1989. Lessons learned about development of public education packages provided a basis for the biodiversity program which has become a public education package model.

BIODIVERSITY EFFORTS—CREATING A MODEL PACKAGE

Programmatically the biodiversity effort began with the naming of an NPS task group of agency interpreters, scientists, administrators, and university researchers. The ultimate objective was to meet the earlier referenced charge given by the Office of Technology Assessment—"formulating the issues in terms that are technically correct yet understandable and convincing."

In the biodiversity program, the task group viewed NPS communicators as the target audience. In turn, these communicators were to deliver biodiversity messages to various NPS audiences. These messages included the impact of loss of biodiversity in national parks as well as loss in the biogeocultural regions where the parks are located. The focus was on appreciation, awareness and knowledge of biodiversity issues and actions necessary to stem the loss of biodiversity.

To fulfill the objective, the extent to which biodiversity was being interpreted, described under the four broad themes of genetic, species, community, and ecosystem diversity, was investigated (Mullins et al. 1989). Interpretive managers reported the extent of related interpretive activities under 23 different categories within these four themes. For example, a high of 62% of the parks reported programs related to edible plants as part of their species diversity interpretation. When asked about interpretation of gene pools (genetic diversity) or origin of crop species (species diversity), less than 30% addressed these issues. Perceived park visitor interest based on visitors' questions about the four themes and 23 categories was also gathered from interpretive managers. Questions about charismatic species such as deer and bear, for example, were found to capture the most attention among NPS visitors.

Also in this study (Mullins et al. 1989), interpretive managers were asked to report their park staff's need for support. Packets of information, training, and additional funding were, in descending order of importance, the most preferred items. Training at the park or regional levels was desired more than national training activities. The same interpretive managers provided a set of responses about their park staff's perception of biodiversity interpretation (Table 37.1).

In a follow-up study, randomly selected NPS interpreters were asked the same set of questions, except about interpreting critical resources issues in general (Mullins and Watson 1992). Acid deposition and air quality, biodiversity, and global climate change are the critical resource issues referred to in this study. Table 37.1 provides a comparison between interpreters' and managers' responses to the questions. Because the differences in the respondents studied and the fact that data are from two different types of studies, statistical analysis was not carried out.

The data from Table 37.1 illustrate that interpretive managers and interpreters are relatively close in their perceptions of the situation. The common perception of lack of time, personnel, and funding to interpret adequately the critical resource issues should be noted. The "neutral" to "disagree" range of responses relating to having access to park science and scientists is also an issue for consideration. Both sets of perceptions are potential barriers to effective communications. Without resources and full access to park science data, interpreters may be unwilling to commit the extra effort required to interpret critical resource issues.

With these data the NPS Task Group on Biological Diversity Interpretation set priorities for the effort. First and foremost the effort was framed as an ongoing attempt to help prepare NPS communicators to communicate critical resource issues related to parks. The extent to which parks would interpret biodiversity was deemed to be primarily the prerogative of each park based on its overall themes, goals, and objectives. Theoretically those objectives reflect the park audience's needs and desires for biodiversity information.

Care was taken to convey to park employees that biodiversity interpretation was not an annual interpretive "initiative" for high profiling one subject and then dropping it. Special efforts were made, primarily because of the misconceptions about the "annual initiative image" of the acid deposition and air quality effort, to help staff understand the loss of biodiversity, like acid deposition, which is an on-going critical resource issue highly relevant to the management of national parks. The concept was conveyed as an opportunity for interpretation to expand its role in the interpretation of critical resource issues. Priorities were set, and a communicator's handbook, a curriculum guide, a slide presentation/videotape, and training, all relating to biodiversity, were planned and produced. Various teams worked on different components.

Table 37.1. Interpreters' and Interpretive Managers' Perception of Interpreting Critical Resource Issues[1]

Interpreters' Mean Score[2] (n=251) 1992	Question	Interpretive Managers' Mean Score[2] (n=225) 1989
3.99	I have sufficient time to develop materials	3.57
3.76	I feel our park staff is of adequate size to effectively conduct programming	3.72
2.27	I feel I have enough to do already	2.06
1.91	I feel I should be interpreting these issues	2.29
3.63	I feel I already adequately interpret these issues	3.32
3.64	I feel I have access to adequate current research on these issues	3.54
3.35	I feel researchers are unwilling to share findings with me	2.53
3.09	I feel research findings available to me are outdated	2.93
2.98	I feel I am uninformed about current issues and research within my park	2.95
3.06	Resource management and interpretation work closely together in my park to interpret these issues	——
4.06	I feel these issues are too complex for visitors in my park to understand	3.80
3.59	I feel visitors to my park do not care about such issues	2.97
3.20	I feel most interpretation in national parks is too elementary	2.10
1.92	I feel the related issues are valuable program areas for our visitors	2.57
3.81	I feel these issues are not applicable to my park	3.64
2.05	I feel interpreting such issues is appropriate for fulfilling the NPS mission	1.84
2.08	I feel that appropriate funding for interpreting such issues is not available	2.09

[1]Under the heading *Question* "issues" is used for illustrative purposes in this table. In Mullins, Fortner, and Palmer's (1989) survey interpretive managers were asked specifically about biological diversity issues. In the 1992 study (Mullins and Watson) interpreters were asked about critical resource issues in general.

[2]The mean scores were calculated from five levels of responses with corresponding values shown — "Strongly Agree" (1), "Agree" (2), "Neutral" (3), "Disagree" (4), and "Strongly Disagree" (5).

A "NPS Handbook for Biological Diversity Communicators" (Mullins 1990) was developed with extensive input from all levels of NPS. Resource managers, scientists, and interpreters in the parks contributed extensively, as did others from within and outside the Service.

Content-wise the handbook begins with an introduction, a letter from former NPS Director William Penn Mott and a charge from Richard L. Cunningham (Regional Chief of Interpretation, Western Region), chairperson of the effort, entitled "Hope is *NOT* an Endangered Species." This is followed by a series of summaries of related technical reports from the federal government and NPS. The purpose of these is to demonstrate that the NPS biodiversity program is part of a larger national effort. A set of concise informational papers is provided to place biodiversity issues within the context of the NPS mission. Following are fact sheets on 26 different concepts among four levels of biodiversity—genetic, species, community, and ecosystem diversity. Each fact sheet provides NPS communicators with several pages of information on each topic such as "Biological Diversity as a Global Concept" and "Evolutionary Processes." A set of select articles reprinted from various journals and magazines is provided for future reference.

Fourteen suggested programs or program segments are also included as examples for integrating biodiversity into parks' existing themes, goals, and objectives. Example programs, such as the "Lyceum Series" of guest scientist lectures at Cuyahoga Valley National Recreation Area and George Washington Memorial Parkway, are highlighted. The intent is to provide a model for programs; the samples were not intended to be prescriptive.

Parks are then encouraged to plan their own set of biodiversity programs. A detailed bibliography and glossary, as well as a "how-to" paper, are provided to aid this task.

Mullins and Watson (1992), in a random survey of NPS interpreters, addressed interpreters' perception of this biodiversity effort. Among those (n=251) rating the utility of the handbook, most found it useful. Twenty-seven percent found it very useful, whereas 54% rated it somewhat useful. Sixteen percent found it seldom useful and 3% had no use for the document. One hundred and twelve respondents found the handbook quality to be good to very good; six respondents rated it less than satisfactory. Persons who were engaged in biodiversity interpretation were twice as likely to have knowledge of the biodiversity handbook as were persons not engaged in such interpretation.

Additionally a biodiversity curriculum guide, targeted to students, was developed by the Minnesota Environmental Education Association (1990). The curriculum guide was funded and distributed inside and outside the Service by the National Park and Conservation Association (NPCA). Currently it is being sold to the general public by NPCA.

Designed for children grades four—six, the curriculum guide presents ten major units:

- Biodiversity: The Spice of Life
- Extinction is Forever
- The Role of National Parks in Maintaining and Preserving Biodiversity
- Biodiversity: Its Value in $$$ and Sense
- Biodiversity: Considering the Global Connection
- Habitat and Niche
- Organisms/Populations of Organisms
- Adaptation/Change
- Communities
- Biodiversity at Historic and Cultural Sites

Evaluation of the curriculum guide by interpreters (Mullins and Watson 1992) indicated that 59% of those interpreters participating in the study were familiar with the curriculum guide. Utility-wise they found the curriculum guide somewhat useful and the quality rating was good. Eighty percent of the reporting interpreters actively engaged in biodiversity interpretation had knowledge of the curriculum guide.

A slide/tape presentation entitled "Biological Diversity: It Makes All the Difference in the World" (Havel and Mullins 1991) was developed based on the concept originally presented by Carol J. Spears, Chief of Interpretation, Channel Islands National Park, in the NPS biological diversity handbook. The slide presentation, later converted to video, centered on three levels of biodiversity—genetic, species, and ecosystem diversity. Biodiversity is also explained utilizing the concepts of ecological, economic, esthetic, and ethical values of maintaining biodiversity. Overall ratings were good although budgetary constraints limited wide distribution of the slide tape program and video.

NPS provided a moderate level of training opportunities relating to biodiversity. A variety of national, regional, and local non-NPS training was also utilized. Twenty-five percent of the respondents reported having engaged in some level of biodiversity training between 1988 and 1991. Most of these individuals had been to one training session (Mullins and Watson 1992). Among the 251 respondents in the Mullins and Watson study (1992), 34% reported being extensively or regularly involved with "interpreting, providing support for, or supervising the interpretation of biodiversity." Another 23% reported being occasionally involved with interpreting this critical resource issue.

DISCUSSION

No single formula exists for the content of a public education package. Each organization must clearly understand its mission and role in biodiversity protection, identify clientele and client needs, and match the mission of the agency to its client needs. Out of this market-oriented process will evolve the appropriate content and technologies for the package.

Clearly at issue in the messages are the immediate "hot topics" relating to biodiversity, such as spotted owl issues, as well as the long-term issues relating to maintenance of ecological, economic, educational, ethical, and esthetic values (E-Values of Biodiversity) associated with biodiversity. These issues are bound up in missions of resource management organizations, scientific studies, legalities, agendas of special interest groups, and the positions taken by various governing bodies.

The complex nature of this evolving web lends itself well to the use of all forms of ecological communications. Various general public and agency constituency groups, as well as the staff of the resource management organizations, often have an unclear understanding of the issues. A public education package is one technique to help all parties better understand what they know, or think they know, about a critical resource issue such as biodiversity.

In essence, there is a blending of sound education and communications concepts with public relations, marketing, and political decision-making to try to bring as much clarity to the subject as possible. The literature relating to dissemination of information and the infusion and acceptance of new ideas, such as found in Rogers' *Diffusion of Innovations* (1983), adds further guidance to such efforts.

As various organizations approach the interpretation of critical resource issues such as the loss of biodiversity, a few caveats are in order. Perhaps out of rethinking why an organization interprets a concept will come a better understanding of philosophical positions and encouragement for organizations to be prepared to accept the consequences of their actions.

First, an organization must look inward at its mission, its capabilities and barriers to carry out successfully the tasks with which it is often charged. A philosophical commitment supported by resources is required of the organization; merely adding a new label without resources or support to existing programs is inadequate.

Although we have reported that NPS is moving forward with its interpretation of critical resource issues, the barriers of time, money, too few personnel, limited training, and uncertainties among interpreters and managers about access to current scientific information continue to impact the program. Certainly the introduction of the National Biological Survey will further uncertainties in some areas. Continually working to overcome the barriers and keeping the "on-the-line" communicators apprised of agency efforts to support them are critical to sustaining a communications effort. Because acid deposition is still present and biodiversity continues to deteriorate, sustained educational efforts over time are necessary.

To sustain these efforts, managers and other decision-makers must recognize that all facets of ecological communications are critical to manage effectively. Such efforts must not be viewed as an optional activity, that is a "nice" add-on. In a pluralistic, democratic society, especially in the present-day context where society is questioning the decision-makers' decisions, it is imperative that a strong two-way flow of information exists. Over time the protection of biodiversity will be greatly affected by the will of the people. Those people must have sound scientific information that is presented to them in such manner that they understand, and in turn, will support efforts to utilize resources wisely.

Often, however, the resources that we seek to protect are not on our managerial turf. Acid deposition, loss of biodiversity, and potential impacts from global climate change are not restricted to a national park's boundary. The power of the pen, the environmental education curriculum, or the public program at the community center or campground might be one of the more powerful weapons for addressing these "threats." Viewing agency "turf" as part of a larger biogeocultural region, similar to what we find in biosphere reserve programs, improves chances of protecting park resources.

The preceding ideas have been voiced in the context of ecological communications as techniques of management, namely, tools to encourage others to follow our lead or to support an agenda. Even if agencies seek not to set agendas, as is many agencies' stated policy, they still exert strong influence over audiences merely by the ecological information chosen to or not to present. Mullins and Neuhauser (1991) argued that resource management agencies and conservation organizations should consider an alternative way of thinking, namely, the citizens right to know. This premise is based on the idea that public education (the right of the public to know) and public participation (the right of the public to be involved) are means of helping citizens become full partners in environmental decision-making.

Public participation is related to the environmental ethics of participants. Resource management agencies and conservation organizations should beware of educational approaches that espouse the "teaching of an environmental ethic." Because ethics are deeply-held psychological constructs, it is probably a fallacy to think they can be taught through ecological communications, and perhaps more importantly, it is inappropriate to force deeply-held values on others. Such a position, however, does not relegate agency activities to a non-adversarial role. It simply encourages members of that agency to be better scientists, communicators, educators, or role models in the area of environmental management. Reasoned people, when presented with well-reasoned arguments and believable scientific data on benefits, costs, and risks, will make improved choices for the environment.

As individuals make choices, different concepts will appeal to different individuals or as Tilden (1957) instructs "somehow relate what is being displayed or described to something within the personality or experience of the visitor...." For example, ecological communicators should use values concepts for biodiversity (ecological, economic, educational, esthetic, and ethical values) as means of psychologically linking with their audiences. Of course, none of the values are truly independent of the other; they are offered merely as a means of formulating issues into understandable segments around which people can clarify their values and make choices.

New data or ways of valuing come from a variety of sources. For example, communicators can find ecological data and messages in *Keeping Options Alive—The Scientific Basis for Conserving Biological Diversity* (Reid and Miller 1989). Herein is reported the why, where, severity of the problem, and tools to address the loss of biodiversity. Economic issues were addressed effectively by McNeely (1988) in a book on developing and using economic incentives for conserving biodiversity. The author provides background data, rationale, and case studies to elucidate economic issues.

The various efforts by the federal resource management agencies, such as NPS and their cooperators, provide an overview of the limited work that has been carried out in biodiversity interpretation. Efforts such as the Smithsonian Institution's exhibit *Seeds of Change* are effective communications tools. Likewise most non-government conservation organizations, such as World Wildlife Fund, provide an array of biodiversity education materials. *On Interpretation -Sociology for Interpreters of Natural and Cultural History* (Machlis and Field 1992) offers views on interpretation for various clientele that are most relevant to biodiversity interpretation.

As one might expect, the literature on esthetics and ethics related to biodiversity issues is not as replete as on the ecological arguments. Wilson's *Biodiversity* (1988) does provide a discussion of these topics, as well as most other topics associated with biodiversity. This book, perhaps more so than others, provides a relatively comprehensive overview of most facets of biodiversity. It is from the research and insights of scientists and authors such as those mentioned in this paper that ecological communicators derive the information, and often the inspiration, to interpret science for agency clientele.

As concepts are interpreted, scientists are finding that simply imparting knowledge, encouraging attitude shifts, and creating an appreciation for a topic such as biodiversity are not enough to promote behavior change. Ecological communicators, such as Hungerford and Volk (1990), argue that our education must provide audiences with the opportunity to learn environmentally responsible behaviors before they can take action.

To continue building public education packages, the National Park Service must seek to overcome the barriers reported herein. By better educating and providing NPS communicators with the tools to reach their audiences, the chances of stemming the tide of biodiversity loss might in some part be affected.

Even though NPS might prepare communicators well, very little is known about how best to communicate critical resource issues to audiences in such a manner as to encourage environmentally responsible behaviors. Much work remains to be done before effectively linking messages, media, and targeted audience in such manner that one can predict the influences of programs on environmental choices and how such choices can reduce the loss of biodiversity.

RECOMMENDATIONS

For those agencies and organizations desiring to engage more actively in ecological communications to help maintain biodiversity the following points are offered for consideration:

- Create within each organization an overall plan for addressing biodiversity.
- Build within the organization a culture that accepts that all facets of communications are intricate components of the program to maintain biodiversity.
- Seek to understand staff needs, perspectives on and understanding of biodiversity before instituting any programs.
- Assess the organization's mission and capabilities as it relates to biodiversity concerns.
- Build on existing programs of fellow agencies and organizations; consider jointly sponsoring efforts such as a biodiversity communications program within the U.S. Department of Interior or U.S. Department of Agriculture.
- Create an effective communications plan that draws upon expert resources from both within and outside the organization; foster an interdisciplinary approach utilizing scientists, resource managers, interpreters/educators, and administrators.
- Identify the clientele; assess their knowledge of, perspectives on, and needs relating to biodiversity.
- Recognize that there are many environmental ethics and that your overall communications program might need to be an agenda cooperatively set between the agency and clientele.
- Seek to go beyond sharing information to interpreting the potential risks associated with loss of biodiversity; encourage audiences to take the action they deem most appropriate based on their environmental ethic.
- Seek to reach persons who do not traditionally visit national parks, for example, or who do not traditionally participate in community conservation activities with biodiversity information relating to their life situations.
- Create packages to address formal, non-formal, and informal education needs so biodiversity messages impact at multiple levels of human activity.
- Provide all agency communicators with packages of materials (rather than various random pieces) that help them fully conceptualize the critical resource issue so they can place this subject in the context of their immediate mission or concerns.
- Package messages around concepts your clientele will relate to and readily understand.
- Identify and target messages to specific audiences, utilizing the most appropriate media (often a familiar media with high reliability is more effective than searching for the most novel technology).

- Recognize that all biodiversity issues are not rooted in fact; accept the appropriateness to interpret scientific uncertainties and to explore ethics and esthetics with an audience.
- Establish systems for both formative and summative-types of evaluation to assess effectiveness of program planning, development and impacts.

Public education packages and the array of other ecological communication tools are much like the concepts we interpret. The effects of acid deposition are still under study, global climate change appears to be occurring but the computer models on impact vary, and the rate of loss of biodiversity is still being calculated. The impact of ecological communications is uncertain, as well. Just as we can ill afford to ignore the risks associated with critical resource issues because of uncertainty, we can ill afford to neglect a strong communications effort to interpret the potential risks we face.

LITERATURE CITED

Dottavio, F. D., P. F. Brussard, and J. D. McCrone, eds. 1990. *Protecting Biological Diversity in the National Parks: Workshop Recommendations.* National Park Service, Washington, D.C.

Havel, C. A., and G. W. Mullins. 1991. *Biological Diversity: It Makes All the Difference In the World.* (a video). National Park Service, Washington, D.C.

Hungerford, H. R., and T. L. Volk. 1990. Changing Learner Behavior Through Environmental Education. *J. Environ. Educ.* 21(3)8-21.

Machlis, G. E., and D. R. Field, eds. 1992. *On Interpretation-Sociology for Interpreters of Natural and Cultural History.* 2nd ed. Oregon State University Press, Corvallis, Ore.

McNeely, J. A. 1988. *Economics and Biological Diversity: Developing and Using Economic Incentives to Conserve Biological Resources.* International Union for Conservation of Nature and Natural Resources, Gland, Switzerland.

Minnesota Environmental Education. 1990. *Biological Diversity: It Makes a World of Difference Curriculum Guide.* National Park and Conservation Association and National Park Service, Washington, D.C.

Mullins, G. W., ed. 1990. *NPS Handbook for Biological Diversity Communicators.* National Park Service, Washington, D.C.

Mullins, G. W., R. W. Fortner, and K. Palmer. 1989. *Report on the Interpreting Biological Diversity in the National Park System Survey.* National Park Service, Washington, D.C.

Mullins, G. W., and H. Neuhauser. 1991. Public Education for Protecting Coastal Barriers. *BioScience* 41(5):326-330.

Mullins, G. W., and M. D. Watson. 1992. *Interpreting Critical Resource Issues in National Parks: An Assessment.* National Park Service, Washington, D.C.

Office of Technology Assessment (OTA). 1987. *Technologies to Maintain Biological Diversity.* OTA-F-330, Government Printing Office, Washington, D.C.

Reid, W. V., and K. R. Miller. 1989. *Keeping Options Alive—The Scientific Basis for Conserving Biological Diversity.* World Resources Institute, Washington, D.C.

Rogers, E. M. 1983. *Diffusion of Innovations.* The Free Press, New York.

Schoenwald-Cox, C., ed. 1987. *The Role of the National Park Service in Protecting Biological Diversity.* National Park Service, Washington, D.C.

Tassier, K. S., G. W. Mullins, and J. D. Peine. 1988. *Our Backyard Biosphere Reserve—Education Modules on Great Smoky Mountains National Park Biosphere Reserve and the Southern*

Appalachian Region. Great Smoky Mountains National Park/Southern Appalachian Regional Man and Biosphere Program, Gatlinburg, Tenn.

Tilden, F. 1957. *Interpreting Our Heritage*. The University of North Carolina Press, Chapel Hill, N.C.

Wilson, E.O. 1988. *Biodiversity*. National Academy Press, Washington, D.C.

38

CONSERVATION AND ZONES OF HUMAN ACTIVITY: THE SPREAD OF HUMAN DISTURBANCE ACROSS A PROTECTED LANDSCAPE

Marybeth Buechner

Biology Department
Cosumnes River College
Sacramento, California
U.S.A.

Ray Sauvajot

Ecology Graduate Group
Division of Environmental Studies
University of California, Davis
Davis, California
U.S.A.

Natural habitats face stresses as the result of the imposition of areas of human-altered habitat between sections of a formerly continuous community or ecosystem (Forman and Godron 1986; Harris 1988; Yahner 1988; Pickett et al. 1992). Protected habitats are not immune to the impacts of contiguous disturbance; even areas as large and significant as U.S. National parks face increasing stress from adjacent development (Stottlemyer 1987; Schonewald-Cox 1988; Schonewald-Cox et al. 1992; Buechner et al. 1992; Slocombe and Nelson 1992). Impacts also arise from human activities occurring within the boundaries of the protected habitat (e.g., Machlis and Tichnell 1987). For example, landscape elements such as roads or recreational areas occur within most parks and reserves and may affect a substantial part of even large parks (Schonewald-Cox and Buechner 1992).

Two issues that must be considered when assessing the impact of intrusive disturbance on natural areas are the distance to which the effects of the disturbance move into natural habitat from source areas and the degree to which these generated edges stress native species. We are concerned that sensitivity to the stresses generated by intrusive human activities might decrease the regional persistence of native plant and wildlife populations. The challenge to conservationists is to develop effective strategies for the maintenance of diverse natural communities within an increasingly modified and fragmented landscape. This includes an understanding of the ways in which parks and reserves are influenced by anthropogenic disturbance affecting the integrity of the protected landscape.

Many types of human-associated disturbance can intrude into nearby natural areas and create zones of disturbance within protected habitats (Table 38.1). Intrusive disturbance can increase the fragmentation of natural habitats, resulting in numerous deleterious effects on sensitive systems (Table 38.2). The consequences of habitat fragmentation have been described as some of the most serious threats facing the conservation of biological diversity in human-altered landscapes (e.g., Wilcox and Murphy 1985; Harris and Silva-Lopez 1992).

Intrusive effects include both diffuse impacts arising from distant, multiple, or non-point sources (e.g., air pollution, acid rain) and more localized intrusions resulting from linear or point sources immediately adjacent to or within a park or reserve (e.g., trampling, littering, the activities of domestic animals). In this paper we focus on localized effects resulting from the spread of human-associated disturbance across a protected landscape. Such localized intrusive effects include both effect intruding from the edges of the parks and those effects arising within park boundaries. Machlis and Tichnell (1987) reported that of the surveyed threats to world parks 24% arose exclusively outside of the park, 34% had sources within the park, and 35% came from a combination of internal and external sources.

How seriously do intrusive effects from areas of intense human activity affect adjacent protected habitats? How might reserve managers best mitigate impacts in fragmented reserves? How can we best delineate the zones of disturbance associated with spreading human activity? How can measurements of these effects be used to make recommendations about reserve management? Whereas an analysis of the issue from the standpoint of U.S. National Parks has suggested some strategies for coping with cross-boundary effects (Schonewald-Cox et al. 1992; Buechner et al. 1992), the answers to these questions remain unclear, partly because few detailed community-level analyses are available (Yahner 1988; Soulé and Kohm 1989; Schonewald-Cox et al. 1992).

In this paper, we focus on one aspect of the problem: the delineation of the zones of disturbance that result from relatively intense and localized human activity within or immediately adjacent to a protected landscape. We begin by discussing, in general terms, the flow of human disturbance from areas of human activity into nearby protected habitats. We then examine a case study of these effects, reporting on the methods and preliminary results of our study of intrusive edge effects in the Santa Monica Mountains near Los Angeles, California. We conclude with a discussion of management and policy issues.

Table 38.1. Summary of the effects of landscape fragmentation on sensitive species

1. Decrease in the size of landscape units due to removal or disturbance of habitat
2. Increase in the perimeter:area ratios of remnant patches and edge effects
3. Increase in the isolation of "core area" species
4. Increase in the influx of material and species associated with human-altered habitat
5. Increased flow of materials and species from core areas into surrounding habitats
6. Possible production of a metapopulation structure for patch-interior species
7. Possible increase in the rate of further landscape fragmentation in the area

Table 38.2. Examples of sources of disturbance associated with human activity which affect parks and reserves (Machlis and Tichnell 1987; Stottlemyer 1987; Schonewald-Cox et al. 1992)

1. Sources External to Park Boundaries
   (a) industrial development
   (b) commercial development
   (c) residential development
   (d) roads and railroads
   (e) logging
   (f) grazing and crop agriculture
   (g) energy extraction/energy production/mining
   (h) military facilities
   (i) recreation/hunting
   (j) other air pollutant sources
   (k) other water pollutant and siltation sources
   (l) other sources altering water flow
   (m) other sources of exotic and pest species and diseases
2. Sources Internal to Park Boundaries
   (a) visitor use/recreation
   (b) utility access corridors
   (c) roads
   (d) trails
   (e) off road vehicles
   (f) other sources of exotic and pest species and diseases
   (g) park operations
   (h) illegal removal of plants or animals
   (i) illegal fires
   (j) illegal dumping of garbage

THE SPREAD OF HUMAN DISTURBANCE ACROSS PROTECTED AREAS: ZONES OF DISTURBANCE AND HUMAN ACTIVITY

The types of human activity that impact parks and reserves are complex and numerous (Table 38.1). In spite of this complexity, however, the ways in which human activity and associated disturbance move across a protected landscape can be grouped into relatively few categories. We have identified three common routes by which human activity moves across a protected landscape and creates zones of varying intensity of disturbance (Figure 38.1, Table 38.3):

(1) Boundary processes—the intrusion of human activity and associated disturbance along broad interfaces between developed areas and protected reserve habitats.

(2) Linear elements—the movement of activity and disturbance along linear landscape elements which intersect protected habitats.

(3) Hot spots—the development of concentrated areas of activity and disturbance within reserve boundaries.

Some recent studies involving these types of intrusive disturbance are given in Table 38.3. Information for this table was gathered by reviewing the last four years

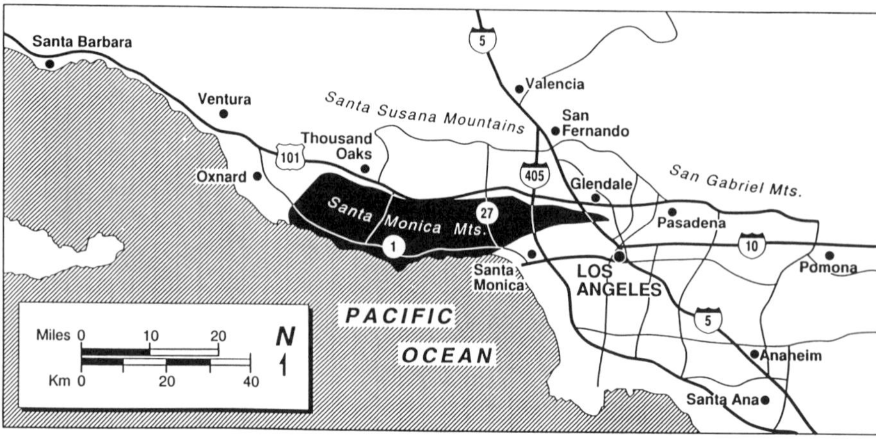

Figure 38.1. Vicinity map showing the Santa Monica Mountains. Our study area is in the section of the mountains between roads 27 and 405 at the northern boundary of the mountains.

(1990-1993) of the following journals: *Biological Conservation, Conservation Biology, Environmental Management, Landscape and Urban Planning, Landscape Ecology*, and *Park Science*. As the table illustrates, even this brief review of recent material provides numerous examples of each type of intrusive disturbance route. Various reviews of less recent material (i.e., prior to 1990) are available and are cited in the discussion below.

Boundary Processes

Human-induced stresses that enter natural habitat from adjacent areas represent one example of a category of phenomena referred to as "edge effects" or "boundary effects" (Forman and Godron 1986; Lovejoy et al. 1986; Schonewald-Cox 1988; Yahner 1988; Harris 1988; Harris and Silva-Lopez 1992; Schonewald-Cox et al. 1992). In the context of this paper, boundary effects are broadly defined as any ecological changes that occur in previously undisturbed communities as the result of the intrusion of disturbance into natural habitat from adjacent human-modified areas (e.g., urban areas). Harris and Silva-Lopez (1992) further categorize various types of habitat fragmentation which may lead to boundary effects. They define "regressive fragmentation" in which habitat clearing and accompanying disturbance occurs from a single direction and one side of the natural habitat is pushed back, "enveloping" fragmentation in which disturbance surrounds the perimeter of an area of natural habitat and impact it from all directions, and " encroaching" fragmentation in which disturbance moves in from either side of a linear corridor of habitat leaving the ends of the corridor connected to larger areas of natural habitat. Buechner (1987), Buechner et al. (1992), Janzen (1986), Machlis and Tichnell (1987), Schonewald-Cox (1988), Schonewald-Cox et al. (1992), Stottlemyer (1987), Temple and Cary (1988), and Yahner (1988) provide reviews of the impacts of boundary processes on areas of natural habitat and wildlife populations.

The importance of intrusive disturbance occurring at the boundary of reserves is recognized by a number of researchers and boundary effects are being increasingly recognized as of potential importance for wildlife conservation (Janzen 1983; Harris 1988; Schonewald-Cox and Bayless 1986; Noss 1987; Schonewald-Cox 1988; Yahner 1988; Buechner et al. 1992; Newmark et al. 1993). Yahner (1988) noted that fragmentation and increased edge are major factors contributing to the reduced distribution and abundance of wildlife species on a broad geographic scale." Diamond et al. (1987) speculated that increased disturbance due to an increase in the accessibility of the area to the public might have been a factor in the extinction of several species of birds within an isolated Javan woodland. Janzen (1983) pointed out that "no park is an island" and that influences from surrounding areas often exert detrimental effects on reserve habitats. Schonewald-Cox suggested that boundary processes and external influences may function as the primary determinants of successful conservation in reserves (Schonewald-Cox and Bayless 1986; Schonewald-Cox 1988). Cortner et al. (1990) noted that "A growing consensus has emerged that urban-wildland issues have reached a threshold whereby they have become among the most contentious and problematic issues for forest managers."

Many intrusive boundary effects are the direct or indirect result of the movement of people (e.g., poaching, trampling, noise, littering). In the Wheatbelt Region of southwestern Australia, 86% of the 90 reserves sampled have been used illegally by people "to the detriment of the reserve" for gravel or sand extraction, rubbish dumping, or timber cutting, with 48% of the reserves experiencing these impacts within the 12 months preceding the study (Wallace and Moore 1987). The impacts of such activities might depend on the number of access points for humans which occur at the boundary of a reserve. Habitat destruction due to humans seeking food, fuel and other essentials has been suggested to be a major problem for reserves in developing countries (Shafer 1990; Eaton 1991; Jeffery and Chooye 1991; Newmark 1993; Shackleton 1993) and in hinterland areas of more developed countries (Slocombe and Nelson 1992).

Exotic and weedy species can invade native vegetation at the edges of reserves (review in Shafer 1990). Timmins and Williams (1987) reviewed information on the ability of weed to invade reserves in New Zealand. They noted that humans have been instrumental in dispersing many weeds into reserves and that areas where the native vegetation has been disturbed are particularly susceptible to invasion by exotic weeds. On Barrow Island, Australia, two plants, capeweed (*Arctotheca calendula*) and black berry nightshade (*Solanum nigrum*) were introduced in the footwear of workmen but were successfully eradicated by hand weeding (Butler 1987). Backhouse (1987) noted that one of the major problems for reserves with high perimeter:area ratios if the degree to which weeds and feral vertebrates can invade the reserve. Dramatic changes in vegetation might occur at the edge of reserves (Taylor 1987). Taylor (1987) described the vegetation which develops at the edge of a reserve as "human-induced vegetation with a degraded native canopy layer and an adventive (weed) native/exotic understory."

Animals associated with human dwellings (e.g., cats, rats, or dogs), other introduced or exotic species (e.g., feral pigs) or species associated with agricultural areas (e.g.,

Table 38.3. Examples of recent publications relating to how anthropogenic intrusive disturbances can affect natural landscapes.

Location	Notes	Reference
	Boundary Processes	
Royal Chitwan National Park, Nepal	In one year, 554 people were fined for stealing firewood or grasses and 1306 were fined for taking cattle or buffaloes into the park	Sharma 1990
Vwaza Marsh Game Reserve, Malawi	Wildlife from within the reserve is used by local people	McShane 1990
Waza National Park, Cameroon	Local resource usage conflicts with park goals	Drijver 1991
Santa Rita Experimental Range, Arizona	Introduced grass planted outside of the area has displaced native spp.	Anable et al. 1992
Donana National Park, Spain	41.7 % of Iberian lynx mortality inside the protected area is due to illegal trapping and 16.7 % is due to hunting with dogs. Adult annual mortality on the edge of the park is many times greater than in the center	Ferreras, et al. 1992
Several remote areas of northern and western Scotland	Feral domestic cats interbreed with European wildcats	Hubbard, et al. 1992
Kapiti Island, New Zealand	Introduced brushtail possums threaten native vegetation and wildlife	Cowan 1992
Las Amoladeras Reserve, Spain	Feral dogs are nest predators for several passerine birds	Suarez et al. 1993
Haleakala National Park, Hawaii	Feral pigs disrupt native ecosystems	Anderson and Stone, 1993
Rocky Mountain National Park, Colorado	Black bears do not use areas on the edge of the park because of human use of areas outside of the park boundary and due to hunting	McCutchen 1993
	Linear Elements	
Grand Canyon National Park, Arizona	Cryptogamic soil crusts are damaged by trampling by hikers	Cole 1990
Royal Chitwan National Park, Nepal	Canoes cause minor disturbance disturbances of ruddy shelducks	Hulbert 1990
Hubbard Brook Experimental Forest, New Hampshire	An experimental approach has detailed the impacts of trampling damage on trails	Kuss and Hall 1991
Hacienda Loma Linda, Canas Gordas de Coto Brus, Costa Rica	An experimental approach studied recovery of vegetation following trampling damage along trails	Boucher et al. 1991
Most of the Netherlands including nature reserves	Badger decline is most closely related to the number of roads	van der Zee et al. 1992
Wilson's Creek National Battlefield, Missouri	An interpretive trail passes through a population of endangered plants causing trampling damage	Thomas and Willson 1992
Donana National Park, Spain	Iberian lynx are killed by road traffic (16.7 % of mortality)	Ferreras et al. 1992
Everglades National Park, Florida	Snakes have high mortality rates on roads	Bernardino and Dalrymple 1992

continued

Table 38.3.—*continued*

Location	Notes	Reference
Kakadu National Park, Australia	In relatively undisturbed natural areas, the most severe infestations of invasive exotic plants occur along roadways and near streams used by feral Asian water buffalo	Cowie and Werner 1993
Rocky Mountain National Park, Colorado	Visitor use of roads and trails is intense on the east side of the park; bears avoid this area in spite of good habitat	McCutchen 1993
Region of First near Grindelwald, Switzerland	Hikers, especially those with dogs, disturb marmots	Mainini et al., 1993
C. Hot Spots		
Grand Canyon National Park, Arizona	Helicopter overflights disturb desert bighorn sheep in some areas	Stockwell et al., 1991
Sakumo and Djange lagoons, Ghana	Sites where fishing is heavy show impacts on some fish species	Ntiamoa-Baidu 1991
Peak District National Park, England	Anglers and other visitors at a reservoir disturb breeding birds	Yalden 1992
Kakudu National Park, Australia	Many invasive exotic plants are found in areas associated with high levels of human activity	Cowie and Werner 1993
Raiganj Wildlife Sanctuary, India	Visitors disturb nesting storks in areas of high human activity	Datta and Pal 1993
Mount Rainier National Park, Washington	Almost 3/4 of all visitors go to Paradise meadow resulting in severe erosion problems; 913 human-caused bare-ground impacts were noted in 1986-87	Rochefort and Gibbons 1993

some mesopredators, cowbirds, white-tailed deer) may also move into areas of natural habitat and cause intrusive disturbances. Several studies have documented the impact of increased predation and nest parasitism on bird communities near the edges of forested habitats (reviews in Temple and Cary 1988; Yahner 1988; Harris and Silva-Lopez 1992). The predators involved include medium-sized mammals such as foxes (*Vulpes fulva*, *Urocyon* sp.), skunks (*Mephitis* spp. and *Spilogale* spp.), domestic dogs (*Canis familiaris*) and domestic cats (*Felis domesticus*), as well as various predatory birds (family Corvidae); nest parasitism is generally caused by the brown-headed cowbird (*Molothrus ater*) (Harris and Silva-Lopez 1992). The detrimental effects of increased predation and nest parasitism can extend up to 600 m into a forest (Yahner 1988). The effects of species such as white-tailed deer (*Odocoileus virginianus*), can extend even further into natural habitats. Bratton (1979) estimated that intensive deer impacts could extend up to 1 km from openings into forests in Great Smoky Mountains National Park. Alverson et al. (1988) concluded that 200 to 400 km sq. reserves within U.S. National Forests would be necessary to retain species sensitive to deer browsing.

The degree to which human activity and associated disturbance affect the natural habitat can be influenced by a number of factors including: (1) the type and density of development present, (2) the "resistance" of the natural landscape to the intrusion of disturbance, (e.g., the ease with which people can move into the natural habitat; Forman and Godron 1986), and (3) the sensitivity of the natural habitat to disruption. Although the first factor can be relatively easily measured from land-use maps, the other factors are much more difficult to measure in the field.

Linear Elements

Linear landscape elements, such as roads, trails, power line corridors or waterways, can facilitate the spread of anthropogenic disturbance into protected areas (Schonewald-Cox et al. 1992; Thomas and Willson 1992). Harris and Silva-Lopez defined the type of habitat fragmentation associated with linear elements as "divisive" fragmentation. They noted that this occurs "when an intrusive force such as a turnpike, a road with attendant strip development, powerlines, or railroads bisects an expansive tract... ." The most obvious linear elements are man-made travel corridors like railroads, trails, or roads; however, linear elements might also be natural features of the landscape. For example, ridges or canyon bottoms can be used by people as travel routes even when no developed trails exist in those locations.

A number of impacts to native vegetation can be caused by intrusive disturbance occurring along linear elements within protected habitats. For example, trampling of vegetation along trails and at visitor resting or sight-seeing spots can be a major concern for parks and reserves (Kus and Hall 1991; Thomas and Willson 1992; Taylor et al. 1993). Invasive exotic or native weedy species may move into protected habitats along linear elements such as roads or waterways (e.g., Cowie and Werner 1993). Power-line corridors both remove vegetation and result in changes in the vegetation along the edges of the corridor (e.g., Luken et al. 1991). Rail corridors have been suggested to have impacts on natural habitats, including both short-term impacts, e.g., those associated with construction activites, and long-term impacts such as habitat fragmentation (DeSanto and Smith 1993). New linear elements can develop in the landscape when off-road vehicles move away from official or established routes (e.g., Slaughter et al. 1990). Backhouse (1987) noted that long narrow reserves (such as riparian reserves) might be especially susceptible to disturbance resulting from human access routes such as trails. Linear elements, such as roads or trails, often occur in conjunction with the boundary of a park or reserve. Roads often start near the boundary of a protected area and carry traffic to sites within the reserve. In some cases, a road might be put through the park in order to connect development on either side, rather than because of the needs of the park. A review of some recent studies of intrusive disturbances which follow linear landscape elements are given in Table 38.3B.

Roads are the most dramatic example of a linear landscape element which conducts human activity through protected habitat. Roads can have major influences on wildlife communities (Singer et al. 1986). We have identified six specific road impact issues which may be important for protected wildlife populations. Roads can (1) remove habitat, (2) result in intrusive biotic changes in habitats along their edges, (3) act as sources of pollution and other abiotic impacts, (4) become barriers to animal dispersal, (5) act as sources of mortality, and (6) function as access routes for people, animals, and other disturbances. The extent to which these effects will influence local wildlife populations will, of course, vary from habitat to habitat.

Numerous studies have outlined the impacts of roads on natural landscapes. The effects of dust from unpaved roads on adjacent vegetation is a major problem in parts of Australia (Butler 1987). Roads, pipelines, and other linear elements can also change the surface runoff patterns of an area (Butler 1987). Mortality from collisions with vehicles can be a major impact on wildlife communities. For example, in Florida collisions with motor vehicles is the primary known source of mortality for several endangered species of large vertebrates (Harris and Silva-Lopez 1992). Roadways and road margins can act as access routes for disturbances associated with the intrusion of pest, predator, and exotic species. This can occur because the altered areas along roadways often represent dispersal corridors or supply food or other resources for such species (Michael et al. 1976; Getz et al. 1978; Vekaar 1988; Baudry and Merriam 1988; Melman et al. 1988). More subtle effects of human activities associated with road access include increased litter accumulations, vegetation trampling, and other direct disturbances (e.g., noise) which can affect some species. The intensity of these disturbances can be expected to reflect the accessibility of natural areas adjacent to roadways. Sauvajot and Buechner (1991) and Schonewald-Cox and Buechner (1992) provide reviews of the effects of roads on parks and reserves.

Various direct impacts associated with roads are known to extend 100 m or more into adjacent habitats (Sauvajot and Buechner 1991; Schonewald-Cox and Buechner 1992). Habitat alteration can facilitate the movement of edge-adapted species near roadways and later the abundance of some animal species along roads. Adams and Geis (1983) in a study of changes in small mammal communities along roadways in northwestern California found that California voles (*Microtus californicus*) and western harvest mice (*Reithrodontomys megalotis*) occurred more often and dusky-footed woodrats (*Neotoma fuscipes*) occurred less often near roads than in adjacent habitats. They found that changes in small mammal species composition and abundance extended 80 to 160 m from road edges. Ferris (1979) investigated the intrusive effects of a highway on birds in northern Maine. He found that breeding bird assemblages contained a larger component of edge species within 100 m of the road and the abundances of some bird species were affected as far as 400 m from the road. Reijnen et al. (1987) suggested that some birds can be affected by road disturbance several hundred meters from roadway edges.

Hot Spots

Areas of intense human activity can develop within the boundaries of the protected natural habitat. Such "hot spots" include visitor centers, trailheads, camping areas, day-use areas, attractive, highly visited sites, such as scenic areas, private inholdings, and the sites of popular recreational activities. For example, campsites can result in substantial impacts on nearby vegetation due to hiking, waste disposal, firewood gathering and other activities (e.g., Cole and Fichtler 1983; Bratton et al. 1982). It is ironic that hot spots can develop precisely because of the attraction of protected resources for visitors. For example, the presence of unusual or interesting (e.g., endangered) animals or plants might attract visitors to the site, creating a local hot spot of disturbance affecting the very species which the area was designed to protect (Dawson 1991). Management activities within the reserve can also create hot spots; ranger housing or park headquarters even scientific field stations are localized areas of intense human activities and can result in disturbance. Hot spots can also develop in areas in which illegal activities, such as illicit off-road-vehicle activity, occur (Slaughter et al. 1990). Private inholdings within protected habitats can function as sources of human disturbance and sites which function in the introduction of weedy species (Harris and Silva-Lopez 1992). Harris and Silva-Lopez (1992) noted that Ocala National Forest in Florida contains approximately 7,000 inholdings and that about 10% of the area within the forest boundary is privately owned (including a U. S. Navy bombing range). In some cases, the animals residing in the reserve can create a hot spot; for example, in Etosha National Park, Namibia, animals congregate in large numbers around artificial waterholes resulting in overgrazing and browsing pressure in some areas (Cloudsley-Thompson 1990). A review of some recent studies of hot spots occurring within protected areas is given in Table 38.3C.

The occurrence of hot spots can be affected by the location of both reserve boundaries and linear travel corridors, such as trails or roads. Hot spots can develop in conjunction with parking spaces or trailheads occurring along roads. Visitor centers, camping and day-use areas, and recreational areas are naturally located at the end of access routes such as roads and trails. In some cases, human-caused intrusive disturbances can "leap-frog" in from the boundary of the protected area, creating a hot spot near, but not on, the boundary of the reserve. Sites along the boundary of a reserve which provide easy access to the natural habitat can result in such hot spots. For example, in our study site hot spots of off-road vehicle (ORV) damage occurs some distance within the park in areas which are reached by access routes from the boundary of the park. As a result, the movement of human activity and associated disturbance in from the edge of a reserve might not be a simple diffusion process, but a more complicated pattern.

ZONES OF HUMAN DISTURBANCE IN THE SANTA MONICA MOUNTAINS: A CASE STUDY OF MULHOLLAND GATEWAY PARK

We have recently been working in a park area adjacent to development. This study serves to illustrate many of the issues discussed above. Because this research is currently in progress, we focus less on definitive results and conclusions and more on a discussion of approaches and preliminary findings regarding the importance of urban proximity, roads, and concentrated areas of human activity as mechanisms for spreading human disturbance across the protected landscapes of this system.

Study Area

Our research is being conducted within the Santa Monica Mountains near Los Angeles, California (Figure 38.1). The numerous park areas that make up much of the Santa Monica Mountains lie next to a tremendous concentration of human development, the Los Angeles metropolitan area. Thus, this are is particularly appropriate for investigating how anthropogenic disturbance spreads across protected landscapes. Throughout the mountains, rapid urbanization has encroached upon remaining natural habitats, and now most protected areas are situated immediately adjacent to massive development. We are interested in how these two contrasting landscapes, protected park lands and adjacent urban development, interact to affect the ecological integrity of the remaining natural areas.

In our work, we focus on a specific study area immediately south of the San Fernando Valley (Figure 38.2). Most of the study area is part of Mulholland Gateway Park, a protected area of natural habitat immediately adjacent to the city. This area includes approximately 400 ha of primarily chaparral habitat and is located adjacent to urbanization to the north and additional natural habitat to the south. Although there is new development in the area, much of the urban area is older development and has been in place for over 20 years. Recently, this region has become the focus of spirited local conservation and planning efforts and, when it is included with the more than 5,000 ha of protected habitat to the south, represents one of the largest areas of natural open space within a city limits anywhere (CDD 1991).

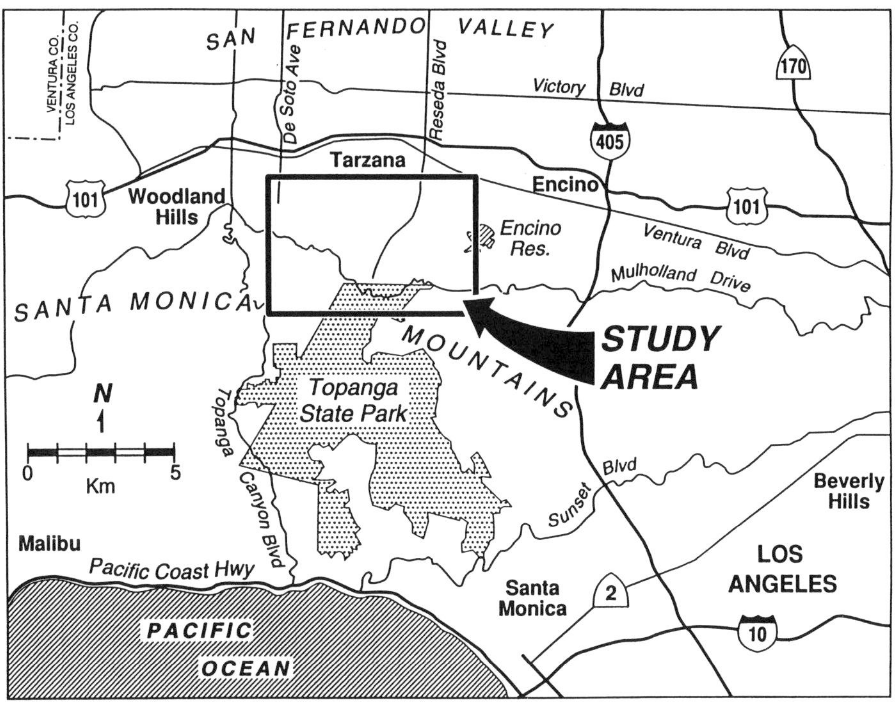

Figure 38.2. Map showing the location of the study area between Topanga State Park and the San Fernando Valley. Much of this area is included in Mulholland Gateway Park.

Boundary Effects and Urban Proximity

During the summer of 1991, we initiated detailed work on the role of urban proximity as a source of disturbance within the study area. Because habitat is immediately adjacent to a very large area of high density development, we felt that this was an excellent location for investigation how anthropogenic disturbances could intrude from development into the nearby protected area. We chose to address this issue by developing a measure of the exposure of natural habitat to urban areas and testing to see if community-level ecological changes in wildlife communities were associated with this exposure.

To assess the effects of proximity to urban area per se, without including other effects (e.g., fire, vegetation clearing) we worked in areas of relatively intact chaparral vegetation and (for this part of the study) avoided areas which show obvious evidence of vegetation damage or removal. We measured the diversity of year-round resident bird species and both diversity and relative abundance for resident small mammal species (Tables 38.4, 38.5) in natural habitat at sampling sites throughout the study area. By sampling this species at sites with varying exposures to the urban areas, we hope to determine if proximity to urban development is an important predictor of community-level changes for these species. We hypothesized that if significant human-associated disturbances are intruding from the developed areas into adjacent natural habitats, we could detect changes for theses species in the sampling sites most exposed to urban areas.

Table 38.4. Presence-absence matrix for resident bird species in the urban exposure transect, 1991. Transects are arranged from those with the lowest urban exposure indices to those with the highest urban exposure indices.

	Transect											
	Least urban exposed——————————› Most urban exposed											
Bird Species	1	2	3	4	5	6	7	8	9	10	11	12
California Quail	X		X		X		X	X	X	X	X	X
Mourning Dove[1]	X	X	X	X	X		X		X	X	X	X
Rock Dove[1]									X			X
Anna's Hummingbird		X	X	X			X	X	X	X	X	X
Scrub Jay	X	X	X	X	X	X	X	X	X	X	X	X
Common Raven		X			X				X		X	X
Plain Titmouse						X						
Bushtit	X		X	X	X				X			
Bewick's Wren	X	X	X	X	X	X		X	X	X	X	X
Wrentit	X	X	X	X	X	X	X	X	X	X	X	X
Northern Mockingbird[1]	X	X	X		X		X		X		X	X
California Thrasher		X	X	X	X	X	X	X			X	
Rufous-sided Towhee	X	X	X	X	X	X	X	X	X	X	X	X
Brown Towhee[1]	X	X	X	X	X	X	X	X		X	X	
House Finch[1]	X	X	X	X	X	X	X	X	X	X	X	X
Total Species	10	11	12	10	12	8	10	9	12	9	12	11

[1] Bird species often associated with urban environments.

Table 38.5. Presence-absence matrix for small mammals trapped in the urban exposure transect, 1991. Transects are arranged from those with the lowest urban exposure indices to those with the highest urban exposure indices.

	Transect											
	Least urban exposed——————————————› Most urban exposed											
Mammal Species	1	2	3	4	5	6	7	8	9	10	11	12
Neotoma fuscipes	X	X	X	X	X	X	X		X	X	X	X
Neotoma lepida		X	X				X	X			X	
Peromyscus boylii	X	X	X	X			X	X		X	X	
Peromyscus californicus		X	X	X				X				X
Peromyscus truei										X		
Peromyscus maniculatus										X		
Perognathus californicus						X						X
Dipodomys agilis					X							
Juvenile sp.	X		X				X	X	X		X	X
Total Species	2	4	4	3	2	2	3	3	1	4	3	3

Our particular design involves sampling birds and mammals along twelve 150-m long transects in relatively undisturbed habitat which varied in exposure to urban areas. Possible transect sites were identified using aerial photographs, land-use maps, and field surveys. Sampling sites (transect locations) were chosen: (1) to include a wide range of urban exposure, (2) with similar habitat to control for habitat differences between transects, and (3) based on site accessibility (Figure 38.3). The distance from the transects to the edge of the nearest urban development ranged from 0 m - 820 m.

To quantify the degree of exposure of each site to urban areas, we used an index of exposure to urban area which was based on measures of all urban area within concentric 250 m wide bands around each sampling site (Figure 38.4; a similar measure was used by Pearson 1993). This approach allowed a more detailed examination of urban exposure than would a simple "distance to urban edge" measure. For example, consider the following two hypothetical sampling sites: (a) a site at the tip of a peninsula of natural habitat extending into the city, and (b) a site near the tip of a finger of development which extends into the natural habitat. Even if the distance to the nearest city edge is the same for both sites, exposure to the intrusive effects of development is clearly higher for the first site and our index would indicate this. Our index also allowed us to use various weightings of urban area at different distances from the sampling sites. The results reported here are based on a linear weighting function with the greatest importance given to the nearest developed area; however, we also explored both uniform weighting and an exponential decay function. The results for all three weighting functions were very similar.

Birds were sampled along each transect using visual and auditory counts (Emlen 1971), with each transect visited on five mornings during the 1991 field season. Small mammals were sampled using 50 traps per night set on five consecutive nights at along each transect (250 trap-nights per transect). In addition, we collected vegetation information, including the percent cover for perennial plant species and other structural

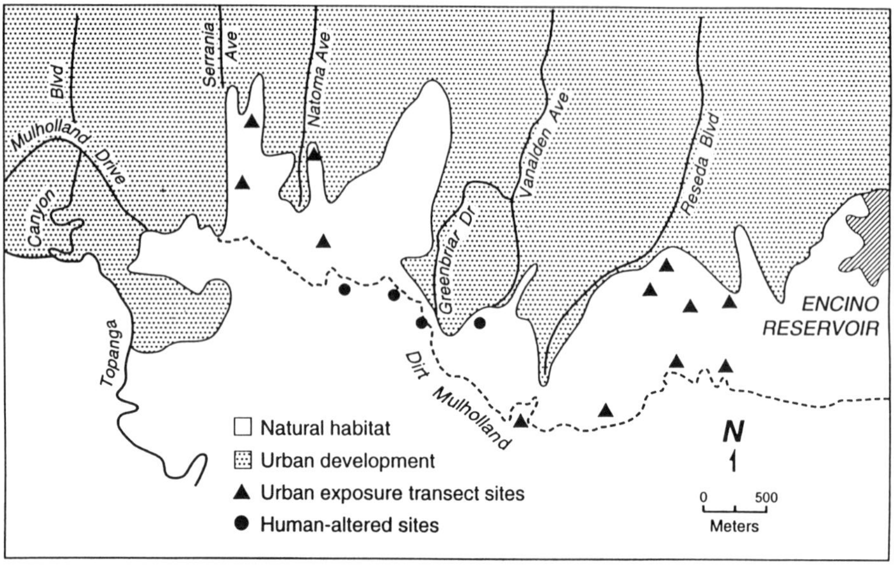

Figure 38.3. Locations of transects within the study area. Approximate locations of the urban exposure transects and human-altered (hot spots) transects are shown.

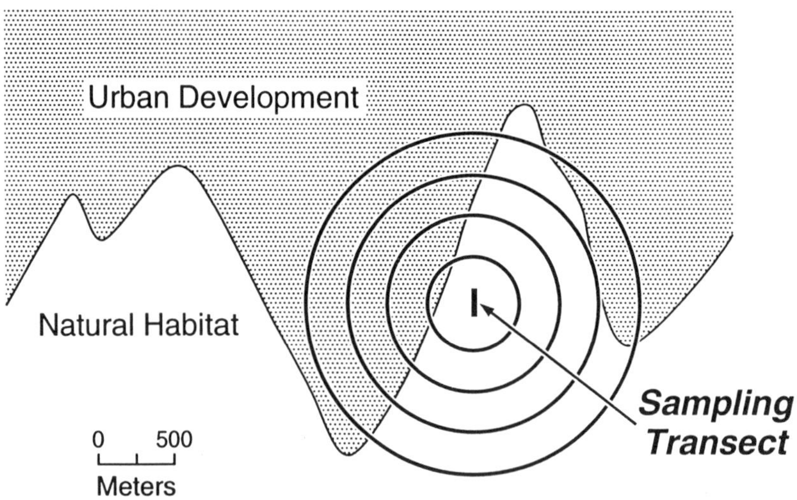

Figure 38.4. Schematic showing how the urban exposure index was calculated. We measured the area of urban development within each of four concentric 250-m bands around the center of the transect. The resulting areas were weighted in various ways. Results reported here use a linearly declining weighting function which places the most importance on the amount of urban area in the innermost band (within 250 m of the transect). Other weighting functions produced similar results. (Although the areas measurements are in square meters, units are not given on this graph because the weighting changes the value so that the numerical value of the index is no longer directly in square meters).

characteristics of the vegetation at each transect using the point-line method. These methods resulted in a total of 3,000 small mammal trap-nights, 60 morning bird samples, and 1,200 vegetation point measurements along the twelve transects.

The data from the summer of 1991 suggest that urban proximity alone does not account for variations in resident bird and mammal communities in our study area. In particular, we have not found a relationship between bird and small mammal species richness or relative abundance and indices of urban exposure (Figures 38.5, 38.6). The species richness of the bird communities varied somewhat between sampling sites but there was no obvious effect of urban exposure Figure 38.5, Table 38.4). All sampling sites contained many resident bird species, including both habitat-specific chaparral birds and urban-associated species (Table 38.4). It is possible that we measured no effect of urban exposure for birds because for these species the study area is effectively all "edge." This would mean that the effects of urbanization extend undiminished at least 1 km into the natural habitat. However, it is important to note that we found a full complement of chaparral species within the study area at all distances from the urban edge.

Although variation was high between sites, differences in small mammal species richness or relative abundance between sites did not appear to be related to urban exposure Figure 38.6, Table 38.5). One of the most surprising findings was that all of the small mammals captured at the sampling sites were native chaparral species—no *Rattus* or *Mus* were captured at any of the sites, even those that were nearly in people's back yards.

Although these results appear to be surprising, it is possible that many mechanisms associated with the intrusive spread of disturbance from urban development are not operating in this system. By design, direct habitat alteration is minimal on the sites we

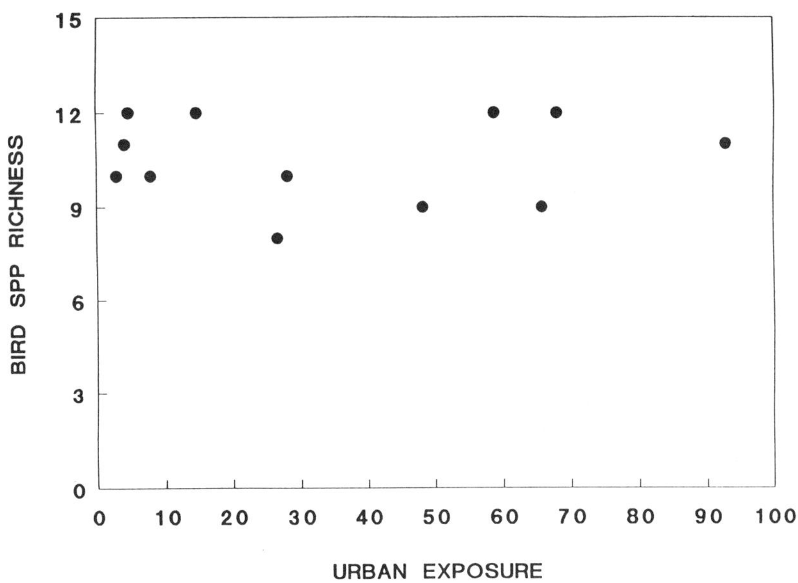

Figure 38.5. Relationship between bird species richness and urban exposure for the species studied.

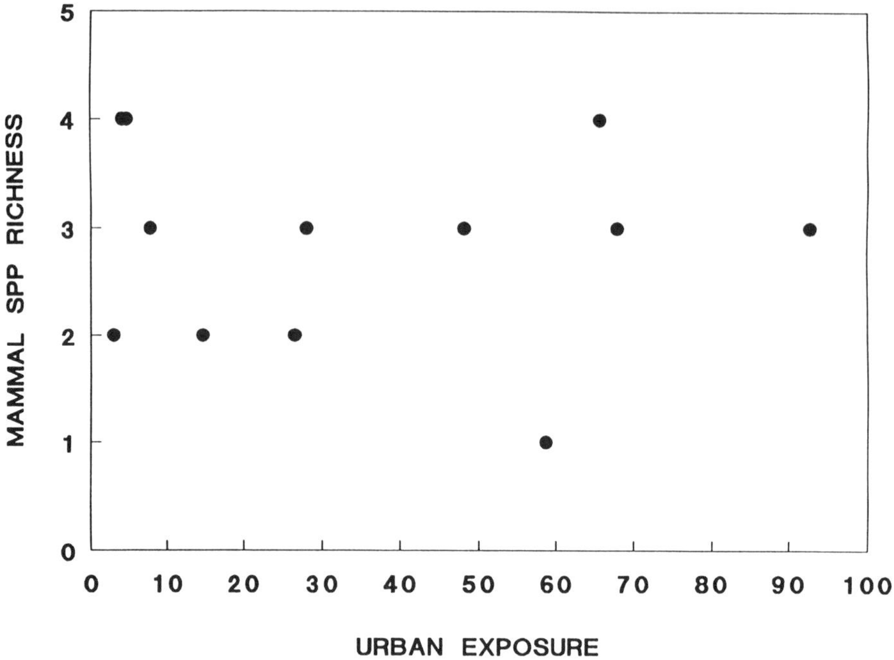

Figure 38.6. Relationship between small mammal species richness and urban exposure for the species studies.

sampled. In addition, however, our observations suggest that few (if any) urban-associated predators (e.g., domestic cats and dogs) and competitors (e.g., *Rattus* or *Mus*) of the small mammals were invading sites near development. Furthermore, human activities in sampling sites near development were minor, (e.g., almost no litter was found within the chaparral habitat), perhaps due to the great difficulty of moving through the chaparral vegetation. Given these factors, we feel that these preliminary results are not unexpected. Further field work is in progress to gather more extensive and detailed information on the resident bird and small mammal communities of the study area.

Linear Elements—Roads

Within the study area, roads and trails might be important in the spread of human disturbance. Although we have not conducted specific field studies on roads as disturbance sources, we have completed landscape analyses that indicate that public roadways are prevalent throughout the study area and could impact wildlife in a variety of ways (Sauvajot and Buechner 1991).

To assess potential impacts of intrusive road effects within the study area, we estimated how much of the park's area occurs close enough to roads to be exposed to intrusive effects. Within the study area, approximately 30 ha (8% of the area) lies within 50 m, 60 ha (15%) lies within 100 m and 300 ha (75%) lies within 500 m of public roads.

A large number of restricted access roads (e.g., fire and service roads) and trails occur within protected habitat. Many other roads technically within urban areas are, in fact, very close to the natural habitat, in some cases immediately adjoining it on one side. Clearly, a substantial portion of the area is very close to roads, and any direct impacts resulting from roadways could be significant.

The potential for road-associated impacts has become an even greater concern because of several recent roadway development proposals for the area. If approved, these projects would significantly increase the already large amount of road surface within the protected habitat. We have reviewed several different road development scenarios suggested for the area, and have estimated that the proposed development could substantially increase the potential for direct roadway impacts (Sauvajot and Buechner 1991). The long-term viability of some species could be impacted, particularly if intrusive disturbance associated with roads reduced the area of undisturbed habitat area between roads (Sauvajot and Buechner 1991). In addition, those species which tend to avoid road developments might decline in the study area if road development is increased (Sauvajot and Buechner 1991). The potential for road-associated disturbances is clearly very high in this area, and we believe that many protected habitats near urban areas experience similar situations. However, to evaluate clearly the role of roadway development as a disturbance source will require more detailed field work in our study area. As our work on this project continues, we hope to examine the role of roads as disturbance conduits in more detail.

Concentrated Areas of Human Activity—Hot Spots

During the summer of 1991, we collected data on year-round resident birds and small mammals in four sites which represented hot spots within the protected habitat (Figure 38.3). These were areas of concentrated human activity in which the structure of the chaparral vegetation had been altered by illegal off-road-vehicle activity, vegetation clearing, littering and dumping, trampling, and other human activities. None of the hot spots were currently being seriously disturbed by these activities as enforcement of park rules has recently been dramatically increased (e.g., barriers to ORV activity have been installed, signs detailing park rules have been installed, and the area now has an on-site ranger present). In terms of habitat structure, these sites exhibited significantly less vegetation cover, less structural diversity, shorter vegetation, and more forbs, grasses, and introduced species compared to the 12 sites described above in which the habitat structure had not be altered by hot spots of disturbance (these will be referred to as the "urban proximity" sites; see section on boundary effects). The four sites sampled were all at least 100 m from the edge of the urban area. Additional hot spots occur throughout the study area.

Each of the four disturbed sites were sampled for birds, mammals, and vegetation during the summer of 1991 using the same sampling methods described above for the urban proximity sites. We found a different small mammal species composition and significantly fewer chaparral-requiring small mammals in the hot spot sites compared to the urban proximity sites ($t = 3.12$; $P = 0.008$) (Figures 38.7, 38.8). In particular, the disturbed areas included many more Pacific Kangaroo Rats (*Dipodomys agilis*), a species

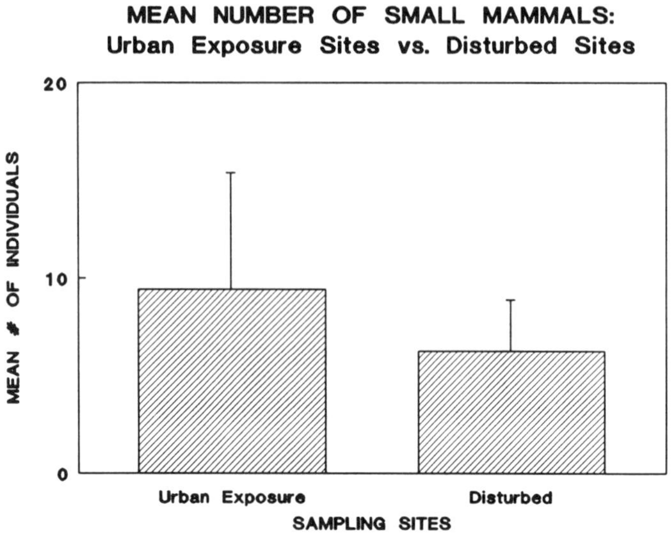

Figure 38.7. Mean number of all individual small mammals captured in urban exposure sites compared to disturbed (hot spot) sites.

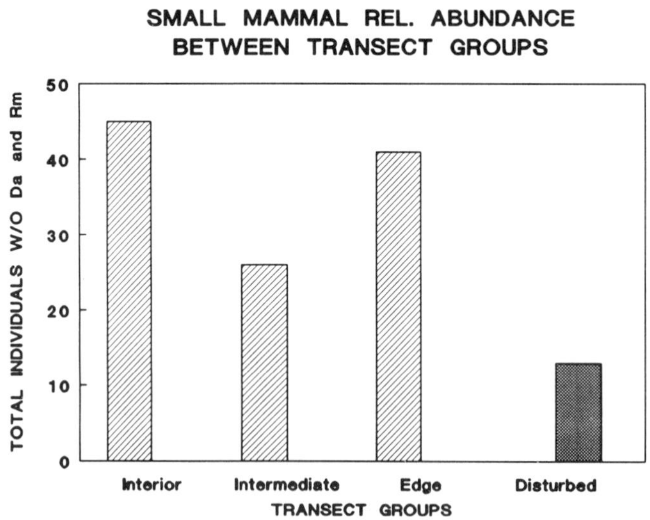

Figure 38.8. Relative abundance of small chaparral-requiring mammals (*Neotoma* and *Peromyscus* spp.) for three sets of urban exposure transects and for the disturbed (hot spot) transects. "Interior" transects are those urban exposure transects with the lowest urban exposure indices, "intermediate" transects are those urban exposure transects with moderate urban exposure indices, and "edge" transects are those urban exposure transects with the highest urban exposure indices. "Disturbed" transects are those in which the vegetation structure had been altered by human activity (hot spot transects). The total number of individual *Neotoma* and *Peromyscus* captured are shown. *Dipodomys agilis* is excluded here because this species prefers open habitats rather than chaparral.

which prefers more open habitats, and harbored fewer of the chaparral mammal species (*Neotoma* and *Peromyscus* spp.) that were common on the urban proximity sites (Figure 38.9). Significant differences between the hot spots and the urban proximity sites were not observed for bird species. Because all of the hot spots sampled were no longer being directly disturbed by human activities, it is possible that the high dispersal abilities of the birds allowed them to utilize the hot spots while small mammals move in more slowly as the habitat returns to a more undisturbed state.

These data, combined with the results from the urban proximity sites, suggest that direct habitat alteration in areas of concentrated human activity might be an important factor to consider in this system. When the chaparral vegetation is modified, important sources of food and cover disappear for many species, particularly small mammals, leading to community-level ecological changes for these species. It is not surprising that alteration of the habitat structure results in changes in the resident small mammal community. What is unexpected, is that unless habitat structure is altered, proximity to urbanization produced no measurable effect. These results have led us to plan more detailed studies of the hot spots.

Conclusions from the Case Study

Overall, this case study illustrates an important point regarding the spread of human disturbance across protected landscapes: each of the three mechanisms investigated can vary in importance, depending on the area and species being investigated. In this case, we believe that direct habitat alteration at hot spots might currently be the most importance

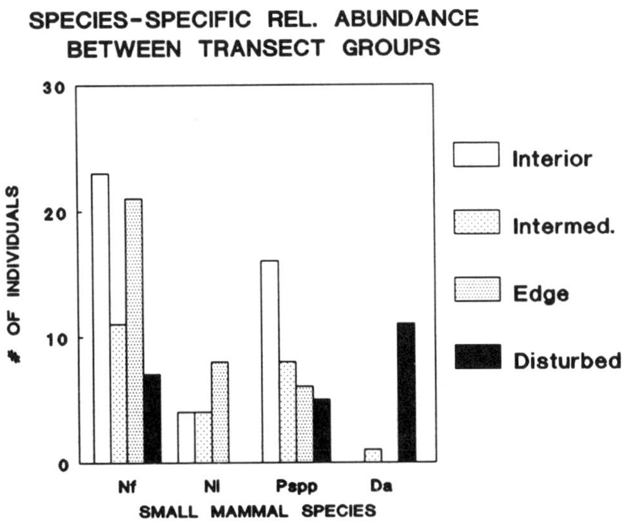

Figure 38.9. Species-specific relative abundance of small mammals in three sets of urban exposure transects and the disturbed (hot spot) transects. Interior, intermediate, and edge transects are defined as in Figure 38.8. Nf = *Neotoma fuscipes,* Nl = *Neotoma lepida,* Pspp = *Peromyscus* spp., and Da = *Dipodomys agilis.*

source of ecological disturbance, at least for the small mammal communities. Exposure to urbanization *per se* (in areas of intact chaparral vegetation) had no measurable effect on the small bird and mammal communities in this study. Because the location of the hot spots and the extent of direct habitat alteration is not associated with our measures of urban proximity (i.e., the urban exposure index), urban proximity by itself is a less important variable.

For this study area, adjacent urbanization probably results in fewer direct impacts to the nearby chaparral habitat than it would in habitats which were less resistant to the influx of local human associated disturbances or more sensitive to the impacts of that disturbance. The effects of people moving into the natural habitat might be reduced in this habitat because the chaparral vegetation presents a formidable barrier to human movements (as the researchers can attest). The difficulty of walking in this habitat no doubt greatly reduces off-trail activity by the people using the park. Chaparral might also be relatively difficult to invade by human-associated species requiring more benign habitat conditions (e.g., no *Mus* or *Rattus* occurred in the chaparral vegetation).

For this system, roads have a potential for being both direct sources of disturbance and serving as access routes creating hot spots. For example, the off-road-vehicles that cause some of the hot spots within the park gain initial access to these sites via roads occurring within the park. Roads and other human access routes (i.e., trails) might act as conduits for the spread of disturbance, allowing human impacts to leap-frog from urban areas into the natural habitat

POLICY AND MANAGEMENT STRATEGIES

Our work suggests that it would be useful to manage the spread of human disturbance through the identification and monitoring of boundary effects, linear landscape elements, and concentrated areas of human activity (hot spots). When sensitive areas can be identified, it might be necessary to protect them from the effects of disturbance. It might also be necessary to act to mitigate disturbance associated with the spread of human activity by re-routing the flow of disturbance and restoring damaged sites.

Our earlier review of the cross-boundary management techniques used by U.S. National Parks (Schonewald-Cox et al. 1992; Buechner et al. 1992) suggests that some strategies can be more useful than others in situations involving boundary effects. These same techniques might be useful in dealing with the entire range of issues arising from the spread of human activity across protected habitats. The techniques identified as most useful by park personnel include: (1) the collection of basic resource data focused on identifying sensitive areas, (2) the inventorying and monitoring of park resources in order to evaluate and document trends, (3) the use of GIS technology to map both disturbance and natural habitat, (4) the use of resource management techniques to manage natural resources and mitigate for disturbance, (5) the development of personal contacts, including the use of cooperative and advisory groups, to maintain communication with neighboring people and other users of park lands, (6) the use of educational and community outreach programs and training programs for park personnel, and (7) land acquisition in areas critical to the integrity of park resources. The first three of these techniques appear to be valuable in gathering information on the flow and impacts of human disturbance across the protected landscape. The last four methods might be of use in the modification or control of the flow of anthropogenic disturbance.

We suggest that, whenever feasible, a geographical study of the spread of human activity through the protected habitat be conducted. The following steps might be used:

(1) Identify and map reserve interfaces with developed areas, major linear elements and localized hot spots. These represent sources of human activity and disturbance.

(2) Delineate zones of varying distance from boundaries, linear elements and hot spots. In some cases, this might be most easily accomplished using GIS technology.

(3) Calculate indices of exposure to these sources of disturbance. The weighting of factors in the indices will depend on the resistance of the habitat to the flow of disturbance and the sensitivity of the natural habitat to human activities.

(4) Monitor those areas with the highest exposure scores for any evidence of community-level impacts. This will require field work involving inventorying and monitoring techniques; the needed field work could be extensive if little information is available on the areas to be studied.

(5) Develop an appropriate strategy to modify the flow of disturbance and/or to mitigate damage. This is likely to involve resource management, communication with local citizens groups, enforcement of park rules, the use of education programs, or land acquisition in areas adjacent to the park.

CONCLUSIONS

A review of the literature and our work at a case study site suggest that boundary effects, linear elements, and areas of intense human activity can all play a role in the development of zones of human disturbance within protected areas. The importance of the different mechanisms by which intrusive anthropogenic disturbance moves across a natural landscape can vary on a case-by-case basis. Unfortunately, this means that it is difficult to write prescriptions for action in any given case in the absence of substantial data on local field conditions. The management of intrusive disturbance will depend on both the gathering of sufficient information on the flow and impacts of human disturbance within the particular habitat of interest, and the development of specifically tailored efforts to modify the flow of human activities over sensitive areas within the local landscape.

LITERATURE CITED

Alverson, W. S., D. M. Waller, and S. L. Solheim. 1988. Forests too deer: Edge effects in northern Wisconsin. *Conserv. Biol.* 2:348-358.

Anable, M. E., M. P. McClaran, and G. B. Ruyle. 1992. Spread of introduced Lehmann lovegrass *Eragrotis lehmanniana* Nees. in Southern Arizona, USA. *Biol.Conserv.* 61:181-188.

Anderson, S. J., and C. P. Stone. 1993. Snaring to control feral pigs *Sus scrofa* in a remote Hawaiian rain forest. *Biol. Conserv.* 63:195-201.

Backhouse, G. N. 1987. Management of remnant habitat for conservation of the Helmeted Honeyeater *Lichenostomus melanops cassidix*. In *Nature Conservation: The Role of Remnants of Native Vegetation*, eds. D. A. Saunders, G. W. Arnold, A. A. Burbidge, and A. J. M. Hopkins, pp. 287-94. Surrey Beatty and Sons Pty Limited, Chipping Norton, Australia.

Baudry, J., and H. G. Merriam. 1988. Connectivity and connectedness: Functional versus structural patterns in landscapes. In *Connectivity in Landscape Ecology, Proceedings of the 2nd International Association for Landscape Ecology (IALE)*, Munster, 1987, ed. K. F. Schreiber, pp. 23-28. Munstersche Geographische Arbeiten 29, 1988.

Bernardino, F. S., and G. H. Dalrymple. 1992. Seasonal activity and road mortality of the snakes of the Pa-hay-okee wetlands of Everglades National Park, USA. *Biol. Conserv.* 62:71-75.

Beissinger, S. R. 1982. Effects of urbanization on avian community organization. *Condor* 84:75-83.

Boucher, D. H., J. Aviles, and R. Chepote. 1991. Recovery of trailside vegetation from trampling in a tropical rain forest. *Environ. Manage.* 15:257-262.

Bratton, S. P. 1979. Impacts of white-tailed deer on the vegetation of Cades Cove, Great Smoky Mountains National Park. *Proc. Ann. Conf. Southeastern Assoc. Fish and Wildl. Agencies* 33:305-312.

Bratton, S. P., L. L. Stromberg, and M. E. Harmon. 1982. Firewood-gathering impacts in backcountry campsites in Great Smoky Mountains National Park. *Environ. Manage.* 6:63-71.

Buechner, M. 1987. Conservation in insular parks: simulation models of factors affecting the movement of animals across park boundaries. *Biol. Conserv.* 41:57-76.

Buechner, M., C. M. Schonewald-Cox, R. Sauvajot, and B. A. Wilcox. 1992. Cross-boundary issues for National Parks: what works "on the ground." *Environ. Manage.* 16:799-809.

Butler, W. H. 1987. Management of Disturbance in an arid remnant: The Barrow Island experience. In *Nature Conservation: The Role of Remnants of Native Vegetation*, eds. D. A. Saunders, G. W. Arnold, A. A. Burbidge, and A. J. M. Hopkins, pp. 279-285. Surrey Beatty and Sons Pty Limited, Chipping Norton, Australia.

Cloudsley-Thompson, J. L. 1990. Etosha and the Kaokoveld: Problems of conservation in Namibia. *Environ. Conserv.* 17:351-354.

Cole, D. N. 1990. Trampling disturbance and recovery of cryptogamic soil crusts in Grand Canyon National Park. *Environ. Manage.* 14:50-57.

Cole, D. N., and R. K. Fichtler. 1983. Campsite impact on three western wilderness areas. *Environ. Manage.* 7:275-288.

Community Development by Design (CDD). 1991. *Mulholland Gateway Park Draft Master Plan.* Unpublished report prepared for the Santa Monica Mountains Conservancy Board and Advisory Committee. Community Development by Design, 2707 Matthew's Street, Berkeley, Calif.

Cortner, H. J., P. D. Gardner, and J. G. Taylor. 1990. Fire hazards at the urban-wildland interface: What the public expects. *Environ. Manage.* 14:57-62.

Cowan, P. E. 1992. The eradication of introduced Australian brushtail possums, *Trichosurus vulpecula*, from Kapiti Island, a New Zealand nature reserve. *Biol. Conserv.* 61:217-226.

Cowie, I. D., and P. A. Werner. 1993. Alien plant species invasive in Kakadu National Park, tropical northern Australia. *Biol. Conserv.* 63:127-135.

Datta, T., and B. C. Pal. 1993. The effect of human interference on the nesting of the openbill stork *Anastomus oscitans* at the Raiganj Wildlife Sanctuary, India. *Biol. Conserv.* 64:149-154.

Dawson, K. J. 1991. Human predation on isolated nature reserves. *Nat. Inst. Urban Wildl. Symp. Series*: 2:83-88.

DeSanto, R. S., and D. G. Smith. 1993. An introduction to issues of habitat fragmentation relative to transportation corridors with special reference to high-speed rail (HSR). *Environ. Manage.* 17:111-114.

Diamond, J. M., K. D. Bishop, and S. Van Balen. 1987. Bird survival in an isolated Javan woodland: island or mirror. *Conserv. Biol.* 1:132-142.

Drijver, C. A. 1991. People's participation in environmental projects in developing countries. *Landscape and Urban Planning* 20:129-139.

Eaton, P. 1991. Wetland management—traditional rights and recreational uses: The Tonda Wildlife Management Area of Papua New Guinea. *Landscape and Urban Planning* 20:67-72.

Emlen, J. T. 1971. Population densities of birds derived from transect counts. *Auk* 88:323-342.

Ferreras, P., J. J. Aldama, J. F. Beltran, and M. Delibes. 1992. Rates and causes of mortality in a fragmented population of Iberian lynx *Felis pardina* Temminck, 1824. *Biol. Conserv.* 61: 197-202.

Ferris, C. R. 1979. Effects of Interstate 95 on breeding birds in Northern Maine. *J. Wlldl. Manage.* 50:497-503.

Forman, R. T. T., and M. Godron. 1986. *Landscape Ecology*. John Wiley and Sons, New York.

Getz, L., F.R. Cole, and D. L. Gates. 1978. Interstate roadsides as dispersal routes for *Microtus pennsylvanicus*. *J. Mamm.* 59:208-212.

Guthrie, D. A. 1974. Suburban bird populations in Southern California. *Am. Midl. Nat.* 92:461-466.

Harris, L. D. 1988. Edge effects and conservation of biotic diversity. *Conserv. Biol.* 2:330-332.

Harris, L. D., and G. Silva-Lopez. 1992. Forest fragmentation and the conservation of biological diversity. In *Conservation Biology: The Theory and Practice of Nature Conservation, Preservation and Management*, ed. P. L. Fiedler and S. Jain, pp. 197-238. Chapman and Hall, New York.

Hubbard, A. L., S. McOrist, T. W. Jones, R. Boid, R. Scott, and N. Easterbeck. 1992. Is survival of European wildcats *Felis silvestris* in Britain threatened by interbreeding with domestic cats? *Biol. Conserv.* 61:203-208.

Hulbert, I. A. R. 1990. The response of ruddy shelduck *Tadorna ferruginea* to tourist activity in the Royal Chitwan National Park of Nepal. *Biol. Conserv.* 52:113-123.

Janzen, D. H. 1983. No park is an island: Increase in interference from outside as park size decreases. *Oikos* 41: 402-410.

Janzen, D. H. 1986. The eternal external threat. In *Conservation Biology: The Science of Scarcity and Diversity*, ed. M. E. Soulé, pp. 286-303. Sinauer Associates, Inc., Sunderland, Mass.

Jeffrey. R. C. V., and P. M. Chooye. 1991. The people's role in wetlands management: the Zambian initiative. *Landscape and Urban Planning* 20:73-79.

Kuss, F. R., and C. N. Hall. 1991. Ground flora trampling studies: Five years after closure. *Environ. Manage.* 15:715-727.

Lovejoy, T. E., R. O. Bierregaard, A. B. Rylands, J. R. Malcolm, C. E. Quintela, L. H. Harper, K. S. Brown, A. H. Powell, G. V. N. Powell, H. O. R. Schubart, and M. B. Hays. 1986. Edge and other effects of isolation on Amazon forest fragments. In *Conservation Biology: The Science of Scarcity and Diversity*, ed. M. E. Soulé, pp. 257-285. Sinauer Associates, Inc., Sunderland, Mass.

Luken, J. O., A. C. Hinton, and D. G. Baker. 1991. Forest edges associated with power-line corridors and implications for corridor siting. *Landscape and Urban Planning* 20:315-324.

Machlis, G. E., and D. L. Tichnell. 1987. Economic development and threats to National Parks: A preliminary analysis. *Environ. Conserv.* 14:151-156.

Mainini, B., P. Neuhaus, and P. Ingold. 1993. Behavior of marmots *Marmota marmota* under the influence of different hiking activities. *Biol. Conserv.* 64:161-164.

McCutchen, H. E. 1993. Ecology of high mountain black bear population in relation to land use at Rocky Mountain National Park. *Park Sci.* 13:25-27.

McShane, T. O. 1990. Wildlands and Human Needs: resource use in an African protected area. *Landscape and Urban Planning* 19: 145-158.

Melman, P. J. M., H. J. Verkaar, and H. Heemsbergen. 1988. The maintenance of road verges as possible ecological corridors of grassland plants. In *Connectivity in Landscape Ecology, Proceedings of the 2nd International Association for Landscape Ecology (IALE)*, Munster, 1987, ed. K. F. Schreiber, pp. 131-133. Munstersche Geographische Arbeiten 29, 1988.

Michael, E. D., C. R. Ferris, and E. G. Haverlack. 1976. Effects of highway rights-of-way on bird populations. *Symp. Environ. Concerns of Rights-of-way Manage.* 1:253-261.

Newmark, W. D., N. L. Leonard, H. I. Sariko, and D. M. Gamassa. 1993. Conservation attitudes of local people living adjacent to five protected areas in Tanzania. *Biol. Conserv.* 63:177-183.

Noss, R. F. 1987. Protecting natural areas in fragmented landscapes. *Natural Areas J.* 7:2-13.

Ntiamoa-Baidu, Y. 1991. Conservation of coastal lagoons in Ghana: the traditional approach. *Landscape and Urban Planning* 20:41-46.

Pearson, S. M. 1993. The spatial extent and relative influence of landscape-level factors on wintering bird populations. *Landscape Ecol.* 8:3-18.

Pickett, S. T. A., V. T. Parker, and P. L. Fiedler. 1992. The new paradigm in ecology: Implications for conservation biology above the species level. In *Conservation Biology: The Theory and Practice of Nature Conservation, Preservation and Management*, eds. P. L. Fiedler and S. Jain, pp. 65-90. Chapman and Hall, New York.

Reijnen, M. J. S. M., J. B. M. Thissen, and G. J. Bekker. 1987. Effects of road traffic on woodland breeding bird populations. *Acta Ecologica* 8:312-313.

Rochefort, R. M., and S. T. Gibbons. 1993. Impact monitoring and restoration in Mount Rainier National Park. *Park Sci.* 13:29-30.

Sauvajot, R. M., and M. Buechner. 1991. *Mulholland Gateway Park Wildlife Report: Evaluation of Possible Roadway Impacts on Wildlife with a Discussion of Mitigation Measures, Wildlife Linkages, and Sensitive Habitats for Mulholland Gateway Park.* Unpublished report prepared for Community Development by Design. Division of Environmental Studies, University of California Davis, Davis, Calif.

Schonewald-Cox, C. M. 1988. Boundaries in the protection of nature reserves. *BioScience* 38:480-486.

Schonewald-Cox, C. M., and J. W. Bayless. 1986. The boundary model: a geographical analysis of design and conservation of nature reserves. *Biol. Conserv.* 38:305-322.

Schonewald-Cox, C. M., M. Buechner, R. Sauvajot, and B. W. Wilcox. 1992. Cross-boundary management between National Parks and surrounding lands: A review and discussion. *Environ. Manage.* 16:273-282.

Schonewald-Cox, C. M., and M. Buechner. 1992. Park protection and public roads. In *Conservation Biology: The Theory and Practice of Nature Conservation, Preservation and Management*, eds. P. L. Fiedler and S. Jain, pp. 373-396. Chapman and Hall, New York.

Shackleton, C. M. 1993. Fuelwood harvesting and sustainable utilization in a communal grazing land and protected area of the eastern transvaal lowveld. *Biol. Conserv.* 63:247-454.

Shafer, C. L. 1990. *Nature Reserves: Island Theory and Conservation Practice.* Smithsonian Institution Press, Washington, D.C.

Sharma, U. R. 1990. An overview of park-people interactions Royal Chitwan National Park, Nepal. *Landscape and Urban Planning* 19:131-144.

Singer, F. J., J. B. Beattie, and R. C. Cunningham. 1986. Large mammals and roads in national parks—a review. *Conference on Science in the National Parks* 4:134. (abstract only).

Slaughter, C. W., C. H. Racine, D. A. Walker, L. A. Johnson, and G. Abele. 1990. Use of off-road vehicles and mitigation of effects in Alaska permafrost environments: a review. *Environ. Manage.* 14:63-72.

Slocombe, D. S., and J. G. Nelson. 1992. Management issues in hinterland national parks: A human ecological approach. *Natural Areas J.* 12:206-214.

Soulé, M. E., and K. A. Kohm. 1989. *Research Priorities for Conservation Biology.* Island Press, Washington, D.C.

Stockwell, C. A., G. C. Bateman, and J. Berger. 1991. Conflicts in national parks: A case study of helicopters and bighorn sheep time budgets at the Grand Canyon. *Biol. Conserv.* 56:317-328.

Stottlemyer, R. 1987. External threats to ecosystems of U. S. National Parks. *Environ. Manage.* 11:91-97.

Suarez, F., M. Yanes, and J. Herranz. 1993. Nature reserves and the conservation of Iberian shrubsteppe passerines: The paradox of nest predation. *Biol. Conserv.* 64:77-81.

Taylor, K. C., R. J. Reader, and D. W. Larson. 1993. Scale-dependent inconsistencies in the effects of trampling on a forest understory community. *Environ. Manage.* 17:239-248.

Taylor, S. G. 1987. Conservation strategies for human-dominated land areas: The South Australian example. In *Nature Conservation: The Role of Remnants of Native Vegetation*, eds. D. A. Saunders, G. W. Arnold, A. A. Burbidge, and A. J. M. Hopkins, pp. 313-322. Surrey Beatty and Sons Pty Limited, Chipping Norton, Australia.

Temple, S. A., and J. R. Cary. 1988. Modeling dynamics of habitat-interior bird populations in fragmented landscapes. *Conserv. Biol.* 2:340-347.

Thomas, L. P., and G. D. Willson. 1992. Effect of experimental trampling on the federally endangered species, *Lesquerella filiformis* Rollins, at Wilson's Creek National Battlefield, Missouri. *Natural Areas J.* 12:101-105.

Timmins, S. M., and P. A. Williams. 1987. Characteristics of problem weeds in New Zealand's protected natural areas. In *Nature Conservation: The Role of Remnants of Native Vegetation*, eds. D. A. Saunders, G. W. Arnold, A. A. Burbidge, and A. J. M. Hopkins 241-247. Surrey Beatty and Sons Pty Limited, Chipping Norton, Australia.

van der Zee, F. F., J. Wiertz, C. J. F. Ter Braak, R. C. van Apledoorn, and J. Vink. 1992. Landscape change as a possible cause of the badger *Meles meles* L. decline in The Netherlands. *Biol. Conserv.* 61:17-22.

Vekaar, H. J. 1988. The possible role of road verges and river dikes as corridors for the exchange of plant species between natural habitats. In *Connectivity in Landscape Ecology, Proceedings of the 2nd International Association for Landscape Ecology (IALE)*, Munster, 1987, ed. K. F. Schreiber, pp. 79-84. Munstersche Geographische Arbeiten 29, 1988.

Wallace, K. J., and S. A. Moore. 1987. Management of remnant bush land for nature conservation in agricultural areas of southwestern Australia—Operational and planning perspectives. In *Nature Conservation: The Role of Remnants of Native Vegetation*, eds. D. A. Saunders, G. W. Arnold, A. A. Burbidge, and A. J. M. Hopkins, pp. 259-268. Surrey Beatty and Sons Pty Limited, Chipping Norton, Australia.

Wilcox, B. A., and D. D. Murphy. 1985. Conservation strategy: The effects of fragmentation on extinction. *Am. Nat.* 125:879-887.

Yahner, R. H. 1988. Changes in wildlife communities near edges. *Conserv. Biol.*2:333-339.

Yalden, D. W. 1992. The influence of recreation disturbance on common sandpipers *Actitis hypoleucos* breeding by an upland reservoir, in England. *Biol. Conserv.* 61:41-49.

39

BALANCING SCIENTIFIC, SOCIAL, AND REGULATORY CONCERNS IN MANAGING BIODIVERSITY

Tim W. Clark

Northern Rockies Conservation Cooperative
Jackson, Wyoming
and School of Forestry and Environmental Studies
Yale University
New Haven, Connecticut
U.S.A.

Dennis Glick

Greater Yellowstone Coalition
Bozeman, Montana
U.S.A.

John D. Varley

Yellowstone National Park, Wyoming
U.S.A.

Managing biodiversity successfully is a complex scientific, social, and regulatory challenge. The scientific challenge is to bring more rigorous and reliable knowledge to bear on management, to broaden the scope from single species to larger landscape scales, and to develop more effective management tools. The social challenge is to identify how and why society values diversity, and to incorporate these values into responsive management schemes. The regulatory challenge is to formulate useful policies and establish effective programs. The uneasy mix of these three concerns is being played out in hundreds of day-to-day management decisions affecting biodiversity all across the United States. The northern spotted owl (*Strix occidentalis caurina*) case in the ancient forests of the Pacific Northwest illustrates the high stakes involved and the collision of science, values, and regulatory processes (Yaffee in press). Systematically sorting out key interacting variables and setting management direction in such complex cases are possible only by using appropriate analytical concepts and applied tools (Clark and Kellert

1988). No ready formula exists for a successful balance of scientific, social, and regulatory concerns, and no simple prescriptions will be offered in this paper. Instead, we focus on learning how to improve our individual and collective performance so we can do a better job of finding the balance that works for a given situation at a given time. Specifically, this paper (1) examines scientific, social, and regulatory dimensions of biodiversity conservation from a general standpoint, (2) identifies and illustrates the interplay of these three concerns in three actual biodiversity management cases in the Greater Yellowstone Ecosystem, and (3) introduces a conceptual frame-work to analyze and understand these three concerns in actual practice and to learn how to participate more effectively in practical conservation efforts.

SCIENTIFIC, SOCIAL, AND REGULATORY CONCERNS IN MANAGING BIODIVERSITY

Integrating scientific, social, and regulatory concerns is nearly always difficult (Varley 1988; Glick et al. 1991; Reading 1993). Among the fundamental difficulties are the ambiguity of the biodiversity concept, the mixed values of the public towards biodiversity, and a lack of useful means by which regulatory agencies can apply large scale management easily. Predicting the future outcomes of today's attempts to balance these three concerns is problematic at best. On what scientific basis, according to whose values, and under what regulatory guidance should the balance be struck?

We know from experience that ignoring the difficulties, complexities, and uncertainties of balancing these three concerns, for example by focusing only on the biophysical dimensions (i.e., only the biology of the plants and animals), results in inadequate management practices (Clark 1993). Simply recognizing the web of interacting scientific, social, and regulatory forces is a first step toward more effective implementation of existing biodiversity policies. It is also necessary to the development of more successful policies, increasing opportunities for greater professional effectiveness, and a sense of understanding and control in the biodiversity management process.

Scientific Concerns in Managing Biodiversity

Biodiversity "refers to the variety and variability among living organisms and ecological complexes in which they occur" (NSF 1989:1). The scientific basis for managing biodiversity is developing rapidly (Noss 1991). In science, reliable observation and measurement is prized as the primary means to explain phenomena, explanations that in turn enable valid predictions in comparable circumstances (Brewer 1981). Many biological researchers call for more use of scientific research in management and in decision-making (Gill 1986). Unfortunately, some of what passes for scientific fact or theory in natural resource management today is not based on adequate science (see Romesburg 1981, 1991; Peters 1991 for critiques).

Management itself is a science, especially in this age of declining populations of native species when management action must often precede scientific explanation. As Walters (1986:9) points out: "[M]anagement should be viewed as an adaptive process:

we learn about the potential of natural populations to sustain harvesting mainly through experience with management itself, rather than through basic research or the development of general ecological theory. We keep running into questions that only hard experience can answer, and a basic issue becomes whether to use management policies that will deliberately enhance that experience."

This is a valuable perspective to adopt in all efforts at biodiversity management, but it is little used at present.

Attention is being focused on varying genetic, population, species, ecosystem, and landscape scales of biodiversity management. Data available to aid management differ at each scale. But much current management of biodiversity is focused on individual species, especially high profile endangered species—a fact that reflects the history of the wildlife profession and its emphasis on economically valuable species. The last decade, however, has brought growing awareness that the major unit of conservation attention should be the ecosystem or landscape because it subsumes all other levels of biodiversity conservation (Noss 1991; Hunter 1991; Minta and Kareiva in press). Recent developments promise to improve scientific management of biodiversity because they identify rare or unrepresented habitats and communities that need protection, and they provide better means to manage and analyze large data sets with computers (Scott et al. 1991).

Thus, the scientific and technical challenge in conserving biodiversity is, first, to define the concept usefully so that it can be applied easily. This will be necessary not only for future conservation-oriented research, but also as a basis for developing of practical policies and programs. A meaningful definition of the challenge is also necessary as a means to build broad public support for this major environmental issue, which is currently little known and underappreciated in comparison to other global environmental problems. The challenge is also to bring more rigorous and sophisticated techniques to bear for inventorying, monitoring, restoring, and conserving biological diversity. A final challenge is to broaden the scope of understanding, management, and research from single species to larger, ecosystem and landscape scales.

Social Concerns in Managing Biodiversity

The social dimension of biodiversity management is an active area of study (Reading 1993). This concern refers to the worth or importance of biodiversity to people and society (Kellert 1980). Unfortunately, only recently has attention been given to measuring public understanding and valuation of biodiversity (S. R. Kellert pers. comm.). Three major classes of values toward nature are often recognized: ecological, social-psychological, and economic (Steinhoff 1980). Understanding public values is essential in management because it permits clarification of policy and program goals and makes implementation much easier. However, simply having this kind of knowledge is not an automatic recipe for improved or successful biodiversity management. In some instances where public valuation of single species has been measured, people may negatively value it (e.g., wolves, *Canis lupus*) or so positively value it that they consume it to extinction (e.g., passenger pigeon, *Ectopistes migratorius*).

To date, measuring social valuation of biodiversity has primarily focused on single species, sometimes on various taxonomic groups, and rarely on biodiversity, ecosystems, or landscapes. This is largely because people do not relate to biodiversity, ecosystems, or landscapes. For example, Reading and Kellert (1993) studied values of people in the region of Montana where endangered black-footed ferrets (*Mustela nigripes*) are likely to be reintroduced, and Kellert (1991) studied social valuation of marine mammals in Canada. On larger and more "abstract" scales, Reading et al. (in press) investigated public knowledge and attitudes towards both economically important species and threatened and endangered species in the Greater Yellowstone Ecosystem. Most people in the ecosystem were aware of these species, their status, or plight, but they were especially concerned with what improved biodiversity management would mean economically and politically to them and to the region. And S. R. Kellert (pers. comm.), in concert with Defenders of Wildlife, carried out a national survey on biodiversity and public knowledge and attitudes. Conclusions showed that the American public is aware we are losing species, such as threatened and endangered species, but is unaware of the magnitude of the losses or potential consequences. Only 20% of Americans say that they ever heard of an issue called "the loss of biological diversity," whereas about one-half of the members of conservation organizations were aware of the issue.

Having this information about the public's knowledge of biodiversity at several biological scales (i.e., individual species, ecosystems, landscapes) is necessary to devise appropriate management strategies. Simply having this information about the public's understanding will not guarantee biodiversity conservation, however, but it will facilitate biodiversity management. Additional studies could significantly clarify how, if, and why society values plants, animals, and the structure and functioning of natural ecosystems.

Social valuation is essential to conserving biodiversity because without positive public support, successful management is not possible. Simply put, the reason that so many plants and animals are currently threatened is the lack of value society has historically attributed to all living resources (Reading 1993).

Regulatory Concerns in Managing Biodiversity

The regulatory basis of biodiversity includes the programs and activities evident in all branches of government—legislative, judiciary, and executive. A variety of laws have been produced that address various aspects of conserving biodiversity. Although the Endangered Species Act (ESA) of 1973, as amended, and the National Forest Management Act (NFMA) of 1976 are two major recent laws mandating biodiversity protection, much earlier organic acts addressed the issue as well. The Yellowstone Act of 1872, for instance, established the Park "for the preservation, from injury or spoilation, of all timber, mineral deposits, natural curiosities, or wonders" and their retention in their natural condition. The National Park Service Act (NPSA) of 1916 set up the agency "to conserve the scenery and the natural and historic objects and the wild life therein...in such manner and by such means as will leave them unimpaired for the enjoyment of future generations." Over 20 other federal laws also relate to biodiversity (see Table 1.3 in OTA 1987:9), but none of them address "diversity" directly or comprehensively. In the judicial branch, the courts have "firmly entrenched" the Endangered Species Act (Coggins 1991:67). One conflict over the snail darter has even reached the Supreme Court, which upheld full species protection.

In the executive branch, the structure and behavior of regulatory agencies and appointed officials are important variables in how biodiversity is managed (Kellert and Clark 1991). In fact, "whether policies succeed or fail in their objectives is largely dependent upon the nature of the organization mandated to carry out those policies" (Clarke and McCool 1985:2). The need to understand explicitly the structure and behavior of agencies that implement policies and programs has been underscored by many policy researchers, biologists, and managers (Yaffee 1982; Tobin 1990).

Implementation of national regulatory or prohibitive policies like the ESA has been stymied by a host of recurring weaknesses (e.g., Clark et al. in press). Several professional assessments have evaluated whether the Act has been successfully carried out (Yaffee 1982; Tobin 1990; Bean 1991). For example, most of the 27 authors in *Balancing on the Brink of Extinction: The Endangered Species Act and Lessons for the Future* (Kohm 1991) noted a significant implementation "gap" (T. W. Clark, R. L. Wallace, and S. R. Kellert, unpubl. data). In part because of these implementation problems, new national legislation concerning the management of biodiversity has been drafted and is being considered, including a Biodiversity Bill (D. Blockstein, pers. comm.) and an Endangered Ecosystem Act (Hunt 1988). These bills seek to enhance biodiversity management, but special attention is needed to ensure adequate definitions and goals, specify means and organizations by which management will be carried out, and provide for rigorous, periodic evaluation.

The regulatory challenge, then, is to improve implementation and evaluation of existing policy, to formulate new policy as needed, and to ensure that public institutions are committed to and capable of meeting policy prescriptions. Participants at all levels must understand how the policy process works and how to configure institutional arrangements and behavior to meet the challenges of biodiversity conservation.

Taken individually or together, scientific, social, and regulatory concerns are all vitally important dimensions of any biodiversity management effort. These three concerns and their interplay are evident in the following three case studies (see Clark 1986).

CASE STUDIES FROM THE GREATER YELLOWSTONE ECOSYSTEM

The following three case studies are brief, but illuminating examples of various scientific, social, and regulatory concerns in the management of biodiversity in the Greater Yellowstone Ecosystem (GYE). The cases were chosen because of their high profile in the region and the authors first-hand knowledge of them. The cases differentially identify the interplay of the scientific, social, and regulatory concerns, either implicitly and explicitly.

Case studies serve as powerful educational tools in wildlife conservation (Clark 1986). Case studies can be constructed in a variety of ways for different purposes. The three case studies below vary in their structure and detail. The cutthroat trout and trumpeter swan cases are normative and brief. In contrast, the forest biodiversity case study is more complex and is based on a model that guided systematic analysis of the many issues involved. Many variables are investigated, many are not necessarily normative, but they are essential to understanding the case. These cases lead up to the final section of this paper which presents a frame-work for analyzing any wildlife conservation case. It is general and identifies basic variables at play in all cases. The section guides the reader to additional readings about this frame-work and its practical application. It is our contention that the analytical approach called for in this frame-work could improved biodiversity conservation. This case study approach was not used in the three cases that follow, because of space limitations, but the forest case study utilizes some of the wording of the frame-work and so serves as an introduction to it. The model has been used successfully in a variety of cases (Kellert and Clark 1991; Reading et al. 1991; Reading 1993).

All three cases come from the GYE, which is a precedent setting area for the management of wildlife and public lands. Just as the creation of Yellowstone National Park in 1872 set a precedent for the establishment of national parks worldwide, so too does the region's current management serve as a model for balancing scientific, social, and regulatory concerns regarding biodiversity management (Clark and Minta 1994). The case of cutthroat trout (*Onchorhyncus clarki bouvieri*) within Yellowstone National Park has been judged successful by most immediate measures, whereas the case of trumpeter swans (*Cygnus buccinator*) in the GYE has been inconclusive to date. The third case we look at, efforts by the national forests in the GYE to conserve biodiversity, shows that many problems persist. Other recently published literature also focuses on biodiversity management in the GYE (Keiter and Boyce 1991; Mattson and Reid 1991).

Before examining the three cases, a brief introduction to the GYE is in order. The GYE encompasses about 18 million acres of federal, state, and private lands located in Wyoming, Montana, and Idaho. Yellowstone National Park lies at its heart. The GYE represents a microcosm of challenges facing managers worldwide who are attempting to conserve biodiversity over a mix of parks, forests, and private lands. The scientific information available to managers is extensive; numerous studies have been carried out in the region or are underway. Society's demands on the GYE for various mixes of commodity and noncommodity uses are intensifying (Jobes 1991, 1993; Power 1991; Patten 1991; Rasker 1993; Johnson and Snepenger 1993). Regulatory direction comes largely from federal and state mandates; the GYE is managed by more than 28 federal, state, and local entities, all of which operate very differently. There is no over-arching policy or effective management body to coordinate agency management for biodiversity (or for any other regionally-distributed resource) in this highly fragmented bureaucratic and policy setting (Clark et al. 1991).

The magnitude of the value clashes between commodity and noncommodity interests is partially captured in statistics of current and planned land uses (Glick et al. 1991). About 6 million acres of the GYE are open to oil and gas development, 150,000 acres of timber are slated for harvest in the next decade, recreation and tourism are booming and growing businesses, 7,500 miles of roads exist on public lands and 800 more miles are planned, over 12,000 active and abandoned mining claims dot the region, over 1 million acres are in private subdivided lands, and 17 more major water projects are proposed. These commodity activities have direct, indirect, and cumulative affects on the biodiversity of the region.

The material covered to this point serves as context for the case studies that follow. The three GYE case studies illustrate the interplay of scientific, social, and regulatory concerns.

Cutthroat Trout Management in the Yellowstone River Drainage

The Yellowstone cutthroat trout is a special Pleistocene presence in the GYE with immense ecological and social value. It is very likely a keystone species, and it is very popular with fishermen and the Park's wildlife viewers. Once the most abundant and widely dispersed subspecies of inland cutthroat trout, it currently survives only in about 10% of its original range (Varley and Gresswell 1988). The watershed that includes Yellowstone Lake is the last major stronghold of the species, but at various times over the past century human activities have conspired to place it at risk: hatchery and spawn-taking operations, introduction of non-native species, and commercial and sport fishing (Gresswell and Varley 1988).

Exploitation of the Yellowstone cutthroat trout fishery began in earnest in the 1860s and grew steadily with the popularity of the Park. Attitudes toward the harvest of trout were extremely liberal in the early years, and excessive harvest, with accompanying waste of killed fish, was common. The first declines in the quality of the fishery, reflecting declines in the abundance of trout, were noted shortly after the turn of the century. The species decline continued until it completely collapsed in the late 1950s and 1960s (Gresswell and Varley 1988).

The decline and collapse were monitored, researched, and documented. Desperate for a simple solution, managers tried numerous regulatory changes, primarily creel limit reductions and prohibition of spawn exploitation. But because these actions were counteracted by annual increases in angler use and effort, the effort failed. Other alternatives for restoring the fishery that were proposed by scientists were foreign to most constituencies, including other fishery biologists, and ran counter to widely held public values. The quickest remedy to restore the trout population would have been a total closure on all fishing, but because the fishery was a cornerstone of the recreation-based regional economy, this option was heretical and thus socially unacceptable. Instead, new, innovative experiments, such as trout sanctuaries, areas of limited human entry, juvenile-only exploitation, catch-and-release-only fishing, and a bait fishing prohibition, were the regulatory mechanisms eased into place in the late 1960s and early 1970s.

Public acceptance of the new program ranged from venomous hostility and outright banditry to puzzled compliance. Among the many affected constituencies, organized fly fishermen were the only segment of the public to support the plans. Park and fishery managers have always been nervous about high-visibility experimentation and were clearly anxious about constituency acceptance of the new program. Nevertheless, they launched a multifaceted, "we've-got-to-hold-the-high-ground" program that focused on public education, including constituency nurturing, continuing reappraisal of the efficacy of the experiment, and, after a grace period of several years, strict law enforcement.

After a decade or so, the restoration program proved a clear success with the limited and regulated boundaries of the park. As a resource issue, the problem followed the classic interplay of constituencies, forces, and time as described in the final section of this paper. Perseverance, education, and enforcement played important roles in its resolution, but each paled beside the power of the new, restored fishery to sell itself. In a relatively few years, the cutthroat trout breeding population had increased about ten-fold. By all accounts, the sport fishing was as good as it had been near the turn of the century, and 82% of today's anglers have satisfactory outings (Jones 1991). Areas that were once "population sinks," such as Fishing Bridge and LeHardy's Rapids, developed a new constituency in the form of "trout watchers." In what may be one of today's fastest growing activities in the Park, more people in Yellowstone National Park now watch fish than catch fish (Varley 1989).

The restoration of the Yellowstone cutthroat trout created important ecological "ripples" that are only now becoming apparent (Varley and Schullery in press). Threatened and endangered species and species of special concern seem to be the primary beneficiaries, although scores of plants and animals have benefitted. Use and consumption of spawning trout by grizzly bears in 59 tributaries of Yellowstone Lake has grown from less than 10% of the streams in the 1950s and 1960s to about two-thirds of all streams in 1990. Restoration of the trout population is also believed to have contributed to the improved status of populations of obligate fish-eaters (e.g., osprey, bald eagles, black bears) (Varley and Schullery in press).

Currently, however, the Yellowstone River north of the Park is subject to degradation from sedimentation from logging and road building, toxic runoff from hard rock mining, dewatering for agriculture, pesticide runoff from fields, and contaminated municipal sewers and rural septic systems. The consequences for biodiversity are likely to be harmful and long-term survival prospects for cutthroat trout in particular are reduced by the species inability to achieve restored numbers over a larger area beyond the park. In short, the park is limited in its ability to addressed threats external to it, but which directly affect it.

Trumpeter Swans: Coordinating For Conservation

Trumpeter swans, the largest of the world's waterfowl species, occurred historically over much of North America. Overhunting for meat and down and the destruction of wetlands for agriculture reduced the species nearly to extinction by about 1920 (Banko 1960). This photogenic and popular bird became a focus of the fledgling United States conservation movement early in the century. The result of this interest by individuals, non-governmental organizations, and federal agencies was the passage of the Migratory Bird Act in 1918. However, surveys in the next decade indicated that fewer than 100 trumpeters remained in the United States, most of which were found in the Red Rocks Lakes area and Yellowstone National Park. In 1919, a breeding pair was discovered in the Park, confirming for the first time in this century that swans still nested in the United States (Skinner 1925). Studies of the range and size of this small breeding population continued into the early 1930s and culminated in 1935 with the establishment of the Red Rocks Lakes National Wildlife Refuge in the Centennial Valley of Montana.

Because of public interest in the swans, protection of habitat in the GYE, winter feeding, and regulatory changes had an immediate effect on the status of the species (Gale 1989): the number of breeding adults doubled between 1935 and 1945. The adult population continued to increase until in the mid-1950s it reached about 450, at which size it stabilized (Gale 1989).

With the five-fold increase in the GYE population and the restoration of populations in the Pacific Coast, Alaska, and Canada (Gale et al. 1987), the trumpeter swan was delisted in the U.S. Fish and Wildlife Service "Redbook" in 1968, although it remained rare. This act was hailed by the management agencies as a classic example of how rare and declining species could be restored to normal, functioning populations. The public and the swan's advocates were pleased.

But then the swans entered a period characterized by declines in both total numbers and breeding adults, and by the 1980s they had very poor cygnet production (Gale et al. 1987). Thus, the swan's saga entered a new phase typified by scientific wrangling over the proximate causes for the decline, agency separatism, turf protection, and predictably, clashing public values. Because swans concentrate on a few small lakes and open water in the GYE, they are particularly vulnerable to stress from a variety of sources, including human and animal disturbance on wintering grounds and at nest sites, parasites, flooding, and low water flows, all of which can lead to death or reproductive failure. The swan's situation, by now elevated to the political realm, was brought into sharp focus during the severe winter of 1988-89, when up to 100 swans died on their winter range.

It is reasonable to conclude from the previous description that the caretakers of the trumpeter swan failed to foresee and meet the needs of the bird. In fact, the case illustrates the best and the worst of conservation science, institutional inertia, and the power of interested constituencies. Just the administrative and jurisdictional complexity among the species primary managers is a formidable obstacle: trumpeter swans in the GYE are managed by three state wildlife departments and two different offices of the U.S. Fish and Wildlife Service, while its habitat is managed by private entities as well as nine agencies. Clearly, these jurisdictions do not coincide with the biological needs of the species.

In the face of an obviously stressed and declining swan population, the constituencies lobbied for immediate listing as "threatened" or "endangered" under the federal Endangered Species Act. This brought about considerable political posturing and the collision between agricultural, conservation, agency, and academic interests.

Although the conflict continues, coordination has accelerated since the 1988-89 die-off, giving some interests hope that some of the problems can be resolved. In many ways, the problems the swan faces in its second crisis of the 20th century are more complicated than those of the 1920s and 1930s. Improved management will depend on minimizing clashes among resource users in the region and eliminating contradictory agency mandates and operating procedures. Whether its advocates can meet the challenge a second time remains to be seen.

Analysis of the Forest Service's Conservation of Biodiversity in the Yellowstone Region

A major study is currently underway to describe and analyze efforts by the U.S. Forest Service to conserve biological diversity on the seven national forests surrounding Yellowstone and Grand Teton National Parks, as mandated under the 1976 NFMA, Whitfield 1993). As such, and because the case deals with large scale biodiversity and not just a single species, this case is different from the two preceding cases. It differs in content and style and includes more of the real life complexity of balancing scientific, social, and regulatory concerns in biodiversity management. As such, it serves as a transition and introduction to the case study frame-work presented in the last section of this paper.

NFMA directs the Forest Service to protect all renewable resources, including wetlands, to maintain diversity, and to protect habitat needed to maintain viable populations of all existing native and desirable non-native vertebrate species (Norse et al. 1986). This comprehensive approach is necessary to move beyond single species management on a case-by-case basis (Scott et al. 1991).

History has shown, however, that substantive change is never achieved by creation of legislation alone (Bolle 1987), and in the 15 years since its enactment, NFMA has not lived up to its promise. Although the National Forest System might be one of the most important reservoirs of biological diversity in the United States, existing planning efforts and management guidelines are inadequate to conserve this biological wealth (Wilcove and Flamm 1986; Wilcove 1988). Within the GYE, a dominance in forest management of extractive uses, primarily timber harvest, continues to threaten biological diversity (Marston and Anderson 1991).

There is growing and informed recognition that the Forest Service is not meeting the law in conserving biodiversity (Wilcove 1988; Scott et al. 1991). Several authors have documented the Forest Service's implementation of its mandate under NFMA (Norse et al. 1986; Scott et al. 1991). Grumbine (1991:27) concluded that the "Forest Service has done little to implement the biodiversity mandates." And, according to Wilcove (1988:v), getting the Forest Service "to fulfill its mandate under the NFMA to protect diversity has been difficult. For example, some of the proposed forest management plans, twisting the intent of Congress in this area, have justified logging on the grounds it will enhance or maintain biological diversity. However, the net effect of commercial logging in these forests is to increase populations of more common and less threatened species at the expense of rarer or more sensitive species. The Forest Service incorrectly regards these changes as benefits to biological diversity." Other problems include the way the Forest Service monitors and inventories wildlife and fragments habitats. If NFMA is ineffective at conserving biodiversity in the highly visible GYE national forests, it is probably failing elsewhere.

Whitfield (1993) examined forest plans and their implementation, conducted interviews, reviewed forest plan appeals, and reviewed extensive literature. He disaggregated the implementation process of NFMA into three classes of variables, viz., the tractability of the problem of maintaining biological diversity in the GYE, the ability of NFMA to structure the implementation process adequately, and the net effect of nonstatutory variables on the accomplishment of NFMA objectives.

He identified several specific problems. First, it is clear that the concept of biodiversity is not always easy to apply because of incomplete scientific theory. Resolution of technical difficulties associated with understanding, measuring, and monitoring diversity remains a major hurdle to successful implementation of NFMA. Second, comprehensive management of biodiversity is also impeded by an inadequate causal theory, that is, the linkage between forest activities and the conservation of biodiversity. Third, there is inadequate allocation of resources needed to implement NFMA's biodiversity conservation objectives. Fourth, many agency officials resist efforts to accomplish biodiversity objectives, particularly where these objectives are viewed as conflicting with traditional agency goals. Fifth, Forest Service inertia is supported by attitudes and actions of several key constituencies (e.g., logging companies, grazing and mining interests). Sixth, biodiversity advocates are not focused on the issue. D. Whittemore (pers. comm.) noted that the conservation community, through its lack of in-depth understanding of biodiversity concerns and its failure to interpret them for the public, is as much an obstacle as a vehicle for progress. Seventh, there are no adequate statutory or decision rule incentives in place to encourage agency officials to conserve biodiversity. Eighth, formal access to the NFMA implementation process by outsiders, both from the public and from other agencies, is impeded by a lack of ecosystem coordination among agency units. Ninth, GYE managers lack the necessary regional (or ecosystem) perspective. And last, a lack of cumulative effects assessments, in part because of inadequate intra- and inter-agency coordination, prevents a landscape level assessment of the effects of forest activities on the conservation of biological diversity.

A FRAME-WORK FOR LEARNING ABOUT THE INTERACTION OF SCIENTIFIC, SOCIAL, AND REGULATORY CONCERNS

These case studies introduce the science, social, and regulatory concerns and their interaction in real management programs. In addition to the normative case study approach to observing, understanding, and managing for scientific, social, and regulatory concerns in widespread professional usage today, more powerful conceptual frame-works have been developed that guide an observer's attention to these and other key variables in biodiversity conservation. The following frame-work to analyze, understand, and improve ongoing and future biodiversity management in the GYE and elsewhere could be used on a regular basis.

How can managers and others understand the complex interplay of scientific, social, and regulatory factors and forces in such a way that lessons can be learned, communicated, and applied? The frame-work of policy analysis outlined by Clark and Kellert (1988), Kellert and Clark (1991), and Clark (1992) offers a conceptual means whereby managers and other people can investigate the cases they become involved in, compare their assessments with each other, and potentially make professional, organizational, and policy improvements as needed. This frame-work is currently being used in some wildlife programs (Reading et al. 1991; Backhouse et al. in press).

Our treatment is only the briefest introduction to the frame-work and the rationale behind it. The frame-work is in large part based on the policy sciences which constitute a rigorous and systematic approach to complex public issues, such as biodiversity conservation (Clark 1993; Brewer and Clark in press).

The frame-work is an analytical guide to constituencies, forces, and time elements in any situation and emphasizes the interactive relationship of various actors or constituencies as they exchange information and values, and attempt to control resources and one another. The frame-work guides its users to appreciate more of the relevant details in a case than they may tend to do on their own by using traditional frames of reference. Thus, insights are fostered that translate into increased professional and organizational effectiveness. Because the frame-work focuses on constituencies, forces, and time, these three variables are introduced below.

Constituencies

Constituencies are the actors in biodiversity issues, whether individual or organizational, consumptive or nonconsumptive, whose lives are affected by some aspect of biodiversity. The actors include the general public, nongovernmental organizations, legislative bodies, regulatory agencies at local, state, and federal levels, and the judiciary, each of which can contain many groups and subgroups. Implementation of any program or policy is the product of interaction among these constituencies. Ideally, any biodiversity program should allow sufficient opportunity for the equitable participation of various actors and interests, but in reality, some constituencies are antagonistic to each other or to the overall program and all perspectives may not be represented. Certain actors may be excluded for various reasons, crass power may be used to drive decision-making, reliable scientific knowledge may be ignored, and the entire process may be carried out to meet the values of narrow, partisan interests. The outcomes of these interactions may benefit or harm biodiversity.

Forces

Four forces are evident in actual conservation efforts: biophysical, social, valuational, and regulatory. Biophysical factors include the limits or bounds on possible management programs, and indeed on policy itself, because of constraints in the availability, condition, or utility of biodiversity. These limits are set by attributes such as the scale of biodiversity (genes, species, populations). Or, if organisms are the focus of management, their morphology, physiology, behavior, life history, population structure and dynamics, and community relationships set limits. The rarity of a genome, population, species, or ecosystem is one example of a biophysical factor that might guide conservation efforts.

Social factors include social forces and constituency influences, including the role of power and property relationships. These mirror the distribution of rights and privileges to use and control elements of biodiversity. The history of natural resource management in the United States has included development of the public trust doctrine legitimizing government control over resources. In many conservation efforts, there is disagreement, conflict, and competition over social or constituency roles, power, and property relationships. In some programs, these variables dominate interactions and influence the direction and pace of the program and, therefore, its ultimate success or failure.

Valuational factors include the worth of biodiversity elements for individuals, organized and unorganized interests, and the public. Underlying values toward nature are a set of base values that individuals hold, such as the desire power, which they may seek to maximize in their interpersonal relationships. Some combination of these values may be established in fixed ideologies that dominate personal views and participation styles in decision and policy processes. Valuations figure prominently in all aspects of biodiversity management.

Regulatory factors are the organizational variables evident in programs and policies of the legislative, executive, and judicial branches of government. Nongovernmental organizations also figure prominently in many natural resource issues; they are often structured and operated quite differently from agency bureaucracies. The kinds of organizational arrangements used to manage biodiversity can either facilitate or hinder the conservation process. Organizations are driven by their histories, cultural outlooks, policy directives, standard operating procedures, self interest, and numerous other factors. An appreciation of organizational variables and knowledge of what can be done to encourage the chances of successful biodiversity management are needed.

Time

This dimension of the analytical frame-work acknowledges that all policies and programs take place over time and undergo six more or less predictable phases (Brewer and deLeon 1983; Ascher and Healy 1990; Table 39.1): initiation, estimation, selection, implementation, evaluation, and termination. Programs are dynamic, changing in response to the needs and perceptions of participants as well as the limitations and opportunities afforded by numerous forces, factors, and influences.

Each phase in the life-span of a program or policy, regardless of the particulars of specific cases, shows similar characteristics. A host of weaknesses can be identified in the phases, and these need to be managed specifically to avoid or ameliorate their negative affects on conservation efforts.

Analytic frame-works such as this one are rarely used to examine past and present biodiversity conservation efforts. Instead, simple normative approaches are used, which cannot provide the systematic depth, insight, and understanding needed for improved biodiversity conservation across policy, organizational, programmatic, and professional scales (Clark 1993). There is every reason to believe that improved biodiversity conservation would result, if analytical frame-works like this one were used commonly to provide systematic learning about our individual and collective performance in the management of living resources.

CONCLUSIONS

All efforts to manage biodiversity require a balance of scientific, social, and regulatory concerns. The dynamic interplay caused by the interaction of these three concerns in any one conservation effort can lead to conflict and competition as the three cases from

Table 39.1. Six phases through which nearly all polices and programs move (after Brewer and deLeon 1983; Ascher and Healy 1990).

Phase	Characteristics
Initiation	Problem identification, construction, and agenda setting
Estimation	Expert analysis and technical considerations of the evolving picture of the problem
Selection	Debate and authorization of goals and a programmatic solution, policy formulation
Implementation	Application of the policy solution through a program which specifies and carries out need actions
Evaluation	Ex post appraisal of the all preceding phases
Termination	Discontinuation or revision of the policy and its programmatic dimensions

the GYE illustrate. But for long-term, sustainable management of biodiversity to become reality, the means to manage their interaction most productively is needed.

As a basis for improvements, we clearly need exemplars of successful biodiversity management. It is possible to improve science and management at the biodiversity and ecosystem levels in places such as the GYE by systematically appreciating contexts, people, agencies, and policy processes at play there. Possibly, a region as important as Yellowstone can serve as a global exemplar for successful conservation of biodiversity. Nevertheless, exemplars should be developed for different types of biodiversity problems, geographic regions, and organizational and policy contexts.

Clearly, many aspects of managing biodiversity must be understood better and dealt with more effectively. In considering the problem of balancing interests and perspectives for biodiversity management, the balance must include an understanding of the valuation of biodiversity, that is, by whom, how, and when biodiversity is valued. Also to be considered are the varying threats to biodiversity, the awareness that only some diversity can or will be saved in all probability, because the necessary resources, public commitment, and available time are limited.

The conceptual frame-work we offer from the policy sciences can guide biologists, managers, and administrators in a systematic examination of the full, interactive complexity of these three concerns. Such a model encourages explicit appreciation of the context of management and administrative actions, as well as thorough orientation to the problem at hand and the institutional arrangements in place. It fosters a necessary self-evaluation at both professional and organizational levels, and it promises to increase understanding of the entire decision and policy process that embodies and impinges on biodiversity conservation. More importantly, it helps chart direction and options for participants in management efforts. Because of its applied utility, the frame-work can potentially result in a significantly better "balance" among scientific, social, and regulatory concerns in biodiversity conservation.

ACKNOWLEDGMENTS

Denise Casey, Peyton Curlee, and one anonymous reviewer commented on the manuscript and we appreciate their time and suggestions.

LITERATURE CITED

Ascher, W., and R. Healy. 1990. *Natural Resource Policy Making in Developing Countries: Environment, Economic Growth, and Income Distribution.* Duke University Press, Durham, N.C.

Backhouse, G. N., T. W. Clark, and R. P. Reading. In Press. The Australian eastern barred bandicoot recovery program: Evaluation and reorganization. In *Endangered Species Recovery: Learning the Lessons, Facilitating the Process*, eds. T. W. Clark, R. P. Reading, and A. L. Clarke. Island Press, Washington, D.C.

Banko, W. E. 1960. The trumpeter swan, its history, habits, and population in the United States. *N. Am. Fauna* 63.

Bean, M. J. 1991. Looking back over the first fifteen years. In *Balancing on the Brink of Extinction: The Endangered Species Act and Lessons for the Future*, ed. K. A. Kohm, pp. 37-42. Island Press, Washington, D.C.

Bolle, A. W. 1987. Forward. In *Land and Resource Planning in the National Forests*, eds. C. F. Wilkinson and H. M. Anderson, pp. 1-6. Island Press, Washington, D.C.

Brewer, G. D. 1981. Where the twain meet: Reconciling science and politics in analysis. *Policy Sci.* 13:269-279.

Brewer, G. D., and P. deLeon. 1983. *The Foundations of Policy Analysis*. Dorsey Press, Homewood, Ill.

Brewer, G. D., and T. W. Clark. In Press. A policy sciences perspective: Improving implementation of the Endangered Species Act. In *Endangered Species Recovery: Learning the Lessons, Facilitating the Process*, eds. T. W. Clark, R. P. Reading, and A. L. Clarke. Island Press, Washington, D.C.

Clark, T. W. 1986. Case studies in wildlife policy education. *Renewable Resources J.* Autumn:11-17.

Clark, T. W. 1992. Practicing natural resource management with a policy orientation. *Environ. Manage.* 16:423-433.

Clark, T. W. 1993. Creating and using knowledge for species and ecosystem conservation: Science, organizations, and policy. *Perspectives in Science and Medicine* 36:497-525, appendices.

Clark, T. W., and S. R. Kellert. 1988. Toward a policy paradigm of the wildlife sciences. *Renewable Resources J.* 6:7-16.

Clark, T. W., E. D. Amato, D. G. Whittemore, and A. H. Harvey. 1991. Policy and programs for ecosystem management in the Greater Yellowstone Ecosystem: An analysis. *Conserv. Biol.* 5:412-422.

Clark, T. W., and S. C. Minta. 1994. *Greater Yellowstone's Future: Prospects for Ecosystem Science, Management, and Policy.* Homestead Press, Moose, Wyo.

Clark, T. W., R. P. Reading, and A. L. Clarke, eds. In Press. *Endangered Species Recovery: Learning the Lessons, Facilitating the Process.* Island Press, Washington, D.C.

Clarke, J. N., and D. McCool. 1985. *Staking out the Terrain: Power Differentials Among Natural Resource Management Agencies.* State University of New York Press, Albany, N.Y.

Coggins, G. C. 1991. Snail darters and pork barrels revisited: Reflections on endangered species and land use in America. In *Balancing on The Brink of Extinction: The Endangered Species Act and Lessons for the Future*, ed. K. A. Kohm, pp. 62-75. Island Press, Washington, D.C.

Gale, R. E. 1989. Trumpeter swan. In *Rare, Sensitive, and Threatened Species of the Greater Yellowstone Ecosystem*, eds. T. W. Clark, A. H. Harvey, D. Genter, T. Wolf, and C. R. Groves, pp. 66-67. Northern Rockies Conservation Cooperative, Montana Natural Heritage Program, The Nature Conservancy, and Mountain West Environmental Services, Jackson, Wyo.

Gale, R. S., E. O. Garten, and I. J. Ball. 1987. The history, ecology and management of the Rocky Mountain population of trumpeter swans. Unpubl. report, Mont. Coop. Wildl. Research Unit, University of Montana, Missoula, Mont.

Gill, R.B. 1986. Wildlife management agency image and professionalism: An insiders view. Paper presented at the Annual Meeting of the Colorado Chapter of The Wildlife Society, Jan. 29, 1988, Colorado Springs, Colo.

Glick, D., M. Carr, and B. Hastings, eds. 1991. *An Environmental Profile of the Greater Yellowstone Ecosystem*. Greater Yellowstone Coalition, Bozeman, Mont.

Gresswell, R. E, and J. D. Varley. 1988. Effects of a century of human influence on the cutthroat trout of Yellowstone Lake. *Am. Fish. Soc. Symp.* 4:45-52.

Grumbine, R. E. 1991. Cooperation or conflict: Interagency relationships and the future of biodiversity for US parks and forests. *Environ. Manage.* 15:27-37.

Hunt, C. E. 1988. Creating an endangered ecosystem act. *Endangered Species Update* 6(3&4):1-4.

Hunter, Jr., M. L. 1991. Coping with ignorance: The coarse-filter strategy for maintaining biodiversity. In *Balancing on The Brink of Extinction: The Endangered Species Act and Lessons for the Future*, ed. K. A. Kohm, pp. 266-281. Island Press, Washington, D.C.

Jobes, P. C. 1991. The Greater Yellowstone social system. *Conserv. Biol.* 5:387-394.

Jobes, P. C. 1993. Population and social characteristics in the Greater Yellowstone Ecosystem. *Soc. Nat. Res.* 6:149-163.

Johnson, J. D., and D. J. Snepenger. 1993. Application of the tourism life cycle concept in the Greater Yellowstone Region. *Soc. Nat. Res.* 6:127-148.

Jones, R. D. 1991. Fishery and aquatic management program. Annual project technical report for 1990. U.S. Fish and Wildlife Service, Yellowstone National Park, Wyo.

Keiter, R. B., and M. S. Boyce, eds. 1991. *The Greater Yellowstone Ecosystem: Redefining America's Wilderness Heritage.* Yale University Press, New Haven, Conn.

Kellert, S. R. 1980. *Knowledge, Affection and Basic Attitudes toward Animals in American Society.* Government Printing Office, Washington, D.C.

Kellert, S. R. 1991. Canadian perceptions of marine mammal conservation and management in the northwest Atlantic. *Internat. Marine Mamm. Assn. Tech. Rep.* #91-04, Guelph, Ont.

Kellert, S. R., and T. W. Clark. 1991. The theory and application of a wildlife policy framework. In *Public Policy Issues in Wildlife Management*, ed. W. R. Mangun, pp. 17-36. Greenwood Press, New York.

Kohm, K. A., ed. 1991. *Balancing on the Brink of Extinction: The Endangered Species Act and Lessons for the Future.* Island Press, Washington, D.C.

Marston, R. A., and J. E. Anderson. 1991. Watersheds and vegetation of the Greater Yellowstone Ecosystem. *Conserv. Biol.* 5:338-346.

Mattson, D. J., and M. M. Reid. 1991. Conservation of the Yellowstone grizzly bear. *Conserv. Biol.* 5:364-372.

Minta, S. C., and P. M. Kareiva. In Press. A conservation science perspective: Conceptual and experimental improvements. In *Endangered Species Recovery: Learning the Lessons, Facilitating the Process*, eds. T. W. Clark, R. P. Reading, and A. L. Clarke. Island Press, Washington, D.C.

National Science Foundation (NSF). 1989. Loss of biological diversity: A global crisis requiring international solutions. Committee on International Science's Task Force on Global Biodiversity, NSB 89-171, Washington, D.C.

Norse, E., K. L. Rosenbaum, D. S. Wilcove, B. A. Wilcox, W. H. Romme, D. W. Johnston, and M. L. Stout. 1986. *Conserving Biological Diversity in our National Forests.* The Wilderness Society, Washington, D.C.

Noss, R. F. 1991. From endangered species to biodiversity. In *Balancing on The Brink of Extinction: The Endangered Species Act and Lessons for the Future*, ed. K. A. Kohm, pp. 227-246. Island Press, Washington, D.C.

Office of Technology Assessment (OTA). 1987. *Technologies to Maintain Biological Diversity.* OTA-F-330. Government Printing Office, Washington, D.C.

Patten, D. T. 1991. Human impacts in the Greater Yellowstone Ecosystem: Evaluating sustainability goals and eco-redevelopment. *Conserv. Biol.* 5:405-411.

Peters, P. H. 1991. *A Critique for Ecology.* Cambridge University Press, New York.

Power, T. M. 1991. Ecosystem preservation and the economy in the Greater Yellowstone area. *Conserv. Biol.* 5:395-404.

Rasker, R. 1993. Rural development, conservation, and public policy in the Greater Yellowstone Ecosystem. *Soc. Nat. Res.* 6:109-126.

Reading, R. P. 1993. Toward an endangered species reintroduction paradigm: A case study of the black-footed ferret. Ph.D. Dissertation, Yale University, New Haven, Conn.

Reading, R. P., and S. R. Kellert. 1993. Attitudes of Montanans toward a proposed black-footed ferret (Mustela nigripes) reintroduction, with special reference to ranchers. *Conserv. Biol.* 7:569-580.

Reading, R. P., T. W. Clark, and S. R. Kellert. 1991. An endangered species reintroduction paradigm. *Endangered Species Update* 8(11):1-3,4.

Reading, R. P., Clark, T. W., and S. R. Kellert. In Press. Attitudes and knowledge of people living in the Greater Yellowstone Ecosystem. *Soc. Nat. Res.*

Romesburg, H.C. 1981. Wildlife science: Gaining reliable knowledge. *J. Wildl. Manage.* 45:293-313.

Romesburg, H.C. 1991. On improving the natural resources and environmental sciences. *J. Wildl. Manage.* 55:744-756.

Scott, J. M., B. Csuti, K. Smith, J. E. Estes, and S. Caicco. 1991. Gap analysis of species richness and vegetation cover: An integrated biodiversity conservation strategy. In *Balancing on The Brink of Extinction: The Endangered Species Act and Lessons for the Future*, ed. K. A. Kohm, pp. 282-297. Island Press, Washington, D.C.

Skinner, M. P. 1925. The birds of Yellowstone National park. *Roosevelt Wild Life Bull.* 3(1):153-157.

Steinhoff, H. W. 1980. Analysis of major conceptual systems for understanding and measuring wildlife values. In *Wildlife Values*, eds. W. W. Shaw and E. H. Zube, pp. 11-22. U.S. Forest Service, Fort Collins, Colo.

Tobin, R. J. 1990. *The Expendable Future: U.S. Politics and the Protection of Biological Diversity.* Duke University Press, Durham, N.C.

Varley, J. D. 1988. Managing Yellowstone National Park into the twenty-first century: The park as an aquarium. In *Ecosystem Management for Parks and Wilderness*, eds, J. K. Agee and D. R. Johnson, pp. 216-225. University of Washington, Seattle, Wash.

Varley, J. D. 1989. *Cultivate National Park Diversity.* St. Paul Pioneer Press, St. Paul, Minn.

Varley, J. D., and R. E. Gresswell. 1988. Ecology, status, and management of the Yellowstone cutthroat trout. *Am. Fish. Soc. Symp.* 4:13-24.

Varley, J. D., and P. Schullery. In Press. Long-term research and monitoring on Yellowstone Lake: A study of the "ripple effect" in ecosystems and their management. In *Efficacy of Long-term Research in U.S. National Parks*, eds, W. L. Halvorson and G. E. Davis. AAAS and NPS, Washington, D.C.

Walters, C. J. 1986. *Adaptive Management of Renewable Resources.* MacGraw-Hill, New York.

Whitfield, M. B. 1993. Conserving biological diversity in the new Forest service. *Northern Rockies Conserv. Coop. News* 6:12-13.

Wilcove, D. S. 1988. *National Forests: Policies for the Future. Protecting Biological Diversity.* The Wilderness Society, Washington, D.C.

Wilcove, D. S., and B. R. Flamm. 1986. Conserving biological diversity: A challenge for national forest management. Paper presented at the IV International Congress of Ecology, Syracuse, N.Y.

Yaffee, S. L. 1982. *Prohibitive Policy: Implementing the Endangered Species Act.* MIT Press, Cambridge, Mass.

Yaffee, S. L. In Press. The northern spotted owl recovery program: An indicator of the importance of the sociolpolitical context of endangered species management. In *Endangered Species Recovery: Learning the Lessons, Facilitating the Process*, eds. T. W. Clark, R. P. Reading, and A. L. Clarke. Island Press, Washington, D.C.

40

ECONOMIC INCENTIVES FOR CONSERVING BIODIVERSITY

Jeffrey A. McNeely

IUCN-The World Conservation Union
Gland
Switzerland

The world is being impoverished by the loss and degradation of its most fundamental capital stock—its genes, species, habitats, and ecosystems. Everyone knows about forests, wetlands, and tigers, but other losses are going almost unnoticed; as many as 30,000 varieties of rice flourished in India a decade ago, but today no more than a dozen dominate three-quarters of the country. Such losses are common in all parts of the planet.

This loss of the living richness of the planet has profound implications for human welfare. Biological resources are renewable: forests, fisheries, wildlife, and crops reproduce themselves and can even increase when managed appropriately. Natural habitats, those that maintain their productivity without significant management inputs from people, have long provided local people with the means for survival, supplying food (meat, nuts, fruits, vegetables), fodder, firewood, construction materials, medicinal plants, and wild genes for domestic plants and animals. Furthermore, the highly diverse natural ecosystems that support this wealth of species also provide important ecological services, including maintaining hydrologic cycles, regulating climate, contributing to the processes of soil formation and maturation, sequestering carbon, storing and cycling essential nutrients, absorbing and breaking down pollutants, and providing sites for tourism, recreation, and research.

It is now widely recognized that biodiversity is valuable because future practical values are unpredictable, because variety is inherently interesting and more attractive, and because our understanding of ecosystems is insufficient to be certain of the role and the impact of removing any component (McNeely et al. 1990; WRI, IUCN, UNEP 1992). Conservationists, and good sense, tell us that when forests are over-exploited, when industrial pollution causes forests to die, when poor farming practices lead to soil erosion, and when over-consumption leads to mountains of waste, our economies must suffer. But economists say that this is not necessarily so. In fact, more food, timber, industrial output, and consumption leads to growing GNP, which is seen by many economists as the primary indicator of human welfare.

So, instead of conserving the rich resources of forest, wetland, and sea, current processes of development are depleting many biological resources at such a rate that they are rendered essentially non-renewable, leading to forms of development that are not sustainable. The roots of this problem are found in the distribution of costs and benefits of both over-exploitation and conservation.

THE DISTRIBUTION OF COSTS AND BENEFITS: THE MAJOR PROBLEM

The distribution of costs and benefits is a topic whose complexities would require many volumes to unravel, so I will not attempt a systematic overview. Instead, I will address just a few issues that affect biodiversity, and which may indicate how economics can be harnessed to support conservation of biodiversity.

Who Benefits From Over-exploitation?

A considerable body of information is now available on how living natural resources are being over-exploited (that is, harvested at rates far in excess of what can be sustained, leading to reduced productivity). Note that over-exploitation is quite different from conversion; if forest is converted into agricultural uses which can be sustained, productivity for humans can often be greatly increased (which of course is one of the objectives of development). So, some conversion of natural ecosystems is probably inevitable, even beneficial.

But the available evidence suggests that current rates and patterns of conversion of natural habitats are in fact not sustainable, and that many species are being lost as a result. For example, the most recent estimates suggest that more than 142,200 km^2 of tropical forests were lost in 1989, a rate of about 27 ha per minute (Myers 1989). Because tropical forests are thought to hold well over half the world's species, and many of the species are confined to relatively small areas, the number of species being lost is certain to be very high indeed; some experts suggest that at least several hundred species are being lost every year (mostly invertebrates). Where these species are lost as a result of over-exploitation (non-sustainable use), the public has great cause for concern.

Many forces are blamed for this over-exploitation, but it is useful to ask who has benefited from it. The short answer is that most of the benefits have flowed into relatively few pockets, and that most of the profits are earned by the wealthier sectors of the population. In the case of tropical forests, most are under public ownership, a management approach that was introduced in colonial times and carried over into independence. Harvesting rights are typically let to timber concessionaires who earn most of the benefits while paying few of the costs (Repetto and Gillis 1988). And economic theory suggests that over-exploitation may be in the interest of the individual seeking to maximize profits (Clark 1976).

Consumers, too, benefit from over-exploitation because the prices they pay for goods are cheaper if the environmental costs are not included ("externalized," in economic jargon). As just one extreme example, the costs of driving that are not paid by motorists and truckers (in terms of air pollution, roads, health costs, traffic police) amounts to some $300 billion per year in the United States, not counting the costs related to the

47,000 people killed per year in motor vehicle accidents (Kenzie et al. 1992). Such figures bring the benefits of over-exploitation and externalization of environmental costs into stark relief, suggesting that everyone "benefits" from over-exploitation.

Who Pays the Costs of Over-exploitation?

Of course, that is only part of the picture, because over-exploitation also carries considerable costs. Some conservationists suggest that everybody pays costs of over-exploitation, to the extent that everybody benefits from biodiversity and, therefore, suffers when it is reduced. This is not a totally satisfactory answer. Although nature certainly has some built-in redundancy, and some species could disappear (indeed are disappearing) without anybody missing them, little data are available on which species are particularly important in the functioning of ecosystems; so we cannot really determine the extent to which the general public is suffering from the loss of biodiversity. But a much more specific answer is often available. In the case of forests, the people who pay are very often the people who live closest to the forest, and who had long earned sustainable benefits from harvesting the goods and services from the natural productivity of the system. In many places, farmers are up in arms over government forestry policies that enable outside concessionaires to deplete the forests which had long been the source of their irrigation water, construction materials, medicinal plants, and game animals.

In the United States, for example, some 125 million acres of national forests, two-thirds of the total system, consistently record below-cost sales (Repetto and Gillis 1988). In other words, the American public is subsidizing the over-exploitation of its own forests, effectively paying twice for the loss of an important resource: once by having to subsidize sales, and again by losing the environmental services and other benefits of forest uses other than logging.

It is also informative to consider who does not pay the costs. Very often, the individuals who are earning excessive rents from resource exploitation do so because they are able to externalize the environmental costs. When logging activities affect downstream agricultural systems, as by reducing the quality or supply of irrigation water or reducing the storage capacity of a reservoir through increased siltation rates, the affected farmers pay. The timber concessionaire typically externalizes such environmental costs, if indeed he is even conscious of them.

Who Earns the Benefits From Conserving Biodiversity?

Two broad kinds of benefits are derived from conservation. Direct benefits are obvious from sustainable harvests of biological resources, contributions to ecological functions, tourism, and other measurable goods and services, including the maintenance of ecological functions. It is less easy to define or measure the benefits from existence value (the value placed by some people on just knowing that a resource exists), option value (the value placed on reserving a resource for future use), and bequest values (the values of passing resources onto future generations), although many economists are attempting to do so.

It is clear that conserving biodiversity, that is, using biological resources in a sustainable manner that does not reduce the overall diversity of the ecosystem involved, can bring very considerable benefits to those who directly harvest the biological resources. Such conservation (often called effective management for sustainable yields) has been a common approach of rural people since time immemorial; and indeed, a precondition for the survival of societies throughout history has been the capacity to manage their resources for a sustainable yield. But in recent times, beginning in the colonial period but greatly accelerating after 1950, the traditional systems for managing a sustainable yield have been replaced by government resource management agencies which have tended to fragment responsibility along sectoral lines and have made insufficient investments to be as successful as they should be.

For example, when protected areas are managed for tourism, insufficient benefits have gone to the local people who live closest to the over 20,000 protected areas that have been established in the past hundred years or so. They may, however, have earned benefits from the conservation of ecological functions such as watershed protection; it is said, for example, that some of India's well-protected tiger reserves are now providing year-round water instead of seasonal flows.

Many of the benefits earned from existence value are being earned by "free riders," who obtain them without having to pay much of the costs involved. Conserving the tiger in India is of much greater importance to comfortable urban-dwellers than to the local people who lose cattle or family to tigers. Farmers (and consumers) in North America benefit when India conserves wild relatives of rice, or the many pharmaceuticals that have been derived from Ayuravedic medicine. The wealthy nations of the north, therefore, earn a considerable consumer surplus when India conserves a natural area containing a high degree of biodiversity.

Who Pays the Opportunity Costs of Conserving Biodiversity?

The opportunity costs of conserving biodiversity are paid disproportionately by the people who live closest to the greatest biodiversity. In one of conservation's greatest ironies, the individuals who live amidst the greatest biological wealth tend to be the poorest of the poor. In another of conservation's ironies, when a central government establishes a national park because of its outstanding value to the nation and to humanity, it simultaneously prohibits access to any marketable value to the local people who once benefitted from the goods and services that are formerly provided, and which the national park was established to protect.

So, the opportunity costs of modern conservation programmes that restrict access to resources are falling disproportionately upon the very communities that development projects are designed to assist. If conservation programmes are to be socially accepted, then new and more appropriate means of apportioning opportunity costs, or providing compensation for them, need to be sought. Care needs to be taken to ensure that the compensation for opportunity costs does not become a subsidy of inefficiency, but those who bear the costs of conserving biodiversity for the good of the greater society could reasonably expect to be justly compensated (Sanders et al. 1990).

DEVELOPING AND USING ECONOMIC INCENTIVES FOR CONSERVING BIODIVERSITY

If the mal-distribution of costs and benefits of over-exploitation and conservation is a major problem facing biodiversity, then solutions should be sought through the better distribution of these costs and benefits. Economic policy can play an important role, at several levels: developing improved macro-economic policies, developing sector-specific policies at national or local levels, and developing ways to integrate economic thinking into programmes for managing biological resources. Economic incentives can be especially useful.

In the context of biodiversity, an incentive for conservation is any inducement that is specifically intended to incite or motivate governments, local people, and organizations to conserve biodiversity. A perverse incentive is one which induces behaviour which depletes biodiversity (such as the subsidies to logging U.S. National Forests cited above). A disincentive is any inducement or mechanism designed to discourage depleting of biodiversity; law enforcement will always need to be a part of the mix because even U.S. National Parks are subject to poaching (see for example, the recent arrest of 48 guides who were leading hunters into Yellowstone National Park to shoot bighorn sheep and elk illegally; Charrington 1989). Together, incentives and disincentives provide the carrot and the stick for motivating behaviour that will conserve biodiversity, and correct the problems which have been caused by perverse incentives.

What are the economic tools which can be used to ensure that the benefits of conserving biodiversity are delivered? I suggest six approaches that show considerable promise, although many more are certainly worth exploring (and indeed are being used in many parts of the United States and elsewhere).

1. *Estimate the economic contribution that biological resources make to the national economy.* Some have said that if current trends continue, we willprogressively lose everything of value in the countryside that does not have a price. Further, it is most unlikely that we will ever be able to assign comprehensive values to all the subtleties of biodiversity. The aesthetic and intangible dimensions are important to all of us, and the ultimate justification for conserving biodiversity is ethical. Even so, much greater attention needs to be devoted to finding ways of assigning prices to biodiversity, including both market and non-market values. Being able to place a financial value on at least some of the attributes of biodiversity will help encourage appropriate investment in its conservation, and help in the design of packages of appropriate economic incentives. This requires:

- implementing methods for ensuring that full environmental costs are included in the price of commodities (in the jargon of economists, "internalizing the externalities");
- developing methods for ensuring that the national accounting system makes explicit the trade-offs and value judgements regarding impacts on biological resources that may not be measured in monetary terms;
- conducting research on methodologies for assessing the cross-sectoral impacts (positive and negative) of resource utilization;

- developing methodologies for assigning values to non-marketed biological resources, appropriate to the needs of the country;
- estimating the economic productivity of various ecosystems, with various types of inputs; and
- developing methods for demonstrating how to assess the true costs of allowing the depletion of biological resources to continue, and devising alternative paths toward sustainable development.

2. *Establish national policies for managing biological resources.* The incentives required to conserve biological resources at the community level usually also require commensurate policies at the national level. A national or regional conservation strategy can be an effective means of reviewing such policies, and determining what shifts are required to achieve national objectives for conserving biological resources. The following steps will help ensure that the economic perspectives are built into such strategies:

- *Quantify* and bring to the attention of policy makers the many economic and financial benefits of integrated rural development linked with conservation of biological resources.
- *Identify* both conflicts and potential for cooperation between the various activities of agriculture, fisheries, forestry, conservation, and rehabilitation of degraded lands.
- *Formulate* legislation consonant with the socio-economic patterns of the target group and the natural resources being utilized, both to institute disincentives and to ensure that incentives carry the power of law.
- *Review* policies and legislation in other sectors for possible application to conservation of biological resources and community involvement in such work.
- *Devise* effective incentives to accelerate integrated development to close any gap between what the individual sees as an investment benefit and what the government considers to be in the national interest.

All government sectors depending on biological resources should be encouraged to develop policies to encourage the sustainable use of those resources. In addition, other sectors with major impacts on biological resources, such as agriculture, commerce, transport, and even the military, could ensure that their policies are consistent with conserving biodiversity, or at least do not unnecessarily deplete biological resources.

Drawing on the latest advances in genetics, population dynamics, and conservation biology, the relevant government policy bodies should state, as a matter of public record, what proportion of the current land and water area is intended to be legally protected for conserving biological resources. Such policy objectives can often be incorporated as part of a national protected area system plan or a national conservation strategy. On the basis of such national objectives, the government can measure the costs and benefits of implementing conservation programmes most effectively.

3. *Establish new administrative structures for managing biological resources.* Even though those resources contained within strictly protected areas, such as national parks, are a government monopoly, biological resources in buffer zones, wildlife sanctuaries, national forests, and communal properties are often "open access goods" and need to be brought into a form of resource-use control which is appropriate to the particular conditions of the region. Such products as firewood, medicinal plants, and meat can often be made available to local communities more effectively through direct harvesting

than through intermediaries, and usage rights can often provide economically disadvantaged communities with highly valued resources.

Incentives can be used to create an institutional setting in which the property rights to specific populations of species of plants or animals are held by a single decision-making unit. This may involve strengthening community-based resource management systems which are functioning well, with the intention of ensuring that the local people steadily enhance their capacity to utilize biological resources in an optimal and sustainable manner.

4. *Enable local communities to participate in conservation efforts.* The moral justification for a system of protected areas or any other programme to conserve biodiversity is strengthened by the flow of human benefits that it generates. The more people benefit directly from the proper conservation and use of wild resources, the greater the incentive for them to protect the resource and the lower the cost to government of doing so. If a government decides that an area is important to be conserved, and this involves opportunity costs to local communities, then it would seem only appropriate that these local communities be compensated for their opportunity costs through a series of economic incentives (McNeely 1988). In fact, one could argue that the cost-benefit ratio of conservation must ultimately be positive for the local people if the resources are to survive. What mechanisms are available to encourage such a positive balance?

At the policy level, methods need to be developed to demonstrate to governments that such allocation of benefits makes economic sense; or, rather, help define the conditions under which such allocation of benefits makes sense (see McNeely and Dobias 1991, for an example).

The foundation of any incentives package is community support, and such support is gained only through involvement. The following elements are part of the package:

- Ensuring that communities are made part of the project planning process from the earliest stages, and making them the leading actors throughout the programme.
- Developing information about incentives packages, including costs and benefits, and any accompanying disincentives.
- Assisting the rural people to develop the skills required to implement the activities stimulated by the incentives package, which implies technical assistance and training as well as education in the broad sense.

5. *Provide technical advice to resource management agencies on how to design realistic packages of economic incentives.* Few resource management agencies have the technical capacity to design realistic packages of economic incentives. The relevant agencies, especially at the state level, would ensure that protected area policies include an economic justification for conserving the areas, provision for comprehensive planning and management to ensure the sustained profitability of the resource, and linkages between protected areas and other relevant sectors (such as agriculture, tourism, communications, community development, forestry, and water resources development). The management authority would be provided with advice on how to specify what each protected area will provide to the national economy in terms of employment, construction costs, cost of food for picnics, fishing and camping equipment, transportation, watershed protection, and genetic resources. Such values can be considerable; one study recently found that the total net economic value of outdoor recreation activities in the United States is $122 billion per year (Bergstrom and Cordell 1991).

6. *Develop innovative funding mechanisms that will enable the public to support conservation of biodiversity.* Because government conservation agencies seldom have sufficient funding to carry out their mandates effectively, innovative funding mechanisms need to be sought outside the traditional government sector. Possible mechanisms include: charging entry and other usage fees for protected areas, including national forests; returning profits from exploiting biological resources to the people living in the region; implementing water use charges for the water produced by a protected area; establishing linkages with major development projects; building conditionality into extractive concession agreements; and considering "conservation concessions," similar to those for forestry (see McNeely 1989 for further details).

Judging from the contributions made to voluntary conservation organizations, the American public is willing to make substantial payments for the existence value of biodiversity. The willingness to pay can be taken as an indication of the demand function for biodiversity. How can conservation activities capture more of this willingness to pay? How can the consumer surplus be reduced, or captured? And perhaps more important, how can we convert this willingness to pay into increased amounts of biodiversity?

Voluntary organizations like The Nature Conservancy, the National Wildlife Federation, Sierra Club, and World Wildlife Fund are one possibility, but the activities supported tend to be research for public awareness rather than recurrent costs or major investments in packages of economic incentives for people living around protected areas. Perhaps a special investment programme needs to be established which would appeal to investors, demonstrating the economic benefits of investing in conserving biodiversity.

CONCLUSIONS

In general, those establishing the economic policies governing the use of biological resources have three broad options: to manage for a sustainable yield and therefore maintain the future option of depleting a resources (when prices may be higher); to draw down supplies slowly, in an effort to spread earnings over many years; or to deplete existing stocks quickly, taking a quick profit which can then be reinvested elsewhere. Most conservationists would suggest that the first option is far preferable, but the third option is in fact the one most often being taken in the case of forests, fisheries, and wildlife.

Determining the full costs of each option will have a major influence on determining which is most economically viable. Careful economic analysis could well demonstrate that rapid depletion is only profitable if the environmental costs are externalized, which implies that the exploitation is unsustainable. Such policies also need to consider the flow of benefits, and it would be useful to provide economic tools which will enable conservation benefits to be delivered to those who live closest to the greatest biodiversity and, therefore, will determine how much of that diversity will survive.

LITERATURE CITED

Bergstrom, J. C., and H. K. Cordell. 1991. An analysis of the demand for and value of outdoor recreation in the United States. *J. Leisure Research* 23(1):67-86.

Charrington, M. 1989. Poaching: Beating swords into plowshares. *Earthwatch* 10:22-29.

Clark, C. W. 1976. *Mathematical Bioeconomics: The Optimal Management of Renewable Resources*. John Wiley, New York.

Kenzie, J. J., R. C. Dower, and D. D. T. Chen. 1992. *The Going Rate: What it Really Costs to Drive*. World Resources Institute, Washington D.C.

McNeely, J. A. 1988. *Economics and Biological Diversity: Developing and Using Economic Incentives to Conserve Biological Resources*. IUCN, Gland, Switzerland.

McNeely, J. A. 1989. How to pay for conserving biological diversity. *Ambio* 18(6): 308-313.

McNeely, J. A., and R. Dobias. 1991. Economic Incentives for Conserving Biological Diversity in Thailand. *Ambio* 20(2): 86-90.

McNeely, J. A., K. R. Miller, W. V. Reid, R. A. Mittermeier, and T. B. Werner. 1990. *Conserving the World's Biological Diversity*. IUCN, World Bank, WRI, Conservation International, and WWF, Gland, Switzerland, and Washington D.C.

Myers, N. 1989. *Deforestation Rates in Tropical Forests and Their Climatic Implications. A Friends of the Earth Report*. London, England.

Repetto, R., and M. Gillis, eds. 1988. *Public Policies and the Misuse of Forest Resources*. Cambridge University Press, New York.

Sanders, L. D., R. G. Walsh, and J. D. Loomis. 1990. Toward Imperical Estimation of the "Total" Value of Protecting Rivers. *Water Resources Research* 26(7): 1345-1357.

WRI, IUCN, and UNEP. 1992. *Global Biodiversity Strategy*. World Resources Institute, International Union for Conservation of Nature and Natural Resources, and United Nations Environment Programme, Washington D.C.

41

THE ROLE OF PRIVATE INDUSTRIAL FOREST LANDS IN THE MANAGEMENT OF BIODIVERSITY

James A. Rochelle

Weyerhauser Company
Corporate Headquarters
Takoma, Washington
U.S.A.

Lorin L. Hicks

Plum Creek
Seattle, Washington
U.S.A.

Several definitions of biodiversity exist, as discussed elsewhere in this volume. All of these definitions are comprehensive, differing primarily in the level of detail and spatial scales at which they address levels of biological organization. Regardless of the specific definition, it is clear that the subject of biodiversity is currently of high interest to the scientific community as well as the general public.

A number of factors are operating to raise collectively the level of concern about biodiversity which, in itself, is a relatively new issue. Clearly the rapid rate of growth in the human population, especially in developing countries, has raised the general level of concern for the environment. The accelerated rate at which species are being lost worldwide is a key factor. Closer to home, regional increases in population, in some cases, are in conflict with maintenance of the environmental amenities that make an area desirable in the first place. Increased affluence, less direct economic dependence on the land, and the influence of the media contribute to changing public expectations for managed forests, including the provision of both commodity and amenity values from both public and private lands. An expanded scientific data base has pointed out a need for change in some forest practices. In addition, widely publicized reductions in abundance of highly-visible organisms, such as some stocks of Pacific salmon, have raised the level of public awareness and concern for biodiversity. Finally, because of the all-encompassing view of what biodiversity is, it has become the umbrella for other issues such as old-growth forest preservation, or the cumulative effects of land use.

Some numbers illustrate the importance of industrial forest lands in the context of biodiversity. The total area of forest land in the United States is 727 million acres, or about one-third of the total land area of the country. About 480 million of these acres are classed as timberland, defined as forest land that is producing, or is capable of producing, crops of merchantable timber and is not administratively or by statute withdrawn from timber utilization. Private industrial timberlands constitute about 71 million acres, or about 15% of this total. Of note is that an area equal to about one-half of this industrial acreage, or about 35 million acres, of productive timberland is withdrawn by law under the National Wilderness System (American Forest Council 1991). An additional 6+ million acres, the exact amount as yet undefined, of productive timberland is currently being withdrawn for the protection and recovery of the northern spotted owl.

Private industrial land ownership patterns range from highly intermingled with public lands to significant blocks of contiguous ownership. In the western United States in particular, private forest lands tend to occur at lower elevation and have higher productive capacity than much of the public land under federal management. These factors, and the specific objectives for which they are managed, result in private industrial lands having an important and somewhat different role in the management of biodiversity than do public lands. When private nonindustrial lands with their wide range of management objectives are included, the range of productive capacities and management approaches represented on the differing ownerships result in a combination of features that, on a regional basis, provide significant contributions to, as well as opportunities for, addressing biodiversity.

All of the factors discussed above lead us quickly to the view that managed landscapes must be part of the solution to managing effectively for biodiversity. As substantial as the 35+ million acres of set-aside forest land may be, it is wholly inadequate to protect and provide the biodiversity our forest lands are capable of sustaining. Clearly a strategy bent on "preserving" all examples of biodiversity will fail, simply because as a society we cannot "lock up" all the landscape necessary to achieve this objective. Consequently, a successful strategy for conserving, protecting, and restoring biodiversity must include managed landscapes and recognize the role and needs of humans in the process (Keystone Center 1991). An example of an approach that recognizes this broader perspective is in the Pacific Northwest where the U.S. Forest Service has initiated an effort to develop a regional approach to biodiversity management that is sensitive to the role of different ownerships (J. Lowe pers. comm.).

Other papers in this volume have thoroughly discussed the many elements of biodiversity and their intrinsic as well as current and future importance from a practical standpoint. The forest industry also recognizes and concurs with those values. To the extent that maintenance of biodiversity and sustained productivity including the production of timber are linked, then it clearly is in the best interest of landowners to understand and find ways to address biodiversity in their overall management plans. The challenge is one of defining the role of industrial lands in the context of the overall land ownership and management objective patterns. Furthermore, this must be done in a way that also allows for cost effective production of the many products of the forest that are in increasing demand by a growing human population.

Evidence of recognition of the importance of biodiversity by the forest products industry includes the participation by representatives of several companies in the Keystone National Biodiversity Dialogue (Keystone Center 1991), in which a wide range of interested parties developed policy-level guidance for the maintenance of biodiversity on federal lands in the United States. Industry representatives also participated in the development of the Society of American Foresters position on biodiversity (Society of American Foresters 1991). The latter effort, in particular, devoted significant effort to understanding and communicating the role of private lands in the conservation of biodiversity. More recently, a number of industrial forest companies have joined with the President's Commission on Environmental Quality (PCEQ 1992) to address the issue of biodiversity on private lands. In this effort, projects encompassing more than a half million acres of private forest land in which maintenance or enhancement of biodiversity is emphasized will be carried out.

THE ROLE OF CONVENTIONAL FOREST MANAGEMENT PRACTICES IN THE CONSERVATION OF BIODIVERSITY

Silvicultural activities that reduce, fragment, or isolate forest habitats, simplify stand structure across a landscape, or eliminate individual species or structural and functional attributes can reduce biodiversity. On the other hand, thoughtful forest management can increase biodiversity (Society of American Foresters 1991).

Existing forest characteristics on private industrial forest lands influence both current and future patterns of biodiversity. Throughout the United States, private forests are almost entirely second- or third-growth; old growth is minimally represented. As a function of past natural disturbance, logging, and other land uses such as agriculture in the southeast, stands occur both in mosaics of different-aged stands and in extensive areas of relatively uniform age.

The continued existence and growth of the land area in forests is in itself a significant contributor to biodiversity. Although timber production prior to the early twentieth century occurred primarily in previously unlogged forests, since the turn of this century, the United States has experienced a net gain in forested landscape. This increase resulted from both the regrowth of harvested areas and the conversion of abandoned agricultural land to forests. Clawson (1979) reported that there were 737 million acres of forest in the conterminous United States in 1977, an increase of 192 million acres over that estimated to exist in 1908.

The association of different groups of animal and plant species with different stages of development of forest stands is well documented (Brown 1985). Post-disturbance patterns of stand development are similar, but natural disturbance maintains a level of structural complexity that can be missing or reduced in intensively managed forests (Hansen et al. 1991). Thus, a landscape supporting a range of age classes will contain a greater diversity of species than a uniform-aged landscape. To the extent that forest management practices result in a landscape mosaic of age classes, biodiversity is enhanced. The establishment of maximum harvest unit sizes and time delays on the harvest of adjacent stands through regulation and internal guidelines is resulting in a wider range of age classes being present on a particular landscape.

The provision of extensive areas of late-successional stands is economically unfeasible on industrial forest lands, but significant efforts are being made to provide some of the structural components of older forests. Through forest practice regulations, voluntary best management practices, internal environmental guidelines, and other actions by landowners, the retention of snags, green trees, and down wood is commonplace in some states. Operational and safety considerations often dictate specific locations in which these features can be retained, and they are frequently left in association with streamside or wetland management zones. In addition to providing older forest components, streamside and wetlands management zones typically support diverse vegetation, and are subject to only limited management intervention. The result is the development of a vegetation complex different both in composition and age from the surrounding managed stands. The proximity to water and the vegetative complexity of riparian and wetland areas often result in the presence of a diverse wildlife community (Oakley et al. 1985).

In addition to the variation in soil types and levels of site productivity found on any particular portion of the landscape, a wide variety of silvicultural practices also affects the vegetative and structural characteristics of forest stands. Methods of harvest, post-harvest site preparation techniques, choice of species for regeneration, and regeneration method, including timing and density of replanting, all affect characteristics of both the forest stand and the associated vegetation. The frequency of silvicultural burning and fertilization, as well as the level and frequency of thinning influence the amount and composition of vegetation throughout the life of the stand. The method of final harvest affects both immediate post-harvest conditions as well as the characteristics of the replacement stand. The faunal and floral communities present will reflect the characteristics of the individual stands and those of the combination of stands and special management areas on the landscape.

EVOLVING ALTERNATIVE SILVICULTURAL APPROACHES

Most will agree that modern forestry is in the midst of significant change. Traditional forest management is being challenged by often mutually exclusive public expectations of limited development and expanded provision for both commodity and amenity resources (Brooks and Grant 1992). Both public and private landowners are experimenting with silvicultural approaches that go beyond insuring replacement of the stand of desired crop trees. Whether these programs are called "New Forestry," "New Perspectives," "Adaptive Forestry" or "Environmental Forestry," their objectives are the perpetuation of ecosystem diversity through a series of management techniques at both the stand and landscape levels (Fiedler 1992).

The exact form that these "alternative silviculture" programs take is variable and evolving. An operational definition (COPE 1992) includes such elements as:

- A level of commodity extraction to meet landowner objectives and public and shareholder expectations
- Innovative treatments guided by a knowledge of biological needs and expected responses
- An appreciation for the social (aesthetic) perception of forestry practices.

Adams (1992) noted a number of implications associated with alternative silviculture including: reduced use of clearcutting, modified management systems and equipment, additional care and expertise in logging, and new concerns for worker safety and impacts on stand growth and yield.

Alternative silviculture in the northwestern and southeastern United States typically involves some level of departure from conventional even-aged management. Emphasis has changed to address both the ecological value of what is retained, as well as the economic value of what is harvested (Franklin 1989). A principal objective is to increase the level of structural diversity remaining on the site, using techniques such as retention of snags, down logs and defective trees, retention of large green trees, and logging methods and cutting patterns that encourage the development of multi-layered canopies. Retention of structural features from the existing stands is an attempt to mimic natural patterns of disturbance. Concentration of harvest units, rather than dispersing them, leads to reduced forest fragmentation (Franklin and Forman 1987).

The degree to which these departures from traditional management are feasible is obviously linked to the objectives for which the land is being managed. With the requirement that industrial forest lands provide adequate shareholder returns, clearly there are limits on the extent to which some alternative treatments, such as the retention of extensive areas of mature timber, can be applied. At the same time, retention of some level of structural elements such as snags, down wood, and green trees, and modification of harvesting patterns are feasible and being done. As we discuss later, a significant number of research efforts and operational trials are underway to understand better both the operational and economic implications and the ecologic benefits of further silvicultural modifications.

As alternative approaches to silviculture evolve, it is clear that is no single approach is appropriate for all ownerships. It is this range of approaches that integrates ecological goals and landowner objectives that broadly constitute "New Forestry" (J. Franklin pers. comm.). It is also important to note that even though these alternative practices address some ecological concerns, they raise others such as aesthetics, need for multiple re-entries to a site, and limitations on future management treatments (Hicks 1991; Hopwood 1991).

OPERATIONAL TRIALS AND RESEARCH EFFORTS IN ALTERNATIVE SILVICULTURE INVOLVING PRIVATE INDUSTRIAL LANDS

The focus of research and operational efforts to date has been the evaluation of the operational, economic, and ecologic aspects of modified silvicultural approaches, as well as measuring the benefits of existing practices aimed at conserving biodiversity. These efforts involve evaluating patterns of structural retention in upland sites, determining wildlife responses to varied riparian zone and upland leave area configurations, and examining options for, and benefits of, landscape level management approaches.

Plum Creek Timber Company began experiments in 1990 by implementing New Forestry prescriptions on 20% of the harvest units in its Cascade and Rocky Mountain regions (Hicks 1991). Prescriptions varied by site, but had the common objective of maintaining structural diversity by leaving representative green overstory and understory trees, defective trees and snags, and large down logs. Resulting post-harvest retention patterns range from uniform distribution to patches of timber in wedge-shaped blocks, contoured patches, and strips. From an operational perspective, there were differences in cost, feasibility related to slope steepness and logging system, and worker safety with the different configurations. Diversity of retained material, aesthetic appearance, ability to regenerate by planting, and potential for immediate use by wildlife also varied between treatments.

A series of operational trials recently installed by the Weyerhaeuser Company in its coastal Washington region examined operational feasibility and will assess wildlife habitat benefits of a range of structural retention treatments. All treatments involved second-growth stands containing both conifer and hardwood trees. Retention patterns included uniform distribution of trees across harvest units, clumping of leave trees in different-sized groups and in different locations within harvest units and distributing trees in strips or clumps adjacent to or connected to streamside management zones. Whereas all treatments could be economically applied, additional costs were incurred, and addressing the human safety hazard increased the difficulty in implementing some treatments. In addition, some configurations might preclude later application of some silvicultural treatments to the regenerated stand. Harvesting equipment used and topography were other major factors that affected the feasibility of these treatments. Biological values are currently being assessed, with emphasis on understanding the current and future contribution to habitat diversity of the retention areas in both the unit in which they occur and the surrounding landscape.

Future directions for alternative silviculture must also include rigorous scientific documentation of the degree to which perceived benefits for wildlife and other resources are achieved. A number of industry sponsored cooperative efforts with universities and other interested parties are addressing that need, and range from evaluations of responses of key groups of wildlife, such as cavity dwellers or interior forest species to alternative treatments, to studies addressing wildlife response to management of riparian areas and forested wetlands. Implications for other key wildlife groups, such as neotropical migratory birds, will be an outcome of all of these studies.

In Washington State, the Timber, Fish and Wildlife (TFW) program is also funding an ambitious research project, being conducted by the University of Washington, Washington State University, and the U.S. Forest Service, to examine wildlife diversity and habitat use patterns in managed forests throughout the state. When combined with results of another TFW study evaluating wildlife use of riparian management zones, a comprehensive view of wildlife diversity at the landscape level in managed forests should emerge.

LANDSCAPE-LEVEL EXPERIMENTS

Geomorphic processes, natural disturbances, variations in site conditions and productivity, and human influences combine to create varied spatial patterns in the forest. Increasingly, landscape ecology, which focuses on the ecological structure, functions, and patterns of change of a landscape, is becoming a vital element of any biodiversity strategy. Examination of the current patterns of diversity, as was done by Hughes (1992) for a managed forest landscape in North Carolina, provides insight as to future conditions and can help direct future efforts to address biodiversity.

Understanding the biodiversity characteristics of managed landscapes is facilitated through the examination of natural forests and their disturbance regimes. Studies within managed landscapes can additionally contribute to understanding landscape ecology, because they provide the opportunity to learn from management experiments. Recognizing that many of our wildlife landscape assemblages occurred incidental to other land management objectives, it would appear that major opportunities exist to improve on them by design. Franklin and Forman (1987) employed a series of geometric models to examine the ecological consequences of landscape patterns created by timber harvest. Walters and Holling (1990), in discussing the concept of adaptive management, pointed out the benefits of combining science and management in large scale experiments that provide both an enhanced scientific understanding and an economic product.

An example of a large-scale experimental management plan comes from our experience with the northern spotted owl. Drawing from his research into the dynamics of spotted owl habitat across disturbed landscapes, Irwin (1990) proposed an active owl management strategy using a network of managed forests and unmanaged reserves. In addition to addressing owl habitat needs, the reserved lands provide a stable, relatively predictable habitat network for other sensitive species. Spotted owls in the managed forest matrix between the reserves should contribute to overall population viability by interacting with the denser sub-populations in the reserves. The managed forests provide a more "risky" environment supporting a sub-population of lower density, with suitable habitat "moving" across the landscape as forest management plans are implemented.

Fundamental to the success of such an active strategy is a research and monitoring program, with an ultimate goal of providing information to help resolve the conflict between preservation and management of the habitats in question. The operative question for investigation becomes "how can timber management be made more compatible with the conservation of biological diversity," rather than the more negative question "what is wrong with the way forests are managed?" (Irwin 1990).

BIODIVERSITY ON INDUSTRIAL FOREST LANDS—A FUTURE VIEW

The above discussion attempts to outline the past and current role of industrial forest lands with regard to the biodiversity currently occurring on these and adjacent lands; and we have suggested some approaches and outlined some activities industry is conducting for addressing it in the future. In looking ahead we can identify some conditions that are critical to successful achievement of our collective biodiversity goals:

- Useful, scientifically valid indicators of biodiversity
- Measurable biodiversity goals and objectives
- The involvement of both working and reserved landscapes
- A large-scale approach with coordination between landowners

- Differing roles for different ownerships
- Validation of the outcome of experimental strategies
- Cost-effective management approaches, or landowner incentives or compensation where economic values foregone are significant

Trends in the management of industrial forest lands are evident, as a reflection of both internal and external forces. Increasingly, the industry recognizes and responds to changing public expectations; conservation of biodiversity is one of those expectations. Sustained productivity of forest land is recognized by the industry as a necessity for long-term success; practices to sustain productivity also contribute to maintenance of biodiversity. Regulations, Best Management Practices, and internal guidelines continue to change incorporating means to address cumulative effects and other landscape scale issues including wildlife protection. In response to increasing demands for both forest products and ecological values, there is a growing recognition of the need to emphasize specific outputs from different portions of the forest landscape. As an example, a recent workshop attended by a broad range of forest industry representatives defined the ecological role of private industrial forest lands in Western Washington to include the maintenance of site productivity, the protection of riparian zones and wetlands, the provision of early- and mid-successional habitats and assistance to public land managers in meeting the needs of species requiring extensive areas of late-successional habitat (ONRC 1993). The concept of a triad (Hunter 1990), in which portions of the landscape are designated for preservation, multiple use, and intensive forest production, respectively, is another model that may apply. Finally, there is an awareness on the part of industry of the need to re-examine some management practices in light of conserving biodiversity, and an attempt to develop new approaches that are also compatible with economic viability. On a global scale, the development of forest management strategies that combine conservation of biodiversity with sustainability of forest products in the United States will help avoid the exportation of environmental problems by avoiding the shift of wood products demands to third-world countries where environmental protection is limited.

The level of concern for the maintenance of biodiversity continues to expand, fueled in large part by the increase in listings or organisms, including members of such familiar groups as Pacific salmon stocks, as threatened and endangered. The introduction of new approaches to the conservation of forest resources, such as the concept of ecosystem management, which received high levels of media attention as a result of the Clinton Forest Summit, further contributes to public awareness of and concern for biodiversity. The development and application of a range of management approaches tailored specifically for different land ownerships will ultimately determine overall patterns of biodiversity on America's forest lands.

LITERATURE CITED

Adams, D. L. 1992. New Forestry in the Inland Northwest. University of Idaho College of Forestry, Wildlife and Range Sciences, Misc. Publ. 16.

American Forest Council. 1991. The American Forest: Facts and Figures 1991. The American Forest and Paper Association, Washington, D.C.

Brooks, D. J., and G. E. Grant. 1992. New approaches to forest management. Background, science issues and research agenda. *J. For.* 90:25-28.

Brown, E. R., ed. 1985. Management of wildlife and fish habitats in forests of western Oregon and Washington. USDA Forest Service Publ. No. R-6 F&WL-1 92. Portland, Ore.

Clawson, M. 1979. Forests in the long sweep of American history. *Science* 204:1168-1175.

Coastal Oregon Productivity Enhancement (COPE) Program and Olympic Natural Resources Center. 1992. Proceedings of First Biannual Regional Alternative Silviculture Conf. Vancouver, Wash. Nov. 25-26, 1991.

Fiedler, C. 1992. New Forestry: concepts and applications. *Western Wildlands* 17:2-7.

Franklin, J. 1989. Toward a new forestry. *Am. For.* Nov./Dec: 37-44.

Franklin, J. F., and R. T. T. Forman. 1987. Creating landscape patterns for forest cutting: ecological consequences and principles. *Landscape Ecol.* 1(1):5-18.

Hansen, A. J., T. A. Spies, F. J. Swanson, and J. L. Ohmann. 1991. Conserving biodiversity in managed forests: lessons from natural forests. *BioScience* 41(6):382-392.

Hicks, L. L. 1991. Plum Creek Timber Company's new forestry experiments: lessons learned and future directions. *For. Perspectives* 1(3):9-10. World Forestry Center, Portland, Ore.

Hopwood, D. 1991. Principles and practices of new forestry. British Columbia Ministry of Forests, Land Management Rept. 71:1-95.

Hughes, J. 1992. Commercial forest management on a landscape scale in coastal North Carolina: Weyerhaeuser's Coastal Carolina tree farm. In *Proceedings Appalachian Section, Society of American Foresters Annual Meeting*, Asheville, NC. Feb. 14, 1992. Pp. 100-105.

Hunter, M. L., Jr. 1990. *Wildlife, Forests and Forestry. Principles of Managing Forests for Biological Diversity.* Prentice Hall, Englewood Cliffs, N.J.

Irwin, L. L. 1990. An active process for conservation of northern spotted owls. National Council for Air and Stream Improvement, Inc., Corvallis, Ore.

Keystone Center. 1991. Biological Diversity on Federal Lands. The Keystone Center, Keystone, Colo.

Oakley, A. L., J. A. Collins, L. B. Everson, D. A. Heller, J. C. Howerton, and R. E. Vincent. 1985. Riparian zones and freshwater wetlands. In *Management of Wildlife and Fish Habitats in Forests of Western Oregon and Washington*, ed. E. R. Brown, pp. 58-79. USDA Forest Service Publ. No. R-6 F&WL-192, Portland, Ore.

Olympic Natural Resources Center. 1993. The role of industrial forestlands in the management of Western Washington's forest ecosystems. A discussion paper. From a workshop sponsored by the College of Forest Resources, University of Washington, Seattle, Wash. March 1-3, 1993.

President's Commission on Environmental Quality (PCEQ). 1992. Biological diversity on private lands initiative. Project Summary. Council on Environmental Quality. Washington, D.C.

Society of American Foresters. 1991. Task Force report on biological diversity in forested ecosystems. Society of American Foresters, Bethesda, Md.

Walters, C. J., and C. S. Holling. 1990. Large-scale management experiments and learning by doing. *Ecology* 71(6):2060-2068.

42

MANAGING BIODIVERSITY ON PRIVATE LANDS

Michael A. O'Connell[1]

World Wildlife Fund
Washington, D.C.
U.S.A.

Scientists and policy-makers have recently focused much attention on management of public lands for biodiversity (Keystone Center 1991, for example). Public lands, however, do not include many important elements of ecosystem diversity, particularly in the eastern United States and comprise only one-third of the land base of the continental United States (Thomas and Salwasser 1989). In regions such as the southern United States, private ownerships comprise up to 90% of the land base (Yarrow 1990). Furthermore, despite considerable dedication and expertise of public land managers, stewardship of wildlife refuges, national forests and other multiple-use areas for biodiversity is frequently constrained by politics, contrasting mandates, and a host of conflicting considerations. In contrast, private lands are rarely bound by those restrictions. Although many private landowners are motivated by profit or discouraged by lack of incentives, the experience of this study has shown that many are receptive to conservation concerns.

In an overall strategy to conserve biodiversity, private lands clearly must play a large role. They contain many important elements of biodiversity not found on public lands and management options are often limited by little more than the interest and willingness of the owner and the ability to reconcile the conservation of biodiversity with other land-use goals. Another area in which the private sector can play a crucial role is leading by example in land-use management. Individual landowners can be compelling models for their cohorts in similar business, and the private sector as a whole can demonstrate to the public and policy makers the potential for harmonizing conservation of biodiversity with sustainable land management. Engagement of the energy, resources, and expertise of the private sector in conserving biodiversity is, thus, essential.

The described considerations notwithstanding, conservation of biodiversity presents many challenges to private land owners and managers. Important among these is the lack of guidance and consensus about how best to manage private lands in a way that includes biodiversity considerations without radically limiting other objectives. Also notable is the large degree of uncertainty among private landowners and managers about what conservation of biodiversity means to practical management decisions and its effect on other responsibilities and goals.

---

[1] Present address: The Nature Conservancy, 2699 Lee Road, Winter Park, Florida.

Private land managers express concern that protection of biodiversity can be a surrogate for eliminating selected uses and, in some cases, all active use. As a result, the growing acceptance of the need for the conservation of biodiversity among scientists and many citizens has not translated into clear management direction for, or commitment to, biodiversity on private lands. In fact, this ambiguity creates several barriers to improving conservation of biodiversity by the private sector. These technical and philosophical hurdles to progress must be addressed if management of private lands for biodiversity is to advance.

This paper describes the project initiated by World Wildlife Fund (WWF) to find solutions to the technical and ideological barriers for biodiversity conservation on private lands. It examines the reasons for such a project, and describes the development of a proposed model and suggested goals for private land management. Finally, the paper examines a case study where a similar strategy was used successfully by a major corporation, and concludes with some lessons learned in pursuit of the WWF project.

OVERCOMING BARRIERS IN THEORY AND PRACTICE

Some authors have explored the need and rationale for conserving biodiversity on private lands, whereas others have offered theoretical principles for managing biodiversity in agriculture, forest, and other managed systems in general (Hansen et al. 1991; Pimentel et al. 1992). These principles have rarely been translated into practical programs. To date, there have been no broad, systematic attempts based on conservation theory to encourage private landowners to maintain biodiversity. Some policies do exist to encourage private landowners to adopt limited conservation measures. For example, the federal government's Conservation Reserve Program under the 1985 Farm Bill subsidizes farmers to remove some wildlife habitat from production for a ten-year period (Yarrow 1990). To begin engaging private landowners in the process of biodiversity conservation, World Wildlife Fund initiated a project in 1989 to address both technical and philosophical challenges. The project has several components, each addressing specific barriers to improving management for biodiversity on private lands. The first of these is a scientifically accurate model frame-work for conservation of biodiversity (O'Connell and Noss 1992).

The model frame-work employs the theory and experience of conservation biology and land-use management and applies them to a stepwise protocol for assessing the elements of biodiversity on a property in their global and regional contexts. This frame-work is linked to an overall goal of maintaining viable populations and natural distributions of native species in the regional landscape. The model is intended to serve as a general strategy concerning biodiversity for private landowners. It allows biodiversity considerations to be part of the land-use decision-making *process* rather than a postscript to the list of objectives an owner desires or is obligated to consider.

The term biodiversity has frequently been discounted as fashionable jargon. It is often misunderstood as either justification for inflexible restrictions on land use by the conservation and regulatory establishments, or, conversely, as an excuse to continue land-use practices that cause habitat loss and fragmentation (Noss 1990). Only when private landowners understand its meaning, its implications for on-the-ground management and other objectives, and the long-term significance of maintaining biodiversity for economic and resource sustainability will this concept be expressly integrated into land use. The second element of the WWF project is, therefore, equally as crucial in changing the attitudes of many private landowners toward biodiversity conservation. It is a two-step process of field testing the model management framework with willing private landowners and communicating the findings of this exercise to the private sector through workshops and other extension activities.

The final element in the WWF effort to promote conservation of biodiversity by private landowners is linking the process to a system of both economic and non-economic incentives and education. Currently, few direct economic incentives exist for private landowners to conserve biodiversity. These include the federal Agricultural Conservation and Conservation Reserve Programs, as well as cost-sharing programs in some states. Some private landowners have begun considering biodiversity because of aesthetics and moral views and a long-term view of the effect of their land-use on natural resources. Many, however, are faithful to traditional views of their responsibility to land—land must be used to be useful—and can only be enticed with economic stimuli. Education is an important way to instill a more effective land ethic (Leopold 1949); thus, education in workshops and other forums completes the approach.

A GOAL FOR MANAGEMENT OF BIODIVERSITY

The first task in describing a strategy for private land management is to establish an overall goal. This goal must be both demanding enough to realize conservation of biodiversity defined as "the variety of life and the processes that support it" (Keystone Center 1991) and sufficiently practical and flexible to accommodate the range of objectives a private owner has for a property. Such a pragmatic goal recognizes that most land-use options affect biodiversity in one way or another, but tries to avoid overall adverse effects at all levels of biodiversity—genetic, species, ecosystem, and landscape.

In the simplest terms, a broad, flexible goal means that responsible private land management does not prevent habitat alteration or even loss of certain species from a property. Correct decisions about land-use activities instead depend on considering their effects in a broader context. For example, the disappearance of white-tailed deer (*Odocoelius virginiana*) from a site in the eastern United States has only local implications, because the species occurs commonly throughout most of the region. When examined from a regional perspective, the species is abundant. Conversely, restoration of a population of globally significant plants or protection of an endemic invertebrate would receive high priority under such a goal.

After examining several potential goals, including such ambitious targets as the National Wetlands Policy Forum's "no net loss" standard (NWPF 1988), O'Connell and Noss (1992) recommended the goal: *maintenance of viable populations and natural distributions of native species and communities in the regional landscape* for private land managers. This standard is broader and more complicated than the scope of traditional site-specific plans for wildlife management. It encourages managers to examine elements of biodiversity on a property from a wider perspective and as part of a landscape of communities and ecosystems on a number of scales. At the same time, it allows more flexibility on individual sites and upholds management actions more appropriate to conservation of biodiversity from a regional and global perspective. For example, a private landowner who chooses to maintain rangeland through prescribed fire instead of mowing would contribute to regional conservation goals.

Under this strategy, appropriate mitigation and restoration become tools that allow certain exploitive uses of land to continue while minimizing or potentially offsetting their harmful effects on biodiversity. Species can be reintroduced, habitat restored, and deleterious factors such as exotic species or excessive predation, controlled. These mitigative actions might produce gains that from a regional or global perspective offset other losses from development or intensive resource use. For example, reintroduction and enhancement of a globally-imperiled plant species on part of a site might more than offset the effects of land clearing on more common native species elsewhere on the property from a global perspective.

THE MANAGEMENT MODEL

O'Connell and Noss (1992) proposed a systematic, stepwise process for integrating biodiversity into private land-use decisions (Figure 42.1). This frame-work allows a manager to examine the significance of a property in its appropriate context and suggests suitable objectives for several potential site conditions. A detailed treatment of the land-use model is beyond the scope of this paper. However, it is the core of the World Wildlife Fund project and so is summarized here.

It is worth noting that steps in the O'Connell and Noss frame-work are by no means rigid, especially on properties that approach landscape size (i.e., tens of thousands of hectares or larger). Managed landscapes are by definition a mosaic of communities with different characteristics and in various states of disturbance (Wilcove 1989). Land managers can collect and evaluate biodiversity data on part of a landscape while simultaneously planning management, carrying out land-use activities and evaluating the effects of land-use in other areas.

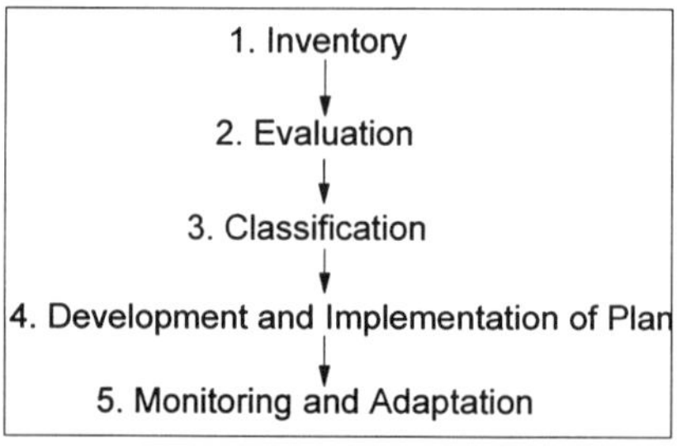

Figure 42.1. Five-step method for the management of biodiversity.

Inventory

The initial procedure in developing land-use plans that consider conservation of biodiversity is a site inventory. Ideally, this activity should progress from general site information to data about the occurrence of specific species and communities. Managers should gather information from aerial photos, from topographic and soils maps, from state Natural Heritage Inventory (NHI) databases, and from government and local sources. A brief survey might be necessary at this stage to identify relevant species and vegetative communities. Ecological processes at both local and landscape levels that are important to maintenance of the site should also be a component of the survey.

Many parcels, such as isolated and degraded fragments that contribute neither native species nor landscape processes, can be classified for management purposes without further effort and expense at this stage. These sites have less to offer global biodiversity conservation than those important to ecological processes or harboring significant elements of biodiversity. They can still contribute to an overall landowner policy of biodiversity conservation, however, as potential restoration sites for native species, thus offsetting losses elsewhere. If the initial evaluation discloses evidence of rare or globally significant elements of biodiversity, a more detailed survey might be necessary to categorize a site further. The results of detailed field study about the occurrence and characteristics of such elements will help plan land-use to minimize and mitigate its effects.

Evaluation

The next step in developing a land-use plan is to evaluate the results of the biodiversity inventory in an appropriate context. No definitive source or system exists to evaluate the importance of a site to regional and global biodiversity, although several are being explored at the national policy level, including the proposed National Biological Survey and National Institute for the Environment. This deficiency can be a serious stumbling block to effective management. However, several tools contribute some guidance.

At the species level, federal and state endangered species lists provide initial direction. These lists also include candidate species that may be no less imperiled, but are not yet legally protected. The Red Data Books of the International Union for the Conservation of Nature are also a useful reference at the species level and in foreign countries. Species in these volumes are not always legally protected.

Simple lists of imperiled flora and fauna do not include many important aspects of biodiversity. The most significant among these missing items are ecological communities, and species that, although declining, have not reached the biologically arbitrary threshold of legal endangerment. These lists also neglect elements of genetic and ecosystem diversity and ecological processes vital to regional and global biodiversity. The ability to evaluate the significance of ecological communities is crucial to a comprehensive strategy for conservation of biodiversity at all levels.

The Natural Heritage Inventory Database (NHI) of The Nature Conservancy is the most accurate system available to the private land manager for evaluating elements of biodiversity. The NHI is maintained through cooperative partnerships and includes globally and regionally important flora and fauna, and many declining species. It frequently includes populations of species and ecological communities not found in other references. The NHI, however, is incomplete and the meaning of negative data is obscure. Lack of records of rare species from a site can either mean that none exist there or that the area has not been surveyed.

The NHI assigns each community and species in the database a ranking code corresponding to its global and regional status. "G" delineates global status, and "S" state (regional) status. A species or community both globally and regionally at high risk of extinction would be coded G1S1. G2S1 denotes an element less rare globally, but very rare in a state. The lowest rank, G5S5, corresponds to an element common regionally and globally. The NHI generally does not assign a rank to exotic species, degraded communities, or species thought not to be in need of conservation.

Site Classifications

O'Connell and Noss (1992) suggested a set of classifications based on the biological character of a property. Each class provides a management objective consistent with the overall goal of maintaining biodiversity from a regional and global perspective. These classes result in a frame-work to help develop land-use plans that include conservation of biodiversity. Some sites might not fit neatly into a specific category, but the following classes encompass most scenarios and provide guidance on the rest. Using these classifications, managers can more easily develop and implement the biodiversity component of site plans:

Class 1—*Sites Supporting Globally Significant Elements*

- Federally Protected Species
- NHI G1 through G3 elements and high-quality examples of lower-ranked communities
- IUCN Endangered and Vulnerable Species

Management Objective: Preservation and Recovery of Significant Elements

These parcels contain elements most in danger of loss. They dictate management with the goal of preserving and recovering significant elements. Such elements are so imperiled that their persistence must be of paramount interest in land-use plans. Managers should attempt to enhance the survival prospects of those elements by, for example, increasing the number of individuals of a certain species, or improving the health of a community by minimizing and mitigating adverse effects, such as logging, grazing or other intensive use.

Class 1 sites occur in many forms. They can encompass an entire endangered or threatened community, or comprise only a part of the community. They might harbor one of the last known populations of a plant species, or be a migration stopover for an endangered bird species. Various scenarios require different management for net benefit. Regulation, such as under state and federal endangered species acts, might also either mandate or prohibit some land-use options and might include both on and off-site management, such as relocation of a species or restoration of a community.

As part of an overall strategy of maintenance of biodiversity from a regional and global perspective, preservation and recovery of species and other elements in Class 1 might appropriately occur at the expense of other, more common species. For example, if a rare plant species occurs on site and is harmed by deer browsing, control of deer numbers is entirely appropriate.

Class 2—*Sites Supporting Regionally Significant Elements*

- State Endangered, Threatened and Candidate Species
- NHI S1 through S3 Species and Communities

Management Objective: No Decline or Degradation of Significant Elements

Elements supported by Class 2 sites are only slightly less imperiled than those in the previous class. This standard often includes species or communities at risk of extirpation from a broad geographic region, but not as close to extinction as G1-3 or federally endangered elements. In the absence of higher priority elements, communities, species or populations on Class 2 sites should receive preference in management.

The suggested goal for Class 2 sites is no decline or degradation of significant elements. Land-use should not decrease populations of a species or deteriorate native communities. As in Class 1, regulation might restrict or limit land-use.

Class 3—*Sites Larger than 500 Hectares Supporting Naturally-Occurring Species Assemblages*

Management Objective: Maintain Native Species Composition and Ecological Process

Flexibility of land-use alternatives increases under the third management classification. This class is not based on the occurrence of critical species or communities, but such elements might occupy parts of these sites. Class 3 sites are instead categorized by their size. They are generally large enough to encompass groups of communities and regional ecological processes like hydrology or fire. Often, areas important to significant elements can be managed expressly, and the remainder of the site managed using the general principles in Table 42.1.

Conservation strategies depend on site-specific attributes and landscape context, such as connection to other habitats (Forman 1983; Noss 1983, 1987a; Harris 1984). Whereas intensive land management might artificially augment local habitat variety and species number, this does not necessarily enhance biodiversity. Loss of rarer elements as weedy species increase or alteration of landscape context might result in a net loss of biodiversity (Noss 1983). The manager must continually keep in mind the broad conservation perspective discussed earlier.

Ecological restoration lends an additional measure of flexibility to management of larger properties such as those in Class 3. Restoration of a degraded community or reintroduction of a rare species might provide benefits that offset impacts from intensive use elsewhere (Jordan et al. 1988). Restoration should parallel natural processes of recovery from disturbance and use only native species (Jordan 1991).

Table 42.1. Principles of Stewardship Common to All Private Land Biodiversity Management (from O'Connell and Noss 1992).

- Maintain native species.
- Coordinate site management with surrounding landscape.
- Protect factors and processes critical to biodiversity (e.g. topography, natural disturbance and succession).
- Maintain landscape connectivity.
- Identify management alternatives mandated or prohibited by law.
- Minimize or eliminate artificial edge effects.
- Control non-native and weedy species using natural or mechanical processes where feasible.
- Minimize use of chemicals.
- Limit disturbance in preserved areas to activities compatible with site qualities.
- Protect important or sensitive habitat features such as wetlands, certain forest cover, unique topography, caves and riparian zones.
- Explore habitat enhancement or restoration to offset impacts.
- Acquire additional acreage to offset unavoidable losses or land-use incompatible with biodiversity conservation.
- Retain an ecologist when necessary. Work with federal, state, local and non-profit biologists when possible.

Class 4—*Relatively Isolated Sites Less Than 500 Hectares with Some Natural Communities*

Management Objective: Maintain Species Composition and Ecological Processes

This class includes parcels of lesser value to biodiversity due to small size or lack of landscape connectivity. Sites in this class might be important habitats for declining or sensitive elements, such as forest interior birds or endemic plants, but do not usually necessitate intensive stewardship for specific elements. The recommended objective for management is perpetuation of native species composition, ecological processes, and other significant site attributes.

Accomplishing the objective for this class often demands more intensive management to achieve conservation objectives. Managers might need to encourage ecological processes, such as fire, important to maintaining site qualities or control those, such as natural succession, that can alter a site. For example, prescribed burning might be necessary to perpetuate a fire-adapted community. Restoration might also allow these sites to play a more significant role in conservation of biodiversity conservation by increasing their value in the regional landscape (Noss 1987b).

Class 5—*Isolated Sites Less Than 100 Hectares Supporting No Elements of Special Significance*

Management Objective: Maintain Native Species

Although by definition these parcels do not contribute greatly to regional or global diversity, they can serve important local functions. They provide opportunities for open space and recreational opportunities, and accommodate intensive human use that might be inappropriate for biodiversity on more sensitive sites. They can also become important in the future because of loss of currently common elements elsewhere.

Stewardship should prevent introduction of exotic and invasive species. Manipulation of habitats is a possible, and in some cases desirable, option for management. Installation of feeding stations, nest boxes, brush piles, artificial ponds, and diminishment of other limiting factors for native species abundance might benefit species of recreational or aesthetic interest such as songbirds or game. Site plans can also include nature trails, birdwatching blinds, and bike paths that can provide significant public relations benefits to landowners desiring them.

Class 6—*Isolated, Ecologically Degraded Parcels with Few Native Species and No Natural Community*

Management Objective: No Further Degradation; Restoration Only if Feasible

Degraded sites with few native species and no natural communities isolated from the surrounding landscape by development, agriculture or other human habitats offer little to regional or global biodiversity. However, the potential for restoration with the predisturbance community as a target makes them attractive alternatives to offset impacts elsewhere. True ecological restorations are difficult and intensive processes (Josselyn et al. 1990). Restorations that seek to achieve a socially or aesthetically acceptable level of revegetation accomplish little to conserve biodiversity and might further damage biodiversity by introducing invasive species (Jordan et al. 1988). Management should, therefore, simply avoid additional loss of native species or deterioration of native communities.

Monitoring

Monitoring is essential to land-use planning that includes conservation of biodiversity (Noss 1990). It is the only way a manager can determine the success or failure of a plan. Criteria to judge whether activities accomplish objectives are vital. For example, if increasing the local population of a globally significant plant species is the goal, the monitoring gauge might be a population size and survivorship rate that assures a high probability of long-term survival (Noss 1990). On the other hand, if the maintenance of a complex of ecological communities and processes on a large site is the objective, the continued presence of several species indicative of those communities might be the best standard.

A detailed, ten-step monitoring process was outlined by Noss (1990) which O'Connell and Noss (1992) adapted for the private-land context. It recommends carefully selected indicators, a well-designed sampling scheme, and methods to evaluate trends and make necessary adjustments in management treatments where necessary. The reader is asked to consult these references for a more detailed treatment of monitoring.

CASE STUDY: THE BAY CHECKERSPOT BUTTERFLY (*Euphydryas editha bayensis*)

The bay checkerspot butterfly was once widespread in central California (Murphy and Ehrlich 1980). Habitat conversion and loss of host plants have reduced numbers to the point that it has been federally listed as threatened since 1987 (Murphy and Weiss 1988). Before this legal protection was adopted, however, the most extensive known habitat of the insect was also the planned site of one of the largest landfills in North America, on 1,200 hectares owned by Waste Management, Inc. (Murphy 1988). The butterfly was restricted to only a few other fragmented habitat parcels throughout its former range. Clearly, siting the landfill without consideration of the butterfly would impair its chances to survive.

Its characteristics place the landfill site in Class 1 of the frame-work outlined by O'Connell and Noss (1992). Initial surveys showed the landfill would disturb 15 % of the habitat on site; about 75 hectares would be altered at any one time. Nearly all existing habitat on the property was disturbed by nonnative species invasions and other limiting factors. The objective for Class 1 dictates that persistence of the butterfly species should take priority in management of the site.

Anticipating legal protection for the butterfly, but eager to conserve it for other benefits, such as public relations, that the action would provide, Waste Management prepared a conservation plan that allowed development of the landfill without diminishing the likelihood of survival and recovery of the butterfly. The U.S. Fish and Wildlife Service accepted and approved the document in 1986 as providing adequate conservation for the subspecies.

The plan contained provisions to identify and protect or enhance significant habitat areas on the site and a trust fund for acquisition, restoration and studies, to which Waste Management, Inc. contributes $50,000 annually (Murphy 1988). As an initial step, the landfill was sited away from high quality habitat. The company leased approximately 75 additional hectares of prime habitat adjacent to the landfill site and instituted a management scheme there. Management included controlled grazing of the habitat to repress nonnative grasses.

The closed cells of the landfill became part of an experiment to test the potential for restoration of the plant community that supports the butterfly. The trust fund provided money to test irrigation, mechanical disturbance, fertilizer, artificial propagation and seeding of host plants to enhance existing habitat value. Consultants conducted studies of the requirements for a viable population of butterflies and regional potential for recovery including mapping of habitat, habitat quality assessments of other parcels in the area and evaluation of potential sites for reintroduction of the insect (Murphy 1988).

Monitoring indicates that conservation measures implemented so far have offset the effect of development of the landfill on the butterfly (D. D. Murphy pers. comm.). In return, Waste Management avoided regulatory complications and gained significant public relations benefit. In fact, the sides of garbage trucks streaming to and from the landfill since 1987 conspicuously display a picture of the butterfly.

Clearly, had the butterfly not existed on the site, management plans for the landfill would have been different. Under Class 3 of O'Connell and Noss (1992) for example, consideration of biodiversity might have taken the form of maintaining species composition and ecological processes in unimpacted areas and compensating for permanent alteration, rather than focusing on a particular species. This case study demonstrates a fundamental theme of managing private lands to conserve biodiversity: economic activity can be reconciled with biodiversity when management is conducted from the proper perspective and when the landowner appreciates the need for conservation.

FIELD TESTING

Lack of understanding about the implications of conservation of biodiversity for on-the-ground management is a significant barrier to adoptions of a mechanism, such as the one outlined above by private landowners. Therefore, a crucial component of any strategy to encourage such management is education through demonstration of the model. The case study of the bay checkerspot butterfly is a good example of the solution adopted by one private landowner for a specific parcel. Nevertheless, a range of scenarios and land-use situations is necessary to illustrate the model's applicability.

Corporations manage extensive landholdings in the United States. Accordingly, corporations were selected as the audience to test the utility of the O'Connell and Noss (1992) model by World Wildlife Fund. Corporations offer several attractive elements for this analysis including their hierarchical nature of decision making, a wide range of land types and uses, funds (generally), internal expertise (occasionally), and a sensitivity to the perceptions of the public and regulatory agencies.

World Wildlife Fund is field testing the model management frame-work by recruiting companies whose landholdings represent a variety of land-use patterns. This ongoing process will soon result in several test cases. The goal of this first approach is to promote adoption of the overall goal suggested by O'Connell and Noss (1992) as the company policy toward biodiversity, and the model conservation frame-work as the way to implement that charge. Lack of understanding by corporate upper management about the definition of biodiversity and rationale for its protection has hindered that process. Simple education of corporate management about the meaning of biodiversity has, therefore, become a secondary objective of the project.

The O'Connell and Noss (1992) model also contributed to the work of the President's Commission on Environmental Quality (PCEQ). PCEQ adopted the World Wildlife Fund private lands biodiversity project as part of its natural resources agenda in late 1991. Fifteen major companies maintain PCEQ membership, offering an excellent opportunity to reach a broad corporate audience. PCEQ members also represent a range of land-use scenarios including intensive forestry, waste disposal, investment real estate, and manufacturing (PCEQ 1993)

Through a series of pilot studies, the PCEQ examined a number of alternative strategies to conserve biodiversity, among them the O'Connell and Noss (1992) model, on individual properties offered by member companies. This effort provided additional case studies of conservation of biodiversity on specific sites.

The World Wildlife Fund and PCEQ projects demonstrated some difficult hurdles that must be overcome before conservation of biodiversity becomes a conventional element of private land-use planning. These will be discussed further in the final section of this paper.

A NOTE ON INCENTIVES

For most private landowners, especially those whose success is influenced by public perceptions and marketing or those engaged in business use of their property, incentives are an important reason to include conservation of biodiversity in their land-use decisions. Incentives can be either economic or non-economic. Direct economic incentives tend to be the most attractive, yet least available alternative. It is, therefore, difficult to convince landowners who demand that direct costs associated with biodiversity management be balanced by direct financial benefits of the value of conserving biodiversity. In addition, whereas many of the broad benefits are public, the costs are private. Some owners simply might not be able to accomplish such actions even if they believe in their value. The lack of a system of economic incentives is a fundamental barrier that requires significant attention.

Occasionally, tax benefits associated with options, such as conservation easements, donation, or fee simple acquisition can encourage activities beneficial to biodiversity. In other instances, fee systems, such as leasing of access rights, private permit programs and federal or state cost-sharing programs such as the Conservation Reserve Program provide some advantages (Yarrow 1990). Yet, although these options are useful mechanisms for achieving conservation in some instances, they do not begin to instill a new biodiversity land ethic, such as described by Leopold (1949).

For some private landowners, non-economic or indirect economic incentives might be worthwhile sufficient. At a workshop for private land managers held by World Wildlife Fund in November 1991, participants developed a list of 26 incentives to conserve biodiversity (WWF 1991). They were in three major categories: ethical, political, and indirectly financial. Incentives ranged from the moral perception that biodiversity conservation is an action demanded by society to the practical issue of getting ahead of the regulatory curve. Others felt that the benefits from public relations were not only measurable, but highly valuable. However, the group concluded that although the standard is changing, non-economic incentives have traditionally been a poor motivation in the face of measurable direct economic costs.

LESSONS LEARNED

Efforts by World Wildlife Fund to date have been particularly instructive about the difficulties in influencing conservation of biodiversity on private lands. It is clear that the changes site managers must make to institute this and other strategies are often in perspective and philosophy, rather than fundamental shifts in the way they conduct business. Yet, common land management ideology is deeply rooted in tradition and can only be altered through education and demonstration. More such ventures are needed to begin to change the perception of private landowners about biodiversity and its meaning for on-the-ground management.

A primary concern of private landowners and managers involves the cost of conservation of biodiversity. Participants in the World Wildlife Fund workshop listed a number of ways to overcome those costs, including conservation easements, selling or donating property, zoning changes, registry programs such as the Land Legacy program of the U.S. Forest Service or The Nature Conservancy Registry, or even analyzing and calculating actual public relations benefits (WWF 1991). These solutions are far from comprehensive. There is an obvious need for more research and policy direction in this area.

Another obstacle to changes in land management is tradition. Conservation theory has taken quantum leaps in recent years, yet the complex concept of biodiversity conservation has not been translated to implementation on the ground. Land managers need involvement and assistance from the research and academic sectors to change old habits. Some private companies have made significant progress toward reforming traditional practices. One corporation provides fiscal incentives for managers by making biodiversity conservation a performance evaluation criterion (C. T. Grigalauski pers. comm.). Another forest products company conducts short courses for managers about the issue and emerging theory (W. Ticknor pers. comm.).

A basic problem for private land management to date has been the lack of technical and biological expertise on the part of many private landowners. Some, such as large forest product companies, employ many biologists and land managers. Others, such as individual landowners or most private companies, have difficulty integrating the specialized concept of biodiversity into everyday decisions. Participants at the World Wildlife Fund workshop suggested several ways for private landowners to overcome lack of technical expertise. They proposed partnerships with non-profit organizations and agencies, working with university biologists and other researchers, examining and effectively using lay expertise among employees, taking advantage of services such as the Natural Heritage Inventory, and hiring consultants (WWF 1991).

A final lesson from the efforts by World Wildlife Fund to engage private landowners in conservation is one of ethics. Many landowners are concerned that sincere attempts to integrate conservation of biodiversity in land-use will place them at a disadvantage with competitors who choose not to do so. Under the current system of incentives, this fear can be justified. Others are eager to profit from public perceptions of environmentalism without committing to genuine actions. This has led to public and agency mistrust of private landowners, especially large corporations. Private landowners and managers recommended a record of positive action, partnerships, dialogues, and case studies as the long-term solution to this essential issue (WWF 1991).

CONCLUSIONS

It is clear from the initial work of World Wildlife Fund that there is considerable progress yet to be made in conservation of biodiversity on private lands. A strategy such as outlined in this paper, and detailed in O'Connell and Noss (1992), is one possible mechanism. Some private landowners, including major corporations, are beginning to institute such management plans for their activities and land holdings and participate in dialogues intended to improve the atmosphere for their efforts. The greatest need for improvement, however, is in the education and public policy arena, to change traditional stereotypes, create incentives and build partnerships that foster innovative land management. Without concentrated effort, the opportunity to include private lands as part of an overall strategy to conserve biodiversity will be missed.

LITERATURE CITED

Forman, R. T. T. 1983. Corridors in a landscape: Their ecological structure and function. *Ekologiya* (CSSR) 2:375-387.

Hansen, A. J., T. A. Spies, F. J. Swanson, and J. L. Ohmann. 1991. Conserving biodiversity in managed forests. *BioScience* 41(5):382-392.

Harris, L. D. 1984. *The Fragmented Forest: Island Biogeography Theory and the Preservation of Biotic Diversity*. University of Chicago Press, Chicago, Ill.

Jordan, W. R., III. 1991. Ecological restoration and the reintegration of ecological systems. In *Proceedings of the Workshop on Bioscience and Society*. Dahlem Conference, Berlin, Germany.

Jordan, W. R., III, R. L. Peters, and E. B. Allen. 1988. Ecological restoration as a strategy for conserving biological diversity. *Environ. Manage.* 12(1):55-72.

Josselyn, M., J. Zedler, and T. Griswold. 1990. Wetland mitigation along the Pacific coast of the United States. In *Wetland Creation and Restoration: The Status of the Science*, eds. J. A. Kusler and M. E. Kentula, pp. 3-36. Island Press, Washington, D.C.

Keystone Center. 1991. The Keystone report on biological diversity on federal lands. The Keystone Policy Center, Keystone, Colo.

Leopold, A. 1949. *A Sand County Almanac*. Oxford University Press, Oxford.

Murphy, D. D. 1988. The Kirby Canyon conservation agreement: a model for the resolution of land-use conflicts involving threatened invertebrates. *Environ. Conserv.* 15(1):45-48.

Murphy, D. D., and P. R. Ehrlich. 1980. Two California checkerspot subspecies: one new, the other on the verge of extinction. *J. Lepidopterists' Soc.* 34:316-320.

Murphy, D. D., and S. B. Weiss. 1988. Ecological studies and conservation of the threatened Bay checkerspot butterfly, *Euphydryas editha bayensis*. *Biol. Conserv.* 46:183-200.

Noss, R. F. 1983. A regional landscape approach to maintain biodiversity. *BioScience* 33(11):700-706.

Noss, R. F. 1987a. Corridors in real landscapes: a reply to Simberloff and Cox. *Conserv. Biol.* 1:159-164.

Noss, R. F. 1987b. Protecting natural areas in fragmented landscapes. *Natural Areas J.* 7:2-13.

Noss, R. F. 1990. Indicators for monitoring biodiversity: a hierarchical approach. *Conserv. Biol.* 4(3):355-364.

National Wetlands Policy Forum (NWPF). 1988. *Protecting America's Wetlands: An Action Agenda*. The Conservation Foundation, Washington, D.C.

O'Connell, M. A., and R. F. Noss. 1992. Private land management for biodiversity conservation. *Environ. Manage.* 16(4):135-151.

President's Commission on Environmental Quality (PCEQ). 1993. Partnerships to progress: The report of the President's commission on environmental quality. Council on Environmental Quality, Washington, D.C.

Pimentel, D., U. Stachow, D. A. Takacs, H. W. Brubaker, A. R. Dumas, J. J. Meaney, J. A. S. O'Neil, D. E. Onsi, and D. B. Cornelius. 1992. Conserving biological diversity in agricultural/forested systems. *BioScience* 42(5):354-362.

Thomas, J. W., and H. Salwasser. 1989. Bringing conservation biology into a position of influence on natural resource management. *Conserv. Biol.* 3(2):123-127.

Wilcove, D. S. 1989. Protecting biological diversity in multiple-use lands: Lessons from the U.S. Forest Service. *Trends Ecol. Evol.* 4:385-388.

World Wildlife Fund (WWF). 1991. Workshop summary: managing private lands to conserve biodiversity. World Wildlife Fund, Washington, D.C.

Yarrow, G. K. 1990. Wildlife management: incentives for the private landowner. Cooperative Extension Service, Clemson University, Clemson, S.C.

43

CREATING POLICY ON GENETIC DIVERSITY

Gene Namkoong

Department of Forest Science
Faculty of Forestry
Vancouver, British Columbia
Canada

Those who attended this conference represent widely varying interests, so it would be surprising if we agreed on all aspects of a genetic diversity policy for land management agencies. However, we have many points of agreement. We agree that genetic diversity is desirable, and that it can be influenced by management. We, therefore, agree that a genetic diversity policy is needed to define goals, procedures, and priorities.

My objective in this paper is to explore the meaning of diversity and to propose some goals and priorities that should be included in a management policy. The underlying situation is complex and involves very significant moral and practical dilemmas. Like Norton (1987), therefore, I have chosen to justify the preservation of genetic diversity and the conservation of species primarily on utilitarian grounds. In doing so, I do not mean to suggest that there are no other significant bases. I do so to simplify the case sufficiently for current analysis and to achieve the broadest possible agreement. I show that a robust genetic diversity policy can be based on human utility values, and it is one that we can all support. I suggest that by accepting such a program now we can make significant progress while we address moral questions that lead to disagreements.

BACKGROUND

Species diversity and genetic diversity are not synonymous. The goals and means for managing genetic diversity are not the same as those for general species diversity or for community or landscape diversity. Each level has elements and dynamics that, while inter-related, are distinct and require different methods of management. For genetic diversity, management options are wide. Gene management can be intensive (as in crop and animal breeding), nonobtrusive (as in monitoring reserves for ecologically important organisms), or both (as in programs for rescuing rare and endangered species). *In situ* and *ex situ* technologies can be applied, and both economic production and species preservation can be goals.

Public agencies like the Forest Service have many constituencies with different goals and priorities. If those constituencies agree to disagree on some issues, it might still be impossible to develop single, integrated policy objectives. Otherwise, the agency might have to settle for a collection of nonintegrated policies for different kinds of species and different levels of biotic organization and hope that some hidden hand will ensure that the sum of all activities will satisfy all goals.

I am hopeful for sufficient temporary agreement to permit progress on an integrated approach. Despite areas of potential disagreement, I think we can all agree that general diversity and genetic diversity are important. We agree that diversity at all levels of biotic organization is a necessary part of a functioning biota. Diversity or variation is necessary for biotic systems to evolve. At the genetic level, variance is a key parameter that determines the rate of adaptation to selective pressures. Enshrined in Fisher's Fundamental Theorem of Natural Selection is the warning that there is no evolution without genetic variance. Furthermore, differences and redundancy of types are needed if losses due to either deterministic or stochastic effects are to be replaced. Thus, there is little disagreement that diversity is a general necessity.

Part of the problem in defining a policy is that diversity is often seen as a goal rather than a method for achieving a goal. If diversity is the designated goal, every variation and every instance become important and the loss of any element, with the possible exception of human beings, threatens large-scale collapse. Rare species are often endowed with high inherent value, and the genetic diversity of useful but common species are considered less important to conserve. If conserving rare species is our goal, then we must define what kinds of rarity are significant, what measures can be used, and what methods of management are preferred. Many of these questions are explored in depth in other volumes (e.g., Falk and Holsinger 1991), but rarity itself is clearly not a simple concept that leads to simple policies.

The broad inclusiveness of a general concept of diversity also generates conflicts between human uses of the biota and its conservation. Humans are destroying whole ecosystems and simplifying others, increasing risks of species extinction and whole **biota** losses. However, if elimination of human activity becomes a goal, one must ask who benefits from management for diversity. In this paper I try to distinguish between goals and the means of achieving goals, and to offer an approach that could assuage some of the unnecessary conflicts that have arisen. Without a general concept of diversity that can encompass different conservation values, a public agency's policy degrades into moderating conflicts rather than maximizing some good.

If we can enunciate a general policy on the management of genetic diversity, we may be able to generate the support needed for multiple methods of reaching goals, to reconcile differences in management methods, and even to integrate methods for achieving those goals. In the search for a single policy, several approaches have been proposed. One is based on the existence of a strong hierarchy of biological organization and, therefore, also of management effects. The biological hierarchy, if accepted, implies that managing the genes of some target organisms will entrain a sufficient management system for all other levels of biotic organization and all other organisms. Focusing on management at the gene level, Frankel (1983) would target genetic management to species that are critical to the stability of communities or ecosystems. With that system, it would be possible to determine if genetic intervention or genetic management is needed to maintain the viability of target species, and hence the viability of ecosystems and all the organisms included in them would be ensured.

An alternative approach that focuses on management at the ecosystem level is based on the possibility that managing targeted and diverse landscapes will entrain a sufficient system for all organisms. An extension of this concept of establishing reserves, but managing them for ecological values, as proposed by Schonewald-Cox (1983), might be to assess certain targeted species for their potential as indicators for entire reserves. Conclusions about the capacities of reserves to protect the genetic diversity of all species would be based on these indicator or flagship species. Both of these approaches could be valid, but any method would have to be assessed on the basis of whether the goals of management are inclusive of all values and if the methods lead to feasible management priorities.

In this paper, I attempt to base policy goals on the narrow basis of utility of genetic diversity to humans. I largely ignore values to other species or to ecosystems and communities themselves. I follow the approach taken by Norton (1987) to justify species conservation on the basis of utility, but I focus on genetic diversity and, on that basis, derive different priority systems. I assume that if a program for managing genetic diversity among all species can be supported on this narrow basis, any other broader basis of justification would differ only in the priority ordering of management efforts. Basing policy on utility also permits an ordering of management priorities by value of the genetic resource.

To guide management, we must establish priorities according to the probable success of management systems in achieving defined goals. Although goals might be very inclusive, there are always limitations on funds, efforts, and personnel. The more inclusive the policy goals are, the more need there is to establish priorities. Without such guidance, management may be unduly influenced by factors that are irrelevant or even inimical to policy goals. The following discussion first considers policy goals, then considers the implications for setting priorities.

POLICY GOALS

From previous papers given in this conference, it is obvious that a complex evolutionary dynamic gives rise to our present endowment of genetic diversity. It is also obvious that genetic diversity can be affected by management actions or inactions, either by intent or neglect. Therefore, we are forced to consider what goals can be reasonably enunciated and what management priorities can be deduced for their attainment.

We know that human activities have profoundly affected the evolution of most species and ecosystems for thousands of years, and that human effects are as natural as any other factor of the physical and biotic environment. Human effects may be stronger and more devastating than most, and different in that they are purposeful and to some extent controllable by will towards some end. The policy question is how to direct those activities among alternative courses of action to achieve objectives that would fulfill our goals.

To some people, the goals for genetic diversity or general biodiversity may seem too obvious to need definition. Consider that for some agencies, including the Forest Service, endangerment of a species affects its value and justifies the use of heroic rescue measures. As a management method however, species rescue would not be permissible in agencies that carry nonintervention as a goal. Either rescue or non-intervention might conflict with an agency goal to pursue human welfare. If policy goals differ, surely the choice of management method and the way we measure management impacts will also differ. Thus, once past the basic assumption that there is a good in diversity, we immediately face problems in defining whose good is served and how any particular measure of diversity is better or more fitting than any other.

A major problem is to define the level of biotic organization that is the subject of concern. The gene is at the most reduced level of potential concern, but the conservation of particular genes is seldom seen as the purpose of management. We do not value genes in the same sense as we do entire organisms or groups of organisms. Genes are complex molecules without consciousness or capacities for independent function. The genes themselves, or any special combination of them, obviously do not serve as ends of conservation in themselves. Rather, they are necessary elements for serving some larger good. Furthermore, because genetic diversity is an ensemble property of a group, policy goals are designed for the good of the group rather than of the individuals making up the group. It must be the good of the ensemble or the good of an entity of which the ensemble is a part that is the policy goal. Thus, even though the good of individual people or organisms may be an end served by genetic diversity, the genetic diversity that we seek to manage is a populational property whose benefits are indirect. Therefore, the policy goals for genetic diversity can include the welfare of individuals, populations, and species, but there is nothing about the level or structure of diversity itself that is of inherent value and which would inhibit how we manage it. Furthermore, as well noted by Brown and Briggs (1991), saving all alleles or allelic combinations that exist now or existed in the past is not feasible and, therefore, management for such an illusive goal is impossible. In addition, the present levels and structures of genetic variation cannot provide normative guidelines for management goals. The present status of diversity provides evidence of what is evolutionarily feasible and can indicate what might not be feasible, but it cannot define for us what should exist. Therefore, the present state of genetic diversity serves as a model, but achieving, maintaining or stabilizing the present state would not satisfy any deeper good. For policy goals, we must look beyond what exists.

The present structure of genetic variation has been the result of deterministic as well as stochastic processes, and has been influenced by the unique history of each species. The balances among selection, population size, and migration have been studied intensively, and the effects of any one factor alone are generally inadequate to understand or predict evolution. The immediate implication for defining goals of management is that the current state of genetic variation cannot be expected to reflect only selection effects. Therefore, the present structure of genetic variation does not reflect a perfect adaptation. Populations carry mutational, inbreeding, and outbreeding loads; they lose alleles and genotypes by stochastic sampling; and local selections are swamped by matings among neighborhoods.

Selection can strongly affect the amount and structure of genetic variation. Long-term selection can build continuous responses on the actions of new mutations, and divergent selection can easily and vastly increase the total genetic variance. Selection responses can, therefore, be the result of action at different gene loci and can use different gene actions. Hence, selection for population differentiation can be a powerful means of managing genetic diversity. Both directed breeding or the erection of mating barriers and the segregation of natural reproduction into mating demes might often be feasible, and would be a far easier means to increase genetic variance than by randomly increasing the population size of a single intermating population. Thus, when genetic management is feasible, it can be orders of magnitude more effective in maintaining or increasing the genetic variance in any set of traits and in capturing alleles than maintaining large, random mating single populations. Maintaining large populations in reserves can often be a preferred method of management for many reasons, but genetic management efficiency is not one of them. The extant structures of interpopulational diversity might well be useful starting points for managing genetic diversity, but they too are the particular results of mixed evolutionary forces and stochasticities and are not in themselves goals to be sought.

Policy goals for genetic diversity can be most easily derived from the more general goal of securing the welfare of present and future generations of humans. Norton (1987) showed that the values of species to humans can include both utility and existence. Even on the narrow basis of human utility alone, all other species can be considered to have some value. Utility values include not only direct consumption and nonconsumption market value but also may include amenity, recreation, scenic, and other noncommodity values. Enhancing existence values might also be a policy goal, but it would be difficult to adjudicate. In species that have human utility, genetic variation is necessary to their survival and value. In many species, genetic variation may be of no direct value to humans, but it may be indirectly valuable because it is necessary for survival and evolution of directly valuable species.

Despite our human utility standard, therefore, we probably would not be manipulating genetic diversity for specific human ends in most species. Our management system would include reserves for scientific research, for recreation, and for scenic values, all of which are human goals. Among species with no known human utility, managing genetic diversity for more than existence would exceed our stated goal. Because the goal of managing genetic diversity cannot be for the genes or individuals involved, the question revolves around whether a species can serve as an end in itself, have a good of its own above that of its constituent individuals, and whether the good of the species *per se*, can be understood well enough that it can be a goal of genetic management, independent of its value to other individuals or species. If a species can have inherent value as a species and the species' good can be a goal of genetic management, the only implication for management that is implied is that the future evolution of the species not be unnecessarily constrained. Actions to maintain genetic diversity in its present form or structure are not implied.

We might, therefore, conclude that for genetic diversity, policy goals would certainly provide for a wide range of species, some with commodity and many with noncommodity values. All species might be included, but intensive gene management would be reserved for species that can possibly affect the good of present and future humans. A policy goal could include the welfare of other species, populations, and individuals, but could not include the welfare of individuals of other species unless those individuals have existence values that are affected by genetic diversity. Several authors (e.g., Taylor 1986) have begun such a justification, but goals have not been defined clearly enough for use in genetic management. The welfare of individual organisms can constrain management practices, but they are not a goal of genetic management. We are generally free to manipulate the genetic diversity of other organisms for the broadly defined good of present and future humans. Policy goals on genetic diversity are limited to not unnecessarily constraining opportunities for the evolution of species. We may provide for their adaptability to present and future environmental variations in multiple populations of useful species and of other coevolving species, but policy on genetic diversity should not have the goal of maintaining the present structure of that diversity.

Because managing the evolutionary possibilities of all species is not practical, more precise and more limited goals are needed. Industrial, governmental, and non-governmental conservation agencies necessarily have different goals. For agencies that have strictly human-based and immediate production goals, present needs for genetic management can be based on species that have proven to be adaptable to human culture and on traits that have proven to be malleable by known forms of genetic management. The management of genetic diversity in the useful and manageable species has been the subject of other symposia and is under active debate and development. A policy goal for such agencies might be to maximize the useful genetic variation of ecologically and economically valuable species. Other species might be ignored completely or assured only of some minimal levels of genetic variation.

For agencies with concern for future human welfare, the values of potentially useful populations, species, and traits are somewhat less clearly defined. It might be feasible to use present needs and culture only as a basis for predicting future values and to include potentially useful species that might have no presently known value. Potential use may extend the scope of concern very widely, but for species with a low probability of significant utility, the probability that management would effectively increase value would be low. Therefore, even if all species are included as having some potential value, the necessity of action on most species would be limited merely to ensuring their evolutionary potential.

A further implication for these kinds of human utility goals is that rarity or endangerment of a species is not a value- determining factor unless rarity itself increases utility. The threatened status of a species can determine the relative urgency with which it is managed but would not determine which species should be managed for their inherent value. Rarity may increase in esthetic or transformative value of a species to humans. Rarity can also increase the utility of a species or population if it is associated with unique alleles or unique adaptabilities which so increase the variance of the taxon that its contribution to future value can be increased by its management. Managing a few individuals of such species may have a much greater marginal effect on total value than managing an equivalent number of a more abundant taxon. Although rarity itself does not influence the inherent value of a species, it can influence the probability that management can be effective.

Even among agencies that include the existence of nonhumans as a goal, all species may not have inherently equal value or be given equal consideration. For example, the argument in terms of animal rights implies that only certain animals can possess rights (Regan 1976) and, therefore, other species have only utility value to them and to us. There is still a question as to whether the species *per se* has value independent of the value of its individuals that may have inherent value and rights as individuals. If it does, then a goal of conservation could be its survival and continued evolution. Even then, however, there is an ordering of the kinds of species that we must consider. There are those whose welfare is a goal and those whose welfare is a utility. Again, there is no inherent good that is attributable to rarity or endangerment of species of utility value unless rarity increases their utility. Hence, even if all species have existence value, it is not clear that the status of threat would be a factor in defining the goals of genetic management.

The goals of management with respect to genetic diversity can, therefore, be limited to providing for or maximizing opportunities for evolution of "valuable" species. We should not specify a state for that diversity, in particular the present state, or the status of threat or endangerment. It may well be an objective of management to pay particular attention to threatened and endangered species or populations and to ensure that populational differences are maintained or increased. The purpose, however, is to increase the opportunities for evolution of the species of value. The state of rare species or populations is not in itself a factor in determining policy goals. Rarity and taxonomic diversity are not goals or value- determining factors, but can be features of populations, species or ecosystems, that increase probabilities of evolution of subject species or may increase the effectiveness of management. Altering the present structure of genetic diversity and even allowing the extinction of some populations and species can sometimes be useful (e.g., if more species can be saved by allowing a weed that excludes many species over broad areas to become extinct).

Each agency should define which species are to be considered to have inherent value and which are of value only to ensuring the viability and evolution of the inherently valuable ones. The latter have value and may be managed but only for the good of others, and hence have instrumental value only. Although the subject species and management priorities would differ, human values would be primary goals, management would support those goals, and genetic diversity would be modified according to some value priorities. Within that context, a management plan would be chosen based on the present state of genetic diversity and the effects of management alternatives. The difference between intensive rescue efforts and nonintervention, and between human ends and those of other species is less one of conflicting goals that one of means.

Among the means of achieving goals, the main factors of evolution that can be easily controlled are the effective population size, the number of populations, and the selective environment. The techniques, of course, have to be adapted to the condition of the species, its population dynamics, and the life histories of the individuals available for management. Increasing or decreasing the genetically effective migration rate between populations can have positive effects on the genetic variance, depending on the structure and levels of variance that exist within and between population. Fragmentation or disruption of populational distributions can simultaneously affect genetic migration and the boundaries of selection demes. Divergent selection can vastly increase interpopulational genetic variance, and genetic variance within selected populations can be managed to reach equilibrium levels between losses due to drift and gain due to mutation. Therefore, by managing population sizes and mating patterns, and by sampling a number of populations and their selective environments, we can increase or decrease the available genetic variance. Furthermore, we have a range of techniques to influence strongly the direction of evolution and the levels of genetic variance, without using highly intrusive management.

PRIORITIES

The lack of sufficient biological knowledge about most species and the costs of management limit intensive genetic management to relatively few species. Hence, with a species-by- species approach, we will never complete a gene management program on only a small fraction of species, and the ones chosen might not be the most effective ones to manage for any goals. Even systems of less intensive management, employing natural reproduction and simple genetic monitoring, could be extended to no more than a few thousand species. Therefore, genetic management will often reduce to ecosystem or landscape management techniques and to systems that would allow for self-sustaining evolution to proceed with minimal intervention (Frankel 1983; Norton 1987). Thus, even though the goals of genetic management and ecosystem management are not the same, the chosen methods of management may turn out to be the same. The distinctions between goals, intermediate objectives, and techniques are important to recognize, however. If different management techniques can be used to achieve a goal, we must recognize the choices we are making. Many substantial, procedural questions can be posed. Here, however, I wish to focus attention on choosing priorities so that management can be optimized for some utility value.

The genetic goal of maintaining or increasing opportunities for evolution is a necessary part of the broader goal of managing biodiversity but satisfying the genetic goal is not sufficient to satisfy all of society's broader goals for biological diversity. For example, genetic diversity within and among species could be maximized, but communities could be all inclusive and landscape level diversity thereby reduced. Genetic diversity might be maximized by segregating populations into non-intermating demes, but such fragmentation might reduce the probability of long-term species survival. Therefore, higher level diversities of species and ecosystems require management at those levels of biotic organization. However, we can also ask the obvious question: is managing diversity at the community, ecosystem, or landscape level sufficient for maintaining opportunities for evolution? A diversity of ecosystems could be set aside in a diversity of reserves that had reduced genetic variation in many species within and among those ecosystems and without interpopulational variance. Therefore, no one level of management is sufficient. All levels of diversity must be managed. In an optimum management system, some ecosystems would be given priority because they contain more genetic diversity than other sets. Conversely, some genetic variations in host-pathogen or mutualistic interactions could be given priority because they were most effective in maintaining species coexistence and even community diversity. In such an optimum system, total diversity would most efficiently be served by an integrated, multilevel management system.

Until a gene-to-landscape integration can be used, Norton's (1987) ecosystem protection system for species conservation can be extended to gene management. He proposed ranking priorities not by relative value but at least partially on the basis of the probability that management actions would effectively conserve target species or large numbers of species. The probable effectiveness of management efforts would help to determine priorities for action or inaction. By this criterion, Norton sought to conserve a maximum number of species and thereby converts species management to conservation of species-rich habitats. Conservation efforts are allocated to priority areas that contain and can continue to support the largest total number of species and taxonomic diversity. This targeting of habitat management is similar to the system suggested by Schonewald-Cox (1983) except that she used target species as the criteria to evaluate reserves established to maintain genetic diversity. The effectiveness of the Schonewald-Cox system requires the additional step of relating genetic variance in the target species to total species genetic diversity. Appropriate target species must be chosen. Conservation of target species will conserve a maximum of other species if the target species selected are those requiring the most inclusive set of environments.

Alternatively, if species can be prioritized by values that are scaled either by human uses or their importance in some ecosystems, then management can target and allocate funds on an ordered basis of species and ecosystems that satisfy the evolutionary needs of priority species. This approach would require multidimensional evaluations because no single scale of value can adequately reflect the multiple values placed on species welfare and genetic diversity. The management systems for priority species would include support of the genetic variation of other species and ecosystems. If enough information became available, the efficiencies of simultaneous gene and ecosystem management would be instituted when feasible. Rare species can claim values that give them a special priority based on importance to scientific research, esthetics, long-term evolution, unique alleles, and taxonomic diversity. They may also be given priority because conserving them at similar cost to conserving populations of a more common species would be more effective. Otherwise, however, neither rarity nor endangerment would endow a species or population with any special priority. This result differs from that derived by Norton, who suggests that, because all species have value, we use endangerment and taxonomic diversity for policy priorities and that practical management of species requires ecosystem or habitat preservation.

Here, value to humans in its broadest sense is suggested as a useful criterion. Targeting genetic variation, populations, and species by value, although controversial, can include wide genetic and species diversity and can be applied by conservation agencies on the basis of present knowledge. Ecosystem and habitat conservation might be especially useful to maintain genetic diversity for species of uncertain value, but it is suggested that when we know more of the biota and can manage genetic diversity better, the management of ecosystem diversity and genetic diversity will become less distinct.

LITERATURE CITED

Brown, A. H. D., and J. D. Briggs. 1991. Sampling strategies for genetic variation in *ex situ* collections of endangered plant species. In *Genetics and Conservation of Rare Plants*, ed. D. A. Falk and K. E. Holsinger, pp. 99-119. Oxford University Press, New York.

Falk, D. A., and K. E. Holsinger. 1991. *Genetics and Conservation of Rare Plants*. Oxford University Press, New York.

Frankel, O. H. 1983. The place of management in conservation. In *Genetics and Conservation*, eds. C. M. Schonewald-Cox, S. M. Chambers, B. MacBryde, and W. L. Thomas, pp. 1-14. Benjamin/Cummings Publishing Co., Menlo Park, Calif.

Norton, B. G. 1987. *Why Preserve Natural Diversity?*. Princeton University Press, Princeton, N.J.

Regan, T. 1976. Do animals have a right to life? In *Animal Rights and Human Obligations*, eds. T. Regan and P. Singer, pp. 197-204. Prentice-Hall Inc., Englewood Cliffs, N.J.

Schonewald-Cox, C. M. 1983. Conclusions: Guidelines to management: A beginning attempt. In *Genetics and Conservation*, eds. C. M. Schonewald-Cox, S. M. Chambers, B. MacBryde, and W. L. Thomas, pp. 414-445. Benjamin/Cummings Publishing Co., Menlo Park, Calif.

Taylor, P. W. 1986. *Respect for Nature: A Theory of Environmental Ethics*. Princeton University Press, Princeton, N.J.

44

CREATING POLICY ON SPECIES DIVERSITY

Michael J. Bean

Environmental Defense Fund
Washington, D.C.
U.S.A.

This paper explores the subject of policy creation with respect to species diversity. In one sense, that topic is easy because we have some twenty years of experience doing it. On the other hand, the policies we have created have not been particularly effective, as evidenced by the continued decline of many of the very species our policies were intended to conserve. They have also become increasingly controversial as they collide with other policies or objectives that many deem equally, if not more, important. Hence, merely to describe the policies we have and how they came about will not suffice. It is necessary, instead, to suggest how our existing policies need to be altered to make them more effective and to make them more compatible with other competing objectives.

HISTORY

In the United States, species conservation policies have a fairly long history. That history is one of steadily expanding concern for a broader and broader array of living organisms. It is also one in which the tools utilized in the conservation effort have also steadily expanded as a result of the recognition of the diversity of threats affecting species well being.

Until the latter part of the present century, species conservation policies were narrowly limited to a very few species. The original policies of the various American colonies, later embodied in the laws of the states, sought to assure the sustainable utilization of "beneficial" species. Animals deemed valuable for their fur, hide, meat, or other products were the exclusive targets of these earliest policies. Later, the recreational values of hunting and fishing became paramount in determining what species would be the object of conservation policies.

Wildlife conservation was almost exclusively a matter of state concern until this century (Bean 1986). In 1918, the federal government assumed responsibility for the conservation of migratory birds when Congress passed the Migratory Bird Treaty Act. Two strands of concern are evident in the congressional deliberations that produced this seminal law. First, Congress recognized that, as a result of the migratory nature of the birds, their conservation could not be assured by the individual states; national action was needed to protect a resource that knew no state, or even international, boundaries. A second concern was with the ecological role of birds. Birds were seen by at least some in Congress as more than just targets for hunters or meat for the table. They were also protectors of our crops, nature's line of defense against a horde of insect pests. By today's standards, this was pretty simple-minded science. In comparison to the myopic justifications for conservation that preceded it, however, this was ground breaking stuff.

Later laws broadened the ambit of federal conservation concern to encompass selected other species. Bald eagles, which were not originally protected by the 1918 Migratory Bird Treaty Act, were the beneficiaries of a federal law enacted in 1940. Marine Mammals were given extensive federal protection with enactment of the Marine Mammal Protection Act in 1972. The most far-reaching and significant of the species conservation laws enacted by Congress, however, were the series of endangered species laws that began in 1966 and culminated with enactment in 1973 of the modern Endangered Species Act. With the Marine Mammal and Endangered Species Acts, the focus of conservation policy had come full circle, from protecting species solely for their value as meat or objects of sporting pleasure, to protecting them for their full array of aesthetic, ecological, utilitarian, and intrinsic values.

Just as the progression of species conservation laws reflected an ever-broadening scope of concern, so too did the tools provided by these laws expand. The early laws relied almost entirely upon limiting harvest to achieve their goals. Soon, the importance of protecting habitats was recognized. The Migratory Bird Treaty Act was supplemented by enactment of the Migratory Bird Conservation Act, which provided authority to acquire bird habitat. Eventually, it was recognized that public acquisition alone could not assure effective conservation. The management of public lands could be inimical to species conservation objectives without some means of restraining and guiding that management. Hence, laws narrowing the discretion of public land managers were enacted. Most significant, and most controversial, was the recognition that for a great many species, private land-use decisions will decide their fate. Fashioning workable means of influencing private land-use decisions became the most formidable challenge with which modern species conservation policy-makers had to grapple.

At the international level, closely parallel developments can be traced. The earliest international conservation initiatives focused on controlling the exploitation of such commercially valuable species as whales, northern fur seals, and ocean fish. In 1972, all potentially vulnerable plant and animal species were the object of the Convention on International Trade in Endangered Species of Wild Fauna and Flora ("CITES"). However, the only tool that CITES provided was the regulation of international commercial trade; the ubiquitous threat of habitat destruction was completely beyond its reach. That gaping omission fueled recognition of the need for a more truly comprehensive treaty for the protection of biodiversity. Efforts to negotiate such a treaty for the 1992 United Nations Conference on Environment and Development were highly contentious as a result of the recognition that species conservation produces both costs and benefits, and that the distribution of these may not always appear equitable from the point of view of those who bear the costs.

CURRENT ISSUES

Species conservation policy in the United States has been dominated over the past two decades by the Endangered Species Act. Although much progress has been made under that Act in reversing the decline of many species and avoiding the extinction of many others, there are nevertheless many noteworthy failures. Those failures, and the sometimes significant impact of that Act upon other interests, have given rise to intense scrutiny of the Act and its future. This section examines some of the key issues receiving that scrutiny, and attempts to offer some useful guidance as to the direction that future policy-making should take in the resolution of these issues.

What Species Shall We Conserve?

The core of current federal policy is to conserve any plant and animal species threatened with extinction, now or in the foreseeable future. Even under the most pessimistic estimates of what fraction of this country's native species may be at risk of extinction, it is clear that most species are not, and that, therefore, there is no explicit species conservation policy for the great bulk of our species. That fact is not necessarily bad, because we may not need a policy for everything, in particular for those things capable of taking care of themselves.

In 1980, Congress enacted a law, commonly known as the "Non-Game Act," that was intended to provide federal grants to enable the states to establish and implement conservation programs for species not at risk of extinction and not already addressed by state game conservation programs. The theory was that such programs would respond to the perceived demand for non-consumptive wildlife recreation and at the same time tend to reduce the likelihood that the species targeted in these programs might someday need the protection of the Endangered Species Act. In fact, twelve years later, the authorized federal grants program has never been established. Many states, without federal support, have established modest non-game conservation programs. These programs do not necessarily fit the model contemplated by Congress, however, in that some encompass endangered species. Without exception, these are very modest programs in comparison to the much more substantial game conservation programs that every state maintains. Thus, more than a decade after Congress sought to broaden the scope of species for which conservation programs and policies would be developed, most species conservation policy is still focused on species that are either at risk of extinction or so abundant that they support a "surplus" for recreational harvest.

Looking more closely at species in peril of extinction, it is evident that the "policy" to conserve all of these is not as comprehensive as it might at first appear. First, authority to protect species under the Endangered Species Act extends only to members of the plant and animal kingdoms. In 1973, Congress may have thought it was including more than it did by referencing these two kingdoms. At the time, many in Congress probably thought that every living species was either plant or animal; thus, any species could be protected by the Act, if necessary. Today, however, scientists recognize other kingdoms of life, and species in these other kingdoms presumably fall outside the scope of the Endangered Species Act. As a practical matter, this is likely to be of little consequence, except possibly in the case of fungi, certain species of which may warrant conservation attention.

Of more consequence is the fact that even within the plant and animal kingdoms there are many more species known to be in some peril of extinction than the administrators of the Act can at present address. Currently, more than 800 species in the United States and its possessions are formally listed as threatened or endangered. Between four and five times that number have been identified as likely to be eligible for listing. These figures encompass all "species" in the unique sense in which that term is used in the Endangered Species Act. As used there, the term includes full species, subspecies, and, in the case of vertebrates only, distinct geographic populations. Thus, this definition is broad enough to allow for the development of both species and genetic diversity policies, at least for vertebrates. For invertebrates and plants, the Act confers no authority to protect individual populations. This does not reflect any principled distinction between vertebrate and other species; rather, it simply reflects a deeply entrenched attitude that vertebrate conservation is more important than invertebrate conservation. In fact, however, authority to protect invertebrate populations could be useful in addressing a number of significant environmental problems, including the decline of coral reefs. Despite its hypothetical reach, the authority to list and protect populations as though they were species has been used quite sparingly; the authority to list and protect subspecies, on the other hand, has been much more liberally utilized (Wilcove et al. 1993).

At present levels of available resources and at current listing rates, the existing backlog of identified species needing protection will not get that protection for several decades. Thus, the necessity exists to choose which of the many species that might be protected should be protected first. In short, the real policy issue is not so much what species to conserve beyond those already in danger of extinction, but rather which of the many species already in danger of extinction should receive priority attention. The logic of protecting endangered species rather than non-game species not yet endangered suggests the conclusion that scarce conservation resources should be allocated first to those species most imminently in peril. And in fact, the published criteria of the Fish and Wildlife Service to guide listing and recovery priorities do give substantial weight to the imminence and gravity of threat affecting a species. The published criteria, however, appear to reflect agency behavior better with respect to listing decisions than with respect to recovery efforts. The latter appear to be driven in large measure by unstated criteria, including the degree of conflict between the conservation needs of a species and politically or economically significant interests, and the degree of public support for conserving a particular species. In the most recent year for which data are available, seven species (one percent of the total listed) were the intended beneficiaries of more than 50 percent of the total state and federal expenditures for all listed species (Fish and Wildlife Service 1992). At least three of these (bald eagle, peregrine falcon, and northern spotted owl) were clearly not among the most imperiled species the Act protects. Thus, actual practice reflects the realpolitik of implementing a species conservation not in a vacuum, but in a cauldron of often very politicized competing interests.

Even if one could limit political "interference" with a species conservation program, there may be good reason to reconsider the wisdom of giving primacy to the imminence and gravity of threat to particular species in deciding which species to protect first. Other considerations that would seem important in determining which species to list and protect first include the following: whether the species is likely to play a key role in its ecosystem, such that its loss may lead to significant disruption of that ecosystem; whether the species is known to have been used in traditional medicine, or is closely related to others that were, such that its conservation might preserve options for future medicinal utilization; whether a species is closely related to crop species, such that its conservation might preserve options for future agricultural application; and whether the species serves as an indicator of the overall health of an ecosystem in which numerous other species may soon also be at risk. The first and last of these suggested criteria have the salutary purpose of linking species conservation concerns with broader ecosystem conservation concerns. The other two seek to tie species conservation priorities more directly to human utilitarian justifications for species preservation.

When to Act?

The conceptual premise of the Endangered Species Act is that there is some point in the decline of a species (or perhaps in the manifestation of threats to that species) at which the species should be considered "endangered" or "threatened." When a species reaches that point, current stated policy is that the government is to intervene, list the species, and carry out those actions necessary to move the species back across the threshold into non-threatened status. What is that threshold?

The Endangered Species Act itself gives only the most general guidance on this question. It defines an "endangered" species as one that is currently in danger of extinction throughout all or a significant portion of its range and a "threatened" species as one that is likely to become endangered in the foreseeable future. Critics of the Act contend that the looseness of these definitions invites arbitrary decision-making by those charged with the Act's administration. They clamor either for clearly stated, measurable criteria (such as, absolute abundance, rate of decline, percentage of original range currently occupied) or for independent "peer review" of agency listing (and recovery) decisions.

As currently written, the Act requires that listing decisions be based solely upon scientific information. Defenders of the Act vigorously resist any suggestion that economic or other non-biological factors should be allowed to influence decisions to list or not list particular species. Such extraneous considerations, they argue, will only result in delaying or withholding the protection that, as a purely scientific matter, a species warrants.

In practice, the implicit contention of the Act's harshest critics, that the administrators of the Act use its nebulous definitions to list species undeserving of protected status, has little evidence to support it. Indeed, much stronger evidence exists to show that protection is withheld from deserving species for unduly long periods (Wilcove et al. 1993). Fully one-half of the plants listed in the past seven years (and plants were the most commonly listed group in that period) had known populations at the time of listing of 120 or fewer individuals. The median known population size of animals listed during the same period was about 1,000 individuals. Nevertheless, it is difficult to discern from actual experience what standards are implicitly being utilized to guide listing and recovery decisions. While many species are being listed when their numbers have been reduced to a few dozen or less, implying a relatively high likelihood of extinction within the near future; others, like the Mojave population of desert tortoises have been listed when their numbers were thought to exceed 1 million, a fact that would seem consistent with a much lower likelihood of extinction within the same time period.

Judgments about which species to protect must be based on some implicit understanding of what probability of extinction within some time frame should trigger protective action. Unfortunately, the administrators of the Act have never specified either the time frame or the probability of extinction within the time frame that guides their actions. Because these parameters have never been articulated, it is likely that they vary, depending on the political, economic, and other circumstances of each case. Species likely to generate significant conflict with powerful political or economic interests may, all other things being equal, require a higher likelihood of extinction in a shorter time frame than other species to be listed. On the other hand, species for which there is an active conservation constituency, particularly one capable of backing up its interest with litigation, may gain protection with a lower probability of extinction over a longer time period than species lacking such constituencies. Admittedly, this is speculation, but it is speculation made possible by the absence of detailed criteria governing listing decisions.

Ultimately, any endangered species conservation program is probably better off if its administrators provide as much detail as possible about the criteria that will guide their decisions. Developing such criteria would have the salutary benefit of forcing administrators to try to develop meaningful standards, capable of general application, rather than relying upon a series of *ad hoc* determinations. Such criteria would also provide outside parties, both conservationists and regulated interests, with a better understanding of how key decisions that trigger a range of significant consequences are supposed to be made. That improved public understanding, while not likely to remove controversy from species listing decisions, should at least tend to focus the controversy upon appropriate issues.

How to Reconcile Competing Interests.

Species conservation often entails changing personal, business, or land-use practices that have brought a species to the point where its conservation has become an urgent policy matter. Effecting such changes increasingly embroils conservation efforts in controversy because of the conflicting desires of those who seek to conserve a species and those who seek to continue the practices that are now called into question. At present, a chorus of critics of the Endangered Species Act are calling for changes that would force the Act's administrators to "balance" species conservation goals with other social and economic goals.

In assessing these proposals, and in carrying out any balancing mechanism, it is extremely important to be mindful of issues of scale, both geographic and temporal. Job losses and other economic impacts that may be severe when viewed from a local and short-term perspective may be of relatively little significance when viewed from a broader geographic and longer-term perspective. This point can be illustrated with a current example.

Trade liberalization between the United States and Mexico is a key objective of current Administration policy. A "North American Free Trade Agreement" is being actively debated. Many American businesses, however, did not wait for the agreement to be finalized. Instead, in anticipation of the expected agreement, they began relocating manufacturing and other facilities to Mexico. In the first five years of this decade, some have estimated that as many as 112,000 American jobs will be lost as result of these shifts (Hufbauer and Schott 1992). For the workers whose jobs will be lost, and the communities in which they live, these consequences are catastrophic. On the other hand, advocates of trade liberalization argue that it will create other jobs as American companies supply goods demanded by an increasingly affluent Mexican market. Hence, whether liberalizing trade with Mexico is a good or bad thing depends entirely on whether one's perspective is local and short-term, or national and long-term.

The same conclusion likely applies to many species conservation controversies. Calls for adding "balance" to the Endangered Species Act stem primarily from the conflict between spotted owl conservation needs and the old-growth logging in the Pacific Northwest. The Bush Administration contended that if it implemented a recovery plan that fulfilled the requirements of the Endangered Species Act, the result would be the loss of over 32,000 jobs in the Northwest over a ten-year period (Sample and Le Master 1992). While substantial, that figure is still dwarfed by the local job losses expected from trade liberalization in only a five-year period. And like the losses associated with trade liberalization, the job losses associated with logging restrictions in the Northwest cannot be meaningfully evaluated without reference to larger areas and longer time-frames. To the extent that logging restrictions preserve otherwise vulnerable jobs in the commercial fishing sector, create opportunities for enhanced employment in tourism and related industries, cause a shift in employment to other regions of the country, or preserve the possibility of beneficially using potential future resources associated with old growth forests, the apparent conflict between species conservation and economic well-being will be less than it otherwise appears.

Unfortunately, few of the calls being heard today for integrating economic concerns into species conservation decisions clearly articulate a need to examine other than short-term, local tradeoffs. The objectives of species conservation are long-term ones, and the beneficiaries are all the citizens of the Nation, not just those in the vicinity of the species. Therefore, any balancing mechanism that looks solely at local, short-term tradeoffs will be tilted heavily against conservation interests. If species conservation efforts are to survive the current clamor for "balance" in decision-making, the balancing mechanisms will have to be ones capable of taking into account both long-term and geographically comprehensive consequences. It also seems important to recognize the desirability of spreading the risks of species conservation more broadly, so that local communities and landowners do not bear the sometimes heavy burden of conservation requirements intended to benefit the entire nation.

Should "Ecosystems" Replace "Species" as the Focus of Concern?

A number of observers, noting the ever increasing number of species on the endangered list and critical of the limited progress made in recovering them, have urged that the "one-at-a-time" species approach to conservation give way to broader conservation efforts aimed at restoring and maintaining whole ecosystems (Scott et al. 1987; Tear et al. 1993). By protecting large areas of habitat, all the species within them will benefit; the challenging tasks of determining the status of each species and designing conservation measures to meet its needs can thus be avoided.

Undoubtedly, protecting large contiguous areas of habitat will have major conservation benefits for the many species that utilize those habitats. That, after all, has been part of the rationale for the system of National Parks and National Wildlife Refuges that has been developed over this century. But it seems doubtful that such a strategy can ever fully replace the need for species-oriented programs like the Endangered Species Act. In the first place, even the very large national parks of the West have been too small to prevent the loss of some species (Newmark 1987). Second, if "protection" of large, contiguous blocks of habitat is tantamount to excluding significant commodity uses of those lands, the political and economic conflicts likely to attend efforts to achieve that protection will be even larger than those that have attended efforts to protect individual species.

If exclusion of significant commodity uses of large blocks of habitat is not the goal, then the question arises as to how such uses and areas are to be managed. The answer, according to many, is "ecosystem management." Exactly what this means is still unclear, although some of the definitions being offered are disturbingly familiar. For example, George Frampton, the Assistant Secretary of Interior for Fish, Wildlife and Parks, recently offered the following: "What we mean by ecosystem management is good science, looking at a range of species in communities and trying to plan for some optimization of resources. Not just one resource at a time" (Environmental Law Institute 1993). What is remarkable about Frampton's definition is how strikingly similar it sounds to "multiple use," the philosophy of public land management that is believed by many to have contributed significantly to the species conservation crises of today. Thus, "ecosystem management" as an alternative to vigorous species conservation policies may prove to be a chimera.

More fruitful might be to integrate ecosystem restoration and maintenance strategies into species recovery efforts. Among politicians and others, there is often an almost reflexive response to any endangered species controversy in favor of relocating a species or taking it into captivity for captive breeding as a way of avoiding conflict with some favored development activity. Bills currently pending in Congress would give even greater emphasis to captive breeding than it receives now. The problem with strategies that rely heavily on intensive, artificial manipulation of a species or its habitat is that they are almost always risky, expensive, and open-ended with respect to the commitments they require. Even if successful, these strategies are unlikely to prevent other species associated with the same ecosystems from themselves becoming in danger of extinction. Thus, the important lesson from the attention now being given to ecosystems may be that recovery strategies for endangered species that emphasize the restoration and maintenance of ecological relationships and functions are likely to be much better long-term strategies than are strategies that emphasize intensive, artificial manipulation. Not only can they achieve recovery for today's endangered species, but make it less likely that other species within their ecosystems will become tomorrow's endangered species.

CONCLUSIONS

The process of creating species conservation policy is both extremely important and suddenly very controversial. Species conservation has been a cornerstone of many conservation programs in the United States for most of a century. The results achieved leave many dissatisfied. Conservation interests fault existing policies for failing to give adequate attention to genetic diversity within species and to community or ecosystem diversity. At whatever level, however, conservation efforts will confront enormous challenges in deciding which of the nearly infinite constellation of natural resources merit our limited attention, when our intervention is necessary, and how we are to make the tradeoffs between conservation and other socially important objectives. The experience gained in addressing those challenges in the context of species conservation efforts should aid in developing effective conservation policies at other levels of biological complexity.

LITERATURE CITED

Bean, M. 1986. Federal legislation and historical perspectives on a national biological survey. In *Foundations for a National Biological Survey*, K. Kim, and L. Knutson, pp. 167-175. Association of Systematics Collections, Lawrence, Kans.

Environmental Law Institute, 1993. Forum profile: George T. Frampton, Jr. *Environ. Forum* 10 (Nov./Dec.):28-33.

Fish and Wildlife Service. 1992. Federal and state endangered species expenditures: fiscal year 1991. Unpublished report, U.S. Department of Interior, Washington, D.C.

Hufbauer, G., and J. Schott. 1992. North American free trade issues and recommendations. Institute for International Economics, Washington D.C.

Newmark, W. D. 1987. A land-bridge island perspective on mammalian extinctions in western North American parks. *Nature* 325:430-432.

Sample, A., and D. Le Master. 1992. Economic effects of northern spotted owl protection: an examination of four studies. *J. For.* 90 (Aug.):31-35.

Scott, J., B. Csuti, J. Jacobi, and J. Estes. 1987. Species richness: a geographic approach to protecting future biological diversity. *BioScience* 37:782-788.

Tear, T., J. Scott, P. Hayward, and B. Griffith. 1993. Status and prospects for success of the Endangered Species Act: a look at recovery plans. *Science* 162:976-977.

Wilcove, D., M. McMillan, and K. Winston. 1993. What exactly is an endangered species? An analysis of the U.S. endangered species list: 1985-1991. *Conserv. Biol.* 7:87-93.

45

CREATING POLICY ON COMMUNITY DIVERSITY

Kenneth A. Dahlberg

Department of Political Science
Western Michigan University
Kalamazoo, Michigan
U.S.A.

If this chapter were part of a book on urban affairs, the title would suggest concern with creating communities that are diverse in the ethnicity, age, and gender of its members as well as diverse in their housing, retail, and industrial patterns. As part of a book on biodiversity, the title suggests other themes. Before pursuing them, it is important to reflect further on the relative nature of all social and biological concepts. They are: (1) relative to the real world distribution of the systems under examination; and (2) relative to the cultural and disciplinary filters and values of those doing the examining.

Ramon Margalef has nicely shown how concepts of "ecosystems" are relative in these two ways (Margalef 1968: 26-27):

> Ecosystems reflect the physical environment in which they have developed, and ecologists reflect the properties of the ecosystems in which they have grown and matured. All schools of ecology are strongly influenced by genius loci that goes back to the local landscape. . . . The mosaic-like vegetation of Mediterranean and Alpine countries, subjected to millennia of human interference, has assisted at the birth of the plant sociology school . . . Scandinavia, with a poor flora, has produced ecologists who count every shoot and sprout. It is a pity that the tropical rain forest, the most complete and complex model of an ecosystem, is not a very suitable place for the breeding of ecologists. And it is only natural that the vast spaces and smooth transitions of North America and Russia have suggested a dynamic approach in ecology and the theory of climax. In such areas, the concept of succession, one of the great and fruitful ideas of classical ecology, was best formulated.

The relative nature of different fields is also implied above, namely, that ecologists are primarily interested in natural systems. Equally, most social scientists are interested primarily in social systems. Unfortunately, few seek to fill the large gaps between the two. For example, only a few anthropologists have sought to analyze the relationships between biodiversity and cultural diversity—and this largely in terms of indigenous or traditional peoples. Stephen B. Brush (1981) has shown that the diversity of potato germplasm is integrally linked to the cultural diversity of the different Andean tribal groups that cultivate different varieties at different altitudes. Even fewer analysts seek to understand the different dimensions of diversity as they interact with, and are affected by modern industrial societies, something that requires one to include in the analysis the importance of technological systems and their constituent infrastructures and institutions (Bennett and Dahlberg 1990).

This volume (and the emerging field of conservation biology more generally) seeks to have ecologists and biologists address issues of biodiversity in managed systems, rather than in the traditional "nature only" setting. Yet there is a tension: the basic organization of the volume still largely reflects traditional ecological concepts and categories while the topics, data sets, concepts, and issues found in the management of such renewable resource-based systems as agriculture, forestry, grazing, and fisheries are only secondarily represented.

More generally, a need exists to try to develop ways to synthesize the data and concepts that a multiplicity of other approaches have generated. For example, in dealing with the analysis and management of sub-national regions, landscape architect/ecologist Ian McHarg identified a long list of different types of regions (each deriving from a different analytic approach) that need to be included: regions based on climate, geology, river basins, groundwater basins, biogeography, human settlement patterns, agriculture, and energy patterns among others (McHarg 1992).

We need broader and more inclusive approaches to be able to integrate such considerations, both within a given level and between levels. Hierarchy theory in ecology has helped significantly to clarify the relationship and linkages between systems operating at different levels. However, hierarch theory still largely focuses on natural systems. As indicated above, social and technological systems also need to be included at each level to pursue what I have termed "contextual analysis" (Dahlberg 1979). In this chapter, I outline the broad dimensions of this approach as applied to biodiversity in managed landscapes.

THE CONTEXTUAL PARAMETERS OF BIODIVERSITY IN MANAGED LANDSCAPES

Global Parameters

Four major global threats face humankind: (1) nuclear war, (2) atmospheric changes that might lead to global warming and climate change, (3) the loss of biodiversity, and (4) the global spread of industrial society. The different bodies of literature addressing these threats reflect not only the concerns of citizens and analysts regarding their possible consequences, but also the different organizing paradigms and the institutions from which they emanate.

Whereas the threat of nuclear war thankfully appears somewhat reduced now, the ongoing economic, development, resource, and environmental costs of the various military-industrial complexes around the world are still staggering and increasing. The industrial countries, suffering from recessions, have sold increasing amounts of older weapons to the Third World, thus reducing the funds the latter have available for development, much less sustainable development. The military represents the most extreme embodiment of those highly centralized, capital-, and energy-intensive systems which Amory Lovins (1977) has termed "hard path" systems. The infrastructural, institutional, and ideological patterns that are a part of such hard path systems are largely antithetical to the more decentralized and flexible soft path systems required both for sustainable (or regenerative) development and the preservation or enhancement biodiversity.

The literature on global change has emerged primarily out of the concerns of atmospheric scientists regarding possible climate changes. The basic definitions of the problems to be addressed (both scientific and political) are largely theirs. The life sciences and social sciences have been latecomers and have had to work largely within these confines. Among the many other weaknesses of the global change literature, one that is particularly relevant is the analytic neglect of systems at other levels. Little concerted attention is given either to natural or regional systems, particularly the renewable resource regimes that encompass the interactions of agriculture, grazing, forestry, and fisheries in the different regions of the world.

The literature on biodiversity reflects the deep concerns of ecologists and biologists over the loss of the ultimate source of the regenerative capacity and resilience of the multitudinous living systems of the biosphere (Wilson 1988; Pimm 1991). Sadly, these biological scientists do not have anywhere near the same degree of support from various industrial, political, and academic constituencies as the national security and atmospheric science establishments. Not only do they have less support from powerful groups, but their message often goes against a number of reigning ideologies, such as, economic growth and free trade. Also, the literature on biodiversity has only slowly begun to include the economic, social, political, and technological issues and parameters that are an integral part of maintaining or enhancing biodiversity. For example, those involved in biodiversity issues have been slow to get involved in the issues and debates surrounding the Brundtland Commission report, *Our Common Future* (WCED 1987), and the UN Conference on Environment and Development, both of which revolve around issues of sustainable development. Equally, they have been slow to assess other global threats to biodiversity. On the physical science side, few studies have focused on the potential impact of acid rain, increased ultraviolet radiation, or increased climate variability upon global or regional biodiversity. On the socio-economic side, few studies have examined either the general way in which the demand for tropical products in the industrial countries leads to losses of biodiversity in the developing countries or the role of specific institutions, corporations, trade patterns, or consumer habits in such losses.

Various approaches have been employed to understand and assess the spread of industrial civilization world-wide, a process which has been accelerated in recent decades by the establishment of instantaneous global communication networks and accompanying mass media and financial systems. One approach, the world systems approach developed by Imanuel Wallerstein (1974, 1980), is essentially neo-Marxist in its emphasis upon the historic spread of capitalist institutions. Others of us see the spread more in terms of the diffusion of industrial institutions and culture, whether in their capitalist or socialist versions (Bennett and Dahlberg 1990). Another approach is that of global computer modelers. *The Limits to Growth* (Meadows et al. 1972) was the first of a series of global models that sought to portray the interactions of population, resources, food, industrial growth, and pollution. Finally, a number of alternative formulations are emerging around James Lovelock's (1979) concept of Gaia. These include explorations of how more sustainable, less patriarchal, and more equitable societies might be promoted in a post fossil-fuel era.

Several things are suggested by the above brief review. First, attempts to understand what are the key *global* processes and parameters are still in their early stages. Second, as in other areas, each field or discipline tends simply to project its specialized concepts and categories to a larger scale, not realizing that new concepts and categories are needed. Two types of value issues underlie these analyses. One involves disputes over the relative priority given to existing goals (national security vs. development; economic growth vs. environment; research vs. action). The other involves fundamental divisions that are emerging between the old industrial order and the transition to a post fossil-fuel era. These divisions include differences in worldview, basic assumptions about progress and the role of science and technology as well as about gender, spirituality, and the value of other species. Those concerned about biodiversity will need to become more familiar with these different bodies of literature, their different assumptions, and the different values underlying them.

International Parameters

Unlike global parameters, which are based upon conceptual categories, international parameters are based more upon historic organizational and jurisdictional categories and institutions. These emerged along with the modern state system, a system formalized with the adoption of the Treaty of Westphalia in 1648. At the international and all lower levels, great discontinuities are found between states (and their legal and administrative structures) and a variety of natural, social, and technological phenomena. Many of these discontinuities are the source of great political conflicts, especially those between the group loyalties of various nations and the state boundaries which can either divide them (North and South Korea) or make them national minorities within a larger empire (the former Soviet Union).

Although the history of the modern state system has primarily been one of the dominance of states, a set of international institutions, laws, regulations, and informal networks has emerged since World War II which in various ways transcend individual states. These include the prime promoters of industrial civilization, that is, multinational corporations, various trading regimes, and agencies such as the World Bank and International Monetary Fund. Also, many non-governmental organizations have grown up to represent the concerns of all major social groupings, including environmentalists and conservationists. Many of these have an influence upon biodiversity and its management, although state and sub-state policy is still the most significant arena for action.

Regional Parameters

It is at the regional level that we can begin to speak about communities, or perhaps more accurately, recognizable collections of communities. These include both biotic communities (biomes and biogeographic regions) and economic communities (the European Communities). Elsewhere, I have argued that we need to develop new concepts and categories to deal with the renewable resource systems and regimes that appear to operate at these scales (Dahlberg 1992a). These systems include complex interactions between natural, social, and technological sub-systems. At this level agriculture, forestry, grazing, and fisheries are largely intertwined. These renewable resource systems are also overlaid with trading systems and monopolies, national development policies, and different cultural preferences for certain types of foods and diets.

We have few effective conceptual tools or data to map the dimensions of these regimes because most of our current ones derive from national data sets organized according to the specific economic sectors that have grown up with the modern industrial state. These difficulties are compounded by the academic specialization and compartmentalization between fields and disciplines that has become increasingly predominant since the creation of the modern research university in Germany in the 19th century.

National and Sub-national Parameters

These are generally the key areas for biodiversity management. Arrangements will vary according to the type of government involved: confederal, federal, or unitary. Also, significant differences exist between industrial and developing countries. Generally, industrial countries have gradually added new layers of industrial infrastructures, institutions, and organizations on top of pre-existing medieval patterns. Hoskins' book, *The Making of the English Landscape* (1955), provides a classic description of this in the United Kingdom. The functional specialization of the Industrial Revolution has been applied not only to academia, but also to government, yielding that necessary, but cumbersome institution, bureaucracy.

The process of imposing a mixture of social, economic, technological, and conceptual patterns over various natural systems creates not only confusion, but great political and managerial problems. Those seeking to develop policies to maintain or enhance biodiversity have to deal with landscapes that are physically, institutionally, and conceptually fragmented. The political, legal, and administrative challenges are significant. Politically, one faces the opposition not only of traditional users, such as ranchers and farmers, but increasingly that of developers. Legally, there are also challenges. In terms of conservation, new legal doctrines which would give natural areas legal standing in the courts are required. This would facilitate their protection. Such doctrines would also help to counter recent attempts by developers and their conservative legal supporters to define environmental protection measures as a "taking" of property which would require compensation. The U.S. Supreme Court, in David H. Lucas v South Carolina Coastal Council (120 L. Ed 2d 798), recently reversed and remanded for a rehearing a South Carolina Supreme Court decision rejecting such a doctrine. This case involved a Coastal Council prohibition on additional development on a barrier sea island, an action which was challenged by Mr. Lucas who had purchased property there for nearly $1 million in the hopes of developing it.

Some creative efforts have facilitated administrative cooperation among U.S. federal agencies and between them and state agencies. The U.S. Man and the Biosphere Program has encouraged such programs in the Smoky Mountains and has begun a case study to identify the reasons for the success of efforts to preserve the New Jersey Pinelands. Major legal obstacles to any significant progress persist. The basic legislation and general legal frameworks and doctrines relating to mining, water rights, forestry, fisheries, and agriculture are all based on 18th and 19th century principles of easy extraction and exploitation of resources. Efforts to add any sort of significant conservation components to these legal regimes have been modest. Provisions in the tax code relating to depletion allowances reinforce these extractive tendencies.

CREATING POLICIES FOR ENHANCING SUSTAINABILITY AND BIODIVERSITY

As suggested earlier, natural scientists have tended to focus their policy efforts on specific threatened ecosystems. These include a whole panoply of protected areas that have evolved over the past century in many countries. The United States, for example, has national parks, wilderness areas, national forests, wildlife refuges, national rivers, national seashores, state parks and forests, and many more categories of protected areas with specific policies and management approaches. These policies and management approaches will not be reviewed here because of an extensive and evolving literature on them. However, the need to continue to improve them is basic, both in terms of their effectiveness and in terms of their being founded on the principles of conservation biology. They are the first line of defense against the loss of biodiversity.

Although such policies are a necessary condition for the maintenance and enhancement of biodiversity, they are not a sufficient condition. What is required more broadly is a transformation towards sustainability of the larger economic, political, organizational, and cultural forces that shape and constrain specific biodiversity policies. Any movement of society towards sustainability must involve all sectors of society and must include not only ecological requirements, but requirements relating to ethics, equity, and economics (Dahlberg 1991). Among these are issues of intergenerational ethics, interspecies ethics, equity between the rich and poor groups within and among the countries of the world, and the need to rethink economics so that it is supportive of the other aspects of sustainability.

System-transforming Policies

To determine whether or not these larger context-setting forces and institutions are moving in sustainable and biodiversity-enhancing directions, social science analyses are needed at all relevant levels. Here, I will briefly review some international-level system-transforming policies, and then review in more detail U.S. national and sub-national policies.

International Policies.— Many more policies and practices are emerging at this level that threaten biodiversity than enhance it. One set of threats revolves around the trade negotiations going on in the General Agreement on Tariffs and Trade (GATT) and the North American Free Trade Agreement. GATT and the doctrines of "free" trade have been the traditional preserve of the major trading countries and multinational corporations. Both seek to "open" markets to "free" trade. What this means in practice is removing physical and legal barriers to the exploitation of other countries' social and natural systems. Many of these "barriers" have protected workers and environments from large-scale exploitation.

In addition to removing "barriers," these trade negotiations also seek to establish an international property regime to complement the already existing international financial regime that is enforced by the International Monetary Fund. The latest GATT agreement, if signed, will require Third World countries to provide patent and copyright protection equivalent to that of their industrial trading partners. This "commodification of the world" will also include the patenting of germplasm. In this regard, GATT is seeking to overcome the efforts of the Third World countries to protect their crop germplasm and to guarantee that they receive a reasonable portion of the profits from its genetic engineering (Fowler and Mooney 1990). These efforts, carried on through the Food and Agriculture Organization, also seek assistance for developing the capacity of Third World countries to carry out their own genetic engineering.

More broadly, GATT and other free trade regimes generally facilitate the global threat noted above, namely, the demand in industrial countries for tropical products. Besides a "commodification" of the world, GATT trade rules seriously threaten efforts of environmentalists to use trade sanctions to protect species. Recently, a GATT tribunal (meeting in secret) ruled in a case brought by Mexico that U.S. legislation prohibiting the sale of tuna caught without dolphin-protecting techniques violates GATT free trade rules and that Mexico is entitled to retaliate. If unchanged, this precedent could threaten one of the most innovative and valuable system-transforming treaties created to protect biodiversity, namely, the Convention on International Trade in Endangered Species of Wild Fauna and Flora (CITES).

Another important international arena is that involving the various negotiations and agreements on global warming. Acid rain, ozone depletion, and increased climate and weather variability all pose significant threats to biodiversity. Unfortunately, as indicated above, both the concepts and the negotiations on global change and warming are dominated by physical scientists and industrial interests. Life scientists and environmentalists need to become much more involved in both this arena and in the trade arena.

National and Sub-national Policies.—The types of contributions that social scientists can make in developing system-transforming policies are illustrated in the work of Lynton K. Caldwell, who played a key role in conceptualizing and developing the National Environmental Policy Act (NEPA) of 1969. This act has radically changed governmental decision making, even though several of its more ambitious provisions were never fully implemented. As part of these larger efforts to encourage long-range integrated planning by all federal agencies, NEPA established the environmental impact statement (EIS) procedure, which broke the information monopoly of the individual agencies and forced their proposed projects into both the public and judicial arenas.

Another system-transforming policy is the Endangered Species Act of 1973, which also has required a fundamental shift in the planning procedures used in federally-funded projects. It has enabled those concerned about species and habitat protection to organize prior to the time that construction or logging begins. It gives groups access to the courts to challenge agency actions. The provisions of the Endangered Species Act, which require significant state participation, have also permitted groups like The Nature Conservancy to develop innovative state programs for habitat preservation. They have set up state offices to work with state officials in identifying, surveying, and preserving habitats containing endangered species.

However, as with federal wetlands protection provisions, the Act is increasingly under challenge, both politically and legally. The spotted owl-old growth forest controversy has been one of the major causes of this challenge. Unfortunately, the media focus has been on the owl, the Act, and the plight of the loggers. Little analysis has been done of the increasing mechanization of the industry and the various trade policies that are at the root of the problem.

In addition to maintaining and strengthening specific acts and policies, institutions need to be built that can gather the data, conduct the research, and integrate the relevant approaches and fields to provide both integrated basic science and policy-relevant studies. The President's Council on Environmental Quality (CEQ), which was created by NEPA, served this function during its early years. Under the Carter Administration, it organized and coordinated the *Global 2000 Report to the President* (1980), which was an effort to get individual federal agencies to broaden their horizons. However, the CEQ was severely cut back at the beginning of the Reagan Administration, thus showing the down-side of having such an agency located in the Office of the President.

The proposed National Institute for the Environment (NIE) is a recent initiative. It would be further removed from the political pressures of any given administration. It would not only provide the type of studies referred to above, but over the long term would help to develop both a Washington and a national constituency for environmentally-relevant research. Potentially, it could help to coordinate the environmental research of the various federal agencies, although the bureaucratic politics of such inter-agency coordination are daunting.

A number of public interest environmental groups have also hired staff scientists to identify and bring together policy- relevant science. Examples, among many others, include the National Resources Defense Council, the Environmental Defense Fund, the Union of Concerned Scientists, and Conservation International. These groups work to develop specific policies and political coalitions to try to "level the playing field" between the environmental policy arena and the economic growth and national security policy arenas.

A final area that relates to system-transforming policies involves values. These permeate policy debates, although they are often couched in legal or scientific language. Basic shifts in values and worldview can be seen in a variety of alternative groups such as the greens, feminists, environmentalists, appropriate technologists, solar energy advocates, alternative agriculturalists and defenders of indigenous peoples. These alternative values and visions should receive more public discussion in the media and in schools. Innovative curricula are needed to provide children with a greater awareness of ecological systems and how they interact with humans. On the life sciences side, there are creative garden-based curricula for K-6 science programs (Life Lab and Grow Lab).

Specific Policies

Specific policies (as well as system-transforming policies) will have to be adapted to the natural, social, and economic conditions of the relevant region or country. This is impeded by the destructive tendency at the international level to "transfer" what are seen to be "neutral" technologies, policies, and concepts. Bilateral and multilateral aid programs do this, avoiding the fact that such "transfers" really involve Western technologies, policies and concepts that very often are inappropriate (Dahlberg 1990). The "transfer" of industrial agriculture through the green revolution has been particularly destructive of both traditional agriculture and its environments, whereas the transfer of public health measures without their being integrally linked to effective development programs has led to the population explosion.

Another general problem is that similar types of natural regions or communities can be governed differently depending on where they are located. Some will be governed by sub-national units, others by national governments, and a few by international bodies. Keeping all these various caveats in mind, let us review the types of policies needed to enhance community sustainability and biodiversity.

Lightly-managed Systems.—For lack of better terms, I will distinguish between lightly- and heavily-managed systems. The former are those primarily addressed by this volume and include riparian systems, marine systems, old growth forests, tropical forests, and grazing lands. Of course, some tropical forests and grazing lands are also heavily managed. The difference is in the degree to which modern industrial approaches are used rather than indigenous or traditional approaches. Other heavily-managed systems include modern agriculture and fisheries.

Making this distinction helps us focus on policies to encourage and strengthen indigenous and traditional approaches. Generally speaking, these are less threatening to environments and biodiversity than industrial approaches, although clearly they can become threatening as populations increase. An example is found in slash-and-burn agriculture, where population increases have led to an increase in plots being cut and to a decrease in rotation and fallow times. The same is true of the collection by villagers of fuelwood from forests.

As mentioned earlier, protecting cultural diversity must be an integral part of any effort to protect biodiversity. This also means collecting and understanding better the wealth of knowledge that indigenous peoples have of their local soils, climate patterns, and local flora and fauna, as well as how they manage these "common property systems" in complex multiple-use patterns (NRC 1986). Also, as is now recognized by many conservation groups, local peoples need to have an economic stake in protecting their local resources. In many ways, achieving this might be easier than trying to block the efforts of those, like large timber cutters, who seek to exploit the natural resources of large areas (often having bribed corrupt officials to gain the necessary permits).

Heavily-managed Systems.—In addition to empowering local peoples to maintain traditional light-management practices, it is also important to develop policies that move heavily-managed systems in the direction of sustainability. Besides seeking national policies that encourage sustainable development, specific sectoral policies are needed for sustainable rural development and sustainable agriculture, forestry, grazing, and fisheries. These should be based on principles of biodiversity conservation and a commitment to shift towards greater reliance upon renewable resources (Wilson and Morren 1990).

An extensive literature exists on both sustainable agriculture and forestry. Much of it, however, still tends to focus only on making production more environmentally sound. What is needed is a systems approach that includes not only production, but processing, distribution, use, and waste recycling, and that considers and integrates economic, social justice, and institutional issues (Allen and Van Dusen 1990). In agriculture, such a shift means moving towards a food systems approach (see Dahlberg 1993, for a full discussion). In forestry, terms like "social forestry" have been used to encompass the notion of the total system. Another dimension neglected in the literature is the informal sector or economy. In the United States, the roughly $18 billion annual value of garden produce (NGA 1989) rivals that of the corn crop. Gardens, however, are of little interest to the US Department of Agriculture because they are not part of the formal cash economy, are highly decentralized, and include significant urban production. The importance of informal sector activities involving gardens, woodlots and firewood gathering, grazing, and fisheries is even greater in the Third World.

In agriculture, a sample of the types of policy changes needed to move in more sustainable directions include the following (moving from the more general to the more specific). Although politically difficult, we should shift away from the current crop support system, which by its focus on particular commodities encourages both individual farm and regional specialization rather than diversified farming (GAO 1990). To offer family farmers some protection against the vagaries of weather and to encourage them in the general maintenance of the rural landscape, a negative income tax could be provided to them. This also would help to counter current trends toward further concentration of farm ownership. However, it should be complemented with vigorous anti-trust enforcement against the oligopolies found among major input suppliers and output purchasers. Finally, many tax code provisions relating to resource depletion and equipment depreciation allowances need to be changed to encourage more sustainable practices.

Changes are essential in federal, and especially state rules which currently do little to discourage the filling in of on-farm wetlands. This points to the desirability of some sort of regional land-use planning. Although very difficult to achieve politically, it is important to have general mechanisms to regulate rural land use. The difference the presence of such regulation can make is seen in the much smaller degree of urban sprawl found in Europe as compared to the United States. Changes are desirable in the definitions of crop acreage bases (the number of acres eligible for crop supports). These should include pastures, something that will encourage farmers to include pastures and rotations in their cropping patterns.

A set of changes is necessary to minimize pesticide use in agriculture. The widespread use of pesticides in agriculture and forestry often damages wildlife and aquatic systems, thus threatening biodiversity. Integrated pest management (IPM) programs, originally developed on the basis of ecosystem approaches, have been touted for a long time, but have in large part been co-opted and re-interpreted to mean integrated pesticide management. Pesticides are an integral part of the currently dominant production paradigm and are supported by powerful interests both within the agricultural establishment and the agrichemical industry (Dahlberg 1992b). The "pesticides complex" is one of the most difficult obstacles facing the movement towards sustainable agriculture.

Some have expressed optimism that the move towards sustainable agriculture can be facilitated by developments in biotechnology. This ignores two basic difficulties. One is that biotechnology is simply the latest, and most powerful elaboration of the reductionist and functionally specialized approaches that helped to create the difficulties we find ourselves in today. Ecosystem-based approaches like integrated pest management (expanded to included social, economic, and technological factors) offer much more potential than reductionist approaches for integrating systems and making them sustainable (see Conway 1985, for a discussion of Third World examples).

The second basic problem relates to the non-neutrality of all technologies, which clearly reflect their design principles, their scale, and their originating environment (Dahlberg 1990). Biotechnologies are currently moving in the predictable directions dictated by the production paradigm, their scale, and the interests of the agrichemical companies. That is, towards an emphasis on high-value specialty crops (strawberries) or increasing production (bovine growth hormone), even where it is not needed. Furthermore, rather than working on nitrogen-fixing and pest-resistant seeds, the agrichemical companies are seeking to develop seeds which will respond to their herbicides and fertilizers. The development of biotechnologies that genuinely support sustainability can be expected only if the dominant production paradigm and the institutions reinforcing it are restructured away from production to sustainability.

Specific habitat support programs should also be considered. The idea of supporting farmers to engage in practices beneficial to the larger community has been accepted in Switzerland, where there are financial supports for maintaining high mountain pastures and grazing as flood protection and anti-avalanche measures. Such support programs could encourage use of hedgerows, maintenance of woodlots, and protection of wetlands, all of which would encourage landscape and biodiversity (Graham and Staton 1989).

A number of other natural resource issues overlap with biodiversity. The literature here has been developed primarily by environmentalists and conservationists. One recent volume, *Natural Resources for the 21st Century* (Sampson and Hair 1990) reviews land, agricultural, forest, fishery, wetland and other natural resource issues and problems. Although not specifically focusing on biodiversity, the policies they suggest for the long-term conservation of natural resources are quite relevant for biodiversity conservation (also, see Western and Pearl 1989).

CONCLUSIONS

The broad underlying argument here has been that much greater conceptual and practical integration is needed between natural scientists, social scientists, humanists, and those managing and regulating renewable natural resources. It has also been stressed that while maintenance and enhancement of biodiversity is a necessary condition for sustainability (or regeneration), it is not sufficient. There must also be a general movement of other sectors and societies in general towards sustainability. In this regard, those concerned about biodiversity must be concerned not only with the policies, rules, and regulations affecting protected areas and species, the first line of defense in the protection of biodiversity, but must also encourage policies which will transform the larger systems, sectors, and societies in more sustainable directions.

Of particular importance are heavily-managed renewable resource systems—agriculture, forestry, fisheries, and grazing. Their current industrial structures, which are fossil-fuel and capital intensive, are not sustainable and threaten the biodiversity base upon which they ultimately depend. They, along with larger urban and industrial processes, also threaten biodiversity through global warming, acid rain, deforestation, and pollution. A hierarchy of management and policy strategies is needed to promote sustainability generally and the maintenance and enhancement of biodiversity in particular. It is not enough to try to protect the biological "roots" of our natural communities; we must revive or find new cultural "roots" that will enable communities of humans to live on a sustained basis with the communities of their fellow species.

ACKNOWLEDGMENTS

I would like to thank Daniel J. Decker and David W. Orr for their helpful comments and suggestions for improving this chapter.

LITERATURE CITED

Allen, P., and D. Van Dusen. 1990. Sustainability in the Balance: Raising Fundamental Issues. Agroecology Program. University of California, Santa Cruz, Calif.

Bennett, J. W., and K. A. Dahlberg. 1990. Institutions, social organization, and cultural values. In *The Earth as Transformed by Human Action*, ed. B. Turner II, pp. 69-86. Cambridge University Press, Cambridge.

Brush, S. B. 1981. The environment and native Andean agriculture. *Am. Indigena* 40: 161-172.

Conway, G. 1985. Agroecosystem analysis. *Agric. Administration* 20: 31-55.

Council on Environmental Quality (CEQ). 1980. *The Global 2000 Report to the President.* U.S. Government Printing Office, Washington, D.C.

Dahlberg, K. A. 1979. *Beyond the Green Revolution: The Ecology and Politics of Global Agricultural Development.* Plenum, New York.

Dahlberg, K. A. 1990. The value content of agricultural technologies. *Agric. Ethics* 2 (2): 87-96.

Dahlberg, K. A. 1991. Sustainable agriculture: fad or harbinger? *BioScience* 41 (5): 337-340.

Dahlberg, K. A. 1992a. Renewable resource systems and regimes: key missing links in global change studies. *Global Environ. Change* 2 (2): 128-152.

Dahlberg, K. A. 1992b. Policies that encourage pesticide use in the United States. In *The Pesticide Question: Environment, Economics and Ethics*, eds. D. Pimentel and H. Lehman, pp. 281-306. Chapman and Hall, New York.

Dahlberg, K. A. 1993. Regenerative food systems: broadening the scope and agenda of sustainability. In *Food for the Future: Conditions and Contradictions of Sustainability*, ed. P. Allen, pp. 75-102. John Wiley, New York.

Fowler, C., and P. Mooney. 1990. *Shattering: Food, Politics, and the Loss of Genetic Diversity.* University of Arizona Press, Tucson, Ariz.

General Accounting Office (GAO). 1990. *Alternative Agriculture: Federal Incentives and Farmers' Options.* GAO/PEMD-90-12. General Accounting Office, Washington, D.C.

Graham, A. E., and M. Staton. 1989. *Integrating Conservation Biology and Agricultural Production: Executive Summary of International Workshops.* Public Service Research and Dissemination Program. University of California, Davis, Calif.

Hoskins, W. G. 1955. *The Making of the English Landscape.* Hodder and Stoughton, London.

Lovelock, J. E. 1979. *Gaia: A New Look at Life on Earth.* Oxford University Press, New York.

Lovins, A. 1977. *Soft Energy Paths.* Harper, New York.

McHarg, I. 1992. Talk presented to the Small City Conference, Western Michigan University, Kalamazoo, Mich., April 1992.

Margalef, R. 1968. *Perspectives in Ecological Theory.* University of Chicago Press, Chicago, Ill.

Meadows, D. H., D. L. Meadows, J. Randers, and W. W. Behrens III. 1972. *The Limits to Growth.* Universe Books, New York.

National Gardening Association (NGA). 1989. National Gardening Fact Sheet. National Gardening Association, Burlington, Vt.

National Research Council. 1986. *Proceedings of the Conference on Common Property Resource Management.* National Academy Press, Washington, D.C.

Pimm, S. L. 1991. *The Balance of Nature: Ecological Issues in the Conservation of Species and Communities.* University of Chicago Press, Chicago, Ill.

Sampson, R. N., and D. Hair, eds. 1990. *Natural Resources for the 21st Century.* Island Press, Washington, D.C.

Wallerstein, I. 1974. *The Modern World System: Capitalist Agriculture and the Origins of the European World-Economy in the Sixteenth Century.* Academic Press, New York.

Wallerstein, I. 1980. *The Modern World System II: Mercantilism and the Consolidation of the European World-Economy, 1600-1750.* Academic Press, New York.

Western, D. C., and M. C. Pearl. 1989. *Conservation for the Twenty-First Century.* Oxford University Press, New York.

Wilson, E. O., ed. 1988. *Biodiversity.* National Academy Press, Washington, D.C.

Wilson, K., and G. E. Morren, Jr. 1990. *Systems Approaches for Improvement in Agriculture and Resource Management.* Macmillan, New York.

World Commission on Environment and Development (WCED). 1987. *Our Common Future.* Oxford University Press, Oxford.

Young, O. R. 1989. *International Cooperation: Building Regimes for Natural Resources and the Environment.* Cornell University Press, Ithaca, N.Y.

46

CREATING POLICY ON LANDSCAPE DIVERSITY

Peter Bridgewater, Dan W. Walton, and John R. Busby

Australian National Parks and Wildlife Service
Canberra ACT
Australia

> "By and large, our present problem is one of attitudes and implements. We are remodeling the Alhambra with a steamshovel, and we are proud of our yardage. We shall hardly relinquish the shovel, which after all has many good points, but we are in need of gentler and more objective criteria for its successful use." *A Sand County Almanac* (Aldo Leopold 1949)

The landscape scale is the forgotten aspect of biodiversity. Public attention and, to a large extent, professional biological attention have been fixed at the species level. Although that is understandable, only at the landscape scale can effective biodiversity conservation be achieved. Conservation of biological diversity is a human activity, policy is a human creation, and landscapes are increasingly anthropogenic.

A statement released recently by The Royal Society (London) and the National Academy of Sciences (Washington, D.C.) (TRSNAS 1992) pointed out that the current rate of reduction in biodiversity is unparalleled over the past 65 million years and that this loss is one of the fastest moving aspects of global change.

Policy for the conservation of landscape diversity is the most essential, but also most neglected of all biodiversity conservation concerns. This situation is due, in part, to the complexity of the landscape scale. National legislation related to biodiversity has been largely directed at protection at the species and genetic levels [Endangered Species Acts, Wildlife Protection (trade) Acts]. International treaties and conventions similarly have embraced the species and genetic levels: Convention on International Trade in Endangered Species of Wild Flora and Fauna (CITES), Convention on Wetlands of International Importance Especially as Waterfowl Habitat (Ramsar Convention), Convention on the Conservation of Migratory Species of Wild Animals (Bonn Convention), International Convention on the Regulation of Whaling, Convention for the Conservation of Antarctic Marine Living Resources (CCAMLR), Convention for the Conservation of Antarctic Seals, United Nations Convention on Biological Diversity. Some efforts to include local communities have been made, but with the exception of heritage coasts or protected landscapes of Europe where the focus was on anthropogenic aspects, landscape scale conservation of biodiversity has remained isolated from mainstream conservation efforts.

Is it all just too hard? Or, are there real reasons why landscape has been ignored or left behind? How can a policy frame-work help? This paper attempts to answer those questions and draws on a scientific frame-work to help set policy goals. Some of these themes have been treated in Bridgewater (1993), Walton et al. (1992) and Bridgewater et al. (1992a). A most important point to make here is that landscape must include seascape. Marine environments have been neglected in many ways and some argue that they are quite distinct systems. Although there is some truth in that perspective, the linkages between land and sea are important to the conservation of biodiversity. Every action that takes place in a catchment has a potential downstream effect, ultimately in the sea. There are increasing concerns that problems in marine biodiversity at all levels originate on land.

THE LANDSCAPE SCALE

Conservation of biodiversity in the last decade has concentrated on the population and community level. Indeed, the early issues of the journal *Conservation Biology* have a monotonous theme of population studies. For the future, population studies, community ecology, and autecological studies will still be important, but will be most useful when set in a context of appropriate spatial and temporal scales. Conservation management has been concerned chiefly with the rare and colorful, rather than the common and perhaps the less colorful, a situation reflecting aesthetic appeal and public perceptions (Wilson 1987). Conservation research effort has inevitably followed available funding, as well as the expertise and personal interests of researchers. Over the coming years a much broader approach to the conservation of biodiversity must prevail, at a landscape scale. And why? Large systems tend to have greater stability than their component parts and this stability tends to dampen chaotic activity of those parts. Focus on landscapes or biocultural regions, therefore, is essential if we are serious about sustainability (Odum 1992). Processes that endanger species arise from landscape degradation, e.g., fragmentation, physiographic alteration, vegetation removal or replacement, or pollution.

Thus, by shifting our study focus from the species and population level to examine the community/ecosystem/landscape level, we can start to make progress. Certainly, the approaches taken by landscape ecologists in the last decade (see *Landscape Ecology*, vols. 1-7) have made major strides, such as emphasizing the importance of metapopulations, which takes population study and embeds it in a landscape framework (Opdam 1991). Recent work on wildlife habitat in Canada also underlines the importance of a landscape approach (Wildlife Habitat Canada 1991).

As Slocombe (1993) observes, the use (misuse or abuse?) of the term "ecosystem" is a problem among ecologists. For those charged with the responsibility for policy, the term often appears to be definable on a sliding scale more attune to the whims of the ecologists that to any consistent conceptual entity useful for policy. The term "habitat" has also acquired whimsical usage and is of decreasing value.

LANDSCAPE INFRASTRUCTURE

Integral to the conservation of biodiversity is the connectivity of landscapes. The use of landscape here follows Forman and Godron (1986): "kilometers-wide area where a cluster of interacting stands or ecosystems is repeated in similar form." Protected areas, within the concept of biosphere reserves, and elements of remnant vegetation connected by corridors or "greenways" can provide the reservoirs of biodiversity and avenues for adaptive responses to environmental change. If, as noted below, protected areas are isolated, their purpose is largely defeated. Also, living associations and their environments change. Corridors or "greenways" reduce landscape fragmentation and assist in the rehabilitation of degraded areas.

Conservation outside reserves not only conserves biodiversity, but has significant influence on the landscape matrix. Erosion control, catchment security, land reclamation, soil fertility improvement, modifications to air temperature and humidity, control of soil salts, water table levels—all are affected by properly placed and designed landscape infrastructure, which then needs careful management (Forman and Godron 1986).

Conservation outside reserves cannot be a haphazard affair; when done properly, it will be education—and information—intensive and require appropriate resources for better and increased levels of management. As Pearce (1993) points out, more carrots rather than more sticks. Resilience provided by adequate *ex* reserve conservation will alleviate the problems which might spring from global change.

The establishment or revitalization of extension services should be given serious consideration (Bennett 1990). Extension services can provide the linkage for information transfer from research scientists and area managers to local communities and the needs of the local communities to research scientists and area managers. Without the goodwill, cooperation, and participation of local communities, there can be little hope of achieving ecological sustainability (Saunders 1990).

SOME EXAMPLES

Implications of landscape infrastructure for wildlife conservation can be seen with a particular species—the Koala (*Phascolarctos cinereus*). Two of the largest challenges for Koala survival are adequate food reserves and habitat through which Koalas can move safely from one area of food trees to another. Koalas prefer foliage from trees growing in those areas with higher nutrient soils, areas now largely cleared for development. Consequently, Koalas are confined to small patches or strips of food trees scattered across open country. The hazards are greatly increased, including attacks from introduced species, to Koalas moving across open country between the patches or strips. Strong territorial behavior and annual recruitment mean Koalas must seek new areas, but their choices are limited by the removal of vegetation associated with development.

The future of Koalas depends heavily on the retention of ample tree cover, as well as planting more of the appropriate species in landscape rehabilitation. Development of landscape infrastructure which provides opportunity for dispersal and decreases vulnerability to predators will assist the survival of Koalas, but there are further hidden advantages. At the landscape level, the increase in vegetation resulting from planting reduces fragmentation, decreases salinization and other land degradation problems are ameliorated.

A landscape ecological approach would have been helpful also in resolving political controversy which surrounded the nomination of the cool rainforests of Tasmania and the warm rainforests of Queensland for inclusion on the UNESCO World Heritage List. In both cases, large areas of forest have now been inscribed. Some arguments against listing originate from a simplistic legal view that "not all the area is needed for World Heritage Listing" or, more precisely, that not all the area "fits" the World Heritage criteria. Briefly, these criteria are as follows. The area must:

(a) be outstanding examples representing the major stages of the earth's evolutionary history; or

(b) be outstanding examples representing significant on-going geological processes, biological evolution and man's interaction with his natural environment; or

(c) contain superlative natural phenomena, formations or features or areas of exceptional natural beauty; or

(d) contain the foremost natural habitats where threatened species of animals or plants of outstanding universal value still survive.

In addition to the above criteria, the sites should also fulfill the conditions of integrity.

Naturally, the criteria are relatively "fuzzy" to allow for the expression of the maximum World Heritage value. The important feature is site integrity. Even though some areas may be viewed as fragmented nodes within a matrix, it is the holistic nature of the area which is critical. Landscape ecological principles relating to the ecological importance of linkages between infrastructure elements help clarify the need for integrity of area, allowing for connectivity between the high quality nodes dispersed through a (possibly) lower quality matrix. From a nature conservation perspective, the nodes of high quality are the vital elements. Landscape ecology theory implies the nodes will rapidly degrade if the surrounding matrix is not included as part of the comprehensive management area.

MARINE AND COASTAL LINKAGES

A major international research thrust on boundaries in the last three to four years has come from UNESCO-MAB and SCOPE activity on land/freshwater ecotones (cf. di Castri et al. 1988). Terrestrial/freshwater boundaries were seen as important because of the hypothesis that greater biodiversity occurred in such boundary zones (Brabyn and Hadley 1988). Nowhere are boundaries more significant than between land and sea. Hayden et al. (1984) proposed a classification of coastal and marine environments. Bunt (1987) and Wilson and Allen (1987) deal with the Australian marine environment and fauna in greater detail. Whether these classifications prove useful for the purposes of the maintenance and management of Australia's biodiversity, several points are relevant. Coastal landforms associated with the drainage of fresh waters into oceans and seas are of special physical and biological significance. Estuaries and wetlands have their own unique characteristics and processes.

Much is often made of the differences between land and sea ecological systems, and real differences are known. Yet there are also similarities, which we can exploit in attempting system modeling and the establishment of geographic information systems. Coastal and marine areas have landscape and seascape characteristics which do not align easily with terrestrial landscapes; demands on and content of geographic information systems for landscapes and seascapes, therefore, will differ. For example, marine, and also migratory bird, species need to viewed as components of four-dimensional corridors (three spatial components and a temporal one).

Marine and coastal zones are already of great significance. In Australia, 80% of the human population lives and works within 50 km of the coastline. Most of the global human population live in and the majority of all marine living resources are harvested from the coastal zone. As the interiors of continental land masses are increasingly degraded, there will be increasing movement of people to the coastal zone. Coastal zone infrastructure, presently under severe stress, will be subjected to even further stress (cf. Nierenberg 1988; Carlton 1989; Ray and McCormick-Ray 1992). That people cannot change the seas and their biodiversity, that marine resources will not be damaged or depleted by exploitation and that the seas can absorb infinite amounts of human waste are views which are untenable (Lasserre 1992).

THREATS TO LANDSCAPE INTEGRITY

Hobbs and Hopkins (1990) consider the problems and opportunities of restoration or reconstruction of Australian plant communities. The reversal of history and a return of the biosphere to some earlier real or imagined state is impossible (Soulé 1985). Ecological rehabilitation techniques, however, might be of great value for maintaining biodiversity in the face of global (and local) change; the possibility exists that entire protected associations can be relocated (Jordan et al. 1988) and, if the scenarios for global change are accurate, will have to be relocated.

Many species require disturbed and degraded areas (Green 1989). New associations undoubtedly have developed in areas which have experienced long-term disturbance or areas which have been subject to extensive invasion (Fox 1990; Green 1990). These synthetic communities (Bridgewater 1990) must be assessed within a landscape context. Opportunities must be seized to explore what Soulé (1990) terms recombinant (or mixo-) ecology while recognizing that human-induced disturbances should be minimized. Our current practice of conducting large-scale uncontrolled experiments on natural and seminatural ecosystems must be recognized as inherently risky. Were comparable experiments to be conducted in other fields, e.g., public health, they would be widely regarded as at least unethical, if not illegal.

Human-induced global change might produce massive ecological disruptions (Soulé 1990). Fragmentation of ecosystems by European settlers in Australia and New Zealand, as well as Polynesian migration to New Zealand and the much earlier migrations of Aborigines to Australia, certainly have paved the way for extinction combined with invasion. Colonization and invasion might be the dominant ecological processes and new associations and combinations the dominant ecological characteristics of the future. Too easily we forget that change is an inherent characteristic of living systems, that what we observe are not static displays of nature.

Global change and landscape are inextricably linked through feedback mechanisms. Global changes come from a variety of sources. Removal of vegetation, atmospheric warming and increased pollution loads are some of the most obvious. Changes in agricultural practices, quite sharp and sudden swings in the pattern of land use, are of tremendous importance in conservation practice. Our management strategies will be developed most effectively if we look at the components of landscape and at the impact of possible changes in a landscape ecological frame-work. This implies an understan ding of the relationship between keystone elements in the landscape, the surrounding matrix, and the need for connectivity.

On the one hand, there is evidence that most of those global changes outlined above tend to reduce species diversity, community diversity, and probably landscape diversity. On the other hand, global warming certainly presents the opportunity for the development of new communities. It is, after all, quite likely that in previous geological eras new communities arose spontaneously as a result of environmental changes brought on by warming or cooling of atmospheres, changes in the amount of available water, in precipitation patterns (Delcourt and Delcourt 1988).

Neither can we ignore the potential impact of non-native species being added to the species mixtures which form our landscapes. Hobbs and Hopkins (1990) give some examples of the scale of the problem in Western Australia. Exotic species, both plant and animal, can alter significantly the physical parameters of the area they invade. In Australia, plant species, for example, can change the fire frequency and/or intensity, totally replace native vegetation (e.g., *Mimosa pigra*, *Eichornia crassipes*) by changing light penetration patterns (e.g., *Cryptostegia grandiflora*) or change the soil chemistry by litter accumulation (e.g., *Tamarix aphylla*) (Longmore 1992).

National nature conservation systems using the bioregional management model, connecting vegetation remnants and landscapes by corridors or "greenways" and embarking on a program of nature development, will be essential to the maintenance and management of biodiversity, especially genetic diversity (Vida 1978; Noss 1983). Bioregional management provides the only setting for dealing with many of the problems of genetics and demography (cf. Lande 1988). It must also encompass environmental monitoring, including research into past environments, and long-term ecological research, which can provide information critical to sound management within a landscape perspective. Human impact on most Australian landscapes is at a level where active intervention is necessary (Goodman 1987).

The message is clear—we need research at the landscape scale to look at system dynamics, species properties, and how we can attempt rehabilitation of degraded areas and replacement of vanished systems. We need ecological technology in the hands of people with the vision and expertise of a latter-day Capability Brown. After all, he created much of the landscape of 18th Century England, and in so doing created wildlife opportunities. Indeed, in Britain, much of the so-called "wildwoods" of today are of quite recent origin (Pederken 1981).

Conservation managers need to be able to distinguish landscape infrastructure from boundaries and pay attention to these boundaries in management plans. This is an area where considerable research appears necessary to characterize the structure and, especially, function of such landscape elements and how they relate to ecological system processes.

Boundaries exist in time and space. Human disturbance creates variations on this theme; boundaries are created and maintained over time and space by human intervention. The "marginalisation" of lands is, in fact, the creation of new margins. Despite the fact that we have imposed boundaries with reckless abandon, we know very little about them. The diversity of flora and fauna which is associated with our imposed boundaries also is known poorly and largely unappreciated.

RESTORATION AND REHABILITATION

Disturbed ecosystems are now the norm (Cairns 1989). We live in an increasingly anthropogenic biosphere whose long-term carrying capacity might well have been exceeded (Ehrlich 1982). The world human population, according to United Nations Population Fund (UNFPA) estimates, will rise from the present level of 5.4 billion to 10 billion by the year 2050. In this scenario, environmental stresses of human origin are unlikely to diminish, while science and technology may offer no hope for avoiding irreversible environmental damage and great human poverty (Kumar 1993).

Most efforts at restoration or rehabilitation have been directed towards endangered species and harvested species. Endangered species are fundamental indicators of disturbance. Because extinction is a process, not a simple event, the recognition that a species is endangered is little more than a snapshot of a moving vehicle. Attempts at therapy most often address symptoms rather than causes. This situation equally applies in large measure to harvested species. Any modern industrial society with even marginally arable land can produce, with the application of appropriate technology, more food and fibre than that society can consume. Small wonder (Costanza and Daly 1991) that many economists and others, equally naive environmentally, view restoration and rehabilitation as a relatively trivial issue comprised of "greenies" saving species and harvesters who simply need more additives, pesticides or "new technology."

We have failed to communicate successfully why rehabilitation and restoration, beyond the narrow focus of the endangered and harvested, are essential. The environmental variables which affect the health and welfare of all the flora and fauna also affect people: water and air quality, availability and concentration of nutrients, available habitat, recycling of organic and inorganic substances, microclimate. Loss of biodiversity means loss of ecological services and options for the future. The cost of replacing ecological services, already great, will increase to staggering proportions. The real and potential wealth represented by conserved biodiversity cannot be replaced.

The joint statement from The Royal Society and the National Academy of Sciences (TRSNAS 1992) emphasized the need to halt irreversible degradation (Schlesinger 1990). The reversible can be addressed. The morbidity of landscapes can be treated and the prevention of further degradation, we believe, will be less costly than intensive care. There are no quick fixes or permanent solutions. Living systems, of which people are part, are dynamic and no place is unimportant. The greater the need to protect living systems and the greater the need to rehabilitate, supplement, and otherwise support living systems, the fewer the options for sustainability.

SOME GUIDING PRINCIPLES FOR POLICY FORMULATION

The challenge lies in finding the will to develop policies which establish priorities, create the efficiency, carry out the necessary institutional reform, discourage a worsening of conditions, provide incentives for improvement, and create the wealth to pay for the conservation of biological diversity. Do we have the will? Is the conservation of biological diversity as important as we say it is or are we merely chasing next year's budget or grant funds? We must find the will and dispel the myth that we are serving self-interest.

As Grubb and May (1991) recommend for research, we must link ecology with economics, sociology, and politics and ensure that good policy intentions do not yield inappropriate results (McNeely 1992; Mott and Bridgewater 1992). Agency and institutional missions must be clearly defined by tasks, not by infra-organizational or professional subculture, so that performance and achievement can be evaluated on a regular basis as part of a greater whole.

In the development of policies for the conservation of biological diversity, the strategic intent of those policies must be clear and unambiguous. That intent should:

(a) capture the essence of success; the strategic intent of the policy (reason for, aim of and relationship to other policies) should be easily understood and elicit a popular enthusiasm;

(b) be stable over time; and

(c) set a target that deserves personal effort and commitment.

Policy initiatives should emphasize existing perceptions of biocultural regions. All countries have them—The Murray-Darling River System and the Western Australia Wheat Belt in Australia; the Tennessee Valley Authority region, Appalachia of the United States, the "Ecologically Sensitive Areas" of the European Community. Research biologists will wish to establish purely "biotic provinces", but conservation is a human endeavor and local communities must "own" the endeavor. Without public enthusiasm and public support, conservation of biological diversity will remain little more than a dream. We have little time in which to gather this support, to assist in the guidance of appropriate policies, and to formulate the necessary actions.

The nature conservation estate (areas with statutory protection) is among the most valuable assets for the maintenance and management of genetic, species, community and landscape diversity, as well as the various ecological processes of importance to humanity (Reid and Miller 1989). As the *Global Biodiversity Strategy* notes: "...protected areas linked by corridors become means of maintaining functioning natural or near-natural ecosystems over large regions" (Anon. 1992a). For the future, conservation of biodiversity MUST focus on management at the landscape level, i.e., bioregional management. Significantly, the recently completed review of the U.S. National Wildlife Refuge System (Weeden and Cutler 1992) recommends this approach, as does the draft National Biodiversity Strategy for Australia (Anon. 1992b).

Similar scientific and politico-administrative responses to the themes of the conservation of biodiversity and global change are emerging in many countries. Lubchenco et al. (1991) consider the needs for different and re-oriented approaches to achieving the "sustainable biosphere," what they regard as a top priority "in the closing decade of the twentieth century". A major thrust of Lubchenco et al. (1991) is the critical importance and interlinkage of the triumvirate of global change, biodiversity, and sustainable ecological systems. All three of these issues have been taken up by the Australian Government in a nationally focused effort attempting to advance to ecologically sustainable development. How these issues impact on biodiversity conservation and where effort should go over the next decade is worth examining.

If countries establish a national nature conservation system, with the aim of maximizing the conservation of biodiversity, and if nature conservation reserves are viewed as part of the landscape matrix, rather than as islands, and are placed within an integrated management frame-work for that landscape, a real beginning will be made.

The concept of nature conservation reserves within an integrated landscape management frame-work is not new. The biosphere reserve (Batisse 1982), a concept developed and championed under the UNESCO Man and the Biosphere Programme, was formulated to integrate conservation and development. Most of the action under the Biosphere Reserve Program, unfortunately, has been misplaced and/or misunderstood and the valuable aims of the program have fallen into disrepair. The concept has undergone considerable refinement since its inception. Fortunately, under the pressure of the numerous international initiatives underway for 1992, supported by United Nations agencies such as UNESCO and UNEP, the concept is being re-vitalized. Batisse (1993) provides a brief historical summary of the MAB Programme, describes its present vitality and the strengths of the biosphere reserve model in relation to the aim of an ecologically sustainable biosphere.

A biosphere reserve is first and foremost a representative ecological area. The concept combines nature conservation with scientific research, environmental monitoring (Herrmann 1990), training, demonstration, and local participation. A strictly protected core area is surrounded by a buffer zone in which research, environmental education, training, and recreation takes place. An outer transition area surrounds the buffer zone. In the transition area, close cooperation can be developed between researchers, managers, and the local population to ensure sustainable development of the natural resources of the area (Brabyn and Hadley 1988). In other words, it is management at a landscape/bioregional level. Biosphere Reserves have an equal or greater role in the coastal/marine environment, a management model for the very complex land-sea interface, as in wholly terrestrial environments (Batisse 1990; Ray and Gregg 1991). The International Network of Biosphere Reserves is the only network of protected areas at the intergovernmental level which combines the *in situ* conservation of representative examples of natural and seminatural ecosystems with scientific research and monitoring, consideration of rural development and local resource use problems, as well as with environmental research and training (Solbrig 1991). The concept continues to evolve and has proved to be useful in overcoming barriers to integrated area management (Ishwaran 1992).

Without invoking biosphere reserves, the Netherlands Government (1990) has produced a *Nature Policy Plan*, which echoes the main points outlined above. That policy proposes a network of national ecological infrastructure, including corridors, nature development areas and endangered species recovery plans. It also recognizes that conservation on land, fresh water, and sea are highly interdependent. Research areas under the plan cover biogeography and metapopulation studies for corridors, fixing of ecological standards, research on reference systems, inventory, monitoring, and ecohydrology.

Above all, we must become more focused (cf. Western 1991). Conservation depends upon the implementation of what we know by managers with bold vision and capable of bold action. Conservation biologists must take a leadership role and accept that mistakes will occur. By imposing differing orthodoxies on research scientists and area managers, we are over-selling and under-utilizing research. The debates in which scientists love to engage (too often the heat of the debate is inversely proportional to what is at stake) must be more judicious and worthy, for they only confuse potential supporters and confound managers. Research, whether applied or curiosity-driven, is fundamentally and finally part of management. Dynamic systems require dynamic management.

The goal of conservation biology (cf. Soulé 1985) and the tasks of the conservation of biodiversity are sufficiently great that it will be the poor institution, agency, or individual unable to contribute. Reductionist approaches have their place, and an important one, but alone are inadequate for the situations presented by today's environment-development issues (di Castri and Hadley 1992).

Conservation by legislation needs more thought and less emotion (Burbidge 1990). We must be cautious that the heady atmosphere of a political victory is not a prelude to a nightmare of legal entanglements which squander badly needed resources and distort issues (see Freeman 1990). We must maintain a perspective on a scale that discourages quick, easy and temporary solutions, and not be demeaned by "the tyranny of small decisions" (Odum 1982).

Entirely too much of the discussion relating to biodiversity has been taken up with the search for whom to blame for the present state of the environment, a search predicated on the premise that "nature" and "people" are unrelated. Inescapably, we all are responsible. That being said, we recognize that we, people, are part of the biosphere. Callicott (1990) made the point most effectively:

> Now that man is completing his tragic conquest of nature, we conservationists are inclined to reverse the roles of hero and villain. But in so doing we only perpetuate a false dualism.

The conservation of biodiversity must not fracture into an endless series of independent skirmishes over research priorities, new wings for museums, endangered species, national parks, what "used to be here", wilderness. Human health, environmental health and economic health are intimately linked (Bridgewater et al. 1992b). The main thrust will have to be management: how we apply what we know, how we determine what we need to know and how the changes in human behavior and institutions can be achieved. The failure to put adequate management in place is a failure to face problems and opportunities (Peterson 1988).

Rackham (1986) expressed concern for four kinds of losses. This concern is borrowed and expressed here as what must be conserved: the beauty of the landscapes, its surprises and marvels of scale; the freedom of our cultural pathways through an environment which has shaped us and which we have shaped; historic vegetation and wildlife; and meaning, the landscape record of our culture and civilization.

CONCLUSIONS

For the ecological scientist, setting a policy frame-work for conservation of biodiversity at a landscape scale is both easy and confusing. It is easy because the scale is best adapted to solving the major issues of biodiversity conservation, but confusing because the complexities operating at that scale do not have the same intuitive appeal of "Let's save the Potoroo/Spotted Owl/Siberian Tiger." For the politician, the challenge is to integrate an array of highly local concerns of constituents into a regional frame-work that can be tied to the national legislative process. For the bureaucrat, the challenge is to provide the essential feed-back on the evaluation of administration. The recognition that people are part of the environment and that change is a characteristic of living systems must be central to policy formulation.

Successful policy development will require an understanding of bioregional approaches, including the definition, design and understanding of bioregions. A bioregional approach, embracing whole catchments and crossing the land/sea interface where appropriate, is the key. Recognition that biodiversity comprises more than "native" species, and that restoration/rehabilitation are vital tools at the landscape scale is critical. Nesting of conservation plans from farms or individual land units, through bioregions to nations and continents is the only viable direction to take.

Without a policy frame-work at the landscape level, all other forms of biodiversity conservation can be caricatured as simply creating zoos, botanic gardens, or, worst of all, museums. Only a living, fully functional landscape will allow us to conserve and manage the maximum biodiversity at global scale. Endangered species led the raising of political consciousness regarding biodiversity in the 1970s. Now we need to raise consciousness about endangered spaces and to develop appropriate policies.

LITERATURE CITED

Anonymous. 1992a. Strengthening protected areas. In *Global Biodiversity Strategy*, ed. K. Courrier, pp. 117-132. World Resources Institute, World Conservation Union, United Nations Environment Programme, Washington, D.C.

Anonymous. 1992b. *A National Strategy for the Conservation of Australia's Biological Diversity*. Draft for Public Comment. Department of the Arts, Sport, the Environment and Territories, Canberra, Commonwealth of Australia.

Batisse, M. 1982. The biosphere reserve: a tool for environmental conservation and management. *Environ. Conserv.* 9:101-111.

Batisse, M. 1990. Development and implementation of the biosphere reserve concept and its applicability to coastal regions. *Environ. Conserv.* 17:111-116.

Batisse, M. 1993. The Silver Jubilee of MAB and Its Revival. *Environ. Conserv.* 20:107-112.

Bennett, A. F. 1990. Habitat Corridors. Their Role in Wildlife Management and Conservation. Victoria Dept. of Conservation and Environment, Melbourne.

Brabyn, H., and M. Hadley. 1988. *Man Belongs to the Earth*. UNESCO Paris.

Bridgewater, P. B. 1990. The role of synthetic vegetation in present and future landscapes of Australia. In *Australian Ecosystems: 200 Years of Utilization, Degradation and Reconstruction*, eds. D. A. Saunders, A. J. M. Hopkins, and R. A. How. *Proc. Ecol. Soc. Australia* 16:129-134.

Bridgewater, P. 1993. Conservation strategy and research in Australia—or how to arrive at the 21st Century in good shape. In *Conservation Biology*, eds. C. Moritz, G. G. Grigg, and J. Kikkawa, in press. Surrey Beatty, Chipping Norton.

Bridgewater, P. B., D. W. Walton, J. R. Busby, and B. J. Reville. 1992a. Theory and Practice in Framing a National System for Conservation in Australia. In *Biodiversity: Broadening the Debate*, pp. 3-16. Australian National Parks and Wildlife Service, Canberra.

Bridgewater, P., A. F. Bennett, J. Landsberg, D. F. McMichael, J. J. Mott, I. R. Noble, P. G. Quilty, and D. W. Walton. 1992b. Biodiversity: Its maintenance and management. In *Scientific*

Aspects of Major Environmental Issues: Biodiversity, pp. 8-19. Papers prepared by two independent working groups for consideration by the Prime Minister's Science Council at its sixth meeting, 18 May 1992. Australian Government Publishing Service, Canberra.

Burbidge, A. A. 1990. Comment on Wildlife conservation in Australia State of the Nation. *Australian Zool.* 26:63-66.

Bunt, J. S. 1987. Marine environment. In *Fauna of Australia*. Vol. 1A. General Articles, eds. G. R. Dyne and D. W. Walton, pp. 17-42. Australian Government Publishing Service, Canberra.

Cairns, J., Jr. 1989. Restoring damaged ecosystems: is predisturbance condition a viable option? *The Environ. Professional* 11:152-159.

Callicott, J. B. 1990. Standards of conservation: then and now. *Conserv. Biol.* 4:229-232.

Carlton, J. T. 1989. Man's role in changing the face of the ocean: Biological invasions and implications for conservation of near-shore environments. *Conserv. Biol.* 3:265-273.

Costanza, R., and H. E. Daly. 1992. Natural capital and sustainable development. *Conserv. Biol.* 6:37-46.

Delcourt, H. R., and P. A. Delcourt. 1988. Quaternary landscape ecology: Relevant scales in space and time. *Landscape Ecol.* 2:23-44.

di Castri, F., and M. Hadley. 1992. Interdisciplinarity and environment and the flaming moth principle. *Biol. Internat.* 25:17-21.

di Castri, F., A. Hansen, and M. M. Holland, eds. 1988. A new look at ecotones: emerging international projects on landscape boundaries. *Biol. Internat.* Special Issue 17, International Union of Biological Sciences, Paris.

Ehrlich, P. R. 1982. Human carrying capacity, extinctions, and nature reserves. *BioScience* 32:331-333.

Forman, R. T. T., and M. Godron. 1986. *Landscape Ecology*. John Wiley and Sons, New York.

Fox, M. D. 1990. Interactions of native and introduced species in new habitats. In *Australian Ecosystems: 200 Years of Utilization, Degradation and Reconstruction*, eds. D. A. Saunders, A. J. M. Hopkins, and R. A. How. *Proc. Ecol. Soc. Australia* 16:141-147.

Freeman, P. B. 1990. Opening Remarks from the Chicago Zoological Society. In *Management and Conservation of Small Mammal Populations*, eds. T. W. Clark and J. H. Seebeck, pp. xvii-xix. Chicago Zoological Society, Chicago, Ill.

Goodman, D. 1987. How do any species persist? Lessons for conservation biology. *Conserv. Biol.* 2:391-396.

Green, B. H. 1989. Conservation in cultural landscapes. In *Conservation for the Twenty-first Century*, eds. D. Western and M. D. Pearl, pp. 182-198. Oxford University Press, New York.

Green, D. G. 1990. Landscapes, cataclysms and population explosions. *Math. Computing and Modeling* 13:75-81.

Grubb, P. J., and R. M. May. 1991. Comments on the Sustainable Biosphere Initiative. *Conserv. Biol.* 5:548.

Hayden, B. P., G. C. Ray, and R. Dolan. 1984. Classification of Coastal and Marine Environments. *Environ. Conserv.* 11:199-207.

Herrmann, R. 1990. Biosphere reserve monitoring and research for understanding global pollution issues. *Parks* 1:23-28.

Hobbs, R. J., and A. J. M. Hopkins. 1990. From frontier to fragments: European impact on Australia's vegetation. In *Australian Ecosystems:*
200 Years of Utilization, Degradation and Reconstruction, eds. D. A. Saunders, A. J. M. Hopkins, and R. A. How. *Proc. Ecol. Soc. Australia* 16:93-144.

Ishwaran, N. 1992. Biodiversity, protected areas and sustainable development. *Nature and Resources* 28:18-25.

Jordan, W. R., III, R. L. Peters, and E. B. Allen. 1988. Ecological restoration as a strategy for conserving biological diversity. *Environ. Manage.* 11:249-256.

Kaule, G., and K. Henle. 1991. Überblick über Wissensstand und Forschungsdefizite. In *Arten- und Biotpschutzforschung für Deutschland*, eds. K. Henle and G. Kaule, pp. 2-45. Universität Stuttgart, Stuttgart.

Kumar, S. 1993. Too many people spoil the planet. *New Scientist* 140(1898):8.

Lande, R. 1988. Genetics and demography in biological conservation. *Science* 241:1455-1460.

Lasserre, P. 1992. The role of biodiversity in marine ecosystems. In *Biodiversity and Global Change*, eds. O. T. Solbrig, H. M. van Emden, and P. G. W. J. van Oordt, pp. 105-130. International Union of Biological Sciences, Paris.

Longmore, R. C., ed. 1992. Plant Invasions. The Incidence of Environmental Weeds in Australia. *Kowari.* No. 2. Australian National Parks and Wildlife Service, Canberra.

Lubchenco, J., A. M. Olson, L. B. Brubaker, S. R. Carpenter, M. M. Holland, S. P. Hubbell, S. A. Levin, J. A. MacMahon, P. A. Matson, J. M. Melillo, H. A. Mooney, C. H. Peterson, H. R. Pulliam, L. A. Real, P. J. Regal, and P. G. Risser. 1991. The sustainable biosphere initiative: an ecological research agenda. *Ecology* 72:371-412.

McNeely, J. A. 1992. The biodiversity crisis: challenges for research and management. In *Conservation of Biodiversity for Sustainable Development*, eds. O. T. Sandlund, K. Hindar, and A. H. D. Brown, pp. 15-26. Scandinavian University Press, Oslo.

Mott, J. J., and P. Bridgewater. 1992. Biodiversity conservation and ecologically sustainable development. *Search* 23:284-287.

Nierenberg, W. A. 1988. The oceans. In *Earth '88 Changing Geographic Perspectives*, ed. H. J. De Blij, pp. 284-299. Proceedings of the Centennial Symposium National Geographic Society, Washington, D.C.

Netherlands Government. 1990. *Nature Policy Plan of the Netherlands.* Ministry of Agriculture, Nature Management and Fisheries, The Hague.

Noss, R. F. 1983. A regional landscape approach to maintain diversity. *BioScience* 33:700-706.

Odum, E. P. 1992. Great ideas in ecology for the 1990s. *BioScience* 42:542-545.

Odum, W. E. 1982. Environmental degradation and the tyranny of small decisions. *BioScience* 32:728-729.

Opdam, P. 1991. Metapopulation theory and habitat fragmentation: a review of holarctic breeding bird studies. *Landscape Ecol.* 5:93-106.

Pearce, F. 1993. Conservation rules, OK! *New Scientist* 140(1898):50-51.

Pederken, G. F. 1981. *Woodland Conservation and Management.* Chapman and Hall, London.

Peterson, R. W. 1988. An earth ethic: our choices. In *Earth '88 Changing Geographic Perspectives*, ed. H. J. De Blij, pp. 360-377. Proceedings of the Centennial Symposium National Geographic Society, Washington, D.C.

Rackham, O. 1986. *The History of the Countryside.* J. M. Dent & Sons Ltd., London.

Ray, G. C., and W. P. Gregg, Jr. 1991. Establishing biosphere reserves for coastal barrier ecosystems. *BioScience* 41:301- 309.

Ray, G. C., and M. G. McCormick-Ray. 1992. *Marine and Estuarine Protected Areas: A Strategy of a National Representative Systems within Australian Coastal and Marine Environments.* Australian National Parks and Wildlife Service, Canberra.

Reid, W. V., and K. R. Miller. 1989. *Keeping Options Alive. The Scientific Basis for Conserving Biodiversity.* World Resources Institute, Washington, D.C.

Saunders, D. A. 1990. The landscape approach to conservation: community involvement, the only practical solution. *Australian Zool.* 26:49-53.

Slocombe, D. S. 1993. Implementing ecosystem-based management. Development of theory, practice, and research for planning and managing a region. *BioScience* 43:612-622.

Solbrig, O. T., ed. 1991. *From Genes to Ecosystems: A Research Agenda for Biodiversity.* International Union of Biological Sciences, Cambridge.

Schlesinger, W. H. 1990. Vegetation an unlikely answer. *Nature* 348:679.

Soulé, M. 1985. What is conservation biology? *BioScience* 35:727- 734.

Soulé, M. 1990. The onslaught of alien species, and other challenges in the coming decades. *Conserv. Biol.* 4:233-239.

The Royal Society and National Academy of Sciences (TRSNAS). 1992. Population Growth, Resource Consumption and a Sustainable World. London and Washington, D.C.

Vida, G. 1978. Genetic diversity and environmental future. *Environ. Conserv.* 5:127-132.

Walton, D. W., M. A. Forbes, and R. M. Thackway. 1992. Biological diversity, environmental monitoring and the nature conservation estate in Australia: Relationship to ecologically sustainable development. In *Biodiversity: Broadening the Debate*, pp. 24-40. Australian National Parks and Wildlife Service, Canberra.

Weeden, R. B., and M. R. Cutler. 1992. *Putting Wildlife First.* Recommendations for Reforming Our Troubled Refuge System. Report of the Commission on New Directions for the National Wildlife Refuge System. Defenders of Wildlife, Washington, D.C.

Western, D. 1991. Biology and conservation: making the relevant connection. *Conserv. Biol.* 5:431-433.

Wildlife Habitat Canada. 1991. *The Status of Wildlife Habitat in Canada.* Ottawa.

Wilson, B. R., and G. R. Allen. 1987. Major components and distribution of marine fauna. In *Fauna of Australia.* Vol. 1A, General Articles, eds. G. R. Dyne and D. W. Walton, pp. 43-68. Australian Government Publishing Service, Canberra.

Wilson, G. R. 1987. Cultural values, conservation and management legislation. In *Fauna of Australia.* Vol. 1A, General Articles, eds. G. R. Dyne and D. W. Walton, pp. 250-265. Australian Government Publishing Service, Canberra.

VI

SUMMARY

47

BIODIVERSITY IN MANAGED LANDSCAPES: PRINCIPLES, PRACTICE, AND POLICY

Robert C. Szaro

USDA Forest Service
Washington, D.C.
U.S.A.

The conservation of biodiversity has never been more hotly debated or discussed than in recent times leading up to the "Earth Summit" in June 1992 in Brazil. At the "Earth Summit" and during negotiations for a global "Convention for the Conservation of Biological Diversity," there was almost universal agreement that conservation of biodiversity is a serious global concern. The loss of the world's living richness has profound implications for the future of human welfare (McNeely this volume). There remains, however, intense disagreement on how to balance preservation and conservation with social, economic, and sustainable use factors.

Clearly every effort should be made to conserve biodiversity (Szaro 1990; Szaro 1992). The conservation of biodiversity encompasses genetic diversity in species populations, richness of species in biological communities, processes whereby species interact with one another and with physical attributes within ecological systems, and the abundance of species, communities, and ecosystems at large geographic scales (Harrington et al. 1990). Current programs to protect, maintain, and enhance populations of particular species contribute to the welfare of components of biodiversity but can only deal with a relatively small portion of the ever expanding list of threatened and endangered species (Reid 1992; Miller this volume).

With increasing controversy, traditional uses of biological resources and habitats have come under question (Szaro and Shapiro 1990). The demands and expectations placed on these resources are high and widely varied, calling for new approaches that go beyond merely reacting to resource crises and concerns (Szaro 1990; Szaro and Salwasser 1991). These new approaches must incorporate fundamental shifts in the scale and scope of conservation practices (Miller this volume). These include the shift of focus from the more traditional single-species and stand level management approaches to management of communities and ecosystems (Reynolds et al. 1992).

PRINCIPLES

Geologic Perspective

Any actions to conserve biodiversity cannot be adequately evaluated without the full consideration of temporal implications. To ensure long-term genetic adaptability, phenotypic diversity, and high probabilities of population persistence, an appropriate planning time-frame should be up to a hundred times that of a single generation of a target species (Marcot and Murphy this volume). For insights on appropriate times and possible impacts, the geologic record provides the only historical basis for determining the long-term responses of populations, species, communities, and ecosystems to large-scale environmental perturbations (Erwin this volume). With all the growing concern over increasing extinction rates, it is often overlooked that extinction has been a normal part of the evolutionary process since life first arose. Indeed, well over 99% of all species that ever existed have become extinct, some in massive extinction events (Gentry this volume), but with the great majority of species expirations having occurred as part of the "background extinction" associated with the evolutionary process (Erwin this volume).

The Council on Environmental Quality (1991) noted that ignoring the long-term health of ecosystems will prove disastrous, since, beyond the value of any individual species they contain, healthy ecosystems collectively provide life support services for the planet. Yet, we must not lose sight of the fact that in an evolutionary context (over tens of millions of years), the species extinction and community reorganization caused by environmental perturbations are a source of innumerable evolutionary opportunities (Erwin this volume). Indeed, growing evidence suggests that the reorganization of communities accompanying mass extinctions might trigger further diversification (Erwin this volume).

Species seem to be disappearing at an unnatural rate with estimates of the current global extinction rate as high as 17,500 species per year (Wilson 1989), compared to the natural "background" extinction rate over the past 600 million years of approximately one species per year (Raup and Sepkoski 1984). However, it appears that even though a major spasm of extinction from tropical deforestation might be imminent, it has not yet happened (Gentry this volume). Very few plant species have yet been lost. Moreover, figures compiled by Reid and Miller (1989) and Reid (1992) indicate only 724 recorded species extinctions since 1600 worldwide. This is hardly a biological catastrophe, but the number is likely a gross underestimate (Gentry this volume). Only two major groups of organisms, birds and mammals are well enough known taxonomically to evaluate directly the current worldwide extinction trends with extinctions in both groups clearly increasing in recent years (Gentry this volume). According to the data collated by Reid and Miller (1989), 42 bird species and 27 mammal species have become extinct since 1900, as compared to 32 birds and 16 mammals in the preceding half century, and 12 and 2 species, respectively, in the half century before that. Yet, there seems little doubt that the earth is on the verge of major extinctions of plants as well as animals (Gentry this volume). Nearly 20,000 taxa are globally rare or threatened and as many as 60,000 face extinction by the middle of the next century (IUCN/WWF 1988; Pavlik this volume). Birds provide but one telling example: nearly a thousand species, more than 11% of all bird species, are at some risk of global extinction (Marcot and Murphy this volume).

And, unlike previous periods of mass extinctions caused by a buildup of environmental stresses that push ecosystems beyond their limits (viewed as natural catastrophes) (Jablonski 1991), currently most extinctions are the result of human influences (Erwin this volume). Extinctions now are primarily due to the degradation, destruction or alteration of habitats, ecological communities and even whole ecosystems. It might still be possible to save most of what is left. Even some of the very small remaining habitat fragments often written off by some zoologists who are more concerned with habitat requirements of top predators, might be of utmost conservation significance in conserving plant species (Gentry this volume).

Evolutionary Processes

Evolution is the unifying concept that integrates all levels of biology and sets the context for a deeper understanding of biological systems and such aspects as adaptation, morphological constraints, life history traits, and social behavior (Meffe this volume). Long-term preservation of biodiversity requires understanding not only the demography and genetics of small populations, but also the ecology and evolution of abundant species (Lande this volume). Conservation actions and plans developed and conducted outside of this context are done apart from biological reality (Meffe this volume).

Because of man's influences many species are now geographically isolated from other populations with which they undoubtedly exchanged genetic information in the past (Meffe this volume). This has led to a focus by many recent conservation efforts on genetic events in small populations, such as threatened and endangered species on the verge of extinction (Lande this volume). The long-term perspective on which this is based recognizes that although in the short-run, genetic variability may usually be less important than other factors, in the long-run, it can play a decisive role, allowing a population to adapt and persist in a changing environment (Lande this volume).

Genetic considerations are often of critical importance in management of harvested populations and in the conservation of biodiversity, particularly for endangered species in small populations (Lande this volume). Many fundamental issues in conservation biology rely on an accurate assessment of the geographic distribution of genetic variation, either within a single widespread species, or among several species or higher taxa (Riddle and Jones this volume). In situ conservation of genetic resources must be based on ecogeographical surveys that correlate environmental factors to intra-specific variability (Ingram this volume). Risks of population extinction from genetic factors, including inbreeding depression, fixation of new deleterious mutations, and lack of genetic adaptability in a changing environment, are still not completely understood, especially in terms of their interactions with each other and with demographic factors (Lande this volume). Effective management of living resources and endangered species should be conservative in allowing ample margins for our uncertainties (Lande this volume).

Long-term management goals should involve the restoration and maintenance of healthy ecosystems, rather than maximum exploitation of resources and maintenance of endangered species at minimum viable population sizes (Lande this volume). Yet these goals should consider two important conclusions from Lande (this volume): "(1) With further habitat destruction or fragmentation, the equilibrium population size for a given amount of suitable habitat declines faster than the amount of suitable habitat and (2) An extinction threshold occurs in the form of a minimum amount of suitable territory in a region that is necessary for population persistence. Vacant patches of suitable habitat may be as important as occupied ones for regional persistence of a population undergoing local extinction and recolonization: continued destruction of temporarily unoccupied habitat will lead to regional decline and extinction of the population." The maintenance of certain levels of intra-specific variation is key to a population's fitness (Ingram this volume).

Central to all planning efforts that involve maintaining population viability is provision of interconnected and well-distributed populations and the habitats that support them (Marcot and Murphy this volume). For remnant populations that are now geographically isolated from other populations, it might be necessary to restore natural levels of gene flow among such populations by periodically moving small numbers among selected localities or by other means such as developing corridors to restore former connections that have been severed by human development (Meffe this volume; Templeton this volume). Otherwise, these newly isolated populations will ultimately lose the within-population genetic diversity that was formerly maintained by gene flow from other sites (Meffe this volume). Although some species are naturally patchily distributed, in general the maintenance of distributions and connectivity is key to sustaining genetically diverse and demographically healthy populations (Marcot and Murphy this volume). Any such artificial or re-created gene flow, however, must not jeopardize among-population divergence and should not allow any movement among geographic subgroups (Meffe this volume).

Meffe also suggests that evolutionary thinking with respect to conservation involves adopting a mind-set that at a minimum accepts the following principles: (1) think in terms of a long time scale, (2) evolution is not directed or goal-oriented; it is without foresight, (3) adaptation can lead to a dead end if the environment changes, (4) evolution is more than natural selection, and (5) evolution is dynamic; by definition it means change. Evolutionary thinking forces conservation actions to be considered relative to long-term interests of the species or system of concern, rather than the perhaps well-intended but ultimately misguided goals of human-oriented interests (Meffe this volume).

A recent approach to addressing the geographic structure of populations in an extended historical context incorporates phylogenetic systematics, biogeography, population genetics, and molecular evolution, and has been referred to as the study of phylogeographic population structure (Avise et al. 1987). Single-species phylogeographic patterns have several implications for conservation biology (Riddle and Jones this volume). For example, Dizon et al. (1992) developed a phylogeographically-based protocol to identify "evolutionarily significant units" (ESUs). The identification of ESUs informs management decisions in at least two ways: (1) by suggesting a priority ranking of populations relative to their evolutionary distinctness as well as amount of genetic information present, and (2) by understanding natural phylogeographic breaks among populations, and, therefore, when not to mix populations from different geographic regions in captive or wild breeding programs (Riddle and Jones this volume).

A knowledge about the geographic distribution of population genetic diversity can give important insights into current patterns of biodiversity as well as predicting consequences of disrupting those patterns by analyzing the historical responses of populations, species, and biotas to habitat alterations (Riddle and Jones this volume). Recent attempts to incorporate population genetics, systematics, molecular evolution, ecology, and biogeography into the analysis of phylogeographic population structure offer a promising approach to understanding the causes behind existing patterns of biodiversity (Riddle and Jones this volume).

Ecosystem Pattern, Complexity, and Dynamics

Change has profound implications for land management. Historically, the quest for stability and preventing change in areas where productivity was maintained by dynamic events has led to the declining quality and quantity of many of our most desired habitats (Szaro 1990; Reid this volume). Much prior management of habitats has viewed systems as being immutable and that all that is necessary is to put a fence around an area and it can be saved forever. Yet, there is abundant evidence that this may seldom be so, both within particular mini-systems and on geographic scales (Willson this volume). Ecosystems by their very nature are in a constant state of flux. Environmental change is inevitable and in a sense the one constant under which ecosystems exist; it is unrealistic to embrace the status quo because it does not exist (Lugo and Brown this volume). Whether people cause the change or not most ecosystems are now influenced in some way by human activities or the artifacts of those activities (Salwasser et al. this volume). The effects of land management decisions must be evaluated using ecologically relevant time scales as well as spatial scales (Brooks and Grant 1992). This is a constant challenge to land managers. Changes resulting from periodic, abrupt, and/or catastrophic environmental factors, lead to displacement, replacement, and succession with species composition tied to the frequency and types of disturbance. Change per se is not necessarily something to be avoided. Ultimately it may be the underlying motivating factor in management decisions. We may wish to alter vegetation structure or composition to emphasize rarer or endangered species or communities (Szaro 1990).

The maintenance of natural levels and patterns of biodiversity requires serious consideration of many kinds of ecological interactions and of variation in their intensities and outcomes (Willson this volume). Hierarchical patterns of heterogeneity are a consequence of the complexity within ecological communities (Dayton 1984; Pearson et al. this volume). Dealing with this complex web of interactions within ecological systems requires a different mind-set, one that focuses on multiple and variable interactions (Willson this volume). These interactions form networks of critical importance to biodiversity (Price et al. 1980; Pianka 1987; Price 1992; Willson this volume). Sustaining these networks of complex ecological patterns are crucial to the maintenance of biodiversity (Pearson et al. this volume). This is especially critical for aquatic systems where many natural disturbance events affect the interactions among watersheds, riparian communities, and aquatic habitats (Williams and Davis this volume). The consequences of these interactions often accumulate in the riparian zone, which is one of the most dynamic and productive portions of the landscape (Gregory et al. 1991; Szaro 1989; Williams and Davis this volume).

A focus on process, function, and interaction provides a basis for making the difficult choices among species to which salvage efforts are to be allocated (Willson this volume).

Protection strategies using a single species perspective can result in an array of complex management strategies likely to contain contradictory prescriptions (Pearson et al. this volume). It may be more effective to focus our efforts on species that have strong interactions with several others, because changes in those species will have a greater impact on total community function than those interacting only weakly or with only one other (Willson this volume). Perhaps, the unit of preservation/conservation should be the interactions/processes themselves, rather than single species (Willson this volume). By focusing more on ecological networks, sets of interacting populations in a variable environment, instead of single species, our chances of maintaining biodiversity are likely to improve (Willson this volume). Taking a broader scale perspective and implementing protection at the landscape level is a more robust protection strategy. An intact landscape will preserve the natural processes and patterns of heterogeneity that its ecosystems and species require (Pearson et al. this volume). A landscape-level characteristic such as habitat connectivity can be preserved by controlling the spatial extent and pattern of habitat loss. The challenge to conservation is to recognize critical landscape-level patterns and to minimize human effects on natural patterns and processes (Pearson et al. this volume).

Socio-economic Factors

Why should people care about protecting biodiversity? Why should they support the effort required to sustain and enhance genetic resources, recover endangered species, restore riparian areas, maintain ancient forests, or conserve trees, insects, and marshes? The answer touches on ethics, esthetics, economics, and quality of life (Szaro and Shapiro 1990; Szaro and Salwasser 1990)

The diversity of life benefits us in infinite ways:

- Our homes, air, livestock, vegetables, fruits, and grains all derive from the products of diverse and healthy ecosystems.
- More than one-half of all our medicines today can be traced to wild organisms; one-quarter of all prescriptions written in the United States every year contain chemicals from plants.
- Diverse communities of plants, animals, and microorganisms provide indispensable ecological services: they recycle wastes, maintain the chemical composition of the atmosphere, and play a major role in determining the world's climate.
- Countless people enjoy the special pleasures of hiking in lush forests, visiting scenic mountains and seashores, and pursuing recreational activities that are dependent on biodiversity, such as hunting and fishing.

Moreover, many people believe that we have an ethical obligation to protect the diversity of living things with which we share our planet, whether or not they are known to be useful to us, simply because humans bear a responsibility for the stewardship of all life on Earth (Szaro and Shapiro 1990).

Yet, the full values of biodiversity far exceed our current knowledge and uses (Machlis and Forester this volume). We know only a small fraction of the species on this planet, especially in tropical ecosystems, despite decades of scientific effort. Every year species are lost before we have a chance to learn anything about them. We will never know which potential foods, medicines, and commercial products have disappeared forever with each extinction? A diversity of wild plants, animals, and microorganisms have served humans since they first walked on earth. An astonishing wealth of wild plants, animals, and microorganisms have served humans since we first walked on Earth. Future generations are entitled to expect, and will need, an environment as biologically rich as the one inherited by today's generations (Szaro and Shapiro 1990).

The root causes of biodiversity loss are primarily human actions, imbedded in social systems that are inextricably intertwined with natural systems (Machlis and Forester this volume). And because biodiversity loss is so crucial a problem in human affairs, its treatment will, like most environmental issues, be primarily as a political, economic, and management problem (Machlis and Forester this volume). Hence, understanding the relation between socio-economic factors and biodiversity loss is a prerequisite for wise and effective biodiversity management and protection. Such understanding will largely rest on sound and careful science (Machlis and Forester this volume). The idea that science should be the basis for the resolution of public resource policy issues and conflicts is relatively new and has only emerged over the last 150 years (Gordon 1992). The translation of scientific information to policy and management processes is still not an exact science (Gordon 1993).

Decision-makers are often not scientists by training, yet, rely heavily on the scientific community for facts, predictions, and advice (Machlis and Forester this volume). Models of biodiversity loss are likely to be utilized by such decision-makers and, if the pattern is similar to other environmental issues (such as risk assessment or global climate change), with mixed results (Machlis and Forester this volume). Progress in ensuring the use of science in decision- making has probably been slowed in part because faulty models are frequently used or pursued (Gordon 1993). Beyond first efforts at modeling the relations between socio-economic factors and biodiversity loss, there is a realistic possibility of truly usable models, perhaps within the next ten years (Machlis and Forester this volume).

PRACTICES

Inventory, Monitoring, and Assessment

Inventory.—The diversity of life constitutes the foundation for all human resources, including the future richness of life for our children for generations to come, yet we are still uncertain even of the magnitude of that resource. Estimates of the number of species of plants, animals and microorganisms ranges from 5 to 80 million (May 1992). To date, only a small fraction (approximately 1.4 million species) has been scientifically described and catalogued (Szaro 1993b).

Better inventories and assessments are needed of current conditions, abundances, distributions, and management directions for conserving genetic resources, species populations, biological communities, and ecological systems. The core of any in situ conservation effort is actually the long-term program of inventory and monitoring (Ingram this volume). Even in the United States with the intensity of scientific effort focused on natural resources, it is extremely difficult to assess the overall status of biodiversity. Rarely have studies been done to examine all the vascular plants and vertebrates and their relationships in any given ecosystem, let alone the thousands of other species found in any given system (Szaro 1993b). Most of what we need to know for biodiversity conservation relates to geography, and the importance of co-location of species comes up repeatedly— not only in regard to efficient priority setting but in defining ecological relationships important to the viability of species, communities, and landscape complexes (Jenkins this volume). Mapped information is one of the main needs of the biodiversity manager (Jenkins this volume).

A biodiversity inventory is a formal cataloging of the occurrences at a point in time of defined elements of biodiversity that are contained in a defined geographic unit (Dennis and Ruggiero this volume). To be conducted formally, this cataloging must be done in response to a clearly stated purpose, according to well established field procedures, within statistically valid sampling designs, and using rigorous quality control and data administration practices (Dennis and Ruggiero this volume). An inventory differs from monitoring in that it attempts to characterize or determine components of a community, ecosystem, or area at any given point in time (Jones and Riddle this volume).

The purpose and orientation of an inventory will determine choice of methods which in turn will influence the completeness of the inventory in terms of taxonomy, community/ecosystem representation, geographic space, and seasonal/temporal representation (Dennis and Ruggiero this volume; Solbrig 1991; Stohlgren and Quinn 1991). An inventory can (1) be an end in itself, (2) develop the baseline needed for follow-up monitoring, and (3) develop the capability to predict future biological events (Dennis and Ruggiero this volume). Once the purpose is determined, there are four possible approaches for orienting an inventory: survey of major elements, identification of indicator elements, identification of targeted elements, and comprehensive assessment of all elements (Dennis and Ruggiero this volume).

An inventory that searches out all occurrences of chosen taxonomic groups will be complete taxonomically, but not necessarily either ecologically or spatially (Dennis and Ruggiero this volume). One that uses stratified sampling designs to ensure inventorying each type of major ecological element in the study area will be complete in terms of ecological systems, but not necessarily either taxonomically or spatially. On the other hand, an inventory that conducts samples at each point of an overlaid grid will be complete geographically, but not necessarily either taxonomically or ecologically (Dennis and Ruggiero this volume).

The idea of the "coarse filter" and "fine filter" was invented for Heritage biodiversity information and planning purposes (an effort of the network of Natural Heritage Data Centers in the United States) and represents a hierarchial approach to inventory efforts (Jenkins this volume). It is obvious that there are more species than we can track or even know about. Therefore, it makes sense first to identify the kinds of major communities found in a region, identify the best location of each of these, and, by protecting sites containing selected examples of each, simultaneously protect populations of most species. The species most likely to fall through the "coarse filter" are the rare ones, so these must be identified individually. They are referred to as a "fine filter" because their populations are mingled with others that are also protected when reserves are established for the target species. It is further assumed that many rare species are habitat specialists found together in their own microcommunities with unknown species which are probably rare for similar reasons (Jenkins this volume).

Monitoring and Assessment.—Comprehensive and coordinated monitoring programs are needed to address status and trends in elements of biodiversity at regional and national scales, as well as those of habitats, ecosystems and landscapes (Jones and Riddle this volume). Changes monitored over time allow us to assess the consequences of management practices (Szaro 1992; Szaro 1993a). This will not be easy because even in an area the size and geographic scope of a typical National Forest in the United States, native biodiversity can easily encompass thousands of species of plants and animals, dozens to hundreds of identifiable biological communities, and an incomprehensible number of pathways, processes, and cycles through which all living things are interconnected. Obviously, it is not possible to address each and every aspect of this complexity. Therefore, some sort of risk assessment or prioritization will be necessary in the identification of specific aspects of diversity, such as distinct species, biological communities, or ecological processes that warrant special consideration (Salwasser 1990).

Three general monitoring approaches as summarized by Jones and Riddle (this volume) have been used to address status and trends in biological and ecological resources: (1) those that evaluate broad patterns of diversity to identify geographic areas undergoing rapid change, (2) those that evaluate changes in biodiversity and/or habitats within specific areas (e.g., a watershed), and (3) those that evaluate the effectiveness of a specific management prescription in counteracting decline or enhancing diversity. Rarely have all these approaches been used in a single effort but an effective national scale monitoring program for biodiversity must include elements from all three levels (Jones and Riddle this volume).

Monitoring should provide sufficient information about the abundance of animals or plants targeted for monitoring to assure that current management practices are not threatening the long- term viability of their populations (Verner 1986). But viability concerns have added an additional layer of complexity to our monitoring problems particularly when we try to derive a number for minimum viable population size. This theoretical concept espoused by Soulé (1987) is useful from the standpoint that there probably is some minimum size population threshold that, when crossed will lead to the demise of a species' population. The level of this threshold clearly varies with the species concerned, for passenger pigeons it probably was in the hundreds of thousands, whereas for desert pupfish the number may be as low as 200. But as useful a concept this may seem, in the real world it is nearly impossible to determine minimum viable population size with any degree of certainty. There are too many variables involved, demographic viables such as immigration, emigration, recruitment, birth rates, survivorship, dispersal mechanisms, catastrophic events, habitat loss, and even changing climatic conditions. It is absolutely ludricous to assume that minimum viable population size can be determined for every species. A more subjective approach might be more realistic: larger numbers of a species are obviously better than fewer numbers of the same species. Perhaps the goal for conserving species should be to try to maintain species with as large a population size as practical.

Monitoring efforts are often severely hampered by the lack of prior planning and thought given to the desired results from any given monitoring effort. Without precisely defining the specific purpose and procedure for monitoring, very little useful information will be generated by the money, time, and effort spent (Pavlik this volume). Actually, poorly-designed monitoring probably does more harm than good because crucial questions go unanswered, actions are delayed, impacts begin to compound, and worse, management decisions based on poor information may actually result in greater harm (Pavlik this volume). It is not enough to select a management indicator species, guild, or other monitoring target with the idea that this will allow us to assess the impact of any given management activity (Szaro and Balda 1982; Szaro 1986; Tilghman and Verner 1989). Ideally, the results from monitoring should feed back into the system to correct or fine tune management activities. It is essential that we strive to have our appraisals of these resources give us the ability to forecast the consequences of human- induced environmental changes accurately (Hoekstra and Flather 1986). But we have a long way to go in this process.

A clear understanding of goals and objectives is a first step. The next step is to assess risk and assign priorities. We need to perform a kind of environmental triage in order to determine when it is time to quit. With limited financial and physical resources it may be necessary to make the highly undesirable decision that no further attempt be made to prevent the extinction of a particular species. Once we have determined those species and remember the importance of ecological processes, we need to formulate the questions whose answers will determine that our goals and objectives are being met. It is absolutely critical to ask the right question in the first place. Why monitor shade cover over perennial streams to maintain water temperature for trout when directly monitoring stream temperature is a more appropriate measure? Thus, our efforts will be geared to maintaining as much shade cover as needed to maintain water temperature.

Specifically, several major monitoring concerns are recognized: 1) how to assure that monitoring is adequate in terms of what is being measured; 2) how much data and how to handle the large volume of data being collected; and 3) are the measurements accurate. Monitoring should have a strong element of quality control and assurance, including setting specific levels of accuracy and precision, timing the inventory to encompass the life-cycles of the target species, as much as possible, and standardizing methods and databases especially when monitoring the same species. However, whatever is done must be as cost-effective as possible. Some possibilities are risk analysis, increasing the scope of monitoring efforts, and determining the needs of monitoring objectives. We might want to limit direct monitoring only on high risk species, on a priority basis, while relying on habitat relationships for most other species. Monitoring efforts should be spread over as large a geographical base as possible (and feasible) to reduce the cost per forest and to increase the scope of applicability. Whenever possible, monitoring should only be asked to detect declining trends because of the potential cost savings (almost 90%) (Verner 1986).

Monitoring efforts also need to develop quality control and assurance programs that ensure: (1) objectives are measurable (and thus monitorable); (2) appropriate measurement techniques and procedures are being used; and (3) management thresholds are clearly identified and incorporated into the planning process so that, if crossed, they automatically trigger a reanalysis of the planned activities. Quality assurance is a process whereby data quality is defined, not a process that forces data to meet a particular standard. The data quality needs to be sufficient to meet the purposes defined for its collection. Quality assurance is simply a process of establishing and documenting what that quality is. Quality assurance deals with such concepts as precision, accuracy, comparability, representativeness, and completeness of the data. Environmental issues seem to be ever more contentious in the public arena. This trend is likely to make data quality assurance much more important in the future.

Developing Measures and Techniques to Determine Progress and Set Priorities.—Monitoring biodiversity requires quantitative methods for measuring and comparing diversity (Gove et al. this volume). Even though the concept of community diversity and its measurement have received considerable attention in community ecology studies for more than four decades, much work is still needed to develop measures useful in assessing the success or failure of management activities (Gove et al. this volume). The selection, evaluation, and implementation of indicators and/or measures are critical in establishing a monitoring program for biodiversity (Jones and Riddle this volume). These can be individual measurements or variables or a combination of measurements or variables, such as indices (Jones and Riddle this volume). Three basic, interrelated types of indicators and indices are known: (1) species measures, (2) measures that integrate composition, structure and functions of biodiversity, and (3) ecological indicators applied at landscape scales (Jones and Riddle this volume). Unfortunately, diversity indices and profiles all assume that the population has been enumerated completely but in cases where this is not possible and sampling is the only alternative, jackknifing (multiple sub-samples of the same data) can be applied for bias correction and hypothesis testing (Gove et al this volume).

Several major efforts have tries to broadern the scale of monitoring biodiversity. One of the more useful aspects developed by the Heritage Data Information Network is a formal process for tracking and documenting Element status, known as Element Ranking (Jenkins this volume). This process is well advanced, with the number of "unknown" or questionable ranks for Elements in better known groups rapidly tending toward zero. Element ranking has proved a useful device. The ranks are used to set priorities for both data collection and conservation (resources are concentrated on the highly ranked Elements). The ranks allow Element lists and databases to be sorted by imperilment classes and the generation of Element lists in sequential imperilment order (Master 1991).

Gap analysis seeks to provide a regional conservation planning perspective by using relatively coarse distribution maps to assess mainly the protection status of plant communities and vertebrate species (Scott et al. 1991). "Gaps" are those natural communities or species not adequately represented in existing reserves (or, more broadly, in areas managed for the long-term maintenance of biodiversity) (Stine et al. this volume). One goal of gap analysis is to help identify and prioritize areas, communities, and species in need of more detailed conservation planning. Gap analysis is not recommended as the only tool for assessing the landscape in the process of designing and selecting ecosystem reserves (Stine et al. this volume). This approach does identify those remaining landscapes that will provide the possible foundation for a regional reserve system (Stine et al. this volume).

The goal for the Smithsonian Institution/Man and the Biosphere Biodiversity Program is the establishment and coordination of a global network of long-term biodiversity inventory and monitoring plots, focusing on temperate, subtropical, and tropical forest ecosystems (Dallmeier this volume). Work revolves around what Davis (1993) describes as the driving purpose of monitoring—to suggest potential agents of change and then, through research, document patterns of cause and effect (Dallmeier this volume). The methodology calls for the establishment of a network of small inventory and monitoring plots (one hectare in size) that provide a snapshot of the larger forest landscape. The general objectives at the plots are to (1) inventory woody plants and their distribution, (2) record forest structure, composition, and diversity and monitor natural and human-induced changes in intervals of three to five years, (3) detect patterns of recruitment, mortality, and regeneration of forest trees in a wide range of natural and managed environments, (4) determine how abiotic factors (e.g., soil and precipitation) affect all of the functions just noted, and (5) contribute data for the preparation of vegetation maps.

Strategies for Maintaining Biodiversity

Maintenance and Recovery of Threatened and Endangered Species.—Extinction is part of the natural evolutionary process that has been occurring since life began billions of years ago. As a result, most of the species that once existed on Earth became extinct long before humans evolved. Human beings, however, have accelerated the process of extinction at alarming rates. It is easy to understand, then, why threatened and endangered species have received most of the conservation attention (Bridgewater et al. this volume). Many are large, easily observable, and oftentimes aesthetically pleasing. Focusing attention on threatened and endangered species exists because: 1) they represent the most fragile and potentially vulnerable members of biological communities, and 2) protecting threatened and endangered species from extinction and preventing additional species from becoming threatened or endangered is central to protecting the ecosystem structure and function and maintaining biodiversity. Successful recovery projects will be those that can demonstrate an increase in population size and an ongoing ability of the target taxon to complete its critical life history processes in natural, enhanced and created populations throughout its historic range (Pavlik this volume).

Why do species decline? Natural resource managers do not purposely degrade habitats and eliminate species. Species are lost when we lack the basic understanding of what an organism requires for its existence or when we lack the capability to predict how a particular ecosystem will respond to perturbation or change. Threatened and endangered species are fundamental indicators of environmental disturbance. The recovery of endangered plants and animals, including the restoration of their critical habitats, might be the greatest challenge facing species-level conservation programs (Pavlik this volume).

Because extinction is a process, not a simple event, the recognition that a species is endangered is little more than a snapshot of a moving vehicle. Attempts at therapy most often address symptoms rather than causes. We have failed to communicate successfully why rehabilitation and restoration beyond the narrow focus of the endangered and harvested are essential. The environmental variables that affect the health and welfare of all flora and fauna also affect people: water and air quality, recycling of organic and inorganic substances, microclimate, etc.

The difficult choices posed in the spotted owl (*Strix occidentalis*) case in the Pacific Northwest of the United States typify many future issues as the conservation of forest biodiversity becomes a higher social priority (Thomas et al. 1990). Regardless of the eventual outcome of this issue, an important lesson can be learned: conserving biodiversity may be expensive and controversial. Federal land management agencies in the United States have increasingly come under fire over management decisions that appear to decrease biodiversity. The USDA Forest Service faces numerous appeals and lawsuits on forest plans for insufficient and sometimes conflicting consideration of forest biodiversity in management decisions. The dispute over the spotted owl and old-growth forests is the most visible example of how difficult it is to blend the conservation of biodiversity with other uses and values of public resources. It illustrates the reality of "no free lunch" in resource allocations. Even though parks, reserves, set- asides, and easements are critical components in the mix for the conservation of biodiversity, they will become more difficult to obtain and maintain and ultimately will require an expansion beyond the "reserve mentality" (Brussard et al. 1992). Somehow we have to develop management plans for public lands that will allow both consumptive and non-consumptive uses but will do so in such a way that no net loss of native species will occur.

One tool in the recovery of threatened and endangered species is that of intentional translocations (Templeton this volume). The goals of such translocations "include the reintroduction of a population into an area from which it was extirpated, the introduction of an endangered population into a new area outside its historical range to protect it from global extinction, the augmentation of natural populations to protect them from extinction through demographic and genetic stochasticity, and the re establishment of gene flow and dispersal between natural populations to counteract the effects of habitat fragmentation" (Templeton this volume). But the primary goal of most reintroductions is the establishment of a self-sustaining population in the area of release (Templeton this volume).

Habitat Restoration.—The remediation of past losses can help offset unavoidable future losses. Habitat destruction, often resulting in greater fragmentation of habitats, is considered the most widespread anthropogenic cause of biodiversity loss that (Machlis and Forester this volume). Habitat fragmentation may differentially increase species and habitat vulnerabilities (Lovejoy 1986; Vermeij 1986) with certain habitats and species more at risk (Terborgh and Winter 1980; Slobodkin 1986; Machlis and Forester this volume). Restoration requires an accurate understanding of ecosystem dynamics and the diversity and ecological roles of all successional stages (Spies and Franklin this volume). The restoration of ecosystems and biological communities is one important means of maintaining biodiversity, or at least of slowing its net loss. Habitat restoration is analogous to the recovery of threatened and endangered species but at a broader ecosystem scale. And like recovery efforts for species, the best way to minimize habitat degradation and fragmentation is prevention (Lugo and Brown this volume).

The restoration and maintenance of healthy habitats and ecosystems should be of equal or greater concern than the recovery and maintenance of threatened and endangered species to a mature science of conservation biology (Salwasser 1990; Lande this volume). The best way to minimize species loss is to maintain the integrity of ecosystem function. The important questions, therefore, concern the kinds of biodiversity that are essential to ecosystem functioning. The best focus of our efforts is to establish how much (or how little) redundancy is found in the biological composition of ecosystems. Functional groups with little or no redundancy warrant priority conservation effort (Walker 1992). It is axiomatic that conservation of biodiversity cannot succeed through "crisis management" of an ever expanding number of endangered species. The best time to restore or sustain a species or ecosystem is when it is still common.

And for certain species and biological communities, the pressing concern is perpetuation or enhancement of the genetic variation that provides for long-term productivity, resistance to stress, and adaptability to change. A biologically diverse ecosystem holds a greater variety of potential resource options for a longer period of time than a less diverse ecosystem. It is more likely to be able to respond to environmental stresses and adapt to a rapidly changing climate. And it may be far less costly in the long run to sustain a rich variety of species and biological communities operating under largely natural ecological processes than to resort to the heroic efforts now being employed to recover California condors (*Gymnogyps californianus*), peregrine falcons (*Falco peregrinus*), and grizzly bears (*Ursus horribilis*). Resource managers know from experience that access to resources is greater and less costly when forests and rangelands are sufficiently healthy and diverse.

Biodiversity is threatened not only by a reduction of habitat acreage and of connections between habitats, but also a degradation of quality of the remaining habitats. Restoration activities respond to these problems by restoring eliminated habitat types (e.g., native prairies and wetlands) and enhancing the condition of remaining habitat fragments. By restoring both the extent and quality of important habitats, restoration programs provide refuges for species and genetic resources that might otherwise be lost. Moreover, as surrounding landscapes are disturbed, organisms disperse from these disturbed areas and affect the recovery and renewed diversity of the biota of surrounding areas.

Many techniques are used to restore ecosystems, depending on the ecosystem and impact type being addressed. These include vegetation planting to control erosion, fertilization of existing vegetation to encourage growth, control of exotics, removal of contaminated soils, fencing to exclude cattle, reintroduction of extirpated species, restoration of hydrologic connections to wetlands, and others. In some cases, choosing the location of a new preserve may be based on requirements to restore or recover populations of a species or group of species (Hopper this volume).

However, not all of these restoration strategies are compatible with the goal of maintaining biodiversity. For example, eroding lands can be rapidly restored by introducing some types of aggressively spreading plants (usually exotics), but the same plants can spread beyond the site and imperil the diversity of flora in adjacent areas. Thus, restoration actions can be either a savior or a nemesis for regional biodiversity, depending on their design, application methods, and existing conditions of the landscape.

Just as there are many restoration techniques, similarly many species exist in habitats that themselves are candidates for restoration. Each species has particular environmental requirements, minimum viable population size, and expected recovery rate and pattern—data or information that is/are essential to evaluating restoration potential. Although responses of some species to specific restoration techniques is known, the theoretical basis is weak for grouping species so that results can be extrapolated to other combinations of techniques and species.

Restoration is not a substitute for the preservation or good management and is both time-consuming and expensive. In tropical forests, the incredible diversity and complexity of the ecosystem make restoration of the original vegetation and ecosystems particularly difficult, costly and, time consuming (Gradwohl and Greenberg 1988; Peters and Lovejoy 1992; Lugo and Brown this volume). But restoration ecology is worth serious consideration even though it addresses the symptom of deforestation rather than the causes. It can speed regeneration in managed systems, renew land productivity, relieve pressure on natural forest resources, and protect closed-canopy forest. It is a strategy that focuses on areas of severe erosion and soil compaction where quick action is desperately needed (Gradwohl and Greenberg 1988). There is always an opportunity to return degraded sites to some level of production and ecological health, with the main lesson being that human activity can be as effective in repairing damaged lands as it is in causing damage (Lugo and Brown this volume).

Dealing With Exotic Species.—Human-assisted invasions by exotic organisms are a major cause of global loss of biodiversity (Courtenay and Moyle this volume; Loope and Stone this volume; Templeton this volume). Movement and establishment of exotic organisms at an ever accelerating pace are facilitated by the ever increasing frequency and intensity of habitat modification and fragmentation (D'Antonio and Vitousek 1992). Taxa that evolved in isolation from each other can suddenly find themselves irreversibly colocated, thus initiating a cascading chain of extinctions (Diamond 1989; Loope and Stone this volume).

The loss of biodiversity from invasions is most severe on oceanic and natural habitat islands, and is increasing in continental ecosystems from accelerating habitat fragmentation, transport of species, and the disruption of natural ecological processes such as fire (Loope and Stone this volume). Much irreversible damage has already happened and further anthropogenic invasion is undoubtedly inevitable, but this only emphasizes the need for immediate action to slow the introduction of exotic species and quickly respond to eradicate such species when first discovered (Loope and Stone this volume).

Regardless of how or why a species was introduced, it has the effect of increasing local species richness, at least temporarily (Courtenay and Moyle this volume). But from a broader regional or global perspective, however, the results of these introductions are oftentimes a net loss of species (Courtenay and Moyle this volume). For example, in the short time in which fishes have been widely introduced, in many instances they have become a form of biological pollution — organisms that expand their ranges and populations rapidly, and disrupt, decrease, or destroy the native biota (Courtenay and Moyle this volume). Making or allowing the continued introduction of non-native fishes can pose severe threats to the continuity and future of native species biodiversity (Courtenay and Moyle this volume).

Definite steps can be followed to deal with exotic species (Courtenay and Moyle this volume; Loope and Stone this volume. First, to avoid future invasions there is a need to identify specific pests that have the potential for invading a new area and becoming pests. Next, once a species has arrived, an obvious requisite to any exclusion or eradication program is the identification of the risk that a species or group of species is potentially hazardous. This process can be difficult because species that are pests on one continent may not be pests in a new region and conversely species that are innocuous in their endemic habitat may become pests after introduction to a new area. Many aggressive exotic plant and animal species not yet established, if introduced, exploit and modify habitats not yet threatened by any already established exotic species. Unless combatted with ingenuity and commitment, the insidious threat of exotic species can be expected to proliferate and to inundate all but the most resistant native ecosystems. In view of the difficulty and expense of controlling invasive species after they are established, preventing establishment of new introductions is not only highly cost effective, but essential. Once an introduction or invasion has been discovered, a quick response by research to provide critical information and management to implement eradication programs is essential.

Should eradication fail, or should an introduction go undetected until after an exotic pest becomes established, emphasis turns to controlling it. Control programs can include the breeding of pest resistance, using biotechnology as a more precise means to pursue genetic resistance by inserting resistant genes, and biological control by using natural control agents from the point of origin. No reasonable alternative to biological control exists for some species. In spite of being subjected to much valid criticism, primarily for past mistakes involving invertebrate introductions, biological control has had some remarkable successes worldwide and a safe long-term track record for dealing with plant pest problems.

Establishing Reserves.—The establishment of reserves for nature conservation is one of the cornerstones in the development of strategies for the conservation of global biodiversity (Hopper this volume). Island biogeography theory provides a basis for establishing a frame-work for designing nature reserves to minimize extinction probabilities and maximize the conservation of biodiversity (Diamond and May 1976; Jensen 1987; Pavlik this volume). Models suggest that the amount of preserved habitat should reflect local species richness, distance from immigration sources, critical resource availability, and the probability of disturbance from beyond reserve boundaries (Pavlik this volume). Centers of species and community richness, representativeness, endemism or rarity have been prominent in the identification of areas deemed suitable for the creation of new reserves for nature conservation (Burbidge et al. 1990; Hopper this volume). But preserves must be designed with more than just conserving species in mind; they must also be designed to maintain functioning interactions between species (Willson this volume).

Debates have centered over the value of large reserves with less edge that would potentially support more species and larger populations for longer periods of time than several small reserves of the same total area (Pavlik this volume). There are obvious limitations to this theoretical approach because often no chance is offered to argue scientifically about the size of a reserve in a checkerboard landscape with rapid development and high real estate values (Pavlik this volume). Regardless of the amount of preserved habitat, most harm originates from the surrounding landscape where managers are unlikely to have much influence (Pavlik this volume). Reserves must also be buffered by land-use zoning, connected by landscape corridors or clustered in an area where a high percentage of the surrounding land is zoned for conservation and can act as a medium of exchange between reserves (Dawson this volume). "But just how large should a small reserve be for a target species? It should be large enough to (1) allow for significant propagule dispersal within, rather than beyond, the reserve, (2) maintain critical ecosystem processes, (3) contain suitable but presently unoccupied habitat, and (4) accommodate population growth (or at least regeneration) of the target species in the future" (Pavlik this volume).

Many existing refuges tend to try to freeze present conditions and act more as museums (often not successfully maintaining the organisms/systems they were set up to protect) than as functional communities that interact with dynamic landscapes or seascapes (Moyle and Sato 1991; Williams and Davis this volume). Yet, not only do individual nature reserves occur in a matrix of changing landscapes and ecosystem processes, but also species exist as dynamic metapopulations undergoing constant evolution (Hopper this volume). Moreover, most parks, refuges, and wilderness areas are designed to protect terrestrial species and landforms but not aquatic ones (Williams and Davis this volume). This results from the fact that because of their linear nature and extensive reaches, river systems are particularly difficult to protect within most existing preserves (Moyle and Sato 1991, Williams and Davis this volume).

Informed management requires sound biological knowledge at various levels of the organizational hierarchy and an ability to integrate such knowledge in a way that leads to maximum conservation outcomes (Hopper this volume). Although it may be convenient to focus policy and management at the species and community levels, it is biologically naive and fails to recognize the need to understand the detail and complexity of population genetic architecture (Hopper this volume). Current networks of protected areas, which support populations of genetic resources, are largely unorganized, ad hoc for those specific purposes, and will require careful review and coordination for adequate germplasm conservation (Ingram this volume). What is needed is ecosystem-level management which is specific to the ecological and political constraints of the reserve (Pavlik this volume). This approach is based on the possibility that managing targeted and diverse landscapes will entrain a sufficient system for all organisms (Namkoong this volume). The effectiveness of such management would be measured by assessing certain targeted species (previously assessed for their potential as indicators for entire reserves) and conclusions about the capacities of reserves to protect the genetic diversity of all species would be based on these indicator or flagship species (Namkoong this volume). Long-term, ecosystem management plans must also be adapted to maximize the value of small reserves because many plant taxa have life history traits that decrease the probability of extinction in small populations (Pavlik this volume).

Maintaining Range Ecosystems.—Many ecologists are recommending that reserves be complemented by surrounding and connecting semi-natural lands where ecological principles are used to manage for a combination of commodity production and resource conservation purposes (Noss and Harris 1986). On an area basis, rangelands constitute the largest category of generally non-tilled, but extensively used land across the world (about 47%) and occur in grassland, shrub steppe, savanna, open woodland, and semi-desert regions (Williams et al. 1968; West this volume). The challenge is how to plan and carry out management on such multipurpose lands to minimize loss, maintain, and even enhance biodiversity (West this volume).

For the much larger fraction of the rangelands that are still dominated by native plant species, we should continue to apply our increasing knowledge of individual species responses and community dynamics in devising low input ways to direct succession to desired outcomes (West this volume). Prescribed burning, behavioral modification of animals, improved ways of distributing animals without fences, and livestock grazing systems compatible with wildlife, recreational use, watershed and soil protection should be further developed (West this volume). It is much cheaper to prevent such semi-natural areas from slipping toward desertification than trying to rehabilitate or restore areas that have already been seriously degraded (West this volume).

Maintaining Tropical Forest Ecosystems.—Many of the most serious problems for conservation are centered in the tropics, where biodiversity is highest and species and whole ecosystems are being lost most rapidly (Raven 1987). Understanding the factors that regulate species richness in the tropics from an evolutionary perspective that strives to determine the origin of high species diversity over millennia is not as relevant to developing management alternatives as one that focuses on understanding how high species densities are maintained in ecosystems over centuries (Lugo and Brown this Volume). Recent tropical deforestation is associated with a pervasive cycle of initial timber extraction followed by shifting cultivation, land acquisition, and subsequent conversion to pasture (Partridge 1984), all of which leads to loss of forest resources, reduction of biodiversity, and impoverishment of rural people (Gomez-Pompa and Kaus 1990). The roots of the problem of resource use in tropical countries are social, compounded by the lack of adequate institutional infrastructure to deal with resource management and the scarcity of trained human resources. Their solution requires greater attention to socioeconomic factors (Lugo and Brown this volume). In developing countries, the issues are most intense, because for hundreds of millions of people, the struggle is for survival (Repetto 1988).

Strict reserves are needed for many rare and endangered species that live exclusively in old-growth habitats. Yet, reserves of virgin forest in the tropics are often difficult to protect, because the management of wildlife in the reserves and their use by humans in the societies surrounding the reserve usually are antagonistic (Gajaseni et al. this volume). Replacing shifting cultivation, which destroys forest reserves with sustainable agroforestry systems that form a buffer around the reserve can help ease conflicts. Although agroforestry systems cannot provide suitable habitat for many rare and endangered species, they can provide patches of structurally more complex vegetation in comparison to the more traditionly used agricultural monocultures, thereby effectively increasing the functional size of the reserve (Gajaseni et al. this volume). Agroforestry systems also could act as "landbridges" between fragments of primary forest (Gajaseni et al. this volume). Moreover, agroforestry systems are better able to sustain a supply of vegetables, grains, and protein for the local farmer, thus relieving the pressure to advance farther into the forest (Gajaseni et al. this volume).

Maintaining Temperate Forest Ecosystems.—Silvicultural activities that reduce, fragment, or isolate forest habitats resulting in the simplification of stand structure or the elimination of individual species can reduce biodiversity (Rochelle and Hicks this volume). Particularly since changes in canopy structure has been shown to be fundamental driver of various ecological properties, including species diversity (Hansen and Hounihan this volume). But well planned forest management can maintain and/or restore biodiversity (SAF 1991). A landscape supporting a range of successional stages will contain a greater diversity of species than a uniform-aged landscape. And to the extent that management practices result in a landscape mosaic of age classes that mimics one caused by natural disturbance, biodiversity is enhanced (Rochelle and Hicks this volume). The individual responses of species to various management practices emphasize the importance of setting specific management objectives (Hansen and Hounihan this volume). Managing for "overall biodiversity" is an impossible goal for any given tract of land because any management action will benefit some species and inhibit others (Hansen and Hounihan this volume). Managers should carefully evaluate which species or community attributes are of most concern, and design silvicultural strategies accordingly (Hansen et al. 1993).

Large tracts of unmanipulated forests are needed to protect fully all ecological values associated with forests (Hunter and Calhoun this volume). The contribution of old-growth forests to biodiversity has only recently become widely recognized (Spies and Franklin this volume). On a global scale, the area of old-growth forests has probably steeply declined over the last 2,000 years as indicated by the fact that the area of "primary" forest or original forest cover, of which old-growth forests would make up a large part, is estimated to be only 24% of amount prior to agriculture settlement (Postel and Ryan 1991; Spies and Franklin this volume). Maintenance or restoration of old growth requires a good understanding of forest dynamics and the diversity and ecological roles of all successional stages (Spies and Franklin this volume). In most biomes, adequate information about the ecological characteristics of old growth does not exist, especially information about how old growth is distinguished from other forest successional stages and how it varies across landscapes and vegetation types (Spies and Franklin this volume).

Understanding the ecology of old growth and maintaining its distinctive biodiversity require a hierarchical perspective using the frame-work comprised of genes, species and populations, ecosystems, and landscapes (Noss 1990; Spies and Franklin this volume)). Placing old growth in such a hierarchy provides a more unified concept of old growth and demonstrates how human manipulations at one biological level can affect old-growth values at other levels (Spies and Franklin this volume). The maintenance and management of old-growth forests require consideration of both patch (stand) and landscape (mosaic) level perspectives of old growth that recognizes the inherent dynamic nature of landscape patterns (Spies and Franklin this volume). The maintenance of old growth within large areas is dependent on the rate, size, pattern, and intensity of disturbances. If the frequency of disturbance is too high, some types of old growth, such as Douglas-fir or ponderosa pine, will not develop in a landscape and if the frequency of disturbance is too low, other types, such as aspen, will be lost from landscapes. If disturbances are too small, then shade intolerant types of old growth, such as alder or long-leaf pine, will not occur (Spies and Franklin this volume).

The diversity of old-growth types means that conservation, management, and restoration strategies must be tailored to the particular type of old growth. Long-term strategies for old- growth conservation require identifying areas that will provide future old-growth patches and landscapes and identifying losses of old growth (Spies and Franklin this volume). Moreover, patch level and landscape level plans and management practices that work for one type might not work well for another. For example, growth types maintained by frequent burning, such as ponderosa pine, will require frequent prescribed burning or thinning to maintain their structure and function (Moir and Dieterich 1988).

Spies and Franklin (this volume) emphasize that "old growth cannot be maintained without a patch dynamics (landscape) perspective. The rates and sizes of disturbances will determine potential types of old growth in an area. Sustaining elements of biological diversity associated with old growth can be facilitated by: (1) maintaining current areas of old growth; (2) providing areas for replacement of old growth lost to disturbances; (3) facilitating organism dispersal among old-growth areas in landscapes; (4) providing old-growth habitat elements in managed forests; and (5) using manipulations to reduce the time needed to develop old-growth characteristics."

Maintaining Aquatic Ecosystems.—The biodiversity of aquatic ecosystems worldwide has been drastically altered or reduced (Rinne et al. this volume; Williams and Davis this volume). This has resulted both from a reduction of the primary medium of these systems (water) and the purposeful introduction and stocking of exotics (particularly fish species). Demands for water for agricultural irrigation, power generation, industrial development, and human consumption have resulted in reduced water flows to many aquatic ecosystems (i.e., the Everglades in the United States) and increased loads of industrial and agricultural pollutants (Rinne et al. this volume). Species loss and rarity in freshwater ecosystems of North America are particularly dramatic (Williams and Davis this volume) with the extinction of 40 fish taxa from North America during the past century, 21 of these since 1960 (Miller et al. 1989). In proportion to the entire fauna, fish species loss and rarity in much of western North America are as great or greater than in tropical regions (Moyle and Williams 1990; Williams and Neves 1992). Moreover, in some river systems habitat loss is so pervasive and non-native species so widely established, that a majority of native fishes are threatened or endangered (Williams and Davis this volume). Existing systems of protected areas have been inadequate to protect North America's native fish diversity (Williams and Davis this volume). A successful strategy for fish conservation requires improvements to our network of protected areas as well as the development of a management framework that maintains aquatic community integrity across managed landscapes (Williams and Davis this volume). Maintaining native fish populations in a protected river, for example, requires compatible upstream agricultural and forestry practices (Reid this volume).

Maintaining Marine Ecosystems.—Conservation efforts for marine environments have lagged far behind those for terrestrial and freshwater systems and are not as well understood (Williams and Davis this volume). Yet, issues in marine ecology and marine biodiversity present an especially important case. Particularly because the volume of the ocean biosphere is at least two orders of magnitude greater than that of the land, any attempt to conserve global biodiversity must not ignore the marine environment. Conservation of marine biodiversity has to contend with the wider variability that exists within the seas and ecosystem boundaries that are less clearly defined as well as biological differences between organisms. Moreover, dramatic differences exist between estuarine, coastal and sea water systems and between surface and deep water systems. Efforts are further confounded by jurisdictional problems arising from differing and sometimes conflicting national and international laws and policies (Wolff this volume). Where marine resource conservation policies and activities exist, they have focused on species-specific measures. There is no marine counterpart to the "Red Data Book."

Less than 1% of the world's marine area is included in an established protected area, and nearly all of the more than 1,000 established marine protected areas allow extractive uses (Salm and Clarke 1984; Kelleher and Bleakley 1992). Truly protected marine areas large enough to assure biological integrity are exceedingly rare, in spite of growing concern about human impacts on the oceans and the worldwide collapse of many major fisheries from using simplistic single-species fisheries models and overfishing (Williams and Davis this volume). However, the management authority of the Great Barrier Reef Marine Park seeks to balance human needs and environmental protection through the application of a broadly based management strategy which caters to use, appreciation, and enjoyment, while protecting a range of specific areas and organisms (Kelleher and Craik this volume). The Great Barrier Reef is managed as a single ecosystem including all the waters of the Great Barrier Reef Region and the 900 islands within its outer boundaries.

Reef fisheries are particularly difficult to manage on a species-specific basis because of the high number of interacting species and the wide variety of gear and fishery objectives (Williams and Davis this volume). Intense fishing reduces abundance, alters population age structure, and can alter fish community structure, all of which are symptoms of overfishing (Munro and Williams 1985; Polovina and Ralston 1987). Protection of entire communities offers many ecological and economic advantages over species-level conservation efforts (Williams and Davis this volume). A community approach to marine fisheries management seems to offer the best opportunity to sustain fishery yields and protect biodiversity, but such an approach has not yet been thoroughly tested (Roberts and Polunin 1991) although it appears to work well in the few cases where it has been examined (Buxton and Smale 1989; Ballantine 1991; Wiliams and Davis this volume).

Coastal zone management has to deal with the interface between terrestrial and marine environments. Species and communities are affected by exploitation, pollution, and degradation of habitats (Wolff this volume). The major processes affecting coastal zones are: (1) development and its influence on geomorphological and hydrological processes and in relation to sea level rise, and (2) biogeographical processes leading to the establishment and extinction of species (Wolff this volume).

An integrated approach to the sustainable use of the world's marine and coastal zones is imperative as marine resources continue to be exhausted (Kelleher and Craik this volume; Williams and Davis this volume). The sustainability of both biodiversity and productivity are inseparable and it needs to be the highest goal of marine resources management as food production in the sea shifts from a "hunt and gather" mode to mariculture (Williams and Davis this volume).

Maintaining Biodiversity on a Landscape Scale

Scale and Scope.—Spatial scale whether local, regional, or global, greatly influences our perceptions of biodiversity (Crow 1989, Spies and Franklin this volume). Understanding the importance of scale is critical to assess accurately the impact of land management practices on biodiversity. The scale of a conservation endeavor affects the strategy involved, the determination of realistic goals, and the probability of success (Pearson et al. this volume). Many significant biological responses and cumulative management effects occur and are only discernible at the landscape level. For example, homogeneity is scale-dependent—what appears heterogeneous over small areas might appear homogeneous over large areas. For example, on a small watershed there could be two very different "forest stands" based on vegetation structure and development, but over a larger area those two stands can be part of a "homogenous" repeating mosaic of vegetation phases (Spies and Franklin this volume). Planners and managers are increasingly aware that adequate decisions cannot be made solely at the stand level. This is particularly obvious because land-use patterns characteristic of a human-dominated landscape are ones in which large, continuous tracts of natural habitat become increasingly fragmented and isolated by a network of developed lands. Thus, regardless what the primary management objective may be, whether producing timber, creating wildlife habitat, protecting watersheds, or providing wilderness experiences, assessing these opportunities clearly requires consideration beyond the boundaries of a particular planning unit.

A strategy to maximize species diversity at the local level does not necessarily add to regional diversity. In fact, often in our haste to "enhance" habitats for wildlife "edge" preferring species have been emphasized at the expense of "area" sensitive ones, and consequently regional diversity might have been decreased. It is important to realize that principles that apply at smaller scales of time and space do not necessarily apply to longer time and larger spatial scales (Crow 1989). Long-term conservation of species and their genetic variation, will require cooperative efforts across entire landscapes (Miller this volume). This is consistent with the growing scientific sentiment that biodiversity should be dealt with at the scale of habitat/ecosystem/landscape levels rather than at the species level (Hunter et al. 1988).

If context (i.e. uses in surrounding areas) is ignored in conservation decisions and the surrounding landscape changes radically in pattern and structure, patch content will be also altered by edge effects and other external influences (Noss this volume). For example, landscape connectivity is a direct consequence of the abundance of suitable habitat, its spatial patterns in the landscape, and the organism's scale of resource utilization (Pearson et al. this volume). Rather than emphasizing homogeneity, as the stand or site concept does, a landscape perspective emphasizes spatial patterns and heterogeneity (Spies and Franklin this volume). Mosaic characteristics, such as kind and proportion of patch types, their sizes, spatial pattern, and stability, affect the biodiversity of the landscape. In fine-grained mosaics, contacts between patches are frequent because of the high edge-to-patch area ratio, whereas in coarser-grained mosaics which occur over larger spatial areas, contacts between differing patch types are less frequent as a proportion of the mosaic area than in fine-grained mosaics (Spies and Franklin this volume).

Ecosystem Management.—The demands and expectations placed on biological resources are high and variable, calling for new approaches that go beyond merely reacting to resource crises and concerns (Szaro and Salwasser 1991; Szaro 1993b; Szaro et al. 1994). New approaches must incorporate fundamental shifts in the scale and scope of conservation practice (Miller this volume; Salwasswer et al. this volume)). These include the shift of focus from the more traditional single-species and stand-level management approach to management of communities and ecosystems (Reynolds et al. 1992). An ecosystem perspective on multiple-use management expands the sustained yield concept beyond single resources to sustaining the processes and component parts of whole ecosystems and the multiple-use concept to a wider array of values and uses (Salwasser et al. this volume).

Public concern about the environment, together with new thinking by scientists and resource managers, has led to a new philosophy about land and resource management. This philosophy espouses managing ecosystems to sustain the full array of values and uses for current and future generations. It also recognizes that humans are an integral part of ecosystems, and promotes a new objective that calls for changes in the traditional ways of managing resources. (Reid this volume). It is an attitude, an approach, and a philosophy about effective multiple-use, sustained-yield management. It is an approach that strives to protect sustainability of ecosystems, while providing for reasonable use of natural resources (Kelleher and Craik this volume). In short, it is environmentally responsible resource management.

To respond to this challenge, it is necessary to look at the dominant social trends affecting future protection, management, and use of natural resources, and evaluate how the trends might affect the content and conduct of our natural resource and research programs. Three trends are the most important in determining future direction: (1) expanding world population and associated demographic changes, (2) increasing competition for the many uses of natural resources from a finite resource base, and (3) increasing public awareness and concern for natural resources and for national and global environmental issues. In addition to traditional natural resource products—timber, big game animals, fish and water—society is now interested in: (1) how ecosystems and climate affect each other, (2) loss of biodiversity especially reflected in threatened and endangered species, (3) growing demand for natural resources integrated with the protection of other resources values, and (4) maintaining ecosystem health.

Intensity of land use varies in a continuum from areas with no human manipulation to practices so intensive that natural ecosystems are replaced by artificial or cultivated ones (Hunter and Calhoun this volume). Hunter and Calhoun (this volume) conceptualized this continuum "as a triad model in which land uses are characterized as: (1) intensive commodity production areas, (2) areas with little or no resource use by people, or (3) areas in which modest resource use occurs while ecological values are carefully protected." All three types of land use have validity and finding the proper balance among these uses is the key to improve resource conservation (Hunter and Calhoun this volume).

An ecosystem management approach, as discussed above, uses as basic principles those derived from conservation biology theory for conserving biodiversity and maintaining ecological systems (Ingram this volume; Soulé and Wilcox 1980; Soulé 1986, 1987; Salwasser et al. this volume). These principles include:

- Recover and conserve threatened and endangered species.
- Provide for viable populations of native plant and animal species.
- Maintain a viable network of native biological communities and ecosystems.
- Maintain structural diversity.
- Sustain genetic diversity.
- Produce and conserve resources needed by people.
- Protect ecosystem integrity—soils, waters, biota, and ecological processes.
- Restore and renew degraded ecosystems.

Ecosystem management should respond to significant shifts in social values, scientific understanding, and land management interests from those of the past. For practical purposes it is generally synonymous with sustainable development, sustainable management, sustainable forestry, and a number of other terms used to identify an ecological approach to land and resource management. Ecosystem management is a goal-driven approach to restoring and sustaining healthy ecosystems and their functions and values. It is based on a collaboratively developed vision of desired future ecosystem conditions and integrates ecological, economic, and social factors. One implication of the large scale required for ecosystem management is that the management of "unprotected" land is as important for the conservation and maintenance of biodiversity and its important ecological processes as the management of protected areas (Reid this volume).

Four basic operating tenets provide an "umbrella" for an ecosystem management approach. Under this umbrella are components driven or related in some degree to participation, collaboration, and using the best science. These tenets are:

- **Partnerships**: Sharing responsibility for land management is fundamental. Ecosystems often cross political boundaries, making the need for cooperation, coordination, and partnerships essential to managing the entire ecosystem.
- **Participation:** Get people involved in all aspects of public resource decision-making so that managers will know their needs and views. A highly participatory process is essential, from beginning to end, before deciding on a course of action, by involving those interested in formulating alternatives, evaluating those alternatives, and developing a process select a desired outcome. The focus should be on end results—desired future ecological and social conditions and the land-use classes and management actions that will best attain them.
- **Scientific Knowledge:** Use the most appropriate scientific information and technologies available to understand the range of choices of actions and the consequences of each. Integrate information and technology, such as ecological classifications, inventories, data management systems, and predictive models, and use them routinely in landscape-scale analyses and conservation strategies. This includes strengthening teamwork between researchers and resource managers to improve the scientific basis of ecosystem management (Soulé and Kohm 1989; Solbrig 1991; Szaro 1994).
- **Ecological Approach:** This means in the simplest terms looking at many factors, across a broad landscape, using several scales, addressing linkages between landscape elements and ecological processes, and a number of other activities. The science of ecology will be applied to multiple-use management recognizing that people are part of the ecosystems we manage. Landscapes should be used as the basic unit for planning and managing ecosystems to meet specific objectives — both desired future ecological conditions and desired economic and social goals while reconciling conflicts between competing uses and values.

Evolving from these four principles are a set of methods and tools that compose the basic elements of any ecosystem management approach (Table 47.1). Independent and unique decisions must be made on individual projects and plans. Many of the tools and methods require sharing and cooperation across administrative boundaries. Much of the information needed at each level to conduct ecosystem management, especially information at the higher geographic scales, is useful to many management units, and many other organizations also interested in ecosystem management.

Adaptive Management.—A formal process of adaptive management will be required to maximize the benefits of any option for land and natural resource management and to achieve the long-term objective of ecosystem management (Szaro et al. 1994). Adaptive management is a continuing process of action-based planning, monitoring, researching and adjusting with the objective of improving the implementation and achieving the desired goals and outcomes (ROD 1994). The process itself is straightforward and simple: new information is identified, evaluated, and a determination is made whether to adjust strategy or goals. The core of any regional program for in situ conservation is a long-term inventory, research, and monitoring effort that in turn informs adaptive management (Ingram this volume). The general planning model illustrated in

Table 47.1. Key methods and tools that compose the basic elements of any ecosystem management approach (Szaro et al. 1994; Szaro and Sexton 1994; Salwasser et al. this volume).

1. Characterize existing site conditions
   (a) Conduct information collection, analyses and planning across administrative and jurisdiction borders to coincide with useful ecological boundaries.
   (b) Develop, seek out, utilize and transfer the very best available scientific knowledge.
   (c) Develop information about a variety of species habitat needs.
   (d) Develop knowledge about the human dimensions of ecosystem management.
   (e) Develop, monitor, and evaluate vital signs of ecosystem health.
2. Characterize ecosystem cycles and limitations
   (a) Define major disturbance factors and their range of historic variation.
   (b) Develop information about ecological processes; carbon cycle, nutrient cycle, hydrologic cycle, succession, biological diversity, population dynamics.
   (c) Develop knowledge of linkages within and between systems and processes.
3. Characterize landscape patterns and linkages
   (a) Develop common approaches to ecological classification.
   (b) Develop and use methods to recognize and address patterns and change over time and space for key elements at multiple scales.
   (c) Address biotic information across levels of biological organization (cell, organism, population, community, ecosystem, landscape, biome, biosphere).
   (d) Conduct analyses over large geographic areas that encompass smaller project areas.
4. Develop mechanisms for information management
   (a) Develop approaches to share information across many borders, including integrated resource inventories and information provided for national uses.
   (b) Develop decision support technologies and methods to support the complexities of ecosystem management. Build recognition of uncertainty into those processes including the fact that most questions will probably never be answered and major mistakes can take a long time to heal.
   (c) Integrate information and technology, such as ecological classifications, inventories, data management systems, and predictive models, and use them routinely in landscape-scale analyses and conservation strategies.
   (d) Adopt means to deal with the complexity that comes with using multiple scales and multiple boundaries across scales for organizing and using information necessary for sound analyses.
5. Develop options for action
   (a) Address activities and information across several geographic scales.
   (b) Address effects at the project level and at least at one scale above and below.
   (c) Cooperatively develop desired conditions.
   (d) Focus on end results—desired future ecological and social conditions and the land-use classes and management actions that will best attain them.
6. Implement an adaptive management process
   (a) Use landscapes as a basic unit for planning and managing ecosystems to meet specific objectives—both desired future ecological conditions and desired economic and social goals while reconciling conflicts between competing uses and values.
   (b) Select scales/boundaries appropriate for highly mobile species.
   (c) Work within the scope of natural processes that shape landscape and ecosystem conditions.
   (d) Use highly participatory process, involve all the publics that want to be from basic data collection through monitoring.
   (e) Seek and form as many partnerships as possible in doing ecosystem management, with federal, state, local and other organizations.
   (f) Use an adaptive management process as an integral part of monitoring and evaluation.

Figure 47.1 illustrates the adaptive approach to management (FEMAT 1993). This approach (Walters and Holling 1990) provides a basis for immediate implementation of ecosystem management.

In an adaptive management approach, Walters and Holling (1990) offer challenges for justifying and designing experimental management programs. These are:

- To demonstrate that a substantial, deliberate change in policy should even be considered, given the alternative of pretending certainty and waiting for nature to expose any gaps in understanding.
- To expose uncertainties (in the form of alternative working hypotheses) and management decision choices in a format that will promote both intelligent choice and a search for imaginative and safe experimental options, by using tools of statistical decision analysis.
- To identify experimental designs that distinguish clearly between localized and large-scale effects, and hence, make the best possible use of opportunities for replication and comparison.
- To develop designs that will permit unambiguous assessment of transient responses to policy changes, in the face of uncontrolled environmental factors that may affect treated and reference experimental units differently.
- To develop imaginative ways to set priorities for investments in research, management, and monitoring, and for design of institutional arrangements that will be in place long enough to measure large-scale responses that may take several decades to unfold.

Whereas the concept of adaptive management is relatively straightforward, applying it to complex management strategies requires answers to several critical questions (ROD 1994). What new information should compel an adjustment to the management strategy? What threshold should trigger this adjustment? Who decides when and how to make adjustments? What are the definitions and thresholds of acceptable results? Adaptive ecosystem management depends on a continually evolving understanding of cause-and-effect relationships in both biological and social systems (Everett et al. 1994). The key features in an adaptive approach are (C. McConnaha, pers. comm.):

1. An *experimental design* for implementation—Adaptive management demands taking a rigorous scientific approach to management. Holling (1978) noted that the "heart of adaptive environmental management" is "an interactive process using techniques that not only reduce uncertainty but also benefit from it. The goal is to develop more resilient policies." Experiments designed to produce information should be an integral part of the actual management activities. Adaptive management can only be effective if monitoring and evaluation procedures are an integral part of the design and are incorporated at the beginning of the process and not simply added post hoc after implementation (Holling 1978).

2. An explicit *description* of the system—To answer critical questions related to resource management, there is a need to understand the current ecological conditions of ecosystems of interest, changes in the ecosystem components over time, and likely ecological trends within the ecosystems. Therefore, it is necessary to develop an assessment to put the social, biological, chemical, and physical components of the management area for which we have stewardship into the larger ecological context.

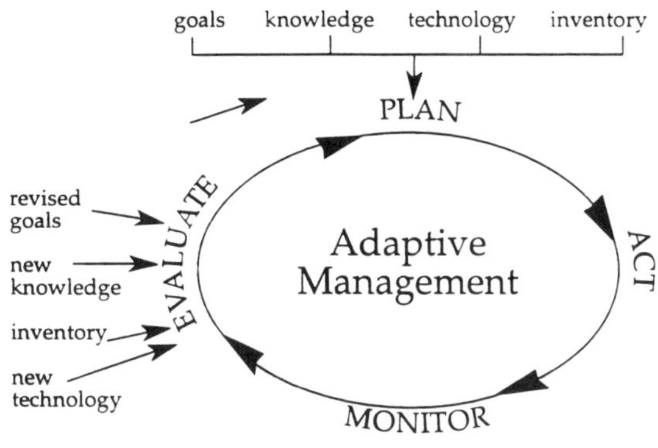

Figure 47.1. Adaptive approach to management (FEMAT 1993).

3. Well defined *goals and objectives*—The planning process, a collaborative and cooperative approach, uses the assessment to assign values to the current condition and describes the "desired future condition" of the resources. An inappropriate decision is likely to occur without widespread understanding, acceptance, and support for the desired future condition. Two relevant working hypotheses formulated by Everett et al. (1994) are: (1) Human values and expectations can be incorporated into ecosystem management by identifying landscape patterns that are representative of these values, and (2) sustainable ecosystems can be achieved by integrating people's expectations with the ecological capacities of ecosystems.

4. Identification of *critical uncertainties*—Because information might be imperfect information a level of uncertainty exists in any decision. The concept of adaptive management acknowledges the need to manage resources under circumstances that contain varying degrees of uncertainty, and the need to adjust to new information (ROD 1994). Even the best decisions may have unacceptable outcomes. To reduce the level and impact of unacceptable outcomes the decision-maker needs to focus on critical uncertainties. Obtaining the best information on these uncertainties and designing a monitoring and evaluation process to track decisions is therefore an essential step in the process.

5. A *monitoring and evaluation program*—What new information should trigger an adjustment of strategy or direction? New information can come from monitoring, statutory or regulatory changes, organizational or process assessments, or a variety of other sources (ROD 1994). There are two distinct phases in a monitoring and evaluation program. The first phase, traditionally, monitors and evaluates the consequences of our management actions. It constantly measures our progress towards our stated desired future ecological condition and provides for mitigation when our actions steer off course. During the evaluation process, information is analyzed to determine the nature, scope, and importance of that information (ROD 1994). Adaptive management depends on both negative and positive feedback in the reiterative evaluation of both the continued desirability of previously selected management goals and progress toward their achievement (Everett et al. 1994). If the impacts detected during monitoring and

evaluation are of little significance, then the management regime can be refined without adjusting the long-term plan. If, however, the impacts are significant, the plan will need revision. This can be done directly or, if needed, by modifying the assessment by incorporating the new knowledge, changed social values, impacted environments, etc...

The second, and less traditional, phase monitors and evaluates social needs in relation to changing societal values and settings. Social values and needs are constantly assessed in relation to the validity of the desired future ecological condition, recognizing that societal values often change more rapidly than ecological conditions.

6. An aggressive approach to *learning*—As noted above the less traditional approach to monitoring and evaluation requires active learning in all aspects of ecosystem management. Resource managers and scientists must learn more about the social, biological, and physical attributes of ecosystems and adapt more quickly to new knowledge. Technology transfer must be an inherent and valued part of ecosystem management.

7. An *adaptable* structure—Lastly, the adaptive management structure itself must be inherently adaptive. This includes the organization responsible for implementing the process. The key element in this flexibility is the development of a management system that operates strategically and not functionally. This system emphasizes functional skills and eliminates functional organizational barriers.

POLICIES

The Need for Science in Policy Development

Our ability to make policy has far exceeded our ability to implement it (Pavlik this volume). Sound science for sound policy is currently a common theme held by many organizations and governments. In resolving issues such as biodiversity, the approach needed is to have both sound science and the "right science" to address the critical policy issues that national and international natural resource management organizations will face. A strong scientific underpinning is needed for major policy decisions to understand their potential effects on the environment. The lack of research data/information underlying the development, implementation, and assessment of most policies and programs can often be a major impediment to effective government (Chelimsky 1991). This is particularly true as management decisions become increasingly complex and are increasingly challenged (Chelimsky 1991). The competition between commodity and environmental interests for priority access to natural resources has intensified. Because problems are not constrained by state, regional, or national boundaries, threats to biodiversity involve global influences and require coordinated global action. And an increasingly involved and well-informed public is challenging the technical credibility of conservation plans and decisions.

The need for scientific information as a foundation for resource management decisions continues to increase dramatically. Even a decision to not manage is essentially a management decision. The need to assemble and format new and existing research results into packages that can be used by managers is great. As such, efforts to synthesize science, and in so doing identify information gaps and set priorities, should be increased.

Long-term observations of biological systems are invaluable in providing information on their response to change and in providing the understanding needed for predictions regarding the future. Research on computer-based decision support models and expert systems hold promise of providing an effective interface between management and research to allow us to predict what kinds of organisms will be present under a given scenario and how these organisms will interact with their new environment. These systems should also help us to predict the future condition of resources based on the management systems used.

The interface between social, physical-biological and ecological models must be improved (Machlis and Forester this volume). An essential task is to improve the ability to quantify social demands for both consumptive and non-consumptive goods. These demands must then be weighed against the need to maintain ecosystems and their attributes. Innovative ecosystem management systems and technologies need to be developed to provide for these demands while maintaining healthy ecosystem functioning.

The integration of science into the decision-making process has been slowed by two primary reasons: (1) decision-makers have not always been aware of how or when research might be useful to them and (2) decision-makers in the past have shown a reluctance to call upon researchers for help, because it meant acknowledging uncertainty, and worse, potentially giving up some power by reducing the range for their discretion. Yet, science can provide two considerable advantages to the decision-making process. First, the decision can be based on past experience and knowledge and second, the fact of ignorance can be confronted because it has been explicitly recognized. This second advantage is especially important because not knowing something in a key area normally means that a policy needs to be cautious and flexible (maybe even reversible) with respect to that unknown and that a program flowing from the policy could be specifically designed to capture the knowledge that is needed (Chelimsky 1991). Thus, policies relating to such a difficult concept as biodiversity need to be designed using an iterative adaptive management strategy that allows for feedback and changes in that policy. Incorporating the evaluation mechanisms that allow policies, programs, and performance to be assessed promotes prudence and responsiveness on the part of government officials.

Developing Policies for Biodiversity Conservation

Gene Conservation Policy.—Part of the problem in defining a policy for genetic diversity is that it is often seen as a goal with every variation and every instance important, rather than as method for achieving a goal (Namkoong this volume). Consequently, rare species are often viewed as having high inherent value while the genetic diversity of useful but common species is considered of less importance (Namkoong this volume). Yet, species diversity and genetic diversity are not synonymous. Each level has elements and dynamics that, while inter-related, are distinct and require different methods of management (Namkoong this volume).

The gene is at the most basic level of potential conservation concern, but the conservation of particular genes is seldom seen as the purpose of management. Genes, or any special combination of them, are not valued in the same sense as are entire organisms or groups of organisms and do not serve as ends of conservation in themselves.

Rather, they are necessary elements for serving some larger goal. Furthermore, because genetic diversity is an ensemble property of a group, policy goals should be designed for the good of the group rather than of the individuals making up the group. Saving all alleles or allelic combinations that exist now or existed in the past is not feasible and, therefore, management for such an illusive goal is impossible. The present status of genetic diversity provides evidence of what is evolutionarily feasible and can indicate what might not be feasible, but it cannot define what should exist in the future (Namkoong this volume).

Genetic diversity can be affected by management actions or inactions, either by intent or neglect. Priorities must be established according to the probable success of management systems in achieving defined goals, because although in theory those goals might be very inclusive, there are always limitations on funds, efforts, and personnel to achieve them (Namkoong this volume). Policy goals for genetic diversity might be most easily derived from the more general goal of securing the welfare of present and future generations of humans. Even on the narrow basis of human utility alone, all other species can be considered to have some value. In many species, genetic variation may be of no direct value to humans, but it may be indirectly valuable because it is necessary for survival and evolution of directly valuable species. Policy goals should certainly provide for a wide range of species, some with commodity and many with noncommodity values, but intensive gene management might then be reserved for species that can possibly affect the welfare of present and future humans. These goals should not unnecessarily constrain opportunities for the continued evolution of species but should provide for their adaptability to present and future environmental variations (Namkoong this volume).

Because managing the evolutionary possibilities of all species is not practical, more precise and more limited goals are needed. A policy goal might be to maximize the useful genetic variation of ecologically and economically valuable species. Other species might be ignored completely or assured only of some minimal levels of genetic variation. The goals of management with respect to genetic diversity could, therefore, be limited to providing for or maximizing opportunities for evolution of "valuable" species. It may well be an objective of management to pay particular attention to threatened and endangered species or populations and to ensure that population differences are maintained or increased (Namkoong this volume).

The lack of sufficient biological knowledge about most species and the costs of management limit intensive genetic management to relatively few species (Namkoong this volume). Hence, by using a species-by-species approach, gene management programs can only be completed for only a small number of species. In these cases, genetic management should be targeted to species that are critical to the stability of communities or ecosystems, and hence the viability of ecosystems and all the organisms included in them would be ensured (Namkoong this volume). Even systems of less intensive management, employing natural reproduction and simple genetic monitoring, can probably only be extended to no more than a few thousand species. Therefore, genetic management will often have to be accomplished through ecosystem or landscape management techniques and systems that allow for self-sustaining evolutionary processes (Frankel 1983; Norton 1987). Thus, even though the goals of genetic management and ecosystem management are not the same, they might be accomplished by the same means (Namkoong this volume).

Species Conservation Policy.—Species conservation policy in the United States has been dominated over the past two decades by the Endangered Species Act (Bean this volume). This Act is often held as the standard upon which the conservation efforts of other countries are evaluated (de Klemm and Shine 1993). Yet, although much progress has been made under that Act in reversing the decline of many species and avoiding the extinction of many others (Rinne et al. this volume), there are still several noteworthy failures (Bean this volume). Of more immediate consequence is the fact that there are many more species known to be in some peril of extinction than can be addressed under the auspices the Act (Bean this volume). Currently, more than 800 species in the United States and its possessions are formally listed as threatened or endangered with between four and five times that number having been identified as likely canidates for listing. With current levels of available resources and at prevailing listing rates, the existing backlog of identified species needing protection will not be cleared for several decades. Thus, the necessity exists to choose which of the many species that might be protected should be protected first. In short, the real policy issue is which of the many species already in danger of extinction should receive priority attention. Judgments about which species to protect must be based on an understanding of its probability of extinction within a given time frame (Bean this volume).

The core of current policy under the Act is to conserve any plant and animal species threatened with extinction, now or in the foreseeable future (Bean this volume). The designation of critical habitat under the Act met with no serious problems as long as the areas involved remained small. However, the spotted owl controversy in the Pacific Northwest of the United States covers a much larger area and has raised serious questions about the implementation of the act and the potential role of other considerations in the decision process (de Klemm and Shine 1993).

Yet, even if one could limit political "interference" with a species conservation program, there may be good reasons to reconsider giving primacy to the imminence and gravity of threat to particular species in deciding which species to protect first (Bean this volume). Other considerations might include: whether the species plays a key role in its ecosystem, such that its loss may lead to significant disruption to that ecosystem; whether the species can serve as an indicator of overall ecosystem health and thus indicate which other species may soon also be at risk; whether the species has been used in traditional medicine, or is closely related to others that were, such that its conservation might preserve options for future medicinal utilization; and whether a species is closely related to crop or domesticated animal species, such that its conservation might preserve options for future agricultural application (Bean this volume). The first two of these suggested criteria would also link species conservation concerns with broader ecosystem conservation concerns. The other two seek to tie species conservation priorities more directly to their human utilitarian values.

As in genetic conservation efforts, species conservation by a "one-at-time" approach to conservation will have to give way to broader conservation efforts aimed at restoring and maintaining whole ecosystems (Scott et al. 1987; Tear et al. 1993). By protecting large areas of habitat, all the species within them will benefit; the challenging tasks of determining the status of each species and designing conservation measures to meet its needs can potentially be avoided. Undoubtedly, protecting large contiguous areas of habitat will have major conservation benefits for the many species that utilize those habitats (Bean this volume).

However, if "protection" of large, contiguous blocks of habitat excludes significant commodity uses of those lands, the political and economic conflicts likely to result from those efforts will be even larger than those that have attended efforts to protect individual species (Bean this volume). If significant commodity uses are not excluded, then the question arises as to how such uses and areas are to be managed (Bean this volume). A more productive approach might be to integrate ecosystem restoration and maintenance strategies into species recovery efforts (Bean, this volume). Recovery strategies for endangered species that emphasize the restoration and maintenance of ecological relationships and functions are likely to be much better long-term strategies than are strategies that emphasize intensive, artificial manipulation (Bean this volume). Not only can they achieve recovery for today's endangered species, but make it less likely that other species within their ecosystems will become endangered in the future.

Unfortunately, many of the arguments being used today for the integration of economic concerns into species conservation decisions focus only on short-term and local tradeoffs (Bean this volume). This is counter to the long-term objectives of species conservation which benefit all citizens of the Nation, not just those living in the vicinity of the species (Bean this volume). If species conservation efforts are to survive the current clamor for "balance" in decision-making, the balancing mechanisms will have to be ones capable of taking into account both long-term and geographically comprehensive consequences. It also seems important to spread the risks of species conservation more broadly, so that local communities and landowners do not incur the sometimes heavy burden of conservation requirements intended to benefit the entire nation (Bean this volume). Trying to find this "balance" behooves us to consider and adopt innovative and progressive approaches in future conservation efforts (Rinne et al. this volume).

Community/Ecosystem/Landscape Conservation Policy.—For most of this century, the focus of many international treaties and much national legislation related to biodiversity have been largely directed at protection at the species and genetic levels (Bridgewater et al. this volume). In contrast, the focus over the last decade has concentrated on a broader scale approach to conservation at the ecosystem and landscape levels. Taking a broader spatial perspective will lead to the evaluation of issues, such as cumulative impact and regional context (Hunter and Calhoun this volume), and ultimately the development of effective strategies for maintaining diverse natural communities within an increasingly modified and fragmented landscape (Buechner and Sauvajot this volume). In balancing interests and perspectives for biodiversity management, the balance must include an understanding of the valuation of biodiversity, that is, by whom, how, and when biodiversity is valued (Clark et al. this volume). Also to be considered are the risks posed by the varied threats to biodiversity and an awareness that only some diversity can or will be saved, because the necessary resources, public commitment, and available time are limited (Clark et al. this volume).

In reality the intensity of land use is a continuum from no use to very intensive human use (Hunter and Calhoun this volume). Removal of vegetation, atmospheric warming, increased pollution loads, and quite sharp and sudden shifts in the pattern of land use are of tremendous importance in conservation practice (Bridgewater et al. this volume). Management strategies should be developed to look at the components of landscape and at the impact of possible changes in a landscape ecological frame-work. This implies an understanding of the relationship between keystone elements in the landscape, the surrounding matrix, and the need for connectivity (Bridgewater et al. this volume). Protected areas, within the concept of biosphere reserves, and elements of remnant vegetation connected by corridors or "greenways" can provide the reservoirs of biodiversity and avenues for adaptive responses to environmental change (Bridgewater et al. this volume).

Successful policy development will require an understanding of ecoregional approaches, including the definition, design, and understanding of ecoregions (Bridgewater et al. this volume). A ecoregional approach, embracing whole catchments and crossing the land/sea interface where appropriate, is the key to biodiversity conservation. Recognition that biodiversity comprises more than "native" species, and that restoration/rehabilitation are vital tools at the landscape scale is critical (Bridgewater et al. this volume). We cannot ignore the potential impact of non-native species being added to the species mixtures which form our landscapes (Bridgewater et al. this volume).

Hunter and Calhoun (this volume) proposed an approach that uses a three-part system called a *triad* that recognizes "that all forms of land use have validity: (1) intensive commodity production areas, (2) areas with little or no resource use by people except low-intensity recreation, and (3) areas in which modest resource use is allowed while ecological values are carefully protected." They argued that all three types of land use have validity and that a different balance among uses could improve resource conservation. For example, in a region where multiple use forest management is dominant, wood production could be intensified on some areas, allowing other areas to be set aside as reserves with no net effect on production (Hunter and Calhoun this volume). Examining current land-use allocation patterns from the triad perspective can reveal imbalances that, if rectified, could improve land use from both an ecological and economic perspective. An implicit assumption of these potential tradeoffs between extensive management and intensive management is that the level of commodity production would remain about the same; it would just be obtained from a limited area through intensive management versus over a larger area through dispersed management (Hunter and Calhoun this volume). The recommendation that intensive but sustainable management has a role in wise land use (Hunter and Calhoun this volume).

Scientists and policy-makers have recently focused much attention on management of public lands for biodiversity (Keystone Center 1991, for example). Public lands, however, do not include many important elements of ecosystem diversity, particularly in regions such as the southern United States, where private ownerships comprise up to 90% of the land base (Yarrow 1990; O'Connell this volume). In an overall strategy to conserve biodiversity, private lands clearly must play a large role (O'Connell this volume). They contain many important elements of biodiversity not found on public lands, and management options are often limited by little more than the interest and willingness of the owner and the ability to reconcile the conservation of biodiversity with other land-use goals (O'Connell this volume).

A recognition of the role of mixed ownerships and multiple-uses within ecoregions is important in the development of any comprehensive approach. The process of imposing a mixture of social, economic, technological, and conceptual patterns over various natural systems creates not only confusion, but great political and managerial problems (Dahlberg this volume). The development of policies to maintain or enhance biodiversity have to be developed with the recognition that landscapes are physically, institutionally, and conceptually fragmented (Dahlberg this volume). Managed landscapes must be part of the solution to managing effectively for biodiversity and a successful strategy for its conservation must recognize the role and needs of humans in the process (Rochelle and Hicks this volume). The development and application of a range of management approaches tailored specifically for different land ownerships will ultimately determine overall patterns of biodiversity across the globe (Rochelle and Hicks this volume).

A primary concern of private landowners and managers involves the cost of conservation of biodiversity (O'Connell this volume). Many landowners are concerned that sincere attempts to integrate conservation of biodiversity in land use will place them at a disadvantage with competitors who choose not to do so (O'Connell this volume). Of particular importance are heavily-managed renewable resource systems—agriculture, forestry, fisheries, and grazing (Dahlberg this volume). A hierarchy of management and policy strategies is needed to promote sustainability generally and the maintenance and enhancement of biodiversity in particular. What is required is a transformation towards sustainability of the larger economic, political, organizational, and cultural forces that shape and constrain specific biodiversity policies (Dahlberg this volume). Any movement of society towards sustainability must involve all sectors of society and must include not only ecological requirements, but requirements relating to ethics, equity, and economics (Dahlberg 1991).

Obviously the development of economic policies that encourage biodiversity conservation is crucial to any long-term strategy. Economic policy plays an important role at several levels: developing improved macro-economic policies, developing sector-specific policies at national or local levels, and developing ways to integrate economic thinking into programs for managing biological resources (McNeely this volume). In general, those establishing the economic policies governing the use of biological resources have three broad options: to manage for a sustainable yield and therefore maintain the future option of depleting a resources (when prices may be higher); to draw down supplies slowly, in an effort to spread earnings over many years; or to deplete existing stocks quickly, taking a quick profit which can then be reinvested elsewhere. Most conservationists prefer the first option, but the third option is the one most often used in the case of forests, fisheries, and wildlife (McNeely this volume).

RESEARCH NEEDS

Our knowledge of biodiversity is often surprisingly superficial, and we have hardly begun to understand the extraordinary complexity of ecological systems (Soulé and Kohm 1989). Several recent efforts have examined research needs for biodiversity (Reid and Miller 1989; Soulé and Kohm 1989; Lubchenco et al. 1991; Solbrig 1991; WRI/IUCN/UNEP 1992). Much overlap is found in all the above efforts but the focus and scope of the effort depends on the perceived needs of the individuals and groups involved.

To provide the scientific basis for conserving biodiversity requires a combination of:

- basic long-term scientific inquiry;
- issue-focused, problem solving interdisciplinary research and technology development;
- scientific support to the ongoing management of lands and resources, such as

through ecological classifications, inventories, ecological analyses, and monitoring and evaluation; and innovative management systems and technologies;
- development of interdisciplinary models; and
- integration of science into the policy-making process.

Soulé and Kohm (1989) identified the following most pressing and important initiatives and research needs:
- A crash program to carry out extensive surveys and mapping to identify critical areas for protection;
- A coordinated research program at selected sites in the tropics for comparative research on populations, communities, and ecosystems in relatively undisturbed and secure situations;
- Enhanced support for research on fundamental species level processes, such as physiology, reproduction, behavior, and viability of individuals, especially with regard to species of critical ecological or economic importance;
- Studies at all spatial scales to assess the kinds, mechanisms, and magnitudes of impacts on ecological systems; and
- Training for scientists and natural resource managers, particularly in tropical developing countries.

These same themes were identified in most other assessments of research priorities. Reid and Miller (1989) added a critical component: "Integrate the study of cultural diversity into biodiversity research." Solbrig (1991) took a hierarchial approach by looking at many detailed hypotheses at varying levels of organization from genes to ecosystems. One additional component also identified in the later effort, was the inclusion of monitoring and how it should be done and the role it should play in future international programs. Lubchenco et al. (1991) in the Sustainable Biosphere Initiative had three broad research recommendations, but stressed a new integrated program of research on the sustainability of ecological systems that should focus on understanding the underlying ecological processes in natural and human-dominated ecosystems.

One critical problem in these efforts is the perception by managers that scientists are often too distant from management issues to focus effectively research on critical information needs. Working with land-managers, the USDA Forest Service identified six critical biodiversity research questions for program emphasis to provide the tools and information needed to support management goals:

- How do we integrate resource uses and ecological values across a range of spatial and temporal scales?
- How can we protect, maintain, and enhance biodiversity and productivity while managing resources for human uses?
- How do we account for and respond to societal needs and economic considerations?
- What can we use as measurable indicators for monitoring changes in biodiversity?
- How can networks of reserved areas, buffer zones, and high use areas be effectively intermixed across the landscape to maintain biodiversity?
- How can we maintain those native species whose range and populations are threatened by introduced insects, diseases, and other exotic organisms?.

Because of the limited state of knowledge, the relationship between human action and biodiversity is a critical component of most biodiversity research agendas (Soulé and Kohm 1989; Reid and Miller 1989; Machlis and Forester this volume). The importance of involving the human component in any research agenda was stressed by the two of the four research action items in the Global Biodiversity Strategy (WRI/IUCN/UNEP 1992):

> Action 81: Strengthen social science research on the connections between biological and social processes.
> Action 82: Strengthen research on ethical, cultural, and religious concerns related to conserving biodiversity.

Research must link ecology with economics, sociology, and politics to ensure that good policy intentions do not yield inappropriate results (Bridewater et al. this volume; Grubb and May 1991; McNeely 1992; Mott and Bridgewater 1992). A necessary step is the construction of specific, testable models that integrate biological and socio-economic variables (Machlis and Forester this volume). Thus, several of the social sciences (such as sociology, economics and geography) can potentially contribute to the development and testing of models that attempt to organize our understanding about the causes of biodiversity loss (Machlis and Forester this volume).

The role of science in the conservation of biodiversity is critical. But that science must be interdisciplinary and theory-grounded research, with a "common framework of biological and social science research" (Machlis and Forester this volume; Soulé and Kohm 1989). More research designed to improve methodologies, distributional and status information, and strategies based on sound information will ultimately provide the basis for sound policy and management decisions.

FUTURE OPTIONS

Biodiversity losses are primarily a consequence of human actions (Machlis and Forester this volume). But how can land managers react to the often painful dilemmas they face on an almost daily basis when making management decisions that can have potentially devastating impacts on ecosystem stability? The discipline of Conservation Biology has been described as a "crisis discipline, where limited information is applied in an uncertain environment to make urgent decisions with sometimes irrevocable consequences" (Maquire 1991). This really speaks to the heart of all land managers. They find themselves trying to find the balance between maintaining and sustaining ecosystems while still providing the commodities products needed by people. Trade-offs will be inevitable and will necessitate formulating and using alternative land management strategies to provide an acceptable mix of commodity production, amenity use, protection of environmental and ecological values, and conserving biodiversity. Conserving diversity now is likely to alter immediate access to resources currently in demand in exchange for increasing the likelihood that long-term productivity, availability, and access are assured.

Never before have there been so many humans on earth using its resources. It is hardly surprising then, that global awareness and concerns for conserving biodiversity are continually increasing. As arguably the most intelligent species on earth, humankind has a responsibility to try as much as possible for the continuance of all forms of life

(Szaro 1995). But how can this be accomplished? The first step is trying to determine the amount, variety, and distribution of species, ecosystems, and landscapes. This will require more comprehensive inventories which must be followed by monitoring efforts to determine the impacts of management activities.

Next, strategies must be developed and implemented for the preservation, maintenance, and restoration of ecosystems. Conservation of entire communities allows species interactions and evolution to continue because management focuses on ecological processes and functions that continue to shape communities, thus promoting dynamic conservation (Williams and Davis this volume). It is essential in an adaptive management strategy that incorporates the of monitoring critical indicators of diversity, particularly those related to management objectives, and those variables that can be used to explain results from management actions and predict future trends. Based on these trends, adaptive management should be used to adjust and refine management practices (Noss this volume). These efforts should also incorporate strategies for the sustainable use of forest resources including more efficient utilization, recycling programs, and forest plantations to meet human needs.

The protection and maintenance of biodiversity is a long-term issue, which will create problems in political systems that deal primarily with the short-term goals and objectives (Szaro 1993b). The most obvious conflicts will be political and economic. There are inherent biases in a market economy that tend to result in environmental degradation. The environmental costs are generally passed to the public at large. Attempts to internalize such costs are resisted strongly by private industry. For example, installation of scrubbers in smokestacks and catalytic converters on cars in the United States were not welcomed by industry. Similar results are likely in actions to protect biodiversity. Construction of Tellico Dam in the United States was given priority over possible extinction of a small fish, the snail darter. Today many in the timber industry are against setting aside old-growth forests because of impacts on local jobs and economies. The actual monetary costs may not be as large as the political costs. For every action taken, there will be winners and losers. The loss of diversity will have long-term effects as we lose what are essentially building blocks for human survival. Using a economic analogy, biological resources should be managed so that we can live off the interest. Living off the principal eventually leads to bankruptcy.

Implementing biodiversity goals will require resources and knowledge. Current scientific understanding of ecological processes is far from perfect. Better inventories and assessments are needed of current conditions, abundances, distributions, and management direction for genetic resources, species populations, biological communities, and ecological systems. Changes monitored over time allow us to assess the consequences of management practices. It is time to make the adaptive management model standard operating procedure for sustaining diverse, healthy, and productive ecosystems (Salwasser et al. this volume). Existing financial resources could be reallocated, but additional resources will be necessary for improving efforts in inventory, monitoring, and basic research. Ecosystem-level management will require new approaches in planning, monitoring, coordination, and administration. A new paradigm is needed, one that balances all uses in the management process and looks beyond the immediate benefits.

The challenge of preserving biodiversity in managed landscapes is to incorporate natural levels of spatial and temporal heterogeneity into management schemes (Pearson et al. this volume). One advantage implied by this approach is that large systems tend to have greater stability than their component parts and this stability tends to dampen chaotic activity of those parts (Bridgewater et al. this volume). Imposing coarse-scale structure on the landscape, like protecting large blocks of habitat or the creation of regional-scale corridors, can preserve connectivity in the face of some acceptable amount of habitat destruction (Pearson et al. this volume).

Future cònservation at larger scales will always be confounded by the potentially large number of political authorities that conduct land management practices on watershed, basin, or even landscape scales. Biodiversity cannot be managed in isolation from political and social realities. Population growth and its resulting impact on resource demands is the most important factor in the fate of forest biodiversity. Thus, maintaining the integrity of the remaining natural ecosystems is closely linked with resource and social issues.

The potential effects of a diversity mandate on other resource uses must be viewed from both a short- and long-term perspective. Conserving biodiversity involves restoring, protecting, conserving, or enhancing the variety of life in an area so that the abundances and distributions of species and communities provide for continued existence and normal ecological functioning, including adaptation and extinction. This does not mean all things must occur in all areas, but that all things must be cared for at some appropriate geographic scale. People must be challenged to "think big"—to expand their thinking from individual timber sales, grazing allotments, and single species management over short time frames to conservation of multiple species, metapopulations, ecosystems, and landscapes over long time spans (Brussard et al. 1992). Processes from which endangered species arise are primarily degradation at the landscape scale, e.g., fragmentation, physiographic alteration, vegetation removal or replacement, and pollution (Bridgewater et al. this volume). So, by shifting the focus of conservation activities from the species and population level to the landscape level, progress towards biodiversity conservation will be started (Bridgewater et al. this volume).

Maintaining biodiversity requires attention to a wider array of components and larger landscape units in determining management options. This is clearly a major thrust of the Sustainable Biosphere Initiative which recognizes the critical importance and interlinkage of global change, biodiversity, and sustainable ecological systems (Lubchenco et al. 1991). The political process of developing a biodiversity policy should be guided by three central principles: (1) efforts at maintaining biodiversity should be directed at maintaining the total diversity of the landscape for many generations; (2) diversity must be understood dynamically, in terms of healthy processes, rather than merely as a maintenance of current elements of the system; and (3) economic activities that complement and enhance, rather than oppose and degrade, ecological processes are to be preferred and encouraged (Norton and Ulanowicz 1992). A variety of approaches must be developed and implemented for the preservation, maintenance, restoration, and sustainable use of ecosystems. All efforts to manage biodiversity require a balance of scientific, social, and regulatory concerns (Clark et al. this volume). The dynamic interplay caused by the interaction of these three concerns in any one conservation effort can lead to conflict and competition, but for long-term, sustainable management of biodiversity to become reality, the means to manage their interaction is needed (Clark et al. this volume). There will be trade-offs, commodity production may decline in the short-term, but in the long-term these trade-offs will result in gains in sustained productivity while maintaining biodiversity with its complete range of ecological processes.

LITERATURE CITED

Avise, J. C., J. Arnold, R. M. Ball, E. Bermingham, T. Lamb, J. E. Neigel, C. A. Reeb, and N.C. Saunders. 1987. Intraspecific phylogeography: the mitochondrial DNA bridge between population genetics and systematics. Ann. Rev. Ecol. Syst. 18:489-522.

Ballantine, B. 1991. Marine reserves for New Zealand. University of Auckland. Leigh Lab. Bull. No. 25.

Brooks, D. J., and G. E. Grant. 1992. New approaches to forest management: background, science issues, and research agenda. J. For. 90(1):25-28, 90(2):21-24.

Brussard, P. F., D. D. Murphy, and R. F. Noss. 1992. Strategy and tactics for conserving biological diversity in the United States. Conserv. Biol. 6:157-159.

Burbidge, A. A., S. D. Hopper, and S. van Leeuwen. 1990. Nature conservation, landscape and recreation values of the Lesueur area. Environ. Protection Authority Bull. 424 (Perth).

Buxton, C. D., and M. J. Smale. 1989. Abundance and distribution patterns of three temperate marine reef fish (Teleostei: Sparidae) in exploited and unexploited areas off the southern cape coast. J. Appl. Ecol. 26:441-451.

Chelimsky, E. 1991. On the social science contribution to governmental decision-making. Science 254:226-231.

Crow, T. R. 1989. Biological diversity and silvicultural systems. In Proceedings of the National Silvicultural Workshop: Silvicultural Challenges and Opportunities in the 1990's, pp. 180-185. [Petersburg, Alaska, July 10-13, 1989] USDA Forest Service, Timber Management. Washington, D.C. 216 pp.

D'Antonio, C. M., and P. M. Vitousek. 1992. Biological invasions by exotic grasses, the fire cycle, and global change. Ann. Rev. Ecol. Syst. 23:

Dahlberg, K. A. 1991. Sustainable agriculture: fad or harbinger? BioScience 41 (5): 337-340.

Davis, G. E. 1993. Design elements of monitoring programs: the necessary ingredients for success. Environ. Mon. Assess. 26:99-105.

Dayton, P. 1984. Processes structuring some marine communities: are they general? In Ecological communities: Conceptual Issues and the Evidence, eds. D. R. Strong, Jr., D. Simberloff, L. G. Abele, and A. B. Thistle, pp. 181-195. Princeton University Press, Princeton, N. J.

de Klemm, C., and C. Shine. 1993. Biological Diversity Conservation and the Law: Legal Mechanisms for Conserving Species and Ecosystems. IUCN Environmental Policy and Law Paper 29:1-292.

Diamond, J. 1989. Quaternary megafaunal extinctions: variations on a theme by Paganini. J. Archaeol. Sci. 16:167-175.

Diamond, J. M., and R. M. May. 1976. Island biogeography and the design of nature reserves. In Theoretical Ecology: Principles and Applications, ed. R. M. May, pp. 163-186. W. B. Saunders, Philadelphia, Pa.

Dizon, A. E., C. Lockyer, W. F. Perrin, D. P. Demaster, and J. Sisson. 1992. Rethinking the stock concept: a phylogeographic approach. Conserv. Biol. 6:24-36.

Everett, R., C. Oliver, J. Saveland, P. Hessburg, N. Diaz, and L. Irwin. 1994. Adaptive Ecosystem Management. In Eastside Forest Health Assessment, Volume II: Ecosystem Management: Principles and Applications, tech. eds. M. E. Jensen and P. S. Bourgeron, pp. 340-354. Gen. Tech. Rep. PNW-GTR-318. USDA Forest Service, Pacific Northwest Research Station, Portland, Ore. Forest Ecosystem Management Assessment Team (FEMAT). 1993.

Forest Ecosystem Management: An Ecological, Economic, and Social Assessment. U.S. Government, Portland, Ore.

Frankel, O. H. 1983. The place of management in conservation. In Genetics and Conservation, eds. C. M. Schonewald-Cox, S. M. Chambers, B. MacBryde, and W. L. Thomas, pp. 1-14. Benjamin/Cummings Publishing Co., Menlo Park, Calif.

Gomez-Pompa A., and A. Kaus. 1990. Traditional management of tropical forests in Mexico. In Alternatives to Deforestation: Steps Toward Sustainable Use of the Amazon Rain Forest, ed. A. B. Anderson, pp. 45-64. Columbia University Press, New York.

Gordon, J. 1992. The role of science in resolving key natural resource issues. In Seeking Common Ground, ed. T. Nygren. A Forum on Pacific Northwest Natural Resources, Feb. 24-25, 1992. Gordon, J. 1993. Assessments I have known: Toward science-based forest policy? Starker Lecture, Yale School of Forestry and Environmental Studies, New Haven, Conn., Oct. 28, 1993.

Gradwohl, J., and R. Greenberg. 1988. Saving the Tropical Forests, Island Press, Washington, D.C.

Gregory, S. V., F. J. Swanson, W. A. McKee, and K. W. Cummins. 1991. An ecosystem perspective of riparian zones. BioScience 41:540-551.

Grubb, P. J., and R. M. May. 1991. Comments on the Sustainable Biosphere Initiative. Conserv. Biol. 5:548.

Hansen, A. J., S. L. Garman, D. L. Urban, and B. Marks. 1993. An approach for managing vertebrate diversity in multiple-use landscapes. Ecol. Appl. 3:481-496.

Harrington, C., D. Debell, M. Raphael, K. Aubry, A. Carey, R. Curtis, J. Lehmkuhl, and R. Miller. 1990. Stand-level information needs related to new perspectives. Pacific Forest and Range Experiment Station, Olympia Forestry Sciences Laboratory, Olympia, Wash.

Hoekstra, T. W., and C. H. Flather. 1986. Theoretical basis for integrating wildlife in renewable resource inventories. J. Environ. Manage. 24:95-110.

Holling, C. S., ed. 1978. Adaptive Environmental Assessment and Management. John Wiley and Sons, New York.

Hunter, M. L., Jr., G. L. Jacobson, Jr., and T. Webb. 1988. Paleoecology and the coarse-filter approach to maintaining biological diversity. Conserv. Biol. 2:375-385.

International Union for the Conservation of Nature/World Wildlife Fund (IUCN/WWF). 1988. Plant Conservation Programme. IUCN, Gland, Switzerland.

Jablonski, D. 1991. Extinctions: a paleontological perspective. Science 253:754-757.

Jensen, D. B. 1987. Concepts of preserve design: What we have learned. In Conservation and Management of Rare and Endangered Plants, ed. T. Elias, pp. 595-603. California Native Plant Society, Sacramento, Calif.

Kelleher, G. K., and C. Bleakley. 1992. Marine and coastal. In Regional Reviews, pp. 1.1-1.16. IVth, World Congress on National Park and Protected Areas, Caracas, Venezuela. IUCN, Gland, Switzerland.

Keystone Center. 1991. Final Consensus Report of the Keystone Policy Dialogue on Biological Diversity on Federal Lands. The Keystone Center, Keystone, Colo.

Lovejoy, T. E. 1986. Species leave the ark one by one. In The Preservation of Species, ed. B. G. Norton, pp. 13-27. Princeton University Press, Princeton, N.J.

Lubchenco, J., A. M. Olsen, L. B. Brubaker, S. R. Carpenter, M. M. Holland, S. P. Hubbell, S. A. Levin, J. A. MacMahon, P. A. Matson, J. M. Melillo, H. A. Mooney, C. H. Peterson, H. R. Pulliam, L. A. Real, P. J. Regal, and P. G. Risser. 1991. The sustainable biosphere initiative: an ecological research agenda. Ecology 72: 371-412.

Maquire, L. A. 1991. Risk analysis for conservation biologists. Conserv. Biol. 5(1):123-125.

Master, L. L. 1991. Assessing threats and setting priorities for conservation. Conserv. Biol. 5(4):559-563.

May, R. M. 1992. Past efforts and future prospects towards understanding how many species there are. In Biodiversity and global change, eds. O. T. Solbrig, H. M. van Emden, and P. G. W. J. van Oordt, pp. 71-82. International Union of Biological Sciences, Paris. Monograph No. 8.

McNeely, J. A. 1992. The biodiversity crisis: challenges for research and management. In Conservation of Biodiversity for Sustainable Development, eds. O. T. Sandlund, K. Hindar, and A. H. D. Brown, pp. 15-26. Scandinavian University Press, Oslo.

Miller, R. R., J. D. Williams, and J. E. Williams. 1989. Extinctions of North American fishes during the past century. Fisheries 14(6):22-38.

Moir, W. H., and J. H. Dieterich. 1988. Old-growth ponderosa pine from succession in pine-bunchgrass forests in Arizona and New Mexico. Natural Areas J. 8:17-24.

Mott, J. J., and P. Bridgewater. 1992. Biodiversity conservation and ecologically sustainable development. Search 23:284-287.

Moyle, P. B., and G. M. Sato. 1991. On the design of preserves to protect native fishes. In Battle Against Extinction: Native Fish Management in the American West, eds. W. L. Minckley and J. E. Deacon, pp. 155-169. University of Arizona Press, Tucson, Ariz.

Moyle, P. B., and J. E. Williams. 1990. Biodiversity loss in the temperate zone: Decline of the native fish fauna of California. Conserv. Biol. 4:275-284.

Munro, J. L., and D. M. Williams. 1985. Assessment and management of coral reef fisheries: biological, environmental, and socio-economic aspects. Proc. Fifth Internat. Coral Reef Congress, Tahiti 4:544-578.

Norton, B. G. 1987. Why Preserve Natural Diversity?. Princeton University Press, Princeton, N. J. Norton, B. G., and R. E. Ulanowicz. 1992. Scale and biodiversity policy: A hierarchical approach. Ambio 21:244-249.

Noss, R. F. 1990. Can we maintain biological and ecological integrity? Conserv. Biol. 4:241-243.

Noss, R. F., and L. D. Harris. 1986. Nodes, networks, and MUMs: Preserving diversity at all scales. Environ. Manage. 10:299-309.

Partridge, W. L. 1984. The humid tropics cattle ranching complex: Cases from Panama reviewed. Human Organization 43:76-80.

Peters, R. L., and T. E. Lovejoy. 1992. Global Warming and Biological Diversity. Yale University Press, New Haven, Conn.

Pianka, E. R. 1987. The subtlety, complexity and importance of population interactions when more than two species are involved. Rev. Chilena Hist. Nat. 60:351-361.

Polovina, J. J., and S. Ralston, eds. 1987. Tropical Snappers and Groupers: Biology and Fisheries Management. Westview Press, Boulder, Colo.

Postel, S., and J. C. Ryan. 1991. Reforming Forestry. In State of the World, 1991, ed. L. R. Brown, pp. 74-92. W. W. Norton, New York.

Price, P. W. 1992. The resource-based organization of communities. Biotropica 24(2b):273-282.

Price, P. W., C. E. Bouton, P. Gross, B. A. McPheron, J. N. Thompson, and A. E. Weis. 1980. Interactions among three trophic levels: Influence of plants on interactions between insect herbivores and natural enemies. Ann. Rev. Ecol. Syst. 11:41-65.

Raup, D. M., and J. J., Sepkoski, Jr. 1984. Periodicity of extinctions in the geologic past. Proc. Nat. Acad. Sci. (USA) 81:801-805.

Raven, P.H. 1987. We're killing our world: Preservation of Biological Diversity. Vital Speeches of the Day, May 15, 1987, pp. 472-478.

Reid, W. V. 1992. The United States needs a national biodiversity policy. In Issues and Ideas. World Resources Institute, Washington, D.C.

Reid, W. V., and K. R. Miller. 1989. Keeping Options Alive: The Scientific Basis for Conserving Biodiversity. World Resources Institute, Washington, D.C.

Repetto, R. 1988. The forest for the trees? Government policies and the misuse of forest resources. World Resources Institute, Washington D.C.

Reynolds, R.T., R.T. Graham, M.H. Reiser, R.L. Bassett, P.L. Kennedy, D.A. Boyce, Jr., G. Goodwin, R. Smith, and E.L. Fisher. 1992. Management recommendations for the northern goshawk in the southwestern United States. Gen. Tech. Rep. RM-217. Fort Collins, CO: U.S.D.A. Forest Service, Rocky Mountain Forest and Range Experiment Station.

Roberts, C. M., and N. V. C. Polunin. 1991. Are marine reserves effective in management of reef fisheries? Rev. Fish Biol. and Fish. 1:65-91.

Record Of Decision (ROD). 1994. Record of Decision for Amendments to the Forest Service and Bureau of Land Management Planning Documents within the Range of the Northern Spotted Owl. Standards and Guidelines for Management of Habitat for Late-successional and Old-growth Forest Related Species Within the Range of the Northern Spotted Owl. Portland, Ore.

Salm, R. V., and J. R. Clarke. 1984. Marine and coastal protected areas: A guide for planners and managers. IUCN. Gland, Switzerland.

Salwasser, H., 1990. Conserving biological diversity: A perspective on scope and approaches. Forest Ecol. Manage. 35:79-90.

Scott, J., B. Csuti, J. Jacobi, and J. Estes. 1987. Species richness: a geographic approach to protecting future biological diversity. BioScience 37:782-788.

Scott, J. M., B. Csuti, and S. Caicco. 1991. Gap analysis: assessing protection needs. In Landscape Linkages and Biodiversity, ed. W. E. Hudson, pp. 15-26. Island Press, Washington, D.C.

Slobodkin, L. B. 1986. On the susceptibility of different species to extinction: Elementary instructions for owners of a world. In The Preservation of Species, ed. B. G. Norton, pp. 226-242. Princeton University Press, Princeton, N.J.

Society of American Foresters. 1991. Biological Diversity in Forested Ecosystems: A Position Statement of the Society of American Foresters. Bethesda, Md.

Solbrig, O. T., ed. 1991. From Genes to Ecosystems: A Research Agenda for Biodiversity. International Union of Biological Sciences, Cambridge.

Soulé, M. E. 1986. Conservation biology and the "real world." In Conservation Biology: The Science of Scarcity and Diversity, ed. M. E. Soulé, pp. 1-12. Sinauer Associates, Sunderland, Mass.

Soulé, M. E., ed. 1987. Viable Populations for Conservation. Cambridge University Press, New York.

Soulé, M. E., and K.A. Kohm, eds. 1989. Research priorities for conservation biology. Island Press, Washington, D.C.

Soulé, M. E., and B. A. Wilcox. eds. 1980. Conservation Biology: an Evolutionary-ecological Perspective. Sinauer Associates, Inc., Sunderland, Mass.

Stohlgren, T. J., and J. F. Quinn. 1991. Status of natural resources databases in national parks: western region. National Park Service, Cooperative Park Studies Unit, Technical Report 44, University of California, Davis, Calif.

Szaro, R. C. 1986. Guild management: An evaluation of avian guilds as a predictive tool. Environ. Manage. 20:681-688.

Szaro, R. C. 1989. Riparian forest and scrubland types of Arizona and New Mexico. Desert Plants 9:66-139.

Szaro, R. C. 1990. Management of dynamic ecosystems: Concluding remarks. In Management of Dynamic Ecosystems, ed. J. Sweeney, pp 173-180. North Cent. Sect., The Wildl. Soc., West Lafayette, Ind.

Szaro, R.C. 1992. Biodiversity inventory and monitoring. In Proceedings—Northern Region Biodiversity Workshop, compiler A. G. Evenden, pp. 57-62 [Missoula, Montana, September 1990], USDA Forest Service, Northern Region, Missoula, Mont.

Szaro, R.C. 1993a. North American boreal and temperate forests: Section 9—Biodiversity. In UNEP/GEMS Report of the UNEP/FAO Expert Consultation on Environmental Parameters in Future Global Forest Assessments. GEMS Report No. 17:62-65.

Szaro, R. C. 1993b. The status of forest biodiversity in North America. J. Trop. For. Science 5:173-200.

Szaro, R. C. 1994. Research Needs and Opportunities: the Response of Forest Biodiversity to

Global Change. In Biodiversity, Temperate Ecosystems and Global Change, eds. T. J. B. Boyle and C. E. B. Boyle, pp. 399-416. [NATO Advanced Science Workshop, August 15-19, 1993, Montebello, Canada] Springer-Verlag, Berlin.

Szaro, R. C. 1995. Biodiversity Maintenance. In Encyclopedia of Energy Technology and the Environment, eds. A. Bisio and S. G. Boots, pp. 423-433. Wiley-Interscience, John Wiley & Sons, Inc., New York.

Szaro, R. C., and R. P. Balda. 1982. Selection and monitoring of avian indicator species: An example from a ponderosa pine forest in the southwest. USDA For. Serv. GTR-RM-89, 8 p., Rocky Mtn. For. & Range Exp. Stn., Ft. Collins, Colo.

Szaro, R. C., G. A. Lessard, and W. T. Sexton. 1994. Ecosystem Management: An Approach for Conserving Biodiversity. In Biodiversity: Science and Development—Towards a New Partnership: Panel N˚5—Biodiversity and Agriculture/Pastures/Forestry, [International Union of Biological Sciences' International Forum, September 5-9, 1994, Paris, France] Biology International (Special Issue) (In press).

Szaro, R. C., and H. Salwasser. 1991. The management context for conserving biological diversity. 10th World Forestry Congress [Paris, France, September 1991]. Revue Forestiere Francaise, Actes Proceedings 2:530-535.

Szaro, R. C., and W. T. Sexton. 1994. Biodiversity conservation in the United States. In Preservation of Natural Diversity in Transboundary Protected Areas: Research needs/management options, [National Academy of Sciences and Polish Academy of Sciences Workshop, May 16-24, 1994, Bieszczady and Tatry Biosphere Reserves, Poland]. National Academy Press, Washington D.C. (In Press).

Szaro, R., and B. Shapiro. 1990. Conserving Our Heritage: America's Biodiversity. The Nature Conservancy, Arlington, Va.

Tear, T., J. Scott, P. Hayward, and B. Griffith. 1993. Status and prospects for success of the Endangered Species Act: a look at recovery plans. Science 162:976-977.

Terborgh, J., and B. Winter. 1980. Some causes of extinction. In Conservation Biology: An Evolutionary-Ecological Perspective, ed. M. E. Soulé and B. A. Wilcox, pp. 119-133. Sinauer Associates, Sunderland, Mass.

Thomas, J. W., E. D. Forsman, J. B. Lint, E. C. Meslow, B. R. Noon, and J. Verner. 1990. A conservation strategy for the northern spotted owl. Interagency Scientific Committee, Portland, Ore.

Tilghman, N. G., and J. Verner. 1989. Featured species, guilds, and species richness—three approaches to wildlife habitat management in northeastern forests. In Timber Management and its Effects on Wildlife, eds. J. C. Finley and M. C. Brittingham, pp. 66-83. Penn. State Cooperative Extension Service, University Park, Pa.

Council on Environmental Quality. 1991. Environmental Quality: Annual Report: 135-150.

Vermeij, G. J. 1986. The biology of human-caused extinction. In The Preservation of Species, ed. B. G. Norton, pp. 28-49. Princeton University Press, Princeton, N.J.

Verner, J. 1986. Future Trends in Management of Nongame Wildlife: A Researchers Viewpoint. In Management of Nongame Birds in the Midwest— A Developing Art, eds. J. B. Hale, L. B. Best and R. L. Clawson, pp 149-171. North Central Section, Wildlife Society, Madison, Wisc.

Walker, B. H. 1992. Biodiversity and ecological redundancy. Conserv. Biol. 6:18-23.

Walters, C. J., and C. S. Holling. 1990. Large-scale management experiments and learning by doing. Ecology 71(6):2060-2068.

Williams, J. E., and R. J. Neves. 1992. Introducing the elements of biological diversity in the aquatic environment. Trans. 57th N. Am. Wildl. Nat. Res. Conf. 57:345-354.

Williams, R. E., B. W. Allred, R. M. Denio, and H. A. Paulsen, Jr. 1968. Conservation, development and use of the world's rangeland. J. Range Manage. 21:355-360.

Wilson, E. O. 1989. Conservation: the next hundred years. In Conservation for the Twenty-first Century, eds. D. Western and M. C. Pearl, pp. 3-7. Oxford University Press, New York.

World Resources Institute, International Union for the Conservation of Nature and Natural Resources, United Nations Environment Programme (WRI/IUCN/UNEP). 1992. Global Biodiversity Strategy: Guidelines for Action to Save, Study and Use Earth's Biotic Wealth Sustainably and Equitably. World Resources Institute, Washington, D.C.

Yarrow, G. K. 1990. Wildlife management: incentives for the private landowner. Cooperative Extension Service, Clemson University, Clemson, S.C.

INDEX